Reading Terrestrial Planet Evolution in Isotopes and Element Measurements

Helmut Lammer • Bernard Marty • Aubrey Zerkle •
Michel Blanc • Hugh O'Neill • Thorsten Kleine
Editors

Reading Terrestrial Planet Evolution in Isotopes and Element Measurements

Previously published in *Space Science Reviews* in the Topical Collection "Reading Terrestrial Planet Evolution in Isotopes and Element Measurements"

Springer

Editors
Helmut Lammer
Space Research Institute
Austrian Academy of Sciences
Graz, Steiermark, Austria

Michel Blanc
Observatoire Midi-Pyrénées
UPS-CNRS
Toulouse Cedex 4, France

Bernard Marty
CRPG/CNRS-Université de Lorraine
Vandoeuvre les Nancy Cedex, France

Hugh O'Neill
Research School of Earth Sciences
Australian National University
Acton, ACT, Australia

Aubrey Zerkle
School of Earth & Environmental Sciences
University of St Andrews
St Andrews, UK

Thorsten Kleine
Institut für Planetologie
University of Münster
Münster, Nordrhein-Westfalen, Germany

ISSN 1385-7525 Space Sciences Series of ISSI
ISBN 978-94-024-2093-7

Cover image credit: NASA's Goddard Space Flight Center/F. Reddy

Printed on acid-free paper

This Springer imprint is published by the registered company Springer Nature B.V.
The registered company address is: Van Godewijckstraat 30, 3311 GX Dordrecht, The Netherlands

Contents

Preface: Topical Collection to "Reading Terrestrial Planet Evolution in Isotopes and Element Measurements"

Helmut Lammer[1] · Bernard Marty[2] · Aubrey L. Zerkle[3] · Manuel Scherf[1] · Hugh O'Neill[4] · Michel Blanc[5] · Thorsten Kleine[6]

Terrestrial planets evolve from primitive material of which remnants are now found in primitive meteorites called chondrites that originated from parent bodies that did not go through the cycle of melting and differentiation. Two broad categories have been defined based on their elemental and isotopic abundances, the carbonaceous chondrites (CC) and the non-carbonaceous chondrites (NC). Using geochemical observations of noble gases and other major volatiles, one can determine what the present-day inventory tells us on its initial building blocks, their sources, accretion processes and the early differentiation of planetary bodies such as the Earth. While many of the present-day volatile reservoirs have chondritic isotopic ratios, their relative abundances are certainly not chondritic, suggesting volatile losses from planetary embryos and growing proto- and early planets during and after their accretion phases. The interdisciplinary research fields discussed contain different processes that produce chemical compositional changes after the solar nebula/chondrite stage. These, for instance, contain collisional erosion, post-nebula volatilization and the escape of noble gases and volatile elements that were dragged by hydrodynamically escaping hydrogen atoms originating from captured primordial atmospheres or magma ocean-related catastrophic outgassed steam atmospheres.

The above-mentioned losses and their efficiencies are dependent on the disk lifetime, the X-ray, and EUV flux evolution of a system's young host star and the accretion history of planets in general. Isotopic and elemental ratios can also be used for constraining the timescale of accretion, the volatile delivery and the origin of the Moon. The amount of radioactive heat producing elements (i.e., U, Th, K) for instance determines the planet's subsequent tectonic regimes and, therefore, its possible evolution to an Earth-like habitat.

Since the discovery of small rocky exoplanets, many speculations on "Earth-like habitats" buzz around within the exoplanet research community and hence the media and the

[1] Space Research Institute, Austrian Academy of Sciences, Schmiedlstr. 6, 8042 Graz, Austria

[2] Université de Lorraine, CNRS, CRPG, 54000 Nancy, France

[3] School of Earth and Environmental Sciences and Centre for Exoplanet Science, University of St. Andrews, St. Andrews, UK

[4] College of Physical & Mathematical Sciences Research School of Earth Sciences, Australian National University Canberra, Canberra, Australia

[5] Institut de Recherche en Astrophysique et Planétologie, IRAP, Toulouse, France

[6] Institut für Planetologie, Universität Münster, Münster, Germany

public. That is why it is very important to understand how Earth obtained its main atmospheric species, N_2 and, later on, O_2. Terrestrial planet evolution in relation to the formation of modern plate tectonics and the influence of life on the origin of Earth's N_2-dominated atmosphere, which can be seen as an atmospheric geo-signature and together with O_2 as a geo-biosignature, is also addressed in this review volume. Future space missions that are important for enhancing our understanding on the evolution of terrestrial planets, including exoplanetary Earth-like habitats, are also discussed. Two special papers, one related to the deep nitrogen and carbon cycles on Earth and a second one, which summarizes the up-to-date data on the relative atomic Solar System abundances, mass fractions, atomic masses of the elements and their isotopes, the composition of the solar photosphere and of the major chondritic meteorite groups finish the collection.

1 The Sun Through Time

This review volume starts with a paper by *M. Güdel* on the importance and role of the young Sun and its activity evolution through time by forcing the evolution of terrestrial planets, especially their atmospheres. The review chapter focuses on the activity evolution of Sun-like G-type stars and discusses direct and indirect observations related to the spin down, of X-ray activities and wind mass loss rates as a function of stellar mass and ages. From these observations, one finds that G-type stars that were born with different rotational periods, from slow, via moderate to fast rotators, have initial EUV fluxes of about 400 to 500 times higher than the present-day Sun during their saturation phase. This phase, however, lasts much longer for fast rotators, than for moderate, and particularly slow rotators. After these saturation phases, the winds and the stellar EUV radiation decline and converge after about 1 billion years (Gyr) after the G star's origin. These relations are particularly important for studying atmospheric evolution since the absorption of the host star's EUV flux heats the upper atmosphere, thereby enhancing thermal and non-thermal escape of particles in many complex ways. EUV-driven atmospheric loss processes also modify volatile element abundances and the ratios of atmospheric isotopes mass-dependently over time, which can tell us something about the rotational history of the planet's host star such as our Sun.

2 Chemical and Isotopic Evolution of the Early Solar System

The significance of the chemical and isotopic compositions of early planetary building blocks in the protoplanetary disk, is reviewed by *Bermingham et al.* The authors examine constraints on the chemical evolution of the disk provided by elemental and nucleosynthetic isotopic compositions of meteorites and discuss if and how the compositional evolution of the disk to the planetary system can be derived by synthesizing astrophysical models and datasets from meteorites that directly sample the disk. A fundamental question regarding the formation of the planets in the solar system concerns the nature and location of the non-carbonaceous (NC) and carbonaceous chondrite (CC) reservoirs, and possible interactions between these reservoirs. Evidence that CC meteorites formed in the outer protoplanetary disk beyond Jupiter's orbit, the contribution of NC and CC reservoirs to the formation of the Earth and its water, and challenges associated with the practice of inferring (radial) accretion locations and movement within the disk based on the volatile element and nucleosynthetic isotope compositions of meteorites are also discussed. Because several papers in this review volume have a strong focus on N_2 and the evolution to Earth's N_2-O_2 dominated present-day atmosphere, this review also discusses the origin and reservoirs of nitrogen based on isotopic constraints provided by carbonaceous and enstatite chondrites. Overall, Earth's hydrogen, oxygen, nitrogen, and nucleosynthetic isotope signatures require building blocks from both the NC and CC reservoirs.

3 Formation of Venus, Earth and Mars: Constrained by Isotopes

The third review by *Lammer et al.* first discusses the current state of knowledge of terrestrial planet formation by reviewing different planet formation models based on the classical, annulus, grand-tack, depleted disk, pebble accretion and early instability hypotheses. A very new aspect of this review chapter is that the authors summarized the current state of knowledge of terrestrial planet formation not only from the aspects of different planet formation models but also from isotopic data of ^{182}Hf-^{182}W geochronology, U-Pb chronology, lithophile-siderophile elements, ^{48}Ca/^{44}Ca isotope samples from planetary building blocks, recent reproduction attempts from Ar and Ne isotope ratios in Venus' and Earth's atmospheres, the expected solar ^{3}He abundance in Earth's deep mantle and Earth's D/H sea water ratios. By discussing the analysis of these different isotope data and chronometrical dating methods, one can conclude that terrestrial planets in the Solar System favored most likely fast accretion scenarios with $\approx$80% of Earth's mass most likely accreted within $\approx$10–30 Myr, while Venus could have even accreted up to its final mass until the end of the disk lifetime. For classical terrestrial planet formation models, however, it is a challenge to accrete large planetary embryos so fast from the dusty materials within the typical disk lifetimes. The authors of this chapter, therefore, conclude that pebble accretion and streaming instabilities could solve this timescale controversy. The later stage of accretion can be explained well with the Grand-Tack model and partly also by the annulus and depleted disk models. More studies, however, are needed to homogenise these studies with data that are provided by chemical, chronological, isotopical and dynamical constraints.

4 Geochemical Constraints on the Origin of the Moon and the Preservation of Ancient Heterogeneities

The latest hypotheses on the origin of the Moon by addressing geochemical constraints and ancient heterogeneities are discussed in the review by *Lock et al.* Although the giant impact theory, with a big impactor named Theia vaporizing much of the early Earth, is the accepted mainstream formation model since about 40 years, details of the Moon's formation remain not well understood until today. Studies show that giant impacts with angular momenta as low as necessary for the Moon-forming event would be a rare one, related to the lowest energy accretionary impact scenario between two massive bodies involved. Therefore, high-energy impacts are now proposed in addition since they provide new environments for the formation of the Moon by reproducing a new class of bodies exceeding the corotation limit, called "synestias", a dynamical, donut shaped, structure of vaporized rock. Regarding the timing of the Moon-forming event, various studies that analyzed isotopic measurements of Lunar, terrestrial and meteoritic samples that constrain a most likely formation age between $\approx$50–200 Myr after the origin of the solar system, with most dating systems favoring ages closer to the younger period. One can argue that the isotopic similarity between the Earth and Moon requires that either the proto-Earth or its large impactor Theia had a very similar isotopic composition, or that the efficiency of mixing during the collision was high. However, the impact between proto-Earth and Theia should not have homogenized early Earth's whole mantle since isotopic signatures of pre-impact heterogeneities are preserved.

5 On the Distribution and Variation of Radioactive Heat Producing Elements Within Meteorites, the Earth, and Planets

The relevance of the distribution and variation of radioactive heat producing elements (HPEs) within meteorites and terrestrial planets as a main factor in controlling a rocky planet's thermal evolution, tectonics, and even its likelihood for habitability is reviewed by O'Neill et al. This type of research and findings are often ignored in habitability studies within the Astrobiology/exoplanet community. The state of knowledge of various HPEs in planetary building blocks, terrestrial planets, and in present Earth models from emerging constraints, and their role related to the tectonic history is summarized. The primordial sources of heat production within terrestrial planets such as the Earth include heat from the accretion process, core formation and heating due to ancient radioactive isotopes in the planet's history. The authors also present a heuristic way to investigate the sensitivity of Earth's current tectonic mode as an example for rocky planets in general to the planet's HPE inventory. They point out that the specifics of tectonic transitions are subject to ongoing debates in the geodynamic research community. There is, however, a common denominator; the internal heating rates of terrestrial planets and their HPE budget are fundamental in controlling the evolutionary paths to Earth-like habitats. Because HPEs and their related processes affect volcanic outgassing and tectonics-related in-gassing, the major atmospheric species associated to these processes, CO_2, SO_2 and H_2O, may be detectable by future spectroscopic measurements on terrestrial exoplanets.

6 Loss and Fractionation of Noble Gas Isotopes and Moderately Volatile Elements from Planetary Embryos and Early Venus, Earth and Mars

The latest knowledge on how various atmospheric escape processes can fractionate noble gas isotopes and moderately volatile rock-forming elements that populate accreting and primordial atmospheres as well as catastrophically outgassed steam atmospheres is reviewed by *Lammer et al.* The authors review and summarize the measured isotopes and their ratios in the atmospheres of Venus, Earth, and Mars. Especially for Venus' case, it is pointed out that several isotope measurements have large error bars or were not measured by previous spacecraft, therefore, it is highly recommended that new precise measurements should be carried out in future missions. Recent discoveries of H_2-dominated primordial atmospheres on most detected low mass exoplanets with known radii and masses are also discussed. Most of these exoplanets have very low densities which indicate that the transit radius is not related to the rocky surface of these bodies, and only extended H_2-dominated gas envelopes can explain the observations. From these discoveries, one can expect that many low mass exoplanets may have Earth-like density cores but a large number of them still host an H_2-dominated primordial atmosphere that was captured from the circumstellar disk – favoring fast accretion scenarios during the lifetime of the gas disk. Finally, it is pointed out that a complex interplay between the young Sun's EUV flux, the accreted protoplanetary mass within the disk lifetime, a possible accumulated primordial H_2-dominated atmosphere, the impacting bodies that were involved in the accretion before and after disk dispersal, as well as the depletion of volatile elements from planetary embryos and the final protoplanets eventually led to today's measured ratios.

7 Nitrogen Atmospheres of the Icy Bodies in the Solar System

The current knowledge on the origin and evolution of N_2-dominated atmospheres on icy bodies such as Titan, Triton and Pluto is reviewed by *Scherf et al.* The authors point out that an important information for understanding the origin and evolution of these atmospheres can be found in various atmospheric isotopic signatures such as $^{14}N/^{15}N$, $^{12}C/^{13}C$ and Ar/N ratios. These isotope ratios in N_2-dominated atmospheres serve as a footprint of the building blocks from which Titan, Triton and Pluto originated and by studying different fractionation processes one can figure out how their atmospheres evolved during their entire evolution. Studies that find that hydrodynamic escape, sputtering and Jeans escape cannot explain the strong enrichment of Titan's atmospheric ^{15}N isotopes over its history if it would have the same origin as the Earth's, which is believed to be chondritic are addressed. Instead, the outgassing of N_2 produced from cometary NH_3 might have played a crucial role in Titan's history pointing to an endogenic origin, which is also backed by the existence of outgassed radiogenic ^{40}Ar isotopes in the atmosphere. The origin and evolution of the N_2-dominated atmospheres of Neptune's large satellite Triton and the dwarf planet Pluto, even after the New Horizon flyby, remain unclear, although it might be likely that these atmospheres originated in a similar way as Titan's. This review also includes a discussion on the diverse nitrogen isotope measurement techniques since an accurate retrieval of isotopic ratios depends on different model and measurement assumptions.

8 Perspectives on Atmospheric Evolution from Xe and Nitrogen Isotopes on Earth, Mars and Venus

The latest knowledge on the evolution of Earth's, Venus' and Mars' atmospheres that can be constrained by Xe and nitrogen isotopes is reviewed by *Avice and Marty.* A major finding for the Earth's paleo-atmosphere is that xenon, the heaviest noble gas, is fractionated in Earth's atmosphere. They point out that the isotopic fractionation of Xe did not take place during or just after Earth's formation but occurred over Gyr timescales. Escape models that try to explain the Xe paradox and summarize the most likely ones, which is dragging of Xe^+ ions by EUV-powered hydrodynamic escape of H^+ atoms during the Archean, and the preferential trapping of heavier Xe^+ isotopes into organic hazes, which are believed to be abundant in the Archean atmosphere are reviewed and discussed. Analyses of Martian meteorites show that the isotopic fractionation and escape of Xe at Mars happened earlier (>4 Ga ago) compared to Earth. For Venus, none of the past space missions managed to detect Xe in its atmosphere. Besides Xe, atmospheric nitrogen and its isotopes can also be used as a tracer of past atmospheric pressure and escape. The authors themselves, for instance, found from data correlations of $N_2/^{36}Ar$ vs. $^{40}Ar/^{36}Ar$ ratios that the partial pressure of atmospheric N_2 was in the range of 0.5–1.0 bar ≥ 3.0 Gyr and most likely lower than the modern one at 3.3 Gyr. Contrary to Earth's, Mars' atmosphere is strongly enriched in the heavy ^{15}N isotope, which is an indication for atmospheric escape of nitrogen during the past $\approx$4 Gyr. The isotopic composition of N_2 in Venus' atmosphere remains largely unknown with estimates of an Earth-like value within an uncertainty of $\approx$20%.

9 Mission to Planet Earth: The First Two Billion Years

The evolution of early Earth with its focus on the first 2 Gyr after the origin of our home planet is reviewed by *Stüeken et al.* They identify five crucial main phases in early Earth's evolution: i) the redox state and volatile content of Earth's building blocks; ii) the chemical properties of the resulting degassed mantle which are important in its ability to recycle volatiles and nutrients via plate tectonics; iii) the emergence of life and life's capabilities to effect Earth's atmosphere (e.g. biological CO_2 and N_2 fixation, oxygenic photosynthesis, etc.); iv) the young Sun's higher EUV flux; and v) evidence of strong photochemical effects on Earth's sulfur (S) cycle, preserved in the form of mass-independent S isotope fractionation, and potentially linked to fractionation in organic C-isotopes. Regarding Earth's atmospheric N_2 build-up, the latest hypotheses and evidence backed by isotope and ancient rock-related air pressure studies are discussed. It is shown that photochemical and geological processes set the stage for habitable conditions and how the origin of life likely had a significant impact on volatile inventories and the atmosphere with major implications for global climate and habitability. Relevant for the search of N_2 atmospheres on terrestrial exoplanets, atmospheric biosignatures on the early Earth are also addressed and explain why N_2-O_2-dominated atmospheres most likely represent geo-biosignatures. This review shows very well that early Earth itself presents an exoplanet analogue that can be investigated through the existing rock record.

10 The Isotopic Imprint of Life on an Evolving Planet

The biological information recorded by stable isotopic compositions observed on Earth in the context of evolving metabolisms on an evolving planet is discussed by the review of *Lloyd et al.* They discuss the isotope biogeochemistry of the biologically essential elements C, N and S, and how these elements are distributed on the modern and early Earth, which is the closest analogue for an extraterrestrial Earth-like habitat. They describe how lifeforms exist in thermodynamic disequilibrium with their environment, and how this is reflected in the stable isotope ratios of materials that they produce. Isotopic analyses can thus be used to study life where direct observations are not possible. The authors also speculate on how the isotopic record would look if the biogeochemical C, N, and S cycles on Earth would operate in the absence of life. This comparison is important in the search for life on other planets. On Earth, by sequestering reduced carbon, nitrogen, and sulfur in minerals and recalcitrant kerogens, life separates electrons from their previous owners, which allows O_2 to persist in Earth's present-day atmosphere. In the absence of life, the burial fluxes of organic matter, the composition of the atmosphere, and the oxidation state of the exosphere likely would have looked very different than today. Microbes would not have mediated a plethora of kinetically slow reactions, such as sulfate reduction or atmospheric N_2 fixation, with important implications for the chemical and isotopic facets of geochemical cycles.

11 Future Missions Related to the Determination of the Elemental and Isotopic Composition of Earth, Moon and Planets

Dandouras et al. review the contribution of future space missions to the determination of the elemental and isotopic composition of Earth, the Moon, the Solar System and exoplanets. Several planned missions to the Earth-Moon system will study the formation history of the Moon, the structure and dynamics of its exosphere, and its interaction with the radiation, solar wind and Earth's escaping atmosphere which might in the past has even been implanted into the Lunar surface. The authors also review the current plans for missions in near Earth orbit that will help better quantify atmospheric escape processes and Earth's upper atmosphere and magnetosphere. Future missions to Venus and Mars will further be key to establish a comparative view of the evolution of the climates and habitability of Earth, Venus, and Mars. The authors point out that sample return is the main task of future exploration, not only for these planets but also for comets and asteroids, which will give an insight into planetary building blocks. Returning samples from the atmosphere or surface of Venus would be the most challenging mission to fly but it would deliver a huge scientific return helping to solve the open questions on why Earth's sister planet is so different. Future missions to Mercury are further expected to provide a test bed for science of hot rocky exoplanets that orbit in close-in stellar environments. These authors conclude their review with a focus on the relevance of the spectroscopic detection of N_2-O_2-dominated atmospheres for the search of extraterrestrial life by future space observatories.

12 Supplementary Information on Earth's N and C-Cycles and the Latest Collection of Relative Atomic Solar System Abundances, Atomic Masses and Solar Photosphere and Major Meteoritic Compositions

For a better understanding how the N- and C-cycles work on Earth and most likely also on terrestrial exoplanets with comparable functioning carbon-silicate cycles, a special paper in the Supplementary Information by *Halama and Bebout* is also included in this topical collection that briefly reviews for an interdisciplinary readership how nitrogen and carbon are converted into the biogeochemical cycle by which these elements are exchanged among the biosphere, pedosphere, geosphere, hydrosphere, and atmosphere of the Earth.

Finally, the latest data for atomic abundances and mass fraction for the elemental and isotopic solar system composition, the atomic masses of the elements and their isotopes, the composition of the solar photosphere, and the compositions of the major chondritic meteorite groups is given in a second Supplementary Information by *Lodders*. The data which are relevant for all reviews and their interested readers are summarized in comprehensive tables that finish this topical collection of relevant reviews related to "Reading Terrestrial Planet Evolution in Isotopes and Element Measurements".

This collection is based on an interdisciplinary international workshop which took place at ISSI Bern during October 22. and 26., where about 48 leading scientists discussed the issues presented here.

Group picture of the workshop "Reading Terrestrial Planet Evolution in Isotopes and Element Measurements" that was organized and hosted by the International Space Science Institute (ISSI) in Bern, Switzerland during October 22 and 26, 2018, and supported by Europlanet 2020.

Acknowledgements The organizers thank ISSI for hosting us in Bern during the workshop and are extremely grateful to all the referees for helping to improve the quality of the review papers in the book. We further acknowledge the editorial staff at Springer Nature and the staff at ISSI, in particular Andrea Fischer, Alexandra Lehmann, Saliba F. Saliba and the ISSI Directors. Without their help, the workshop and the book would not have been possible. Finally, the financial support by ISSI, and Europlanet 2020 Research Infrastructure is also gratefully acknowledged.

Space Sci Rev (2020) 216:143
https://doi.org/10.1007/s11214-020-00773-9

The Sun Through Time

Manuel Güdel[1]

Received: 24 December 2019 / Accepted: 17 November 2020 / Published online: 8 December 2020

Abstract Magnetic activity of stars like the Sun evolves in time because of spin-down owing to angular momentum removal by a magnetized stellar wind. These magnetic fields are generated by an internal dynamo driven by convection and differential rotation. Spin-down therefore converges at an age of about 700 Myr for solar-mass stars to values uniquely determined by the stellar mass and age. Before that time, however, rotation periods and their evolution depend on the initial rotation period of a star after it has lost its protostellar/protoplanetary disk. This non-unique rotational evolution implies similar non-unique evolutions for stellar winds and for the stellar high-energy output. I present a summary of evolutionary trends for stellar rotation, stellar wind mass loss and stellar high-energy output based on observations and models.

Keywords Stellar activity evolution · Stellar wind evolution · Stellar mass-loss evolution · Stellar rotation · Stellar spin-down

1 Introduction

Solar activity was recognized as an important property of the Sun in observations of magnetic fields in sunspots (Hale 1908), the occurrence of giant outbursts initially in white light (Carrington 1859; Hodgson 1859), the presence of a very hot (millions of K) corona initially in optical lines (Grotrian 1939; Edlén 1942), and later in radio waves (Hey 1946) and X-rays (Burnight 1949). We now know that magnetic fields are at the origin of all these phenomena, as they can store energy derived from convective motions at and below the photospheric level, but can also continuously or episodically release energy in higher atmospheric layers (the chromosphere, the transition region, the corona, and the solar-wind region).

Reading Terrestrial Planet Evolution in Isotopes and Element Measurements
Edited by Helmut Lammer, Bernard Marty, Aubrey L. Zerkle, Michel Blanc, Hugh O'Neill and Thorsten Kleine

✉ M. Güdel
manuel.guedel@univie.ac.at

1 Department of Astrophysics, University of Vienna, Türkenschanzstr. 17, 1180 Vienna, Austria

Magnetic fields originate in the dynamo deep inside the convection zone; differential rotation and convective motion act together to amplify magnetic fields that rise buoyantly to the surface where they form magnetic active regions with "magnetic loops" reaching out into the corona. A solar coronal wind was inferred initially from observations of comets (Biermann 1951) and explained theoretically by Parker (1958), followed by *in situ* observations by space probes in the early 1960s.

In what sense solar/stellar rotation itself matters for "magnetic activity" was less clear; it was stellar astronomy that made important contributions to this question. Kraft (1967) reported that the average rotational velocities of stars with strong Ca II emission are higher than velocities for weak Ca emitters. This pointed to an evolutionary spin-down of stars because stronger Ca II emitters are younger; Kraft (1967) suggested that magnetically coupled winds like the solar wind are responsible for the angular momentum loss and associated this process with outer convective layers in cool stars. The angular momentum loss process was theoretically explained by Weber and Davis (1967) via magnetic wind torques. In another seminal paper, Skumanich (1972) laid the foundations of subsequent stellar statistical studies of age, evolution of activity, and rotation. Skumanich suggested that Ca II emission declines with the inverse square root of the age, and the same decay law should also hold for the rotation velocity and the surface magnetic field strength. These observations were the foundation of age-rotation-activity relations that have been quantitatively refined until the present time.

Apart from the role it plays for stellar structure, atmospheres and spin-down, stellar activity has been recognized as a key parameter for the evolution of planets. Short-wavelength radiation (ultraviolet [UV], extreme ultraviolet [EUV], and X-rays) affects the thermal structure of planetary atmospheres, can drive chemical reactions, and can lead to escape and erosion of planetary atmospheres. But also the stellar wind, shock fronts and high-energy particles it transports interact with planetary outer atmospheres and – if present – magnetospheres; a variety of interactions can again lead to chemical reactions, heating, and atmospheric loss.

Understanding the evolution of stellar magnetic fields and magnetic activity as expressed in high-energy radiation and winds is therefore pivotal if we are to understand the long-term evolution of planets. In fact, an observation of an exoplanet provides a snapshot in time but cannot by itself explain the composition and structure of an atmosphere. Comprehensive evolutionary models of the host star are required to trace the evolutionary processing of atmospheres, for rocky planets complemented with evolutionary models of the planetary interiors (convection, cooling, outgassing and ingassing).

The following subsections discuss the long-term evolution of stellar rotation and spin-down, winds, and high-energy (UV/EUV/X-ray = XUV) radiation with the aim of summarizing trends in the evolution of this stellar output. Stellar ages themselves come therefore into the discussion, but the intent here is not to provide means to estimate stellar ages (see Mamajek and Hillenbrand 2008 for a discussion of stellar ages and their relation with other stellar parameters). Many of the properties discussed below in fact evolve non-uniquely within an age range that is crucial for the setup of planetary atmospheres and their potential habitability, i.e., up to several 100 Myr after star/planet formation.

2 Stellar Rotation

Stellar rotational evolution has been the subject of many publications aiming at studying basic physical concepts and testing them with observations (e.g., Gallet and Bouvier 2013).

We briefly discuss here a model that includes all key ingredients for a semi-empirical description of stellar spin-down (Johnstone et al. 2015).

Observationally, young open clusters (i.e., co-eval samples of stars) show a large spread in rotation periods for any given stellar mass. Most likely, these stars were starting their rotational evolution with different initial conditions (which we take to be the rotational state after the dispersal of protostellar/protoplanetary disks a few million years after the formation of a protostar). In the pre-main sequence stages, stars still contract and therefore spin up until they reach the zero-age main sequence. After that time, stars can only spin down as a magnetized wind removes angular momentum. Interestingly, after about 700 Myr of evolution for solar-mass stars, they all end up at a rotation period nearly uniquely determined by the stellar mass and age (Soderblom et al. 1993). This *convergence* is ascribed to a feedback between angular momentum removal by a magnetized coronal wind and the rotationally induced operation of an internal dynamo that generates the magnetic fields. The close relation between magnetized and ionized winds and stellar coronae suggests that the wind mass-loss rate also declines with time.

To understand evolutionary spin-down, we therefore need knowledge on the initial stellar rotation rate, Ω_0, the internal structure of the star (from stellar-structure models), and the rate at which angular momentum is removed from the star by the wind. To discuss the basic physical ingredients, I use the example of Johnstone et al. (2015) although other models have been presented in the literature (e.g., Gallet and Bouvier 2013, see also the physical wind model by Airapetian and Usmanov 2016). The model of Johnstone et al. (2015) consists of a system of equations with some free parameters that need to be fitted to the observed rotational distributions of stellar clusters. It was initially applied to stars with a radiative core and a convective envelope in the mass range of 0.4–1.1 $M_\odot$ including core-envelope coupling torques, but was recently extended to include also fully convective stars down to 0.1 $M_\odot$ using a different torque formula (Johnstone et al. 2020).

We denote with $\Omega_*(t)$ the angular rotation rate, $\dot{M}(t)$ the wind mass-loss rate and $B(t)$ the surface magnetic field strength. The magnetic torque acting on the star, τ, is a function of stellar mass M_*, radius R_*, $\dot{M}$, B, and Ω_*,

$$\tau = f(B, \dot{M}, M_*, R_*, \Omega_*) \approx B^{0.87} \dot{M}^{0.56} R_*^{2.87} \Omega_* \tag{1}$$

where the numerical example on the right-hand side is simplified from a torque formula derived from MHD simulations by Matt et al. (2012). The spin-down rate then is,

$$\frac{\mathrm{d}\Omega_*}{\mathrm{d}t} = \frac{1}{I_*}\left(\tau - \frac{\mathrm{d}I_*}{\mathrm{d}t}\Omega_*\right), \tag{2}$$

where I_* is the star's moment of inertia from stellar structure models. We next assume

$$\dot{M} = \dot{M}_\odot \left(\frac{R_*}{R_\odot}\right)^2 \left(\frac{\Omega_*}{\Omega_\odot}\right)^a \left(\frac{M_*}{M_\odot}\right)^b \tag{3}$$

where the exponents a and b are fitted to observational constraints. Vidotto et al. (2014) derived the equatorial magnetic field strength of the dominant dipole component from observations as

$$B = B_\odot \left(\frac{\Omega_* \tau_*}{\Omega_\odot \tau_\odot}\right)^{1.32} \tag{4}$$

where τ_* is the convective turnover time of the star. B "saturates" for rapidly rotating stars, i.e., is no longer a function of Ω_*, which means that $\dot{M}$ should also saturate. The rotation rate where saturation sets in can be described by

$$\Omega_{*,\mathrm{sat}}(M_*) = \Omega_{\mathrm{sat}}(M_\odot)\left(\frac{M_*}{M_\odot}\right)^c \tag{5}$$

where c is a further fit parameter, and $\Omega_{*,\mathrm{sat}}(M_\odot) \approx 15\,\Omega_\odot$.

The above equations can be solved once the free parameters a, b, and c have been fitted to rotation distributions for different ages, and initial conditions for $\Omega(t_0)$ are assumed, using observed Ω_* distributions of clusters in star-forming regions.

Fig. 1 shows examples for the spin-down evolution of a solar-mass star with different initial rotation rates, as well as the required wind mass-loss history from the same model. As required by the observations, the rotation rates converge to a unique age-dependent value after ~700 Myr (where the width of the Ω distribution between the 10th and the 90th percentiles is ~0.2 dex; Johnstone et al. 2015; Soderblom et al. 1993), and the same must apply for $\dot{M}$ as the primary cause for spin-down. After ~ 700 Myr, the mass-loss rate declines in time t roughly as

$$\dot{M} \propto t^{-0.75}. \tag{6}$$

It relates to the rotation period as

$$\dot{M} \propto R_*^2\, M_*^{-3.36}\, \Omega_*^{1.33} \tag{7}$$

i.e., $\dot{M}$ decreases strongly with stellar mass but increases with rotation velocity; the saturation mass-loss rate is

$$\dot{M}_{\mathrm{sat}} = 37\dot{M}_\odot \left(\frac{M_*}{M_\odot}\right)^{1.3}, \tag{8}$$

showing that higher-mass stars can achieve higher mass-loss rates. Saturation of $\dot{M}$ limits the mass-loss rates of young, very rapid rotators, as shown in Fig. 1.

Because of the widely distributed initial rotation rates after star formation, the $\dot{M}$ and Ω_* histories are non-unique. We see that $\dot{M}$ of solar analogs is distributed over nearly an order of magnitude (10th to 90th percentile, green area) near the Zero-Age Main-Sequence (ZAMS) age (Fig. 1-right; Johnstone et al. 2015).

3 Stellar Winds

Stellar winds have already been described above, together with stellar magnetic fields, as the key agent for stellar spin-down. What is the observational situation? Stellar winds are very difficult to detect directly owing to their weak emission; however, a few observational methods have been developed to infer the mass-loss rate $\dot{M}$ from other features that are controlled by the stellar wind. Observational wind detection/measurement methods can therefore be classified as *direct* and *indirect.*

Direct wind measurements make use of electromagnetic emission of the wind plasma itself, or optical-depth related absorption or attenuation in the wind. *Indirect wind measurements* use features for which the wind is responsible but that themselves are distinct from the wind.

We briefly summarize methods described in the literature and provide some key results for each one.

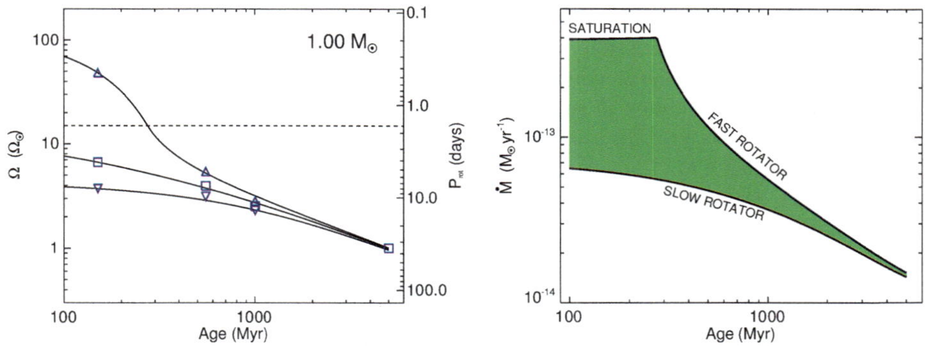

Fig. 1 Left: Modeled rotational evolution tracks of a 1 $M_\odot$ star at the 10th, 50th, and 90th percentiles of the rotational distribution, shown from 100 Myr to 5 Gyr. The blue symbols mark the same percentiles from observational distributions. The dashed line indicates the limiting rotation rate above which the wind mass-loss rate and the surface magnetic field saturate. – Right: The evolution of the solar wind mass-loss rate with age on the main sequence. The green area includes all possible $\dot{M}$ evolutionary tracks for different initial rotation periods at the 10th percentile (bottom edge) and the 90th percentile (top edge). The fastest rotators remain saturated at a maximum $\dot{M}$ level during the first 300 Myr. (From Johnstone et al. 2015.)

3.1 Direct Wind-Detection Methods

Bremsstrahlung at Radio Wavelengths Because winds from cool stars are ionized, they emit bremsstrahlung as a consequence of accelerating/decelerating interactions between electrons and ions. The bremsstrahlung emissivity is a function of electron temperature, T, and number density, n_e. The radio free-free flux spectrum for an optically thick, constant-velocity, fully ionized isothermal spherical (isotropic) wind follows from

$$S_\nu = 9 \times 10^{10} \left(\frac{\dot{M}}{v}\right)^{4/3} T^{0.1}\, \nu^{0.6}\, d^{-2}\ \mathrm{mJy}, \tag{9}$$

(Panagia and Felli 1975; Wright and Barlow 1975; Olnon 1975). Radio observations over a wide range of frequencies can therefore provide detections or upper limits for $\dot{M}$ given a wind temperature and a terminal wind velocity, v. Searches best involve high radio frequencies or millimeter wavelengths (Doyle and Mathioudakis 1991; Mullan et al. 1992), but interesting upper limits have been reported also for centimetric wavelengths, such as an upper limit for the evolved late F-type star Procyon translating to $\dot{M} < 2 \times 10^{-11}\ M_\odot\,\mathrm{yr}^{-1}$ (Drake et al. 1993). Initially, the method was often used for radio-loud nearby M dwarfs, for which Lim and White (1996), Lim et al. (1996), and van den Oord and Doyle (1997) derived $\dot{M} <$ a few times $10^{-10}\ M_\odot\,\mathrm{yr}^{-1}$, down to $\lesssim 7 \times 10^{-12}\ M_\odot\,\mathrm{yr}^{-1}$ for a wind temperature of 10^4 K or 10^6 K, respectively. Turning to solar-mass stars, Gaidos et al. (2000) and Fichtinger et al. (2017) studied an age sequence of solar analogs at 6–14 GHz but found only upper limits for the radio flux corresponding to wind mass-loss rates of $\dot{M} < 3 \times 10^{-11}\ M_\odot\,\mathrm{yr}^{-1}$. The latter authors constrained the ZAMS mass of the Sun to no more than 1.02 times the present mass, which rejects a popular explanation of the Faint Young Sun Paradox (FYSP). The FYSP states that geological evidence implies mild climates on early Earth and Mars even though the solar luminosity was only $\approx 70-75\%$ of the present level. A slightly more massive Sun would have been more luminous in the early main-sequence stages, and therefore the Sun could have been brighter than in the standard evolution model. To solve the FYSP using an Earth-like or CO_2 atmosphere, a mass excess of $> 2\%$ is required at the ZAMS.

Radio Free-Free Optical Depth A sufficiently dense wind can become optically thick to underlying stellar radio emission. If the optically thick surface is located above the lower coronal level, then radio flares would not be seen on active stars, in contrast to observational evidence. From the condition of small optical depth, Lim and White (1996) obtained $\dot{M} \lesssim 5 \times 10^{-14} - 10^{-12}\ M_{\odot}\ \mathrm{yr}^{-1}$ for $v = 300 - 600\ \mathrm{km\,s}^{-1}$ and wind temperatures of $10^4 - 10^6$ K for the dMe star YZ CMi. For solar analogs we are focusing on here, Fichtinger et al. (2017) derived upper limits at different ages confirming their upper limits from bremsstrahlung emission (see above).

Radio Wave Propagation Radio waves can propagate only if their frequencies exceed the local electron plasma frequency that itself depends on $n_{\mathrm{e}}^{1/2}$ (n_{e} being the electron number density). If a planet is supposedly a magnetospheric radio source, non-detections of this radiation can set upper limits to the mass-loss rate given a wind velocity. Vidotto and Donati (2017) estimated $\dot{M} \lesssim 10^{-10}\ M_{\odot}\ \mathrm{yr}^{-1}$ for the T Tauri star V830 Tau and a close-in hot Jupiter.

Radio Waves from Shocks in Winds Coronal eruptions during large magnetic reconnection events can produce *coronal mass ejections (CMEs)* that form shocks which travel through the interplanetary space. Particles accelerated in these shock fronts induce radio radiation at the plasma frequency (or its harmonic) and could therefore be detected. This radiation is well known as Type-II radio emission in the solar corona. In very active stars, the CME mass loss rate maybe an important contributor to the total mass loss rate (Khodachenko et al. 2007), but observationally stellar Type-II bursts remain elusive (Crosley and Osten 2018).

X-Ray Emission A hot ($\approx 10^6$ K) ionized wind produces continuum and line emission in the X-ray regime, like stellar coronae but presumably at much lower flux levels. Because thermal X-ray emission scales with $n_{\mathrm{e}}^2 V$ where V is an emitting volume, the observable emission should be stronger for higher wind mass-loss rates (given a fixed velocity), and the emission should come from close to the star. A detection is difficult because the emission from active coronal regions outshines wind emission. Lim and White (1996) estimate $\dot{M} \lesssim 6 \times 10^{-11}\ M_{\odot}\ \mathrm{yr}^{-1}$ for the active mid-M dwarf YZ CMi, but generalizing such estimates to other active stars including detected solar-mass stars (taking the detected X-ray luminosity as an upper limit) results in upper limits in the range of $\dot{M} < 10^{-12} - 10^{-10}\ M_{\odot}\ \mathrm{yr}^{-1}$.

Astrospheric Charge Exchange Neutral hydrogen from the interstellar medium can penetrate the stellar wind region (the "astrosphere") and induce charge exchange (CX) with the heavy wind ions. Such reactions can lead to line emission in the X-rays. This method is very challenging, requiring a nearby star and a very low general X-ray background; it has been applied only once, to the M dwarf Proxima Centauri; Wargelin and Drake (2001) and Wargelin and Drake (2002) estimated the expected emission and specifically investigated high-resolution X-ray images around Prox. Cen for excess CX emission from the astrosphere. They found none, setting an upper limit of $\dot{M} < 3 \times 10^{-13}\ M_{\odot}\ \mathrm{yr}^{1}$ for the total wind mass loss rate.

3.2 Indirect Wind-Detection Methods

Lyα Absorption in the Astrospheric Hydrogen Wall The interaction between the wind-blown ionized "astrosphere" around a star (equivalent to the heliosphere, the latter with a radius of order 120 au) with a partially neutral interstellar medium leads to charge exchange interactions and a pile-up of hot neutral hydrogen in the heliopause region. This "hydrogen wall" produces excess absorption in the stellar Lyα profile that can be related to a wind mass-loss rate based on complex numerical simulation models. More Lyα absorption essentially means higher $\dot{M}$. This method was pioneered by B. Wood an colleagues (see review by Wood 2004) and has succeeded in obtaining $\dot{M}$ values for nearly 20 stars, many of them having masses similar to the Sun's although some M dwarfs were also included. They infer a correlation with the average coronal X-ray surface flux of the form $\dot{M} \propto F_X^{1.34\pm0.18}$ and a wind decay law

$$\dot{M}(t) \propto t^{-2.33\pm0.55}, \tag{10}$$

for G and K-type stars. Interestingly, however, they find only low upper limits for $\dot{M}$ for active, young stars ($\lesssim 700$ Myr for solar analogs). It is unclear if this is related to specific magnetic configurations on the star that lead to wind suppression (e.g., large dipolar fields), or polar outflows instead of spherical winds; the breakdown of winds is problematic for the observed spin-down, however (Johnstone et al. 2015).

Wind-Planet Interactions Stellar winds interact with planetary outer atmospheres and magnetospheres in diverse processes (ion escape, sputtering, charge exchange etc, see Lammer et al. 2003). They also transport shocks and high-energy particles that can ionize and chemically modify upper atmospheres (e.g., Airapetian et al. 2016). The collision between the wind and a planetary magnetosphere can form a detectable shock in front of the planet (Vidotto et al. 2011a). Several estimates have been made for the wind density or mass-loss rates based on observed spectral features. Kislyakova et al. (2014) estimated, from excess Lyα absorption during a transit of the hot Jupiter, $\dot{M} = 4 \times 10^{-14}\ M_\odot\ \mathrm{yr}^{-1}$ for the near-solar analog (F9 V-type) HD 209458. Lecavelier des Etangs et al. (2012) similarly found $\dot{M} \approx 5 \times 10^{-15}\ M_\odot\ \mathrm{yr}^{-1}$ for the cooler K2 V star HD 189733, and Vidotto and Bourrier (2017) derived $\dot{M} \approx 1.2^{+1.3}_{-0.75} \times 10^{-15}\ M_\odot\ \mathrm{yr}^{-1}$ for the M2.5 dwarf GJ 436.

Slingshot Prominences Jardine and Collier Cameron (2019) argued that prominences seen in absorption features when they transit in front of an active star can be signs of episodic mass loss like a wind. These prominences predominantly form at magnetic loop tops located around the co-rotation radius. Depending on the relative location of the Alfvén radius, these cooling loop-top condensations can break open and leave the corona as a wind. Extrapolation of the observed features to the entire star leads to estimates of $\dot{M}$ that seem to continue the trend for H wall estimates (see above), reaching $\dot{M}$ per unit surface area up to 3000 times the solar value. This study included three solar-type stars with masses of 0.82, 1.0, and 1.16 $M_\odot$ and two lower-mass M dwarfs.

Accretion Contamination in White Dwarf Atmospheres This method is mentioned for completeness although it has focused on M dwarf winds. While white dwarf (WD) upper atmospheres should be made of pure H and He, contamination by heavier elements detected spectroscopically has been ascribed to accretion of interstellar gas. For close WD+M star binaries, Debes (2006) argued that the M dwarf wind accretes onto the WD, producing an equilibrium excess abundance of elements like Ca. Observations of several WD+M pairs would imply $\dot{M} \approx 10^{-16} - 6 \times 10^{-15}\ M_\odot\ \mathrm{yr}^{-1}$ for the M dwarfs.

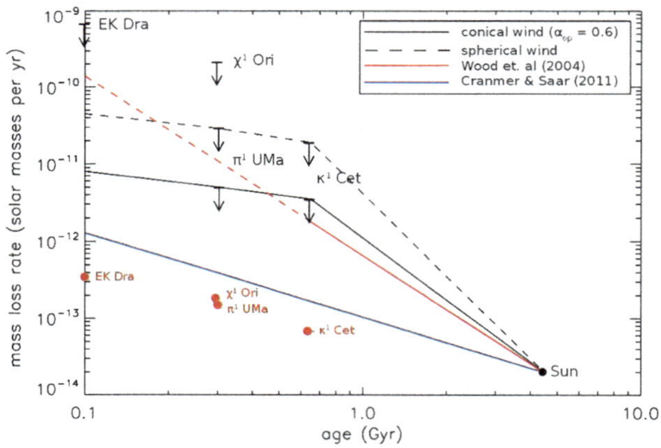

Fig. 2 Mass-loss evolution for solar-type stars. The two black solid lines are upper limit estimates for the $\dot{M}$ evolution based on non-detections of stellar wind radio bremsstrahlung (Sect. 3.1; arrows indicate corresponding $\dot{M}$ upper limits). The upper black dashed line refers to a spherical wind, the lower solid black line to a conical wind with an opening angle of 40 degrees. The red circles are mass-loss estimates from the spin-down model of Johnstone et al. (2015) (Sect. 2), while the red line shows the fit from the Lyα absorption model of Wood (2018) (Sect. 3.2; the dashed part is where Wood 2018 report a breakdown of the relation following observations). The blue solid line relates to the theoretical model of Cranmer and Saar (2011). (From Fichtinger et al. 2017.)

3.3 Conclusions for Wind Mass-Loss Rates

Direct wind-detection methods use straightforward interpretation of observational data of emission of and absorption by the ionized stellar wind. Such effects tend to be weak so that essentially all observations have provided upper limits to $\dot{M}$. The most severe problem is competing and much brighter similar emission from the star itself. The upper limits from such observations are mostly in the range of $\dot{M} \approx 10^{-13} - 10^{-10}\ M_\odot\ \mathrm{yr}^{-1}$, and this includes solar-mass stars.

Indirect wind measurements use features that often "amplify" the evidence of a wind. The disadvantage of these methods is their need for complex models relating the observed features to the putative stellar winds. Such measurements have nevertheless been more successful in providing estimates for $\dot{M}$, mostly in the range of $\dot{M} \approx 10^{-16} - 10^{-12}\ M_\odot\ \mathrm{yr}^{-1}$.

Collectively, the $\dot{M}$ estimates suggest a correlation between $\dot{M}$ and stellar activity, and therefore also rotation and age. The value of $\dot{M}$ most likely decays by 1–2 orders of magnitude during the evolution of a solar analog from a very active, early "Sun" with rapid rotation to a star like the present-day Sun (Fig. 2).

4 Stellar High-Energy Radiation

All late-type main-sequence stars are X-ray emitters because of the presence of a hot, million-degree tenuous corona akin to the solar corona. The field of stellar coronal X-ray astronomy was opened after the first detailed study of the solar X-ray corona with Skylab, and the first stellar X-ray surveys with the *Einstein* satellite (Catura et al. 1975). It was soon realized that the Sun was among the weakest late-type stellar X-ray emitters in absolute terms, and that X-ray luminosities comprised at least the range of $10^{27} - 10^{30}\ \mathrm{erg\,s^{-1}}$ and temperatures from the solar value, ~ 2 MK, to at least 10–20 MK.

As already mentioned in Sect. 1, activity is correlated with rotation and therefore also with age. This is due to the operation of a dynamo that amplifies magnetic fields in a process involving differential rotation and convection. These magnetic fields, in turn, can build up energy that they release in coronal heating, coronal mass ejections, and probably in wind heating and acceleration. This feedback loop leads to a convergence of the rotation period, and hence activity, about 700 Myr after the arrival on the ZAMS for a solar analog, regardless of the initial conditions. Before that age, activity evolution tracks are non-unique. The key relations are the following.

1) *Activity-rotation relation.* X-ray activity indicators such as the X-ray luminosity $L_{\rm X}$ (for a given stellar mass M_*), the X-ray surface flux, $F_{\rm X}$, or the ratio between total X-ray luminosity and bolometric luminosity $L_{\rm bol}$ scale with the stellar rotation period P_* roughly as

$$L_{\rm X,M_*}, F_X, \frac{L_X}{L_{\rm bol}} \propto \Omega_*^2 \propto P_*^{-2} \tag{11}$$

(e.g., Pallavicini et al. 1981; Pizzolato et al. 2003; Wright et al. 2011) but this holds only up to a threshold Ω_* above which these activity indicators saturate, i.e., become independent of Ω_*. For solar analogs, this occurs at $\Omega_{\rm sat} \approx 15\ \Omega_\odot$ (Johnstone et al. 2020).

2) *Rotation-age relation.* This was already discussed in Sect. 2. Skumanich (1972) proposed $\Omega \propto t^{-1/2}$ while Ayres (1997) found, for the equatorial rotation velocity v of solar analogs,

$$v \propto t^{-0.6} \tag{12}$$

and Johnstone et al. (2015) derived

$$\Omega_* \propto t^{-0.57} \tag{13}$$

for solar-type stars.

3) *Activity-age relation.* This follows from the above two relations as, for example, $L_{\rm X}/L_{\rm bol} \propto t^{-1.2}$. Statistics from observations imply, for solar analogs,

$$L_{\rm X} \approx (3 \pm 1) \times 10^{28} t^{-1.5\pm0.3}\ {\rm erg\,s^{-1}} \tag{14}$$

(Maggio et al. 1987; Güdel et al. 1997). Such relations would be important for research on the evolution of planetary atmospheres given the influence of X-rays on upper-atmospheric heating, ionization, chemistry, and loss into space. However, there are two important caveats to mention: i) Eq. (14) does not take saturation into account. ii) While approximately correct for ages exceeding ∼700 Myr for solar analogs, the extrapolation to younger stars is invalid given the wide range of initial Ω_* values observed after star formation. Eq. (14) must therefore not be used for solar-type stars with ages less than about 700 Myrs. Similar constraints apply to K and M dwarfs (Johnstone et al. 2015).

By combining rotational evolution models with Eq. (11), we find a range of evolutionary tracks for $L_{\rm X}$ (or the EUV luminosity $L_{\rm EUV}$), as illustrated in Fig. 3. The main points to emphasize are these: i) There is a wide spread of X-ray luminosities between the 10th and the 90th percentile of the initial rotation period distribution, starting at ages of 10–20 Myr and ending only at ages of ∼1 Gyr, when rotation converges. ii) The fastest rotators reside at an $L_{\rm X}$ saturation level for several 100 Myr, while the slowest rotators reach moderate X-ray luminosities already at a few tens of Myr. iii) Within the first few 100 Myr, $L_{\rm X}$ values are distributed across a factor of ~ 40 in $L_{\rm X}$, implying very different atmospheric evolution scenarios for planetary atmospheres due to the different $L_{\rm X}$ evolutionary tracks.

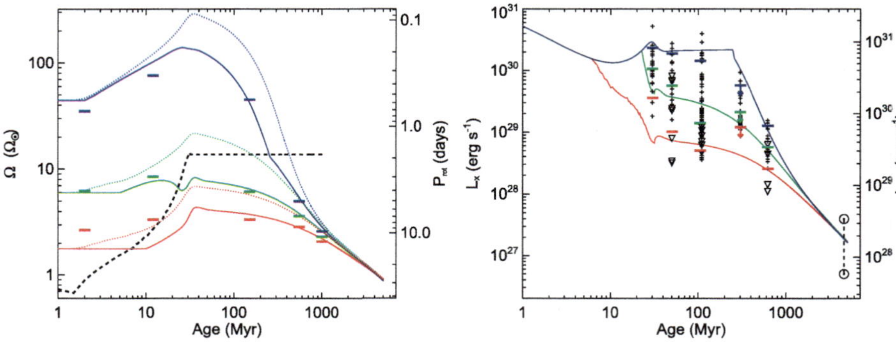

Fig. 3 Left (**a**) Predicted rotational evolution tracks for stars at the 10th (red), 50th (green), and 90th (blue) percentiles of the rotational distribution. The solid and dotted lines show the envelope and core rotational evolution, respectively, and the horizontal solid bars show the observational constraints on the percentiles. The dashed black line shows the time dependent saturation threshold for $\dot{M}$, $B_{\rm dip}$, and $L_{\rm X}$. Right (**b**): Predicted $L_{\rm X}$ along each of the rotation tracks and comparisons to observed $L_{\rm X}$ values of single stars in several clusters with ages $\geq$ 30 Myr (for observational samples at younger ages, see Johnstone et al. 2020), with upper limits shown by $\triangledown$ symbols. The solid horizontal bars show the 10th, 50th, and 90th percentiles of the observed distributions of $L_{\rm X}$ at each age, calculated by counting upper limits as detections. The two solar symbols at 4.5 Gyr show the range of $L_{\rm X}$ for the Sun over the course of the solar cycle. The scale on the right y-axis shows the associated $L_{\rm EUV}$.

Tu et al. (2015) analyzed a large sample of $L_{\rm X}$ measurements from open clusters and compared it with the results implied from rotational evolution (Johnstone et al. 2015). They found, for the 10th (slowest), 50th, and 90th (fastest) percentile in the initial Ω_* distribution, corresponding to $\Omega_0 = 1.8, 6.2$, and $45.6\ \Omega_\odot$, respectively, the evolutionary trends

$$L_{\rm X} = \begin{cases} 2.0 \times 10^{31} t^{-1.12} \\ 2.6 \times 10^{32} t^{-1.42} \\ 2.3 \times 10^{36} t^{-2.50} \end{cases} \quad L_{\rm EUV} = \begin{cases} 7.4 \times 10^{31} t^{-0.96} & \text{10th percentile} \\ 4.8 \times 10^{32} t^{-1.22} & \text{50th percentile} \\ 1.2 \times 10^{36} t^{-2.15} & \text{90th percentile} \end{cases} \tag{15}$$

(t in Myrs) where these approximations apply after the end of saturation at ages of $t_{\rm sat} = 5.7, 23$, and 226 Myrs, respectively. These evolutionary tracks are shown in Fig. 3 together with the corresponding rotation evolution tracks.

5 Summary and Conclusions

The long-term evolution of magnetic fields, magnetic activity, high-energy emission and winds of stars on the main sequence are all intimately related to stellar spin-down through an internal magnetic dynamo. Since spin-down itself is due to removal of angular momentum in a magnetized wind, a feedback loop leads to convergence of rotation periods several hundred Myr after star formation. By implication, the winds and the high-energy radiation also decline with age and also converge to values almost only determined by age for a given stellar mass and ages beyond $\sim$1 Gyr. Some of these trends are observationally well studied while others are inferred from indirect observations. For the rotation rate Ω_*, the wind mass-loss rate $\dot{M}$, and the X-ray luminosity $L_{\rm X}$, the following trends are found (mostly for solar-type stars):

$$\Omega(t) \propto t^{-0.57} \quad \text{for } t > 700 \text{ Myr, from spin-down}$$
$$\dot{M}(t) \propto t^{-1.75} \quad \text{for } t > 700 \text{ Myr, from spin-down}$$
$$\dot{M}(t) \propto t^{-2.33} \quad \text{for } t > 700 \text{ Myr, from Ly}\alpha \text{ absorption}$$
$$L_X(t) \propto t^{-1.12} \quad \text{for } t > 5.7 \text{ Myr, for slow rotator}$$
$$L_X(t) \propto t^{-1.42} \quad \text{for } t > 23 \text{ Myr, for intermediate rotator}$$
$$L_X(t) \propto t^{-2.50} \quad \text{for } t > 226 \text{ Myr, for fast rotator.}$$

It should be understood that these are best-fit relations from various methods and observing data sets; the uncertainties in the above exponents may typically be of order several times 0.1. As indicated earlier, for a solar analog the age dependence for ages $\lesssim$700 Myr is non-unique as indicated above for L_X; this is equally true for Ω and $\dot{M}$ as described in Sect. 2 and 3 but the relations above are given only for the ages $\gtrsim$700 Myr.

These relations are important to study atmospheric evolution of planets, but even more important is a correct treatment of the evolution at ages $\lesssim$700 Myr for a solar analog when evolutionary tracks are non-unique and the initial conditions after the dispersal of the protoplanetary disk need to be taken into account as well.

Further high-energy properties of a star may matter. In particular, high-energy flares add considerably to the X-ray output of a star or may even be the principal contributors to such emission. Audard et al. (2000) found that the extreme-ultraviolet flare rate correlates linearly with the activity level represented by the average L_X. Also, coronal mass ejections (CMEs) can complement if not dominate the mass-loss rate of a star although extrapolation of solar trends lead to excessive CME mass-loss rates, suggesting that CMEs must be suppressed at high activity levels (Drake et al. 2013).

Acknowledgement I would like to thank the organizers of the ISSI workshop and the staff of ISSI for a pleasant meeting in a relaxed atmosphere, and for giving me the opportunity to present this overview. The article profited from helpful comments made by two anonymous referees. This research has been supported by the Austrian Science Fund FWF through project S116 *Pathways to Habitability: From Disks to Active Stars, Planets and Life* and the related subproject S11604 *Radiation & Wind Evolution from the T Tauri Phase to ZAMS and Beyond.*

Funding Note Open access funding provided by Austrian Science Fund (FWF).

References

V.S. Airapetian, A.V. Usmanov, Astrophys. J. **817**, L24 (2016)

V.S. Airapetian, A. Glocer, G. Gronoff et al., Nat. Geosci. **9**, 452 (2016)
M. Audard, M. Güdel, J.J. Drake, V.L. Kashyap, Astrophys. J. **541**, 396 (2000)
T.R. Ayres, J. Geophys. Res. **102**, 1641 (1997)
L. Biermann, Z. Astrophys. **29**, 274 (1951)
T.R. Burnight, Phys. Rev. A **76**, 165 (1949)
R.C. Carrington, Mon. Not. R. Astron. Soc. **20**, 13 (1859)
R.C. Catura, L.W. Acton, H.M. Johnson, Astrophys. J. **196**, L47 (1975)
S.R. Cranmer, S.H. Saar, Astrophys. J. **741**, 54 (2011)
M.K. Crosley, R.A. Osten, Astrophys. J. **862**, 113 (2018)
J.H. Debes, Astrophys. J. **652**, 636 (2006)
J.G. Doyle, M. Mathioudakis, Astron. Astrophys. **241**, L41 (1991)
S.A. Drake, T. Simon, A. Brown, Astrophys. J. **406**, 247 (1993)
J.J. Drake, O. Cohen, S. Yashiro, N. Gopalswamy, Astrophys. J. **764**, 170 (2013)
B. Edlén, Z. Astrophys. **22**, 30 (1942)
B. Fichtinger, M. Güdel, R.L. Mutel et al., Astron. Astrophys. **599**, A127 (2017)
E.J. Gaidos, M. Güdel, G.A. Blake, Geophys. Res. Lett. **27**, 501 (2000)
F. Gallet, J. Bouvier, Astron. Astrophys. **556**, A36 (2013)
W. Grotrian, Naturwissenschaften **27**, 214 (1939)
M. Güdel, E.F. Guinan, S.L. Skinner, Astrophys. J. **483**, 947 (1997)
G.E. Hale, Astrophys. J. **28**, 315 (1908)
J.S. Hey, Nature **157**, 47 (1946)
R. Hodgson, Mon. Not. R. Astron. Soc. **20**, 15 (1859)
M. Jardine, A. Collier Cameron, Mon. Not. R. Astron. Soc. **482**, 2853 (2019)
C.P. Johnstone, M. Güdel, I. Brott, T. Lüftinger, Astron. Astrophys. **577**, A28 (2015)
C.P. Johnstone, M. Bartel, M. Güdel (2020, in press). Astron. Astrophys. arXiv:2009.07695
M.L. Khodachenko, I. Ribas, H. Lammer et al., Astrobiology **7**, 167 (2007)
K.G. Kislyakova, M. Holmström, H. Lammer et al., Science **346**, 981 (2014)
R.P. Kraft, Astrophys. J. **150**, 551 (1967)
H. Lammer, H.I.M. Lichtenegger, C. Kolb et al., Icarus **165**, 9 (2003)
A.Lecavelier des Etangs, V. Bourrier, P.J. Wheatley et al., Astron. Astrophys. **543**, L4 (2012)
J. Lim, S.M. White, Astrophys. J. **462**, L91 (1996)
J. Lim, S.M. White, O.B. Slee, Astrophys. J. **460**, 976 (1996)
A. Maggio, S. Sciortino, G.S. Vaiana et al., Astrophys. J. **315**, 687 (1987)
E.E. Mamajek, L.A. Hillenbrand, Astrophys. J. **687**, 1264 (2008)
S.P. Matt, K.B. MacGregor, M.H. Pinsonneault, T.P. Greene, Astrophys. J. **754**, L26 (2012)
D.J. Mullan, J.G. Doyle, R.O. Redman, M. Mathioudakis, Astrophys. J. **397**, 225 (1992)
F.M. Olnon, Astron. Astrophys. **39**, 217 (1975)
R. Pallavicini, L. Golub, R. Rosner et al., Astrophys. J. **248**, 279 (1981)
N. Panagia, M. Felli, Astron. Astrophys. **39**, 1 (1975)
E.N. Parker, Astrophys. J. **128**, 664 (1958)
N. Pizzolato, A. Maggio, G. Micela et al., Astron. Astrophys. **397**, 147 (2003)
A. Skumanich, Astrophys. J. **171**, 565 (1972)
D.R. Soderblom, J.R. Stauffer, K.B. MacGregor, B.F. Jones, Astrophys. J. **409**, 624 (1993)
L. Tu, C.P. Johnstone, M. Güdel, H. Lammer, Astron. Astrophys. **577**, L3 (2015)
G.H.J. van den Oord, J.G. Doyle, Astron. Astrophys. **319**, 578 (1997)
A.A. Vidotto, V. Bourrier, Mon. Not. R. Astron. Soc. **470**, 4026 (2017)
A.A. Vidotto, J.-F. Donati, Astron. Astrophys. **602**, A39 (2017)
A.A. Vidotto, M. Jardine, C. Helling, Mon. Not. R. Astron. Soc. **414**, 1573 (2011a)
A.A. Vidotto, S.G. Gregory, M. Jardine et al., Mon. Not. R. Astron. Soc. **441**, 2361 (2014)
B.J. Wargelin, J.J. Drake, Astrophys. J. **546**, L57 (2001)
B.J. Wargelin, J.J. Drake, Astrophys. J. **578**, 503 (2002)
E.J. Weber, L. Davis Jr., Astrophys. J. **148**, 217 (1967)
B.E. Wood, Liv. Rev. Solar Phys. **1**, 2 (2004)
B.E. Wood, Dissipative and heating processes in collisionless plasma: the solar corona, the solar wind, and the interstellar medium. J. Phys. Conf. Ser. **1100**, 012028 (2018)
A.E. Wright, M.J. Barlow, Mon. Not. R. Astron. Soc. **170**, 41 (1975)
N.J. Wright, J.J. Drake, E.E. Mamajek, G.W. Henry, Astrophys. J. **743**, 48 (2011)

Space Sci Rev (2020) 216:133
https://doi.org/10.1007/s11214-020-00748-w

The NC-CC Isotope Dichotomy: Implications for the Chemical and Isotopic Evolution of the Early Solar System

Katherine R. Bermingham[1,2] · Evelyn Füri[3] · Katharina Lodders[4] · Bernard Marty[3]

Received: 14 October 2019 / Accepted: 4 October 2020 / Published online: 17 November 2020

Abstract Understanding the formation of our planetary system requires identification of the materials from which it originated and the accretion processes that produced the planets. The compositional evolution of the solar system can be constrained by synthesizing astronomical datasets and numerical models with elemental and isotopic compositions from objects that directly sampled the disk: meteorites and their constituents (chondrules, refractory inclusions, and matrix). This contribution reviews constraints on early solar system evolution provided by the so-called non-carbonaceous (NC) and carbonaceous chondrite (CC) groups and their relationship to the volatile element characteristics of chondritic meteorites. In previous work, the NC or CC character of a parent body was used to infer its accretion location in the protoplanetary disk. The NC groups purportedly originated in the inner disk, and the CC groups were derived from the outer disk, where the NC and CC regions of the disk may have been separated early on by proto-Jupiter, a pressure maximum, or a dust trap in the disk. The tenet that all CC parent bodies accreted in the outer disk is, in part, based on evidence that a handful of CC meteorites are enriched in volatile species compared to NC meteorites. Here, it is reviewed if and how the volatile element and nucleosynthetic isotope compositions of meteorites can be linked to accretion locations within the disk. The nucleosynthetic isotope compositions of whole rock meteorite samples contrast the trends found for their major volatile element compositions (i.e., C, N, and O). Although there may be an increase in volatile abundances when comparing some stony NC and CC meteorites and their inferred accretion locations within the disk, this is not necessarily a general rule. The difficulties with inferring parent body accretion locations are discussed. It is found that it cannot always be

Reading Terrestrial Planet Evolution in Isotopes and Element Measurements
Edited by Helmut Lammer, Bernard Marty, Aubrey L. Zerkle, Michel Blanc, Hugh O'Neill and Thorsten Kleine

✉ K.R. Bermingham

1 Department of Earth and Planetary Sciences, Rutgers University, Piscataway, NJ 08854, USA

2 Department of Geology, University of Maryland, College Park, MD 20742, USA

3 Université de Lorraine, CNRS, CRPG, 54000 Nancy, France

4 Department of Earth and Planetary Sciences and McDonnell Center for the Space Sciences, Washington University in St. Louis, St. Louis, MO 63130, USA

assumed that parent bodies which formed in the CC reservoir are "volatile-rich" relative to those that formed in the NC reservoir which are "volatile-poor". Consequently, tracing the origin of terrestrial volatiles using the NC-CC isotope dichotomy remains challenging.

Keywords NC-CC · Volatiles · Nucleosynthetic isotope anomalies · Meteorites · Protoplanetary disk evolution

1 Introduction to Solar System Formation

The solar system formed from the solar nebula, which was a segment of a molecular cloud made of interstellar gas and dust particles that had spun into a disk around the forming Sun. This includes dust from stellar sources that was added to the interstellar medium at least $\sim 3 \pm 2$ Ga[1] before the birth of the solar system (e.g., Heck et al. 2020). The solar nebula separated from the main molecular cloud mass due to gravitational collapse of the cloud which was caused by a sufficient weakening of magnetic field support following ambipolar diffusion (Shu et al. 1987) or by shockwaves from nearby supernova (Cameron and Truran 1977; Boss 1995). Due to conservation of angular momentum, the solar nebula formed a rotating disk (generally referred to as the protoplanetary disk) around the central protostar. During the earliest stage of the protoplanetary disk, angular momentum was transported outward, while significant mass was transported inward and accreted onto the protostar or was lost by FUV photoevaporation (e.g., Williams and Cieza 2011).

The oldest preserved solids thought to have condensed first in the protoplanetary disk are calcium-aluminum-rich inclusions (CAIs), which have an average absolute U-corrected Pb-Pb age of $4{,}567.30 \pm 0.16$ Ma (Connelly et al. 2012). This age is the cosmochemical start (t_0) of the solar system. Within <1 Ma of CAI formation, planetesimals began to accrete throughout the protoplanetary disk (e.g., Kruijer et al. 2014; Sugiura and Fujiya 2014). Some planetesimals had sufficient concentrations of heat-producing radioactive isotopes (primarily ^{26}Al) to cause melting, differentiation, and formation of a metallic core and a silicate mantle and crust (Lee et al. 1977; Shukolyukov and Lugmair 1993a, 1993b; Hevey and Sanders 2006; Sahijpal et al. 2007; Moskovitz and Gaidos 2011). Collisional accretion and subsequent oligarchic growth of planetesimals gave rise to the evolution of planetesimals into a planetary system (for a recent review, see Walsh and Levison 2019).

Presently, a gross zonation in chemistry and mass is exhibited by the planets, where the outer giant planets (Jupiter, Saturn, Uranus, Neptune) are gas- and ice-rich,[2] whereas the inner rocky planets (Mercury, Venus, Earth, Mars) are relatively depleted in highly-volatile and volatile elements. In part, the chemical gradient is considered to be a consequence of clearing gas from the inner solar system by early solar activity, rapid growth of the giant planets, and incomplete condensation from the protoplanetary disk (for a review, see Taylor 2001; Lodders and Fegley 2011; Palme et al. 2014; Cieza et al. 2016, and references therein). The ice to rock ratio favors a larger mass of ice in cooler regions in planetary accretion disks

[1]The IUPAC-IUGS convention for expressing dates following recommendation by Holden et al. (2011) is that annus (a) is used for one year as a unit of time, both for absolute time and time differences. Hence, considering the declensions, it is Ga = Giga annis (ablative for age; used as appellative) or Ga = Giga annos (pl. accusative, for how long ago).

[2]The enrichments of C, N, and O (relative to solar) observed in outer planet atmospheres were caused by the accretion of ices bearing these highly volatile elements. Highly volatile elements have 50% condensation temperatures below 371 K, volatile elements below 665 K, moderately volatiles between 1335 and 665 K, and refractories above 1335 K (for a gas of solar composition at a total pressure of 10^{-4} bar; Lodders 2003). See last chapter in this edition by K. Lodders for elemental data for the Sun and meteorites.

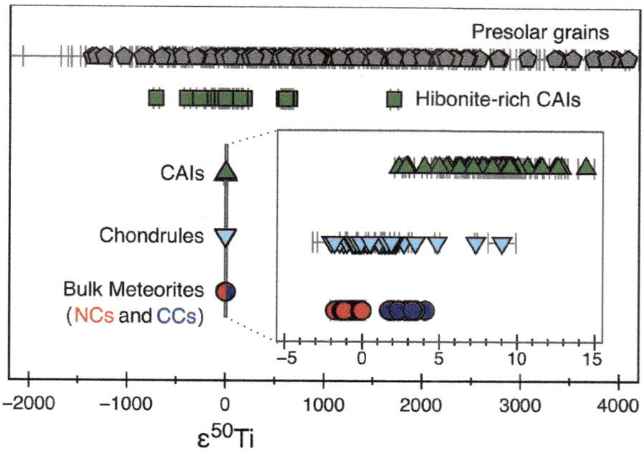

Fig. 1 Variation in $^{50}Ti/^{47}Ti$, expressed in ε units (parts per 10,000; see Sect. 3 for details), in presolar grains, CAIs, chondrules, and "bulk" CC- and NC-type meteorites (chondrites and achondrites). The inset illustrates that CAIs, chondrules, and bulk meteorites record small $\varepsilon^{50}Ti$ variations (<20 ε units) compared to presolar grains ($\geq$5000 ε units, corresponding to 500 ‰ in δ units). Error bars are smaller than symbol sizes in most cases, except for presolar grains. Modified from Torrano et al. (2019) with data from the literature (Alexander and Nittler 1999; Amari et al. 2001; Davis et al. 2018; Gerber et al. 2017; Gyngard et al. 2018; Hoppe et al. 1994; Huss and Smith 2007; Ireland et al. 1991; Kööp et al. 2016; Niemeyer 1988; Render et al. 2019; Torrano et al. 2019; Trinquier et al. 2009; Zhang et al. 2012; Zinner et al. 2007)

that enables rapid growth of gas-giant planets and provides the critical mass necessary to allow efficient gravitational capture of nebular gas (e.g., Wasson 1988). This provides an explanation for the H- and He-rich atmospheres of the outer planets and their much larger masses than the terrestrial planets. Observations of nearby planetary systems indicate that these planetary forming processes are not unique to our solar system (e.g., Cieza et al. 2016; Kudo et al. 2018).

Despite understanding the generalities of planet formation, important questions remain. For example, what was the composition of the disk and how did it evolve? To what extent are the different stages of disk evolution and planetary accretion recorded by the composition of the planets, asteroids, and meteorites? Can these cosmochemical materials be used to accurately reconstruct the formation of the solar system? What was the volatile element distribution throughout the disk and how did it change with planet growth? What was the source of water and other volatiles to Earth?

Insights to these questions come from analyzing material that directly sampled the protoplanetary disk: chondritic, achondritic, and iron meteorites. Recent developments in analytical instrumentation have made it possible to determine the subtle (on the parts per million scale) isotopic variations between bulk meteorites and their individual components (Figs. 1 and 2).

Here, isotopic variations are defined as isotopic deviations from a terrestrial composition.[3] In parallel, attempts have been made to relate meteorite isotopic and volatile composi-

[3]Typically, isotope *ratios* are measured, and meteorite samples with an excess or depletion of a given isotope will have a higher or lower (respectively) isotope ratio than the terrestrial reference value. Positive or negative isotope anomalies arise from the definition of the isotope notation, which is the difference between the measured sample ratio from that of the terrestrial standard value in percent (%), permil (‰), on a scale of parts per 10,000 (ε units), or in parts per 1,000,000 (μ units). The advantage of using the differences in the ratios is that these scales are easier to work with, especially for samples with small isotope anomalies (bulk meteorites, chondrules, CAIs). Terrestrial materials are generally defined as 'normal' (or equal to zero) in isotopic composition.

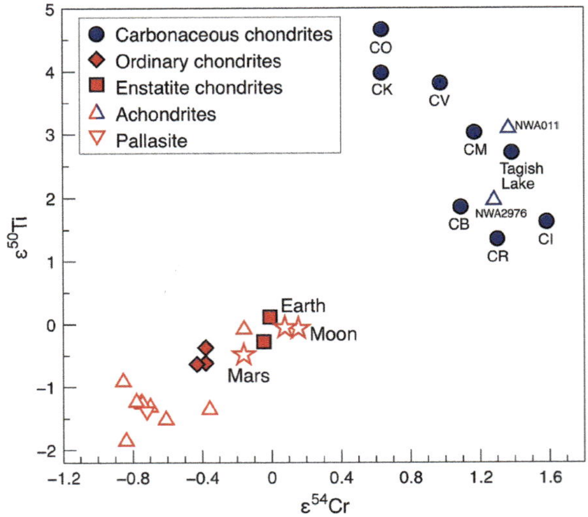

Fig. 2 Variation in $^{50}Ti/^{47}Ti$ and $^{54}Cr/^{52}Cr$, expressed in ε units (see Sect. 3 for details), between undifferentiated (filled symbols) and differentiated meteorites (open symbols) in comparison to Earth, Moon, and Mars (star symbols). CC- and NC-type bodies are represented in blue and red, respectively. Modified from Warren (2011) and Scott et al. (2018), and data sources therein

tions to their parent body's accretion location in the protoplanetary disk, and to use models to reconstruct the architecture of the evolving protoplanetary disk (e.g., Warren 2011; Kruijer et al. 2017). This approach relies on the assumptions that: (1) the terrestrial meteorite collection is representative of the entire or a substantial population of planetesimals present during accretion; (2) the elemental and isotope composition of a meteorite group faithfully represents the composition of the disk region from which its parent body(ies) accreted; (3) the volatile content of meteorites accurately records the volatile content of the parent body and the disk region it sampled; (4) the combination of a meteorite's isotope composition and the volatile element content can trace the communication between different regions of the disk.

This contribution reviews work that uses the composition of meteorites to deduce the evolution of the protoplanetary disk from dust to the planetary system observed today. It begins with a brief review of meteorites and their components. This is followed by an outline of isotopic variations in meteorites, meteorite volatile content, parent body accretion locations, and the applicability of these data to reconstructing the evolution of the disk and, finally, the accretion history of Earth. We present several old and new ideas without claiming to be comprehensive. Our aim is to make the reader aware that, despite the recognition of the presence of a dichotomy in isotopic composition, a synthesis of perspectives and data to produce a unique explanation of the origin of all elemental and isotopic peculiarities is still under development. This state is reflected in the primary literature, which often focuses on subsets of specific isotopic systems and/or particular theoretical approaches but does not always include a rigorous assessment of existing and/or other new ideas. In time, a clearer picture will likely emerge when sample-based data are more completely understood and integrated with astronomical datasets and numerical models.

2 Meteorites and Their Components

Most meteorites originate from the asteroid belt,[4] which is a low mass region (representing $\sim$4% of the mass of the Moon) of the solar system between the orbits of Mars and Jupiter.

[4] Additional sources of achondrites to Earth include Mars and the Moon. The meteorites referred to here do not include micrometeorites.

The asteroid belt provides the most direct sampling of the protoplanetary disk via meteorites delivered to Earth. The belt has a gross chemical stratification, with an inner region that is dominated by silicaceous (S-type) asteroids and an outer region dominated by dark carbonaceous (C-type) asteroids. While some material in the asteroid belt presumably accreted *in situ*, the diversity of meteorite compositions (ranging from dry, igneous rocks to 'comet-like' ice-rock mixtures) indicates that the asteroid belt received material originating from the "inner" and "outer" solar system. Here, the material inboard of Jupiter is defined to come from the inner solar system and material outboard of Jupiter originated from the outer solar system (following Warren 2011). It has been proposed that some asteroids were scattered inwards and outwards during the growth and/or migration of the gas giants (Walsh et al. 2011; Warren 2011; Morbidelli et al. 2015; for a recent review, see Raymond et al. 2020). Hence, the present distribution of asteroids is unlikely reflective of their original accretion location.

Meteorites are from either melted (achondrites, stony irons, and irons) or unmelted, 'primitive' (chondrites) parent bodies. 'Parent body' refers to either a planet, moon, or asteroid. Achondrites and iron meteorites sample the silicate-dominated crust/mantle and metallic core, respectively, of bodies that experienced large-scale melting and differentiation. It should be noted that iron meteorites are not chemically or isotopically linked to known achondrites, which suggests that the silicate mantles of iron meteorite parent bodies are "missing" and might have been destroyed by impacts in the early solar system. Also, there is a lack of achondrites with the expected mineralogy and chemical composition of mantles of differentiated planetesimals (except for martian and lunar meteorites).

Main group pallasites (so-called stony irons, composed of olivine and Fe-Ni-rich metal) had long been thought to represent the transition zone between the core and mantle of differentiated asteroids (Scott 1977). The highly siderophile element abundances of pallasites are consistent with late crystallization of a metallic core, as expected near the core-mantle boundary (Scott 1977; Davis 1977). Updated metallographic cooling rates and chemical compositions, however, indicate an impact origin for these meteorites (Yang et al. 2007, 2010). Recent high strain-rate deformation experiments were used to identify the process of pallasite formation, and results indicated that pallasites could form via two-stages: (1) inefficient core-mantle differentiation, and (2) an impact event (Walte et al. 2020).

Chondrites sample planetesimals that have not experienced wholesale melting and differentiation. As such, they are often referred to as "primitive" meteorites, and their compositions may reflect the composition of the disk material from which they accreted. This does not mean they record the exact state of the primitive condensable materials that were present in the solar nebula because all chondrites underwent metamorphic (aqueous and/or thermal) processing on their assembled parent bodies. Nonetheless, compared to planets, satellites, and achondrites, chondrites represent the least altered materials from the solar nebula which can be studied.

Based on mineralogical, chemical, and isotopic differences, chondrites are divided into three major classes: carbonaceous, ordinary, and enstatite chondrites (Fig. 2), and these classes are further divided into chondritic groups (Fig. 3). There are substantial mineralogical and petrographic characteristics as well as whole-rock chemical and O-isotopic compositions which support the interpretation that chondrite groups sample separate parent bodies (for review, see Weisberg et al. 2006). Spectroscopic observations link enstatite/ordinary and carbonaceous chondrites to asteroids in the inner and outer belt, respectively, (e.g., Morbidelli et al. 2012, and references therein). Vernazza et al. (2014, 2016) and Patzer and Schultz (2002), however, questioned if each chondrite group represents a single distinct parent body.

Chondritic meteorites consist of three main components: refractory inclusions (CAIs, hibonite-rich CAIs, and amoeboid olivine aggregates), chondrules, and matrix (e.g., Hezel

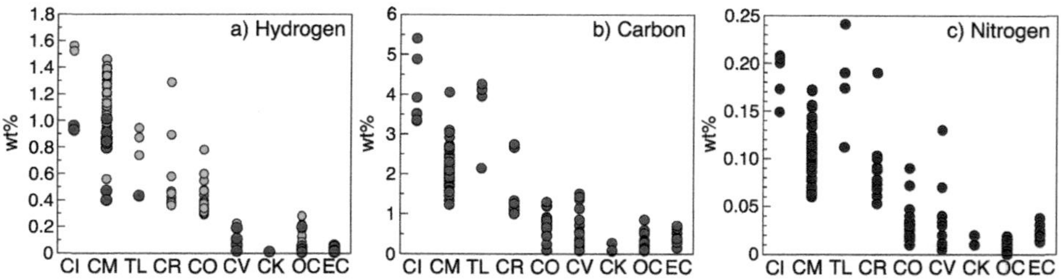

Fig. 3 (**a**) Hydrogen, (**b**) carbon, and (**c**) nitrogen content (in wt%) of carbonaceous (CI, CM, Tagish Lake (TL), CR, CO, CV, CK), ordinary (OC), and enstatite chondrites (EC). Hydrogen abundances from Vacher et al. (2020) (dark green) are lower, in most cases, than those from Alexander et al. (2012, 2018) (light green) as a result of pre-degassing and removal of atmospheric contamination under vacuum at 120 °C for 48 h before measurement. Hydrogen abundances of ECs are from Piani et al. (2020). Carbon and nitrogen concentrations are from Alexander et al. (2012, 2018), Grady et al. (1986), Hashizume and Sugiura (1995), and Pearson et al. (2006)

et al. 2019; Krot 2019). Recent advances have established a chronology of formation of CAIs and chondrules in the context of plausible formation models that account for their chemical and isotopic compositions. Evidence from short- and long-lived isotopes, oxygen isotopes, nucleosynthetic isotopic anomalies, and petrologic studies suggest that CAIs were the first solids to have formed in the protosolar disk (for a review, see MacPherson 2014). The oxygen isotopic compositions of CAIs suggest that they formed in regions that were predominantly solar-like and ^{16}O-rich compared to Earth (e.g., MacPherson 2014). Thermochemical modelling of the condensation of refractory inclusions indicates that the main minerals in CAIs condensed in a low pressure, high temperature (>1300 K) environment from a solar composition gas (which had an oxygen fugacity several orders of magnitude below the iron-wüstite buffer, IW) (Grossman and Larimer 1974; Lodders 2003). There are, however, indications from trace element abundance patterns that the oxygen fugacity during condensation was variable (e.g., Fegley and Palme 1985; Rubin et al. 1988).

Chondrules, the dominant component in most chondrites, are up to millimeter-sized silicate-rich spherules that had been partially or completely molten prior to incorporation into a planetesimal (Scott and Krot 2005). They preserve mineral compositions and textures that indicate crystallization within minutes to hours (Krot et al. 2018 and references therein). ^{26}Al-^{26}Mg systematics indicate that most chondrules formed 1-3 Ma after CAIs (e.g., Villeneuve et al. 2009; Pape et al. 2019). This conclusion is contrary to Pb-Pb ages that indicate some chondrules formed contemporaneously with CAIs which would imply that chondrules record a heterogeneous distribution of the short-lived ^{26}Al nuclide in the protoplanetary disk (Connelly and Bizzarro 2017). The chemical and mineralogical characteristics of chondrules indicate that they formed under diverse physico-chemical conditions in the nebula that were generally more oxidizing, at higher total pressures and/or higher dust:gas ratios than the CAI-forming region.

Chondrite matrices are volatile-rich compared to refractory inclusions and chondrules. Matrices are characterized by fine-grained mixtures of materials with diverse origins in the protoplanetary disk. This fine-grained mixture can be annealed during parent body thermal metamorphism. In objects that have not undergone thermal metamorphism, the matrix generally shows signs of aqueous alteration, e.g., from the presence of secondary mineral phases such as carbonates, sulfates, and phyllosilicates (for a review, see Brearley 2006). Matrices also house primary constituents that formed in the disk, including broken fragments of high-temperature components (e.g., CAIs and chondrules) and low-temperature phases (e.g., organic materials).

Chondrite matrices can also preserve presolar grains and insoluble/refractory organics (e.g., Lodders and Amari 2005; Zinner 2014; Floss and Haenecour 2016; Nittler and Ciesla 2016; Takigawa et al. 2018). Presolar grains formed in stellar outflows or ejecta and remained intact throughout their journey into the solar system where they were preserved in the disk (Lodders and Amari 2005). Since presolar grains incorporate the dominant isotope products of nucleosynthetic reactions operating in their parent star, their isotope compositions are highly variable and well outside the range found in solar system materials (Fig. 1). Presolar grains exhibit diverse mineralogy and include SiC, graphite, carbides or metal grains in graphite, Si_3N_4, corundum, spinel, hibonite, TiO_2, and amorphous and crystalline silicates (for review, see Zinner 2014; Floss and Haenecour 2016; Nittler and Ciesla 2016). Preservation of these grains in some meteorites is a testament to their chemical robustness and ability to withstand the process of disk formation and evolution, and parent body accretion and alteration. Differentiated planetesimals presumably formed from material which included presolar grains; however, these grains are not preserved in achondrites and irons because the extensive processes of heating and melting (and potentially the low oxygen fugacity) destroyed them.

3 Nucleosynthetic Isotope Variations in the Protoplanetary Disk

Historically, the protoplanetary disk was considered to be a hot, compositionally homogeneous disk of dust and gas (e.g., Urey 1952). The understanding of the composition of the disk changed considerably with the study of meteorites. The discovery of isotopic variations in xenon (Reynolds and Turner 1964), neon (Black and Pepin 1969), N, Si, C (Lewis et al. 1987; Zinner et al. 1987), and SiC grains (Bernatowicz et al. 1987) in meteorites led to the realization that products of precursor stars were preserved in the solar system. These observations required that the protoplanetary disk did not reach temperatures high enough to vaporize (and homogenize) all presolar grains. Following this and the discovery of oxygen isotopic variations in meteorites[5] (Clayton et al. 1973), the concept of a cooler, isotopically heterogeneous disk that retained a cosmic chemical memory of the solar system's parental molecular cloud was advanced (e.g., Clayton 1982).

Subsequent isotope measurements of meteorites revealed variations in individual components and bulk samples that could not be explained by mass-dependent (physical or chemical) processes, radioactive decay, cosmic ray exposure effects, or nuclear field shift effects. Generally, bulk sample isotopic compositions of meteorites are less anomalous than those found in individual presolar grains, which is likely due to dilution of the presolar grain compositions by disk dust (Fig. 1). By contrasting isotope compositions of bulk meteorite samples with the much larger isotopic anomalies recorded by presolar grains, it was proposed that bulk sample isotopic variations were caused by the heterogeneous distribution of presolar grains in the protoplanetary disk (for review, see Qin and Carlson 2016). These isotopic variations were named nucleosynthetic isotope anomalies.

To date, meteoritic nucleosynthetic anomalies have been recorded in isotopes of Ne, Kr, Xe, Ca, Ti, Cr, Ni, Sr, Zr, Mo, Ru, Pd, Ba, W, Nd, Sm (e.g., Warren 2011; Ott 2014; Mayer et al. 2015; Bermingham et al. 2016; Dauphas and Schauble 2016; Qin and Carlson 2016;

[5] Proposals for the origin of meteoritic oxygen isotopic variations now include photochemical effects (Clayton 2002) and non-mass dependent isotopic effects (Thiemens and Heidenreich 1983).

Kruijer et al. 2017; Bermingham et al. 2018a, 2018b). Bulk meteorite isotopic variations are generally considered to reflect the bulk (or average) composition of the parent body, and thus the nebular region from which it accreted (e.g., Walker et al. 2015). The small scale of nucleosynthetic isotope anomalies indicates that the protoplanetary disk became reasonably, but not perfectly, homogenized during its evolution. Consensus on the process(es) that led to the heterogeneous distribution of presolar grains in the disk is yet to be reached. Leading theories suggest it is the result of poor mixing of these grains into the protoplanetary disk (Clayton 1982 and references therein) and/or their selective destruction via thermal chemical processing in the protoplanetary disk (Trinquier et al. 2009).

4 Can Meteorite Compositions Be Used to Constrain the Evolution of the Protoplanetary Disk?

As most meteorites are samples of disk material that have been preserved for ~4.56 Ga, their chemical and isotopic compositions can help reconstruct the physico-chemical evolution of the protoplanetary disk. If nucleosynthetic isotope signatures exhibited by bulk meteorites reflect the unique regional characteristics of the disk from which their parent bodies accreted, these isotopic variations could be used to trace heterogeneities in different regions of the disk (e.g., inner *vs.* outer solar system) and communication between these regions. Furthermore, if a robust link can be found between the nucleosynthetic isotope composition of a parent body, its accretion location, and the amount of volatiles (e.g., H, C, N) it accreted, the source of water and other volatiles for Earth and the terrestrial planets could potentially be traced.

4.1 Relating Meteorite Compositions to Accretion Locations in the Disk

It has been suggested that chondrite properties (e.g., bulk sample oxidation state, oxygen isotopic composition, volatile and refractory lithophile element abundance, matrix/chondrule modal abundance ratio) varied with heliocentric distance, and each chondrite group represents a different parent asteroid formed in a distinct nebular region and/or at a particular time (e.g., Wasson 1985, 1988; Rubin and Wasson 1995; Gradie and Tedesco 1982; Wood 2005; Raymond et al. 2009). The early article by Wasson (1988) argued for spatially separated formation locations for the different chondrite groups. The author proposed a model for forming the carbonaceous chondrites at the outer edge of the asteroid belt and the enstatite chondrites closest to the Sun (<1 AU), and the ordinary chondrites in-between. Based on chemical and physical properties, it is now generally assumed that the parent bodies of enstatite chondrites formed closest to the Sun, whereas ordinary and carbonaceous chondrites sample planetesimals that accreted at increasing heliocentric distances (e.g., Rubin and Wasson 1995; Wood 2005). First, the proportion of chondrules in enstatite and ordinary chondrites has been interpreted to indicate that their parent bodies formed in the inner disk, where higher shock velocities than in the outer disk resulted in higher concentrations of chondrules (e.g., Wood 2005). The differences in chondrule chemistry, however, still require spatially separated accretion locations with different oxidation conditions, where resorting and concentrating chondrules alone is not a sufficient explanation. Second, the abundance of water and other volatiles in unequilibrated chondrites has been used as an approximate indicator of the radial distance of parent body accretion, thereby suggesting that carbonaceous chondrites formed in the outer protoplanetary disk (near or beyond the snow line; Wasson 1988; Sect. 5.2), where they were able to accrete abundant organic- and ice-rich components (e.g., Morbidelli et al. 2012;

Trigo-Rodríguez et al. 2019). The formation of carbonaceous chondrites in the outer part of the protoplanetary disk was reinforced by reflection spectroscopy of objects in the outer asteroid belt (see Wasson 1988) and beyond (objects between Jupiter and Uranus, the irregular satellites of the giant planets, comets, and Kuiper belt objects) (e.g., Cloutis et al. 2012a, 2012b). Building on work that integrated disk dynamics and meteorite compositions, Gounelle et al. (2006) estimated the orbit for carbonaceous chondrite Orgueil (Ivuna-type; CI) as similar to the orbits of Jupiter-family comets, and suggested a cometary origin for CI meteorites, as had others before (Lodders and Osborne 1999 and references therein). Considering the uncertainties in the input data (150-year-old visual observations) for the orbital calculations, Gounelle et al. (2006) concluded this was more of a best-informed guess than a certainty; however, there are additional arguments for a connection of CI and CM (Mighei-like; CM) chondrites to comets and asteroids that can be regarded as "extinct comets" (see Lodders and Osborne 1999).

The general placement of where planets and meteorite parent bodies accreted in the disk may be deduced from the 'classical' interpretation of a solar nebula temperature-pressure structure together with the thermal stability of condensed phases from thermodynamic calculations (see Lewis 1974). Most subsequent models also invoked radial gradients in temperature and composition in the disk influencing the composition of the forming planetesimals as functions of location and time. The role dynamical re-distribution of accreted objects played, however, has recently been recognized and complicated the classical interpretation. Numerical models, such as the low-mass asteroid belt, Grand Tack, and early instability models (for review, see Raymond et al. 2020) indicate that the rapid growth and/or gas-driven migration of the giant planets destabilized the orbits of nearby planetesimals, many of which were scattered inward and outwards. This produced an inner asteroid belt dominated by inner solar system objects and an outer belt dominated by outer solar system objects (e.g., Walsh et al. 2011; Raymond and Izidoro 2017). A consequence of dynamical redistribution is that the heliocentric distance at which a planetesimal resides (now) does not necessarily reflect its initial accretion location in the disk.

4.2 Relating Nucleosynthetic Isotope Signatures of Bulk Meteorites to Accretion Locations in the Disk: The NC-CC Isotope Dichotomy

In 2011, a dichotomy was identified among solar system objects based on their nucleosynthetic isotope composition. Building on work by Trinquier et al. (2007, 2009), Warren (2011) compiled Ti, Ni, Cr, and O bulk rock isotope data and observed that carbonaceous chondrites were generally more isotopically anomalous than other chondrites (i.e., ordinary and enstatite chondrites), achondrites, Earth, Moon, and Mars. He termed the carbonaceous chondrite group "carbonaceous chondrite, CC" and the other group "non-carbonaceous chondrite, NC" (Fig. 2). The isotopic bimodality has since been found in siderophile element isotope compositions (e.g., Mo, Ru, W) of magmatic and non-magmatic iron meteorites (e.g., Budde et al. 2016; Kruijer et al. 2017; Poole et al. 2017; Worsham et al. 2017; Bermingham et al. 2018b). The use of carbonaceous *chondrite* (*CC*) and non-carbonaceous *chondrite* (N*C*) to describe iron meteorites is a misnomer because iron meteorites are not chondritic meteorites. It has, however, become convention to use the NC-CC notation to describe the isotopic bimodality regardless of whether or not a meteorite is a chondrite as defined by its mineralogy and petrogenesis.

As the CC and NC groups do not overlap in nucleosynthetic isotope compositions, the dichotomy appears to have a bimodal origin. Warren (2011) proposed a spatial or temporal origin for the dichotomy based on the inferred link between the chemical composition of a given parent body and its general accretion location in the protoplanetary disk (following reasoning by Wood 2005; Gounelle et al. 2006; Gounelle et al. 2008). Warren (2011) speculated that all chondritic CC meteorites formed in the outer (relatively volatile-rich) solar system, whereas the NC parent bodies formed in the inner (relatively volatile-depleted) solar system with the two regions possibly being separated by proto-Jupiter.

What the NC and CC isotopic dichotomy represents is debated. Kruijer et al. (2017) advanced the interpretation by Warren (2011) that proto-Jupiter caused a segregation of the inner (NC) and outer (CC) protoplanetary disk. Using Mo and W isotope compositions of bulk chondritic and iron meteorites, Kruijer et al. (2017) proposed that these groups exhibit distinct accretion ages: NC (<0.4 Ma after CAI formation) and CC ($0.9^{+0.4}_{-0.2}$Ma after CAI formation). Authors concluded the temporal and isotopic constraints implied that the NC and CC reservoirs coexisted in the disk (from ~1 Ma until at least 3–4 Ma), but not in the same accretion region. To obtain spatial segregation without inward drift of material, authors proposed the formation of a gap in the disk caused by the accretion of proto-Jupiter's core within ~1 Ma of CAI formation. Authors conjectured that the cause of the isotopic offset between CC and NC was an addition of r-process-rich material to the CC reservoir which did not infiltrate the coexisting, yet spatially separated, NC reservoir.

Scott et al. (2018) extended the meteorite database and also found that the chemical and isotopic characteristics of CC meteorites could be accounted for by formation of CC parent bodies beyond proto-Jupiter. Authors argued, however, the dichotomy cannot be related to temporal variations in the disk because differentiated bodies that exhibit the NC-CC dichotomy accreted 1–3 Ma before chondrites. Recent ^{182}W-derived accretion ages reported by Hilton et al. (2019) indicate that some CC parent body accretion ages are not resolved from the accretion ages of NC parent bodies. Brasser and Mojzsis (2020) raised doubts about the role Jupiter played in the separation of the NC and CC reservoirs by finding that the separation event may have been caused by a pressure maximum or a dust trap in the disk near the location at which Jupiter later formed.

4.3 Implications of the NC-CC Dichotomy on Disk Evolution Models

Despite the absence of consensus on what caused the isotopic difference between the NC and CC groups, it is a common interpretation that the NC group samples material accreted in the inner disk and the CC group samples material accreted in the outer disk. Using this interpretation, conclusions have been drawn about how the composition of the disk, its condensates, and planets evolved.

Models describing the cause of an isotopic difference between the NC and CC regions include the addition of material to the CC region, but not the NC region. For example, in Mo isotope compositional space (μ^{94}Mo *vs.* μ^{95}Mo), the NC and CC groups lie on two parallel or close to parallel lines (e.g., Budde et al. 2016; Worsham et al. 2017; Poole et al. 2017; Kruijer et al. 2017; Budde et al. 2019; Yokoyama et al. 2019). The lines could be offset by more r-process and/or p-process material in the CC region than the NC region (e.g., Budde et al. 2016; Worsham et al. 2017; Poole et al. 2017) or different s-process components (Stephan and Davis 2019). Yokoyama et al. (2019) proposed the variation in Mo isotope compositions in CC meteorites is related to either local parent-body processing or a heterogeneous distribution of s-process matter in the outer solar system. In the NC group, these authors identified two end-member components (NC-A and NC-B), where the NC-B component reflects the remaining complement of the type B CAIs gaseous source reservoir.

Desch et al. (2018) developed a numerical model explaining why carbonaceous chondrites have an abundance of CAIs (refractory component), which are long-known hosts for nucleosynthetic isotope anomalies. Their model was able to predict where in the disk different meteorite parent bodies formed. Authors found CC and NC bodies formed in the outer and inner disk, respectively, where regions were separated by proto-Jupiter, and that inside Jupiter's orbit, CAIs were depleted by aerodynamic drag. Alexander (2019a, 2019b) performed a comprehensive study of the elemental and isotopic characteristics of CC and NC meteorites. Alexander (2019a) showed that different mixtures of the same four meteorite components can reproduce most bulk elemental and isotopic compositions of CC meteorites, but NC meteorites require distinctly different mixtures (Alexander 2019b). Burkhardt et al. (2019) found that by mixing "CAI-like" material into an NC-like composition, the chemical and isotopic composition of the CC reservoir could be produced.

Nanne et al. (2019) and Burkhardt et al. (2019) proposed similar models detailing the cause of distinct isotope compositions in CAIs, the NC, and CC groups. A variant of the infall model was published by Spitzer et al. (2020). Authors propose that the most straightforward explanation for a change in isotope composition (between NC and CC) is a change in the composition of infalling material from the parental molecular cloud into the disk, coupled with variable mixing and subsequent isolation (by proto-Jupiter) of reservoirs within the disk (e.g., Nanne et al. 2019).

As an example, the model by Nanne et al. (2019) can be distilled into the following: (1) early infall into the disk occurred where molecular cloud material enriched in presolar carriers produced in neutron-rich environments (e.g., ^{58}Ni, ^{50}Ti, ^{54}Cr and *r*-process Mo) was added to the disk; (2) CAIs, which formed first and close to the Sun, inherited this mix of carriers and are the closest compositional match to the initial isotopic composition of the disk; (3) CAIs were transported outwards by rapid radial expansion of the early infalling material; (4) after removal of the early-formed CAIs, the composition of infalling material changed to be more NC-like, which was not enriched in the neutron-rich nuclides, and this period of infall provided most of the mass of the inner disk and it did not reach the outer disk via infall. This scenario produces a dominance of an NC-like composition in the inner solar system, where presumably the NC parent bodies accreted. The outer solar system became more CC-like as a result of inner and outer disk mixing between the initial disk material with the later accreting NC-like material. To avoid homogenization of the disk, the model invokes an early formed proto-Jupiter that inhibited exchange of material between the inner and outer disk.

Although infall models have appealing features, their feasibility depends on central assumptions that are yet to be verified. For example, these models rely on a change in the *isotopic* composition of infalling material. Nanne et al. (2019, pp. 50 and 52, respectively) surmised the change in isotopic composition of infalling material occurred because it came "from different parts of a compositionally heterogeneous parental molecular cloud, or the composition of the cloud itself changed over time" where the "isotopic heterogeneity within the solar system's parental molecular cloud has been maintained within the solar accretion disk because material infalling at different times derives from different parts of the parent cloud". The key questions are what caused the specific change in isotopic composition of the infalling molecular cloud material and over what time period could these changes occur?

It was thought that molecular clouds in star forming regions were *chemically* homogeneous because star clusters (which formed from large molecular clouds) are homogeneous (e.g., Klessen 2011; Bland-Hawthorn and Freeman 2014). It has also been proposed, however, that chemical heterogeneities (e.g., carbon monoxide and water) may persist between the pre-stellar core, the protostar, and circumstellar disk (Visser et al. 2009). Whether or not this chemical homogeneity extends to the specific presolar carriers of nucleosynthetic isotope anomalies is yet to be verified.

After the onset of star formation in a large molecular cloud, contamination of a presolar molecular cloud can occur by supernova ejecta (e.g., Bland-Hawthorn and Freeman 2014). The fast evolution of supernovae in star-forming regions is the basis for the supernova-triggered formation of the Sun, including injection of supernova material into the presolar cloud (e.g., Cameron and Truran 1977; Foster and Boss 1997; Boss 2013), or for hypotheses involving later injection of "fresh" supernova material into the evolving solar nebula (e.g., Chevalier 2000; Hester et al. 2004). The injection of supernova material may provide a large proportion of neutron-rich nuclides and nuclides made by the r-process. To account for the relatively high ^{26}Al and low ^{60}Fe abundances inferred for the early solar system, however, formation of the solar system may have been initiated by a triggered star formation at the edge of a Wolf–Rayet bubble (Dwarkadas et al. 2017). Nanne et al. (2019), however, concludes that the NC-CC isotope dichotomy reflects isotopic heterogeneity within the solar system's parental molecular cloud that cannot be linked to a specific event around t_0.

Additionally, the infall models involve CAI formation close to the Sun and subsequent transportation outward by rapid radial expansion of early-infalling material, a similar process through which the underlying disk was produced (Yang and Ciesla 2012; Nanne et al. 2019). Early in disk evolution, rapid radial expansion may occur as mass from the infalling cloud is added preferentially to the inner regions of the disk, thereby fueling outward flow (Yang and Ciesla 2012). In Yang and Ciesla (2012), the standard case scenario has infall ending at ~0.3 Ma. How the timing and extent of mixing predicted by infall models and rapid radial expansion align with the age and composition constraints of NC-CC meteorites requires further consideration. Also needing more detailed assessment is how rapid radial expansion fits with observations and other models of disk formation (e.g., Williams and Cieza 2011). Such models often describe the presolar cloud material accreting onto the protosun and the disk, and the Sun's growth eventually becoming dominated by accretion of disk material. In these models, during the Sun's growth, the accretion flow is from outside-in, and the disk loses mass via accretion onto the star. Further, the effects of outflow and jet formation from the protosun are yet to be fully considered. Here, it is briefly noted that the X-wind models by Shu (1997), Shu et al. (2001) and Lee et al. (1998) are relevant to infall models (Yang and Ciesla 2012). From Yang and Ciesla (2012), unresolved issues include the role of magnetic fields, supersonic turbulence, rotation in the cloud, and a variable infall rate. X-wind models (or variants thereof) could provide mechanisms for CAI formation near the Sun, and their transport out to larger heliocentric distances. At this stage, the timing and process of infall and transportation of disk material (including CAIs) to the outer disk needs further investigation.

Central to determining the feasibility of infall models will be constraining the cause and timing of the change in isotopic composition and physical location of the infalling molecular cloud material to the disk, and the distribution of disk material in the broader context of what is known about solar nebula evolution. Although important strides are being made, a comprehensive model describing what the NC-CC dichotomy represents, if it reflects inner *vs.* outer solar system compositions, what caused the difference in isotopic composition between and within the two groups, and defining Jupiter's role are topics still under development.

5 Does the NC-CC Dichotomy Extend to Major Highly-Volatile Elements?

In addition to the implications the NC-CC dichotomy has on protoplanetary disk evolution models, the isotope dichotomy is starting to be used to decipher the source(s) of

Earth's water and other volatiles such as C and N. For example, accretion of volatile-rich CC material towards the end of Earth's formation – through the Moon-forming impactor and/or late accretion – has been argued for based on a comparison of the Mo-Ru isotope composition of meteorites and the bulk silicate Earth (Budde et al. 2019; Hopp et al. 2020). This seems consistent with the constraints provided by elemental compositions, which require Earth to have accreted from at least two components, initially a reduced one, which now would be sampled by NC meteorites, and later a more oxidized and volatile-rich component, represented by CC meteorites (e.g., Rubie et al. 2015; for a recent review, see Fegley et al. 2020).

The principle that the NC-CC dichotomy can be used to trace volatile accretion to Earth, however, requires a robust link between the NC-CC isotope signature of meteorite groups, the volatile content of their parent bodies, and the volatile content of the protoplanetary disk from which they accreted. If the NC-CC isotope dichotomy reflects an isotopic heterogeneity between the inner and outer disk, a difference in volatile element abundances may be expected between materials derived from the two reservoirs: non-carbonaceous bodies would be depleted in highly-volatile elements relative to CC bodies, which are expected to be comparatively volatile-rich. This application of the NC-CC dichotomy is discussed in the sections below which consider the distribution of highly-volatile major elements (H, C, and N) in the disk and meteorites.

5.1 Volatile Species and Their Condensation Fronts in the Protoplanetary Disk

During the protoplanetary disk phase, molecular or atomic volatile species (e.g., H_2, H_2O, He, CO, CO_2, CH_4, CH_3OH, N_2, NH_3, HCN, H_2S, etc.) with low condensation temperatures ($\leq$160 K; e.g., Zhang et al. 2015) resided in the gas phase throughout a significant portion of the disk. At sufficiently low temperatures (i.e., below the condensation temperature of water), the solid mass density in the disk increased substantially because of water ice formation, thereby aiding planet accumulation. Condensation of other C- and N-bearing ices could have increased the mass density further. In addition, volatile ices likely played an essential role in planet formation by promoting coagulation of 'sticky' ice-coated dust grains into larger pebbles (Wang et al. 2005). Thus, the initial distribution and physical state of volatile species (solid *vs.* vapor) in the disk was critical in determining the composition and evolution of planetesimals.

To the first order, the thermal structure of the protoplanetary disk controlled where hydrogenated molecules (e.g., H_2O) and other volatile compounds (e.g., carbon- and nitrogen-bearing species) condensed into ice or froze out onto small dust grains that were subsequently incorporated into accreting planetesimals. While H_2 and He remained in the gaseous state throughout the disk, water molecules were able to freeze out into solid ice grains at and beyond the so-called "snow line" (i.e., the H_2O condensation front at $\sim$160 K at the midplane; Hayashi 1981; Lodders 2003, 2004; Krijt et al. 2016) (Fig. 4). Other condensation fronts, defined by the physical state of H-, C-, and N-bearing molecules, included those of ammonia, carbon monoxide, carbon dioxide, methane, methanol, and hydrogen cyanide. Importantly, ice formation depends on the kinetics of the reaction(s) making the condensing molecules. For example, the very stable molecules CO and N_2 essentially contain all the carbon and nitrogen in the solar gas at high temperatures. Only at lower temperatures and high total pressures, will the reaction equilibrium shift to methane and ammonia, i.e., $CO + 3H_2 = CH_4 + H_2O$ and $N_2 + 3H_2 = 2\,NH_3$ (Lewis and Prinn 1980; Fegley and Prinn 1989; Fegley 2000; Lodders 2003). However, at the low total pressure within the outer protoplanetary disk, the efficient conversion of CO to CH_4 and of N_2 to NH_3 is kinetically inhibited.

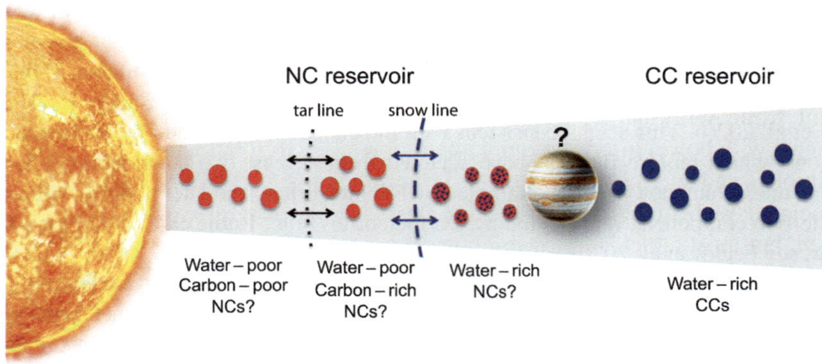

Fig. 4 Schematic cross-section of the protoplanetary disk (not to scale). Planetesimals with NC- and CC-type nucleosynthetic compositions are represented in red and blue, respectively. The two reservoirs are considered to have been spatially separated (e.g., Warren 2011), possibly as a result of the early accretion of proto-Jupiter (Kruijer et al. 2017). Based on the radial gradient in temperature, planetesimals that accreted beyond the snow line (where $T \leq 160$ K at the midplane; Hayashi 1981; Lodders 2004) are expected to be water-rich, whereas 'dry' (i.e., ice-free) planetesimals that formed between the tar- and snow line could have incorporated refractory organics (Lodders 2004; Nakano et al. 2020). The radial position of the mobile condensation fronts relative to the Jupiter barrier likely played a fundamental role in the distribution and transport of inorganic ices (and organics) in the disk during the first few million years of solar system history

Note that this also affects the total amount of water because O is locked up in CO. Given that other potential C- and N-bearing molecules are affected in a similar manner, models of their condensation locations must consider kinetics and dynamical modelling, and it follows that the NH_3, CH_3OH, and H_2O ice lines essentially coincided while CO and CH_4 condensed at much greater heliocentric distances (e.g., Dodson-Robinson et al. 2009; Bergin and Cleeves 2018).

In contrast to these inorganic volatile ices that only exist at $T \leq 160$ K, laboratory heating experiments under high vacuum demonstrate that interstellar organic analogs persist at temperatures above 350 to 450 K (Nakano et al. 2003). Lodders (2004), therefore, argued that the disk also had a "tar line" inward of the snow line, delineating a region where refractory organics (such as poly-aromatic hydrocarbons) were stable but water existed in vapor form only (Fig. 4). Because innermost disk temperatures were above the condensation temperature of all volatile ices, planetary objects that formed sunward of the snow line are not expected to have accreted icy grains, in contrast to bodies formed at greater heliocentric distances. Between the snow and tar lines, however, organics, especially their refractory fractions, could have been incorporated into 'ice-free' planetesimals (Fig. 4); these organics represent a potential source of water for bodies accreted sunward of the snow line (Nakano et al. 2020).

It is important to note that a condensation front is not a straight "line", but rather a surface whose position in the disk and vertical-horizontal structure depends on both temperature and (vapor) pressure in the disk, and, therefore, on the disk mass and heating processes within the disk. Meijerink et al. (2009), for example, pointed out that the heliocentric distance of the water condensation front increases with height above and below the disk midplane due to heating of the disk surface by the central star. Vertical diffusive transport of water vapor across the snow line resulted in condensation and subsequent settling of ice particles and ice-coated grains to the midplane, thereby increasing the dust-to-gas ratio along the midplane and promoting particle growth (e.g., Krijt et al. 2016).

The positions of the snow and tar lines at the midplane are essential when contemplating planetesimal formation and accretion of water ice and organics. Not only were these "lines" 3-dimensional structures, but their location changed with time as the disk's temperature, total pressure, and density evolved (e.g., as a function of the luminosity of the central star, surface density, mass accretion rate, opacity; note that condensation also feeds into the disk because opacity sources from the gas are removed and dust/ice opacity is generated) (Davis 2005; Garaud and Lin 2007; Oka et al. 2011; Morbidelli et al. 2016; Sato et al. 2016). The absolute positions (and relative positions to accreting planets, including Jupiter) and rate of change of these positions during the disk phase, however, are model dependent. Different models place the snow line anywhere between $\sim$1 (e.g., Sasselov and Lecar 2000) and 4 to 7 AU (Cameron 1995; Boss 1996; Dodson-Robinson et al. 2009; see Fegley 2000 for a summary of models of P-T conditions in the solar nebula). According to the disk model of Boss (1996), the snow line never migrated closer than 3 AU of the Sun. In the model of Garaud and Lin (2007), the snow line migrated to $\leq$1 AU as the mass accretion rate onto the Sun decreased within the first few million years of the nascent solar system, implying that water ice could condense at $\sim$1 AU from the residual nebula gas and that water ice-rich planetesimals existed in the terrestrial planet-forming region. However, Morbidelli et al. (2016) presented a speculative argument that even as the temperature decreased below the water condensation threshold, the disk inwards of Jupiter may have remained ice-depleted because proto-Jupiter effectively prevented the inward drift of icy particles once it reached 20 Earth masses; as a result, the snow line could have "fossilized" at $\sim$3 AU. This model implies that planetesimals which formed sunward of Jupiter's orbit would not have accreted any water ice. Evidence from NC chondrites and achondrites, however, contradicts this scenario (see Sect. 5.2).

Overall, based on predictions of a strong heliocentric radial gradient in temperature and total pressure, the volatile content and composition of different planetesimals should have been characterized by sharp transitions between species that were refractory enough to freeze out, and those that were too volatile and remained in the gas phase. Planetesimals that accreted close to the Sun are expected to be water- and carbon-poor, whereas bodies that formed in the outer solar system should have accreted H-C-N-rich ices and organics (Fig. 4). The 3-dimensional distribution of ices and vapor, however, continuously evolved over the lifetime of the protoplanetary disk as the condensation fronts moved inwards and outwards. Currently, it is unresolved how this movement affected the distribution of volatile ices in the feeding zone of planetesimals during the first several Ma of solar system history. In addition, there are factors other than temperature and total pressure that control the volatile characteristics of planetesimals and meteorites (see Sect. 5.2). Thus, linking the volatile content of a meteorite to a (radial) accretion location of a planetary body is not straightforward.

5.2 Major Volatile Element (H, C, N) Abundances in NC and CC Chondrites

The distinct nucleosynthetic isotopic composition of NC- and CC-type meteorites has been interpreted to indicate that their parent bodies accreted in the inner and outer solar system, respectively, separated by Jupiter (Fig. 4). If so, NC and CC meteorites could serve (to the first order) as probes of the volatile element signatures of different reservoirs. Do the volatile contents of these objects follow the expected trend that NC meteorites are more volatile-poor than CC meteorites?

The H-C-N content of achondrites (and presumably larger differentiated planetary building blocks) does not reflect that of the disk or of undifferentiated precursors because melting and differentiation caused decomposition of H-C-N carrier phases (such as organics), variable degrees of degassing, and re-distribution of hydrogen, carbon, and nitrogen between the silicate and metal portions of the parent body (e.g., Mikhail and Füri 2019). Similarly, iron meteorites do not preserve a direct record of the major volatile element abundances of their parent bodies and formation regions because the H-C-N content of planetary cores depends on a complex set of parameters, such as the composition of the Fe-rich alloy (including the abundance of H, C, N, S, and Si), pressure, temperature, and oxygen fugacity during metal-silicate equilibration (e.g., Wood et al. 2013).

In contrast, the preservation of H-C-N in chondrites indicates that these materials may, to an extent, record the (minimum) volatile content of their parent body and presumably the inherited disk material. Although the volatile content of primitive (i.e., those that are least affected by aqueous alteration or thermal metamorphism and are unshocked) planetesimals is fundamentally controlled by several parameters, such as the timing of accretion, location of accretion relative to the position of the ice lines and tar line, redox conditions, as well as the distribution of H-C-N carriers (e.g., presolar grains, organics) in the disk, the H-C-N abundances in different chondrites could reflect the volatile element composition of the region of the disk from which they accreted. Accordingly, can the NC-CC dichotomy be extended to volatile element abundances in chondrites?

Considering examples of NC-type objects, enstatite chondrites are highly reduced, implying that their parent bodies formed in an oxygen-poor environment. They have been presumed to be devoid of water (Mason 1966; Hutson and Ruzicka 2000; Rubin and Ma 2017), until Piani et al. (2020) recently showed that enstatite chondrites contain up to 0.06 wt.% H (i.e., 0.54 wt.% equivalent H_2O). Ordinary chondrites contain slightly larger quantities of water (i.e., $\leq$0.2 wt.% H; Alexander et al. 2012, 2018; Vacher et al. 2020; Fig. 3), indicating that their parent bodies, which formed $\sim$2 Ma after CAIs at $\sim$2 AU (e.g., Henke et al. 2012; Blackburn et al. 2017; Sugiura and Fujiya 2014), accreted some water (ice). Sutton et al. (2017) argued that a significant amount (i.e., up to $\sim$80%) of water was lost on the parent body of Semarkona, the most primitive ordinary chondrite known, as a result of Fe oxidation by water during aqueous alteration and/or metamorphic dehydration. If large amounts of reactive (liquid) water had been present, much of the iron metal in ordinary chondrites would have been converted to magnetite, and sulfides would have been attacked; this, however, is not the case. Nonetheless, the hydrogen content of ordinary chondrites, together with the presence of hydrated minerals (smectite) and secondary phases (calcite) (Alexander et al. 1989), suggests that the temperature had dropped below the condensation temperature of water ice at $\sim$2 AU by the time the ordinary chondrites formed (Alexander et al. 2018). Furthermore, it requires either that some small icy particles and/or organics (inherited, in part, from the interstellar medium and preserved in the outer solar nebula; Lodders 2004; Deloule et al. 1998; Piani et al. 2015) were able to get past Jupiter, or that water-bearing material was trapped sunward of Jupiter's orbit prior to the rapid growth of its core.

Considering CC-type bodies, CI and CM carbonaceous chondrites are significantly more water-rich (containing up to 10 to 20 wt.% equivalent water in the form of hydrous minerals; Kerridge 1985; Pearson et al. 2001; Piani et al. 2020; Robert and Epstein 1982;

Robert 2003; Vacher et al. 2020; Fig. 3)[6], suggesting that their parent bodies formed near or beyond the snow line, where they accreted variable amounts of water-ice grains (Vacher et al. 2020). However, several carbonaceous chondrite groups (e.g., Vigarano-like CV and Karoonda-like CK chondrites) contain only small quantities of water (i.e., $\leq$0.03 to 0.19 wt.% H; Vacher et al. 2020; Fig. 3). Although the water abundances of their parent bodies were modified by thermal metamorphism and/or metal oxidation (Sutton et al. 2017; Vacher et al. 2020), these observations indicate that CC-type bodies cannot systematically be classified as "water-rich". Overall, there is no sharp difference in the water mass fraction of planetesimals inferred to have accreted in the inner (ordinary chondrites) and outer solar system (carbonaceous chondrites).

Other volatile element abundances also do not display systematic radial variations. Some members of both the ordinary and enstatite chondrite classes contain more carbon and nitrogen than certain carbonaceous chondrites (Grady et al. 1986; Wasson and Kallemeyn 1988; Hashizume and Sugiura 1995; Grady and Wright 2003; Rubin and Choi 2009); parent body alteration processes, however, may have significantly affected bulk chondrite carbon and nitrogen abundances (e.g., Pearson et al. 2006). Carbonaceous chondrites themselves record a wide range of carbon ($\leq$0.2 to $\sim$5 wt.%) and nitrogen ($\leq$0.01 to 0.5 wt.%) abundances (e.g., Pearson et al. 2006, and references therein) (Fig. 3). Some (ppm-levels) of the carbon and nitrogen occurs in the form of presolar grains (see Sect. 2) but most is present as organic matter dispersed in the matrix of carbonaceous chondrites, and in the form of reduced species (graphite, carbides, nitrides) in enstatite chondrites (e.g., Grady and Wright 2003). This suggests that the accretion and retention of nominally volatile elements (C, N) in the inner regions of the disk may have been facilitated by different oxidation states. Before this can be concluded, however, details about the evolution of the redox conditions in the disk, including how they affected chemical reactions in the nebula and the dynamics of accretion, need to be further documented.

The key point is that planetesimals which formed in the NC reservoir cannot be generally characterized as "volatile-depleted" with respect to volatile and highly volatile elements compared to CC bodies. Although they are relatively deficient in water and carbon as compared to *some* bodies accreted within the CC reservoir, several volatile elements are comparable in abundance in the most reduced (enstatite chondrites from the NC group) and oxidized (the carbonaceous chondrites from the CC group) chondrites. Thus, with current datasets, the concept of the NC-CC dichotomy cannot yet be extended to all volatile element abundances, and explanations are needed to account for the discrepancies between bulk volatile contents and the NC-CC grouping of meteorites. In addressing these discrepancies, it will be important to determine the extent to which the volatile element composition of a meteorite reflects that of its parent body and the disk region the parent body sampled.

[6]Measurable retention of any originally accreted water (ice) on the parent body requires post-accretion aqueous alteration; the water abundance then also depends on kinetics and the degree of aqueous processing and/or thermal metamorphism on the parent body. It should be noted that important intra- and inter-chondrite variations in water concentrations exist which can be explained, in part, by varying proportions of H-bearing components in a given (sub-)sample (e.g., Pearson et al. 2001). Analytically, not all extraction techniques may clearly distinguish between primordial and (terrestrial) adsorbed water (e.g., Robert and Epstein 1982; Vacher et al. 2020), and terrestrial exposure of "finds" can modify water concentrations (Stephant et al. 2018). Thus, the reported bulk water contents of chondrites cannot directly reflect the original accreted water in all cases.

6 Implications for Volatile Accretion: Building Earth from NC- and CC-Type Precursors

Based on oxygen and nucleosynthetic isotope compositions, Earth is classified as a NC-body (Warren 2011; Fig. 2), although in μ^{94}Mo *vs.* μ^{95}Mo compositional space Earth lies between the NC and CC groups (Budde et al. 2019). This classification implies that early Earth's building blocks were dominated by materials with NC isotopic signatures. Although it has been proposed that in elemental composition Earth is relatively close to carbonaceous chondrites, in particular CV and CK chondrites (e.g., Palme and O'Neill 2003), oxygen isotopes and numerous nucleosynthetic isotope data preclude more than a few percent carbonaceous chondrite material accreting to Earth. These data require that Earth formed from a reservoir that is isotopically similar to enstatite chondrites (or a mixture of bodies that bracket the isotopic composition of Earth) (e.g., Clayton et al. 1984; Lodders 2000; Dauphas 2017). Earth cannot have been constructed solely out of enstatite chondrites because of the resolvable chemical (including oxidation state) and mass–dependent isotopic differences between enstatite chondrites and Earth (e.g., McDonough and Sun 1995; Lodders 2000; Drake and Righter 2002; Fegley et al. 2020). Dauphas (2017) proposed that enstatite meteorites and Earth formed from the same isotopic nebular reservoir, and the reason that the bodies differ in bulk chemical compositions is because their chemical evolution diverged due to fractionation via nebular and planetary processes.

Determining the role that NC- and CC-type bodies (and chondrites *vs.* achondrites) played in the accretionary history of Earth can become complicated when the isotopic and chemical characteristics of an object might be decoupled. Given the absence of meteoritic materials that match Earth's elemental and isotopic compositions, it is likely that Earth was primarily constructed out of building blocks that shared feeding zones with enstatite chondrites, but these materials are not sampled by the current meteorite collection (e.g., Drake and Righter 2002; Dauphas 2017; for a recent review, see Fegley et al. 2020).

6.1 Constraints from Hydrogen Isotopes

The present-day Earth has a significant quantity of water at its surface, and five to ten ocean masses worth of hydrogen as OH and H_2O may reside in the terrestrial mantle (e.g., Marty 2012; Marty et al. 2016; Hirschmann 2018). The source of Earth's water, and the timing of water accretion, is an ongoing investigation in planetary science, with models proposing different timing and means via which water was delivered to Earth (e.g., via nebular ingassing, asteroids, or comets; e.g., Marty 2012; Hallis et al. 2015; Marty et al. 2016; Alexander 2017; Alexander et al. 2018; Wu et al. 2018). Notably, the recent detection of significant quantities of water (i.e., up to 0.54 wt.% equivalent H_2O) in enstatite chondrites represents a paradigm shift in the understanding of the origin of Earth's water, as a significant, even dominant fraction of terrestrial water could be derived from EC-like bodies (Piani et al. 2020).

Whether or not planetesimals from the inner or outer solar system were a dominant source of Earth's water can be further interrogated by comparing the H isotope composition of meteorites with the terrestrial mantle. Hydrogen isotopes (δD = –218 to –125‰ in the deep Earth's mantle; Deloule et al. 1991; Hallis et al. 2015) identify enstatite-type objects ($\delta D_{EC,bulk}$ = –146 to –98‰; Piani et al. 2020) or carbonaceous (CI/CM) chondrite-like material ($\delta D_{CI/CM}$ = –227 to +300‰; e.g., Alexander et al. 2012; Marty et al. 2016;

Alexander et al. 2018; Piani et al. 2020),[7] as the potential sources of Earth's water, with minor (<1%) cometary (Marty et al. 2016) and nebular (Hallis et al. 2015; Wu et al. 2018) contributions. A nebular volatile contribution is independently indicated by the presence of solar-like neon in Earth's mantle (Honda et al. 1993; Yokochi and Marty 2004; Williams and Muhkopadyay 2018), but mass balance considerations suggest that this contribution was minor for major volatiles such as hydrogen, carbon, and nitrogen (e.g., Marty 2012).

The new findings of Piani et al. (2020) identify EC-type bodies as the likely sources of water in Earth's interior. However, Earth's surficial (oceans and atmosphere) water reservoir, which is more enriched in D than the mantle, appears to require a contribution of carbonaceous chondrite-like material (Piani et al. 2020). The most likely explanation for the accretion of this component is that 'wet' (i.e., 'icy') CI/CM-type bodies were scattered inward from greater heliocentric distances as a result of Jupiter's growth and/or gas-driven migration after the first 3 to 4 million years of solar system history (Raymond and Izidoro 2017; Alexander et al. 2018).

6.2 Constraints from Siderophile Element Isotopes

The importance of late accretion (often referred to as the "late veneer") for adding water to Earth remains a matter of debate. Late accretion provided the highly siderophile elements to Earth's mantle after core formation (Chou 1978) and represents a contribution of $\sim$0.5 wt.% (by mass) of chondritic material (Walker 2009). Based on highly siderophile element isotope signatures, this material was inferred to originate from the inner solar system, which may imply that late accretion did not provide a significant amount of water to our planet (Bermingham and Walker 2017; Fischer-Gödde and Kleine 2017; Bermingham et al. 2018b). A recent comparison of the Ru isotope composition of Archean rocks from South West Greenland (Fischer-Gödde et al. 2020) with the modern silicate mantle (Bermingham and Walker 2017), however, indicates that carbonaceous-chondrite-like material may have been delivered to Earth during late accretion. The implications for the origin of Earth's water are yet to be determined because the extent to which the unusual Ru isotope composition of South West Greenland represents the mantle composition before late accretion is as yet unknown (Bermingham 2020). Future work that determines the extent of Ru isotope variability in Earth's mantle will be important for constraining the origin and timing of delivery of Earth's volatiles.

6.3 Constraints from Nitrogen Isotopes

A clear link between nitrogen abundance and the NC-CC classification is yet to be determined (see Sect. 5.2). However, a dichotomy seems to exist for nitrogen isotope ($^{15}N/^{14}N$) ratios, with CC irons being enriched in ^{15}N compared to NC irons (Scott et al. 2018). These observations are consistent with the general trend of increasing ^{15}N-enrichments with radial distance from the Sun (Füri and Marty 2015), i.e., from the Sun ($\delta^{15}N_{Sun} = -383 \pm 8‰$; Marty et al. 2011) to the Earth-Moon system ($\delta^{15}N_{Earth\text{-}Moon} \approx -40$ to $+26‰$; Palot et al. 2012; Cartigny and Marty 2013; Füri et al. 2015; Labidi et al. 2020) and Mars

[7] Individual CI and CM chondrites record a wide range of hydrogen isotope (δD) values, possibly due, in part, to variable degrees of aqueous alteration and sampling biases (Alexander et al. 2012, 2018; Kerridge 1985; Robert 2003). Therefore, Alexander et al. (2018) used average values of $78 \pm 7‰$ and $-53 \pm 130‰$ for the CI and CM chondrite groups, respectively.

($\delta^{15}N_{Mars,mantle} \approx -30‰$; Mathew and Marti 2001) and to comets ($\delta^{15}N_{Comet} \approx +800$ to 1000‰; Bockelée-Morvan et al. 2015, and references therein). In contrast, nitrogen isotope ratios recorded by chondrites do not display the NC-CC dichotomy. Enstatite chondrites are characterized by low $\delta^{15}N$ values ($\delta^{15}N_{EC} = -47$ to 0‰; Kung and Clayton 1978; Grady et al. 1986), but the nitrogen isotopic compositions of ordinary ($\delta^{15}N_{OC} = -4$ to +75‰; Hashizume and Sugiura 1995) and carbonaceous chondrites ($\delta^{15}N_{CI/CM} = -6$ to +56‰, excluding Bells (CM2); Kerridge 1985; Alexander et al. 2012) largely overlap.[8] Since only enstatite chondrites have negative $\delta^{15}N$ values, Javoy et al. (1986) suggested that they are the source of nitrogen in Earth's primitive mantle ($\delta^{15}N \approx -40‰$; Palot et al. 2012; Cartigny and Marty 2013; Dalou et al. 2019). As outlined in Sect. 5.2, enstatite chondrites contain a small but significant amount of water, and they contain important quantities of nitrogen, possibly as a result of enhanced nitrogen solubility under the reducing conditions at which their parent bodies formed (Libourel et al. 2003; Boulliung et al. 2020). Therefore, enstatite chondrite-*like* bodies, presumably derived from the NC reservoir (Warren 2011), are not only strong candidates for Earth's primary building blocks but also for the dominant source of terrestrial water and nitrogen (Javoy et al. 1986; Lodders 2000; Dauphas 2017; Piani et al. 2020).

It is noteworthy that the nitrogen isotopic signature of the modern (upper) terrestrial mantle ($\delta^{15}N \approx -5 \pm 2‰$ in diamonds and mid-ocean ridge basalts; Cartigny and Marty 2013) has previously been interpreted to reflect that CI/CM-like planetesimals were the major sources of nitrogen on Earth (e.g., Alexander 2017). Yet, Li et al. (2016) and Dalou et al. (2019) recently showed that core formation, possibly coupled with degassing, could have increased the $\delta^{15}N_{mantle}$ value from an enstatite chondrite-like signature to the present-day value. The degree of N isotope fractionation, however, depends on the redox conditions and on N speciation. The distinct nitrogen isotope signature of other terrestrial reservoirs such as the atmosphere and crust ($\delta^{15}N \approx 0$ to + 6‰; Cartigny and Marty 2013) suggests that CC-type material contributed some nitrogen to an enstatite chondrite-*like* Earth (Javoy et al. 1986).

Together, Earth's hydrogen, nitrogen, oxygen, and nucleosynthetic isotope signatures of refractory elements such as those shown in Fig. 2 suggest that the terrestrial building blocks may have included both NC- and CC-type bodies. A contribution of CC-type (CI- and/or CM-like) material may also be required to explain the isotopic composition of hydrogen and nitrogen in the interiors of Mars and Earth's Moon (Alexander 2017; Füri et al. 2015, 2017). If so, this would imply that the NC and CC reservoirs communicated with each other during the main stages of terrestrial planet accretion. Given the lack of any (known) samples from Venus and Mercury, however, it remains unclear if terrestrial planets closer to the Sun accreted both NC- and CC-type material, and if so, how much of each type.

Challenges, however, remain in relating meteorite volatile compositions to accretion locations in the disk, some of which have been raised in this chapter. These complicate the interpretation that NC meteorites represent inner disk material that is strictly "volatile-poor" relative to CC meteorites which represent the outer disk and are "volatile-rich"; As such, if and how the NC-CC dichotomy can be used to reconstruct planetary volatile accretion requires further investigation.

[8] The different chondrite groups (e.g., CI, CM, CR, CO, CV, CK) within a given class (e.g., carbonaceous chondrites) record a wide range of $\delta^{15}N$ values (e.g., Pearson et al. 2006). For a given group, the $\delta^{15}N$ value varies with petrologic type, i.e., the extent of thermal metamorphism; $\delta^{15}N$ generally decreases with increasing petrologic type from 1 to 4 (Pearson et al. 2006). In addition, the nitrogen extraction method (combustion *vs.* pyrolysis) can yield distinct $\delta^{15}N$ values (e.g., Grady et al. 1986). Therefore, care should be taken when comparing data from different studies.

7 Conclusions

The NC-CC isotope dichotomy appears to record fundamental chemical and/or physical processes in the disk. Although there is an abundance of models and circumstantial evidence that converge on the idea that the parent bodies of CC meteorites formed in the outer protoplanetary disk beyond Jupiter's orbit and the snow line, irrefutable evidence of their (radial) accretion location is yet to be identified. It remains to be answered what the relationship is between nucleosynthetic isotope signatures in meteorites, their volatile compositions, and their precursor disk composition, and thus if the combination of a nucleosynthetic isotope composition and volatile element content can trace communication between different regions of the disk. Detailed observations of the planet forming process(es) in neighboring systems, synthesized with disk and planetary accretion models and the chemical/isotopic compositions of meteorites, will undoubtedly provide the most comprehensive understanding of the evolution of the protoplanetary disk. These data will be essential to resolving what the NC-CC isotope dichotomy represents. Does the NC-CC isotope dichotomy provide constraints on the temporal, spatial, and chemical evolution of the protoplanetary disk or does it not because of decoupling between the isotopic and chemical characteristics of an object? Resolving this and other issues is important to determining the significance of the NC-CC isotope dichotomy and if the NC-CC character of a parent body permits it to be used to trace the interplay between planetary building blocks.

Acknowledgements We thank the International Space Science Institute (Bern, CH) for hosting and supporting the workshop "Reading Terrestrial Planet Evolution in Isotopes and Element Measurements" and the editors of this book. Our sincere thanks to editor in chief H. Lammer for his guidance and patience. We also thank thorough reviews by A.M. Davis and C.M.O'D. Alexander. KRB was supported by NASA Emerging Worlds grants 80NSSC18K0496 and NNX16AN07G, NASA SSERVI grant NNA14AB07A, and the Department of Earth and Planetary Sciences, Rutgers University. EF and BM were supported by the European Research Council (ERC) under the European Union's Horizon 2020 research and innovation program (grant agreements no. 715028 and no. 695618, respectively). Work by KL was supported in part by NSF grant AST 1517541 and the McDonnell Centre for the Space Sciences. This is CRPG-CNRS contribution 2374.

References

C.M.O'D. Alexander, The origin of inner solar system water. Philos. Trans. R. Soc. A, Math. Phys. Eng. Sci. **375**, 20150384 (2017)

C.M.O'D. Alexander, Quantitative models for the elemental and isotopic fractionations in chondrites: the carbonaceous chondrites. Geochim. Cosmochim. Acta **254**, 277–309 (2019a)

C.M.O'D. Alexander, Quantitative models for the elemental and isotopic fractionations in the chondrites: the non-carbonaceous chondrites. Geochim. Cosmochim. Acta **254**, 246–276 (2019b)

C.M.O'D. Alexander, D.J. Barber, R. Hutchison, The microstructure of Semarkona and Bishunpur. Geochim. Cosmochim. Acta **53**, 3045–3057 (1989)

C.M.O'D. Alexander, L.R. Nittler, The galactic evolution of Si, Ti, and O isotopic ratios. Astrophys. J. **519**, 222–235 (1999)

C.M.O'D. Alexander, R. Bowden, M.L. Fogel, K.T. Howard, C.D. Herd, L.R. Nittler, The provenances of asteroids, and their contributions to the volatile inventories of the terrestrial planets. Science **337**, 721–723 (2012)

C.M.O'D. Alexander, K.D. McKeegan, K. Altwegg, Water reservoirs in small planetary bodies: meteorites, asteroids, and comets. Space Sci. Rev. **214**, 36 (2018)

S. Amari, L.R. Nittler, E. Zinner, K. Lodders, R.S. Lewis, Presolar SiC grains of Type A and B: Their isotopic compositions and stellar origins. Astrophys. J. **559**, 463–483 (2001)

E.A. Bergin, L.I. Cleeves, Chemistry during the gas-rich stage of planet formation, in *Handbook of Exoplanets*, ed. by H. Deeg, J. Belmonte (Springer, Cham, 2018), pp. 2221–2250
K.R. Bermingham, K. Mezger, E.E. Scherer, M.F. Horan, R.W. Carlson, D. Upadhyay, T. Magna, A. Pack, Barium isotope abundances in meteorites and their implications for early Solar System evolution. Geochim. Cosmochim. Acta **175**, 282–298 (2016)
K.R. Bermingham, R.J. Walker, The ruthenium isotopic composition of the oceanic mantle. Earth Planet. Sci. Lett. **474**, 466–473 (2017)
K.R. Bermingham, N. Gussone, K. Mezger, J. Krause, Origins of mass-dependent and mass-independent Ca isotope variations in meteoritic components and meteorites. Geochim. Cosmochim. Acta **226**, 206–223 (2018a)
K.R. Bermingham, E.A. Worsham, R.J. Walker, New insights into Mo and Ru isotope variation in the nebula and terrestrial planet accretionary genetics. Earth Planet. Sci. Lett. **487**, 221–229 (2018b)
K.R. Bermingham, Ancient rock bears isotopic fingerprints of Earth's origins. Nature **579**, 195–196 (2020)
T. Bernatowicz, G. Fraundorf, T. Ming, E. Anders, B. Wopenka, E. Zinner, P. Fraundorf, Evidence for interstellar SiC in the Murray carbonaceous chondrite. Nature **330**, 728–730 (1987)
D.C. Black, R.O. Pepin, Trapped neon in meteorites - II. Earth Planet. Sci. Lett. **6**, 395–405 (1969)
T. Blackburn, C.M.O'D. Alexander, R. Carlson, L.T. Elkins-Tanton, The accretion and impact history of the ordinary chondrite parent bodies. Geochim. Cosmochim. Acta **200**, 201–217 (2017)
J. Bland-Hawthorn, K. Freeman, *The Origin of the Galaxy and Local Group, Saas-Fee Advanced Course*, vol. 37 (Springer, Berlin, 2014), p. 1
D. Bockelée-Morvan, U. Calmonte, S. Charnley, J. Duprat, C. Engrand, A. Gicquel, M. Hässig, E. Jehin, H. Kawakita, B. Marty, S. Milam, A. Morse, P. Rousellot, S. Sheridan, E.S. Wirström, Cometary isotopes measurements. Space Sci. Rev. **197**, 47–83 (2015)
A.P. Boss, Collapse and fragmentation of molecular cloud cores. II. Collapse induced by stellar shock waves. Astrophys. J. **439**, 224–236 (1995)
A.P. Boss, Evolution of the solar nebula. III. Protoplanetary disks undergoing mass accretion. Astrophys. J. **469**, 906–920 (1996)
A.P. Boss, Mixing and transport of short-lived and stable isotopes and refractory grains in protoplanetary disks. Astrophys. J. **773**, 5 (2013)
J. Boulliung, E. Füri, C. Dalou, L. Tissandier, L. Zimmermann, Y. Marrocchi, Oxygen fugacity and melt composition controls on nitrogen solubility in silicate melts. Geochim. Cosmochim. Acta **284**, 120–133 (2020)
A.J. Brearley, The action of water, in *Meteorites and the Early Solar System II*, ed. by D. Lauretta, H.Y. McSween Jr. (University of Arizona Press, Tucson, 2006), pp. 587–624
R. Brasser, S.J. Mojzsis, The partitioning of the inner and outer Solar System by a structured protoplanetary disk. Nat. Astron. **4**, 492–499 (2020)
G. Budde, C. Burkhardt, G.A. Brennecka, M. Fischer-Gödde, T.S. Kruijer, T. Kleine, Molybdenum isotopic evidence for the origin of chondrules and a distinct genetic heritage of carbonaceous and non-carbonaceous meteorites. Earth Planet. Sci. Lett. **454**, 293–303 (2016)
G. Budde, C. Burkhardt, T. Kleine, Molybdenum isotopic evidence for the late accretion of outer Solar System material to Earth. Nat. Astron. **3**, 736–741 (2019)
C. Burkhardt, N. Dauphas, U. Hans, B. Bourdon, T. Kleine, Elemental and isotopic variability in solar system materials by mixing and processing of primordial disk reservoirs. Geochim. Cosmochim. Acta **261**, 145–170 (2019)
A.G.W. Cameron, The first ten million years in the solar nebula. Meteoritics **30**, 133–161 (1995)
A.G.W. Cameron, W. Truran, The supernova trigger for formation of the solar system. Icarus **30**, 447–461 (1977)
P. Cartigny, B. Marty, Nitrogen isotopes and mantle geodynamics: the emergence of life and the atmosphere-crust-mantle connection. Elements **9**, 359–366 (2013)
R.A. Chevalier, Young circumstellar disks near evolved massive stars and supernovae. Astrophys. J. **538**, L151–L154 (2000)
C.-L. Chou, Fractionation of siderophile elements in the Earth's upper mantle, in *Proceedings of the 9th Lunar and Planetary Science Conference* (1978), pp. 219–230
L.A. Cieza, S. Casassus, J. Tobin, S.P. Bos, J.P. Williams, S. Perez, Z. Zhu, C. Caceres, H. Canovas, M.M. Dunham, A. Hales, J.L. Prieto, D.A. Principe, M.R. Schreiber, D. Ruiz-Rodriguez, A. Zurlo, Imaging the water snow-line during a protostellar outburst. Nature **535**, 258–261 (2016)
R.N. Clayton, L. Grossman, T.K. Mayeda, A component of primitive nuclear composition in carbonaceous meteorites. Science **182**, 485–488 (1973)
R.N. Clayton, T.K. Mayeda, A.E. Rubin, Oxygen isotopic compositions of enstatite chondrites and aubrites. J. Geophys. Res. **89**, 245–249 (1984)
D.D. Clayton, Cosmic chemical memory: a new astronomy. Q. J. R. Astron. Soc. **23**, 174–212 (1982)

R.N. Clayton, Solar system: self-shielding in the solar nebula. Nature **415**, 860–861 (2002)
E.A. Cloutis, P. Hudon, T. Hiroi, M.J. Gaffey, P. Mann, J.F. Bell III., Spectral reflectance properties of carbonaceous chondrites: 6. CV chondrites. Icarus **221**, 328–358 (2012a)
E.A. Cloutis, P. Hudon, T. Hiroi, M.J. Gaffey, Spectral reflectance properties of carbonaceous chondrites: 7. CK chondrites. Icarus **221**, 911–924 (2012b)
J.N. Connelly, M. Bizzarro, Pb-Pb chronometry and the early Solar System. Geochim. Cosmochim. Acta **201**, 345–363 (2017)
J.N. Connelly, M. Bizzarro, A.N. Krot, A. Nordlund, D. Wielandt, M.A. Ivanova, The absolute chronology and thermal processing of solids in the solar protoplanetary disk. Science **338**, 651–655 (2012)
C. Dalou, E. Füri, C. Deligny, L. Piani, M.-C. Caumon, M. Laumonier, Redox control on nitrogen isotope fractionation during planetary core formation. Proc. Natl. Acad. Sci. USA **116**, 11485–14494 (2019)
N. Dauphas, The isotopic nature of the Earth's accreting material through time. Nature **541**, 521–524 (2017)
N. Dauphas, E.A. Schauble, Mass fractionation laws, mass-independent effects, and isotopic anomalies. Annu. Rev. Earth Planet. Sci. **44**, 709–783 (2016)
A.M. Davis, The cosmochemical history of the pallasites. PhD dissertation, Yale University (1977)
A.M. Davis, J. Zhang, N.D. Greber, J. Hu, F.L.H. Tissot, N. Dauphas, Titanium isotopes and rare Earth patterns in CAIs: evidence for thermal processing and gas-dust decoupling in the protoplanetary disk. Geochim. Cosmochim. Acta **221**, 275–295 (2018)
S.S. Davis, Condensation front migration in a protoplanetary nebula. Astrophys. J. **620**, 994–1001 (2005)
E. Deloule, F. Albarède, S.M.F. Sheppard, Hydrogen isotope heterogeneities in the mantle from ion probe analysis of amphiboles from ultramafic rocks. Earth Planet. Sci. Lett. **105**, 543–553 (1991)
E. Deloule, F. Robert, J.C. Doukhan, Interstellar hydroxyl in meteoritic chondrules: implications for the origin of water in the inner solar system. Geochim. Cosmochim. Acta **62**, 3367–3378 (1998)
S.J. Desch, A. Kalyaan, A.C.M. O'D, The effect of Jupiter's formation on the distribution of refractory elements and inclusions in meteorites. Astrophys. J. Suppl. Ser. **238**, 11 (2018)
S.E. Dodson-Robinson, K. Willacy, P. Bodenheimer, N.J. Turner, C.A. Beichman, Ice lines, planetesimal composition and solid surface density in the solar nebula. Icarus **200**, 672–693 (2009)
M.J. Drake, K. Righter, Determining the composition of the Earth. Nature **416**, 39–44 (2002)
V.V. Dwarkadas, N. Dauphas, B. Meyer, P. Boyajian, M. Bojazi, Triggered star formation inside the shell of a Wolf–Rayet bubble as the origin of the solar system. Astrophys. J. **851**, 147 (2017)
B. Fegley, Kinetics of gas-grain reactions in the solar nebula. Space Sci. Rev. **92**, 177–200 (2000)
B. Fegley, H. Palme, Evidence for oxidizing conditions in the solar nebula from Mo and W depletions in refractory inclusions in carbonaceous chondrites. Earth Planet. Sci. Lett. **72**, 311–326 (1985)
B. Fegley, K. Lodders, N.S. Jacobson, Volatile element chemistry during accretion of the Earth. Geochemistry **80**, 125594 (2020)
B. Fegley Jr., R.G. Prinn, Solar nebula chemistry: implications for volatiles in the solar system, in *The Formation and Evolution of Planetary Systems*, ed. by H.A. Weaver, L. Danly (Cambridge University Press, Cambridge, 1989), pp. 171–211
M. Fischer-Gödde, T. Kleine, Ruthenium isotopic evidence for an inner Solar System origin of the late veneer. Nature **541**, 525–527 (2017)
M. Fischer-Gödde, B.-M. Magnus, C. Münker, K. Szilas, W.D. Maier, N. Messling, T. Morishita, M. Van Kranendonk, H. Smithies, Ruthenium isotope vestige of Earth's pre-late-veneer mantle preserved in Archaean rocks. Nature **579**, 240–244 (2020)
C. Floss, P. Haenecour, Presolar silicate grains: abundance, isotopic and elemental compositions, and the effects of secondary processing. Geochem. J. **50**, 3–25 (2016)
P.N. Foster, A.P. Boss, Injection of radioactive nuclides from the stellar source that triggered the collapse of the presolar nebula. Astrophys. J. **489**, 346–357 (1997)
E. Füri, B. Marty, Nitrogen isotope variations in the Solar System. Nat. Geosci. **8**, 515–522 (2015)
E. Füri, P.H. Barry, L.A. Taylor, B. Marty, Indigenous nitrogen in the Moon: constraints from coupled nitrogen-noble gas analyses of mare basalts. Earth Planet. Sci. Lett. **431**, 195–205 (2015)
E. Füri, E. Deloule, R. Trappitsch, The production rate of cosmogenic deuterium at the Moon's surface. Earth Planet. Sci. Lett. **474**, 76–82 (2017)
P. Garaud, D.N.C. Lin, The effect of internal dissipation and surface irradiation on the structure of disks and the location of the snow line around Sun-like stars. Astrophys. J. **654**, 606–624 (2007)
S. Gerber, C. Burkhardt, G. Budde, K. Metzler, T. Kleine, Mixing and transport of dust in the early solar nebula as inferred from titanium isotope variations among chondrules. Astrophys. J. Lett. **841**, L17 (2017)
M. Gounelle, A. Morbidelli, P.A. Bland, P. Spurný, E.D. Young, M. Sephton, Meteorites from the outer solar system? in *The Solar System Beyond Neptune*, ed. by M.A. Barucci, H. Boehnhardt, D.P. Cruikshank, A. Morbidelli (University of Arizona Press, Tucson, 2008), pp. 525–541

M. Gounelle, P. Spurný, P.A. Bland, The orbit and atmospheric trajectory of the Orgueil meteorite from historical records. Meteorit. Planet. Sci. **41**, 135–150 (2006)
J. Gradie, E. Tedesco, Compositional structure of the asteroid belt. Science **216**, 1405–1407 (1982)
M.M. Grady, I.P. Wright, Elemental and isotopic abundances of carbon and nitrogen in meteorite. Space Sci. Rev. **106**, 231–248 (2003)
M.M. Grady, I.P. Wright, L.P. Carr, C.T. Pillinger, Compositional differences in enstatite chondrites based on carbon and nitrogen stable isotope measurements. Geochim. Cosmochim. Acta **50**, 2799–2813 (1986)
L. Grossman, J.W. Larimer, Early chemical history of the solar system. Rev. Geophys. **12**, 71–101 (1974)
F. Gyngard, S. Amari, E. Zinner, K.K. Marhas, Correlated silicon and titanium isotopic compositions of presolar SiC grains from Murchison CM2 chondrite. Geochim. Cosmochim. Acta **221**, 154–161 (2018)
L.J. Hallis, G.R. Huss, K. Nagashima, G.J. Taylor, S.A. Halldórsson, D.R. Hilton, M.J. Mottl, K.J. Meech, Evidence for primordial water in Earth' s deep mantle. Science **350**, 795–797 (2015)
K. Hashizume, N. Sugiura, Nitrogen isotopes in bulk ordinary chondrites. Geochim. Cosmochim. Acta **59**, 4057–4069 (1995)
C. Hayashi, Structure of the solar nebula, growth and degay of magnetic fields and effects of magnetic and turbulent viscosities on the nebula. Prog. Theor. Phys. Suppl. **70**, 35–53 (1981)
P.R. Heck, J. Greer, L. Kööp, R. Trappitsch, F. Gyngard, H. Busemann, C. Maden, J.N. Ávila, A.M. Davis, R. Wieler, Lifetimes of interstellar dust from cosmic ray exposure ages of presolar silicon carbide. Proc. Natl. Acad. Sci. **1107**, 1884–1889 (2020)
S. Henke, H.-P. Gail, M. Trielof, W.H. Schwarz, T. Kleine, Thermal history modelling of the H chondrite parent body. Astron. Astrophys. **545**, A135 (2012)
J.J. Hester, S.J. Desch, K.R. Healy, L.A. Leshin, The cradle of the solar system. Science **304**, 1116 (2004)
P.J. Hevey, I.S. Sanders, A model for planetesimal meltdown by ^{26}Al and its implications for meteorite parent bodies. Meteorit. Planet. Sci. **41**, 95–106 (2006)
D.C. Hezel, P.A. Bland, H. Palme, E. Jacquet, J. Bigolski, Compositions of chondrules and matrix and their complementary relationship in chondrites, in *Chondrules: Records of Protoplanetary Disk Processes*, ed. by S.S. Russell, H.C. Connolly Jr., A.N. Krot (Cambridge University Press, Cambridge, 2019), pp. 91–121
C.D. Hilton, K.R. Bermingham, R.J. Walker, T.J. McCoy, Genetics, crystallization sequence, and age of the South Byron Trio iron meteorites: new insights to carbonaceous chondrite (CC) type parent bodies. Geochim. Cosmochim. Acta **251**, 217–228 (2019)
M.M. Hirschmann, Comparative deep Earth volatile cycles: the case for C recycling from exosphere/ mantle fractionation of major (H_2O, C, N) volatiles and from H_2O/Ce, CO_2/Ba, and CO_2/Nb exosphere ratios. Earth Planet. Sci. Lett. **502**, 262–273 (2018)
N.E. Holden, M.L. Bonardi, P. De Bièvre, P.R. Renne, I.M. andVilla, IUPAC-IUGS common definition and convention on the use of the year as a derived unit of time (IUPAC recommendations 2011). Pure Appl. Chem. **83**, 1159–1162 (2011)
M. Honda, I. Mcdougall, D. Patterson, Solar noble gases in the Earth: the systematics of helium-neon isotopes in mantle derived samples. Lithos **30**, 257–265 (1993)
T. Hopp, G. Budde, T. Kleine, Heterogeneous accretion of Earth inferred from Mo-Ru isotope systematics. Earth Planet. Sci. Lett. **534**, 116065 (2020)
P. Hoppe, S. Amari, E. Zinner, T. Ireland, R.S. Lewis, Carbon, nitrogen, magnesium, silicon, and titanium isotopic compositions of single interstellar silicon carbide grains from the Murchison carbonaceous chondrite. Astrophys. J. **430**, 870–890 (1994)
G.R. Huss, J.B. Smith, Titanium isotopic compositions of well-characterized silicon carbide grains from Orgueil (CI): implications for s-process nucleosynthesis. Meteorit. Planet. Sci. **42**, 1055–1075 (2007)
M. Hutson, A. Ruzicka, A multi-step model for the origin of E3 (enstatite) chondrites. Meteorit. Planet. Sci. **35**, 601–608 (2000)
T.R. Ireland, E.K. Zinner, S. Amari, Isotopically anomalous Ti in presolar SiC from the Murchison meteorite. Astrophys. J. **376**, L53–L556 (1991)
M. Javoy, F. Pineau, H. Delorme, Carbon and nitrogen isotopes in the mantle. Chem. Geol. **57**, 41–62 (1986)
J.F. Kerridge, Carbon, hydrogen and nitrogen in carbonaceous chondrites: abundances and isotopic compositions in bulk samples. Geochim. Cosmochim. Acta **49**, 1707–1714 (1985)
R.S. Klessen, in *Star Formation in the Local Universe*. European Astronomical Society Publications Series, vol. 51 (2011), pp. 133–167
L. Kööp, A.M. Davis, D. Nakashima, C. Park, A.N. Krot, K. Nagashima, T.J. Tenner, P.R. Heck, N.T. Kita, A link between oxygen, calcium and titanium isotopes in ^{26}Al-poor hibonite-rich CAIs from Murchison and implications for the heterogeneity of dust reservoirs in the solar nebula. Geochim. Cosmochim. Acta **189**, 70–95 (2016)
S. Krijt, F.J. Ciesla, E.A. Bergin, Tracing water vapor and ice during dust growth. Astrophys. J. **883**, 285 (2016)

A.N. Krot, K. Nagashima, G. Libourel, K.E. Miller, Multiple mechanisms of transient heating events in the protoplanetary disk: evidence from precursors of chondrules and igneous Ca, Al-rich inclusions, in *Chondrules: Records of Protoplanetary Disk Processes*, ed. by S.S. Russell, H.C. Connolly Jr., A.N. Krot (Cambridge University Press, Cambridge, 2018), pp. 11–56

A.N. Krot, Refractory inclusions in carbonaceous chondrites: records of early solar system processes. Meteorit. Planet. Sci. **54**, 1647–1691 (2019)

T.S. Kruijer, M. Touboul, M. Fischer-Gödde, K.R. Bermingham, R.J. Walker, T. Kleine, Protracted core formation and rapid accretion of protoplanets. Science **344**, 1150–1154 (2014)

T.S. Kruijer, C. Burkhardt, G. Budde, T. Kleine, Age of Jupiter inferred from the distinct genetics and formation times of meteorites. Proc. Natl. Acad. Sci. USA **114**, 6712–6716 (2017)

T. Kudo, J. Hashimoto, T. Muto, H.B. Liu, R. Dong, Y. Hasegawa, T. Tsukagoshi, M. Konishi, A spatially resolved au-scale inner disk around DM Tau. Astrophys. J. Lett. **868**, L5 (2018)

C.C. Kung, R.N. Clayton, Nitrogen abundances and isotopic compositions in stony meteorites. Earth Planet. Sci. Lett. **38**, 421–435 (1978)

J. Labidi, P.H. Barry, D.V. Bekaert, M.W. Broadley, B. Marty, T. Giunta, O. Warr, B. Sherwood Lollar, T.P. Fischer, G. Avice, A. Caracausi, C.J. Ballentine, A. Halldórsson Stefánsson, M.D. Kurz, E.D. Young, Hydrothermal $^{15}N^{15}N$ abundances constrain the origins of mantle nitrogen. Nature **580**, 367–371 (2020)

T. Lee, D.A. Papanastassiou, G.J. Wasserburg, Aluminum-26 in the early solar system: fossil or fuel? Astrophys. J. **211**, L107–110 (1977)

T. Lee, F.H. Shu, S. Shang, A.E. Glassgold, K.E. Rehm, Protostellar cosmic rays and extinct radioactivities in meteorites. Astrophys. J. **506**, 898–912 (1998)

J.S. Lewis, The temperature gradient in the solar nebula. Science **186**, 440–443 (1974)

J.S. Lewis, R.G. Prinn, Kinetic inhibition of CO and N_2 reduction in the solar nebula. Astrophys. J. **238**, 357–364 (1980)

R.S. Lewis, T. Ming, J.F. Wacker, E. Anders, E. Steel, Interstellar diamonds in meteorites. Nature **326**, 160–162 (1987)

Y. Li, B. Marty, S. Shcheka, L. Zimmermann, H. Keppler, Nitrogen isotope fractionation during terrestrial core-mantle separation. Geochem. Perspect. Lett. **2**, 138–147 (2016)

G. Libourel, B. Marty, F. Humbert, Nitrogen solubility in basaltic melt. Part I. Effect of oxygen fugacity. Geochim. Cosmochim. Acta **67**, 4123–4135 (2003)

K. Lodders, An oxygen isotope mixing model for the accretion and composition of rocky planets. Space Sci. Rev. **92**, 341–354 (2000)

K. Lodders, Solar system abundances and condensation temperatures of the elements. Astrophys. J. **591**, 1220–1247 (2003)

K. Lodders, Jupiter formed with more tar than ice. Astrophys. J. **611**, 587–597 (2004)

K. Lodders, S. Amari, Presolar grains from meteorites: remnants from early times of the solar system. Chem. Erde **65**, 93–166 (2005)

K. Lodders, B. Fegley, *Chemistry of the Solar System* (Royal Society of Chemistry, Cambridge, 2011). 476 pp.

K. Lodders, R. Osborne, Perspectives on the comet-asteroid-meteorite link. Space Sci. Rev. **90**, 289–297 (1999)

G.J. MacPherson, Calcium-aluminum-rich inclusions in chondritic meteorites, in *Meteorites and Cosmochemical Processes*, vol. 1, ed. by A.M. Davis (Elsevier, Oxford, 2014), pp. 139–179. Treatise on Geochemistry, 2nd edn., ed. by H.D. Holland, K.K. Turekian

B. Marty, The origins and concentrations of water, carbon, nitrogen and noble gases on Earth. Earth Planet. Sci. Lett. **313–314**, 56–66 (2012)

B. Marty, M. Chaussidon, R.C. Wiens, A.J.G. Jurewicz, D.S. Burnett, A ^{15}N-poor isotopic composition for the Solar System as shown by Genesis solar wind samples. Science **332**, 1533–1536 (2011)

B. Marty, G. Avice, Y. Sano, K. Altwegg, H. Balsiger, M. Hässig, A. Morbidelli, O. Mousis, M. Rubin, Origins of volatile elements (H, C, N, noble gases) on Earth and Mars in light of recent results from the ROSETTA cometary mission. Earth Planet. Sci. Lett. **441**, 91–102 (2016)

B. Mason, The enstatite chondrites. Geochim. Cosmochim. Acta **30**, 23–39 (1966)

K.J. Mathew, K. Marti, Early evolution of Martial volatiles: nitrogen and noble gas components in ALH84001 and Chassigny. J. Geophys. Res. **106**, 1401–1422 (2001)

B. Mayer, N. Wittig, M. Humayun, I. Leya, Palladium isotopic evidence for nucleosynthetic and cosmogenic isotope anomalies in IVB iron meteorites. Astrophys. J. **809**, 180–188 (2015)

W.F. McDonough, S.-S. Sun, The composition of the Earth. Chem. Geol. **120**, 223–253 (1995)

R. Meijerink, K.M. Pontoppidan, G.A. Blake, D.R. Poelman, C.P. Dullemond, Radiative transfer models of mid-infrared H_2O lines in the planet-forming region of circumstellar disks. Astrophys. J. **704**, 1471–1481 (2009)

S. Mikhail, E. Füri, On the origin(s) and evolution of Earth's carbon. Elements **15**, 307–312 (2019)
A. Morbidelli, B. Bitsch, A. Crida, M. Gounelle, T. Guillot, S. Jacobson, A. Johansen, M. Lambrechts, E. Lega, Fossilized condensation lines in the Solar System protoplanetary disk. Icarus **267**, 368–376 (2016)
A. Morbidelli, K.J. Walsh, D.P. O'Brien, D.A. Minton, W.F. Bottke, Dynamical evolution of the asteroid belt, in *Asteroids IV*, ed. by P. Michel, F. DeMeo, W.F. Bottke (University of Arizona Press, Tucson, 2015), pp. 493–508
A. Morbidelli, J.I. Lunine, D.P.O. Brien, S.N. Raymond, K.J. Walsh, Building terrestrial planets. Annu. Rev. Earth Planet. Sci. **40**, 251–278 (2012)
N. Moskovitz, E. Gaidos, Differentiation of planetesimals and the thermal consequences of melt migration. Meteorit. Planet. Sci. **46**, 903–918 (2011)
H. Nakano, A. Kouchi, S. Tachibana, A. Tsuchiyama, Evaporation of interstellar organic materials in the solar nebula. Astrophys. J. **592**, 1252–1262 (2003)
H. Nakano, N. Hirakaw, Y. Matsubara, S. Yamashita, T. Okuchi, K. Asahina, R. Tanaka, N. Suzuki, H. Naraoka, Y. Takano, S. Tachibana, T. Hama, Y. Oba, Y. Kimura, N. Watanabe, A. Kouchi, Precometary organic matter: a hidden reservoir of water inside the snow line. Sci. Rep. **10**, 7755 (2020)
J.A.M. Nanne, F. Nimmo, J.N. Cuzzi, T. Kleine, Origin of the non-carbonaceous–carbonaceous meteorite dichotomy. Earth Planet. Sci. Lett. **511**, 44–54 (2019)
S. Niemeyer, Titanium isotope anomalies in chondrules from carbonaceous chondrites. Geochim. Cosmochim. Acta **52**, 309–318 (1988)
L.R. Nittler, F. Ciesla, Astrophysics with extraterrestrial materials. Annu. Rev. Astron. Astrophys. **54**, 53–93 (2016)
A. Oka, T. Nakamoto, S. Ida, Evolution of snow line in optically thick protoplanetary disks: effect of water ice opacity and dust grain size. Astrophys. J. **738**, 141 (2011)
U. Ott, Planetary and pre-solar noble gases in meteorites. Chem. Erde **74**, 519–544 (2014)
H. Palme, H. O'Neill, Cosmochemical estimates of mantle composition, in *The Mantle and Core*, vol. 3, ed. by R.W. Carlson (2003), pp. 1–39. Treatise on Geochemistry, vol. 3, 2nd edn., ed. by H.D. Holland, K.K. Turekian (Elsevier)
H. Palme, K. Lodders, A. Jones, Solar system abundances of the elements, in *Treatise on Geochemistry*, vol. 2, ed. by H.D. Holland, K.K. Turekian 2nd edn. (Elsevier, Oxford, 2014), pp. 15–36
M. Palot, P. Cartigny, J.W. Harris, F.V. Kaminsky, T. Stachel, Evidence for deep mantle convection and primordial heterogeneity from nitrogen and carbon stable isotopes in diamond. Earth Planet. Sci. Lett. **357–358**, 179–193 (2012)
J. Pape, K. Mezger, A.-S. Bouvier, L.P. Baumgartner, Time and duration of chondrule formation: constraints from ^{26}Al-^{26}Mg ages of individual chondrules. Geochim. Cosmochim. Acta **244**, 416–436 (2019)
A. Patzer, L. Schultz, Noble gases in enstatite chondrites II: the trapped component. Meteorit. Planet. Sci. **37**, 601–612 (2002)
V.K. Pearson, M.A. Sephton, I. Gilmour, I. Franchi, Hydrogen isotopic composition of the Tagish Lake meteorite: comparison with other carbonaceous chondrites, in *32nd Annual Lunar and Planetary Science Conference, Abstract 1861* (2001)
V.K. Pearson, M.A. Sephton, I.A. Franchi, J.M. Gibson, I. Gilmour, Carbon and nitrogen in carbonaceous chondrites: elemental abundances and stable isotopic compositions. Meteorit. Planet. Sci. **41**, 1899–1918 (2006)
L. Piani, F. Robert, L. Remusat, Micron-scale D/H heterogeneity in chondrite matrices: a signature of the pristine solar system water? Earth Planet. Sci. Lett. **415**, 154–164 (2015)
L. Piani, Y. Marrocchi, T. Rigaudier, L.G. Vacher, D. Thomassin, B. Marty, Earth's water may have been inherited from material similar to enstatite chondrite meteorites. Science **369**, 1110–1113 (2020).
G.M. Poole, M. Rehkämper, B.J. Coles, T. Goldberg, C.L. Smith, Nucleosynthetic molybdenum isotope anomalies in iron meteorites – new evidence for thermal processing of solar nebula material. Earth Planet. Sci. Lett. **473**, 215–226 (2017)
L. Qin, R.W. Carlson, Nucleosynthetic isotope anomalies and their cosmochemical significance. Geochem. J. **50**, 43–65 (2016)
S.N. Raymond, D.P. O'Brien, A. Morbidelli, N.A. Kaib, Building the terrestrial planets: constrained accretion in the inner solar system. Icarus **203**, 644–662 (2009)
S.N. Raymond, A. Izidoro, Origin of water in the inner solar system: planetesimals scattered inward during Jupiter and Saturn's rapid gas accretion. Icarus **297**, 134–148 (2017)
S.N. Raymond, A. Izidoro, A. Morbidelli, Solar system formation in the context of extrasolar planets, in *Planetary Astrobiology*, ed. by V. Meadows, G.N. Arney, B.E. Schmidt D.J. Marais (University of Arizona Press, Tucson, 2020), pp. 287–324
J. Render, S. Ebert, C. Burkhardt, T. Kleine, G.A. Brennecka, Titanium isotopic evidence for a shared genetic heritage of refractory inclusions from different carbonaceous chondrites. Geochim. Cosmochim. Acta **254**, 40–53 (2019)

J.H. Reynolds, G. Turner, Rare gases in chondrite Renazzo. J. Geophys. Res. **49**, 3263–3281 (1964)

F. Robert, The D/H ratio in chondrites. Space Sci. Rev. **106**, 87–101 (2003)

F. Robert, S. Epstein, The concentration and isotopic composition of hydrogen, carbon and nitrogen in carbonaceous meteorites. Geochim. Cosmochim. Acta **46**, 81–95 (1982)

D.C. Rubie, S.A. Jacobson, A. Morbidelli, D.P. O'Brien, E.D. Young, J. de Vries, F. Nimmo, H. Palme, D.J. Frost, Accretion and differentiation of the terrestrial planets with implications for the composition of early-formed Solar System bodies. Icarus **248**, 89–108 (2015)

A.E. Rubin, B. Fegley, R. Brett, Oxidation state in chondrites, in *Meteorites and the Early Solar System*, ed. by J.F. Kerridge, M.S. Mathews (University of Arizona Press, Tucson, 1988), pp. 488–511

A.E. Rubin, B.-G. Choi, Origin of halogens and nitrogen in enstatite chondrites. Earth Moon Planets **105**, 41–53 (2009)

A.E. Rubin, J.T. Wasson, Variations of chondrite properties with heliocentric distance. Meteoritics **30**, 569 (1995) (Abstr.)

A.E. Rubin, C. Ma, Meteoritic minerals and their origins. Geochemistry **77**, 325–385 (2017)

S. Sahijpal, P. Soni, G. Gupta, Numerical simulations of the differentiation of accreting planetesimals with ^{26}Al and ^{60}Fe as the heat sources. Meteorit. Planet. Sci. **42**, 1529–1548 (2007)

D.D. Sasselov, M. Lecar, On the snow line in dusty protoplanetary disks. Astrophys. J. **528**, 995–998 (2000)

T. Sato, S. Okuzumi, S. Ida, On the water delivery to terrestrial embryos by ice pebble accretion. Astron. Astrophys. **589**, A15 (2016)

E.R.D. Scott, Formation of olivine-metal textures in pallasite meteorites. Geochim. Cosmochim. Acta **41**, 693–710 (1977)

E.R.D. Scott, A.N. Krot, Chondritic meteorites and the high-temperature nebular origins of their components, in *Chondrites and the Protoplanetary Disk*, ed. by A.N. Krot et al. ASP Conference Series, vol. 341 (Astronomical Society of the Pacific, San Francisco, 2005), pp. 15–52

E.R.D. Scott, A.N. Krot, I.S. Sanders, Isotopic dichotomy among meteorites and its bearing on the protoplanetary disk. Astrophys. J. **854**, 164 (2018)

F.H. Shu, F.C. Adams, S. Lizano, Star formation in molecular clouds: observation and theory. Annu. Rev. Astron. Astrophys. **25**, 23–72 (1987)

F.H. Shu, X-rays and fluctuating X-winds from protostars. Science **277**, 1475–1479 (1997)

F.H. Shu, H. Shang, M. Gounelle, A.E. Glassgold, T. Lee, The origin of chondrules and refractory inclusions in chondritic meteorites. Astrophys. J. **548**, 1029–1050 (2001)

A. Shukolyukov, G.W. Lugmair, Live iron-60 in the early solar system. Science **259**, 1138–1142 (1993a)

A. Shukolyukov, G.W. Lugmair, ^{60}Fe in eucrites. Earth Planet. Sci. Lett. **119**, 159–166 (1993b)

F. Spitzer, C. Burkhardt, G. Budde, T.S. Kruijer, A. Morbidelli, T. Kleine, Isotopic evolution of the inner Solar System inferred from molybdenum isotopes in meteorites. Astrophys. J. Lett. **898**, L2 (2020)

T. Stephan, A.M. Davis, Molybdenum in presolar silicon carbide grains reveal details of s-, r-, and p-process nucleosynthesis, in *82nd Annual Meeting of the Meteoritical Society* (2019). LPI Contrib. No. 2157, abstract #6107

A. Stephant, L.A.J. Garvie, P. Mane, R. Hervig, M. Wadhwa, Terrestrial exposure of a fresh Martian meteorite causes rapid changes in hydrogen isotopes and water concentrations. Sci. Rep. **8**, 12385 (2018)

N. Sugiura, W. Fujiya, Correlated accretion ages and ε^{54}Cr of meteorite parent bodies and the evolution of the solar nebula. Meteorit. Planet. Sci. **49**, 772–787 (2014)

S. Sutton, C.M.O'D. Alexander, A. Bryant, A. Lanzirotti, M. Newville, E.A. Cloutis, The bulk valence state of Fe and the origin of water in chondrites. Geochim. Cosmochim. Acta **211**, 115–132 (2017)

S.R. Taylor, *Solar System Evolution*, 2nd edn. (Cambridge University Press, Cambridge, 2001). 484 pp.

A. Takigawa, R.M. Stroud, L.R. Nittler, C.M.O'D. Alexander, A. Miyake, High-temperature dust condensation around an AGB star: evidence from a highly pristine presolar corundum. Astrophys. J. **862**, L13 (2018)

M.H. Thiemens, J.E. Heidenreich, The mass independent fractionation of oxygen: a novel isotope effect and its possible cosmochemical implications. Science **219**, 1073–1075 (1983)

Z.A. Torrano, G.A. Brennecka, C.D. Williams, S.J. Romaniello, V.K. Rai, R.R. Hines, M. Wadhwa, Titanium isotope signatures of calcium-aluminum-rich inclusions from CV and CK chondrites: implications for early Solar System reservoirs and mixing. Geochim. Cosmochim. Acta **263**, 13–30 (2019)

J.M. Trigo-Rodríguez, A. Rimola, S. Tanbakouei, V. Cabedo Soto, M. Lee, Accretion of water in carbonaceous chondrites: current evidence and implications for the delivery of water to early Earth. Space Sci. Rev. **215**, 18 (2019)

A. Trinquier, J.L. Birck, C.J. Allègre, Widespread ^{54}Cr heterogeneity in the inner solar system. Astrophys. J. **655**, 1179–1185 (2007)

A. Trinquier, T. Elliott, D. Ulfbeck, C. Coath, A.N. Krot, M. Bizzarro, Origin of nucleosynthetic isotope heterogeneity in the Solar protoplanetary disk. Science **324**, 374–376 (2009)

H.C. Urey, *The Planets: Their Origin and Development* (Yale University Press, New Haven, 1952), 245 pp.

L. Vacher, L. Piani, T. Rigaudier, D. Thomassin, G. Florin, M. Piralla, Y. Marrocchi, Hydrogen in chondrites: influence of parent body alteration and atmospheric contamination on primordial components. Geochim. Cosmochim. Acta **281**, 53–66 (2020)

P. Vernazza, B. Zanda, R.P. Binzel, T. Hiroi, F.E. DeMeo, M. Birlan, R. Hewins, L. Ricci, P. Barge, M. Lockhart, Multiple and fast: the accretion of ordinary chondrite parent bodies. Astrophys. J. **791**, 120–144 (2014)

P. Vernazza, M. Marsset, P. Beck, R.P. Binzel, M. Birlan, E.A. Cloutis, F.E. DeMeo, C. Dumas, T. Hiroi, Compositional homogeneity of CM parent bodies. Astrophys. J. **152**, 54–64 (2016)

J. Villeneuve, M. Chaussidon, G. Libourel, Homogeneous distribution of ^{26}Al in the Solar System from the Mg isotopic composition of chondrules. Science **325**, 985–988 (2009)

R. Visser, E.F. van Dishoeck, S.D. Doty, C.P. Dullemond, The chemical history of molecules in circumstellar disks. Astron. Astrophys. **495**, 881–897 (2009)

R.J. Walker, Highly siderophile elements in the Earth, Moon and Mars: update and implications for planetary accretion and differentiation. Geochemistry **69**, 101–125 (2009)

R.J. Walker, K. Bermingham, J. Liu, I.S. Puchtel, M. Touboul, E.A. Worsham, In search of late-stage planetary building blocks. Chem. Geol. **411**, 125–142 (2015)

N.P. Walte, G.F.D. Solferino, G.J. Golabek, D.S. Souza, A. Bouvier, Two-stage formation of pallasites and the evolution of their parent bodies revealed by deformation experiments. Earth Planet. Sci. Lett. **546**, 116419 (2020)

K.J. Walsh, A. Morbidelli, S.N. Raymond, D.P. O'Brien, A.M. Mandell, A low mass for Mars from Jupiter's early gas-driven migration. Nature **475**, 206–209 (2011)

K.J. Walsh, H.F. Levison, Planetesimals to terrestrial planets: collisional evolution amidst a dissipating gas disk. Icarus **329**, 88–100 (2019)

H. Wang, R.C. Bell, M.J. Iemeda, A.A. Tsekouras, J.P. Cowin, Sticky ice grains aid planet formation: unusual properties of cryogenic water ice. Astrophys. J. **620**, 1028–1032 (2005)

P.H. Warren, Stable-isotopic anomalies and the accretionary assemblage of the Earth and Mars: a subordinate role for carbonaceous chondrites. Earth Planet. Sci. Lett. **311**, 93–100 (2011)

J.T. Wasson, *Meteorites: Their Record of Early Solar-System History* (W.H. Freeman, New York, 1985)

J.T. Wasson, The building stones of the planets, in *Mercury*, ed. by F. Vilas, C.R. Chapman, M.S. Matthews (University of Arizona Press, Tucson, 1988), pp. 622–650

J.T. Wasson, G.W. Kallemeyn, Compositions of chondrites. Philos. Trans. R. Soc. A, Math. Phys. Eng. Sci. **32**, 391–403 (1988)

M.K. Weisberg, T.J. McCoy, A.N. Krot, Systematics and evaluation of meteorite classification, in *Meteorites and the Early Solar System II*, ed. by D.D. Lauretta, H.Y. McSween (University of Arizona Press, Tucson, 2006), pp. 19–52

C.D. Williams, S. Muhkopadyay, Capture of nebular gases during Earth's accretion is preserved in deep-mantle neon. Nature **565**, 78–81 (2018)

J.P. Williams, L.A. Cieza, Protoplanetary disks and their evolution. Annu. Rev. Astron. Astrophys. **49**, 67–117 (2011)

J.A. Wood, The chondrite types and their origins, in *Chondrites and the Protoplanetary Disk*, ed. by A.N. Krot, E.R.D. Scott, B. Reipurth (Astronomical Society of the Pacific, Hawai'i, 2005), pp. 953–971

B.J. Wood, J. Li, A. Shahar, Carbon in the core: its influence on the properties of core and mantle. Rev. Mineral. Geochem. **75**, 231–250 (2013)

E.A. Worsham, K.R. Bermingham, R.J. Walker, Characterizing cosmochemical materials with genetic affinities to the Earth: genetic and chronological diversity within the IAB iron meteorite complex. Earth Planet. Sci. Lett. **467**, 157–166 (2017)

J. Wu, S.J. Desch, L. Schaefer, L.T. Elkins-Tanton, K. Pahlevan, P.R. Buseck, Origin of Earth's water: chondritic inheritance plus nebular ingassing and storage of hydrogen in the core. J. Geophys. Res., Planets **123**, 2691–2712 (2018)

L. Yang, F.J. Ciesla, The effects of disk building on the distributions of refractory materials in the solar nebula. Meteorit. Planet. Sci. **47**, 99–119 (2012)

J. Yang, J.I. Goldstein, E.R.D. Scott, Iron meteorite evidence for early formation and catastrophic disruption of protoplanets. Nature **446**, 888–891 (2007)

J. Yang, J.I. Goldstein, E.R.D. Scott, Main-group pallasites: thermal history, relationship to IIIAB iron, and origin. Geochim. Cosmochim. Acta **74**, 4471–4492 (2010)

R. Yokochi, B. Marty, A determination of the neon isotopic composition of the deep mantle. Earth Planet. Sci. Lett. **225**, 77–88 (2004)

T. Yokoyama, Y. Nagai, R. Fukai, T. Hirata, Origin and evolution of distinct molybdenum isotopic variabilities within carbonaceous and non-carbonaceous reservoirs. Astrophys. J. **883**, 62 (2019)

J. Zhang, N. Dauphas, A.M. Davis, I. Leya, A. Fedkin, The proto-Earth as a significant source of lunar material. Nat. Geosci. **5**, 251–255 (2012)

K. Zhang, G.A. Blake, E.A. Bergin, Evidence of fast pebble growth near condensation fronts in the HL Tau protoplanetary disk. Astrophys. J. Lett. **806**, L7 (2015)

E. Zinner, Presolar grains, in *Meteorites and Cosmochemical Processes*, ed. by A.M. Davis (Elsevier, New York, 2014), pp. 181–213. Treatise on Geochemistry, vol. 1, 2nd edn., ed. by H.D. Holland, K.K. Turekian

E. Zinner, M. Tang, E. Anders, Nature **330**, 730–732 (1987)

E. Zinner, S. Amari, R. Guinness, C. Jennings, A.F. Mertz, A.N. Nguyen, R. Gallino, P. Hoppe, M. Lugaro, L.R. Nittler, R.S. Lewis, NanoSIMS isotopic analysis of small presolar grains: search for Si_3N_4 grains from AGB stars and Al and Ti isotopic compositions of rare presolar SiC grains. Geochim. Cosmochim. Acta **71**, 4786–4813 (2007)

Space Sci Rev (2021) 217:7
https://doi.org/10.1007/s11214-020-00778-4

Formation of Venus, Earth and Mars: Constrained by Isotopes

Helmut Lammer[1] · Ramon Brasser[2] ·
Anders Johansen[3] · Manuel Scherf[1] · Martin Leitzinger[4]

Received: 26 November 2019 / Accepted: 28 November 2020 / Published online: 22 December 2020

Abstract Here we discuss the current state of knowledge of terrestrial planet formation from the aspects of different planet formation models and isotopic data from ^{182}Hf-^{182}W, U-Pb, lithophile-siderophile elements, ^{48}Ca/^{44}Ca isotope samples from planetary building blocks, recent reproduction attempts from ^{36}Ar/^{38}Ar, ^{20}Ne/^{22}Ne, ^{36}Ar/^{22}Ne isotope ratios in Venus' and Earth's atmospheres, the expected solar ^{3}He abundance in Earth's deep mantle and Earth's D/H sea water ratios that shed light on the accretion time of the early protoplanets. Accretion scenarios that can explain the different isotope ratios, including a Moon-forming event ca. 50 Myr after the formation of the Solar System, support the theory that the bulk of Earth's mass ($\geq$80%) most likely accreted within 10–30 Myr. From a combined analysis of the before mentioned isotopes, one finds that proto-Earth accreted most likely a mass of 0.5–0.6 M_{Earth} within the first $\approx$3–4.5 Myr, the approximate lifetime of the protoplanetary disk. For Venus, the available atmospheric noble gas data are too uncertain for constraining the planet's accretion scenario accurately. However, from the available imprecise Ar and Ne isotope measurements, one finds that proto-Venus could have grown to a mass of up to 0.85–1.0 M_{Venus} before the disk dissipated. Classical terrestrial planet formation models have struggled to grow large planetary embryos, or even cores of giant planets, quickly from the tiniest materials within the typical lifetime of protoplanetary disks. Pebble accretion could solve this long-standing time scale controversy. Pebble accretion and streaming instabilities produce large planetesimals that grow into Mars-sized and larger planetary embryos during this early accretion phase. The later stage of accretion can be explained well

Reading Terrestrial Planet Evolution in Isotopes and Element Measurements
Edited by Helmut Lammer, Bernard Marty, Aubrey L. Zerkle, Michel Blanc, Hugh O'Neill and Thorsten Kleine

✉ H. Lammer

[1] Space Research Institute, Austrian Academy of Sciences, Schmiedlstr. 6, 8042 Graz, Austria

[2] Earth-Life Science Institute, Tokyo Institute of Technology, 2-12-1 Ookayama, Meguro-ku, Tokyo 152-8551, Japan

[3] Lund Observatory, Department of Astronomy and Theoretical Physics, Lund University, Box 43, 22100 Lund, Sweden

[4] Institute of Physics/IGAM, University of Graz, Universitätsplatz 5/II, 8010 Graz, Austria

with the Grand-Tack model as well as the annulus and depleted disk models. The relative roles of pebble accretion and planetesimal accretion/giant impacts are poorly understood and should be investigated with N-body simulations that include pebbles and multiple protoplanets. To summarise, different isotopic dating methods and the latest terrestrial planet formation models indicate that the accretion process from dust settling, planetesimal formation, and growth to large planetary embryos and protoplanets is a fast process that occurred to a great extent in the Solar System within the lifetime of the protoplanetary disk.

Keywords Terrestrial planet formation · Pebble accretion · Noble gases · Isotopes

1 Introduction

Terrestrial planets accrete mass by numerous collisions between small objects that accumulate into larger ones, thereby forming planetesimals, and Moon- and Mars-mass planetary embryos that eventually grow into planets. An illustration of a schematic time scale of the accretion of terrestrial planets is shown in Fig. 1. One can see four main accretion stages: i) dust settling; ii) planetesimal formation; iii) formation of Moon- to Mars-mass planetary embryos; and iv) the accretion of protoplanets up to terrestrial planetary masses by giant impacts (e.g., Morbidelli et al. 2012; Brasser 2013; Johansen et al. 2015; Birnstiel et al. 2016). The bulk composition of the silicate part of terrestrial planets has long been linked to a fraction of carbonaceous chondritic meteorites (CI, CV, CM, etc.), which was mixed with enstatite chondrites (EC) and ordinary chondrites (OC) according to mainly oxygen isotope evidence (Drake and Righter 2002; Dauphas 2017; Bermingham et al. 2020, this journal), and/or ureilite-like material based on recent analysed $^{48}Ca/^{44}Ca$ data in samples that originated from ureilite and angrite meteorites, as well as from Vesta, Mars and Earth (Schiller et al. 2018).

As illustrated in Fig. 1, depending on their speed of growth before or after disk dissipation and their orbital locations of origin/destination, protoplanets will evolve into objects with different volatile and elemental abundances, which results in different elemental ratios compared to their initial ones due to subsequent atmospheric escape and/or impact erosion of atmospheres and crustal/mantle material (e.g., Lammer et al. 2020a; this journal; O'Neill et al. 2020; this journal).

Safronov (1969) proposed the first physical model for the formation of the terrestrial planets. He suggested that the earliest stage of accumulation of dust into planetesimals was caused by gravitational instabilities in a thin dust layer in the midplane of the protoplanetary disk. Due to their gravitational interaction, the ensuing planetesimals are stirred and can encounter each other. Safronov (1969) showed that the relative velocities between planetesimals are typically of the order of their surface escape velocity, so that each planetesimal's gravitational cross section is limited by the geometrical one, restricting the accretion rate. Wetherill (1980) built upon these ideas and suggested that the terrestrial planets coagulated from planetesimals, and that the formation of these planets was linked with the evolution of the asteroid belt. Wetherill and Stewart (1989) subsequently showed that in a disk of planetesimals, some would undergo runaway growth and form a sequence of planetary embryos. These embryos would then further collide to form the terrestrial planets.

The past decades have seen much progress in understanding the formation of the terrestrial planets. Models that fully explain the formation of terrestrial planets in the inner Solar System have to explain how planetesimals grow to $\approx$100 km-scale bodies because of dust coagulation (Johansen and Lambrechts 2017). These planetesimals then fur-

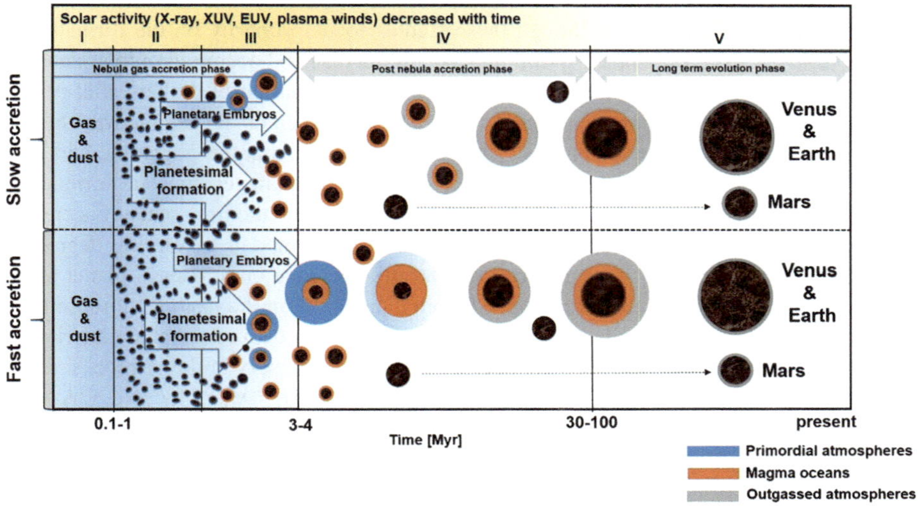

Fig. 1 Illustration of the five main formation stages of the terrestrial planet formation: dust settling (I), planetesimal formation (II), formation of planetary embryos (III), giant impacts including Earth's Moon-forming event (IV). The long-term evolution phase where secondary atmospheres origin (V). The scenarios above the horizontal dashed line illustrate so-called slow accretion scenarios where terrestrial planets accrete most of their mass after the disk dissipated. The scenario below the dashed-horizontal line corresponds to a two-stage accretion scenario according to Jacobsen and Harper (1996), Harper and Jacobsen (1996), and Yu and Jacobsen (2011). Here, terrestrial planets accrete faster so that they can capture primordial atmospheres, which subsequently escape while the proto-planets accrete their final masses. The shaded zone represents the presence of the protoplanetary nebula in the Solar System that dissipated after $\approx$3.3–4.5 Myr (Bollard et al. 2017; Wang et al. 2017). In both scenarios, the planets outgas steam atmospheres when their magma oceans solidify, and secondary atmospheres build up later-on due to volcanic outgassing and tectonic activities

ther coagulate into Mars-mass or even larger planetary embryos, most likely by planetesimal accretion (Kokubo and Ida 1995; Kokubo and Ida 1998). Finally, terrestrial planet formation models that form Venus, Earth and Mars must match constraints related to the involved planetary embryos and asteroids, including their orbital and compositional distributions as well as their isotope and elemental ratios (Morbidelli et al. 2012; Brasser et al. 2017).

At present, there exists a great variety of terrestrial planet formation models: the Classical model (e.g., Wetherill and Stewart 1989; Chambers 2001; Raymond et al. 2006, 2009), Grand Tack model (e.g., Walsh et al. 2011; Brasser et al. 2017, 2018; Woo et al. 2018), annulus model (Agnor et al. 1999; Hansen 2009; Walsh and Levison 2016), depleted disk model (e.g., Izidoro et al. 2014, 2015; Raymond and Izidoro 2017a) and early instability model (Clement et al. 2019). In contrast, giant planet formation in the solar system has not been studied as extensively (Helled et al. 2013). As such, most of the terrestrial planet formation models in the literature presuppose that terrestrial and giant planet formation can be treated separately because they frequently assume that the giant planets have already formed and reside on their current orbits. Only the Grand Tack model includes the migration of the giant planets. The models listed above tend to study the later stages of terrestrial planet formation, usually focused on the giant impact stage (Chambers 2001) that follows the oligarchic growth phase (Kokubo and Ida 1998). However, we would argue that an additional model of terrestrial planet formation can be constructed based on pebble accretion (e.g., Ormel and Klahr 2010; Lambrechts and Johansen 2012; Lambrechts et al. 2014;

Drazkowska et al. 2016). Even though these studies focus primarily on the formation of planets from a stream of pebbles spiralling through the disc and accreting onto a planetary core, Levison et al. (2015b) constructed a model of the formation of the terrestial planets from pebble accretion starting from a planetesimal disk and then ran this all the way through the oligarchic and giant impact phases. Their model suggests that the oligarchic and giant impact phases are a natural outcome irrespective of how the planetesimals are formed and what they subsequently accrete. The only potential difference is the time scale of accretion.

Here we discuss most of the terrestrial planet formation models in varying degrees of depth (Sect. 2). In Sect. 3, we summarise the latest knowledge on terrestrial planet accretion, including time scales that are obtained from data based on the isotopic chronometers ^{182}Hf-^{182}W and U-Pb, lithophile-siderophile elemental isotope data of accreted material, the latest data analysis from ^{48}Ca/^{44}Ca isotope samples from planetary building blocks, recent reproduction attempts of Venus' and Earth's atmospheric ^{36}Ar/^{38}Ar, ^{20}Ne/^{22}Ne, ^{36}Ar/^{22}Ne isotope ratios, the evidence of solar ^{3}He abundance in Earth's deep mantle, and the D/H ratio in Earth's sea water. These findings are then compared with the various terrestrial planet formation models. Finally, Sect. 4 concludes the review.

2 Terrestrial Planet Formation Models

2.1 The Classical Model for Terrestrial Planet Formation

Wetherill and Stewart (1989) demonstrated that in a protoplanetary disk consisting of equal-mass planetesimals, some would undergo rapid collisions with other planetesimals leading to a phase of runaway growth wherein the mass of the planet increases as $\mathrm{d}M/\mathrm{d}t \propto M^{\beta}$, where β is a free parameter. When $\beta > 1$ runaway growth occurs. In contrast, if $\beta < 1$ then the accretion is more egalitarian and no specific body will begin to dominate over the others.

During the earliest stages of planet formation $\beta = 4/3$ due to effective gravitational focusing (Ida and Makino 1993; Kokubo and Ida 1996). Here the planetesimals gravitationally stir each other (Ida and Makino 1993), resulting in their eccentricities and inclinations slowly increasing. Once the eccentricities reach orbit-crossing values, the planetesimals will eventually collide. Under the assumption of perfect mergers, the formation enters a runaway growth phase in which the size-frequency distribution of the bodies in the disk becomes approximately bimodal (Kokubo and Ida 1996). During this phase the solids in the disk consist of approximately lunar- to Mars-sized protoplanets (dubbed planetary embryos) spaced by about 10 mutual Hill radii, which are embedded in a sea of planetesimals (Kokubo and Ida 1998). The Hill radius of a planet is given by $R_{\mathrm{H}} = a(m_{\mathrm{pl}}/3M_{\mathrm{star}})^{1/3}$, where a is the semi-major axis, m_{pl} and M_{star} are the mass of the planet and the star; the mutual Hill radius uses the sum of the masses of two neighbouring planets and the mean of their semi-major axes. During the bimodal phase the planetary embryos gravitationally perturb the planetesimals and each other (Kokubo and Ida 1998). If no gas remains from the protoplanetary disk the system would be unstable, but if any residual gas remains then the eccentricities and inclinations of all bodies are kept low due to damping forces from the gas, and the system is dynamically quasi-stable (Kominami and Ida 2002). The planetary embryos all grow at roughly the same rate as they vie for space and for their ability to accrete planetesimals. This phase is called the oligarchic growth stage, and approximately half of the solid mass within the disk is in the embryos, with the other half being in planetesimals (Kokubo and Ida 1998).

The gas density from the solar nebula decreases with time (e.g. Hartmann et al. 1998) so that the damping forces from the gas disk on the embryos also weaken with time. There comes a point in time when gravitational stirring overcomes the tidal forces from the gas disk (Kominami and Ida 2002, 2004; Kominami et al. 2005). By then, the mutual gravitational perturbations from the embryos cause their eccentricities to rise to orbit-crossing values so that the whole chain eventually becomes dynamically unstable and the system subsequently evolves into a giant impact phase. Here planetary mergers through large impacts are commonplace. Over tens of millions of years these mutual protoplanetary collisions eventually lead to the observed terrestrial planet inventory (e.g. Chambers and Wetherill 1998; Chambers 2001; Raymond et al. 2006, 2009).

The so-called Classical model is based on the evolution described above. Due to limitations in computing power, however, the runaway growth was extremely difficult to model so that simulations of this model began with the Sun being surrounded by a disk of planetesimals and protoplanets that are gravitationally interacting (e.g. Raymond et al. 2006). Typical simulations of the Classical model assume that the gas from the protoplanetary disk has already dissipated (Chambers 2001), and that Jupiter and Saturn have already formed and reside on their current orbits. More recent iterations invoke the presence of gas from the protoplanetary disk due to advances in the theory of planet migration and eccentricity damping (e.g. Tanaka et al. 2002; Tanaka and Ward 2004; Cresswell and Nelson 2008; Paardekooper et al. 2011). In N-body simulations of the model, the protoplanets are usually assumed to have the same size and mass; the same is true for planetesimals. The surface density of material typically has a steep gradient (typically $\Sigma \propto r^{-3/2}$) to match the mass distribution of the planets (Weidenschilling 1977).

Early simulations based on the Classical model of terrestrial planet formation estimates the growth time scale of the terrestrial planets to be tens of millions of years. Overall, numerical simulations indicated that the final terrestrial system would be assembled within 100 Myr (Chambers 2001; Raymond et al. 2006, 2009). In most of these early simulations, however, the resulting planetary systems were found to suffer from excess eccentricity and inclination of the final planets, but the inclusion of a large number of planetesimals–which exert strong dynamical friction on the growing planets–alleviated this concern (O'Brien et al. 2006). Still, a chronic and fundamental shortcoming of the Classical model remained: the output systematically yielded a Mars analogue that was far too massive (e.g. Chambers 2001; Raymond et al. 2006, 2009).

2.2 The Grand Tack Model

The problem of Mars' low mass compared to Earth's has been the focus of several recent models of terrestrial planet formation. The most common of these is the Grand Tack model (Walsh et al. 2011), which relies on the early, gas-driven radial migration of Jupiter. The idea behind the model is that after the gas giant formed it opened an annulus in the gas disk and continued migrating towards the Sun (Lin and Papaloizou 1986) because the torques acting on the planet from the protoplanetary disk are imbalanced. At the same time, Walsh et al. (2011) argue that beyond Jupiter Saturn also accreted its gaseous envelope, albeit more slowly. Once Saturn reached some critical mass (as suggested by Walsh et al. (2011) to be $50 M_{\mathrm{Earth}}$, but it could be lower) it rapidly migrated Sunwards too, caught up with Jupiter and became trapped in the 2:3 mean motion resonance (Masset and Snellgrove 2001). This particular, and fortuitous, configuration of orbital spacing and mass ratio between the gas giants reversed the total torque from the gas disk on both planets (Masset and Snellgrove 2001), thus reversing their migration; the planets 'tacked', a term from sailing reflecting a change of direction. Pierens et al. (2014) showed that outward migration is possible with Jupiter and Saturn in both the 2:3 and 1:2 resonances (cf. Zhang and Zhou 2010), but that dynamically stable evolution with little migration may also happen i.e. the planets could stall once they are caught in a resonance. Thus, the inward-outward migration is a possible outcome, but is not guaranteed, and the amount of outward migration depends sensitively on the disk structure and evolution (D'Angelo and Marzari 2012). To ensure that Mars did not grow beyond its current mass, the location of the 'tack' of Jupiter was set at 1.5 AU (Walsh et al. 2011). Apart from Mars' low mass compared to Earth, the Grand Tack scenario can further account for the semi major axis distribution of the terrestrial planets and the stirred demography of the asteroid belt (DeMeo and Carry 2014; Brasser et al. 2019).

The dynamical consequences of the Grand Tack model on the terrestrial planet region are two-fold. First, as Jupiter migrated towards the Sun, it strongly scattered all protoplanets and planetesimals in its path due to its high mass. This scattering places more than half of the mass of the disk from the inner Solar System to farther than 5 AU from the Sun. Second, Jupiter pushed some protoplanets and planetesimals towards the Sun. This shepherding of material mostly occurred within the 2:1 mean motion resonance with the planet. When Jupiter reversed its migration at 1.5 AU it caused a pileup of solid material within 1 AU. Thus, almost all protoplanets and planetesimals that originally formed beyond 1 AU were scattered outwards by Jupiter, with a few pushed within 1 AU. This evolution provides a plausible explanation for the low amount of mass in the asteroid belt region.

Figure 2 shows an example of the evolution of a Grand Tack simulation taken from Brasser et al. (2016). The top-left panel depicts the initial conditions. Jupiter and Saturn are at 3.5 and 4.5 AU, respectively. The protoplanetary disk resides between 0.7 AU and 8 AU, interrupted by the gas giants. The disk contains planetary embryos of 0.05 Earth masses and two thousand planetesimals.

Closer than 1.5 AU the disk is assumed to consist of material similar to enstatite chondrite (EC; green) (Brasser et al. 2017, 2018), while between 1.5 AU and Jupiter the material is akin to ordinary chondrite (OC; orange). The blue dots correspond to the carbonaceous chondrites (CC). Symbol sizes are proportional to their masses; for Jupiter and Saturn a fixed size was chosen so as not to fill most of the panels. As the gas giants migrate inwards (top-right panel) a significant fraction of OC and CC material is ejected while some OC material is pushed closer to the Sun by Jupiter.

After 5 Myr of simulation evolution the gas giants have reversed their migration and stalled close to their current positions (bottom-left). They have emptied out the asteroid belt region and left behind predominantly EC material in the inner Solar System. Earth and Venus have reached approximately half of their masses. By 30 Myr (middle-right) the terrestrial planets are almost fully formed and not much happens after 60 Myr of simulation evolution (bottom-left) to 150 Myr (bottom-right).

The Classical model and the Grand Tack model predict slightly different outcomes for the bulk compositions of the terrestrial planets (Woo et al. 2018). In the Classical model, the planets tend to accrete mostly locally. In contrast, in the Grand Tack model, all the accretion is confined to an annulus between approximately 0.5 AU and 1 AU, but this region also contains material that originated in the outer parts of the protoplanetary disk that was scattered inwards by the migration of Jupiter. Additional contamination of the inner Solar System with material from the outer Solar System occurs during the growth of the gas giants (Raymond and Izidoro 2017b).

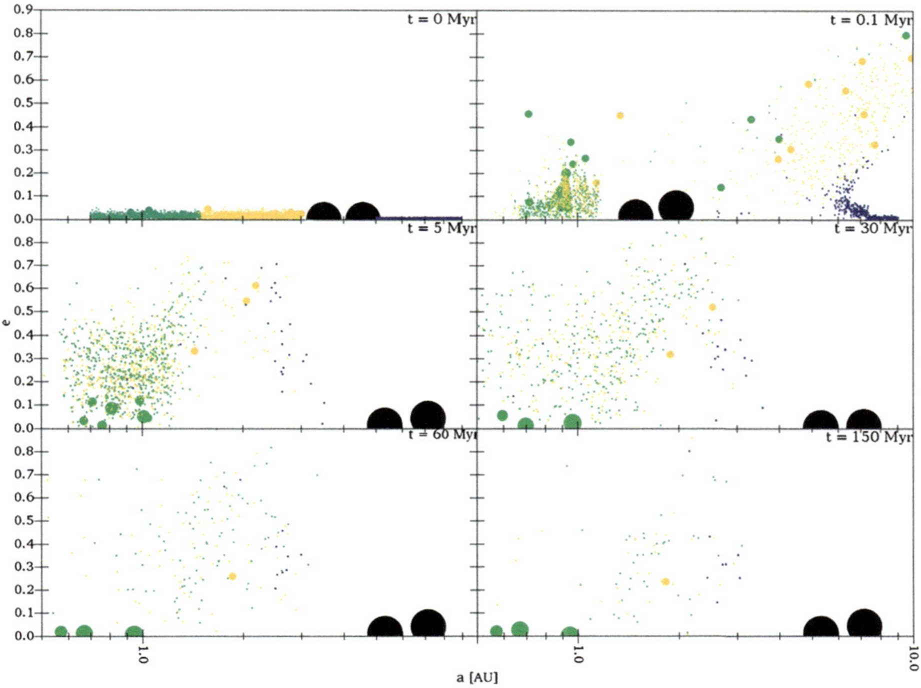

Fig. 2 Snapshots of the evolution of a Grand Tack simulation from Brasser et al. (2016). The colour coding is green for bodies that began closer than 1.5 AU to the Sun, orange for bodies originally between 1.5 AU and 3.0 AU and blue for material beyond that. Jupiter and Saturn are depicted as large black bullets. The symbol sizes are proportional to the mass of the bodies; for Jupiter and Saturn a fixed size was chosen so as not to fill most of the panels (from Brasser et al. 2016)

2.3 Alternative Model Hypotheses

We briefly mention several alternative models of terrestrial planet formation, which have either not (yet) gained much traction or are too new to have been thoroughly explored, especially for their cosmochemical implications. The first is the annulus model (Agnor et al. 1999; Hansen 2009; Walsh and Levison 2016), wherein solid material in the disk is confined to a narrow region around the Sun, typically between 0.5 AU or 0.7 AU to 1 AU or 1.5 AU. The justification for the inner edge of the annulus model is due to the large mass ratio between Mercury and Venus; in a similar vein, the justification for the outer edge of the annulus is to attempt to reproduce the large mass ratio between Earth and Mars. The rationale for the edges is that a low density of solid material will hinder the Mercurian and Martian embryos from gathering it so that they cannot grow into fully fledged planets such as Venus and Earth.

The second is the depleted disk model (Izidoro et al. 2014, 2015) wherein the mass of the asteroid belt is initially much lower than elsewhere in the protoplanetary disk; a variation of this model (Raymond and Izidoro 2017a) fills the asteroid belt region up with the outwards scattering of material from the terrestrial planet formation zone and the inwards scattering of material from the giant planet region during giant planet migration. This variation is, in essence, no different from the annulus model. In this model, Mars' small mass compared to Earth's is caused by the density drop of solid material around 1 to 1.5 AU. As such, the Mars analogue was unable to accrete much material once the Martian embryo had already mostly grown by locally accreting material. The reason for the density drop is left unspoken.

A third model (Clement et al. 2019) invokes a dynamical instability in the outer Solar System wherein the giant planets scatter each other and all material around them. There is ample evidence that such a dynamical instability has occurred (Thommes et al. 1999; Tsiganis et al. 2005; Gomes et al. 2005; Morbidelli et al. 2005; Brasser et al. 2009; Morbidelli et al. 2010; Levison et al. 2011; Nesvorný and Morbidelli 2012; Nesvorný 2015a, 2015b; Brasser and Lee 2015), but thus far it has been assumed that this instability happened after terrestrial planet formation to account for the late heavy bombardment purported to be recorded by the formation of the large lunar impact basins (Tera et al. 1974). In Clement et al. (2019) the authors assume that this dynamical instability occurred early, while the terrestrial planets were still forming. Mars' low mass is then a result of material in its vicinity being removed by perturbations from the migrating giant planets as they evolved through a high-eccentricity phase. Based on evidence from reset ages of meteorites for radiometric systems with different closure temperature Mojzsis et al. (2019) showed that giant planet migration must have commenced before 4480 million years ago (Ma), lending some support to the early instability model of Clement et al. (2019).

A fourth variation is the pebble accretion model (Ormel and Klahr 2010; Lambrechts and Johansen 2012; Johansen and Lambrechts 2017; Johansen et al. 2020) wherein Sun-wards-drifting pebbles accrete onto large planetesimals or small planetary embryos (Johansen et al. 2015). The high growth rate of pebbles compared to planetesimals is due to dissipation of kinetic energy by gas drag as a pebble is scattered by a growing protoplanet. These embryos then rapidly grow into either the giant planets (Levison et al. 2015a) or the terrestrial planets (Levison et al. 2015b). The radial pebble flux through the protoplanetary disc is a key parameter that determines the final masses of the embryos. Lambrechts et al. (2019) demonstrated that low pebble fluxes lead to the formation of Mars-mass embryos that then go on to collide as in the classical terrestrial planet formation picture. Larger pebble fluxes lead instead to the formation of super-Earths with masses between 1 and 10 times the mass of the Earth. The pebble accretion model is actively studied both in the Solar System and for exoplanets (e.g. Johansen and Lambrechts 2017; Matsumura et al. 2017; Izidoro et al. 2019; Johansen et al. 2020). Furthermore, this model also explores the idea of drifting pebbles piling up in the terrestrial planet region, first forming planetesimals and subsequently the terrestrial planets (Drazkowska et al. 2016). This model is then a precursor for the annulus or depleted disk models.

2.4 The Terrestrial Planet Formation Chart

A good model of terrestrial planet formation should match a variety of criteria. The criteria that we consider here are: fast accretion of mass (<30 Myr; e.g., Kleine et al. 2002; Yin et al. 2002; Yu and Jacobsen 2011; Kleine and Walker 2017, and references therein), the mass-semimajor axis (M-a relation) distribution of the planets, the low mass of Mars (closely related to the first criterion), the angular momentum deficit (AMD) of the planets, the radial concentration distribution (S_r), the timing of the Moon-forming event (Lock et al. 2020, this journal), the isotopic composition of the planets, and the lunar and Martian cratering chronology. We have composed a chart wherein we list each of the criteria from relevant literature sources and whether the models listed above match the criteria. When certain criteria have not yet been tested, we write 'U' for 'Unknown'. Table 1 is filled with the results from the studies listed in the previous subsections, as well as those of O'Brien et al. (2006, 2014), Jacobson et al. (2014), Brasser et al. (2018, 2019) and Woo et al. (2018). The

Table 1 Terrestrial planet formation comparison chart related to the different accretion stages I, II, III, and IV illustrated in Fig. 1. The criteria and the symbols are explained in the main text

Accretion stages	I–III	IV						
Model	Fast mass accretion	M-a relation	Mars	AMD	S_r	Moon	Composition	Chronology
Classical	✗	✗	✗	~	✗	✓	✓	U
Annulus	✗	✓	✓	~	✓	✓	U	U
Grand Tack	✗	✓	✓	✓	✓	✓	✓	✓
Depleted disk	✗	✓	✓	✓	✗	U	U	U
Pebble	✓	~	~	U	U	U	U	U
Early instability	✗	~	~	~	✗	✓	U	U

criteria for success vary from one publication to another. Here we have taken the results from literature sources and tallied the successes or failures based on what the authors wrote in their respective publications. Where unclear, we e-mailed the authors with specific requests. An exhaustive list of individual successes and failures is beyond the scope of this work.

We consider the model to work if the success probability for a particular event exceeds 50% in the relevant literature sources (symbol ✓), inconclusive if it falls from 25% to 50% (symbol ~) and a failure if it is below 25% (symbol ✗). *We caution against accepting these results at face value because different studies have different criteria for what is considered a successful semi-major axis relationship, or a mass threshold for a Mars analogue*. This table should be considered as *indicative only* and prompt further investigation into particular aspects of each model. It further has to be noted that the outcome of the simulations and the various models also depend on the behaviour and evolution of the giant planets in the respective simulations. Depending on this evolution, the outcome of the different models can be consistent with each other.

3 Terrestrial Planet Accretion Rates Constrained by Isotopes

As briefly reviewed above, planetesimals and planetary embryos began to form from solid grains originating from the gas and dust in the disk ≈4567 Ma (e.g.; Jacobsen 2003; Bollard et al. 2017). Planetesimals then formed within the first few 10^5 years (Kruijer et al. 2014) followed by planetary embryos, such as Mars, over the next few million years (Dauphas and Pourmand 2011). The terrestrial planets subsequently formed from accretion of Moon- and Mars-sized planetary embryos that originated at various heliocentric locations. During the past decades, the terrestrial planet formation models and scenarios discussed above used different accretion rates, timing of metal-silicate partitioning, and compositions of planetary building blocks. However, the before-mentioned different model approaches and parameter spaces that are applied by numerous planetary formation modellers (e.g., Morbidelli et al. 2012; Raymond et al. 2004, 2009; O'Brien et al. 2006; Johansen et al. 2007; Izidoro and Raymond 2018; Raymond and Morbidelli 2020; and references therein), and core formation models (magma ocean model, two-stage model, local equilibrium model, see: Jacobsen et al. 2008; and references therein) yield different time scales for the formation of large terrestrial planets like Earth or Venus that lie within a range of $\leq$30–100 Myr.

Since measurable variations in isotopic ratios of certain elements in terrestrial planets and their atmospheres, as well as in meteorites, provide records of accretion, formation of the core and crust, we briefly review the so-called isotopic chronometers ^{182}Hf-^{182}W and U-Pb (e.g., Jacobsen and Harper 1996; Yin et al. 2002; Halliday 2004; Kleine et al. 2004; Rudge et al. 2010; Jacobsen et al. 2008; Yu and Jacobsen 2011; Kleine and Walker 2017), and isotopic systems such as ^{50}Ti, ^{54}Cr, ^{92}Mo and ^{100}Ru (e.g., Dauphas 2017; Brasser et al. 2018; Woo et al. 2018; Carlson et al. 2018), ^{48}Ca/^{44}Ca isotope systematics (Schiller et al. 2018), atmospheric ^{36}Ar/^{38}Ar, ^{20}Ne/^{22}Ne, ^{36}Ar/^{22}Ne ratios (Marty 2012; Lammer et al. 2020b), the evidence of solar ^{3}He abundance in Earth's deep mantle and the D/H ratio in Earth's sea water, that can be used to constrain the accretion time scales of the terrestrial planets.

3.1 Radiometric Dating Systems

3.1.1 ^{182}Hf-^{182}W Chronometry for Earth and Mars' Accretion

The fractionation of gas and dust during the stages of disk formation and accretion of planetesimals in the solar nebula (<5 Myr, Bollard et al. 2017) led to a fractionation between volatile and refractory elements (e.g., Wai and Wasson 1977; Jacobsen et al. 2008). At this early time, Hafnium (Hf) and Tungsten (W) isotopes are not significantly fractionated. The differentiation of planetesimals, planetary embryos and protoplanets, however, leads due to the separation of metal-sulphide melt from silicates to the formation of cores resulting in Hf-W fractionation (e.g., Jacobsen et al. 2008; Kleine and Walker 2017). The Hf-W-system therefore provides the most reliable radiometric chronometer of core formation for terrestrial planets such as the Earth (e.g., Jacobsen and Harper 1996; Harper and Jacobsen 1996; Li and Agee 1996; Halliday and Lee 1999; Halliday and Lee 1999; Halliday 2000; Kleine et al. 2002, 2004; Yin et al. 2002; Jacobsen 2005; Wood and Halliday 2005; Jacobsen et al. 2008; Kleine et al. 2009; Yu and Jacobsen 2011; Kleine and Walker 2017).

The elements Hf and W are both highly refractory trace elements. Hafnium is a lithophile element and strongly partitioned into the silicate portion of a planet, while W is moderately siderophile and partitioned preferentially into the coexisting metallic phase. Therefore, $\geq$90% of Earth's W has moved into the core during its formation (Halliday 2000) (see Fig. 3a), making the history of planetary bodies and the timing of accretion, differentiation, and core formation a valuable dating technique in planet formation (Kleine et al. 2009).

The basic principles of the ^{182}Hf-^{182}W chronometry method are briefly described below. ^{182}Hf decays to ^{182}W via ^{182}Ta with a short half-life time of 8.9 Myr. Initially, their parent and trace elements should be in chondritic portions $\approx$1:1 in the early Solar System bodies because both elements are highly refractory (e.g., Jacobsen and Harper 1996; Halliday and Lee 1999; Halliday 2000; Kleine and Walker 2017).

As mentioned above Hf is lithophile and remains in the mantle but W, which exists preferentially into the metallic phase, sinks into the core. Therefore, the parent-to-daughter ratio $f_{\mathrm{Hf/W}}$ is perturbed by the formation of the core (e.g. Kleine and Walker 2017; and references therein). The residual silicate portion in Earth's crust and mantle is commonly described as the bulk silicate Earth (BSE) and has a $f_{\mathrm{Hf/W}}$ in the range of $\approx$10–40 (Newsom et al. 1996). More recent studies that correlated W with other refractory lithophile trace elements such as U, Th, Ta and Ba (McDonough 2003; Dauphas and Pourmand 2011; König et al. 2011) indicate that $f_{\mathrm{Hf/W}}$ of the BSE is 18.8 ± 5 (Kleine and Walker 2017). This value is an order of magnitude higher than the chondritic value (see Fig. 3b). If the fractionation of Hf from W caused by core formation takes place during the lifetime of ^{182}Hf, an excess ^{182}W relative to other isotopes of W should develop in the silicate portion of the planet because of an enhanced $f_{\mathrm{Hf/W}}$ as illustrated in Fig. 3b.

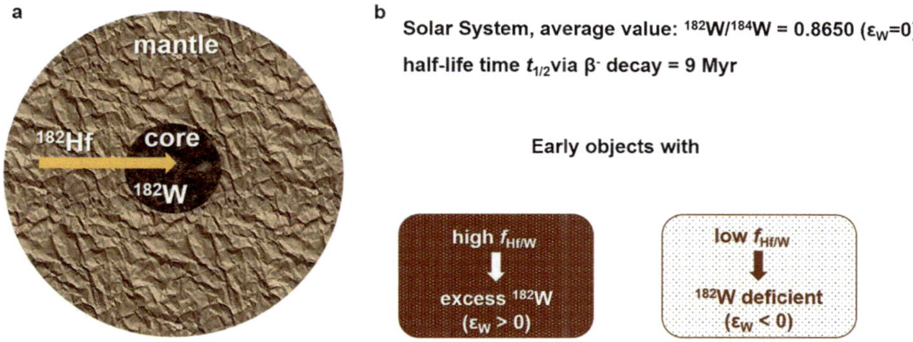

Fig. 3 Panel **a**: The lithophile element ^{182}Hf remains in the mantle while the moderately siderophile element ^{182}W moves preferentially into the metallic core. Because ^{182}Hf decays with a half-life time of only 9 Myr via ^{182}Ta into ^{182}W these elements represent a precision timer of accretion, differentiation, and core formation of terrestrial planets. Panel **b**: Illustration of the essential Hf-W feature (after Halliday 2000)

By using ^{182}Hf-^{182}W chronometry and the ratio $f_{\mathrm{Hf/W}}$ one can now constrain the rate of Earth's accretion and the timing of core formation by geochemical models. Several Moon- to Mars-mass planetary embryo impactors released enough energy that proto-Earth produced temporary magma oceans so that core formation could take place (e.g., Rubie et al. 2007).

The segregation of metal and hence core formation occurs within the early magma ocean state with a metal-silicate equilibrium at very high pressure and temperature (e.g., Rubie et al. 2003; Jacobsen 2003; Kleine and Walker 2017). Core formation models assume that the terrestrial planets accrete dust and planetesimals with a chondritic Hf/W ratio related to an Hf and W-isotopic composition. Jacobsen and Harper (1996) suggested a model that assumes local equilibrium at which the newly added material to the growing protoplanet immediately differentiates into metal and silicate, with the metal sinking to the core without equilibrating with the remaining silicate mantle. However, this local equilibrium model results in a very long mean-time for Earth's core formation of $\approx$108 Myr ($\approx$4460 Ma) with a 90% core-formation time of the order of $\approx$250 Myr ($\approx$4318 Ma) (Jacobsen and Harper 1996; Jacobsen et al. 2008). A reason for this long time-scale is that the newly segregated core material in this model reflects only the average isotopic composition of the delivered material.

This long core formation mean-time is also not in agreement with the strong evidence for an early silicate differentiation at $\approx$4530 Myr ago from measured ^{143}Nd/^{144}Nd and ^{142}Nd/^{144}Nd ratios of the 3800 Myr ago supracrustal rocks of Isua from West Greenland (Harper and Jacobsen 1992; Jacobsen and Harper 1996; Boyet et al. 2003; Caro et al. 2003; Boyet and Carlson 2005; Bennett et al. 2007; Jacobsen et al. 2008). These findings led to a general consensus that the related Sm-Nd-systematics i.e. the fractionation of Sm/Nd taking place within the first 100 Myr after the origin of the Solar System, i.e. before 4468 Ma. This would also indicate that the Moon-forming event, which led to the last magma ocean, occurred within the first 100 Myr (e.g. Touboul et al. 2007; Asphaug 2014; Lock et al. 2020, this journal).

Li and Agee (1996) developed a so-called metal pond core formation model, where the authors assumed that the newly segregated material is equilibrated at the bottom of the magma ocean at roughly 1000 km. Harper and Jacobsen (1996) and Jacobsen (2005) developed the so-called magma ocean differentiation model, where the newly segregated core material reflects the average isotopic composition of the silicate mantle due to full equilibrium in a magma ocean. Today, the magma ocean model is the most widely used core formation model related to the interpretation of early Earth's Hf-W-chronometry studies (e.g., Yin et al. 2002; Wood and Halliday 2005; Jacobsen et al. 2008; Kleine et al. 2009; Yu and Jacobsen 2011; Kleine and Walker 2017). Two-stage accretion scenarios yield results between the magma ocean model with complete equilibration between metal and silicates and the local equilibrium model (e.g., Jacobsen et al. 2008; references therein).

Experimental work and model results indicate that the primitive mantle abundances of the moderately siderophile elements Fe, Ni, Co, Mo, W, P, the moderately siderophile element Ge and the highly siderophile element Re are consistent with a so-called metal-silicate equilibrium at the base of a 700 to 1000 km deep magma ocean in the early Earth, and hence a homogeneous accretion (Righter and Drake 2020; Righter et al. 2011). During the stage of planetary embryo formation surface temperatures can reach up to 2000–3000 K (Ikoma and Genda 2006; Bouhifd and Jephcoat 2011; Stökl et al. 2016), so that deep magma oceans would have been present (Lammer et al. 2020a, this journal; 2020b).

By assuming complete metal-silicate equilibrium the Hf-W systematics yield a fast accretion of the main mass of the Earth within 10 Myr (Wade and Wood 2005; Jacobsen et al. 2008). The Moon-forming event, however, finished Earth's main phase of accretion between $\geq$30 Myr (e.g., Jacobsen 2005; Albarède 2009; Barboni et al. 2017; Thiemens et al. 2019) to $\approx$100 Myr (e.g., Halliday 2008; Halliday and Wood 2009; Boyet and Carlson 2005; Connelly et al. 2019) i.e. between 4538 Myr and 4468 Myr ago. Metal segregation happens faster than accretion (Jacobsen and Harper 1996; Jacobsen et al. 2008, and references therein), so that the time scale of core formation is generally used to constrain Earth's accretion rate (e.g., Rudge et al. 2010; Yu and Jacobsen 2011; Kleine and Walker 2017; and references therein). Equilibrium two-stage model results for Hf-W systematics agree with accretion ages of $\approx 31.0 \pm 4.4$ Myr (Rudge et al. 2010; Kleine and Walker 2017), meaning that core formation was established by 4537 ± 4.4 Myr ago.

According to these studies, the ^{182}Hf-^{182}W chronometry method indicates that at least 80% of the early Earth must have accreted in $\leq$35 Myr, while the final $\approx$10% must have been accreted by 120 Myr. In partial metal-silicate equilibrium models, proto-Earth would have accreted $\approx$0.63 M_{Earth} after a duration of some 10–25 Myr (Jacobsen et al. 2008; Mamajek 2009; Rudge et al. 2010; Kleine and Walker 2017; Morbidelli 2018).

Yu and Jacobsen (2011) showed that one could overcome discrepancies mainly explained by metal-silicate disequilibrium during giant impacts in Hf-W systematics by including constraints from partitioning of refractory siderophile elements Ni, Co, W, V, and Nb during core formation. These authors found that the concentrations of the siderophile elements in the mantle are consistent with high-pressure metal-silicate equilibration in a terrestrial magma ocean (Righter and Drake 2020; Righter et al. 2011). In this study, the earliest time the Moon-forming event could have taken place is at roughly 30 Myr after the origin of the Solar System. Under these assumptions, the model of Yu and Jacobsen (2011) indicate that the timing of the Moon-forming event is inversely correlated with the time scale of Earth's main accretion stage of the first 0.87 M_{Earth}.

If the Moon-forming giant impact occurred later than 50 Myr (i.e. after 4510 Myr ago), the main stage of early Earth's accretion would need to have occurred before $< 10.7 \pm 2.5$ Myr, while an earlier Moon-forming event would increase the time taken for Earth's primary accretion (Yu and Jacobsen 2011). To summarise, the evidence for Earth's (terrestrial planet) accretion from ^{182}Hf-^{182}W chronometry within its uncertainties results in a time scale for the main mass accretion of 8.2–35 Myr and proto-Earth masses at the time when the disk dissipated at 3.5–4 Myr (Bollard et al. 2017; Wang et al. 2017) of 0.5–0.75 M_{Earth}, if one assumes full metal-silicate equilibrium (Yin et al. 2002; Jacobsen et al. 2008; Rudge et al.

2010; Yu and Jacobsen 2011; Kleine and Walker 2017). Yu and Jacobsen (2011) obtain a mass for proto-Earth at roughly 4 Myr that is of the order of 0.56 M_{Earth} with a Hf/W ratio of $\approx$15 and a late giant Moon-forming impact $\geq$52 Myr (i.e. after 4510 Myr ago).

As was shown in Lammer et al. (2020a, this journal; 2020b), if proto-Earth accreted in excess of $>$0.58 M_{Earth} during the lifetime of the solar nebula, the present-day atmospheric ^{36}Ar/^{38}Ar, ^{20}Ne/^{22}Ne and ^{36}Ar/^{22}Ne ratios (see also Sect. 3.2.4) cannot be reproduced through EUV-driven hydrodynamic escape of a thin primordial atmosphere. Moreover, these authors also show that if the proto-Earth accreted $>$0.7 M_{Earth} during the disk lifetime ($\approx$4 Myr), it would need hundreds of Myr or even Gyrs to lose its primordial atmosphere. One can therefore conclude that the upper mass of proto-Earth at 4.0 Myr, i.e. the time when the disk dissipated, should have been $\leq$0.58 M_{Earth} (Lammer et al. 2020b).

The analysis of ^{182}Hf-^{182}W chronometry for dating Mars' accretion is more complex because one cannot measure the Hf/W ratio of bulk silicate Mars (Kleine and Walker 2017). In Mars' case one must compare the abundance of W to other lithophile elements that have the same incompatibility as W. If one assumes that refractory elements such as U, Th, Hf and W occur in chondritic relative portions in bulk Mars, than Hf/W of bulk silicate Mars can be estimated from U/W $\times$ chondritic Hf/U or Th/W $\times$ chondritic Hf/Th of Martian rock samples (Dauphas and Pourmand 2011). By using this assumption, two-stage accretion models yield a core formation age for Mars of 4.1 $\pm$ 2.7 Myr, while exponential growth models for Mars yield $\approx$63% accreted mass at roughly 2.4 Myr and up to $\leq$10 Myr for the remainder of Mars' accretion (Dauphas and Pourmand 2011; Kleine and Walker 2017). Using ^{60}Fe-^{60}Ni chronometry, Tang and Dauphas (2014) independently obtained the same accretion time scale. Thus, one can say that Mars was almost fully formed when the disk dispersed after 3.3–4.5 Myr (Bollard et al. 2017; Wang et al. 2017).

3.1.2 Uranium-Lead (U-Pb) Isotopic Dating System

The U-Pb dating system is one of the earliest radiometric dating schemes that was used for dating rocks that crystallised between millions of years to over 4500 Ma (e.g., McDonough and Sun 1995; Kellogg et al. 2007; Jacobsen et al. 2008; Schaltegger et al. 2015). Because the exact rate at which U decays into Pb is well known, the current ratio of Pb to U in a rock sample can be used to determine its age, provided that in the interim no Pb loss occurred. The U-Pb dating system relies on two separate isotope decay chains: the first one is from ^{238}U to ^{206}Pb, with a half life of 4.47 Gyr, while the second decay chain begins at ^{235}U which decays to ^{207}Pb with a half life of 710 Myr. U-Pb systematics applied to Earth's mantle suggest that the time it takes for the Earth to grow to $\approx$0.63 M_{Earth} is 56–130 Myr for an equilibrium two-stage model, and 21.5–51 Myr for an exponential growth model (Rudge et al. 2010).

Compared to ^{182}Hf-^{182}W systematics of Earth's mantle discussed above, U-Pb systematics yield accretion time scales that are much longer (e.g., Halliday 2004; Kleine et al. 2004; Rudge et al. 2010; Kleine and Walker 2017). Because U-Pb chronometers are long-lived chronometers that are susceptible to geological processes throughout Earth's history, as well as at its beginning, this method does not give as strong an accretion constraint as the Hf-W chronometer (Jacobsen et al. 2008; Kleine and Walker 2017; and references therein).

The blanketing effect of the H_2/He-dominated primordial atmosphere that surrounded growing protoplanets during and, depending on their mass, also some time after the disk lifetime, as well as collisions between Moon- to Mars-mass planetary embryos that were involved in the growth of proto-Earth and Venus, released enough energy to produce a global magma ocean (e.g., Jacobsen et al. 2008; Lammer et al. 2020a, this journal; 2020b), so core formation as discussed before started much earlier than estimated by the U-Pb chronology (e.g., Rubie et al. 2007; Kleine and Walker 2017). Due to this discrepancy, Kamber and Kramers (2006) and Yin and Jacobsen (2006) suggested that U-Pb systematics do not constrain core formation since bulk Earth's U/Pb ratio and the Pb isotopic composition are not known well enough. Moreover, the discrepancy in accretion time scales related to U-Pb systematics may be caused by a late segregation of Pb-bearing sulphides into Earth's core (Wood and Halliday 2005; Halliday and Wood 2007), disequilibrium during core formation (Halliday 2004; Allègre et al. 2008), and the possible addition of Pb during the late veneer (Albarède 2009). To summarise this brief discussion: according to Rudge et al. (2010) and Kleine and Walker (2017), one can conclude that U-Pb systems contain little information on the early accretion, but constrain Earth's late accretion phase very well.

However, the main uncertainties between Hf-W and U-Pb systematics for the determination of the core formation time-scale is the strong dependence on the metal-silicate equilibrium (e.g., Kleine et al. 2002; Allègre et al. 2008) and the uncertainties in knowledge related to the bulk Earth Pb isotopic composition and U/Pb ratio (e.g., Yin and Jacobsen 2006). One can bring the Hf-W and U-Pb age determinations into agreement if one assumes a full metal-silicate equilibrium combined with a very fast early accretion that subsequently slowed down and was followed by a long phase of much slower growth culminating in a late giant impact, i.e., the Moon-forming event (Rudge et al. 2010; Yu and Jacobsen 2011).

Moreover, Kellogg et al. (2007) showed that a mean age of accretion and core formation of $\approx$10 Myr is consistent with the isotopic composition of modern mid-ocean ridge basalts when taking into account recycling of continental and oceanic crusts into the mantle over time. This result agrees with the study of Rudge et al. (2010) that there is no need to infer a late core formation to explain Earth's Pb isotopic composition. According to Jacobsen et al. (2008) it seems that the fractionation of U-Pb in the nebula is a dominant process so that the core-related fractionation has to be very accurately determined compared with the volatile-related fractionation in order to obtain accurate results for the timing of core formation based on the U-Pb-system. Therefore, the U-Pb-isotopic dating system does not place strong constraints on the chronology of accretion and core formation.

3.2 Constrains from Lithophile-Siderophile Isotope Data, ^{48}Ca/^{44}Ca-Systematics and Noble Gases

3.2.1 Earth's Accretion Constrained by Lithophile-Siderophile Elements

For decades, researchers have tried to estimate Earth's bulk chemical composition by comparison with meteorites. Clayton (1970) found that Earth's oxygen composition is different compared to most meteorites apart from the silicon-rich and highly reduced enstatite chondrites. However, the chemical composition of enstatite chondrites is very different compared to rocks on Earth's surface (e.g., Drake and Righter 2002; Carlson 2017), so the community continued to assume that the more oxidised and volatile-rich carbonaceous chondrites were involved in early Earth's accretion.

Recently Dauphas (2017) used the unique isotope content of different types of meteorites to identify those that could have been involved in Earth's accretion over time. He calculated that the isotopic mantle signatures of lithophile elements such as O, Ca, Ti and Nd, and moderate to highly siderophile elements like Cr, Ni, Mo and Ru recorded different accretion stages of the early Earth. Dauphas (2017) showed that these elements point to a large fraction of Earth's building blocks, some 70%, that were isotopically similar to enstatite meteorites. It was found that for the first 60% of its accretion, proto-Earth most likely formed from a mixture of an isotopic composition of 51% enstatite, 40% ordinary chondrites and 9% carbonaceous chondrites (i.e. 5.4% CC in total). For the second accretion stage, i.e. the next 60–99.5% of Earth's mass, and for the final 0.5% accreted during the late veneer, the isotopic record of these elements are mainly akin to enstatite chondrite-like material.

At the same time, Fischer-Gödde and Kleine (2017) concluded from Ru isotopic measurements from a sample of meteorites that material arriving as late accretion was relatively dry and isotopically identical to enstatite chondrites. This finding is also in support of Dauphas (2017) that states enstatite chondrites most likely played an important role in the accretion of the Solar System's terrestrial planets, although in an earlier study Wang and Becker (2013) state that S, Se and Te abundance in the terrestrial mantle requires a volatile-rich late veneer. Carlson et al. (2018) conclude that the composition of the planetary building blocks changes over time and Earth's deep interior should have a different composition compared to its surface layers. However, the findings and three-stage accretion model presented by Dauphas (2017) neither gives a time scale for the two main accretion stages nor describes how much mass was accreted while the planet was embedded within the disk or after its dispersal.

3.2.2 Accretion Rates of Inner Disk Mass Versus Time Obtained from $^{48}Ca/^{44}Ca$ Isotope Data

Current models that study the formation of planetary embryos from dust settling to planetesimal formation via pebbles (that originate from the same flow) struggle to explain the different compositions of Solar System bodies (e.g., Johansen et al. 2015; Morbidelli 2018). Schiller et al. (2018) measured the $^{48}Ca/^{44}Ca$ isotope content in samples from Earth, Mars, Vesta, and the angrite and ureilite parent bodies. Here, $\mu^{48}Ca$ corresponds to the $^{48}Ca/^{44}Ca$ ratio relative to that in the terrestrial reference standard, given in parts per million (p.p.m.). A positive correlation between $^{48}Ca/^{44}Ca$ isotope ratios and planetary masses was identified. To explain these findings, a highly debated and unorthodox view of planetary growth is proposed by these authors. Schiller et al. (2018) suggest that pebble accretion is correct, however, the different bodies were growing at the same rate but stopped growing at different times (Schiller et al. 2018; Morbidelli 2018). This hypothesis is non-conventional because the mainstream theory is that bodies of different masses and at different locations grew at different rates during the lifetime of the disk. Figure 4 illustrates their hypothesis of planetary accretion. In the model approach of Schiller et al. (2018), the gas and dust in the disk surrounding the young Sun initially contained inner-disk material with low $\mu^{48}Ca \approx -150$ ppm, similar to the ureilites, and outer disk particles with high $\mu^{48}Ca$ values of $\approx +200$ ppm, which corresponds to carbonaceous chondrites.

In this scenario, the planetesimals and planetary embryos grew by accreting matter from the inner disk. The inner disk started to be fed with inwards-drifting pebbles and, after some time, the inner-disk material was largely supplemented by the material from the outer disk, resulting in an increase of the mean $\mu^{48}Ca$ value of the inner disk. The planetary embryos continued to grow until they were displaced by perturbations from other bodies. The assumed $\mu^{48}Ca$ mixing from the inner and outer material resulted in values of $\mu^{48}Ca$ ≈ -100 p.p.m. for Vesta, ≈ -20 p.p.m for Mars and 0 p.p.m. for Earth and the Moon. From their analysis, the authors conclude that proto-Earth grew to $\approx 0.8\ M_{\text{Earth}}$ within the disk lifetime and its accreted material consisted of $\approx$60% Ureilite and $\approx$40% CI-chondrite-like material (which are the end-members of the model). This rapid accretion is difficult to reconcile with current dynamical and Hf-W chronometry models (e.g. Carlson et al. 2018; Yu and Jacobsen 2011).

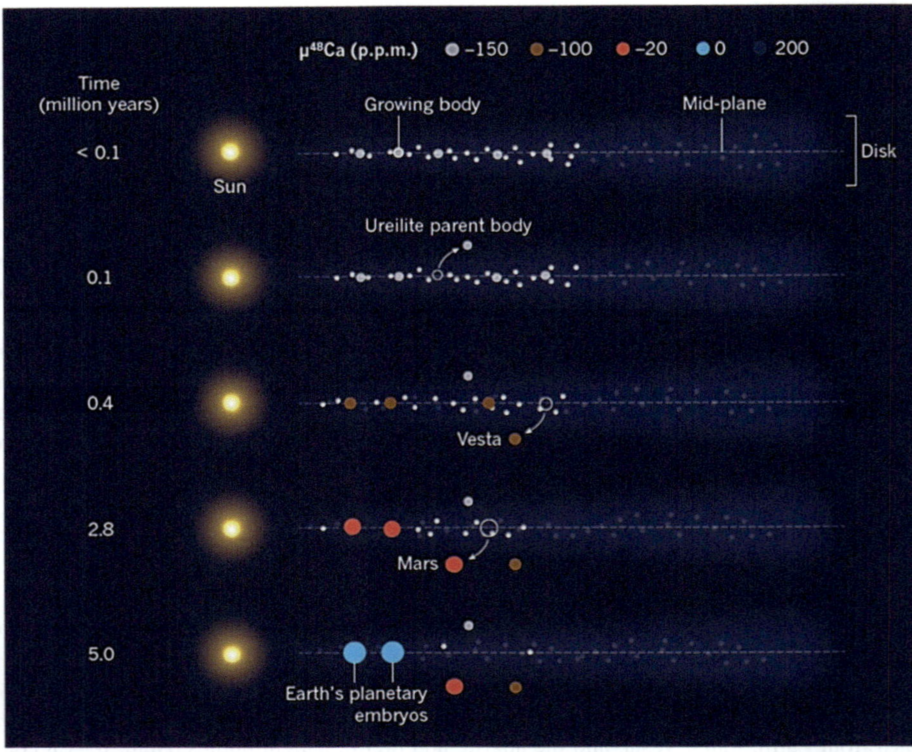

Fig. 4 Illustration of the hypothesis suggested by Schiller et al. (2018) for the accretion of Venus, Earth and Mars based on measured $^{48}Ca/^{44}Ca$ isotope ratios (μ ^{48}Ca [p.p.m.]) that correlate between the masses of the inner Solar System bodies. The times indicate accretion stages during the disk's lifetime in which various bodies originated. In this scenario, the parent bodies of Ureilite-type meteorites, the asteroid Vesta, Mars and planetary embryos that probably formed Earth are also shown (from Morbidelli 2018)

In a brief report, Morbidelli (2018) remarked various inconsistencies within the hypothesis proposed by Schiller et al. (2018). One problem is that, for the Earth, the model of Schiller et al. (2018) requires about 0.4 M_{Earth} to be accreted from the outer disk, which is higher than previous calculations by Dauphas (2017) and Morbidelli et al. (2016). However, the previously mentioned studies did not include ureilite-like material in their models because of the inability from the existing samples of the Ureilites and other achondrites to obtain the bulk elemental concentrations of the parent body that the models require.

The added outer Solar System material is represented isotopically in Ca ratios of CI chondrites (Schiller et al. 2018). The composition of the proto-Earth as inferred by Schiller et al. (2018) is not in agreement with the isotope measurements of the lithophile, moderately and highly siderophile elements of the material out of which Earth's mantle accreted, as determined by Dauphas (2017), since it does not include a high amount of enstatite material (see also Sect. 3.1.2) because Schiller et al. (2018) focused on an end-member model only. Furthermore, it is unlikely that a terrestrial mantle consisting of 40% CI chondrite is consistent with the terrestrial Mg/Si elemental ratio as well as with oxygen, titanium, chromium and other isotopic systems. Recently, Schiller et al. (2020) claim that terrestrial Fe isotopes are nearly identical to those of CI chondrites. From their measurements they imply that the terrestrial mantle consists of 40% CI. It is clear that further studies are needed to resolve this dilemma.

One should also note that Lammer et al. (2020b) concluded that protoplanets that accreted masses of $>0.58\ M_{\text{Earth}}$ within the disk lifetime of 3.3–4.5 Myr (Bollard et al. 2017; Wang et al. 2017) would capture a primordial atmosphere that is too massive to be consistent with the present atmospheric ratios, while a proto-Earth with a mass of $\geq 0.7\ M_{\text{Earth}}$ at 4 Myr likely ends up as a gas dwarf (e.g., Owen and Wu 2016; Fossati et al. 2017; see also Lammer et al. 2020a; this journal).

As shown in Fig. 5, it is possible to reproduce the present-day atmospheric Ar and Ne noble gas ratios of the Earth with a composition as suggested by Schiller et al. (2018) but not by growing up to 80% of its mass within the nebula (that dispersed at 3.3–4.5 Myr). In such a case proto-Earth accretes up to ≈ 0.53–$0.58\ M_{\text{Earth}}$ within the gas disk and grows fast up to $\geq 0.8\ M_{\text{Earth}}$ immediately afterwards and still before the escape of a tiny captured primordial atmosphere (within <7 Myr) through planetesimal accretion. Afterwards mainly CCs or CIs–as suggested by Schiller et al. (2018)–that are depleted in the noble gases and moderately volatile elements finalised Earth's accretion. In such a case, the process after disk dispersal would be a complex interplay between EUV-driven hydrodynamic and impact-induced atmospheric escape combined with mass accretion. Such a fast accretion scenario also fits with the Hf-W chronometry for $f_{\text{Hf/W}} \approx 15$, if full metal-silicate equilibrium is assumed (Rudge et al. 2010; Yu and Jacobsen 2011; Kleine and Walker 2017) (see also Sect. 3.1.1).

To conclude, the controversial hypothesis of Schiller et al. (2018) may give some answers as to why different Solar System bodies have varying compositions if they grew from the accretion of the same flow of pebbles (cf. Morbidelli 2018), but it neither agrees with the isotopic nature of Earth's accreting material, as inferred by Dauphas (2017), (see also Sect. 3.2.1) nor are the implications clear for other isotopic systems and for the elemental abundances within the Earth. Future research will have to show how the μ^{48}Ca isotope data connects with the other isotopical, chemical, chronological and dynamical constraints of terrestrial planet formation.

3.2.3 Constraining Early Earth's and Venus' Accretion by Reproducing Their Atmospheric $^{36}Ar/^{38}Ar$, $^{20}Ne/^{22}Ne$, $^{36}Ar/^{22}Ne$ Isotope Ratios

Besides the before mentioned dating methods, ^{36}Ar/^{38}Ar, ^{20}Ne/^{22}Ne and ^{36}Ar/^{22}Ne noble gas isotope ratios in Earth's and Venus' atmospheres also provide important information on the origin and evolution of these planets (e.g. Pepin 1991, 1997; Gillmann et al. 2009; Odert et al. 2018; Lammer et al. 2020a, this journal; Lammer et al. 2020b). Previous studies (e.g., Pepin 1991, 1997; Gillmann et al. 2009; Odert et al. 2018) that have tried to reproduce the present atmospheric isotope ratios of these noble gases during the past-applied arbitrary assumptions related to the EUV flux of the young Sun, simple atmospheric escape models and initial H_2-rich reservoirs related to primordial and steam atmospheres.

Marty (2012) analysed the Ar and Ne abundances in the present atmosphere of the Earth and found that the ratios of today's atmosphere and concentrations of water, carbon and nitrogen indicate a volatile content equivalent to $\approx 98\%$ non-chondritic depleted/dry proto-Earth with only $\approx 2\%$ contribution of carbonaceous chondritic material. As discussed above, Dauphas (2017) analysed isotopic signatures of lithophile, moderately siderophile and highly siderophile elements from Earth's mantle material and concluded that roughly 5% carbonaceous chondritic material was involved in early Earth's accretion. As discussed above, more recently, Schiller et al. (2018) inferred from the analy-

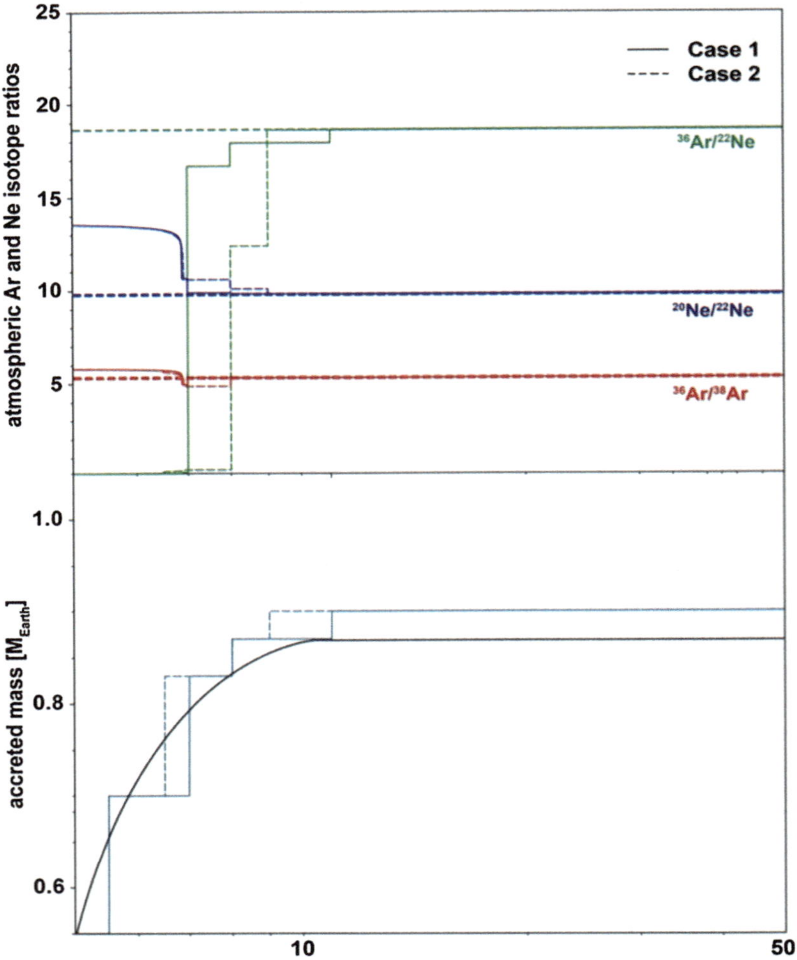

Fig. 5 Evolution scenario that reproduces present Earth's atmospheric ^{20}Ne/^{22}Ne, ^{36}Ar/^{38}Ar and ^{36}Ar/^{22}Ne ratios and the corresponding early mass accretion of proto-Earth as modelled by Lammer et al. (2020a) with a proto-Earth-mass at 4 Myr of 0.55 M_{Earth}, a small captured H_2-dominated primordial atmosphere that is lost within <7 Myr after the origin of the Solar System, and ≈3 Myr after disk dispersal due to impact erosion and EUV-driven hydrodynamic escape related to a young Sun that rotates slower than a moderate rotator. *Case 1*: Reproduction of Earth's atmospheric ^{20}Ne/^{22}Ne, ^{36}Ar/^{38}Ar and ^{36}Ar/^{22}Ne ratios if one assumes the initial composition given in Dauphas (2017), but with a CC contribution of about 5% after a tiny primordial atmosphere escaped (see also Sect. 3.2.1). *Case 2*: Reproduction of Earth's atmospheric ^{20}Ne/^{22}Ne, ^{36}Ar/^{38}Ar and ^{36}Ar/^{22}Ne ratios by assuming the initial composition suggested by Schiller et al. (2018) (see also Sect. 3.2.2) with a contribution of ≈40% of CC. The chondritic post-H_2-envelope impactors are depleted in 70% and 30% in Ne and Ar, respectively. The corresponding accretion scenario agrees well with that of Yu and Jacobsen (2011) that the main stage of Earth's accretion of ≈0.87 M_{Earth} was completed rapidly within 10.7 ± 2.5 Myr, corresponding to a Hf-W determined Moon giant impact time t_{MGI} at ≈53 Myr after Solar System formation (black solid line). (After Lammer et al. 2020a)

sis of $^{48}Ca/^{44}Ca$ isotope ratios in various early Solar System objects, that proto-Earth accreted roughly 40% carbonaceous chondrites (mainly CI chondrites) at the time when the disk dissipated. This value is roughly comparable to that obtained by Warren (2011), who uses the lever method to obtain ≈24–32%. In the case that proto-Earth accreted higher amounts of carbonaceous chondritic material compared to the 2% inferred from today's data (Marty 2012), thermal processing in a possible primordial atmosphere and atmospheric escape from accreting planetary embryos and the growing protoplanet could have been depleted in these volatile elements (Odert et al. 2018; Benedikt et al. 2020; Young et al. 2019; Lammer et al. 2020a, this journal; 2020b), altering the bulk and isotopic composition.

Recently, Lammer et al. (2020b) reproduced Earth's present atmospheric $^{36}Ar/^{38}Ar$, $^{20}Ne/^{22}Ne$ and $^{36}Ar/^{22}Ne$ noble gas isotope ratios within realistic young Sun and initial H_2-reservoir conditions by applying a sequence of sophisticated impact and upper atmosphere escape models.

These authors studied the boil-off phase of primordial H_2-dominated atmospheres that accumulate around protoplanets with different masses during the accretion in the disk phase. They applied a hydrodynamic upper atmosphere model for the escape calculation of possible H_2-envelopes that remained during disk dispersal on various protoplanetary masses, including impact erosion, delivery, and composition/ratio modifications. The atmosphere evolution studies were carried out by including the young Sun's EUV flux as constrained by an activity-rotation relation of young solar-like stars according to Tu et al. (2015).

Finally, Lammer et al. (2020b) could reproduce Earth's present $^{36}Ar/^{38}Ar$, $^{20}Ne/^{22}Ne$ and $^{36}Ar/^{22}Ne$ and Venus' $^{36}Ar/^{38}Ar$, $^{20}Ne/^{22}Ne$ noble gas ratios within the uncertainties of the parameter space, by proto-Venus and proto-Earth growing to masses of up to 0.68–$0.81 M_{\rm Earth}$ and 0.53–0.58 $M_{\rm Earth}$, respectively, at 4 Myr at the time when the disk dissipated.

In early Earth's case, today's atmospheric $^{36}Ar/^{22}Ne$ ratio of 18.8 (Marty and Allé 1994; Marty 2012) can be best reproduced by Lammer et al. (2020a, 2020b) if the post-nebula impactors (of a combined mass of 0.15–0.3 $M_{\rm Earth}$) contain roughly 5% carbonaceous chondritic material (see Fig. 5; dashed lines) if one assumes an initial composition as suggested by Dauphas (2017), and ≥70% (see Fig. 5; solid lines) if they assume the composition as expected by Schiller et al. (2018). The 95% and ≤30% of the accreted post-nebula material has to be depleted in the studied elements and hence originates from volatile-poor material.

Figure 5 shows reproduction attempts of Earth's present atmospheric Ar and Ne noble gas ratios when the proto-Earth was $0.55 M_{\rm Earth}$ at the time when the disk dissipated at 4 Myr. In such a case, a tiny primordial H_2-dominated atmosphere with a mass fraction of about 3.5×10^{-5} $M_{\rm Earth}$ remained after the boil-off phase during disk dispersal (Lammer et al. 2020a, this journal; 2020a). The growing proto-Earth lost this tiny primordial atmosphere during ≤3 Myr afterwards by impact erosion and EUV-driven hydrodynamic escape of H atoms that drag and fractionate the Ar and Ne isotopes.

In Venus' case, the reproduction of its atmospheric $^{36}Ar/^{38}Ar$ and $^{20}Ne/^{22}Ne$ isotope ratios is more uncertain due to the large error bars of the available measurements of previous spacecraft. According to Lammer et al. (2020b) Venus' present atmospheric Ar and Ne isotope ratios can be reproduced if it captured a primordial atmosphere that corresponds to a mass for proto-Venus of ≈0.83–1 $M_{\rm Venus}$ at the time when the disk dissipated. If this were the case this would indicate that Venus accreted faster than Earth. As pointed out by Dandouras et al. (2020, this journal), future Venus missions that will precisely re-measure the atmospheric noble gas ratios are necessary to constrain the initial composition and evolution of Venus.

3.2.4 Constraining Early Earth's Accretion by ^{3}He, and ^{20}Ne/^{22}Ne, Isotope Data from the Deep Mantle

The question of whether the growing proto-Earth was surrounded by at least a low-mass primordial atmosphere has long been debated (e.g., Hayashi et al. 1979; Mizuno et al. 1980; Jacobsen and Harper 1996; Harper and Jacobsen 1996; Porcelli et al. 2001; Dauphas 2003; Yokochi and Marty 2004; Jacobsen et al. 2008; Raquin and Moreira 2009; Hallis et al. 2015; Jaupart et al. 2017; Péron et al. 2016; Williams and Mukhopadhyay 2019). That early Earth most likely accreted partially before and after the nebular disk dispersed was discussed by Harper and Jacobsen (1996), who studied the abundance of ^{3}He and the before mentioned Ne isotopes in the mantle. In such a two-stage accretion scenario hypothesis, proto-Earth accreted enough mass during the disk lifetime to attract an H_2-He-dominated primordial atmosphere directly from the nebular gas. For explaining the evidence of the solar-like Ne and He noble gas abundances, as well as the presence of a large ^{129}Xe excess in the deep Earth, Harper and Jacobsen (1996) and Jacobsen et al. (2008) suggest, that these elements were embedded in the solar-like H_2-He-dominated primordial atmosphere, dissolved in the magma ocean and transported into the deep Earth. Additionally, planetesimals and dust continuously added noble gas components into the growing protoplanet. Although, the presence of a large ^{129}Xe excess, produced in the deep mantle due to a fast radioactive decay of ^{129}I with a half-life of 15.7 Myr, is consistent with a very early atmosphere (Jacobsen and Harper 1996; Jacobsen et al. 2008; Caracausi et al. 2016; Avice et al. 2017), the initially produced ^{129}Xe is contaminated by Xe isotopes that were delivered during the late veneer. A detailed discussion on Xe isotopes and their perspectives on atmospheric evolution on Earth, Venus and Mars can be found in Avice and Marty (2020, this journal).

According to Harper and Jacobsen (1996), the inferred ^{3}He abundance from deep-mantle plumes of Hawaiian volcanos (e.g., Poreda and Craig 1992; Harper and Jacobsen 1996; Timmerman et al. 2019) is more or less comparable to the abundance observed in primitive carbonaceous chondrites (e.g., Mazor et al. 1970; Kellogg and Wasserburg 1990), but is much higher compared to all the other highly volatile elements in the deep Earth. This evidence is thought to be a primitive un-degassed reservoir that was preserved from the early accretion stages, when the planetary body was not large enough to support the energy of the impacts over the degassing threshold (Atreya et al. 1989; Harper and Jacobsen 1996; Jacobsen et al. 2008).

On the other hand, one can expect that Earth's deep mantle should also be depleted in ^{3}He in a similar way as the other volatiles, because of volatile depletion in proto-Earth's nebula source reservoir, metamorphic degassing in planetesimals, as well as degassing from large planetesimals and planetary embryos during accretion (Harper and Jacobsen 1996; Jacobsen et al. 2008; Lichtenberg et al. 2016, 2018; Benedikt et al. 2020; Lammer et al. 2020a, this journal). Short-lived radioactive isotopes such as ^{26}Al, and ^{60}Fe also determine the thermal history and interior structure of the early planetary building blocks additionally to impacts and gravitational energy (Lichtenberg et al. 2016, 2018). The following thermo-mechanical evolution results in internal differentiation, which is related to a rapid volatile outgassing that was recently studied in Benedikt et al. (2020) and also discussed in Lammer et al. (2020a, this journal).

Therefore, the primordial ^{3}He, which was accreted to the Earth through the delivery of carbonaceous chondrites, should have been efficiently lost, similar as the other volatile elements, compared to its initial abundance in the primitive meteorite value. By comparing the ^{3}He abundance with other highly volatile elements that should have been accreted from the same building blocks, Harper and Jacobsen (1996) argued that the inferred ^{3}He abundance in the deep Earth is at least two orders of magnitude higher than that expected from accretion of dust and planetesimals, which was probably most likely $\leq 1\%$ of its initial value (Harper and Jacobsen 1996; Jacobsen et al. 2008). Because of this paradox, these authors concluded that the ^{3}He in Earth's deep mantle cannot be an accreted primitive component but is most likely a remnant of a magma ocean equilibration-related primitive H_2-He-dominated primordial atmosphere.

The preservation of such a primordial helium reservoir over the Earth's history has been questioned on the basis of geophysical evidence of slab subduction into the lower mantle and mantle convection models (e.g., Van der Hilst et al. 1997; Meibom et al. 2003; Class and Goldstein 2005; Parman 2007). Recent studies, however, confirmed such a reservoir of primordial helium from studies of the hottest mantle plumes (Jackson et al. 2017) and from measured He-Sr-Pb isotope ratios in superdeep diamond fluid inclusions from the transition zone in a depth of 410–660 km which is unaffected by degassing and shallow crustal contamination (Timmerman et al. 2019). These authors conclude that their findings are indications that a less degassed, high-^{3}He/^{4}He deep mantle source infiltrates this transition zone, where it interacts with recycled material, creating the diverse compositions recorded in ocean island basalts. This is, furthermore, an indication of a dense high-^{3}He/^{4}He deep-mantle domain that could have remained isolated from the convecting mantle (e.g., Samuel and Farnetani 2003, which explains and supports the preservation of very early (>4500 Ma) geochemical anomalies in lavas (e.g., Harper and Jacobsen 1996; Jacobsen et al. 2008; Mukhopadhyay 2012; Jackson et al. 2017).

Harper and Jacobsen (1996) suggested that the best explanation for the high abundance of ^{3}He isotopes in the deep mantle is that 0.8 $M_{\rm Earth}$ accreted within the presence of a solar nebular-composition atmosphere, which indicates a fast accretion of the early Earth. Such a scenario would be in agreement with the constrains obtained by the atmospheric Ar and Ne noble gas ratios (Lammer et al. 2020b), the Hf-W chronometer (Kleine and Walker 2017, and references therein), and the Ca isotope systematics (Schiller et al. 2018), if proto Earth grew within the disk ($\approx$3.3–4.5 Myr; Bollard et al. 2017; Wang et al. 2017) to a mass of $\approx$0.53–0.58 $M_{\rm Earth}$, and to $\approx$0.8 $M_{\rm Earth}$ immediately after disk dispersal, whilst the growing body was surrounded by the escaping H_2-He-dominated primordial envelope (see Fig. 5).

Besides ^{3}He, there is also an evidence of solar-like Ne isotopes in the deep mantle (e.g., Mizuno et al. 1980; Honda et al. 1991; Porcelli et al. 2001; Yokochi and Marty 2004; Jacobsen et al. 2008; Raquin and Moreira 2009; Jaupart et al. 2017; Péron et al. 2016). Williams and Mukhopadhyay (2019) measured a ^{20}Ne/^{22}Ne isotope ratio from Earth's deep interior of 13.01–13.45, which is higher than a ratio of 12.52–12.75 that could also be produced through implantation of the solar wind onto accreted meteoritic material (Moreira et al. 2013; Moreira and Charnoz 2016). This range is well in accordance with the ^{20}Ne/^{22}Ne isotope ratio of $\approx$13.34 of the solar nebula that was derived by Heber et al. (2012) from Genesis data. In agreement with the hypothesis of Harper and Jacobsen (1996) and Jacobsen et al. (2008), Williams and Mukhopadhyay (2019) suggest that some Ne isotopes that were embedded in a primordial atmosphere were incorporated with the solar ratio into proto-Earth's interior via a magma ocean. Jacobsen et al. (2008) also argued that solar wind related implanted components would most likely be thoroughly outgassed and lost from early differentiated planetesimals due to the before mentioned ^{26}Al heating (Lichtenberg et al. 2016, 2018), which comports with the results of recent studies such as Young et al. (2019), Sossi et al. (2019), Benedikt et al. (2020) and Lammer et al. (2020b).

Slower accretion scenarios with proto-Earth accreting <0.53 $M_{\rm Earth}$ during the disk lifetime of 3.3–4.5 Myr (Bollard et al. 2017; Wang et al. 2017) would need planetary embryos that were extremely depleted in volatile and moderately volatile rock-forming elements (i.e.

K, Na, Si, Mg, etc.) after the disk evaporated and other explanations than a primordial atmospheric origin for the high ^{3}He abundance in the deep mantle would have to be found.

3.2.5 Constraining Early Earth's Accretion from D/H

If proto-Earth grew partly within the gas disk then one can argue that there should be visible evidence in the D/H ratio of Earth's ocean water (e.g. Morbidelli et al. 2000; Meech and Raymond 2019; and references therein; Raymond and Morbidelli 2020), since the D/H ratio of nebular gas of $(21 \pm 5) \times 10^{-6}$ is roughly 7 times lower (Geiss and Gloeckler 1998; Robert et al. 2012) than Earth's seawater D/H ratio of $(150 \pm 10) \times 10^{-6}$ (e.g., Robert et al. 2012), which fits well within the measured D/H ratio in carbonaceous chondrites (e.g., Marty 2012). According to Ikoma and Genda (2006) and Ikoma et al. (2018), the atmospheric H_2 within the primordial H_2-dominated atmosphere can be oxidized by gas-rock interactions between the atmosphere and a magmatic surface with temperatures $\geq$1500 K, thereby producing H_2O. The amount of nebular-related water depends both on the mass of the surrounding primordial atmosphere and on the available ion oxides and fayalite (Fe_2SiO4) in the magma ocean. The ratios of produced water to the primordial atmospheric hydrogen M_{H2O}/M_{H2} are $\approx$0.49, $\approx$24.02, and $\approx$0.88 for the quarz-iron-fayalite-oxygen ($2Fe + SiO_2 + O_2 \leftrightarrow Fe_2SiO_4$), the wüstite-magnetite ($6.696Fe_{0.974}O + O_2 \leftrightarrow 2.174Fe_3O_4$), and the ion-wüstite ($1.894Fe + O_2 \leftrightarrow 2Fe_{0.947}O$) buffers (Robie et al. 1978; Ikoma and Genda 2006). Ikoma and Genda (2006) estimated that under such conditions a protoplanet that captured an H_2-envelope from the disk gas with a mass of $M_{H2} \approx 10^{22}$ kg (0.00167 M_{Earth}) would produce an H_2O amount comparable to the mass of the Earth's current ocean of $M_{H2O} \approx 1.4 \times 10^{21}$ g if such oxygen buffers were available.

By applying the same assumption as Ikoma and Genda (2006) for an accretion scenario that can reproduce the present day atmospheric Ar and Ne isotope ratios based on the simulations of Lammer et al. (2020a), a proto-Earth with a mass of 0.55 M_{Earth} (see also Fig. 5) with a captured primordial H-dominated atmosphere of roughly $M_{H2} \approx 3.5 \times 10^{-5}$ M_{Earth} (or 2.15×10^{20} kg) produces a water reservoir of M_{H2O} of $\approx 3.0 \times 10^{19}$ g from the H_2-envelope corresponding to only $\approx$2% of the current value of Earth's seawater (see Fig. 6). Such a low value lies within the error bars of the D/H seawater-carbonaceous chondrite "match" which is $\approx$10–20% (Pahlevan et al. 2019; and references therein).

If we estimate the hypothetical H_2O production with similar assumptions as before but for a proto-Earth with a mass of 0.75 M_{Earth} at the disk dispersal time of 4 Myr, an H_2-envelope of 0.0077 M_{Earth} (4.59×10^{22} kg) would be captured (Lammer et al. 2020b). In such a case, roughly 4.5 times of Earth's present seawater value, or some 4.5 Earth-oceans, could have been produced due to the nebula gas-magma ocean interaction processes. One can see from Fig. 6, that proto-Earth masses >0.6 M_{Earth} should have left behind D/H ratios that would differ largely from the measured carbonaceous chondritic value.

Another indication that at least a small fraction of Earth's H_2O may be derived from a nebular origin comes from measurements of the D/H ratios in glassy melt inclusions in two basaltic rock samples that originated in a deep mantle reservoir by Hallis et al. (2015). These authors found D/H ratios in their samples that are closer to the nebular than to the carbonaceous chondritic ratio.

However, the contribution of other minor bodies could have also altered the initial D/H ratio of the Earth. Comets have D/H ratios that are on average significantly higher than the terrestrial value (e.g., Alexander et al. 2018); while the Earth has a D/H ratio of $(150 \pm 10) \times 10^{-6}$ (e.g., Robert et al. 2012), the value of comet 67P/Churyumov-Gerasimenko is three times higher ($(530 \pm 10) \times 10^{-6}$; Altwegg et al. 2015). Based on

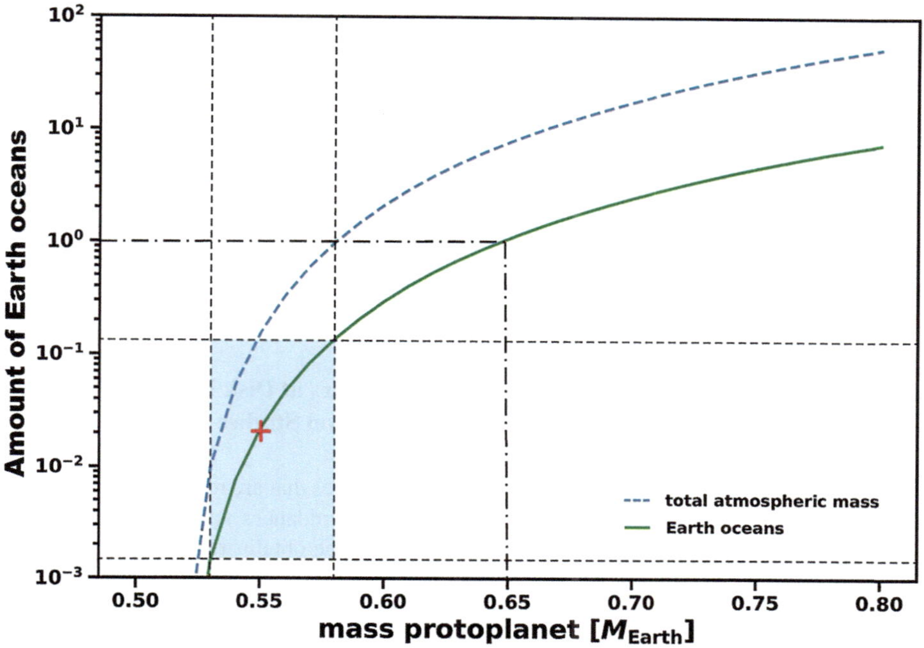

Fig. 6 The amount of Earth oceans (green solid line) that can be produced by the captured H_2-dominated primordial atmospheres (blue dashed line) as modelled in Lammer et al. (2020b) due to gas-rock interactions between the atmosphere and a magma ocean according to Ikoma and Genda (2006). The red "+" corresponds to the accretion scenario that can reproduce Earth's atmospheric Ar and Ne isotope ratios, as shown in Fig. 5. In this scenario, a negligible amount of 2% of an Earth ocean with solar nebula origin would be produced. The blue area corresponds to the proto-Earth masses at the disk dissipation time for which the nebula-based D/H-ratio would only modify the chondritic value of Earth's seawater within its error bars. The dashed-dotted lines mark the production of an Earth ocean from a primordial atmosphere in case of proto-Earth accreting to 0.65 M_{Earth} before the gas disk dissipated

the D/H ratio of 67P/Churyumov-Gerasimenko, Marty et al. (2016) estimated a cometary contribution to the Earth's water reservoir of only $\leq 1\%$. Higher contributions would have significantly altered the present-day value if one assumes a predominantly carbonaceous chondritic origin. Ordinary (OC) chondrites, on the other hand, also show a twice as high D/H ratio compared to the Earth ($(310 \pm 10) \times 10^{-6}$; Alexander et al. 2012). Both chondrite types, ordinary and carbonaceous, but in addition also enstatite chondrites, have their origin from different parent bodies, mainly stemming from the main asteroid belt (e.g., Alexander et al. 2018). If one would assume a larger contribution of ordinary chondrites to the building blocks of the Earth (e.g., Dauphas 2017), or even of comets, the Earth's D/H ratio might potentially also be reproduced by a stronger contribution of disk gas accreted water to counterbalance the higher D/H ratios of these minor bodies. However, this must then also be in-line with other volatile elements such as C, N, and noble gases.

According to the reproduction attempts of Venus' atmospheric Ar, Ne and bulk K/U ratios (Lammer et al. 2020a, this journal; 2020b), more nebula-based water could have been produced compared to early Earth's expected evolution as shown in the blue area of Fig. 6 due to proto-Venus possibly being slightly more massive. This would indicate that Venus' initial D/H ratio was probably closer to the nebula ratio than to carbonaceous chondrites. Since the planet's initial water inventory and its D/H ratio was modified during the planets evolution by thermal escape and various nonthermal atmospheric loss processes as well as water vapor outgassing episodes (see Lammer et al. 2020a, this journal) it is difficult to draw definite conclusions. New future in-situ measurements of Venus atmospheric D/H ratio in combination with heavier noble gas isotopes as discussed in Dandouras et al. (2020; this journal) are essential for constraining and understanding the evolution of early Venus' water inventory and atmosphere.

One should also note that the production of nebula-based water depends on the nebula properties such as the dust grain depletion factor, etc. (Lammer et al. 2014; Kimura and Ikoma 2020), which can be different on other star systems. Recently, Kimura and Ikoma (2020) showed that primordial atmospheres could be highly enriched in water vapor if one does not assume solar abundances for the gas disk properties.

3.3 Protoplanetary Masses of Venus, Earth and Mars at Disk Dispersal Time Constrained from Different Isotope Reproduction Studies

If we summarize the different discussed isotope-systematics that are related to D/H, primordial ^{3}He, ^{36}Ar/^{38}Ar, ^{20}Ne/^{22}Ne, ^{36}Ar/^{22}Ne, ^{48}Ca/^{44}Ca abundances and ratios, and chronometric methods based on ^{182}Hf-^{182}W and U-P isotopes one obtains the accretion scenarios for early Earth illustrated in Fig. 7.

The min. and max. disk lifetime according to Bollard et al. (2017) and Wang et al. (2017) is represented by the two vertical dashed-lines. The lower horizontal dashed-line corresponds to the minimum proto-Earth-mass, which is necessary for a protoplanet at 1 AU to build-up and keep an H_2-He-dominated primordial atmosphere (Lammer et al. 2020b) from the solar nebula for some while. The blueish shaded area above the mass fraction of proto-Earth corresponds to masses for which primordial atmospheres remain too long around the growing planetary nucleus for a distance of 1 AU (Sasaki and Nakazawa 1990; Lammer et al. 2020a, this journal; 2020b). As discussed in Sect. 3.2.4, there is an excess of ^{3}He in the deep mantle, which is two orders of magnitude higher than the ^{3}He amount that was delivered by planetesimals and dust. Because it is expected that this ^{3}He excess originated most likely due to ingassing (equilibration with a global magma ocean) from a primordial H_2-He-dominated atmosphere (e.g., Harper and Jacobsen 1996; Jacobsen et al. 2008; Jackson et al. 2017; Timmerman et al. 2019), one can use this primordial ^{3}He evidence for constraining the minimum proto-Earth-mass at the disk dispersal time, in agreement with the reproduced atmospheric ^{36}Ar/^{38}Ar, ^{20}Ne/^{22}Ne, ^{36}Ar/^{22}Ne, noble gas ratios (see Sect. 3.2.3, Lammer et al. 2020a; this journal; 2020b) of 0.53 M_{Earth}.

The upper mass of proto-Earth at the disk dispersal time can be constrained with the D/H ratio of Earth's seawater in agreement with the reproduction of the atmospheric Ar and Ne isotope ratios (see Sect. 3.2.3; Lammer et al. 2020a; this journal; 2020b) and recent pebble accretion model results of Johansen et al. (2020) of $0.5 < M \leq 0.6\ M_{\text{Earth}}$.

As discussed in Sect. 3.2.5, higher proto-Earth-masses would have resulted in too massive H_2-He-dominated primordial envelopes that would have produced a too high amount of ocean water through the reaction between the primordial atmospheric hydrogen and oxides in the magma ocean (Ikoma and Genda 2006). Such a mixture of water from nebular and carbonaceous chondritic origin would have resulted in different D/H ratios than measured in Earth's seawater, which is close to the D/H value measured in carbonaceous chondrites (e.g., Marty 2012). The dark grey box that is confined between proto-Earth masses of $\approx$0.53–0.6 M_{Earth}, and the min and max disk dispersal time of 3.3 and 4.5 Myr, therefore, corresponds to the range to which the accreting proto-Earth should have evolved to produce the preserved

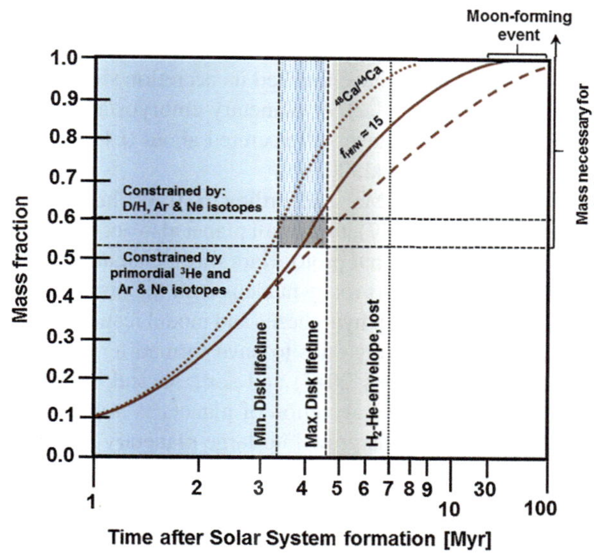

Fig. 7 Illustration of the most likely proto-Earth accretion scenarios as constrained by different isotope-systematics (D-H, atmospheric Ar & Ne, primordial ^{3}He abundance in the deep mantle) and isotopic chronometers (Hf-W, U-P) in dependence on the disk lifetime and activity of the young Sun (dotted line: fast accretion; dashed line: slower accretion scenario). Proto-Earth's mass fraction during disk dispersal most likely was $\approx$0.5–0.6 $M_{\rm Earth}$ (dark grey area), which is also in agreement with recent pebble accretion model results by Johansen et al. (2020)

isotope records. The dotted-vertical line marks the time when the primordial atmosphere was lost (Lammer et al. 2020a).

In case proto-Earth accreted very fast (dotted brownish line), almost to its whole mass within 6–7 Myr. Schiller et al. (2018) inferred it from Ca-isotope systematics (see Sect. 3.2.2), then the disk lifetime should have been close to its minimum value. At this time, the luminosity of the young Sun was yet higher so that a slightly denser primordial atmosphere could have been lost from a more massive proto-Earth of 0.6 $M_{\rm Earth}$ at the time of disk dispersal. A continuous fast growth until about 7 Myr after the origin of the Solar System combined with the escaping primordial atmosphere can reproduce the present atmospheric Ar and Ne isotope ratios including ^{36}Ar/^{22}Ne (see Lammer et al. 2020a for details).

The solid and dashed brownish lines correspond to the dual or two stage accretion scenarios as proposed by Yu and Jacobsen (2011) and Dauphas (2003). Both scenarios can also produce the studied isotope record. Most of the recent studies that used the Hf-W-chronometry method for dating the core formation expect a complete equilibrium or nearly complete equilibrium of metal and silicates favouring Hf/W values between 10–20, thereby indicating a fast accretion of $\geq$0.8 $M_{\rm Earth}$ before $\leq$30 Myr (Rudge et al. 2010; Kleine et al. 2002; Yin and Jacobsen 2006; Yu and Jacobsen 2011; Kleine and Walker 2017; and references therein). There are difficulties to homogenise isotopic data related to the composition of Earth's accreted material between the measurements of lithophile, and moderately and highly siderophile elements (Dauphas 2017) and μ^{48}Ca (Schiller et al. 2018).

From this discussion, one can see that the initial setup of planets that may later evolve to an Earth-like habitat is a complex interplay between the lifetime of the gas-disk, the accretion rate of the terrestrial planet and the EUV evolution path of its young host star. If the Solar System disk lifetime had been about 8 Myr instead of $\approx$4 Myr (Bollard et al. 2017; Wang et al. 2017), Earth would have accreted too much primordial atmospheric mass within the protoplanetary nebula and would have ended as a small sub-Neptune or an ocean planet surrounded by a remnant of its primordial atmosphere that could not escape during its lifetime.

Due to insufficient noble gas isotope data, at present one cannot conclude much about the main composition of proto-Venus. If, however, proto-Venus accumulated a primordial atmosphere, it should have finished its accretion slightly faster than proto-Earth and did not grow further due to a lack of planetary embryos/accreting material at 0.7 AU, while early Earth at 1 AU could have still accreted about 0.4 M_{Earth} after the disk dispersal (Lammer et al. 2020a; this journal; 2020b).

In the case of early Mars, according to Hf-W chronometry (Dauphas and Pourmand 2011; Kleine and Walker 2017), this small planet also accreted most of its mass ($>$0.63–1.0 M_{Mars}) during the disk phase, but proto-Mars was too small to capture a tiny primordial atmosphere that remained around its rocky nucleus after the dissipation of the disk (Lammer et al. 2018).

Out of the formation hypotheses and model results that are compared in Table 1 of Sect. 2, currently the most likely ones to have formed terrestrial planets are pebble accretion (Johansen and Lambrechts 2017) and some Grand Tack models (Brasser et al. 2016); they come the closest to the accretion of planetary embryos with masses of $\approx 0.5 M_{\mathrm{Earth}}$ after a few Myr. Planetesimals grow into large planetary embryos, even giant cores, or gas giants by accreting planetesimals or pebbles, or most likely a combination of both (Lambrechts et al. 2019). One should note that model simulations of planetesimal accretion struggle to grow giant massive planetary cores within the disk lifetime of 3–5 Myr (e.g., Levison et al. 2015a, 2015b). Such fast accretion scenarios may be solved by pebble accretion, but key aspects of pebble accretion remain to be better understood (Lambrechts and Johansen 2012; Johansen et al. 2015; Johansen and Lambrechts 2017; Birnstiel et al. 2016; Bitsch et al. 2019).

According to studies of Johansen et al. (2015), recent simulations indicate that the main growth of planetesimals results from the gas drag-assisted accretion of chondrules. Thus, chondrules represent the building blocks of planetesimals, planetary embryos, protoplanets and finally, terrestrial planets (e.g., Lambrechts and Johansen 2012; Johansen et al. 2015; Birnstiel et al. 2016; Johansen and Lambrechts 2017; Alexander 2019a, 2019b). It is expected that these scenarios evolve from a two-stage process, where the first generation of accreting bodies form very fast by streaming instabilities followed by continuous growth that is dominated by the gas-drag in combination with the accretion of mm-sized particles for planetesimals with radii larger than about 200 km (Johansen et al. 2015). The streaming instability most likely represents a promising mechanism that can explain how dust particles grow to 100 km large planetesimals. The streaming instability, however, requires specific conditions to operate and this implies that planetesimals may form in preferential locations, just beyond the snow line (Izidoro and Raymond 2018). However, at present knowledge, pebble accretion and streaming instabilities most likely solve the first stages in planet accretion (Fig. 1: I and II), as it results in the fast formation of Mars-sized embryos and even more massive protoplanets $\geq$0.5 M_{Earth} over typical disk lifetimes of 3–5 Myr. The later stage of accretion (Fig. 1: III, IV) of the terrestrial planet structure of the inner Solar System can then be quite well explained by the Grand-Tack model and to some extent by the annulus and depleted disk models as well. A clear future step in terrestrial planet formation will be an N-body simulation that also includes the reproduction of isotopic and geochemical constraints.

4 Conclusions

We reviewed the classical scenarios of terrestrial planet formation, including the Grand-Tack model and more recently proposed alternatives. The terrestrial planet formation models were compared with accretion time scales obtained from isotope data of planetary building blocks, the atmospheric Ar and Ne isotope data of Earth and Venus, Earth's D/H isotope ratio in the sea water and the evidence of ^{3}H abundance in the deep mantle, as well as ^{182}Hf-^{182}W chronometry. The analysis of these data indicate that the early terrestrial planets in the Solar System could have accreted most of their mass ($\geq 0.8\ M_{\text{Earth}}$) within the first 10–30 Myr, including a Moon-forming event related to Earth ≥ 50 Myr after the Solar System originated. From the combined evidence of the atmospheric ^{36}Ar/^{38}Ar, ^{20}Ne/^{22}Ne, ^{36}Ar/^{22}Ne isotope ratios, the D/H ratio in Earth's seawater and the estimated ^{3}He abundance that likely originated due to ingassing from a primordial atmosphere, one finds that proto-Earth accreted a mass of 0.5–0.6 M_{Earth} at ≈ 4 Myr, the time when the disk dissipated. The planetesimals grow into planetary embryos and or protoplanets by accreting pebbles, planetesimals, or a combination of both. Pebble accretion most likely solves time scale conflicts of fast accretion, although some aspects of pebble accretion remain to be better understood. For instance, more observations of the radial dependence of pebble sizes in disks, coagulation models that include volatiles, and experiments on the sticking properties of aggregates of a range of compositions are necessary. The late stage accretion of terrestrial planets can than explain the structure of the inner Solar System. Here, the Grand Tack model, the primordial low-mass asteroid belt, and the primordial empty asteroid belt models yield quite accurate results. More studies are needed to homogenise pebble accretion and later stage accretion models with data that are provided by chemical, chronological, isotopical and dynamical constraints. Furthermore, precise measurements of the isotopic composition of Venus are necessary for testing the planet's evolution, including the isotopic composition of its building blocks.

Acknowledgements R. B. acknowledges financial assistance from the Japan Society for the Promotion of Science (JSPS) Shingakujutsu Kobo (JP19H05071). A.J. acknowledges funding from the European Research Foundation (ERC Consolidator Grant 724687-PLANETESYS), the Knut and Alice Wallenberg Foundation (Wallenberg Academy Fellow Grant 2017.0287) and the Swedish Research Council (Project Grant 2018-04867). H.L. and M.S. acknowledge the support of Europlanet 2020 RI. Europlanet 2020 RI has received funding from the European Union's Horizon 2020 research and innovation programme under grant agreement No 654208. We thank Nathan Kaib and Matt Clement for information regarding their 'Early instability' model. H. L. and M. S. acknowledge support from the FWF NFN subproject S11607-N16. M. L. acknowledge support of the Austrian FWF projects P27256-N27 and P30949-N36. The authors also thank the International Space Science Institute (ISSI) in Bern, Switzerland for the support. Finally, we thank the referee S. B. Jacobsen and an anonymous referee for their very helpful comments and suggestions that helped to improve this review article.

References

C.B. Agnor, R.M. Canup, H.F. Levison, On the character and consequences of large impacts in the late stage of terrestrial planet formation. Icarus **142**, 219–237 (1999)

F. Albarède, Volatile accretion history of the terrestrial planets and dynamic implications. Nature **461**, 1227–1233 (2009)

C.M.O. Alexander, Quantitative models for the elemental and isotopic fractionations in chondrites: the carbonaceous chondrites. Geochim. Cosmochim. Acta **254**, 277–309 (2019a)

C.M.O. Alexander, Quantitative models for the elemental and isotopic fractionations in the chondrites: the noncarbonaceous chondrites. Geochim. Cosmochim. Acta **254**, 246–276 (2019b)

C.M.O. Alexander, R. Bowden, M.L. Fogel, K.T. Howard, C.D.K. Herd, L.R. Nittler, The provenances of asteroids, and their contributions to the volatile inventories of the terrestrial planets. Science **337**, 721–723 (2012)

C.M.O. Alexander, K.D. McKeegan, K. Altwegg, Water reservoirs in small planetary bodies: meteorites, asteroids, and comets. Space Sci. Rev. **214**, 36 (2018)

C.J. Allègre, G. Manhès, C. Göpel, The major differentiation of the Earth at ~4.45 Ga. Earth Planet. Sci. Lett. **267**, 386–398 (2008)

K. Altwegg, H. Balsiger, A. Bar-Nun, J.J. Berthelier, A. Bieler, P. Bochsler, C. Briois, U. Calmonte, M. Combi, J. De Keyser, P. Eberhardt, B. Fiethe, S. Fuselier, S. Gasc, T.I. Gombosi, K.C. Hansen, M. Hässig, A. Jäckel, E. Kopp, A. Korth, L. LeRoy, U. Mall, B. Marty, O. Mousis, E. Neefs, T. Owen, H. Rème, M. Rubin, T. Sémon, C.-Y. Tzou, H. Waite, P. Wurz, 67P/Churyumov-Gerasimenko, a Jupiter family comet with a high D/H ratio. Science **347**(6220), 1261952 (2015). https://doi.org/10.1126/science.1261952

E. Asphaug, Impact origin of the Moon. Annu. Rev. Earth Planet. Sci. **42**, 551–578 (2014)

S.K. Atreya, J.B. Pollack, M.S. Matthews, *Origin and evolution of planetary and satellite atmospheres origin and evolution of planetary and satellite atmospheres* (University of Arizona Press, Tucson, 1989). 881 pp. ISBN 978-0-8165-1105-1

G. Avice, B. Marty, Perspectives on atmospheric evolution from noble gas and nitrogen isotopes on Earth, Mars & Venus. Space Sci. Rev. **216**, 36 (2020). https://doi.org/10.1007/s11214-020-00655-0. 2020

G. Avice, B. Marty, R. Burges, The origin and degassing history of the Earth's atmosphere revealed by Archean xenon. Nat. Commun. **8**, 15455 (2017)

M. Barboni, P. Boehnke, B. Keller, I.E. Kohl, B. Schoene, E.D. Young, K.D. McKeegan, Early formation of the Moon 4.51 billion years ago. Sci. Adv. **3**, e1602365 (2017)

M.R. Benedikt, M. Scherf, H. Lammer, E. Marcq, P. Odert, M. Leitzinger, N.V. Erkaev, Escape of rock-forming volatile elements and noble gases from planetary embryos. Icarus **347**, 113772 (2020)

V.C. Bennett, A.D. Brandon, A.P. Nutman, Coupled 142Nd–143Nd isotopic evidence for Hadean mantle dynamics. Science **318**, 1907–1910 (2007)

K. Bermingham, E. Füri, K. Lodders, B. Marty, Chemical and isotopic evolution of the early Solar System. Space Sci. Rev. **216**, 133 (2020). https://doi.org/10.1007/s11214-020-00748-w, this journal

T. Birnstiel, M. Fang, A. Johansen, Dust evolution and the formation of planetesimals. Space Sci. Rev. **205**, 41–45 (2016)

B. Bitsch, A. Izidoro, A. Johansen, S.N. Raymond, A. Morbidelli, M. Lambrechts, S. Jacobson, Formation of planetary systems by pebble accretion and migration: growth f gas giants. Astron. Astrophys. **623**, A88 (2019)

J. Bollard, J.N. Connelly, M.J. Whitehouse, E.A. Pringle, L. Bonal, J.K. Jørgensen, Å. Nordlund, F. Moynier, M. Bizzarro, Early formation of planetary building blocks inferred from Pb isotopic ages of chondrules. Sci. Adv. **3**, 1–9 (2017)

M.A. Bouhifd, A.P. Jephcoat, Convergence of Ni and Co metal-silicate partition co-efficients in the deep magma-ocean and coupled silicon-oxygen solubility in iron melts at high pressures. Earth Planet. Sci. Lett. **307**, 341–348 (2011)

M. Boyet, R.W. Carlson, 142nd evidence for early (>4.53 Ga) global differentiation of the silicate Earth. Science **309**, 576–581 (2005)

M. Boyet, J. Blichert-Toft, M. Rosing, M. Storey, P. Telouk, F. Albarede, ^{142}Nd evidence for early Earth differentiation. Earth Planet. Sci. Lett. **214**, 427–442 (2003)

R. Brasser, The formation of Mars: building blocks and accretion time scale. Space Sci. Rev. **174**, 11–25 (2013)

R. Brasser, M.H. Lee, Tilting Saturn without tilting Jupiter: constraints on giant planet migration. Astron. J. **150**, 157 (2015)

R. Brasser, A. Morbidelli, R. Gomes, K. Tsiganis, H.-F. Levison, Constructing the secular architecture of the solar system II: the terrestrial planets. Astron. Astrophys. **507**, 1053–1065 (2009)

R. Brasser, S. Matsumura, S. Ida, S.J. Mojzsis, S.C. Werner, Analysis of terrestrial planet formation by the grand tack model: system architecture and tack location. Astrophys. J. **821**, 75–93 (2016)

R. Brasser, S.J. Mojzsis, S. Matsumura, S. Ida, The cool and distant formation of Mars. Earth Planet. Sci. Lett. **468**, 85–93 (2017)

R. Brasser, N. Dauphas, S.J. Mojzsis, Jupiter's influence on the building blocks of Mars and Earth. Geophys. Res. Lett. **45**, 5908–5917 (2018)

R. Brasser, S. Matsumura, S. Ida, S.J. Moizsis, S.C. Werner, Analysis of terrestrial planet formation by the Grand Tack model: system architecture and tack location. Astrophys. J. **821**, 75 (2019)

A. Caracausi, G. Avice, P.G. Burnard, E. Füri, B. Marty, Chondritic xenon in the Earth's mantle. Nature **533**, 82–85 (2016)

R.W. Carlson, Earth's building blocks. Nature **541**, 468–469 (2017)

R.W. Carlson, R. Brasser, Q.-Z. Yin, M. Fischer-Gödde, L. Qin, Feedstocks of the terrestrial planets. Space Sci. Rev. **214**, 121 (2018)

G. Caro, B. Bourdon, J.-L. Birck, S. Moorbath, ^{146}Sm–^{142}Nd from Isua metamorphosed sediments for early differentiation of the Earth's mantle. Nature **423**, 428–431 (2003)
J.E. Chambers, Making more terrestrial planets. Icarus **152**, 205–224 (2001)
J.E. Chambers, G.W. Wetherill, Making the terrestrial planets: N-body integrations of planetary embryos in three dimensions. Icarus **136**, 304–327 (1998)
C. Class, S.L. Goldstein, Evolution of helium isotopes in the Earth's mantle. Nature **436**, 1107–1112 (2005)
D.D. Clayton, Nucleosysnthesis in stars, in *Proceed. of the 11th Int. Conf. on Cosmic Rays*, vol. 5 (1970), pp. 21–39
M.S. Clement, N.A. Kaib, S.N. Raymond, J.E. Chambers, K.J. Walsh, The early instability scenario: terrestrial planet formation during the giant planet instability, and the effect of collisional fragmentation. Icarus **321**, 778–790 (2019)
J.N. Connelly, M. Schiller, M. Bizzarro, Pb isotope evidence for rapid accretion and differentiation of planetary embryos. Earth Planet. Sci. Lett. **525**, 115722 (2019)
P. Cresswell, R.P. Nelson, Three-dimensional simulations of multiple protoplanets embedded in a protostellar disc. Astron. Astrophys. **482**, 677–690 (2008)
I. Dandouras, M. Blanc, L. Fossati, M. Gerasimov, E.W. Guenther, K.G. Kislyakova, H. Lammer, Y. Lin, B. Marty, S. Rugheimer, C. Sotin, S. Tachibana, P. Wurz, M. Yamauchi, Future missions related to isotope and element measurements. Space Sci. Rev. **216**, 121 (2020). https://doi.org/10.1007/s11214-020-00736-0, this journal
G. D'Angelo, F. Marzari, Outward migration of Jupiter and Saturn in evolved gaseous disks. Astrophys. J. **757**, 50 (2012)
N. Dauphas, The dual origin of the terrestrial atmosphere. Icarus **165**, 326–339 (2003)
N. Dauphas, The isotopic nature of the Earth's accreting material through time. Nature **541**, 521–524 (2017)
N. Dauphas, A. Pourmand, Hf-W-Th evidence for rapid growth of Mars and its status as a planetary embryo. Nature **473**, 489–492 (2011)
F.E. DeMeo, B. Carry, Solar System evolution from compositional mapping of the asteroid belt. Nature **505**, 629–634 (2014)
M.J. Drake, K. Righter, Determining the composition of the Earth. Nature **416**, 39–44 (2002)
J. Drazkowska, Y. Alibert, B. Moore, Close-in planetesimal formation by pile-up of drifting pebbles. Astron. Astrophys. **594**, A105 (2016)
M. Fischer-Gödde, T. Kleine, Ruthenium isotopic evidence for an inner Solar System origin of the late veneer. Nature **541**, 525–527 (2017)
L. Fossati, N.V. Erkaev, H. Lammer, P.E. Cubillos, P. Odert, I. Juvan, K.G. Kislyakova, M. Lendl, D. Kubyshkina, S.J. Bauer, Aeronomical constraints to the minimum mass and maximum radius of hot low-mass planets. Astron. Astrophys. **598**, 90–99 (2017)
J. Geiss, G. Gloeckler, Abundances of Deuterium and Helium-3 in the protosolar cloud. Space Sci. Rev. **84**, 239–250 (1998)
C. Gillmann, E. Chassefière, P. Lognonné, A consistent picture of early hydrodynamic escape of Venus atmosphere explaining present Ne and Ar isotopic ratios and low oxygen atmospheric content. Earth Planet. Sci. Lett. **286**, 503–513 (2009)
R. Gomes, H.F. Levison, K. Tsiganis, A. Morbidelli, Origin of the cataclysmic Late Heavy Bombardment period of the terrestrial planets. Nature **435**, 466–469 (2005)
A.N. Halliday, Hf-W chronometry and inner solar system accretion rates. Space Sci. Rev. **92**, 355–370 (2000)
A.N. Halliday, Mixing, volatile loss and compositional chance during impact-driven accretion of the Earth. Nature **427**, 505–509 (2004)
A.N. Halliday, A young Moon-forming impact at 70–110 million years accompanied by late-stage mixing, core formation and degassing of the Earth. Philos. Trans. - Royal Soc., A **366**, 4163–4181 (2008)
A.N. Halliday, D.C. Lee, Tungsten isotopes and the early development of the Earth and Moon. Geochim. Cosmochim. Acta **63**, 4157–4179 (1999)
A.N. Halliday, B.J. Wood, The composition and major reservoirs of the Earth around the time of the Moon-forming giant impact, in *Treatise on Geophysics: Evolution of the Earth*, vol. 9, ed. by D. Stevenson (2007), pp. 13–50. Chap. 9.02
A.N. Halliday, B.J. Wood, How did Earth accrete? Science **325**, 44–45 (2009)
L.J. Hallis, G.R. Huss, K. Nagashima, G.J. Taylor, S.A. Halldórsson, D.R. Hilton, M.J. Mottl, K.J. Meech, Evidence for primordial water in Earth's deep mantle. Science **350**, 795–797 (2015)
B.M.S. Hansen, Formation of the terrestrial planets from a narrow annulus. Astrophys. J. **703**, 1131–1140 (2009)
C.L. Harper, S.B. Jacobsen, Evidence from coupled ^{147}Sm-^{143}Nd and ^{146}Sm-^{142}Nd systematics for very early (4.5-Gyr) differentiation of the Earth's mantle. Nature **360**, 728–732 (1992). https://doi.org/10.1038/360728a0
C.L. Harper Jr., S.B. Jacobsen, Noble gases and Earth's accretion. Science **273**, 1814–1818 (1996)

L. Hartmann, N. Calvet, E. Gullbring, P. D'Alessio, Accretion and the evolution of T Tauri disks. Astrophys. J. **495**, 385–400 (1998)

C. Hayashi, K. Nakazawa, H. Mizuno, Earth's melting due to the blanketing effect of the primordial dense atmosphere. Earth Planet. Sci. Lett. **43**, 22–28 (1979)

V.S. Heber, H. Baur, P. Bochsler, K.D. McKeegan, M. Neugebauer, D.B. Reisenfeld, R. Wieler, R.C. Wiens, Isotopic mass fractionation of Solar Wind: Evidence from fast and slow Solar Wind collected by the GENESIS mission. Astrophys. J. **759**, 121 (2012)

R. Helled, P. Bodenheimer, M. Podolak, A. Boley, F. Meru, S. Nayakshin, J.J. Fortney, L. Mayer, Y. Alibert, A.P. Boss, Giant planet formation, evolution, and internal structure, in *Protostars and Planets VI*, ed. by H. Beuther, R.S. Klessen, C.P. Dullemond, T. Henning (University of Arizona Press, Tucson, 2013), pp. 643–665

M. Honda, I. McDougall, D.B. Patterson, A. Doulgeris, D.A. Clague, Possible solar noble-gas component in Hawaiian basalts. Nature **349**, 149–151 (1991)

S. Ida, J. Makino, Scattering of planetesimals by a protoplanet—slowing down the runaway growth. Icarus **106**, 210–227 (1993)

M. Ikoma, H. Genda, Constraints on the mass of a habitable planet with water of nebular origin. Astrophys. J. **648**, 696–706 (2006)

M. Ikoma, L. Elkins-Tanton, K. Hamano, J. Suckale, Water partitioning in planetary embryos and protoplanets with magma oceans. Space Sci. Rev. **214**, 76 (2018)

A. Izidoro, S.N. Raymond, Formation of terrestrial planets, in *Handbook of Exoplanets*, ed. by H. Deeg, J. Belmonte (Springer, New York, 2018), pp. 2365–2423

A. Izidoro, N. Haghighipour, O.C. Winter, M. Tsuchida, Terrestrial planet formation in a protoplanetary disk with a local mass depletion: a successful scenario for the formation of Mars. Astrophys. J. **782**, 31 (2014)

A. Izidoro, S.N. Raymond, A. Morbidelli, O.C. Winter, Terrestrial planet formation constrained by Mars and the structure of the asteroid belt. Mon. Not. R. Astron. Soc. **453**, 3619–3634 (2015)

A. Izidoro, B. Bitsch, S.N. Raymond, A. Johansen, A. Morbidelli, M. Lambrechts, S.A. Jacobson, Formation of planetary systems by pebble accretion and migration: hot super-Earth systems from breaking compact resonant chains. Astron. Astrophys. (2019), submitted. ArXiv e-prints. arXiv:1902.08772

M.G. Jackson, J.G. Konter, T.W. Becker, Primordial helium entrained by the hottest mantle plumes. Nature **542**, 340–343 (2017)

S.B. Jacobsen, How old is planet Earth? Science **300**, 1513–1514 (2003)

S.B. Jacobsen, The Hf-W isotopic system and the origin of the Earth and Moon. Annu. Rev. Earth Planet. Sci. **33**, 531–570 (2005)

S.B. Jacobsen, C.L. Harper Jr., Accretion and early differentiation history of the Earth based on extinct radionuclides. Geophys. Monogr. **95**, 47–74 (1996)

S.B. Jacobsen, M.C. Ranen, M.I. Petaev, J.L. Remo, J.O. O'Connel, D.D. Sasselov, Isotopes as clues to the origin and earliest differentiation history of the Earth. Philos. Trans. - Royal Soc., A **366**, 4129–4162 (2008)

S.A. Jacobson, A. Morbidelli, S.N. Raymond, D.P. O'Brien, K.J. Walsh, D.C. Rubie, Highly siderophile elements in Earth's mantle as a clock for the Moon-forming impact. Nature **508**, 84–87 (2014)

E. Jaupart, S. Charnoz, M. Moreira, Primordial atmosphere incorporation in planetary embryos and the origin of Neon in terrestrial planets. Icarus **293**, 199–205 (2017)

A. Johansen, M. Lambrechts, Forming planets via pebble accretion. Annu. Rev. Earth Planet. Sci. **45**, 359–387 (2017)

A. Johansen, J.S. Oishi, L.M.M. Mac, H. Klahr, T. Henning, A. Youdin, Rapid planetesimal formation in turbulent circumstellar disks. Nature **448**, 1022–1025 (2007)

A. Johansen, M.M. Mac Low, P. Lacerda, M. Bizzarro, Growth of asteroids, planetary embryos, and Kuiper belt objects by chondrule accretion. Sci. Adv. **1**(3), e1500109 (2015)

A. Johansen, T. Ronnet, M. Bizzarro, M. Schiller, M. Lambrechts, Å. Nordlund, H. Lammer, A pebble accretion model for the formation of the terrestrial planets in the Solar System. Sci. Adv. (2020), accepted

B.S. Kamber, J.D. Kramers, How well can Pb isotopes date core formation? Nature **444**, E1–E2 (2006)

L.H. Kellogg, G.J. Wasserburg, The role of plumes in mantle helium fluxes. Earth Planet. Sci. Lett. **99**, 276–289 (1990)

J.B. Kellogg, S.B. Jacobsen, R.J. O'Connell, Modeling lead isotopic heterogeneity in mid-ocean ridge basalts. Earth Planet. Sci. Lett. **262**, 328–342 (2007)

T. Kimura, M. Ikoma, Formation of aqua planets with water of nebular origin: Effects of water enrichment on the structure and mass of captured atmospheres of terrestrial planets. MNRAS **496**(3), 3755–3766 (2020). https://doi.org/10.1093/mnras/staa1778. arXiv:2006.09068

T. Kleine, R.J. Walker, Tungsten isotopes in planets. Annu. Rev. Earth Planet. Sci. **45**, 389–417 (2017)

T. Kleine, C. Munker, K. Mezger, H. Palme, Rapid accretion and early core formation on asteroids and the terrestrial planets from Hf-W chronometry. Nature **418**, 952–955 (2002)
T. Kleine, K. Mezger, H. Palme, C. Münker, The W isotope evolution of the bulk silicate Earth: constraints on the timing and mechanisms of core formation and accretion. Earth Planet. Sci. Lett. **228**, 109–123 (2004)
T. Kleine, M. Touboul, B. Bourdon, F. Nimmo, K. Mezger, H. Palme, S.B. Jacobsen, Q.-Z. Yin, A.N. Halliday, Hf-W chronology of the accretion and early evolution of asteroids and terrestrial planets. Geochim. Cosmochim. Acta **73**, 5150–5188 (2009)
S. König, C. Münker, S. Hohl, H. Paulik, A.R. Barth, M. Lagos, J. Pfänder, A. Regelous, The Earth's tungsten budget during mantle melting and crust formation. Geochim. Cosmochim. Acta **75**, 2119–2136 (2011)
E. Kokubo, S. Ida, Orbital evolution of protoplanets embedded in a swarm of planetesimals. Icarus **106**, 247–257 (1995)
E. Kokubo, S. Ida, On runaway growth of planetesimals. Icarus **123**, 180–191 (1996)
E. Kokubo, S. Ida, Oligarchic growth of planetesimals. Icarus **131**, 171–178 (1998)
J. Kominami, S. Ida, The effect of tidal interaction with a gas disk on formation of terrestrial planets. Icarus **157**, 43–56 (2002)
J. Kominami, S. Ida, Formation of terrestrial planets in a dissipating gas disk with Jupiter and Saturn. Icarus **167**, 231–243 (2004). https://doi.org/10.1016/j.icarus.2003.10.005
J. Kominami, H. Tanaka, S. Ida, Orbital evolution and accretion of protoplanets tidally interacting with a gas disk. I. Effects of interaction with planetesimals and other protoplanets. Icarus **178**, 540–552 (2005)
T.S. Kruijer, M. Touboul, M. Fischer-Gödde, K.R. Bermingham, R.J. Walker, T. Kleine, Protracted core formation and rapid accretion of protoplanets. Science **344**, 1150–1154 (2014)
M. Lambrechts, A. Johansen, Rapid growth of gas-giant cores by pebble accretion. Astron. Astrophys. **544**, A32 (2012)
M. Lambrechts, A. Johansen, A. Morbidelli, Separating gas-giant and ice-giant planets by halting pebble accretion. Astron. Astrophys. **572**, A35 (2014)
M. Lambrechts, A. Morbidelli, S.A. Jacobsen, A. Johansen, B. Bitsch, A. Izodoro, S.N. Raymond, Formation of planetary systems by pebble accretion and migration. How the radial pebble flux determines a terrestrial-planet or super-Earth growth mode. Astron. Astrophys. **627**, A83 (2019)
H. Lammer, A. Stökl, N.V. Erkaev, E.A. Dorfi, P. Odert, M. Güdel, Y.N. Kulikov, K.G. Kislyakova, M. Leitzinger, Origin and loss of nebula-captured hydrogen envelopes from 'sub'- to 'super-Earths' in the habitable zone of Sun-like stars. Mon. Not. R. Astron. Soc. **439**, 3225–3238 (2014)
H. Lammer, A.L. Zerkle, S. Gebauer, N. Tosi, L. Noack, M. Scherf, E. Pilat-Lohinger, M. Güdel, J.L. Grenfell, M. Godolt, A. Nikolaou, Origin and evolution of the atmospheres of early Venus, Earth and Mars. Astron. Astrophys. Rev. **26**, 2 (2018)
H. Lammer, M. Leitzinger, M. Scherf, P. Odert, C. Burger, D. Kubyshkina, C.P. Johnstone, T. Maindl, C.M. Schäfer, M. Güdel, N. Tosi, A. Nikolaou, E. Marcq, N.V. Erkaev, L. Noak, K.G. Kisylakova, L. Fossati, E. Pilat-Lohinger, F. Ragossnig, E.A. Dorfi, Measured atmospheric $^{36}Ar/^{38}Ar$, $^{20}Ne/^{22}Ne$, $^{36}Ar/^{22}Ne$ noble gas isotope and bulk K/U ratios constrain the early evolution of Venus and Earth. Icarus **339**, 11351 (2020b)
H. Lammer, M. Scherf, H. Kurokawa, Y. Ueno, C. Burger, Z. Leinhard, T. Maindl, C.P. Johnstone, M. Leizinger, M. Benedikt, L. Fossati, B. Marty, B. Fegley, P. Odert, K.G. Kislyakova, Loss and fractionation of noble gas isotopes and moderate volatile elements from planetary embryos and Venus, Earth and Mars early evolution. Space Sci. Rev. **216**, 74 (2020a). https://doi.org/10.1007/s11214-020-00701-x, this journal
H.F. Levison, A. Morbidelli, K. Tsiganis, D. Nesvorny, R. Gomes, Late orbital instabilities in the outer planets induced by interaction with a self-gravitating planetesimal disk. Astron. J. **142**, 152 (2011)
H.F. Levison, K.A. Kretke, M.J. Duncan, Growing the gas-giant planets by the gradual accumulation of pebbles. Nature **524**, 322–324 (2015a)
H.F. Levison, K.A. Kretke, K.J. Walsh, W.F. Bottke, Growing the terrestrial planets from the gradual accumulation of submeter-sized objects. Proc. Natl. Acad. Sci. **112**, 14180–14185 (2015b)
J. Li, C. Agee, Geochemistry of mantle–core differentiation at high pressure. Nature **381**, 686–689 (1996)
T. Lichtenberg, G.J. Golabek, T.V. Gerya, M.R. Meyer, The effects of short-lived radionuclides and porosity on the early thermo-mechanical evolution of planetesimals. Icarus **274**, 350–365 (2016)
T. Lichtenberg, G.J. Golabek, C.P. Dullemond, M. Schöonbächler, T.V. Gerya, M.R. Meyer, Impact splash chondrule formation during planetesimal recycling. Icarus **302**, 27–43 (2018)
D.N.C. Lin, J.C.B. Papaloizou, On the tidal interaction between protoplanets and the protoplanetary disk III—orbital migration of protoplanets. Astrophys. J. **309**, 846–857 (1986)
S.J. Lock, K.R. Bermingham, R. Parai, M. Boyet, Geochemical constraints on the origin of the Moon and preservation of ancient terrestrial heterogeneities. Space Sci. Rev. **216**, 109 (2020). https://doi.org/10.1007/s11214-020-00729-z

E.E. Mamajek, Initial conditions of planet formation: lifetimes of primordial disks. AIP Conf. Proc. **1158**, 3–10 (2009)

B. Marty, The origins and concentrations of water, carbon, nitrogen and noble gases on Earth. Earth Planet. Sci. Lett. **313**, 56–66 (2012)

B. Marty, P. Allé, Neon and Argon isotopic constraints on Earth-atmosphere evolution, in *Noble Gas Geochem. Cosmochem.*, ed. by J. Matsuda (Terra Sci. Pub., Tokyo, 1994), pp. 191–204

B. Marty, G. Avice, Y. Sano, K. Altwegg, H. Balsiger, M. Hässig, A. Morbidelli, O. Mousis, M. Rubin, Origins of volatile elements (H, C, N, noble gases) on Earth and Mars in light of recent results from the ROSETTA cometary mission. Earth Planet. Sci. Lett. **441**, 91–102 (2016)

F. Masset, M. Snellgrove, Reversing type II migration: resonance trapping of a lighter giant protoplanet. Mon. Not. R. Astron. Soc. **320**, L55–L59 (2001)

S. Matsumura, R. Brasser, S. Ida, N-body simulations of planet formation via pebble accretion. I. First results. Astron. Astrophys. **607**, A67 (2017)

E. Mazor, D. Heymann, E. Anders, Noble gases in carbonaceous chondrites. Geochim. Cosmochim. Acta **34**, 781–824 (1970)

W.F. McDonough, Compositional model for the Earth's core, in *Treatise on Geochemistry, vol. 2: The Mantle and Core*, ed. by K.K. Turekian H.D. Holland (Elsevier, New York, 2003), pp. 547–568

W.F. McDonough, S.-S. Sun, The composition of the Earth. Chem. Geol. **120**, 223–253 (1995)

K. Meech, S.N. Raymond, Origin of Earth's water: sources and constraints, in *Planetary Astrobiology*, ed. by V. Meadows, G. Arney, D.D. Marais, B. Schmidt (2019), in press. arXiv:1912.04361

A. Meibom, D.L. Anderson, N.H. Sleep, R. Frei, C.P. Chamberlain, M.T. Hren, J.L. Wooden, Are high ^{3}He/^{4}He ratios in oceanic basalts an indicator of deep-mantle plume components? Earth Planet. Sci. Lett. **208**, 197–204 (2003)

H. Mizuno, K. Nakazawa, C. Hayashi, Dissolution of the primordial rare gases into the molten Earth's material. Earth Planet. Sci. Lett. **50**, 202–210 (1980)

S.J. Mojzsis, R. Brasser, N.M. Kelly, O. Abramov, S.C. Werner, Onset of giant planet migration before 4480 million years ago. Astrophys. J. **881**, 44 (2019)

A. Morbidelli, Calcium signals in planetary embryos. Nature **555**, 451–452 (2018)

A. Morbidelli, J. Chambers, J.I. Lunine, J.M. Petit, F. Robert, G.B. Valsecchi, K.E. Cyr, Source regions and time scales for the delivery of water to Earth. Meteorit. Planet. Sci. **35**, 1309–1320 (2000)

A. Morbidelli, H.F. Levison, K. Tsiganis, R. Gomes, Chaotic capture of Jupiter's Trojan asteroids in the early Solar System. Nature **435**, 462–465 (2005)

A. Morbidelli, R. Brasser, R. Gomes, H.F. Levison, K. Tsiganis, Evidence from the asteroid belt for a violent past evolution of Jupiter's orbit. Astron. J. **140**, 1391–1401 (2010)

A. Morbidelli, J.I. Lunine, D.P. O`Brien, S.N. Raymond, K.J. Walsh, Building terrestrial planets. Annu. Rev. Earth Planet. Sci. **40**, 251–275 (2012)

A. Morbidelli, B. Bitsch, A. Crida, M. Gounelle, T. Guillot, S. Jacobson, A. Johansen, M. Lamnrechts, E. Lega, Fossilized condensation lines in the Solar System protoplanetary disk. Icarus **267**, 368–376 (2016). https://doi.org/10.1016/j.icarus.2015.11.027

M.A. Moreira, S. Charnoz, The origin of the neon isotopes in chondrites and on Earth. Earth Planet. Sci. Lett. **433**, 249–256 (2016)

M.A. Moreira, M.D. Kurz, Noble Gases as Tracers of Mantle Processes and Magmatic Degassing, in *The Noble Gases as Geochemical Tracers. Advances in Isotope Geochemistry*, ed. by P. Burnard (Springer, Berlin, 2013). https://doi.org/10.1007/978-3-642-28836-4_12

S. Mukhopadhyay, Early differentiation and volatile accretion recorded in deep-mantle neon and xenon. Nature **486**, 101–104 (2012)

D. Nesvorný, Jumping Neptune can explain the Kuiper belt kernel. Astron. J. **150**, 68 (2015a)

D. Nesvorný, Evidence for slow migration of Neptune from the inclination distribution of Kuiper belt objects. Astron. J. **150**, 73 (2015b)

D. Nesvorný, A. Morbidelli, Statistical study of the early Solar System's instability with four, five, and six giant planets. Astron. J. **144**, 117 (2012)

H.E. Newsom, W/Hf Fractionation in Chondrites and the Earth: Constraints on timing of core formation. Lunar Planet. Sci. **27**, 957 (1996)

D.P. O'Brien, A. Morbidelli, H.F. Levison, Terrestrial planet formation with strong dynamical friction. Icarus **184**, 39–58 (2006)

D.P. O'Brien, K.J. Walsh, A. Morbidelli, S.N. Raymond, A.M. Mandell, Water delivery and giant impacts in the 'Grand Tack' scenario. Icarus **239**, 74–84 (2014)

P. Odert, H. Lammer, N.V. Erkaev, A. Nikolaou, H.I.M. Lichtenegger, C.P. Johnstone, K.G. Kislyakova, M. Leitzinger, N. Tosi, Escape and fractionation of volatiles and noble gases from Mars-sized planetary embryos and growing protoplanets. Icarus **307**, 327–346 (2018)

C. O'Neill, H.St.C. O'Neill, A.M. Jellinek, On the distribution and variation of radioactive heat producing elements within meteorites, the Earth and planets. Space Sci. Rev. (2020). https://doi.org/10.1007/s11214-020-00656-z. this journal
C.W. Ormel, H.H. Klahr, The effect of gas drag on the growth of protoplanets. Analytical expressions for the accretion of small bodies in laminar disks. Astron. Astrophys. **520**, A43 (2010)
J.E. Owen, Y. Wu, Atmospheres of low-mass planets: the "boil-off". Astrophys. J. **817**, 107 (2016)
S.-J. Paardekooper, C. Baruteau, W. Kley, A torque formula for non-isothermal type I planetary migration—II. Unsaturated horseshoe drag. Mon. Not. R. Astr. Soc. **410**, 293–303 (2011)
K. Pahlevan, L. Schaefer, M.M. Hirschmann, Hydrogen isotopic evidence for early oxidation of silicate Earth. Earth Planet. Sci. Lett. **526**, 115770 (2019)
S.W. Parman, Helium isotopic evidence for episodic mantle melting and crustal growth. Nature **446**, 900–903 (2007)
R.O. Pepin, On the origin and early evolution of terrestrial planet atmospheres and meteoritic volatiles. Icarus **92**, 2–79 (1991)
R.O. Pepin, Evolution of Earth's noble gases: consequences of assuming hydrodynamic loss driven by giant impact. Icarus **126**, 148–156 (1997)
S. Péron, M. Moreira, A. Colin, L. Arbaret, B. Putlitz, M.D. Kurz, Neon isotopic composition of the mantle constrained by single vesicle analyses. Earth Planet. Sci. Lett. **449**, 145–154 (2016)
A. Pierens, S.N. Raymond, D. Nesvorny, A. Morbidelli, Outward migration of Jupiter and Saturn in 3:2 or 2:1 resonance in radiative disks: implications for the Grand Tack and Nice models. Astrophys. J. **795**, L11 (2014)
D. Porcelli, D. Woolum, P. Cassen, Deep Earth rare gases: initial inventories, capture from the solar nebula, and losses during Moon formation. Earth Planet. Sci. Lett. **193**, 237–251 (2001)
R.J. Poreda, H. Craig, He and Sr isotopes in the Lau Basin mantle: depleted and primitive mantle components. Earth Planet. Sci. Lett. **113**, 487–493 (1992)
A. Raquin, M. Moreira, Atmospheric $^{38}Ar/^{36}Ar$ in the mantle: implications for the nature of the terrestrial parent bodies. Earth Planet. Sci. Lett. **287**, 551–558 (2009)
S.N. Raymond, A. Izidoro, The empty primordial asteroid belt. Sci. Adv. **3**, 1–6 (2017a)
S.N. Raymond, A. Izidoro, Origin of water in the inner Solar System: planetesimals scattered inward during Jupiter and Saturn's rapid gas accretion. Icarus **297**, 134–148 (2017b)
S.N. Raymond, A. Morbidelli, Planet formation: key mechanisms and global models, in *Lecture Notes of the 3rd Advanced School on Exoplanetary Science*, ed. by M. Biazzo, B. Sozzetti (2020). 100 pp. arXiv:2002.05756
S.N. Raymond, T. Quinn, J.I. Lunine, Making other Earths: dynamical simulations of terrestrial planet formation and water delivery. Icarus **168**, 1–17 (2004)
S.N. Raymond, T. Quinn, J.I. Lunine, High-resolution simulations of the final assembly of Earth-like planets I. Terrestrial accretion and dynamics. Icarus **183**, 265–282 (2006)
S.N. Raymond, D.P. O'Brien, A. Morbidelli, N. Kaib, Building the terrestrial planets: constrained accretion in the inner Solar System. Icarus **203**, 644–662 (2009)
K. Righter, M.J. Drake, Metal/silicate equilibrium in the early Earth—new constraints from the volatile moderately siderophile elements Ga, Cu, P, and Sn. Geochim. Cosmochim. Acta **64**, 3581–3597 (2020)
K. Righter, C. King, L. Danielson, K. Pando, C.T. Lee, Experimental determination of the metal/silicate partition coefficient of Germanium: implications for core and mantle differentiation. Earth Planet. Sci. Lett. **304**, 379–388 (2011)
F. Robert, D. Gautier, B. Dubrulle, The Solar System D/H ratio: observations and theories. Space Sci. Rev. **92**, 201–224 (2012)
R.A. Robie, B.S. Hemmingway, J.R. Fisher, Thermodynamic properties of minerals and related substances at 298.15 K and 1 bar (105 pascals) pressure and at higher temperatures. Geol. Surv. Bull., pp. 1452–1452 (1978)
D.C. Rubie, H.J. Melosh, J.E. Reid, C. Liebske, K. Righter, Mechanisms of metal-silicate equilibration in the terrestrial magma ocean. Earth Planet. Sci. Lett. **205**, 239–255 (2003)
D.C. Rubie, F. Nimmo, H.J. Melosh, Formation of the Earth's core., in *Treatise on Geophysics*, vol. 9 (2007), pp. 51–90. Chap. 9.03
J.F. Rudge, T. Kleine, B. Bourdon, Broad bounds on Earth's accretion and core formation constrained by geochemical models. Nat. Geosci. **3**, 439–443 (2010)
V.S. Safronov, *Evolution of the Protoplanetary Cloud and Formation of the Earth and Planets* (Israel Program for Scientific Translations, Keter Publishing House, Jerusalem, 1969), p. 212. Translated from Russian
H. Samuel, C.G. Farnetani, Thermochemical convection and helium concentrations in mantle plumes. Earth Planet. Sci. Lett. **207**, 9–56 (2003). https://doi.org/10.1016/S0012-821X(02)01125-1
S. Sasaki, K. Nakazawa, Did a primary solar-type atmosphere exist around the proto-Earth? Icarus **85**, 21–42 (1990). https://doi.org/10.1016/0019-1035(90)90101-E

U. Schaltegger, A.K. Schmitt, M.S.A. Horstwood, U-Th-Pb zircon geochronology by ID-TIMS, SIMS, and laser ablation ICP-MS: recipes, interpretations, and opportunities. Chem. Geol. **401**, 89–110 (2015)

M. Schiller, M. Bizzarro, V.A. Fernandes, Isotopic evolution of the protoplanetary disk and the building blocks of Earth and the Moon. Nature **555**, 507–510 (2018)

M. Schiller, M. Bizzarro, J. Siebert, Iron isotope evidence for very rapid accretion and differentiation oft He proto-Earth. Sci. Adv. **6**, 1–7 (2020)

P.A. Sossi, S. Klemme, H.S.C. O'Neill, J. Berndt, F. Moynier, Evaporation of moderately volatile elements from silicate melts: experiments and theory. Geochim. Cosmochim. Acta **260**, 204–231 (2019)

A. Stökl, E.A. Dorfi, C.P. Johnstone, H. Lammer, Dynamical accretion of primordial atmospheres around planets with masses between 0.1 and 5 M_{Earth} in the habitable zone. Astrophys. J. **825**, 86 (2016)

H. Tanaka, W.R. Ward, Three-dimensional interaction between a planet and an isothermal gaseous disk. II. Eccentricity waves and bending waves. Astrophys. J. **602**, 388–395 (2004)

K.K. Tanaka, H. Tanaka, K. Nakazawa, Non-equilibrium condensation in a primordial solar nebula: formation of refractory metal nuggets. Icarus **169**, 197–207 (2002)

H. Tang, N. Dauphas, ^{60}Fe-^{60}Ni Chronology of core formation in Mars. Earth Planet. Sci. Lett. **390**, 264–274 (2014)

F. Tera, D.A. Papanastassiou, G.J. Wasserburg, Isotopic evidence for a terminal lunar cataclysm. Earth Planet. Sci. Lett. **22**, 1–21 (1974)

M.M. Thiemens, P. Sprung, R.O.C. Fonseca, F.P. Leitzke, C. Münker, Early Moon formation inferred from hafnium-tungsten systematics. Nat. Geosci. **12**, 696–700 (2019)

E.W. Thommes, M.J. Duncan, H.F. Levison, The formation of Uranus and Neptune in the Jupiter-Saturn region of the Solar System. Nature **402**, 635–638 (1999)

S. Timmerman, M. Honda, A.D. Burnham, Y. Amelin, S. Woodland, D.G. Pearson, A.L. Jaques, C. Le Losq, V.C. Bennett, G.P. Bulanova, C.B. Smith, J.W. Harris, E. Tohver, Primordial and recycled helium isotope signatures in the mantle transition zone. Science **365**, 692–694 (2019)

M. Touboul, T. Kleine, B. Bourdon, H. Palme, R. Wieler, Late formation and prolonged differentiation of the Moon inferred fromWisotopes in lunar metals. Nature **450**, 1206–1209 (2007)

K. Tsiganis, R. Gomes, A. Morbidelli, H.F. Levison, Origin of the orbital architecture of the giant planets of the Solar System. Nature **435**, 459–461 (2005)

L. Tu, C.P. Johnstone, M. Güdel, H. Lammer, The extreme ultraviolet and X-ray Sun in time: high-energy evolutionary tracks of a solar-like star. Astron. Astrophys. **577**, L3 (2015)

R.D. Van der Hilst, S. Widiyantoro, E.R. Engdahl, Evidence for deep mantle circulation from global tomography. Nature **386**, 578–584 (1997)

J. Wade, B.J. Wood, Core formation and the oxidation state of the Earth. Earth Planet. Sci. Lett. **236**, 78–95 (2005)

C.M. Wai, J.T. Wasson, Nebular condensation of moderately volatile elements and their abundances in ordinary chondrites. Earth Planet. Sci. Lett. **36**, 1–13 (1977). https://doi.org/10.1016/0012-821X(77)90182-0

K.J. Walsh, H.F. Levison, Terrestrial planet formation from an annulus. Astron. J. **152**, 68 (2016)

K.J. Walsh, A. Morbidelli, S.N. Raymond, D.P. O'Brien, A.M. Mandell, A low mass for Mars from Jupiter's early gas-driven migration. Nature **475**, 206–209 (2011)

Z. Wang, H. Becker, Ratios of S, Se and Te in silicate Earth require a volatile-rich late veneer. Nature **499**, 328–331 (2013). https://doi.org/10.1038/nature12285.

H. Wang, B.P. Weiss, X.N. Bai, B.G. Downey, J. Wang, C. Suavet, R.R. Fu, M.E. Zucolotto, Lifetime of the solar nebula constrained by meteorite paleomagnetism. Science **355**, 623–627 (2017)

P.H. Warren, Stable-isotopic anomalies and the accretionary assemblage of the Earth and Mars: a subordinate role for carbonaceous chondrites. Earth Planet. Sci. Lett. **311**, 93–100 (2011)

S.J. Weidenschilling, The distribution of mass in the planetary system and solar nebula. Astrophys. Space Sci. **51**, 153–158 (1977)

G.W. Wetherill, Formation of the terrestrial planets. Annu. Rev. Astron. Astrophys. **18**, 77–113 (1980)

G.W. Wetherill, G.R. Stewart, Accumulation of a swarm of small planetesimals. Icarus **77**, 330–357 (1989)

C.D. Williams, S. Mukhopadhyay, Capture of nebular gases during Earth's accretion is preserved in deep-mantle neon. Nature **565**, 78–81 (2019)

J.M.Y. Woo, R. Brasser, S. Matsumura, S.J. Mojzsis, S. Ida, The curious case of Mars' formation. Astron. Astrophys. **617**, A17 (2018)

B.J. Wood, A.N. Halliday, Cooling of the Earth and core formation after the giant impact. Nature **437**, 1345–1348 (2005)

Q. Yin, S.B. Jacobsen, Does U-Pb date Earth's core formation? Nature **444**, E1 (2006)

Q. Yin, S.B. Jacobsen, K. Yamashita, J. Blichert-Toft, P. Télouk, F. Albarède, A short timescale for terrestrial planet formation from Hf-W chronometry of meteorites. Nature **418**, 949–952 (2002)

R. Yokochi, B. Marty, A determination of the neon isotopic composition of the deep mantle. Earth Planet. Sci. Lett. **225**, 77–88 (2004)

E.D. Young, A. Shahar, F. Nimmo, H.E. Schlichting, E.A. Schauble, H. Tang, J. Labidi, Near-equilibrium isotope fractionation during planetesimal evaporation. Icarus **323**, 1–15 (2019)

G. Yu, S.B. Jacobsen, Fast accretion of the Earth with a late Moon-forming giant impact. Proc. Natl. Acad. Sci. **108**, 17604 (2011)

H. Zhang, J.-L. Zhou, On the orbital evolution of a giant planet pair embedded in a gaseous disk. I. Jupiter-Saturn configuration. Astrophys. J. **714**, 532–548 (2010)

Space Sci Rev (2020) 216:109
https://doi.org/10.1007/s11214-020-00729-z

Geochemical Constraints on the Origin of the Moon and Preservation of Ancient Terrestrial Heterogeneities

Simon J. Lock[1] · Katherine R. Bermingham[2,3] · Rita Parai[4] · Maud Boyet[5]

Received: 26 November 2019 / Accepted: 26 August 2020 / Published online: 10 September 2020

Abstract The Moon forming giant impact marks the end of the main stage of Earth's accretion and sets the stage for the subsequent evolution of our planet. The giant impact theory has been the accepted model of lunar origin for 40 years, but the parameters of the impact and the mechanisms that led to the formation of the Moon are still hotly debated. Here we review the principal geochemical observations that constrain the timing and parameters of the impact, the mechanisms of lunar formation, and the contemporaneous evolution of Earth. We discuss how chemical and isotopic studies on lunar, terrestrial and meteorite samples relate to physical models and how they can be used to differentiate between lunar origin models. In particular, we argue that the efficiency of mixing during the collision is a key test of giant impact models. A high degree of intra-impact mixing is required to explain the isotopic similarity between the Earth and Moon but, at the same time, the impact did not homogenize the whole terrestrial mantle, as isotopic signatures of pre-impact heterogeneity are preserved. We summarize the outlook for the field and highlight the key advances in both measurements and modeling needed to advance our understanding of lunar origin.

Keywords Earth · Moon · Impacts · Formation · Geochemistry · Isotopes

Reading Terrestrial Planet Evolution in Isotopes and Element Measurements
Edited by Helmut Lammer, Bernard Marty, Aubrey L. Zerkle, Michel Blanc, Hugh O'Neill and Thorsten Kleine

✉ S.J. Lock
slock@caltech.edu

1 Division of Geological and Planetary Sciences, California Institute of Technology, Pasadena, CA 91125, USA

2 Department of Earth and Planetary Sciences, Rutgers University, Piscataway, NJ 08854, USA

3 Department of Geology, University of Maryland, College Park, MD 20742, USA

4 Department of Earth and Planetary Sciences, Washington University in St. Louis, Saint Louis, MO 63130, USA

5 Laboratoire Magmas et Volcans, Université Clermont Auvergne, CNRS, IRD, OPGC, 63000 Clermont-Ferrand, France

1 Introduction

The terminal event in the main stage of Earth's accretion is thought to be the Moon-forming giant impact (Hartmann and Davis 1975; Cameron and Ward 1976). A planetary body (often referred to as Theia) collided with the proto-Earth, injecting material into orbit from which the Moon formed. The giant impact melted and vaporized much of Earth and set the initial conditions for the subsequent evolution of our planet. The giant impact theory has been the accepted model of lunar origin for 40 years, yet the parameters of the impact and the mechanisms that led to the formation of the Moon are intensely debated (see reviews by Asphaug 2014; Barr 2016).

Lunar origin is one of the best characterized problems in terrestrial planetary formation as we have comparatively large quantities of samples from each body available for geochemical analysis, and a wealth of geophysical measurements and orbital observations. It is partly for these reasons, however, that the topic of lunar origin remains so controversial. Models that match the first order physical observations (such as the mass of the Moon and Earth) are potentially excluded by other lines of evidence from geochemical studies. Geochemical constraints and geodynamical models must go hand in hand in the effort to elucidate the details of lunar formation.

Here, we provide an overview of the various permutations of the giant impact model and how the Moon is formed in each case. We then review the constraints on lunar origin and how they bear on the different models, focusing on evidence from geochemical measurements of lunar and terrestrial samples.

2 Giant Impact Models for Lunar Origin

Early models of the Moon-forming giant impacts were focused on reproducing physical constraints, e.g., the mass of the Moon. Cameron and Ward (1976) proposed that the Moon-forming giant impact set the angular momentum of the Earth-Moon system, which placed a very strong constraint on the parameters of the impact (e.g., the masses of the colliding bodies, impact velocity, impact angle). Numerical simulations have found that a narrow range of impacts, grazing impacts of roughly Mars-mass impactors near the mutual escape velocity,[1] can possibly inject sufficient mass and angular momentum into orbit to form an iron-poor (see Sect. 4), lunar-mass Moon (Canup and Asphaug 2001; Canup 2004, 2008). The energy of the impact melts or vaporizes much of the colliding bodies and leaves the proto-Earth spinning with a roughly 5 hour period. This so-called canonical scenario has become the de-facto working model for lunar formation and has been the basis of our understanding of the outcome of giant impacts in the solar system.

A canonical Moon-forming impact injects material into orbit, forming a multiphase (i.e., a mixture of liquid and vapor) disk around the proto-Earth (Fig. 1a). Condensed material (i.e., not vapor) that is injected beyond the Roche limit (the point beyond which large objects can survive without being torn apart by tidal forces[2]) accretes to form one or two large satellites on a timescale of order weeks (Ida et al. 1997; Kokubo et al. 2000). Further mass is

[1]The mutual escape velocity is the minimum velocity required to overcome gravity and separate two bodies that are initially touching. Alternatively, the mutual escape velocity is the impact velocity of two bodies drawn together by gravity from infinite distance given zero initial relative velocity.

[2]For the case of the post-impact Earth, the Roche limit is at a distance of about 2.9 R_{Earth} from the center of the body, where R_{Earth} is the radius of the present-day Earth.

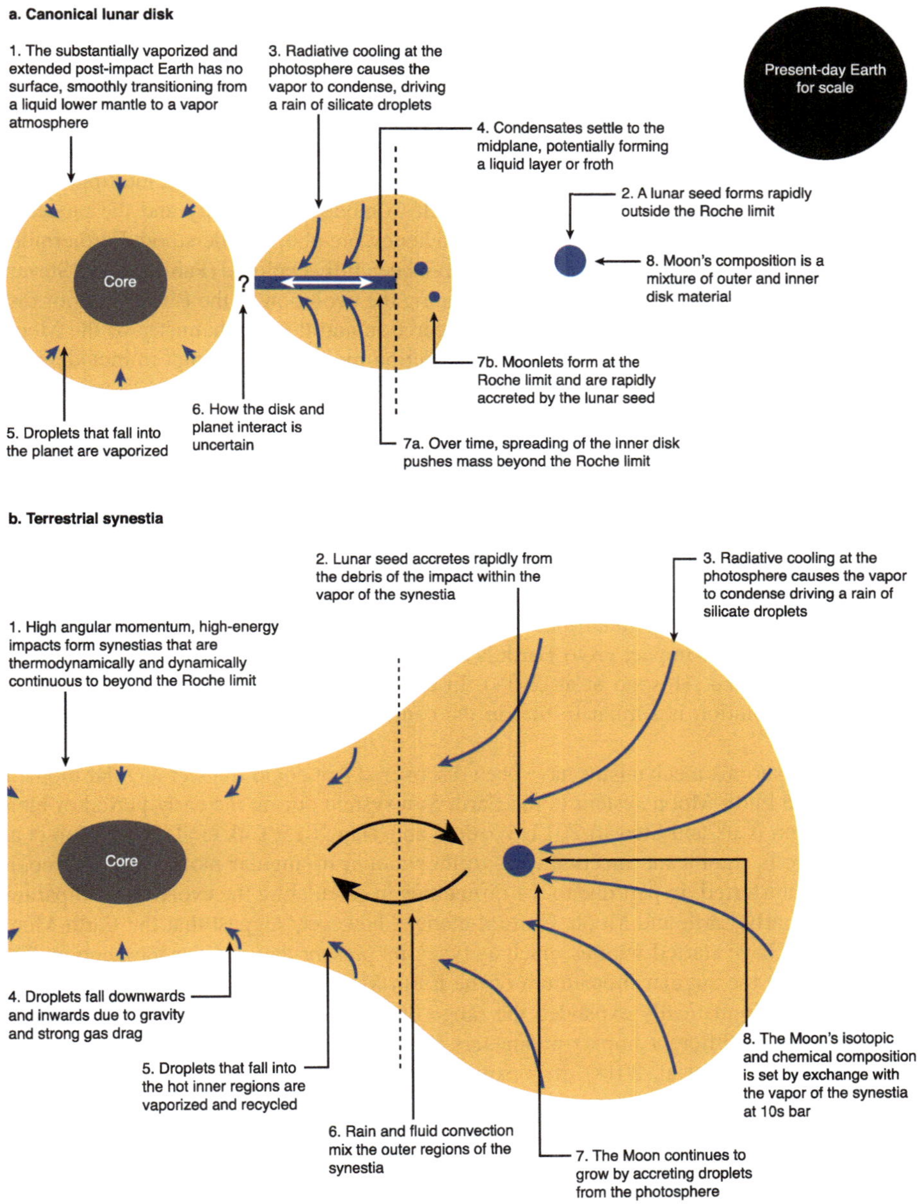

Fig. 1 Schematic of the canonical (**a**) and synestia (**b**) Moon-formation models. Post-impact structures are shown roughly to scale as they would appear a few months after the impact with the depicted surface being the photosphere, the point at which the structure becomes optically thin. Colors are used to differentiate between: the metal core of the bodies (grey); the silicate portion of the bodies (yellow); and the existence and motions of condensates in the low-pressure regions of the structures (blue). Vertical dashed lines indicate the Roche limit, the point beyond which large objects can survive without being torn apart by tidal forces ($\sim 2.9\ R_{\mathrm{Earth}}$).

then added to the proto-Moon over tens to hundreds of years by viscous spreading of material from the Roche-interior disk (Thompson and Stevenson 1988; Salmon and Canup 2012, 2014; Charnoz and Michaut 2015; Ward 2017). The numerical techniques used for giant impact simulations cannot resolve the complex physics and chemistry, or long timescales, of disk evolution (years to hundreds of years). Separate models are required to understand the formation of the Moon and to build scaling laws that can predict the mass of the satellite that would be formed by a disk of a given mass and angular momentum as determined by giant impact simulations. The complex structure and dynamics of the multiscale, multiphase, and multicomponent Roche-interior disk are difficult to model numerically and the processes that could drive viscous spreading, and their efficiency, are poorly understood. Furthermore, how the disk interacts with the proto-Earth has not been fully explored (Pahlevan and Stevenson 2007; Desch and Taylor 2013). Understanding the evolution of the Roche-interior disk is vital for predicting the efficiency of lunar accretion and the geochemistry of the Moon. For example, using scaling laws from lunar accretion models that attempt to include some of the complexities of the Roche-interior disk (Salmon and Canup 2012, 2014), very few of the published canonical Moon-forming impacts would produce a lunar-mass Moon (see discussion in Lock et al. 2018). Determining how the Moon forms following an impact remains a major stumbling block for lunar origin studies.

Although the canonical impact has been the dominant focus for work on lunar formation, other potential impacts have been explored. Reufer et al. (2012) showed that a hit-and-run collision could leave the largest remnant with a circumplanetary disk. It has also been proposed that the Moon was formed by not one but multiple giant impacts, with each impact delivering a fraction of the lunar mass (Rufu et al. 2017). A modification of the canonical impact where the pre-impact proto-Earth is hot enough to have a molten upper mantle has also been suggested (Hosono et al. 2019). In each of these cases, it is assumed that the post-impact evolution is similar to that in the canonical case, with varying silicate-vapor fractions.

Recently, various mechanisms have been discovered that could transfer angular momentum from the Earth-Moon system to the Earth-Sun system during the early period of lunar tidal evolution (Ćuk and Stewart 2012; Wisdom and Tian 2015; Ćuk et al. 2016; Tian et al. 2017). There is significant uncertainty as to the amount of angular momentum that could have been transferred, in part due to the difficulties in calculating the evolving tidal parameters of the early Earth and Moon. Current models, however, suggest that the Earth-Moon system could have started with as much as twice its present-day angular momentum. The possibility that the angular momentum of the Earth-Moon system has evolved since lunar formation has dramatically expanded the range of potential Moon-forming giant impacts and a number of different impact parameters have been explored (Canup 2012; Ćuk and Stewart 2012; Lock et al. 2018). It is expected that most planets will acquire significant angular momentum during accretion as a result of single or multiple giant impacts (Agnor et al. 1999; Kokubo and Genda 2010; Rufu et al. 2017). In fact, studies that have attempted to track the angular momentum of terrestrial planets indicate that terminal impacts with angular momenta as low as the canonical impact would be rare (Kokubo and Genda 2010). It should be noted that these models have neglected several significant processes (e.g., the formation and tidal evolution of moons) that could affect the angular momentum of planets during accretion and models that include these effects are required to confirm the angular momentum distribution of proto-planets. The canonical Moon-forming impact is also a relatively low energy impact and the range of potential impact parameters (e.g., the masses of the colliding bodies, impact velocity, impact angle) is very narrow. There is a much wider array of giant impacts that occur during accretion and the vast majority of accretionary impacts between embryo-sized bodies are of higher specific energy than the canonical scenario (see e.g., Quintana et al. 2016), resulting in hotter and more vaporized post-impact bodies (Lock and Stewart 2017). Therefore, given the expected range of giant impacts that occur during accretion and the potential for the angular momentum of the Earth-Moon system to have been reduced during lunar tidal evolution, it is reasonable to expect that the terminal giant impact for Earth would be of higher angular momentum and energy than the canonical impact.

High-energy, high-angular momentum impacts provide new environments for lunar formation. In particular, it has been found that a range of such impacts produce a new class of planetary bodies called synestias (Lock and Stewart 2017). Synestias are bodies that exceed the corotation limit (CoRoL). As the angular momentum of a corotating planet is increased, it rotates faster and becomes increasingly oblate. At a critical angular momentum the angular velocity at the equator is equal to that of a Keplerian orbit, i.e., the angular velocity at the equator is the same as a body in orbit. This point, at which the angular momentum is high enough that the angular velocity at the equator equals that of a circular Keplerian orbit, is the CoRoL. More generally, the CoRoL is a surface that depends on the properties of the body (e.g., thermal state, angular momentum, total mass, and compositional layering). Post-impact synestias are thermodynamically and dynamically continuous bodies with an inner corotating region transitioning to a significantly sub-Keplerian disk-like region further from the rotation axis. Post-impact synestias are typically many times larger than an equivalent condensed planet and form large biconcave disk-shaped structures. Synestias are predicted to be common during terrestrial planet formation and most final planets would have been transformed into a synestia at least once during their accretion (Lock and Stewart 2017). Note that all previously proposed high-angular momentum impacts (Canup 2012; Ćuk and Stewart 2012) produce synestias but this was not realized at the time.

A synestia is a distinct dynamical structure from a planet and condensate-dominated circumplanetary disk system, and, as a result, different processes dominate the early evolution of a synestia (Fig. 1b). Recently, it was proposed that if a giant impact produced a sufficiently extended synestia the Moon would have formed inside it, surrounded by 10s bar of silicate vapor (Lock et al. 2018). A lunar seed, accreted from large debris injected into orbit during the impact, would have grown over years to tens of years by accreting silicate droplets raining from the photosphere, where radiative cooling leads to condensation of the vapor. The rain of droplets from the photosphere also acts to transport mass and angular momentum in the synestia as the droplets are dragged inwards by the gas, leading to mixing of the outer layers. Over time, as the vapor cools and condenses, the synestia shrinks to within the lunar orbit terminating lunar accretion. Although promising as a model for lunar origin, the formation of the Moon from a synestia is a very recent idea and many pieces of the scenario need to be explored in more depth.

The range of possible high-angular momentum giant impacts, and the corresponding range of post-impact states, are only beginning to be explored. There are many possible scenarios between the end member cases of the canonical disk and the highly extended synestias proposed by Lock et al. (2018) and future work is needed to explore this continuum of possibilities.

3 Timing of Moon Formation

Various different approaches have been used to constrain the timing of the Moon-forming giant impact, resulting in ages that vary from 60 Myr to 250 Myr after the beginning of the solar system, defined as the time of the condensation of the first solids (calcium aluminum inclusions) from the cooling nebular gas of the protoplanetary disk at 4567 ± 0.16 Ma (Connelly et al. 2012). Estimates of the timing of the Moon-forming giant impact are determined from dating lunar samples (Apollo samples and lunar meteorites), calculation of the timing of lunar magma ocean (LMO) crystallization, and numerical simulations. Time constraints on terrestrial events that post-date the last giant impact are derived from terrestrial samples (e.g. the oldest zircons derived from the early crust, samples that record the formation of different reservoirs during crystallization of the terrestrial magma ocean) and provide a limit on how late the Moon could have formed. It is important to note that none of the approaches to dating the Moon-forming giant impact directly dates the impact itself. Each method dates an event that occurred on Earth or the Moon which was tied to the impact, lunar formation, or differentiation of Earth and/or the Moon in some way. There is therefore always an additional uncertainty, on top of any intrinsic uncertainty in the approach, as to what event a given technique is dating and how the timing of that event relates to the timing of the giant impact.

All ages defined by lunar samples are interpreted based on the same assumed evolution after lunar formation. Mafic silicates crystallized from an initially extensively molten Moon and subsequently sank into the mantle to form the source regions of much later mare basalt magmatism (Warren 1985). After some 70-80% crystallization, plagioclase began to crystallize from what was by then a dense iron-rich differentiated magma, causing the plagioclase to float to form the ferroan anorthosite (FAN) suite (FAS) of the lunar highlands rocks (Dowty et al. 1974). The FAS, therefore, likely contains the oldest lunar rocks. Further crystallization resulted in a residual liquid strongly enriched in incompatible elements. This reservoir is the source of KREEP basalts. They were given the name KREEP based on their marked enrichment in potassium (K), rare earth elements (REE) and phosphorus (P), among other incompatible elements (see e.g., Snyder et al. 1995). Studies of short- and long-lived systematics have helped refine the lunar chronology from the formation of the different source reservoirs to lunar magmatism. No method can directly provide the age of the Moon's formation but different systems can date events that are closely linked to the LMO stage. For example, the oldest ages measured for the lunar anorthosite crust are assumed to date the magma ocean crystallization if this process is relatively short. In general, this is a good assumption as models of the LMO agree that a stable plagioclase flotation crust formed in less than a million years. However, it should be noted that several processes could extend the cooling timescale for the LMO (tidal heating, formation of a thick lid, presence of an optically-thick atmosphere: Tian et al. 2017; Elkins-Tanton et al. 2011; Saxena et al. 2017) potentially resulting in the persistence of partially molten regions of the lunar mantle for several million years.

Below we summarize the different methods that have been used to determine the age of the Moon and their results (Fig. 2):

1. Absolute ages for FAS rocks: 100-200 Myr after the beginning of the solar system. Only two samples have produced concordant ages from multiple isotope dating techniques: sample 60025 dated at 4360±3 Ma (Borg et al. 2011) and sample 67075 dated at 4470±70 Ma (Nyquist et al. 2010). Concordant ages (4304±12 Ma) from the ^{147}Sm-^{143}Nd, ^{146}Sm-^{142}Nd, Rb-Sr, and Ar-Ar isotopic systems were obtained on a lithic breccia from the noritic anorthosite 60016 (Marks et al. 2019). However, for this specific sample this age is not an age of crystallization because all these isotope systematics have different closure temperatures; rather it reflects the time of excavation of this sample from the crust by a large impact event (Marks et al. 2019). In contrast, Ar-Ar ages on FAN 60025 and 67075 are significantly younger and around 4.1 Ga (see Borg et al. 2011) showing later disturbances. The troctolite anorthosite 62237 recently measured by Sio

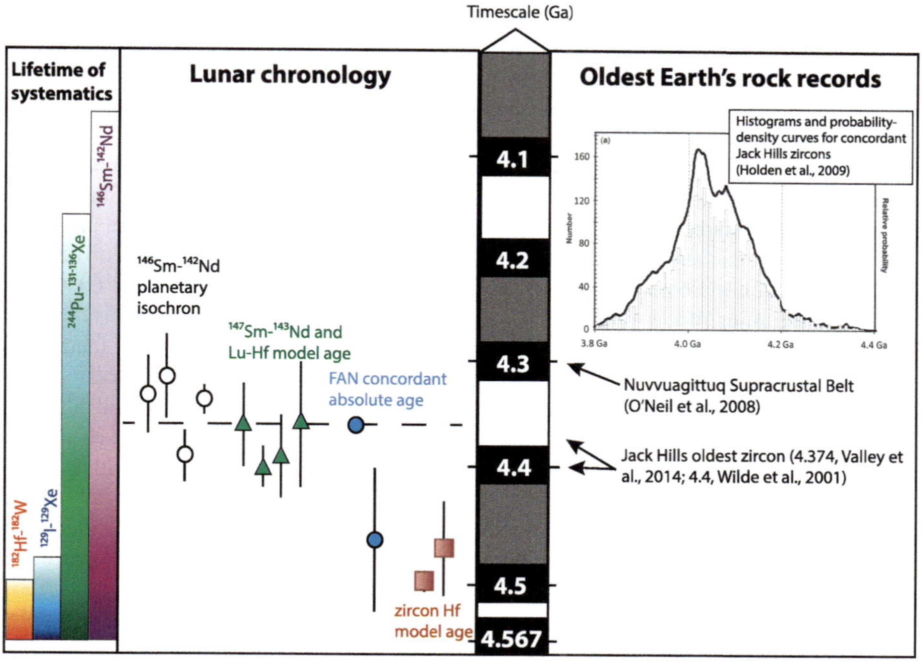

Fig. 2 Age estimates for the crystallization of the lunar magma ocean compared to the oldest terrestrial records. References: ^{146}Sm-^{142}Nd planetary isochron (Nyquist et al. 1995; Boyet and Carlson 2007; McLeod et al. 2014; Borg et al. 2019), ^{147}Sm-^{143}Nd and Lu-Hf model age on bulk-rocks (Lugmair and Carlson 1978; Sprung et al. 2013; Gaffney and Borg 2014), FAN absolute datations (Borg et al. 2011; Nyquist et al. 2010), zircon Hf model age (Taylor et al. 2009; Barboni et al. 2017). The dashed line represents the most concordant estimate for the age of a lunar sample and so our preferred estimate for the timing of the giant impact that formed the Moon. Interpreting these ages (crust crystallization, differentiation of the lunar interior) as those of the formation of the Moon requires a rapid crystallization of the lunar magma ocean (see Sect. 3 for more details). Radioactive parent isotopes for short-lived systematics are alive during five half lives (lifetimes shown on the left).

et al. (2020) plot on the same internal Sm-Nd isochron as 60025 suggesting they formed contemporaneously.

2. U-Pb ages of lunar zircons: the oldest lunar zircon is estimated to have formed 151±6 Myr after the beginning of the solar system. This zircon was located in the matrix of a clast-rich impact melt breccia (Nemchin et al. 2009). Nemchin et al. (2009) interpreted this age as the age of formation of the KREEP source as experiments show that zircon saturation is achieved only at the very end of LMO crystallization (Dickinson and Hess 1982). The significance of zircon ages remains debated and could represent magmatic and/or impact activity on the Moon (Grange et al. 2013).
3. Model ages for the LMO crystallization (KREEP, mare basalts, Mg-suite crustal rocks): 130-200 Myr after the beginning of the solar system (Lugmair and Carlson 1978; Sprung et al. 2013; Gaffney and Borg 2014). Model ages determined using both Sm-Nd and Lu-Hf long-lived systematics give similar results. Here age refers to the formation of the different source reservoirs (i.e., the time at which fractionation of parent/daughter ratios occurred).
4. Planetary isochron age using short-lived ^{146}Sm-^{142}Nd systematics: 180-230 Myr after the beginning of the solar system (Nyquist et al. 1995; Boyet and Carlson 2007; Brandon

et al. 2009; McLeod et al. 2014; Borg et al. 2019). Ages are calculated from the slope of the fossil isochron where the $^{142}Nd/^{144}Nd$ ratio of different samples is plotted against the parent/daughter ratio of their sources. Here assumptions are made for calculating the source parent/daughter ratio (e.g., one or several stage evolution, initial composition of the bulk Moon) which explains at least part of the dispersion in ages.
5. The absence of variation in $^{182}W/^{184}W$ recorded in the lunar samples measured to date precludes global differentiation of the Moon prior to 60-70 Myr after the beginning of the solar system (Touboul et al. 2015; Kruijer and Kleine 2017) (cf., Thiemens et al. 2019).
6. Hafnium isotopic analyses were performed on the same volumes of lunar zircon dated by high-precision U-Pb geochronology and a model Lu-Hf differentiation age of 60 Myr after the beginning of the solar system was obtained (Barboni et al. 2017). Several samples have elevated $^{178}Hf/^{177}Hf$ ratios showing that the Hf isotopic compositions (and hence the inferred ages) are strongly affected by capture of secondary neutrons produced during cosmic ray exposure. Although the robustness of the correction was raised, a sample not affected by this secondary effect gives an age of 60 Myr. Previous Lu-Hf data on lunar zircons dated the crystallization of the lunar magma ocean at 100 Ma with an upper limit of 200 Ma (Taylor et al. 2009).
7. Numerical simulations of planet formation that attempt to match the highly siderophile element (HSE[3]) abundances in the terrestrial mantle: 95±32 Myr after the beginning of the solar system (Jacobson et al. 2014).

Recent isotope studies support a young Moon. Borg et al. (2019) showed that mare basalt sources and ferroan anorthosite suite cumulates define a linear array on a $^{146}Sm/^{144}Nd$ versus $^{142}Nd/^{144}Nd$ isochron plot demonstrating these materials were derived from a common reservoir at 4336^{+31}_{-32} Ma. This age is similar to the age obtained on the FAN 60025 (4360±3 Ma, Borg et al. 2011). The age derived for sample 60025 is the most robust age determined for lunar samples because three different chronometers (^{146}Sm-^{142}Nd, ^{147}Sm-^{143}Nd, ^{207}Pb-^{206}Pb) produce the same age within uncertainty. Isotope studies on highlands rocks enriched in magnesium, known as the Mg-suite and interpreted to be partial melts of cumulates in the lunar interior, also produce similar ages (Carlson et al. 2014). A lunar crust formed contemporaneously with different terrestrial mantle reservoirs suggests a rapid crystallization of the LMO, in agreement with numerical models (< 1 Myr, e.g., Elkins-Tanton et al. 2011). It has been proposed that the young ages of the lunar crust could be due to solidification of the LMO being prolonged by tidal heating (Elkins-Tanton et al. 2011), but if the Moon formed late, then a mechanism to prolong the solidification of the LMO is not required.

The age of sample 60025 is also in agreement with ages proposed for the last magma ocean event on Earth estimated to be around 150-200 Myr after the beginning of the solar system from $^{146,\,147}Sm$-$^{142,\,143}Nd$ systematics measurements on Eoarchean samples (e.g., Rizo et al. 2011; Caro et al. 2017). A major melting event on Earth at this time is consistent with the presence of zircon formed in a terrestrial crust at 4.4 Ga (Wilde et al. 2001; Valley et al. 2014) and the age of 4.3 Ga estimated from the ^{146}Sm-^{142}Nd systematics for samples from the Nuvvuagittuq greenstone belt (northern Quebec, Canada) representing the oldest preserved crustal section on Earth (O'Neil et al. 2008). However, it may be challenging to reconcile outgassing of a late magma ocean with ^{129}I-^{129}Xe systematics in mid-ocean ridge basalts, which exhibit large excesses in ^{129}Xe that are intrinsic to the upper mantle and

[3]HSEs are those elements that have a strong propensity to be in metals over silicates and so are overwhelmingly incorporated into Earth's core during accretion. Typically, HSEs are defined as elements that have metal–silicate partition coefficients (D values, concentration ratio of an element in liquid metal to liquid silicate) >10,000.

not derived from a deeper-seated mantle source (e.g., Mukhopadhyay and Parai 2019; Parai et al. 2019). Mantle excesses in ^{129}Xe indicate that retention of Xe began within the lifetime of ^{129}I (<100 Myr after CAIs: Porcelli et al. 2001; Mukhopadhyay and Parai 2019). If the Moon-forming giant impact generated a terrestrial magma ocean after ^{129}I was extinct, this implies that magma ocean outgassing after the Moon-forming giant impact was inefficient. The efficiency by which noble gases are degassed from a magma ocean is uncertain, and depends on the style of magma ocean solidification (fractional or batch crystallization) and the fraction of melt that is trapped as the mantle freezes which are both poorly understood (see e.g., Solomatov and Stevenson 1993a,b,c; Miyazaki and Korenaga 2019).

Geochronology on the Moon remains a challenging task due to the nature and limited number of samples available. Most lunar samples were collected during the Apollo missions, which landed predominantly in the mare basin areas where basalts are relatively young (< 3.9 Ga) in comparison to the average lunar crust, thereby potentially biasing the sample suite. FANs have been mostly sampled during the Apollo 16 mission and some samples are also available in the lunar meteorite collection. The question of whether samples taken from a small area of the lunar surface constrain the age of the crust persists since lunar highland meteorites and Luna mission anorthosites are more magnesium-rich than Apollo ferroan anorthosites (Gross et al. 2014). Lunar crustal rocks were not all formed during the crystallization of the LMO; some were likely generated via complex crustal processes later in time. Impacts on the lunar surface have also extensively shocked crustal rocks, which can disturb radiogenic isotope signatures. For example, Rb-Sr systematics in FANs are vulnerable to shock resetting that may yield young apparent ages (Norman et al. 2003). The large deficit in ^{142}Nd measured in FAN 62255 may indicate direct crystallization from the LMO 60-125 Myr after the beginning of the solar system (Boyce et al. 2015), but such measurements of ^{142}Nd in FANs with extremely low rare earth element abundances are challenging. Additional samples and further analyses are needed to better understand lunar chronology from the radiogenic isotope record.

4 Bulk Lunar Elemental Composition

Estimates of the Moon's bulk composition (inclusive of the core, mantle, and crust) suggest that it is very similar to Earth's with two principal exceptions: the Moon has a much smaller metal core (Williams et al. 2014; Weber et al. 2011; Garcia et al. 2011); and the Moon is more depleted in volatile and moderately volatile elements (MVEs; e.g., K, Na, Zn, and Cl: Ringwood and Kesson 1977; Wanke et al. 1977b; Morgan et al. 1978; Longhi 2006; Albarède et al. 2015; Taylor 2014; Hauri et al. 2015; Taylor and Wieczorek 2014; Ni et al. 2019, see Fig. 3). The similarity of the Earth and Moon in the major elements (excepting some siderophile elements, those that have a propensity to partition into metals, such as iron) suggests that the two bodies formed from material from a similar reservoir, consistent with the observed similarity in non-mass dependent isotopic composition (Sect. 6). The similarities and differences in composition between the two bodies can give us insight into the mechanisms of lunar formation.

It is estimated that the Moon has a metallic iron core that is only ~2% of its mass (compared to ~30% for Earth) which places a significant constraint on lunar origin models (Williams et al. 2014; Weber et al. 2011; Garcia et al. 2011). This constraint is met in the canonical giant impact model as, in a subset of collisions, only a small quantity of core material from the impactor is injected into orbit, creating an iron-depleted proto-lunar disk.

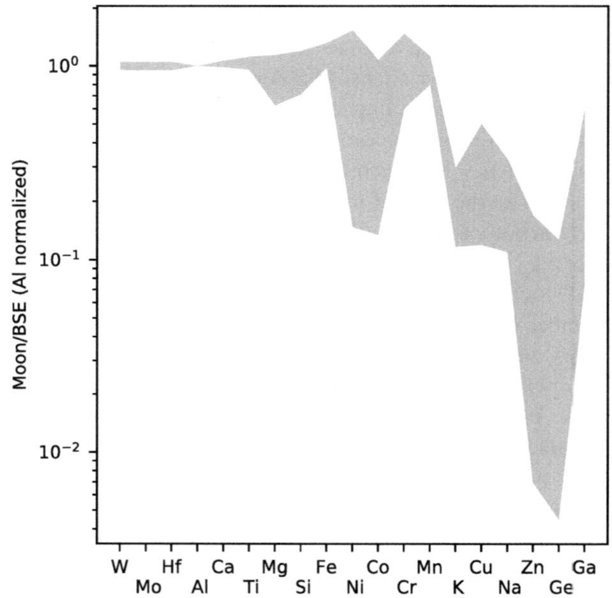

Fig. 3 The lunar composition is depleted in moderately volatile elements (MVEs) relative to the bulk silicate Earth (BSE). The gray band is the range of published estimates for the bulk lunar composition, doubly normalized to the mole fraction of Al and the abundance of each element in the BSE (McDonough and Sun 1995). The range shown is based on estimates and discussions provided by Ringwood and Kesson (1977), Wänke et al. (1977a), Morgan et al. (1978), Ringwood (1979), Taylor (1982), Wänke and Dreibus (1982), Ringwood et al. (1987), Warren (2005), Longhi (2006), Taylor and McLennan (2009), Taylor (2014), and Hauri et al. (2015) (see discussion in Section S5.1 of Lock et al. 2018).

The majority of the impactor's core material merges with the core of the target. The generation of iron-depleted disks has been used as a constraint in other lunar origin scenarios (Reufer et al. 2012; Rufu et al. 2017; Hosono et al. 2019). Lock et al. (2018) proposed an alternative explanation for the small lunar core in the context of the synestia model which may also be relevant in other lunar formation scenarios. They argued that since the bulk composition of Earth was set by the impact and that the Moon-forming region had a composition similar to that of the rest of the silicate portion of the post-impact body (as required by the isotopic constraints, see Sect. 6), the vapor in the outer regions of the synestia would have had close to bulk silicate Earth (BSE) composition. Using physical chemistry models, the authors showed that $\sim$2 wt% metal could be precipitated from a cooling BSE liquid at low to modest pressures (10^{-6}-10^{2} bar). The metal that now constitutes the core of the Moon could have been partially, or completely, sourced from metal that precipitated directly from silicate liquid in moonlets or the Moon itself, and injection of free metal into the disk may not be required. Note that some amount of metal from condensed material injected into orbit during the impact may also contribute to the mass of the lunar core in the synestia model. Further work on the combined thermal and chemical evolution of the lunar disk and the Moon is required to explore the relative contributions to the lunar core from different potential mechanisms.

The MVE depletion of the Moon is a key chemical constraint on lunar origin models. In a series of studies (Ringwood and Kesson 1977; Ringwood 1986), Ringwood and colleagues argued that the lunar composition could be explained if the Moon was a partial condensate of vapor derived from Earth's mantle. Few studies have attempted to combine the dynamics, thermodynamics and chemistry of lunar origin, which is necessary to test the proposed models. Recently, Canup et al. (2015) used the canonical lunar disk models of Salmon and Canup (2012) and physical chemistry calculations to link the dynamics and thermodynamics of accretion from a canonical disk. They also suggested that the lunar volatile element depletion could be explained if the material that formed the observable Moon was a partial condensate of disk material. However, the model presented by Canup et al. (2015) does not quantitatively explain the magnitude, nor the pattern, of moderately volatile element depletion observed for the Moon.

Lock et al. (2018) combined physical and geochemical models to estimate the composition of the Moon formed from a terrestrial synestia and showed that such a scenario can reproduce both the pattern and magnitude of MVE depletion in the Moon. By simulating the evolution of a terrestrial synestia using a modified hydrodynamic code, they demonstrated that the pressure of silicate vapor surrounding the growing Moon could be on the order of 10s bar. They then argued that the Moon would equilibrate with the gas around it in the structure at the pressure of the gas and a temperature set by the onset of vaporization of the major elements (particularly Si). Using physical chemistry calculations, they showed that equilibration of a BSE chemical system at those pressures and temperatures could reproduce the bulk lunar composition. MVEs that were not incorporated into the Moon would remain in the vapor of the synestia. Over time, radiative cooling of the synestia would cause it to contract. Eventually the synestia contracted within the orbit of the Moon, cutting off exchange between the vapor of the synestia and the Moon, and leaving the Moon with its distinctive MVE depleted composition. The MVEs that remained in the vapor of the synestia would have been incorporated into the BSE. The result that the lunar composition can be explained by partial condensation of BSE material at 10s bar and at temperatures tied to the onset of vaporization of the major elements may also place a constraint on the equilibration conditions of lunar material in other lunar origin models. However, the thermodynamic properties of MVEs at the high temperatures of post-impact bodies are not well known, and more experimental studies of silicate vaporization are needed to confirm this result.

Alternatively, it has been proposed that the MVE depletion of the Moon could be explained by evaporative loss from the Moon during the early stages of the LMO (Charnoz et al. 2019; Young and Tang 2019). We will discuss this suggestion in the next section in connection with MVE isotopic fractionation.

Determining the bulk chemical (and isotopic) composition of the Moon is challenging due to the limited spatial coverage of samples and limited constraints on the composition of the core (Steenstra et al. 2016, 2017, 2020; Jing et al. 2014; Antonangeli et al. 2015). Furthermore, determining the composition of the lunar mantle from only crustal samples requires the use of petrological melting models and there are intrinsic degeneracies in the inferred source compositions. As a result, there is still significant variation in different estimates of lunar composition. Future missions that sample new locations and different lithologies, particularly sample return missions, are vital to further constrain the lunar composition.

5 Moderately Volatile Element Isotopic Fractionation

Further insight on the mechanism of MVE depletion of the Moon relative to Earth is provided by isotopic studies. It has been found that lunar samples exhibit mass-dependent isotopic fractionation relative to Earth for various MVEs (K, Rb, Zn, Cl, Sn, Cr, Ga; Herzog et al. 2009; Sharp et al. 2010; Paniello et al. 2012; Boyce et al. 2015, 2018; Kato et al. 2015; Wang and Jacobsen 2016; Day et al. 2017; Kato and Moynier 2017; Pringle and Moynier 2017; Dhaliwal et al. 2018; Sossi et al. 2018; Wang et al. 2019a,b; Stephant et al. 2019; Nie and Dauphas 2019; Wimpenny et al. 2019, see Fig. 4). Different elements show different directions and magnitudes of fractionation and there is considerable variability in the fractionation of different lunar samples for some elements. MVE isotopic fractionation has the potential to be a powerful tool to constrain the formation and later evolution of the Moon, but interpreting the observed fractionations is difficult as numerous processes can cause

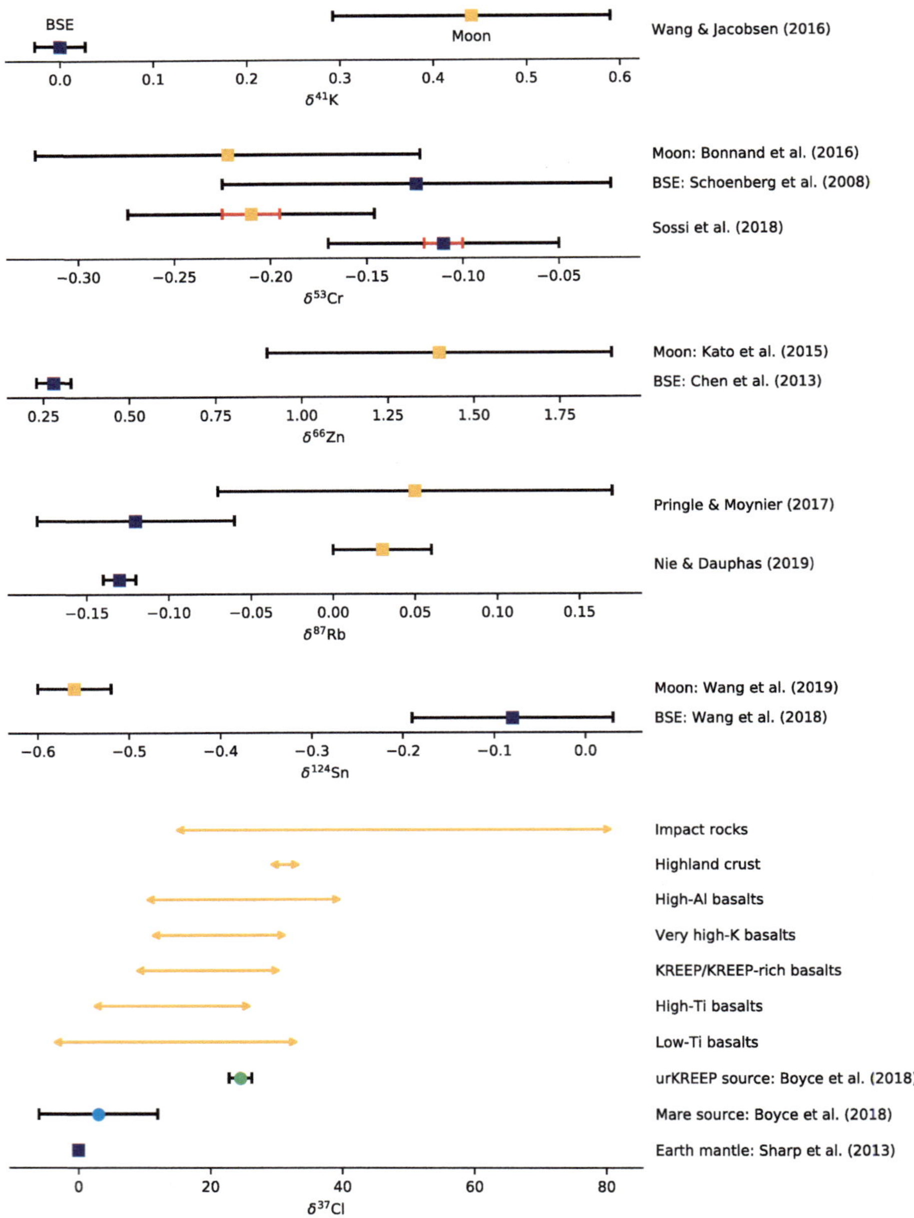

mass-dependent fractionation. Here we will restrict ourselves to the discussion of potential mechanisms for MVE isotopic fractionation during lunar formation or on the newly formed Moon. We will not discuss fractionation due to other processes, such as post-emplacement alteration (e.g., Potts et al. 2018; Barnes et al. 2019), and those acting after the lunar magma ocean had largely solidified (e.g., Barnes et al. 2016), or during the ascent, eruption or emplacement of the melts that formed lunar rocks (e.g., Sharp et al. 2010).

◀ **Fig. 4** Lunar samples exhibit mass-dependent isotopic fractionation relative to Earth for various volatile elements (MVEs). Shown are estimates for the isotopic composition of the bulk silicate Earth (BSE, dark blue squares) and Moon (yellow squares) for a selection of MVEs (K, Cr, Zn, Rb, Sn and Cl). Isotopic compositions are expressed as δ^{41}K, δ^{53}Cr, δ^{66}Zn, δ^{87}Rb, δ^{124}Sn, and δ^{37}Cl, parts per thousand deviation of the isotope ratio of interest in the sample/body from a standard. The relevant isotope ratios for each element plotted here are ^{41}K/^{39}K, ^{53}Cr/^{52}Cr, ^{66}Zn/^{64}Zn, ^{87}Rb/^{85}Rb, ^{124}Sn/^{116}Sn, and ^{37}Cl/^{35}Cl. K: Average of lunar (yellow) and terrestrial (blue) samples measured by Wang and Jacobsen (2016). Error bars are two standard deviations of sample measurements. Cr: Estimates for lunar isotopic composition from Bonnand et al. (2016) and estimates for the lunar and terrestrial isotopic composition from Sossi et al. (2018). Also plotted is the estimate for BSE from Schoenberg et al. (2008) which was used for comparison in the Bonnand et al. (2016) study. Black errors bars are two standard deviations of sample measurements. For Sossi et al. (2018) the standard error of all measurements of samples from each body (17 and 36 for lunar and terrestrial samples respectively) is also shown in red. Sossi et al. (2018) argued that their measurements, combined with literature values, for each individual body were consistent with being drawn from a single normal distribution and thus the standard error of all measurements of samples from each body was a more suitable estimate of the uncertainty in the terrestrial and lunar composition. Note that the Cr isotopic fractionations shown here are mass-dependent unlike the mass-independent variation shown in Figs. 5 and 6. Zn: Estimate of the bulk Moon from Kato et al. (2015) and estimate of the BSE from Chen et al. (2013). Error bars are two standard deviations of sample measurements. Rb: Estimates for lunar and terrestrial isotopic composition from Pringle and Moynier (2017) and Nie and Dauphas (2019). Error bars are two standard deviations of sample measurements and 95% confidence intervals, respectively. Sn: Estimates for lunar and terrestrial isotopic composition from Wang et al. (2019a) and Wang et al. (2018), respectively. Error bars are two standard deviations of sample measurements. Cl: The Cl isotopic compositions of lunar samples are highly variable and the yellow arrows in the third panel show the range of reported values for appetites and melt inclusions in different lunar lithologies from: Sharp et al. (2010), Wang et al. (2019b), Barnes et al. (2016), Tartèse et al. (2014), Treiman et al. (2014), Boyce et al. (2015), Potts et al. (2018), Barnes et al. (2019), and Stephant et al. (2019). Colored circles show estimates of the Cl isotopic composition of the Mare and urKREEP sources based on mixing models constrained by sample measurements (Boyce et al. 2018). Also plotted is the estimate for the composition of the terrestrial mantle from (Sharp et al. 2013).

MVE isotopic fractionation could arise during the formation of the Moon due to processes in the lunar disk or synestia. Wang and Jacobsen (2016) proposed that the observed isotopic fractionation in K could be produced by equilibrium exchange between liquid and vapor at moderate pressures (10s bar) consistent with the conditions expected for the formation of the Moon from a terrestrial synestia (Lock et al. 2018). However, at the high temperatures of liquid-vapor exchange in synestias and in the midplane of the canonical disk at the Roche-limit ($>\sim$3500 K; Lock et al. 2018; Canup et al. 2015) the expected equilibrium isotopic fractionation is too small to explain the observations (Dauphas et al. 2018). To amplify the effect from equilibrium fractionation, it has been suggested, on the basis of the observed light isotopic composition of Sn, that the lunar material experienced repeated stages of equilibration and phase separation at somewhat lower temperatures ($\sim$2500 K; Wang et al. 2019a). Alternatively, a small degree of non-equilibrium fractionation during partial evaporation of MVEs could explain the observed heavy isotopic signature of many MVEs (Nie and Dauphas 2019), but may be in disagreement with light isotopic signatures such as those observed for Sn (Wang et al. 2019a). Assessing the feasibility of each of these arguments and understanding the implications of MVE isotopic fractionations for lunar origin is challenging. Different parcels of lunar material follow varying pressure-temperature-composition-time paths before (and after) being incorporated into the Moon and likely go through multiple steps of vaporization and condensation. Thus, the lunar MVE isotopic composition may reflect the weighted average of the varying signatures from material that experienced different histories. Current models of Moon formation do not track the thermodynamic paths of lunar material to the level of detail needed to access the resulting isotopic fractionation.

The newly-formed Moon would be substantially molten and it has been suggested that the loss of silicate vapor from the surface of the LMO could fractionate MVEs both elementally and isotopically (Paniello et al. 2012; Boyce et al. 2015; Kato et al. 2015; Day et al. 2017; Kato and Moynier 2017; Sossi et al. 2018; Stephant et al. 2019; Charnoz et al. 2019; Young and Tang 2019). Recently, Charnoz et al. (2019) and Young and Tang (2019) have attempted to model this process and found that it is possible to lose significant amounts of MVEs from the Moon by evaporative loss, producing mass-dependent isotopic fractionations. However, these models assume that the loss from the Moon is in steady state, which is not necessarily the case given the surface of the Moon is far out of equilibrium with its surroundings. Processes that may prove important, such as radiative transfer and condensation of the expanding vapor, have also been neglected.

Understanding MVE fractionation between Earth and the Moon is a new and rapidly developing frontier in lunar formation studies, but more work, particularly studies measuring multiple different elements in the same samples, is required to untangle the contributions of different processes to the isotopic fractionation (e.g., Nie and Dauphas 2019). In order to use these observations to place constraints on the origin and evolution of the Moon, coupled chemical, thermal and dynamical models will be required to quantify the isotopic fractionation due to different processes. Models that can couple the elemental depletion and isotopic fractionation of MVEs could prove particularly powerful determinants of lunar origin and evolution. For such models, there is a need for improved experimental constraints on the behavior of MVEs at the very high temperatures expected during lunar formation and magma ocean evolution, a regime that is exceedingly difficult to reach in a lab setting (Sossi et al. 2019).

6 Similarity in Lunar and Terrestrial Non-mass Dependent Isotope Ratios

Increasingly precise isotopic measurements of terrestrial and lunar samples have shown that Earth and the Moon share very similar non-mass dependent stable isotope ratios for a range of elements including Cr, O, Si, and Ti (e.g., Lugmair and Shukolyukov 1998; Wiechert et al. 2001; Armytage et al. 2012; Zhang et al. 2012; Young et al. 2016; Greenwood et al. 2018, see Figs. 5 and 6). Significant non-mass dependent variation in the isotope ratios of these elements are observed among other planetary bodies (Clayton and Mayeda 1996; Yin et al. 2002; Trinquier et al. 2007; Zhang et al. 2012). Such isotopic signatures are thought to reflect the provenance of material that formed different bodies (i.e., the region in the protoplanetary disk from which material was derived). Given the non-mass dependent variation of stable isotope ratios observed among planetary materials, the impactor Theia is generally expected to have had non-mass dependent stable isotopic signatures distinct from the proto-Earth.

Refractory radiogenic isotope variations (e.g., ^{182}W) are also a type of non-mass dependent isotope variation and ^{182}W isotope signatures were likely substantially different on Theia and the proto-Earth. Radioactive decay of lithophile ^{182}Hf generates radiogenic moderately siderophile ^{182}W (half-life of 8.9 Myr, Vockenhuber et al. 2004), such that ^{182}W isotopic ratios are sensitive to the conditions and timing of metal-silicate partitioning associated with core formation on each of the colliding bodies. A difference in ^{182}W isotopic composition in Theia and the proto-Earth is therefore anticipated, as it is improbable that the two bodies had independent differentiation histories that produced identical ^{182}W isotope compositions through the stochastic process of planet formation (Kleine et al. 2009; Kruijer et al. 2015; Touboul et al. 2015; Kruijer and Kleine 2017). However, as for the stable isotopes, the Earth and Moon likely had similar ^{182}W isotope compositions immediately

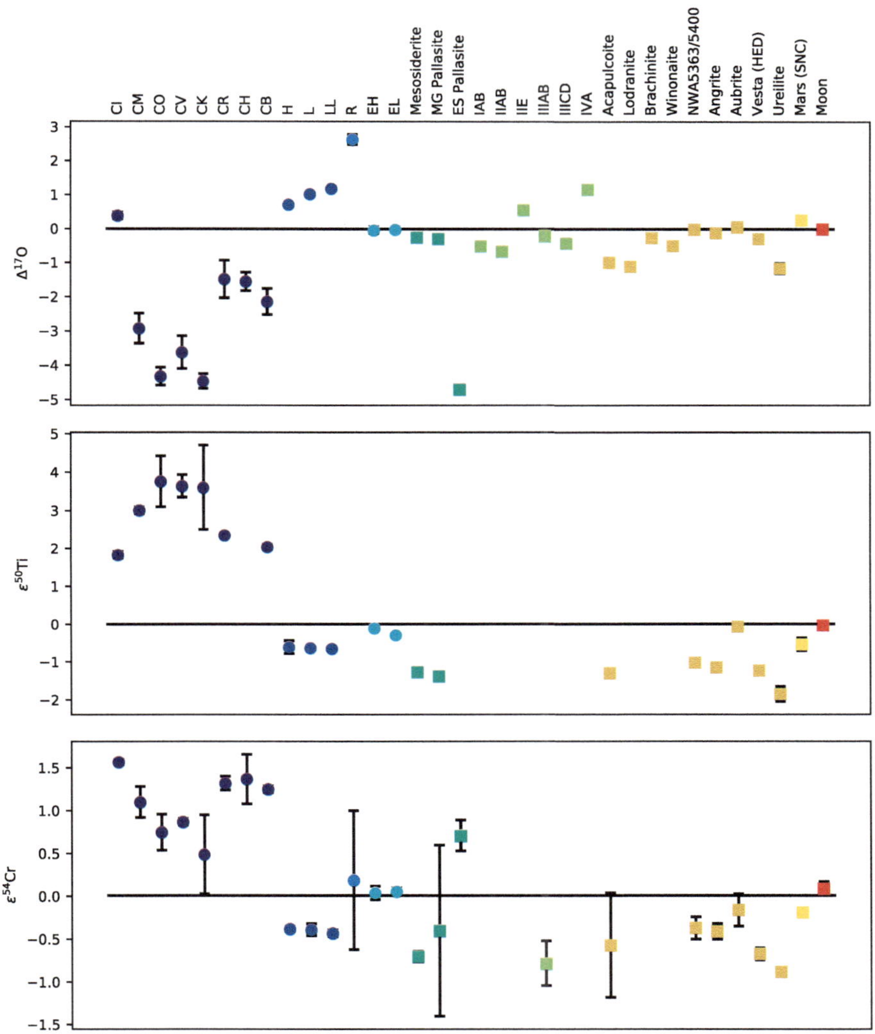

Fig. 5 The Earth and the Moon are very similar in a range of isotopes that trace the origin of planetary materials (non-mass dependent stable isotope ratios that vary due to nucleosynthetic anomalies). Shown are the Δ^{17}O, ϵ^{50}Ti, and ϵ^{54}Cr for a range of planetary materials from both undifferentiated (chondrites; circles) and differentiated (squares) bodies: carbonaceous chondrites (CI, CM, CO, CV, CK, CH, CB); ordinary chondrites (H, L, LL); Rumuruti type chondrites (R); enstatite chondrites (EH, EL); stony-iron meteorites (Mesosiderites, main group Pallasities, and Eagle-Station group Pallasites); iron meteorites (IAB, IIAB, IIE, IIIAB, IIICD, IVA), rocky achondrites (Acapulcoites, Lodranites, Brachinites, Winonaites, NWA5363/5400, Angrites, Aubrites, Vesta (as characterized by the Howardites, Eucrites and Diogenites meteorite groups), Urelites); Mars (as characterized by the Shergottite, Nakhlite, and Chassigny meteorite groups); and the Moon. Earth is at 0 by definition (black horizontal lines). Δ^{17}O is the mass-independent offset in O isotopes relative to Earth, the offset in ^{17}O/^{16}O relative to a given ^{18}O/^{16}O from a reference mass-dependent isotope fractionation line reported in units of parts per thousand (per mil, ‰). ϵ^{50}Ti and ϵ^{54}Cr are the parts per ten thousand deviations of the isotope ratio of interest in the sample/body from a terrestrial standard. The relevant isotope ratios are ^{50}Ti/^{47}Ti, and ^{54}Cr/^{52}Cr. All error bars are two standard deviations of sample measurements or 95% confidence intervals. In some cases the error is smaller than the symbols. Data were taken from a compilation by Dauphas (2017), with the exception that the lunar value for ϵ^{54}Cr was taken from Mougel et al. (2018). Original data sources are given in Appendix A.1.

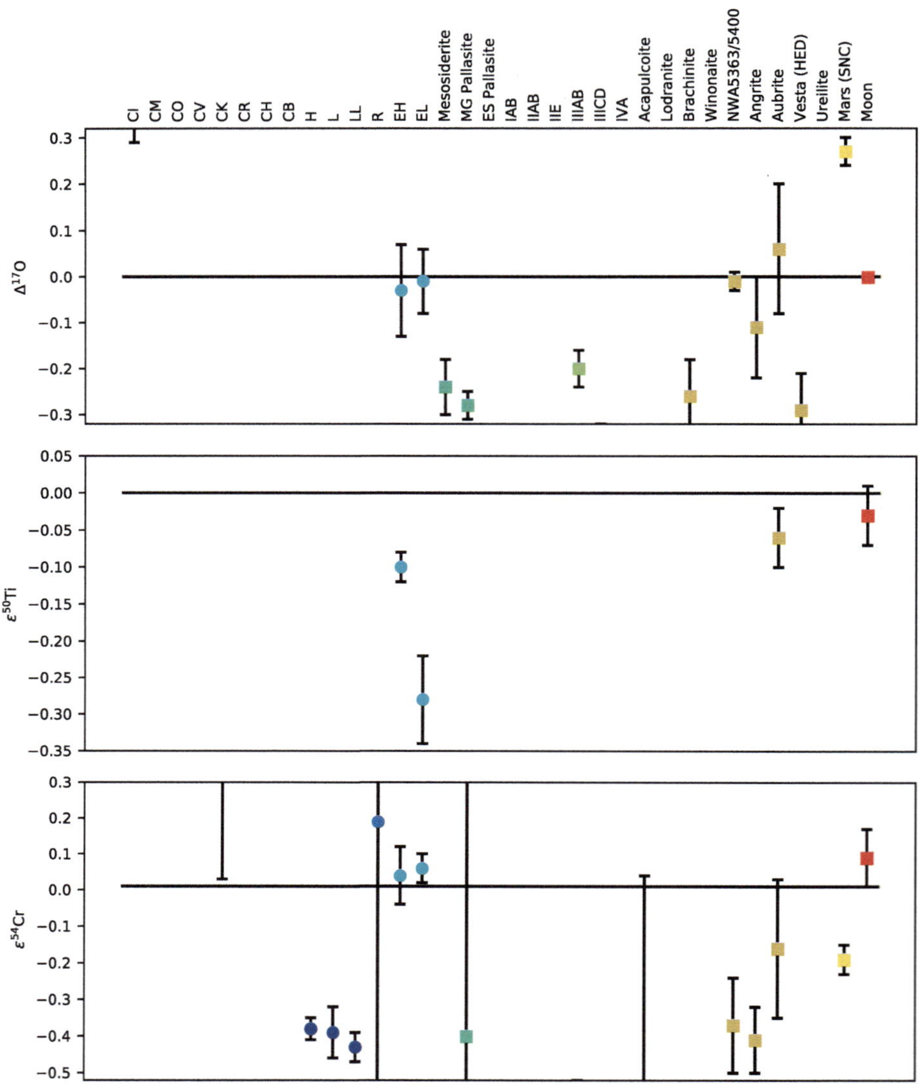

Fig. 6 The Earth and the Moon are very similar in a range of isotopes that trace the origin of planetary materials (non-mass dependent stable isotope ratios called nucleosynthetic anomalies). As Fig. 5 except showing a restricted range of isotopic compositions, focused on those values close to that of the Earth and Moon.

following Moon formation (Kruijer et al. 2015; Touboul et al. 2015). Recent W isotopic studies measured a small ^{182}W excess of $\sim$25 ppm (parts per million) in lunar rocks relative to modern terrestrial samples (Touboul et al. 2015; Kruijer et al. 2015; Kruijer and Kleine 2017). Earth and the Moon have accreted different amounts of material since the Moon-forming impact which has slightly changed their respective ^{182}W isotopic compositions and the observed ^{182}W excess in lunar rocks is consistent with estimates of the disproportionate late accretion to the Earth and Moon. Although alternatives have been proposed (e.g., Thiemens et al. 2019), the most parsimonious conclusion is therefore that Earth and the Moon started with similar ^{182}W isotope compositions.

Numerical simulations of giant impacts predict that the canonical lunar disk is derived primarily from the impactor, and in this case the Moon would be expected to acquire the distinct isotopic signature of the impactor (Canup and Asphaug 2001; Canup 2004, 2008). Such a scenario would likely result in measurable refractory nucleosynthetic and radiogenic isotopic differences between Earth and the Moon (Pahlevan and Stevenson 2007; Melosh 2014; Young et al. 2016; Kruijer and Kleine 2017), which have not been observed. Note that if the impactor and target had significantly different bulk mantle compositions, an elemental compositional difference between Earth and the Moon would also be expected. In particular, if the impactor and target had substantially different core formation histories (as would likely be the case in most proposed scenarios, not just the canonical model) then their mantles would have had different abundances of moderately siderophile elements (MSEs, e.g., V, Cr, Mn). However, the BSE and the Moon have very similar MSE concentrations and it has been argued that it would be hard to reconcile this fact with a scenario in which lunar material was largely sourced from the impactor (Seifert and Ringwood 1988; Drake et al. 1989), although the strength of this argument is dependent on the core formation histories of the impacting bodies, the mass and nature of late veneer etc. The contradiction between the geochemical observations of similarity and dynamical expectations of difference is often referred to as the 'lunar isotopic crisis.'

Three classes of solutions to the crisis of unexpected isotopic similarity have been proposed. First, the impactor and proto-Earth could have formed from the same source material (i.e., they formed from material derived from the same region in the disk) and thus shared nearly identical nucleosynthetic isotopic signatures (Jacobsen et al. 2013; Dauphas et al. 2014; Dauphas 2017). This suggestion has a basis in the observation that some meteorite groups (e.g., enstatite chondrites) have non-mass dependent stable isotope compositions not dissimilar to that of Earth (e.g., Dauphas et al. 2014; Dauphas 2017; Boyet et al. 2018). If the impactor and target accreted the majority of their mass from a common reservoir with a specific mix of nucleosynthetic components, the Earth and Moon would inherit similar non-mass dependent stable isotopic signatures. This explanation for the isotopic similarity between Earth and the Moon, however, relies upon coincidence to explain the inferred similarity of the ^{182}W isotopic compositions of the Earth and Moon before late accretion (Dauphas et al. 2014; Touboul et al. 2015; Kruijer and Kleine 2017) and could be difficult to reconcile with the similar MSE concentrations in the BSE and the Moon (see discussion in Sossi and Moynier 2017). Nevertheless, the possibility of a more homogeneous inner solar system relaxes the isotopic constraint on models of Moon-formation, as a given scenario need only produce similar ^{182}W compositions in the Earth and Moon.

Alternatively, the Moon may have been constructed from material injected into orbit by many smaller impacts (Rufu et al. 2017). Such a process would be a natural outcome of the many impacts Earth experienced during its formation and the Moon would then be a mixture of the compositions of each impactor. However, to reproduce the isotopic similarity between Earth and the Moon it is likely that 10s of such impacts are required, and the satellite from each previous impact must survive and merge with the newly formed moon (Rufu et al. 2017; Rufu and Aharonson 2019; Citron et al. 2018). The likelihood of such a combination of events being able to form our Moon is still being worked out but is potentially quite low.

In the third class of solution, extensive mixing of impactor and target material during the impact or post-impact mixing between different reservoirs could mitigate initial isotopic heterogeneities. The requirement is that there ends up being a similar mass fraction of impactor material in the region of the post-impact body from which the Moon forms (i.e., the disk) as there is in the post-impact body as a whole. Due to the large shearing flows during giant impacts, there is potential for substantial mixing (Nakajima and Stevenson 2015).

However, the shear is highly heterogeneous in the colliding bodies and mixing is inhibited by thermal buoyancy of highly shocked compared to weakly shocked material (Carter et al. 2020). The degree of mixing is likely underestimated in current giant impact models due to the lack of thermal equilibration between parcels of mass in Lagrangian methods and the fact the resolution is too low to capture small-scale instabilities (cf. Deng et al. 2019). Nevertheless, as the canonical impact is a grazing impact, opportunity for mixing of impactor and target material is limited and the lunar disk will mostly be derived from Theia. Pahlevan and Stevenson (2007) suggested that exchange of material between the planet and the disk could erase isotopic heterogeneity between the two reservoirs. However, the extent of mixing required to explain the observations, especially the isotopic similarity for the most refractory elements, is likely to be unfeasible due to the large dynamical and thermodynamic boundary between the planet and the disk (Melosh 2014; Lock and Stewart 2017).

In most high energy and/or high angular momentum impacts there is greater opportunity for mixing between impactor and target during the impact. The higher energy of the impact causes the colliding bodies to be more substantially disrupted than in the canonical impact and the greater torques lead to increased shear. The ratio of impactor to target material in the Moon-forming regions of the post-impact structure (synestia or otherwise) is, therefore, typically more similar to the bulk planet than in the canonical model (Canup 2012; Ćuk and Stewart 2012; Lock et al. 2018). It has been found that in a limited range of scenarios the mixing is thorough enough in current giant impact simulations to produce an isotopic similarity between Earth and the Moon by impact processes alone (Canup 2012; Ćuk and Stewart 2012), but the probability of such specific impact parameters is thought to be low. Lock et al. (2018) argued that silicate rain and thermal convection in post-impact synestias could mix a large mass fraction (10%) of the post-impact structure, diluting any isotopic heterogeneity after the impact and increasing the range of impacts that could produce an isotopically similar Earth and Moon.

However, there are considerable barriers to mixing within all post-impact bodies. Firstly, post-impact bodies are highly thermally stratified. Figure 7 shows the thermal structure of post-impact bodies formed by a canonical impact (Fig. 7c) and a high-energy, high-angular momentum impact that resulted in a synestia (Fig. 7f). Specific entropy is a useful variable with which to access the thermal stratification of a structure made of a single material, as it gives an indication of the thermal state of a parcel of material without a dependency on pressure (unlike temperature). A parcel of higher entropy will always be more buoyant than one of lower entropy if they are of the same composition. The outer regions of post-impact bodies (beyond $\sim 10 \times 10^6$ m in the examples in Fig. 7), from where most lunar material is sourced, are typically near-isentropic and there is only a weak thermal barrier to mixing in the outer regions (Nakajima and Stevenson 2014; Lock and Stewart 2017; Lock et al. 2018). However, the strong increase in entropy with radius (and pressure) in both post-impact bodies between $\sim 5 \times 10^6$ m and $\sim 10 \times 10^6$ m suppresses any mixing between the outer regions and the cooler interior of the post-impact body. Mixing of material from the lower mantle to the outer regions is unfeasible without significant cooling of the outer regions or large-scale instability. Secondly, post-impact bodies are rapidly rotating (Fig. 7a, d) and there is a significant angular momentum gradient with radius (Fig. 7b, e). For a parcel of mass to move outwards in the structure it must gain angular momentum either from shear in the vapor or from drag by infalling condensates. The rapid rotation and angular momentum gradient create a significant barrier to radial mixing and mass exchange perpendicular to the rotation axis in the post-impact body (Pahlevan and Stevenson 2007). Convection parallel to the rotation axis is less inhibited by rapid rotation, and in the isentropic outer regions vertical mixing is rapid (on timescales of order weeks), driven by cooling of the vapor at the photosphere (Pahlevan and Stevenson 2007; Lock et al. 2018). The effect of rapid rotation on

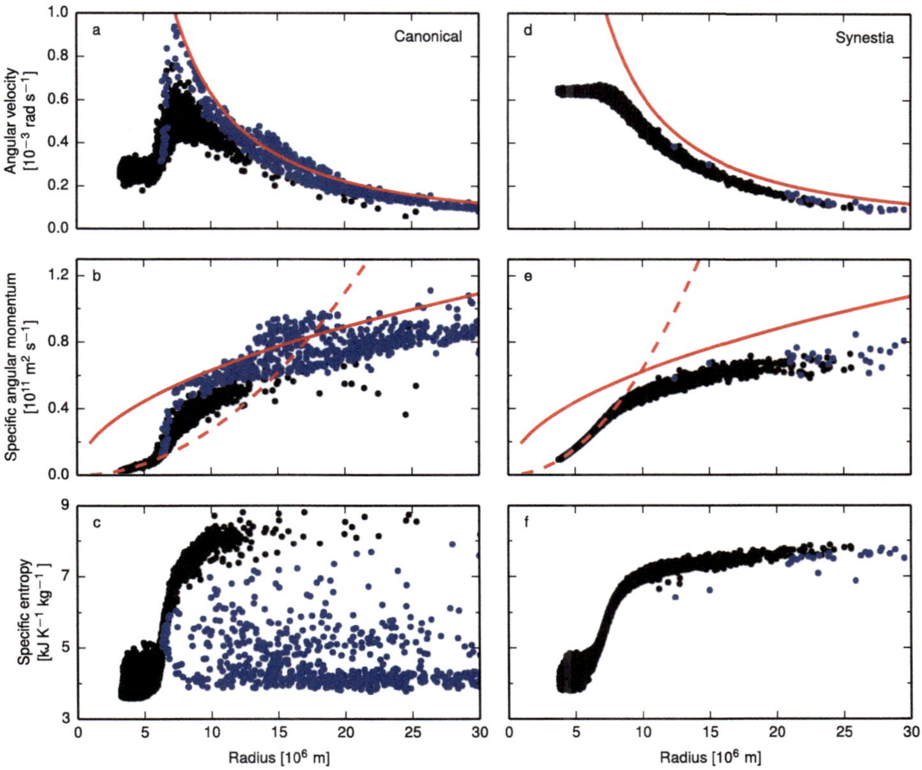

Fig. 7 There are significant thermal and rotational barriers to mixing in post-impact bodies. Shown are the post-impact bodies formed in smoothed-particle hydrodynamics (SPH) impact simulations for example canonical and terrestrial synestia Moon-forming impact scenarios (one in each column). Dots represent silicate SPH particles in the midplane of each post-impact structure 48 hrs after first contact. Solid red lines denote the angular velocity or specific angular momentum of a circular Keplerian orbit around a point mass. Dashed red line indicates the specific angular momentum of material corotating with the inner region. The black particles are either at pressures above the critical point (25.5 kbar) or are pure vapor, and the blue particles are mixtures of liquid and vapor. The super-Keplerian particles in a and b are the remnants of a disrupting moonlet. Modified from Lock and Stewart (2017).

limiting radial mass transport has been demonstrated in simulations of convection in giant planets (Kaspi et al. 2009), rapidly-rotating magma oceans (Maas and Hansen 2019), and protoplanetary disks (Shariff 2009). As a result of the substantial barriers to mixing, post-impact mixing is likely only able to homogenize a fraction of the post-impact body. Lock et al. (2018) argued that, in the case of a synestia, radial transport of silicate rain droplets due to gas drag would be able to overcome the angular momentum barrier and mix regions of the body with specific entropies above the critical point (up to 50% of the body, but typically less). However, even if such large fractions of the post-impact body mixed, a considerable amount of intra-impact mixing is still required if the impactor and target had substantially different isotopic compositions.

To demonstrate the difficulty in producing an Earth and Moon with similar stable isotope compositions, even with significant post-impact mixing, we can use a simple calculation (see Appendix A.2 for full details). For our model, we will use the similarity between Earth and the Moon in O isotopes as our principle constraint (Young et al. 2016; Greenwood

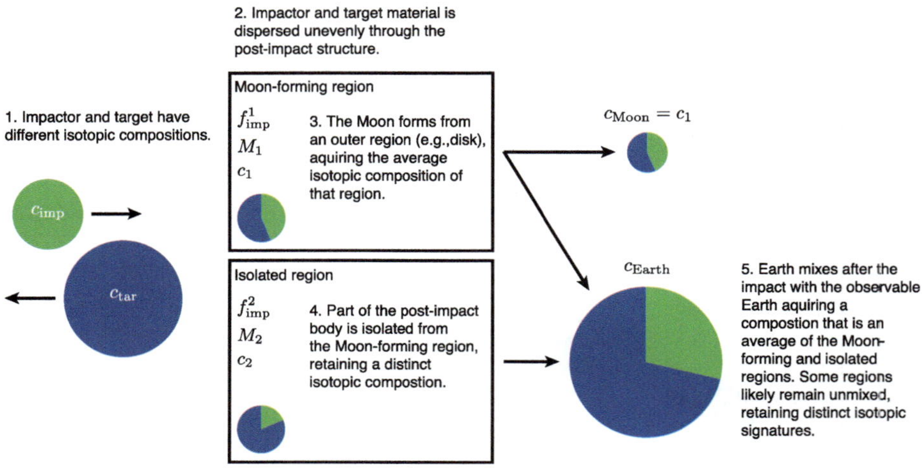

Fig. 8 Schematic of mixing model. For details see Sect. 6 and Appendix A.2

et al. 2018). There is a large amount of non-mass dependent O isotope variability between solar system bodies (Figs. 5 and 6), and it is typically assumed that the bodies that collided to form the Moon would have had somewhat different O isotope compositions. Non-mass dependent variation in O isotopes is typically quantified in terms of $\Delta^{17}O$, the offset in $^{17}O/^{16}O$ relative to a given $^{18}O/^{16}O$ ratio from the mass-dependent isotope fractionation line defined by terrestrial rocks (commonly known as the terrestrial fractionation line, TFL), and are reported in units of parts per million (ppm) or parts per thousand (per mil, ‰). Oxygen is a primary constituent of mantle rocks and only thought to be in relatively low concentrations in planetary cores, making it a good tracer of the bulk silicate material from the impactor and target. Also, as the concentrations of O in impactor and target mantles would have been very similar, mixing between the impactor and the target is linear in isotope space. Using O as a primary constraint avoids the complications that affect modeling of minor elements (e.g., W), such as extreme concentration contrasts between different chemical reservoirs and a strong dependence on the previous accretion histories of the impacting bodies.

In our model, we divide the post-impact body into two regions: an outer region from which the Moon inherits its composition (e.g., the lunar disk or outer synestia) and an inner region which remains isolated from the outer region during the formation of the Moon (Fig. 8). Both regions are made up of a mixture of impactor and target material. After the Moon forms, either the lunar disk collapses back to Earth or the synestia cools and contracts, and the remnants of the outer Moon-forming region and the inner region are combined. The material from the remnant outer and inner regions is then assumed to mix to produce the present-day observable mantle. Note that the Moon-forming region in this model is not descriptive of the physical position of the region, or the role it plays in forming the Moon – it is simply the fraction of the body from which the Moon inherits its composition. We can calculate the fraction of the body that must participate in mixing after the impact (i.e., the mass of the outer, Moon-forming region from which the Moon inherits its composition) to explain the isotopic similarity of the Earth and Moon for a given isotopic difference between the impactor and target and a given degree of intra-impact mixing. We parameterized the degree of intra-impact mixing by the difference in the mass fraction of the two regions that derives from the impactor, i.e., how well impactor material is dispersed through the body during the impact.

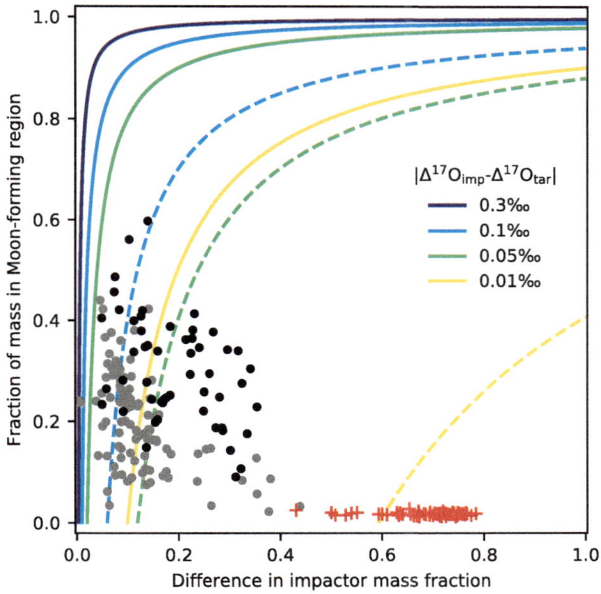

Fig. 9 Producing an isotopically similar Earth and Moon requires substantial intra-impact mixing. Lines show the silicate mass fraction of the post-impact body that must be mixed (y-axis) to reproduce the O isotope similarity between the Earth and Moon given an initial post-impact difference in the mass fraction of impactor in the Moon-forming and isolated reservoirs (x-axis). The x-axis is a measure of the efficiency of mixing during the impact and the y-axis is a measure of the efficiency of post-impact mixing. The more mixing that occurs during the impact (lower values on the x-axis) the less post-impact mixing is required to satisfy the O isotope constraint. Colored lines indicate the results for different $\Delta^{17}O$ offsets between the impactor and target for an Earth-Moon difference of -1 ppm (solid lines, as determined by Young et al. 2016) and -6 ppm (dashed lines, the most extreme value within 2 standard error from Young et al. 2016). The greater the $\Delta^{17}O$ offsets between the impactor and target, or the smaller the isotopic difference between the present-day Earth and Moon, the more intra- or post-impact mixing that is required. For reference, the difference in $\Delta^{17}O$ between Earth and Mars today is 0.3‰ (Franchi et al. 1999) and the average for enstatite chondrites calculated by Dauphas (2017) is 0.01‰. Points show the mass fraction of synestias from the simulations of Lock et al. (2018) and Lock and Stewart (2019) that could be mixed by silicate rain and the corresponding difference in impactor fraction. The gray dots are those simulations in the styles of Ćuk and Stewart (2012) and Canup (2012) that are designed to maximize mixing. The red crosses show the results of successful canonical impacts (Canup 2004, 2008) assuming the Moon is sourced from a perfect mixture of the material implanted in the disk with no-post impact exchange between the planet and disk (cf., Pahlevan and Stevenson 2007).

Figure 9 shows the results of our mixing calculation. The lines mark the silicate mass fraction of the post-impact body that must be mixed to reproduce an Earth and Moon with similar O isotope compositions given an initial difference in the mass fraction of impactor in the Moon-forming and isolated reservoirs. These two variables are indicators of the degree of intra-impact (x-axis) and post-impact mixing (y-axis). The lower the efficiency of mixing during the impact (higher values on the x-axis) the larger the mass fraction of the post-impact body that must be mixed after the impact (y-axis) to explain the isotopic similarity between Earth and the Moon. Different colored lines indicate the results for different $\Delta^{17}O$ offsets between the impactor and target for an Earth-Moon difference of -1 ppm (solid lines, as determined by Young et al. 2016) and -6 ppm (dashed lines, the most extreme value within 2 standard error from Young et al. 2016). For reference, the difference between Earth and

Mars today is 0.3‰ (Franchi et al. 1999), and the average difference between Earth and enstatite chondrites calculated by Dauphas (2017) is 0.01‰ (Figs. 5 and 6). Note that the most recent measurements of oxygen isotopes are in the same range as that of the Young et al. (2016) determination, with a reported difference of -3 ± 3 ppm between terrestrial and lunar samples, and hence a maximum offset within 2 standard error of -6 ppm (Greenwood et al. 2018). Greenwood et al. (2018) attributed the inferred difference between the Earth and Moon to differential accretion of water from the late veneer, requiring the post-impact difference between the two bodies to be even smaller. To explain the given offset between Earth and the Moon, the combined degree of intra- and post-impact mixing must be such that it lies on the colored line in each case. The greater the $\Delta^{17}O$ offsets between the impactor and target, or the smaller the isotopic difference between the present-day Earth and Moon, the more intra- or post-impact mixing that is required.

The red crosses in Fig. 9 show the results of canonical impacts reported by Canup (2004, 2008) assuming the Moon is sourced from a perfect mixture of the material implanted in the disk (which is sourced largely from the impactor) with no-post impact exchange between the planet and disk (cf., Pahlevan and Stevenson 2007). Black and grey points show synestias from the simulations of Lock et al. (2018) and Lock and Stewart (2019). We followed the arguments of Lock et al. (2018) to determine the fraction of the outer regions of the synestia that could be mixed by silicate rain (y-axis) and determined the corresponding difference in impactor fraction between the mixed outer regions and the unmixed inner regions (x-axis). All of these impacts are high-angular momentum and so experience significantly more mixing than in the canonical case, but the gray dots are those simulations in the styles of Ćuk and Stewart (2012) and Canup (2012) that are explicitly designed to maximize mixing. Note, again, that intra-impact mixing is likely underestimated in current numerical simulations and so the x-axis values for impact simulations should be viewed as a weak upper limit.

If the impactor and target were isotopically dissimilar (e.g., the $\Delta^{17}O$ difference was similar to that between Earth and Mars today, 0.3‰; Franchi et al. 1999), the only way to produce isotopic similarity is to extremely efficiently mix impactor and target material during the impact or to mix almost the entire body afterwards. Even if the isotopic difference between the impactor and target are small (e.g., similar to the average for enstatite chondrites calculated by Dauphas 2017, 0.01‰) a significant amount of intra- or post-impact mixing is required unless the true isotope difference between the Earth and Moon today is at the extremes allowed by the uncertainties in observations. Nevertheless, post-impact mixing may still be important for homogenizing smaller-scale isotopic heterogeneities in the Moon-forming region. Given the difficulties in mixing the majority of the post-impact body (see discussion above), it is inescapable that either the impactor and target were nearly identical in O isotopes or extensive intra-impact mixing is required.

Compared to that provided by O isotopes, the constraint offered by ^{182}W is harder to quantify. Predicting the ^{182}W isotopic composition of the Earth and Moon from a given model is difficult due to the uncertainties in the disproportionate late accretion onto the two bodies, the effects of core formation on both the Earth and Moon, and the dependence of the pre-impact history of the colliding bodies on the concentrations of W and isotopic differences. Tungsten is also a trace element for which concentrations can vary significantly between different reservoirs, leading to highly non-linear mixing. Note that the range in likely impactor ^{182}W compositions is larger than for O (Kruijer et al. 2015; Touboul et al. 2015; Kruijer and Kleine 2017; Kleine and Walker 2017; Young et al. 2016). The similarity of the Earth and Moon in ^{182}W isotopic compositions before the addition of late veneer therefore places a weaker constraint on mixing than O, but likely still requires a significant degree of intra-impact mixing or a coincidentally similar composition impactor. Hence, although post-impact mixing can aid in producing an Earth and Moon with similar stable isotopic compositions, a significant degree of intra-impact mixing and/or a somewhat isotopically similar impactor and target is probably also required.

Note that satisfying the constraint of isotopic similarity by intra-impact mixing of impactor and target material does not strictly require homogenization of a significant fraction of the mantle. As long as the Moon-forming region was made up of a similar mass-fraction of impactor and target material as the rest of the body (assuming that the material injected into the Moon-forming region was representative of the bulk silicate composition of both bodies), the material outside the Moon-forming region could have remained poorly mixed with significant heterogeneity. This is in contrast to the case where the constraint of isotopic similarity was satisfied by post-impact mixing, in which a large fraction of the post-impact Earth must have been homogenized to produce a Moon-forming region representative of the composition of the whole body. However, the distribution of impactor and target material produced by an impact is not independent of the degree of internal homogenization of the post-impact structure. Impacts that lead to similar impactor mass fractions in the outer and inner regions of the post-impact body are typically high-energy and high-angular momentum. Such impacts have the potential to homogenize large regions of the body due to the significant shear during the impact (Nakajima and Stevenson 2015), although the efficiency is uncertain (see above). Also, if the impacting bodies had significant internal isotopic heterogeneity, then it is unlikely that a representative sample of the silicate of each impacting body would have been injected into the Moon-forming region without homogenization of a significant fraction of the post-impact body. Regardless, it is important to bear in mind the important distinction between mixing of impactor and target material between the Moon-forming and interior regions of the post-impact body, and the internal homogenization of each region.

Current giant impact simulations suggest that there are a range of high angular momentum giant impacts that could produce synestias that are sufficiently well mixed to meet the constraint of isotopic similarity (Lock et al. 2018). These include the originally proposed scenarios of Canup (2012) and Ćuk and Stewart (2012) but also a number of intermediate scenarios (e.g., 0.3 and 0.75, or 0.4 and 0.65 Earth-mass bodies colliding). Extensive intra-impact mixing is more likely in many high-angular momentum impacts than in the canonical grazing impact (whether of note they produce synestias), making them better candidates for lunar origin. However, accurately quantifying mixing, and therefore the range of possible Moon-forming giant impacts, remains a significant challenge. Significant developments in the codes used for giant impact studies are required to accurately capture the mixing processes during impacts. It may be the case that the required methods and resolution to accurately quantify mixing are computationally unfeasible, in which case more innovative solutions must be found to quantify the degree of mixing. In any case, the isotopic similarity of Earth and the Moon remains a significant constraint that must be satisfied by models of lunar origin.

7 Preservation of Ancient Heterogeneities in Earth

Early terrestrial evolution was intrinsically tied to lunar formation. The giant impact melted and vaporized large fractions of Earth, at least partially mixed the mantle and core, and set the initial conditions for Earth's subsequent evolution (Nakajima and Stevenson 2015; Lock and Stewart 2017, 2019; Lock et al. 2020). The processes that occurred during the formation and early evolution of Earth are recorded in ancient and modern terrestrial samples as varia-

Table 1 The most frequently utilized short-lived radioactive isotope systems for constraining the early evolution of the Earth-Moon system in order of half-life

Short-lived radioactive parent nuclides and daughter nuclides	Half-life ($t_{1/2}$)
$^{182}Hf \rightarrow {}^{182}W$	8.9 Myr (Vockenhuber et al. 2004)
$^{129}I \rightarrow {}^{129}Xe$	15.7 Myr (Eckerman et al. 1993)
$^{244}Pu \rightarrow {}^{131\text{-}136}Xe$	80 Myr (Akovali 2003)
$^{146}Sm \rightarrow {}^{142}Nd$	103 Myr (Friedman et al. 1966; Meissner et al. 1987)

tions in isotopic ratios that are affected by short-lived radioactive decay. These variations are interpreted to reflect parent-daughter element ratio variations that existed during the lifetime of the radioactive parent isotope, i.e., a period of $\sim$ 5-6 half-lives. Radioactive isotopes with relatively short half-lives (e.g., ^{129}I, ^{244}Pu, ^{146}Sm, ^{182}Hf; see Table 1) trace processes that generated radiogenic isotope variations early in Earth's history. Examination of short-lived radiogenic isotope signatures in terrestrial samples can thus provide insights into the timing and mechanism of lunar formation and the Earth's subsequent evolution.

7.1 I-Pu-Xe Systematics

Present-day mantle xenon (Xe) isotopic signatures in mid-ocean ridge basalts, plume-derived samples, and well-gases are consistent with extensive degassing of volatiles within the first 100 Myr of Earth's history (Allègre et al. 1983; Yokochi and Marty 2005; Mukhopadhyay 2012). This is interpreted to reflect strong degassing either throughout accretion or during discrete loss events (e.g., only during giant impacts; Porcelli et al. 2001). A model age for the cessation of terrestrial magma ocean outgassing can be determined using I-Pu-Xe systematics in mantle-derived rocks (a mantle 'closure age'; Wetherill 1975). However, uncertainties in initial I budgets, the partitioning behavior of I during accretion (Armytage et al. 2013; Jackson et al. 2018), and the extent of degassing before, during and after a giant impact all contribute to uncertainty in determining the timing of this last outgassing event (Mukhopadhyay and Parai 2019). To preserve ^{129}Xe excesses from ^{129}I decay, retention of Xe in the mantle must have begun within the lifetime of ^{129}I.

Recent studies have noted heterogeneity in radiogenic ^{129}Xe signatures between the mantle source of mid-ocean ridge basalts (MORBs) and the portion of the mantle tapped by plume-related volcanism (Mukhopadhyay 2012). While both MORB and plume Xe isotopic signatures indicate extensive early degassing, ^{129}I-^{129}Xe signatures are distinct among MORB and plume-derived samples and cannot be related to one another by recent processes. Rather, the plume source must have differentiated from the upper mantle within the lifetime of ^{129}I, and the chemical signature of this separation was not erased by 4.45 Gyr of mantle convection (Mukhopadhyay 2012; Tucker et al. 2012; Parai et al. 2012; Pető et al. 2013; Parai and Mukhopadhyay 2015; Mukhopadhyay and Parai 2019; Parai et al. 2019). The preservation of early-formed I-Xe heterogeneity means that Earth's mantle was never mixed to the point of homogenization after ^{129}I became extinct at $\sim$100 Myr after the start of the Solar System.

7.2 Sm-Nd Systematics

Early-formed chemical heterogeneities in the terrestrial silicate reservoir are evident in the short-lived ^{146}Sm-^{142}Nd system ($t_{1/2} = 103$ Myr). Variations in $^{142}Nd/^{144}Nd$ are commonly expressed as μ^{142}Nd, which refers to deviations in parts per million from a terrestrial standard. The first anomalies in μ^{142}Nd were measured in 3.8 Gyr old Archean samples coming from the Isua Supracrustal Belt in SW Greenland (Harper and Jacobsen 1992; Boyet et al. 2003; Caro et al. 2003). These samples showed ^{142}Nd excesses that reflect a Sm/Nd fractionation produced during the first 500 Ma of Earth's history while ^{146}Sm was alive. Both parent and daughter elements are refractory and lithophile, but are fractionated by partial melting and crystallization (e.g., during magma ocean crystallization and crust formation) as Nd is more incompatible than Sm. Reservoirs characterized by high Sm/Nd ratios (i.e., those that preferentially derive from the residue of partial melts) over time evolve excesses in ^{142}Nd, whereas those characterized by low Sm/Nd ratios (i.e., those that preferentially derive from melt) develop deficits in ^{142}Nd relative to an unfractionated reservoir. The timing of this fractionation can be constrained by combining short-lived and long-lived Sm-Nd systematics (^{146}Sm-^{142}Nd and ^{147}Sm-^{143}Nd, respectively). Models predict that chemical fractionation associated with the crystallization of the magma ocean would have occurred around 100-150 Myr after the beginning of solar system accretion to explain observed short- and long-lived Sm-Nd signatures (e.g., Caro et al. 2003; Rizo et al. 2011).

Both excesses and deficits in ^{142}Nd relative to the average modern mantle have now been measured in samples coming from the oldest known terranes on Earth: the Kaapvaal craton in South Africa, Canada (Superior Province, Labrador, Northwest Territories), the Yilgarn craton in Western Australia and in the Napier complex in East Antarctica (see the recent reviews in Carlson et al. 2019; Guitreau et al. 2019). Very small variations of ^{142}Nd were also reported in modern samples from the Réunion hotspot, and these variations may correlate with helium isotope ratios (Peters et al. 2018). Samples from the Samoa and Hawaii hotspots which have anomalies in ^{182}W/^{184}W have been measured for ^{146}Sm-^{142}Nd systematics Horan et al. (2018). Measured deviations are very small and remain poorly resolved with respect to analytical precisions (μ^{142}Nd$<$3 ppm) even if these samples have high ^{3}He/^{4}He ratios (R/Ra up to 34). Thus, ^{142}Nd variations in terrestrial samples indicate that early-formed chemical reservoirs may have been preserved in the deep mantle despite 4.5 Gyr of mantle convection.

7.3 ^{182}Hf-^{182}W Systematics

Parts per million scale heterogeneity is also observed in ^{182}W/^{184}W signatures among terrestrial rocks ranging in age from $\sim$4 Ga to modern day (for recent reviews, see Rizo et al. 2019; Carlson et al. 2019). In terrestrial studies, because variations in ^{182}W/^{184}W are small they are commonly expressed as μ^{182}W, which refers to deviation in parts per million of the ^{182}W/^{184}W ratio from a high purity W standard. The composition of the W standard is commonly taken to be the composition of the BSE. For reference, chondrites have μ^{182}W $\sim$-190$\pm$10 (Kleine et al. 2009) and mid-ocean ridge basalts have μ^{182}W$\sim$0 (e.g., Rizo et al. 2016b). The different geochemical behavior of the parent (^{182}Hf; lithophile) and daughter (^{182}W; moderately siderophile) nuclides coupled with the short half-life of ^{182}Hf ($t_{1/2} = 8.9$ Ma, Vockenhuber et al. 2004) makes the ^{182}Hf-^{182}W system an effective tracer of metal-silicate differentiation events (e.g., core formation) that occurred during the first $\sim$60 Myr of solar system history (Touboul et al. 2012). Early differentiation events can also be traced using the Hf-W system because W is more incompatible than Hf which leads to fractionation in the Hf/W ratio during silicate differentiation (Shearer and Righter 2003). If any parent-daughter (Hf/W) fractionation occurs during the lifetime of ^{182}Hf, excesses or depletions in ^{182}W/^{184}W can develop in mantle or crustal reservoirs (Touboul et al. 2012).

Most mantle-derived rocks from $\sim$4 to $\sim$2.7 Ga are characterized by excesses in μ^{182}W (+5 to +20, with the majority between +10 and +15). Samples defining this compositional range include ~4 Ga supracrustal rocks from Acasta (Willbold et al. 2015; Reimink et al. 2018), ≥3.8 Ga supracrustal rocks from Nuvvuagittuq (Touboul et al. 2014), ~3.8 Ga supracrustal rocks from Saglek (Liu et al. 2016), 3.8-3.3 Ga supracrustal rocks from Isua (Willbold et al. 2011; Rizo et al. 2016a; Dale et al. 2017), 3.5 Ga basalts from the Pilbara Craton (Rizo et al. 2019), and 2.8-2.7 Ga komatiites from Kostomuksha (Touboul et al. 2012) and Boston Creek (Puchtel et al. 2018). Exceptions to the general enrichment in $\mu^{182}W$ in ancient materials are 3.5 Ga Komati komatiites ($\mu^{182}W$ 0 to +5, unresolved from the W standard composition; Touboul et al. 2012), 3.55 Ga komatiites from Schapenburg ($\mu^{182}W$ -6 to -12; Puchtel et al. 2016), and >2.3 Ga South African glacial diamictites ($\mu^{182}W$ 0 to -15; Mundl et al. 2018).

In modern samples, heterogeneity in $\mu^{182}W$ is also observed. Phanerozoic flood basalts that record positive $\mu^{182}W$ (+10 to +48 ppm) compositions were reported by Rizo et al. (2016b). Kruijer and Kleine (2018), however, proposed a revision to some of these data suggesting that there were no ^{182}W isotope excesses in Phanerozoic rocks from the Ontong Java Plateau. Further investigation is required to resolve this debate. Non-anomalous $\mu^{182}W$ compositions and depletions in $\mu^{182}W$ are also manifest in the modern rock record (+5 to -20; Willbold et al. 2011; Touboul et al. 2012; Liu et al. 2016; Rizo et al. 2016a; Mundl et al. 2017; Rizo et al. 2019). Samples defining the depleted ^{182}W signature are primarily ocean island basalts (OIBs), some of which display a relationship with depletions in ^{182}W being associated with high $^3He/^4He$ (Mundl et al. 2017). Samples with high $^3He/^4He$ are commonly interpreted to represent tapping of a region of the mantle that is less degassed relative to regions of the modern mantle (e.g., mid ocean ridge basalts; Kurz et al. 1982). The general relationship between $\mu^{182}W$ and $^3He/^4He$ in some OIBs suggests that high $^3He/^4He$ plumes either sample reservoirs in the deep mantle that were formed prior to ~60 Ma or sample reservoirs that record core-mantle interaction (Mundl et al. 2017; Rizo et al. 2019; Mundl-Petermeier et al. 2020).

From the current rock record, it is clear that the terrestrial mantle is defined by $\mu^{182}W$ heterogeneity that persisted from the Hadean-Archean to the present day. How the magnitude and variability in ^{182}W developed and why modern samples tend to exhibit normal or depleted ^{182}W signatures compared to Hadean and Archean-derived samples remain a matter of debate. Processes proposed to explain ^{182}W heterogeneity can be divided into three general scenarios: (1) disproportional late accretion, (2) early metal-silicate or silicate-silicate fractionation, or (3) core-mantle interaction, each of which are discussed below.

Disproportional late accretion requires that mantle-derived samples showing excesses in $\mu^{182}W$ be derived from mantle domains which did not receive the full complement of late-accreted material (Willbold et al. 2011; Touboul et al. 2014; Rizo et al. 2016a; Puchtel et al. 2018). Late-accreted material refers to the final ~0.5 to 2 wt% added to Earth after final core segregation (Kimura et al. 1974; Chou 1978; Walker 2009; Morbidelli and Wood 2014; Marchi et al. 2018). This material is considered to be chondritic in composition and, as such, would have been depleted in $\mu^{182}W$ relative to the present-day mantle (chondritic $\mu^{182}W = -190 \pm 10$; Kleine et al. 2009) and relatively enriched in W and HSEs compared to the pre-late veneer mantle (Morgan 1985; Becker et al. 2006; Walker 2009). Addition of late-accreted material into the mantle would have led to a decrease in $\mu^{182}W$ coupled with an enrichment in HSE abundance (Willbold et al. 2011). Consequently, regions isolated from the addition of late accreted material would be comparatively enriched in $\mu^{182}W$ and depleted in HSE. Coupled $\mu^{182}W$-HSE concentration estimates of primordial source regions, however, do not consistently exhibit the predicted relationship between $\mu^{182}W$ and HSE abundance. This may be due to the difficulties in accurately determining the HSE source composition (e.g., supracrustal materials are the product of multi-phase magmatic and metamorphic processes; Willbold et al. 2011). HSE abundances in the mantle can only be confidently determined using mantle peridotites and komatiites and there are few examples of Archean mantle peridotites. The complex μ^{182}W-HSE relationship may also indicate that disproportional late accretion is not the sole mechanism producing μ^{182}W heterogeneity in the mantle (see Carlson et al. 2019, and references therein).

Early silicate differentiation following crystallization of a magma ocean or partial melting during the lifetime of ^{182}Hf would result in μ^{182}W heterogeneity (Touboul et al. 2012). Alternatively, early metal-silicate fractionation followed by the removal of metal from an isolated mantle domain (e.g., a basal magma ocean), may have produced variable μ^{182}W (Touboul et al. 2012) (see summary by Puchtel et al. 2018). These scenarios can be assessed by combining Hf-W systematics with the short-lived ^{146}Sm-^{142}Nd system. The ^{146}Sm-^{142}Nd system comprises lithophile elements, where the daughter is more incompatible than the parent in mantle melting (see Sect. 7.2). As such, this radiometric system is sensitive to silicate-liquid fractionation, and insensitive to metal-silicate fractionation. If silicate differentiation processes were responsible for producing variations in μ^{182}W, these anomalies would be coupled with variations in ^{142}Nd so long as the fractionation event occurred within the lifetimes of both parent species (e.g., Rizo et al. 2019; Carlson et al. 2019). There are, however, limited datasets that report ^{142}Nd-^{182}W data obtained from the same sample (Touboul et al. 2014; Puchtel et al. 2016; Rizo et al. 2016a; Horan et al. 2018). From these coupled datasets, correlated behavior between ^{142}Nd-^{182}W is generally not observed, except in Schapenburg komatiites which are linked to silicate-silicate differentiation (Puchtel et al. 2016; Rizo et al. 2019; Carlson et al. 2019). This general lack of correlation indicates that early silicate differentiation was likely not solely responsible for production of anomalies in μ^{182}W and μ^{142}Nd (e.g., Touboul et al. 2014; Rizo et al. 2016a; Horan et al. 2018; Carlson et al. 2019). Given the very limited number of coupled μ^{142}Nd-μ^{182}W data obtained from the same samples, however, future work combining these systems on the same samples is required to assess the robustness of this interpretation. A few martian meteorites were measured for ^{182}Hf and ^{142}Nd isotopes and even if the martian mantle differentiated during the lifetime of these systematics, no correlation has been identified (Foley et al. 2005).

Core-mantle interaction has recently been reemphasized as a viable process through which μ^{182}W heterogeneity may have formed (Mundl et al. 2017; Rizo et al. 2019; Mundl-Petermeier et al. 2020). Rizo et al. (2019) proposed that the change in μ^{182}W mantle signatures (from excess to normal or depleted) after 2.7 Ga was due to a change in mantle dynamics occurring after the Archean or the onset of inner core crystallization. This model is based on new OIB data interpreted in the context of a compilation of the past near-decade of high precession ^{182}W analysis. Either mechanism proposed by Rizo et al. (2019) could produce deep mantle reservoirs characterized by μ^{182}W depletions that could subsequently be entrained in mantle plumes. Notably, as the core is predicted to be highly enriched in HSEs, the deep mantle reservoirs housing μ^{182}W depletions are expected to be enriched in HSEs. There is, however, no correlation between depleted μ^{182}W and enriched HSE concentrations in studied OIBs, suggesting that there was insignificant siderophile element exchange between the core and the deep mantle reservoir despite pervasive diffusive isotope exchange (Rizo et al. 2019; Mundl-Petermeier et al. 2020). Following Puchtel and Humayun (2000) and Humayun (2011), Mundl-Petermeier et al. (2020) suggested core-mantle interaction via isotopic equilibration as a process by which modification of ^{182}W isotope compositions could occur without modification of HSE abundances in the lowermost mantle. Alternatively, the HSE concentration in deep mantle source regions may not faithfully be recorded in the HSE complement of OIBs due to the complexities in estimating HSE concentrations from the basaltic rock record. It has also been suggested that HSE abundances may become decoupled from W during the low degrees of partial melting involved in the formation of OIBs (Jones et al. 2019). Future work further developing the accuracy of HSE concentration estimates in source regions and experimentally determining the mechanism and efficiency of chemical and isotopic exchange across the core-mantle boundary will aid in determining the role that core-mantle interaction plays in establishing μ^{182}W mantle heterogeneity (e.g., Yoshino et al. 2020).

Because Sm and Nd are lithophile elements, processes associated with core-segregation are not anticipated to significantly affect Sm/Nd ratios or Nd isotope compositions. This may account for the absence of coupled μ^{182}W and μ^{142}Nd compositions in some samples (Rizo et al. 2019). Carlson et al. (2019) extended these authors' interpretation by proposing μ^{182}W variations in the mantle record a predominantly core-mantle segregation process while μ^{142}Nd variations record crust-mantle differentiation events that occurred after ^{182}Hf was extinct. Carlson et al. (2019) proposed that Nd model ages calculated using Archean samples reflect the time of the giant impact, with the requirement that the impact did not homogenize W isotopic signatures imparted during early core segregation. Consequently, they suggest that the μ^{142}Nd variations reflect differentiation of the mantle after the Moon-forming impact which were effectively homogenized by the end of the Archean. Future studies combining μ^{182}W-μ^{142}Nd-HSE concentration datasets, experimentally derived constraints on chemical and isotopic exchange under core-mantle boundary conditions, and numerical models tracing mantle dynamics during different stages of Earth's history will enable these theories to be further vetted.

7.4 Using Terrestrial Geochemical Characteristics to Constrain the Moon Forming Event

In summary, very early formed isotopic heterogeneities have been preserved in the terrestrial mantle to the present day. Generation of these heterogeneities did not necessarily post-date the Moon-forming giant impact; they could have formed earlier and survived, or could have been generated by giant impacts themselves. For example, high ^{3}He/^{22}Ne ratios observed in Earth's mantle compared to precursor compositions (e.g., solar, chondritic) are attributed to ingassing of solar nebular gas into a terrestrial magma ocean early in Earth's history, due to the higher solubility of He relative to Ne in silicate melts (e.g., Graham 2002). The ^{3}He/^{22}Ne ratio of the MORB mantle ($\sim$10) is significantly higher than that observed in plume sources ($\sim$2.5). Tucker and Mukhopadhyay (2014) argued that at least three magma ocean events occurred to raise ^{3}He/^{22}Ne in the upper mantle to $\sim$10, and that the later events did not affect the whole mantle gas budget. Therefore, there is simultaneously geochemical evidence for widespread mantle melting, outgassing, and differentiation, and evidence for inefficient mixing of the early Earth's interior during and after giant impacts. Here we will discuss the constraints the preservation of primordial reservoirs places on the formation of the Moon, and the recovery of Earth immediately after the impact.

The interpretation of the constraints imposed by each radioactive system on Moon formation is critically dependent on the lifetime of the system relative to the timing of the Moon-forming impact (Sect. 3). If the Moon-forming giant impact occurred after a radioactive system went extinct, then variations in radiogenic isotopes produced by that system must either have been produced before the impact and preserved, record heterogeneous late addition of material to the mantle from extraplanetary bodies (such as late accretion), or record exchange between the mantle and core. If the Moon formed at $\sim$4.36 Ga ($\sim$200 Myr after the start of the solar system) as argued in Sect. 3, this requirement would apply to both ^{182}W and ^{129}Xe isotopic anomalies as well as ^{3}He/^{22}Ne ratios. If some of the observed anomalies (such as ^{182}W, ^{129}Xe) were formed in the proto-Earth's mantle before the Moon forming impact (as seems likely based on the arguments above) the Moon-forming impact could not have homogenized the whole mantle. As W and noble gases are trace elements (unlike e.g., O), the concentration of these elements could vary significantly between different reservoirs and pre-Moon forming impact signatures may have been preserved in relatively small volumes of the mantle. The requirement for incomplete homogenization of Earth during the impact is a powerful constraint on lunar formation, particularly given the inference that efficient intra-impact mixing of impactor and target material is required to explain the isotopic similarity between the Earth and Moon (Sect. 6). A successful Moon forming impact must mix the majority of the mantle but leave distinct domains, of uncertain volume, that are not homogenized with the rest of the mantle. The balance between these two requirements could dramatically reduce the number of plausible Moon-forming giant impacts. However, in order to leverage this constraint, significant improvements in giant impact models are required to accurately capture mixing and thermal equilibration (see Sect. 6).

The need to preserve anomalies generated before the Moon-forming giant impact would also place constraints on the recovery of Earth after the Moon-forming impact. After the impact, Earth cooled rapidly by radiation, driving condensation of the outer regions and subsequent freezing of the mantle. In order to preserve pre-impact mantle heterogeneity these processes must not have homogenized the mantle. Preservation of ^{182}W and noble gas anomalies is particularly challenging as these elements are incompatible in solid silicates (at least at upper mantle pressures) and thus are concentrated in the liquid during fractional crystallization. The post-impact evolution is dependent on the parameters of the Moon-forming impact as they determine the initial thermal and rotational state of the planet (Nakajima and Stevenson 2015; Lock and Stewart 2017, 2019; Lock et al. 2020). Lock and Stewart (2019) also showed that internal pressures in the post-impact body are dependent on the parameters of the impact and could have been much lower than in present-day Earth (10s of GPa lower at the core-mantle boundary) with the lowest pressures after high-angular momentum impacts. The pressures would have subsequently increased during cooling and lunar tidal recession as the body condensed and its rotation was slowed. Varying internal pressures in the aftermath of different giant impacts would change the stability of different phases and so how the mantle freezes (Lock and Stewart 2019; Lock et al. 2020). The rate of rotation of the post-impact body also dictates the efficiency of mixing in the magma ocean, with high rotation rates suppressing mixing (Maas and Hansen 2019) and aiding in the preservation of primordial reservoirs. The generation of anomalies from radioactive systems that were still alive at the time of the Moon-forming impact (such as ^{146}Sm-^{142}Nd) also depends on the evolution of the post-impact body and hence the parameters of the Moon-forming impact. Studying the geochemical consequences of different pathways of recovery (condensation, magma ocean freezing and tidal evolution) could offer new and important constraints on the initial conditions of the early Earth and hence the parameters of the Moon-forming giant impact.

8 Conclusions and Future Directions

Compositional and isotopic measurements of both lunar, terrestrial, and meteoritic samples have given valuable insights into the formation of the Moon and the early evolution of Earth. Constraints from various radioactive systems suggest that the Moon formed between about 50 and 200 Myrs after the start of the solar system, with several systems favoring younger ages. Comparison between the bulk composition of the Earth and Moon has revealed that the Moon likely formed in a high temperature, modest pressure environment from an iron-depleted disk or synestia. We also argue that the isotopic similarity between the Earth and Moon requires that either the proto-Earth and impactor had very similar isotopic compositions, or that the efficiency of mixing during the collision had to have been high. However, simultaneously the impact could not have homogenized the whole mantle as isotopic signatures of pre-impact heterogeneity are preserved. Mantle heterogeneities must also be preserved during the recovery of Earth after the impact and survived 4.5 Gyr of mantle convection, providing an important test of mantle evolution models. Together, these observations provide strong constraints with which to test different models of the formation of the Moon and early evolution of Earth.

Our understanding of planet formation and the early Earth is rapidly developing with increasingly sophisticated measurements of lunar, meteoritic and terrestrial samples. Particularly important for elucidating this critical time in Earth's history are studies that combine measurements of the abundances and isotopic ratios for multiple elements from the same samples. For example, to test mechanisms for producing the MVE depletion and isotopic fractionation in the Moon relative to Earth it is important to observe if and how the different elements are coupled. Combining μ^{182}W-μ^{142}Nd-noble gas-HSE concentration measurements of terrestrial samples would also help distinguish the relationship between these different systems, would allow for identification of later-stage alteration, and would provide tests of different models for the origins of chemical and isotopic heterogeneities.

In order to be able to interpret the plethora of geochemical constraints, there is a need for models that couple the physical and geochemical processes that occurred during lunar formation and the early evolution of Earth. In this paper, we have highlighted the importance of developing new techniques to capture mixing and thermal equilibration in giant impact simulations to be able to use the efficiency of mixing to both explain the isotopic similarity between the Earth and Moon and the survival of primordial reservoirs as constraints on lunar origin. In addition, to fully utilize the survival of primordial reservoirs as a constraint on Earth's history, coupled dynamic and chemical models that track the recovery and long-term evolution of Earth are required. To make the connection between processes in the lunar disk or terrestrial synestia and observations, we must also develop more sophisticated models that account for the small-scale multiphase and multicomponent dynamics of these substantially vaporized structures.

Producing accurate models requires laboratory experiments to constrain the properties of materials and behavior of different elements at the extreme conditions of lunar formation and mantle evolution. Developing advanced equations of state for silicates and metals, in particular EOS that include phase transitions, at very high pressures and temperatures are fundamental to improving the accuracy of giant impact and mantle evolution models. Thermodynamic data on how materials behave at high temperatures and modest pressures in the lunar disk and terrestrial synestia are also required to accurately predict the MVE depletion and isotopic fractionation produced by different lunar formation models.

We are in a new age of lunar exploration. Several nations and private companies are planning missions to the Moon, including crewed missions. Most imminently, the China National Space Administration is planning sample return missions from the near (Chang'e 5) and potentially also the far (Chang'e 6) sides of the Moon. Acquisition of high precision measurements from different regions of the lunar surface would allow for more accurate determinations of the Moon's age, composition, and thermal history. In addition, the lunar far side may host samples from the lunar mantle (Li et al. 2019) providing a more complete view of lunar chemistry. These new data, and further studies of existing samples, will continue to elucidate the coupled early history of Earth and its Moon.

Acknowledgements This paper was instigated at the International Space Science Institute (ISSI) workshop 'Reading Terrestrial Planet Evolution in Isotopes and Element Measurements' and the authors would like to thank ISSI and Europlanet for their support. We would also like to thank Paolo Sossi and an anonymous reviewer for comments that helped improve the clarity and completeness of the manuscript, and Helmut Lammer for editorial handling. We thank Jessica Barnes for providing their Cl isotope database. SJL acknowledges funding from NSF (awards EAR-1947614 and EAR-1725349) and the Division of Geological and Planetary Sciences at the California Institute of Technology. KRB acknowledges funding from NASA Emerging Worlds grants 80NSSC18K0496 and NNX16AN07G, NASA SSERVI grant NNA14AB07A, and support from the Department of Earth and Planetary Sciences, Rutgers University. RP acknowledges support from Washington University. MB received funding from the European Research Council (ERC Grant agreement No. 682778 - ISOREE).

Appendix

A.1 Data Sources for Figs. 5 and 6

Data for Figs. 5 and 6 were taken from a compilation by Dauphas (2017) with the exception that the lunar value for ϵ^{54}Cr was taken from Mougel et al. (2018). Below, we give the original data sources for this compilation.

A.1.1 $\Delta^{17}O$

Δ^{17}O data was compiled by Dauphas (2017) with the aid of MetBase (Meteorite Information Database, http://www.metbase.org/). Δ^{17}O data were taken from Bischoff et al. (1991, 1993, 1998), Bischoff (1994), Bridges et al. (1997, 1999), Brearley et al. (1989), Burkhardt et al. (2017), Buchanan et al. (1993), Clayton and Mayeda (1978, 1981, 1984, 1985, 1989, 1990, 1996), Clayton et al. (1976, 1977, 1983, 1984a,b, 1991, 1997a,b), Connolly et al. (2007), Franchi et al. (1992, 1999), Grossman and Zipfel (2001), Grossman et al. (1987), Grossman (1999), Grady et al. (1987), Gooding et al. (1983), Goodrich et al. (1987), Halbout et al. (1984, 1986), Ivanov et al. (1987), Jabeen et al. (1998), Jackel et al. (1996), Kallemeyn et al. (1996), Keller et al. (1994), McCoy et al. (1995, 1996, 1997), Mayeda and Clayton (1980, 1983, 1989a,b), Mayeda et al. (1987, 1995), Moroz et al. (1988), Nehru et al. (1996), Olsen et al. (1987, 1994), Onuma et al. (1978, 1983), Petaev et al. (1988), Prinz et al. (1991), Pun et al. (1991), Romanek et al. (1998), Rowe et al. (1994), Russell et al. (1998, 2005), Ruzicka et al. (1995), Recca et al. (1986), Sears et al. (1990), Simon et al. (1995), Schulze et al. (1994), Stepniewski et al. (1998), Weber et al. (1996, 1997), Weisberg et al. (1991, 1993, 1995, 1996, 1997), Zolensky et al. (1989, 1997); and Young et al. (2016).

A.1.2 $\epsilon^{50}Ti$

ϵ^{50}Ti data were taken from Zhang et al. (2011, 2012), Trinquier et al. (2009); and Burkhardt et al. (2017).

A.1.3 $\epsilon^{54}Cr$

ϵ^{54}Cr data were taken from Trinquier et al. (2007), Qin et al. (2010a,b), Shukolyukov and Lugmair (2006a,b), Shukolyukov et al. (2009), Schiller et al. (2014), Yamashita et al. (2005, 2010), Yamakawa et al. (2010), Larsen et al. (2011), Göpel and Birck (2010), Burkhardt et al. (2017); and Mougel et al. (2018).

A.2 Details of Mixing Model

In Sect. 6 we present the results of a mixing model to investigate the requirements for mixing during and after the Moon-forming giant impact imposed by the isotopic similarity between the Earth and Moon. Here we provide the details of that model.

We divide the post-impact body into two regions: the Moon-forming region and an isolate region that does not communicate with the Moon forming region during the period of Moon formation. The mass of these two regions are M_1 and M_2 respectively, where 1 denotes the Moon-forming region and 2 the isolated region (Fig. 8). The masses of the two regions are such that

$$M_1 + M_2 = M_{\text{Earth}} + M_{\text{Moon}} \,, \tag{1}$$

where M_{Moon} and M_{Earth} are the mass of the present-day Moon and Earth respectively.

The two regions are made up of a different mass fraction of impactor material, f^i_{imp}, and so have different average Δ^{17}O isotopic compositions, c_i. Note that both these regions can be internally heterogeneous and the only requirement is that the Moon inherits the average composition of the Moon-forming region. For simplicity, we assume that the majority of Earth mixes after the impact and that the present-day observable mantle is a mixture of the isolated region and the fraction of the Moon-forming region that was not incorporated into the Moon such that

$$(M_1 - M_{\text{Moon}})c_1 + M_2 c_2 = M_{\text{Earth}} c_{\text{Earth}} = 0 \,, \tag{2}$$

where c_{Earth} is the composition of the present-day Earth's mantle which is zero by definition. The composition of each region is related to the composition of the target and impactor by

$$c_i = f^i_{\text{imp}} c_{\text{imp}} + (1 - f^i_{\text{imp}}) c_{\text{tar}} \,, \tag{3}$$

where c_{imp} and c_{tar} are the Δ^{17}O isotopic composition of the impactor and target respectively.

We can determine an expression for the mass fraction of the post-impact structure that must be mixed in to the Moon-forming region ($F_{\text{mix}} = M_1/(M_1 + M_2) = M_1/(M_{\text{Earth}} + M_{\text{Moon}})$) given a present-day difference in Δ^{17}O between the Moon and Earth ($\Delta c_{\text{M-E}} = c_1$ by definition), a specified Δ^{17}O offset between the impactor and target ($\Delta c_{\text{imp-tar}} = c_{\text{imp}} - c_{\text{tar}}$), and a difference in the mass fraction of impactor in the two regions ($\Delta f_{\text{imp}} = f^1_{\text{imp}} - f^2_{\text{imp}}$). By substituting for M_2 in equation (2) using equation (1) we can find an expression for M_1 as a function of the composition of the two regions:

$$M_1 = \frac{M_{\text{Moon}} c_1 - (M_{\text{Earth}} + M_{\text{Moon}}) c_2}{(c_1 - c_2)} \,. \tag{4}$$

Independently we can re-express equation 2 for the isolated region ($i = 2$) to give an expression for c_2 in terms of the composition of the impactor and target:

$$c_2 = c_{\text{imp}} - (1 - f^2_{\text{imp}}) \Delta c_{\text{imp-tar}} \,. \tag{5}$$

Similarly, rearranging equation (2) for the Moon-forming region ($i = 1$) we obtain an expression for c_{imp}:

$$c_{\text{imp}} = \Delta c_{M-E} + (1 - f^1_{\text{imp}}) \Delta c_{\text{imp-tar}} \,. \tag{6}$$

Substituting for $c_{\rm imp}$ in equation (5) using equation (6) we obtain

$$c_2 = \Delta c_{M-E} - \Delta f_{\rm imp} \Delta c_{\rm imp-tar} \,. \tag{7}$$

Substituting for c_2 in equation (4) using equation (7) and rearranging we find that

$$F_{\rm mix} = \frac{\Delta f_{\rm imp} \Delta c_{\rm imp-tar}(M_{\rm Earth} - M_{\rm Moon}) - \Delta c_{\rm M-E} M_{\rm Earth}}{\Delta f_{\rm imp} \Delta c_{\rm imp-tar}(M_{\rm Earth} + M_{\rm Moon})} \,, \tag{8}$$

or alternatively:

$$F_{\rm mix} = 1 - \frac{\Delta c_{\rm M-E} M_{\rm Earth}}{\Delta f_{\rm imp} \Delta c_{\rm imp-tar}(M_{\rm Earth} + M_{\rm Moon})} \,. \tag{9}$$

The expression given in equation (9) is significant as it relates the degree of post-impact mixing that is required to explain the Earth-Moon isotopic similarity for a given degree of intra-impact mixing and compositional difference between impactor and target. We can therefore use it to understand the trade offs between mixing during and after the impact (Sect. 6).

References

C. Agnor, R.M. Canup, H.F. Levison, On the character and consequences of large impacts in the late stage of terrestrial planet formation. Icarus **142**(1), 219–237 (1999). https://doi.org/10.1006/icar.1999.6201

Y.A. Akovali, Nuclear data sheets for A = 244. Nucl. Data Sheets **99**(1), 197–273 (2003). https://doi.org/10.1006/ndsh.2003.0008

F. Albarède, E. Albalat, C.T.A. Lee, An intrinsic volatility scale relevant to the Earth and Moon and the status of water in the Moon. Meteorit. Planet. Sci. **50**(4), 568–577 (2015). https://doi.org/10.1111/maps.12331

C.J. Allègre, T. Staudacher, P. Sarda, M. Kurz, Constraints on evolution of Earth's mantle from rare gas systematics. Nature **303**(5920), 762–766 (1983). https://doi.org/10.1038/303762a0

D. Antonangeli, G. Morard, N.C. Schmerr, T. Komabayashi, M. Krisch, G. Fiquet, Y. Fei, H.K. Mao, Toward a mineral physics reference model for the Moon's core. Proc. Natl. Acad. Sci. USA **112**(13), 3916–3919 (2015). https://doi.org/10.1073/pnas.1417490112

R. Armytage, R. Georg, H. Williams, A. Halliday, Silicon isotopes in lunar rocks: implications for the Moon's formation and the early history of the Earth. Geochim. Cosmochim. Acta **77**, 504–514 (2012). https://doi.org/10.1016/j.gca.2011.10.032

R.M. Armytage, A.P. Jephcoat, M.A. Bouhifd, D. Porcelli, Metal-silicate partitioning of iodine at high pressures and temperatures: implications for the Earth's core and ^{129*}Xe budgets. Earth Planet. Sci. Lett. **373**, 140–149 (2013). https://doi.org/10.1016/j.epsl.2013.04.031

E. Asphaug, Impact origin of the Moon? Annu. Rev. Earth Planet. Sci. **42**(1), 551–578 (2014). https://doi.org/10.1146/annurev-earth-050212-124057

M. Barboni, P. Boehnke, B. Keller, I.E. Kohl, B. Schoene, E.D. Young, K.D. McKeegan, Early formation of the Moon 4.51 billion years ago. Sci. Adv. **3**(1), e1602,365 (2017). https://doi.org/10.1126/sciadv.1602365

J.J. Barnes, R. Tartèse, M. Anand, F.M. McCubbin, C.R. Neal, I.A. Franchi, Early degassing of lunar urKREEP by crust-breaching impact(s). Earth Planet. Sci. Lett. **447**, 84–94 (2016). https://doi.org/10.1016/j.epsl.2016.04.036

J.J. Barnes, I.A. Franchi, F.M. McCubbin, M. Anand, Multiple reservoirs of volatiles in the Moon revealed by the isotopic composition of chlorine in lunar basalts. Geochim. Cosmochim. Acta **266**, 144–162 (2019). https://doi.org/10.1016/j.gca.2018.12.032

A.C. Barr, On the origin of Earth's Moon. J. Geophys. Res., Planets **121**(9), 1573–1601 (2016). https://doi.org/10.1002/2016JE005098

H. Becker, M.F. Horan, R.J. Walker, S. Gao, J.P. Lorand, R.L. Rudnick, Highly siderophile element composition of the Earth's primitive upper mantle: constraints from new data on peridotite massifs and xenoliths. Geochim. Cosmochim. Acta **70**(17), 4528–4550 (2006). https://doi.org/10.1016/j.gca.2006.06.004

A. Bischoff, Acfer 217 - a new member of the Rumuruti chondrite group (R). Meteoritics **29**(2), 264–274 (1994). https://doi.org/10.1111/j.1945-5100.1994.tb00680.x

A. Bischoff, H. Palme, R.N. Clayton, T.K. Mayeda, T. Grund, B. Spettel, T. Geiger, M. Endreß, W. Beckerling, K. Metzler, New carbonaceous and type 3 ordinary chondrites from the Sahara Desert. Meteoritics **26**, 318 (1991)

A. Bischoff, H. Palme, R.D. Ash, R.N. Clayton, L. Schultz, U. Herpers, D. Stöffler, M.M. Grady, C.T. Pillinger, B. Spettel, H. Weber, T. Grund, M. Endreß, D. Weber, Paired Renazzo-type (CR) carbonaceous chondrites from the Sahara. Geochim. Cosmochim. Acta **57**(7), 1587–1603 (1993). https://doi.org/10.1016/0016-7037(93)90014-N

A. Bischoff, D. Weber, R. Bartoschewitz, R.N. Clayton, T.K. Mayeda, L. Schultz, B. Spettel, H.W. Weber, A. Bischoff, D. Weber, R. Bartoschewitz, R.N. Clayton, T.K. Mayeda, L. Schultz, B. Spettel, H.W. Weber, Characterization of the Rumuruti chondrite regolith breccia Hughes 030 (R3-6) and implications for the occurrence of unequilibrated lithologies on the R-chondrite parent body. Meteorit. Planet. Sci. **33**, A15 (1998)

P. Bonnand, I.J. Parkinson, M. Anand, Mass dependent fractionation of stable chromium isotopes in mare basalts: implications for the formation and the differentiation of the Moon. Geochim. Cosmochim. Acta **175**, 208–221 (2016). https://doi.org/10.1016/j.gca.2015.11.041

L.E. Borg, J.N. Connelly, M. Boyet, R.W. Carlson, Chronological evidence that the Moon is either young or did not have a global magma ocean. Nature **477**(7362), 70–73 (2011). https://doi.org/10.1038/nature10328

L.E. Borg, A.M. Gaffney, T.S. Kruijer, N.A. Marks, C.K. Sio, J. Wimpenny, Isotopic evidence for a young lunar magma ocean. Earth Planet. Sci. Lett. **523**, 115,706 (2019). https://doi.org/10.1016/j.epsl.2019.07.008

J.W. Boyce, A.H. Treiman, Y. Guan, C. Ma, J.M. Eiler, J. Gross, J.P. Greenwood, E.M. Stolper, The chlorine isotope fingerprint of the lunar magma ocean. Sci. Adv. **1**(8), e1500,380 (2015). https://doi.org/10.1126/sciadv.1500380

J.W. Boyce, S.A. Kanee, F.M. McCubbin, J.J. Barnes, H. Bricker, A.H. Treiman, Early loss, fractionation, and redistribution of chlorine in the Moon as revealed by the low-Ti lunar mare basalt suite. Earth Planet. Sci. Lett. **500**, 205–214 (2018). https://doi.org/10.1016/j.epsl.2018.07.042

M. Boyet, R.W. Carlson, A highly depleted moon or a non-magma ocean origin for the lunar crust? Earth Planet. Sci. Lett. **262**(3–4), 505–516 (2007). https://doi.org/10.1016/j.epsl.2007.08.009

M. Boyet, J. Blichert-Toft, M. Rosing, M. Storey, P. Télouk, F. Albarède, ^{142}Nd evidence for early Earth differentiation. Earth Planet. Sci. Lett. **214**(3–4), 427–442 (2003). https://doi.org/10.1016/S0012-821X(03)00423-0

M. Boyet, A. Bouvier, P. Frossard, T. Hammouda, M. Garçon, A. Gannoun, Enstatite chondrites EL3 as building blocks for the Earth: the debate over the ^{146}Sm–^{142}Nd systematics. Earth Planet. Sci. Lett. **488**, 68–78 (2018). https://doi.org/10.1016/j.epsl.2018.02.004

A.D. Brandon, T.J. Lapen, V. Debaille, B.L. Beard, K. Rankenburg, C. Neal, Re-evaluating ^{142}Nd/^{144}Nd in lunar mare basalts with implications for the early evolution and bulk Sm/Nd of the Moon. Geochim. Cosmochim. Acta **73**(20), 6421–6445 (2009). https://doi.org/10.1016/j.gca.2009.07.015

A.J. Brearley, E.R. Scott, K. Keil, R.N. Clayton, T.K. Mayeda, W.V. Boynton, D.H. Hill, Chemical, isotopic and mineralogical evidence for the origin of matrix in ordinary chondrites. Geochim. Cosmochim. Acta **53**(8), 2081–2093 (1989). https://doi.org/10.1016/0016-7037(89)90326-8

J.C. Bridges, I.A. Franchi, M.M. Grady, A.S. Sexton, C.T. Pillinger, The δ^{18}O composition of Feldspar and other minerals in Lafayette. Meteorit. Planet. Sci. **32**(4), A21–A22 (1997)

J.C. Bridges, I.A. Franchi, A.S. Sexton, C.T. Pillinger, Mineralogical controls on the oxygen isotopic compositions of UOCs. Geochim. Cosmochim. Acta **63**(6), 945–951 (1999). https://doi.org/10.1016/S0016-7037(98)00317-2

P.C. Buchanan, M.E. Zolensky, A.M. Reid, Carbonaceous chondrite clasts in the howardites Bholghati and EET87513. Meteoritics **28**(5), 659–682 (1993). https://doi.org/10.1111/j.1945-5100.1993.tb00637.x

C. Burkhardt, N. Dauphas, H. Tang, M. Fischer-Gödde, L. Qin, J.H. Chen, S.S. Rout, A. Pack, P.R. Heck, D.A. Papanastassiou, In search of the Earth-forming reservoir: mineralogical, chemical, and isotopic characterizations of the ungrouped achondrite NWA 5363/NWA 5400 and selected chondrites. Meteorit. Planet. Sci. **52**(5), 807–826 (2017). https://doi.org/10.1111/maps.12834

A.G.W. Cameron, W.R. Ward, The origin of the Moon, in *Lunar and Planetary Science Conference Abstracts*, vol. 7 (1976), p. 120

R.M. Canup, Simulations of a late lunar-forming impact. Icarus **168**(2), 433–456 (2004). https://doi.org/10.1016/j.icarus.2003.09.028

R.M. Canup, Lunar-forming collisions with pre-impact rotation. Icarus **196**(2), 518–538 (2008). https://doi.org/10.1016/j.icarus.2008.03.011

R.M. Canup, Forming a Moon with an Earth-like composition via a giant impact. Science **338**(6110), 1052–1055 (2012). https://doi.org/10.1126/science.1226073

R.M. Canup, E. Asphaug, Origin of the Moon in a giant impact near the end of the Earth's formation. Nature **412**(6848), 708–712 (2001). https://doi.org/10.1038/35089010

R.M. Canup, C. Visscher, J. Salmon, B. Fegley Jr., Lunar volatile depletion due to incomplete accretion within an impact-generated disk. Nat. Geosci. **8**(12), 918–921 (2015). https://doi.org/10.1038/ngeo2574

R.W. Carlson, L.E. Borg, A.M. Gaffney, M. Boyet, Rb-Sr, Sm-Nd and Lu-Hf isotope systematics of the lunar Mg-suite: the age of the lunar crust and its relation to the time of Moon formation. Philos. Trans. R. Soc. A, Math. Phys. Eng. Sci. **372**(2024), 20130,246 (2014). https://doi.org/10.1098/rsta.2013.0246

R.W. Carlson, M. Garçon, J. O'Neil, J. Reimink, H. Rizo, The nature of Earth's first crust. Chem. Geol. **530**, 119,321 (2019). https://doi.org/10.1016/j.chemgeo.2019.119321

G. Caro, B. Bourdon, J.L. Birck, S. Moorbath, ^{146}Sm-^{142}Nd evidence from Isua metamorphosed sediments for early differentiation of the Earth's mantle. Nature **423**(6938), 428–432 (2003). https://doi.org/10.1038/nature01668

G. Caro, P. Morino, S.J. Mojzsis, N.L. Cates, W. Bleeker, Sluggish Hadean geodynamics: evidence from coupled $^{146,\,147}$Sm–$^{142,\,143}$Nd systematics in Eoarchean supracrustal rocks of the Inukjuak domain (Québec). Earth Planet. Sci. Lett. **457**, 23–37 (2017). https://doi.org/10.1016/j.epsl.2016.09.051

P.J. Carter, S.J. Lock, S.T. Stewart, The energy budgets of giant impacts. J. Geophys. Res., Planets **125**(1), e2019JE006042 (2020). https://doi.org/10.1029/2019JE006042

S. Charnoz, C. Michaut, Evolution of the protolunar disk: dynamics, cooling timescale and implantation of volatiles onto the Earth. Icarus **260**, 440–463 (2015). https://doi.org/10.1016/j.icarus.2015.07.018

S. Charnoz, Y. Lee, P. Sossi, L. Allibert, J. Siebert, R. Hyodo, H. Genda, F. Moynier, Efficient early Moon devolatilisation just after its formation, through tidally assisted hydrodynamic escape, in *Lunar and Planetary Science Conference Abstracts*, vol. 50 (2019), p. 2395

H. Chen, P.S. Savage, F.Z. Teng, R.T. Helz, F. Moynier, Zinc isotope fractionation during magmatic differentiation and the isotopic composition of the bulk Earth. Earth Planet. Sci. Lett. **369–370**, 34–42 (2013). https://doi.org/10.1016/J.EPSL.2013.02.037

C.L. Chou, Fractionation of siderophile elements in the Earth's upper mantle, in *Lunar and Planetary Science Conference Abstracts*, vol. 9 (1978), pp. 219–230

R.I. Citron, H.B. Perets, O. Aharonson, The role of multiple giant impacts in the formation of the Earth–Moon system. Astrophys. J. **862**(1), 5 (2018). https://doi.org/10.3847/1538-4357/aaca2d

R.N. Clayton, T.K. Mayeda, Multiple parent bodies of polymict brecciated meteorites. Geochim. Cosmochim. Acta **42**(3), 325–327 (1978). https://doi.org/10.1016/0016-7037(78)90185-0

R.N. Clayton, T.K. Mayeda, Redox processes in chondrules and chondrites, in *Lunar and Planetary Science Conference Abstracts*, vol. 12 (1981), pp. 154–156

R.N. Clayton, T.K. Mayeda, The oxygen isotope record in Murchison and other carbonaceous chondrites. Earth Planet. Sci. Lett. **67**(2), 151–161 (1984). https://doi.org/10.1016/0012-821X(84)90110-9

R.N. Clayton, T.K. Mayeda, Oxygen isotopes in Shergotty, in *Lunar and Planetary Science Conference Abstracts*, vol. 16 (1985), p. 11

R.N. Clayton, T.K. Mayeda, Oxygen isotopic composition of LEW 86010. Meteoritics **24**, 259 (1989)

R.N. Clayton, T.K. Mayeda, Oxygen isotopic composition of Antarctic meteorites, in *Workshop on Differences Between Antarctic and non-Antarctic Meteorites*, vol. 90-01 (1990), pp. 30–31

R.N. Clayton, T.K. Mayeda, Oxygen isotope studies of achondrites. Geochim. Cosmochim. Acta **60**(11), 1999–2017 (1996). https://doi.org/10.1016/0016-7037(96)00074-9

R.N. Clayton, N. Onuma, T.K. Mayeda, A classification of meteorites based on oxygen isotopes. Earth Planet. Sci. Lett. **30**(1), 10–18 (1976). https://doi.org/10.1016/0012-821X(76)90003-0

R.N. Clayton, N. Onuma, L. Grossman, T.K. Mayeda, Distribution of the pre-solar component in Allende and other carbonaceous chondrites. Earth Planet. Sci. Lett. **34**(2), 209–224 (1977). https://doi.org/10.1016/0012-821X(77)90005-X

R.N. Clayton, T.K. Mayeda, E.J. Olsen, M. Prinz, Oxygen isotope relationships in iron meteorites. Earth Planet. Sci. Lett. **65**(2), 229–232 (1983). https://doi.org/10.1016/0012-821X(83)90161-9

R. Clayton, T. Mayeda, K. Yanai, Oxygen isotopic compositions of some Yamato meteorites. Mem. Natl. Inst. Polar Res., Spec. Issue **35**, 267–271 (1984a)

R.N. Clayton, T.K. Mayeda, A.E. Rubin, Oxygen isotopic compositions of enstatite chondrites and aubrites. J. Geophys. Res. **89**(S01), C245 (1984b). https://doi.org/10.1029/jb089is01p0c245

R.N. Clayton, T.K. Mayeda, J.N. Goswami, E.J. Olsen, Oxygen isotope studies of ordinary chondrites. Geochim. Cosmochim. Acta **55**(8), 2317–2337 (1991). https://doi.org/10.1016/0016-7037(91)90107-G

R.N. Clayton, T.K. Mayeda, T. Hiroi, M. Zolensky, M.E. Lipschutz, Oxygen isotopes in laboratory-heated CI and CM chondrites, in *Annual Meteoritical Society Meeting*, vol. 60 (1997a), p. 5255

R.N. Clayton, T.K. Mayeda, H. Kojima, M.K. Weisberg, M. Prinz, Hydration and dehydration in carbonaceous chondrites, in *Lunar and Planetary Science Conference Abstracts*, vol. 28 (1997b), p. 239

J.N. Connelly, M. Bizzarro, A.N. Krot, Å. Nordlund, D. Wielandt, M.A. Ivanova, The absolute chronology and thermal processing of solids in the solar protoplanetary disk. Science **338**(6107), 651–655 (2012). https://doi.org/10.1126/science.1226919

H.C. Connolly, J. Zipfel, L. Folco, C. Smith, R.H. Jones, G. Benedix, K. Righter, A. Yamaguchi, H.C. Aoudjehane, J.N. Grossman, The Meteoritical Bulletin, No. 91, 2007 March. Meteorit. Planet. Sci. **42**(3), 413–466 (2007). https://doi.org/10.1111/j.1945-5100.2007.tb00242.x

M. Ćuk, S.T. Stewart, Making the Moon from a fast-spinning Earth: a giant impact followed by resonant despinning. Science **338**(6110), 1047–1052 (2012). https://doi.org/10.1126/science.1225542

M. Ćuk, D.P. Hamilton, S.J. Lock, S.T. Stewart, Tidal evolution of the Moon from a high-obliquity, high-angular-momentum Earth. Nature **539**, 402–406 (2016). https://doi.org/10.1038/nature19846

C.W. Dale, T.S. Kruijer, K.W. Burton, Highly siderophile element and ^{182}W evidence for a partial late veneer in the source of 3.8 Ga rocks from Isua, Greenland. Earth Planet. Sci. Lett. **458**, 394–404 (2017). https://doi.org/10.1016/j.epsl.2016.11.001

N. Dauphas, The isotopic nature of the Earth's accreting material through time. Nature **541**(7638), 521–524 (2017). https://doi.org/10.1038/nature20830

N. Dauphas, C. Burkhardt, P.H. Warren, T. Fang-Zhen, Geochemical arguments for an Earth-like Moon-forming impactor. Philos. Trans. R. Soc. A, Math. Phys. Eng. Sci. **372**(2024), 20130,244 (2014). https://doi.org/10.1098/rsta.2013.0244

N. Dauphas, M. Meheut, M. Blanchard, H. Zeng, G. Galli, R.N. Canup, C. Visscher, N. Nie, Can lunar formation theories be tested with K isotopes? in *Lunar and Planetary Science Conference Abstracts*, vol. 49 (2018), p. 2481

J.M. Day, F. Moynier, C.K. Shearer, Late-stage magmatic outgassing from a volatile-depleted Moon. Proc. Natl. Acad. Sci. USA **114**(36), 9547–9551 (2017). https://doi.org/10.1073/pnas.1708236114

H. Deng, C. Reinhardt, F. Benitez, L. Mayer, J. Stadel, A.C. Barr, Enhanced mixing in giant impact simulations with a new Lagrangian method. Astrophys. J. **870**(2), 127 (2019). https://doi.org/10.3847/1538-4357/aaf399

S.J. Desch, G.J. Taylor, Isotopic mixing due to the interaction between the protolunar disk and the Earth's atmosphere, in *Lunar and Planetary Science Conference Abstracts*, vol. 44 (2013), p. 2566

J.K. Dhaliwal, J.M. Day, F. Moynier, Volatile element loss during planetary magma ocean phases. Icarus **300**, 249–260 (2018). https://doi.org/10.1016/J.ICARUS.2017.09.002

J.E. Dickinson, P.C. Hess, Zircon saturation in lunar basalts and granites. Earth Planet. Sci. Lett. **57**(2), 336–344 (1982). https://doi.org/10.1016/0012-821X(82)90154-6

E. Dowty, M. Prinz, K. Keil, Ferroan anorthosite: a widespread and distinctive lunar rock type. Earth Planet. Sci. Lett. **24**(1), 15–25 (1974). https://doi.org/10.1016/0012-821X(74)90003-X

M.J. Drake, H.E. Newsom, C.J. Capobianco, V, Cr, and Mn in the Earth, Moon, EPB, and SPB and the origin of the Moon: Experimental studies. Geochim. Cosmochim. Acta **53**(8), 2101–2111 (1989). https://doi.org/10.1016/0016-7037(89)90328-1

K.F. Eckerman, R.J. Westfall, J.C. Ryman, M. Cristy, Nuclear decay data files of the dosimetry research group No. ORNL/TM-12350. Tech. rep. Oak Ridge National Laboratory (ORNL), Oak Ridge, TN (1993). https://doi.org/10.2172/10116928

L.T. Elkins-Tanton, S. Burgess, Q.Z. Yin, The lunar magma ocean: reconciling the solidification process with lunar petrology and geochronology. Earth Planet. Sci. Lett. **304**(3–4), 326–336 (2011). https://doi.org/10.1016/j.epsl.2011.02.004

C.N. Foley, M. Wadhwa, L.E. Borg, P.E. Janney, R. Hines, T.L. Grove, The early differentiation history of Mars from ^{182}W-^{142}Nd isotope systematics in the SNC meteorites. Geochim. Cosmochim. Acta **69**(18), 4557–4571 (2005). https://doi.org/10.1016/j.gca.2005.05.009

I.A. Franchi, T. Akagi, C.T. Pillinger, Laser fluorination of meteorites - small sample analysis of δ^{17}O and δ^{18}O. Meteoritics **27**, 222 (1992)

I.A. Franchi, I.P. Wright, A.S. Sexton, C.T. Pillinger, The oxygen-isotopic composition of Earth and Mars. Meteorit. Planet. Sci. **34**(4), 657–661 (1999). https://doi.org/10.1111/j.1945-5100.1999.tb01371.x

A.M. Friedman, J. Milsted, D. Metta, D. Henderson, J. Lerner, A.L. Harkness, D.J. Rokop, Alpha decay half lives of ^{148}Gd ^{150}Gd, and ^{146}Sm. Radiochim. Acta **5**(4), 192–194 (1966). https://doi.org/10.1524/ract.1966.5.4.192

A.M. Gaffney, L.E. Borg, A young solidification age for the lunar magma ocean. Geochim. Cosmochim. Acta **140**, 227–240 (2014). https://doi.org/10.1016/j.gca.2014.05.028

R.F. Garcia, J. Gagnepain-Beyneix, S. Chevrot, P. Lognonné, Very preliminary reference Moon model. Phys. Earth Planet. Inter. **188**(1–2), 96–113 (2011). https://doi.org/10.1016/j.pepi.2011.06.015

J.L. Gooding, T.K. Mayeda, R.N. Clayton, T. Fukuoka, Oxygen isotopic heterogeneities, their petrological correlations, and implications for melt origins of chondrules in unequilibrated ordinary chondrites. Earth Planet. Sci. Lett. **65**(2), 209–224 (1983). https://doi.org/10.1016/0012-821X(83)90159-0

C.A. Goodrich, K. Keil, J.L. Berkley, J. Laul, M. Smith, J.F. Wacker, R.N. Clayton, T.K. Mayeda, Roosevelt County 027: a low-shock Ureilite with interstitial silicates and high noble gas concentrations. Meteoritics **22**(3), 191–218 (1987). https://doi.org/10.1111/j.1945-5100.1987.tb00619.x

C. Göpel, J. Birck, Mn/Cr systematics: a tool to discriminate the origin of primitive meteorites? in *Goldschmidt Conference Abstracts* (2010), p. A348

M. Grady, D. Aylmer, G. Kurat, T. Ntaflos, U. Ott, H. Palme, B. Spettel, Yamato-82042: an unusual carbonaceous chondrite with CM affinities. Mem. Natl. Inst. Polar Res., Spec. Issue **46**(46), 162–178 (1987)

D.W. Graham, Noble gas isotope geochemistry of mid-ocean ridge and ocean island basalts: characterization of mantle source reservoirs. Rev. Mineral. Geochem. **47**(1), 247–317 (2002). https://doi.org/10.2138/rmg.2002.47.8

M.L. Grange, R.T. Pidgeon, A.A. Nemchin, N.E. Timms, C. Meyer, Interpreting U-Pb data from primary and secondary features in lunar zircon. Geochim. Cosmochim. Acta **101**, 112–132 (2013). https://doi.org/10.1016/j.gca.2012.10.013

R.C. Greenwood, J.A. Barrat, M.F. Miller, M. Anand, N. Dauphas, I.A. Franchi, P. Sillard, N.A. Starkey, Oxygen isotopic evidence for accretion of Earth's water before a high-energy Moon-forming giant impact. Sci. Adv. **4**(3), eaao5928 (2018). https://doi.org/10.1126/sciadv.aao5928

J. Gross, A.H. Treiman, C.N. Mercer, Lunar feldspathic meteorites: constraints on the geology of the lunar highlands, and the origin of the lunar crust. Earth Planet. Sci. Lett. **388**, 318–328 (2014)

J.N. Grossman, The Meteoritical Bulletin, No. 83, 1999 July. Meteorit. Planet. Sci. **34**(S4), A169–A186 (1999). https://doi.org/10.1111/j.1945-5100.1999.tb01762.x

J.N. Grossman, J. Zipfel, The Meteoritical Bulletin, No. 85, 2001 September. Meteorit. Planet. Sci. **36**(S9), A293–A322 (2001). https://doi.org/10.1111/j.1945-5100.2001.tb01542.x

J.N. Grossman, R.N. Clayton, T.K. Mayeda, Oxygen isotopes in the matrix of the Semarkona (LL3.0) chondrite. Meteoritics **22**, 395 (1987)

M. Guitreau, M. Boyet, J.L. Paquette, A. Gannoun, Z. Konc, M. Benbakkar, K. Suchorski, J.M. Hénot, Hadean protocrust reworking at the origin of the Archean Napier Complex (Antarctica). Geochem. Perspect. Lett. **12**, 7–11 (2019). https://doi.org/10.7185/geochemlet.1927

J. Halbout, M. Javoy, F. Robert, Oxygen isotopes in type 3 ordinary chondrites, in *Lunar and Planetary Science Conference Abstracts*, vol. 15 (1984), pp. 339–340

J. Halbout, F. Robert, M. Javoy, Oxygen and hydrogen isotope relations in water and acid residues of carbonaceous chondrites. Geochim. Cosmochim. Acta **50**(8), 1599–1609 (1986). https://doi.org/10.1016/0016-7037(86)90123-7

C.L. Harper, S.B. Jacobsen, Evidence from coupled ^{147}Sm-^{143}Nd and ^{146}Sm-^{142}Nd systematics for very early (4.5 Gyr) differentiation of the Earth's mantle. Nature **360**(6406), 728–732 (1992). https://doi.org/10.1038/360728a0

W.K. Hartmann, D.R. Davis, Satellite-sized planetesimals and lunar origin. Icarus **24**(4), 504–515 (1975). https://doi.org/10.1016/0019-1035(75)90070-6

E.H. Hauri, A.E. Saal, M.J. Rutherford, J.A. Van Orman, Water in the Moon's interior: truth and consequences. Earth Planet. Sci. Lett. **409**, 252–264 (2015). https://doi.org/10.1016/j.epsl.2014.10.053

G.F. Herzog, F. Moynier, F. Albarède, A.A. Berezhnoy, Isotopic and elemental abundances of copper and zinc in lunar samples, Zagami, Pele's hairs, and a terrestrial basalt. Geochim. Cosmochim. Acta **73**(19), 5884–5904 (2009). https://doi.org/10.1016/j.gca.2009.05.067

M.F. Horan, R.W. Carlson, R.J. Walker, M. Jackson, M. Garçon, M. Norman, Tracking Hadean processes in modern basalts with 142-Neodymium. Earth Planet. Sci. Lett. **484**, 184–191 (2018). https://doi.org/10.1016/j.epsl.2017.12.017

N. Hosono, Si. Karato, J. Makino, T.R. Saitoh, Terrestrial magma ocean origin of the Moon. Nat. Geosci. **12**(6), 418–423 (2019). https://doi.org/10.1038/s41561-019-0354-2

M. Humayun, A model for osmium isotopic evolution of metallic solids at the core-mantle boundary. Geochem. Geophys. Geosyst. **Q03**, 007 (2011). https://doi.org/10.1029/2010GC003281

S. Ida, R.M. Canup, G.R. Stewart, Lunar accretion from an impact-generated disk. Nature **389**(6649), 353–357 (1997). https://doi.org/10.1038/38669

A.V. Ivanov, A.A. Ulyanov, V.I. Ustinov, Y.A. Shukolyukov, The Kaidun meteorite: oxygen isotopic composition, in *Lunar and Planetary Science Conference Abstracts*, vol. 18 (1987), p. 453

I. Jabeen, M. Kusakabe, K. Nagao, T. Nakamura, Oxygen isotope study of Tsukuba chondrite, some HED meteorites and Allende chondrules. Antarct. Meteor. Res. **11**, 122–135 (1998)

A. Jackel, A. Bischoff, R.N. Clayton, T.K. Mayeda, Dar AL Gani 013—a new Saharan Rumuruti-chondrite (R3-6) with highly unequilibrated (type 3) fragments, in *Lunar and Planetary Science Conference Abstracts*, vol. 27 (1996), p. 595

C.R. Jackson, N.R. Bennett, Z. Du, E. Cottrell, Y. Fei, Early episodes of high-pressure core formation preserved in plume mantle. Nature **553**(7689), 491–495 (2018). https://doi.org/10.1038/nature25446

S.B. Jacobsen, M.I. Petaev, S. Huang, D.D. Sasselov, An isotopically homogeneous region of the inner terrestrial planet region (Mercury to Earth): evidence from E chondrites and implications for giant Moon-forming impact scenarios, in *Lunar and Planetary Science Conference Abstracts*, vol. 44 (2013), p. 2344

S.A. Jacobson, A. Morbidelli, S.N. Raymond, D.P. O'Brien, K.J. Walsh, D.C. Rubie, Highly siderophile elements in Earth's mantle as a clock for the Moon-forming impact. Nature **508**(1), 84–87 (2014). https://doi.org/10.1038/nature13172

Z. Jing, Y. Wang, Y. Kono, T. Yu, T. Sakamaki, C. Park, M.L. Rivers, S.R. Sutton, G. Shen, Sound velocity of Fe-S liquids at high pressure: implications for the Moon's molten outer core. Earth Planet. Sci. Lett. **396**, 78–87 (2014). https://doi.org/10.1016/j.epsl.2014.04.015

T.D. Jones, D.R. Davies, P.A. Sossi, Tungsten isotopes in mantle plumes: heads it's positive, tails it's negative. Earth Planet. Sci. Lett. **506**, 255–267 (2019). https://doi.org/10.1016/j.epsl.2018.11.008

G.W. Kallemeyn, A.E. Rubin, J.T. Wasson, The compositional classification of chondrites: VII. The R chondrite group. Geochim. Cosmochim. Acta **60**(12), 2243–2256 (1996). https://doi.org/10.1016/0016-7037(96)88430-4

Y. Kaspi, G.R. Flierl, A.P. Showman, The deep wind structure of the giant planets: results from an anelastic general circulation model. Icarus **202**(2), 525–542 (2009). https://doi.org/10.1016/j.icarus.2009.03.026

C. Kato, F. Moynier, Gallium isotopic evidence for extensive volatile loss from the Moon during its formation. Sci. Adv. **3**(7), e1700,571 (2017). https://doi.org/10.1126/sciadv.1700571

C. Kato, F. Moynier, M.C. Valdes, J.K. Dhaliwal, J.M.D. Day, Extensive volatile loss during formation and differentiation of the Moon. Nat. Commun. **6**(1), 7617 (2015). https://doi.org/10.1038/ncomms8617

L.P. Keller, K.L. Thomas, R.N. Clayton, T.K. Mayeda, J.M. DeHart, D.S. McKay, Aqueous alteration of the Bali CV3 chondrite: evidence from mineralogy, mineral chemistry, and oxygen isotopic compositions. Geochim. Cosmochim. Acta **58**(24), 5589–5598 (1994). https://doi.org/10.1016/0016-7037(94)90252-6

K. Kimura, R.S. Lewis, E. Anders, Distribution of gold and rhenium between nickel-iron and silicate melts: implications for the abundance of siderophile elements on the Earth and Moon. Geochim. Cosmochim. Acta **38**(5), 683–701 (1974). https://doi.org/10.1016/0016-7037(74)90144-6

T. Kleine, R.J. Walker, Tungsten isotopes in planets. Annu. Rev. Earth Planet. Sci. **45**(1), 389–417 (2017). https://doi.org/10.1146/annurev-earth-063016-020037

T. Kleine, M. Touboul, B. Bourdon, F. Nimmo, K. Mezger, H. Palme, S.B. Jacobsen, Q.Z. Yin, A.N. Halliday, Hf–W chronology of the accretion and early evolution of asteroids and terrestrial planets. Geochim. Cosmochim. Acta **73**(17), 5150–5188 (2009). https://doi.org/10.1016/j.gca.2008.11.047

E. Kokubo, H. Genda, Formation of terrestrial planets from protoplanets under a realistic accretion condition. Astrophys. J. **714**(1), L21–L25 (2010). https://doi.org/10.1088/2041-8205/714/1/L21

E. Kokubo, S. Ida, J. Makino, Evolution of a circumterrestrial disk and formation of a single moon. Icarus **148**(2), 419–436 (2000). https://doi.org/10.1006/icar.2000.6496

T.S. Kruijer, T. Kleine, Tungsten isotopes and the origin of the Moon. Earth Planet. Sci. Lett. **475**, 15–24 (2017). https://doi.org/10.1016/j.epsl.2017.07.021

T.S. Kruijer, T. Kleine, No ^{182}W excess in the Ontong Java Plateau source. Chem. Geol. **485**, 24–31 (2018). https://doi.org/10.1016/j.chemgeo.2018.03.024

T.S. Kruijer, T. Kleine, M. Fischer-Gödde, P. Sprung, Lunar tungsten isotopic evidence for the late veneer. Nature **520**, 534–537 (2015). https://doi.org/10.1038/nature14360

M.D. Kurz, W.J. Jenkins, S.R. Hart, Helium isotopic systematics of oceanic islands and mantle heterogeneity. Nature **297**(5861), 43–47 (1982). https://doi.org/10.1038/297043a0

K.K. Larsen, A. Trinquier, C. Paton, M. Schiller, D. Wielandt, M.A. Ivanova, J.N. Connelly, Å. Nordlund, A.N. Krot, M. Bizzarro, Evidence for magnesium isotope heterogeneity in the solar protoplanetary disk. Astrophys. J. Lett. **735**(2), L37 (2011). https://doi.org/10.1088/2041-8205/735/2/L37

C. Li, D. Liu, B. Liu, X. Ren, J. Liu, Z. He, W. Zuo, X. Zeng, R. Xu, X. Tan, X. Zhang, W. Chen, R. Shu, W. Wen, Y. Su, H. Zhang, Z. Ouyang, Chang'E-4 initial spectroscopic identification of lunar far-side mantle-derived materials. Nature **569**, 378–382 (2019). https://doi.org/10.1038/s41586-019-1189-0

J. Liu, M. Touboul, A. Ishikawa, R.J. Walker, D. Graham Pearson, Widespread tungsten isotope anomalies and W mobility in crustal and mantle rocks of the Eoarchean Saglek Block, northern Labrador, Canada: implications for early Earth processes and W recycling. Earth Planet. Sci. Lett. **448**, 13–23 (2016). https://doi.org/10.1016/j.epsl.2016.05.001

S.J. Lock, S.T. Stewart, The structure of terrestrial bodies: impact heating, corotation limits, and synestias. J. Geophys. Res., Planets **122**(5), 950–982 (2017). https://doi.org/10.1002/2016JE005239

S.J. Lock, S.T. Stewart, Giant impacts stochastically change the internal pressures of terrestrial planets. Sci. Adv. **5**(9), eaav3746 (2019). https://doi.org/10.1126/sciadv.aav3746

S.J. Lock, S.T. Stewart, M.I. Petaev, Z. Leinhardt, M.T. Mace, S.B. Jacobsen, M. Ćuk, The origin of the moon within a terrestrial synestia. J. Geophys. Res., Planets **123**(4), 910–951 (2018). https://doi.org/10.1002/2017JE005333

S.J. Lock, S.T. Stewart, M. Ćuk, The energy budget and figure of Earth during recovery from the Moon-forming giant impact. Earth Planet. Sci. Lett. **530**, 115,885 (2020). https://doi.org/10.1016/j.epsl.2019.115885

J. Longhi, Petrogenesis of picritic mare magmas: constraints on the extent of early lunar differentiation. Geochim. Cosmochim. Acta **70**(24), 5919–5934 (2006). https://doi.org/10.1016/j.gca.2006.09.023

G.W. Lugmair, R.W. Carlson, The Sm-Nd history of KREEP, in *Lunar and Planetary Science Conference Abstracts*, vol. 9 (1978), pp. 689–704

G. Lugmair, A. Shukolyukov, Early solar system timescales according to ^{53}Mn-^{53}Cr systematics. Geochim. Cosmochim. Acta **62**(16), 2863–2886 (1998). https://doi.org/10.1016/S0016-7037(98)00189-6

C. Maas, U. Hansen, Dynamics of a terrestrial magma ocean under planetary rotation: a study in spherical geometry. Earth Planet. Sci. Lett. **513**, 81–94 (2019). https://doi.org/10.1016/j.epsl.2019.02.016

S. Marchi, R.M. Canup, R.J. Walker, Heterogeneous delivery of silicate and metal to the Earth by large planetesimals. Nat. Geosci. **11**(1), 77–81 (2018). https://doi.org/10.1038/s41561-017-0022-3

N.E. Marks, L.E. Borg, C.K. Shearer, W.S. Cassata, Geochronology of an Apollo 16 clast provides evidence for a basin-forming impact 4.3 billion years ago. J. Geophys. Res., Planets **124**(10), 2465–2481 (2019). https://doi.org/10.1029/2019JE005966

T.K. Mayeda, R.N. Clayton, Oxygen isotopic compositions of Aubrites and some unique meteorites. Geochim. Cosmochim. Acta **14**(2), 1145–1151 (1980)

T.K. Mayeda, R.N. Clayton, Oxygen isotopic composition of ALHA 81005, in *Lunar and Planetary Science Conference Abstracts*, vol. 14 (1983), p. 472

K.T. Mayeda, N.R. Clayton, Oxygen isotopic compositions of unique Antarctic meteorites. Proc. NIPR Symp. Antarct. Meteor. **14**, 172 (1989a)

T.K. Mayeda, R.N. Clayton, Oxygen isotopes in the Bholghati Howardite, in *Lunar and Planetary Science Conference Abstracts*, vol. 20 (1989b), p. 648

T. Mayeda, R. Clayton, K. Yanai, Oxygen isotopic compositions of several Antarctic meteorites. Mem. Natl. Inst. Polar Res., Spec. Issue **46**, 144–150 (1987)

T.K. Mayeda, K. Yanai, R.N. Clayton, Another martian meteorite, in *Lunar and Planetary Science Conference Abstracts*, vol. 26 (1995), pp. 917–918

T.J. McCoy, K. Keil, D.D. Bogard, D.H. Garrison, I. Casanova, M.M. Lindstrom, A.J. Brearley, K. Kehm, R.H. Nichols, C.M. Hohenberg, Origin and history of impact-melt rocks of enstatite chondrite parentage. Geochim. Cosmochim. Acta **59**(1), 161–175 (1995). https://doi.org/10.1016/0016-7037(94)00231-A

T.J. McCoy, K. Keil, R.N. Clayton, T.K. Mayeda, D.D. Bogard, D.H. Garrison, G.R. Huss, I.D. Hutcheon, R. Wieler, A petrologic, chemical, and isotopic study of Monument Draw and comparison with other acapulcoites: evidence for formation by incipient partial melting. Geochim. Cosmochim. Acta **60**(14), 2681–2708 (1996). https://doi.org/10.1016/0016-7037(96)00109-3

T.J. McCoy, K. Keil, R.N. Clayton, T.K. Mayeda, D.D. Bogard, D.H. Garrison, R. Wieler, A petrologic and isotopic study of lodranites: evidence for early formation as partial melt residues from heterogeneous precursors. Geochim. Cosmochim. Acta **61**(3), 623–637 (1997). https://doi.org/10.1016/S0016-7037(96)00359-6

W. McDonough, S. Sun, The composition of the Earth. Chem. Geol. **120**(3–4), 223–253 (1995). https://doi.org/10.1016/0009-2541(94)00140-4

C.L. McLeod, A.D. Brandon, R.M.G. Armytage, Constraints on the formation age and evolution of the Moon from ^{142}Nd-^{143}Nd systematics of Apollo 12 basalts. Earth Planet. Sci. Lett. **396**, 179–189 (2014). https://doi.org/10.1016/j.epsl.2014.04.007

F. Meissner, W.D. Schmidt-Ott, L. Ziegeler, Half-life and α-ray energy of ^{146}Sm. Eur. Phys. J. A **327**(2), 171–174 (1987). https://doi.org/10.1007/BF01292406

H.J. Melosh, New approaches to the Moon's isotopic crisis. Philos. Trans. R. Soc. A, Math. Phys. Eng. Sci. **372**(2024), 20130 (2014). https://doi.org/10.1098/rsta.2013.0168. 168

Y. Miyazaki, J. Korenaga, On the timescale of magma ocean solidification and its chemical consequences: 2. Compositional differentiation under crystal accumulation and matrix compaction. J. Geophys. Res., Solid Earth **124**(4), 3399–3419 (2019). https://doi.org/10.1029/2018JB016928

A. Morbidelli, B. Wood, Late accretion and the late veneer, in *The Early Earth: Accretion and Differentiation*, ed. by J. Badro, M. Walter (John Wiley & Sons, Inc., Hoboken, 2014). Chap. 4. https://doi.org/10.1002/9781118860359.ch4

J.W. Morgan, Osmium isotope constraints on Earth's late accretionary history. Nature **317**(6039), 703–705 (1985). https://doi.org/10.1038/317703a0

J.W. Morgan, J. Hertogen, E. Anders, The Moon: composition determined by nebular processes. Moon Planets **18**(4), 465–478 (1978). https://doi.org/10.1007/BF00897296

L.V. Moroz, V.I. Ustinov, N.N. Kononkova, N.I. Zaslavskaya, Y.A. Shukolyukov, Oxygen isotopes of chromite and chemical composition of the minerals from polymineral nodules in Sikhote-Alin meteorite, in *Lunar and Planetary Science Conference Abstracts*, vol. 19 (1988), pp. 809–810

B. Mougel, F. Moynier, C. Göpel, Chromium isotopic homogeneity between the Moon, the Earth, and enstatite chondrites. Earth Planet. Sci. Lett. **481**, 1–8 (2018). https://doi.org/10.1016/j.epsl.2017.10.018

S. Mukhopadhyay, Early differentiation and volatile accretion recorded in deep-mantle neon and xenon. Nature **486**(7401), 101–104 (2012). https://doi.org/10.1038/nature11141

S. Mukhopadhyay, R. Parai, Noble gases: a record of Earth's evolution and mantle dynamics. Annu. Rev. Earth Planet. Sci. **47**(1), 389–419 (2019). https://doi.org/10.1146/annurev-earth-053018-060238

A. Mundl, M. Touboul, M.G. Jackson, J.M.D. Day, M.D. Kurz, V. Lekic, R.T. Helz, R.J. Walker, Tungsten-182 heterogeneity in modern ocean island basalts. Science **356**(6333), 66–69 (2017). https://doi.org/10.1126/science.aal4179

A. Mundl, R.J. Walker, J.R. Reimink, R.L. Rudnick, R.M. Gaschnig, Tungsten-182 in the upper continental crust: evidence from glacial diamictites. Chem. Geol. **494**, 144–152 (2018). https://doi.org/10.1016/j.chemgeo.2018.07.036

A. Mundl-Petermeier, R.J. Walker, R.A. Fischer, V. Lekic, M.G. Jackson, M.D. Kurz, Anomalous ^{182}W in high $^{3}He/^{4}He$ ocean island basalts: fingerprints of Earth's core? Geochim. Cosmochim. Acta **271**, 194–211 (2020). https://doi.org/10.1016/j.gca.2019.12.020

M. Nakajima, D.J. Stevenson, Investigation of the initial state of the Moon-forming disk: bridging SPH simulations and hydrostatic models. Icarus **233**, 259–267 (2014). https://doi.org/10.1016/j.icarus.2014.01.008

M. Nakajima, D.J. Stevenson, Melting and mixing states of the Earth's mantle after the Moon-forming impact. Earth Planet. Sci. Lett. **427**, 286–295 (2015). https://doi.org/10.1016/j.epsl.2015.06.023

C.E. Nehru, M. Prinz, M.K. Weisberg, M.E. Ebihara, R.N. Clayton, T.K. Mayeda, A new Brachinite and petrogenesis of the group, in *Lunar and Planetary Science Conference Abstracts*, vol. 27 (1996), pp. 943–944

A. Nemchin, N. Timms, R. Pidgeon, T. Geisler, S. Reddy, C. Meyer, Timing of crystallization of the lunar magma ocean constrained by the oldest zircon. Nat. Geosci. **2**(2), 133–136 (2009). https://doi.org/10.1038/ngeo417

P. Ni, Y. Zhang, S. Chen, J. Gagnon, A melt inclusion study on volatile abundances in the lunar mantle. Geochim. Cosmochim. Acta **249**, 17–41 (2019). https://doi.org/10.1016/j.gca.2018.12.034

N.X. Nie, N. Dauphas, Vapor drainage in the protolunar disk as the cause for the depletion in volatile elements of the Moon. Astrophys. J. **884**(2), L48 (2019). https://doi.org/10.3847/2041-8213/ab4a16

M.D. Norman, L.E. Borg, L.E. Nyquist, D.D. Bogard, Chronology, geochemistry, and petrology of a ferroan noritic anorthosite clast from Descartes breccia 67215: clues to the age, origin, structure, and impact history of the lunar crust. Meteorit. Planet. Sci. **38**(4), 645–661 (2003). https://doi.org/10.1111/j.1945-5100.2003.tb00031.x

L.E. Nyquist, H. Wiesmann, B. Bansal, C.Y. Shih, J.E. Keith, C.L. Harper, ^{146}Sm-^{142}Nd formation interval for the lunar mantle. Geochim. Cosmochim. Acta **59**(13), 2817–2837 (1995). https://doi.org/10.1016/0016-7037(95)00175-Y

L.E. Nyquist, C.Y. Shih, Y.D. Reese, J. Park, D.D. Bogard, D.H. Garrison, A. Yamaguchi, Lunar crustal history recorded in lunar anorthosites, in *Lunar and Planetary Science Conference*, vol. 41 (2010), p. 1383

E.J. Olsen, B.D. Dod, R.A. Schmitt, P.P. Sipiera, Monticello: a class-rich Howardite. Meteoritics **22**(1), 81–96 (1987). https://doi.org/10.1111/j.1945-5100.1987.tb00885.x

E. Olsen, A. Davis, R.S. Clarke, L. Schultz, H.W. Weber, R. Clayton, T. Mayeda, E. Jarosewich, P. Sylvester, L. Grossman, M.S. Wang, M.E. Lipschutz, I.M. Steele, J. Schwade, Watson: a new link in the HE iron chain. Meteoritics **29**(2), 200–213 (1994). https://doi.org/10.1111/j.1945-5100.1994.tb00672.x

J. O'Neil, R.W. Carlson, D. Francis, R.K. Stevenson, C.M. Deibel, B. DiGiovine, J.P. Greene, D.J. Henderson, C.L. Jiang, S.T. Marley, T. Nakanishi, R.C. Pardo, K.E. Rehm, D. Robertson, R. Scott, C. Schmitt, X.D. Tang, R. Vondrasek, A. Yokoyama, Neodymium-142 evidence for Hadean mafic crust. Science **321**(5897), 1828–1831 (2008). https://doi.org/10.1126/science.1161925

N. Onuma, R. Clayton, T. Mayeda, K. Yanai, Oxygen isotopes in several Yamato meteorites. Mem. Natl. Inst. Polar Res., Spec. Issue **8**, 220–224 (1978)

N. Onuma, Y. Ikeda, T.K. Mayeda, R.N. Clayton, K. Yanai, Oxygen isotopic composition of photographically described inclusions from Antarctic unequilibrated ordinary chondrites, in *Memoirs of National Institute of Polar Research. Special Issue*, vol. 30 (1983), pp. 306–314

K. Pahlevan, D.J. Stevenson, Equilibration in the aftermath of the lunar-forming giant impact. Earth Planet. Sci. Lett. **262**(3–4), 438–449 (2007). https://doi.org/10.1016/j.epsl.2007.07.055

R.C. Paniello, J.M.D. Day, F. Moynier, Zinc isotopic evidence for the origin of the Moon. Nature **490**(7420), 376–379 (2012). https://doi.org/10.1038/nature11507

R. Parai, S. Mukhopadhyay, The evolution of MORB and plume mantle volatile budgets: constraints from fission Xe isotopes in Southwest Indian Ridge basalts. Geochem. Geophys. Geosyst. **16**(3), 719–735 (2015). https://doi.org/10.1002/2014GC005566

R. Parai, S. Mukhopadhyay, J. Standish, Heterogeneous upper mantle Ne, Ar and Xe isotopic compositions and a possible Dupal noble gas signature recorded in basalts from the Southwest Indian Ridge. Earth Planet. Sci. Lett. **359–360**, 227–239 (2012). https://doi.org/10.1016/j.epsl.2012.10.017

R. Parai, S. Mukhopadhyay, J.M. Tucker, M.K. Pető, The emerging portrait of an ancient, heterogeneous and continuously evolving mantle plume source. Lithos **346–347**, 105,153 (2019). https://doi.org/10.1016/j.lithos.2019.105153

M.I. Petaev, V.I. Ustinov, N.I. Zaslavkaya, E.Y. Gavrilov, A. Shukolyukov, Oxygen isotopes in Pomozdino meteorite, in *Lunar and Planetary Science Conference Abstracts*, vol. 19 (1988), p. 917

B.J. Peters, R.W. Carlson, J.M. Day, M.F. Horan, Hadean silicate differentiation preserved by anomalous ^{142}Nd/^{144}Nd ratios in the Réunion hotspot source. Nature **555**(7694), 89–93 (2018). https://doi.org/10.1038/nature25754

M.K. Pető, S. Mukhopadhyay, K.A. Kelley, Heterogeneities from the first 100 million years recorded in deep mantle noble gases from the Northern Lau Back-arc Basin. Earth Planet. Sci. Lett. **369–370**, 13–23 (2013). https://doi.org/10.1016/j.epsl.2013.02.012

D. Porcelli, D. Woolum, P. Cassen, Deep Earth rare gases: initial inventories, capture from the solar nebula, and losses during moon formation. Earth Planet. Sci. Lett. **193**(1–2), 237–251 (2001). https://doi.org/10.1016/S0012-821X(01)00493-9

N.J. Potts, J.J. Barnes, R. Tartèse, I.A. Franchi, M. Anand, Chlorine isotopic compositions of apatite in Apollo 14 rocks: evidence for widespread vapor-phase metasomatism on the lunar nearside $\sim$ 4 billion years ago. Geochim. Cosmochim. Acta **230**, 46–59 (2018). https://doi.org/10.1016/j.gca.2018.03.022

E.A. Pringle, F. Moynier, Rubidium isotopic composition of the Earth, meteorites, and the Moon: evidence for the origin of volatile loss during planetary accretion. Earth Planet. Sci. Lett. **473**, 62–70 (2017). https://doi.org/10.1016/j.epsl.2017.05.033

M. Prinz, N. Chatterjee, M.K. Weisberg, R.N. Clayton, T.K. Mayeda, MAC88177: a new type of achondrite? in *Lunar and Planetary Science Conference Abstracts*, vol. 22 (1991), p. 1099

I. Puchtel, M. Humayun, Platinum group elements in Kostomuksha komatiites and basalts: implications for oceanic crust recycling and core-mantle interaction. Geochim. Cosmochim. Acta **64**(24), 4227–4242 (2000). https://doi.org/10.1016/S0016-7037(00)00492-0

I.S. Puchtel, J. Blichert-Toft, M. Touboul, M.F. Horan, R.J. Walker, The coupled ^{182}W-^{142}Nd record of early terrestrial mantle differentiation. Geochem. Geophys. Geosyst. **17**(6), 2168–2193 (2016). https://doi.org/10.1002/2016GC006324

I.S. Puchtel, J. Blichert-Toft, M. Touboul, R.J. Walker, ^{182}W and HSE constraints from 2.7 Ga komatiites on the heterogeneous nature of the Archean mantle. Geochim. Cosmochim. Acta **228**, 1–26 (2018). https://doi.org/10.1016/j.gca.2018.02.030

A. Pun, K. Keil, G.J. Taylor, E. King, A unique Eucrite clast from the Kapoeta Howardite, in *Lunar and Planetary Science Conference Abstracts*, vol. 22 (1991), p. 1105

L. Qin, C.M. Alexander, R.W. Carlson, M.F. Horan, T. Yokoyama, Contributors to chromium isotope variation of meteorites. Geochim. Cosmochim. Acta **74**(3), 1122–1145 (2010a). https://doi.org/10.1016/j.gca.2009.11.005

L. Qin, D. Rumble, C.M. Alexander, R.W. Carlson, P. Jenniskens, M.H. Shaddad, The chromium isotopic composition of Almahata Sitta. Meteorit. Planet. Sci. **45**(10–11), 1771–1777 (2010b). https://doi.org/10.1111/j.1945-5100.2010.01109.x

E.V. Quintana, T. Barclay, W.J. Borucki, J.F. Rowe, J.E. Chambers, The frequency of giant impacts on Earth-like worlds. Astrophys. J. **821**(2), 126 (2016). https://doi.org/10.3847/0004-637X/821/2/126

S. Recca, E. Scott, K. Keil, R. Clayton, T. Mayeda, G. Huss, E. Jarosewich, K. Weeks, F. Hasan, D. Sears, R. Wieler, P. Signer, Ragland, an LL3.4 chondrite find from New Mexico. Meteoritics **21**(2), 217–229 (1986). https://doi.org/10.1111/j.1945-5100.1986.tb01243.x

J.R. Reimink, T. Chacko, R.W. Carlson, S.B. Shirey, J. Liu, R.A. Stern, A.M. Bauer, D.G. Pearson, L.M. Heaman, Petrogenesis and tectonics of the Acasta Gneiss Complex derived from integrated petrology and ^{142}Nd and ^{182}W extinct nuclide-geochemistry. Earth Planet. Sci. Lett. **494**, 12–22 (2018). https://doi.org/10.1016/j.epsl.2018.04.047

A. Reufer, M.M. Meier, W. Benz, R. Wieler, A hit-and-run giant impact scenario. Icarus **221**(1), 296–299 (2012). https://doi.org/10.1016/j.icarus.2012.07.021

A.E. Ringwood, *Origin of the Earth and Moon* (Springer, New York, 1979)

A.E. Ringwood, Terrestrial origin of the Moon. Nature **322**(6077), 323–328 (1986). https://doi.org/10.1038/322323a0

A.E. Ringwood, S.E. Kesson, Basaltic magmatism and the bulk composition of the Moon. Moon **16**(4), 425–464 (1977). https://doi.org/10.1007/BF00577902

A. Ringwood, S. Seifert, H. Wänke, A komatiite component in Apollo 16 highland breccias: implications for the nickel-cobalt systematics and bulk composition of the moon. Earth Planet. Sci. Lett. **81**(2–3), 105–117 (1987). https://doi.org/10.1016/0012-821X(87)90149-X
H. Rizo, M. Boyet, J. Blichert-Toft, M. Rosing, Combined Nd and Hf isotope evidence for deep-seated source of Isua lavas. Earth Planet. Sci. Lett. **312**(3–4), 267–279 (2011). https://doi.org/10.1016/j.epsl.2011.10.014
H. Rizo, R. Walker, R. Carlson, M. Touboul, M. Horan, I. Puchtel, M. Boyet, M. Rosing, Early Earth differentiation investigated through ^{142}Nd, ^{182}W, and highly siderophile element abundances in samples from Isua, Greenland. Geochim. Cosmochim. Acta **175**, 319–336 (2016a). https://doi.org/10.1016/j.gca.2015.12.007
H. Rizo, R.J. Walker, R.W. Carlson, M.F. Horan, S. Mukhopadhyay, V. Manthos, D. Francis, M.G. Jackson, Preservation of Earth-forming events in the tungsten isotopic composition of modern flood basalts. Science **352**(6287), 809–812 (2016b). https://doi.org/10.1126/science.aad8563
H. Rizo, D. Andrault, N.R. Bennett, M. Humayun, A. Brandon, I. Vlastelic, B. Moine, A. Poirier, M.A. Bouhifd, D.T. Murphy, ^{182}W evidence for core-mantle interaction in the source of mantle plumes. Geochem. Perspect. Lett. **11**, 6–11 (2019). https://doi.org/10.7185/geochemlet.1917
C.S. Romanek, E.C. Perry, A.H. Treiman, R.A. Socki, J.H. Jones, E.K. Gibson, Oxygen isotopic record of silicate alteration in the Shergotty-Nakhla-Chassigny meteorite Lafayette. Meteorit. Planet. Sci. **33**(4), 775–784 (1998). https://doi.org/10.1111/j.1945-5100.1998.tb01683.x
M.W. Rowe, R.N. Clayton, T.K. Mayeda, Oxygen isotopes in separated components of CI and CM meteorites. Geochim. Cosmochim. Acta **58**(23), 5341–5347 (1994). https://doi.org/10.1016/0016-7037(94)90317-4
R. Rufu, O. Aharonson, Impact dynamics of moons within a planetary potential. J. Geophys. Res., Planets **124**(4), 1008–1019 (2019). https://doi.org/10.1029/2018JE005798
R. Rufu, O. Aharonson, H.B. Perets, A multiple-impact origin for the Moon. Nat. Geosci. **10**, 89–94 (2017). https://doi.org/10.1038/ngeo2866
S.S. Russell, T.J. McCoy, E. Jarosewich, R.D. Ash, The Burnwell, Kentucky, low iron oxide chondrite fall: description, classification and origin. Meteorit. Planet. Sci. **33**(4), 853–856 (1998). https://doi.org/10.1111/j.1945-5100.1998.tb01691.x
S.S. Russell, M.E. Zolensky, K. Righter, L. Folco, R.H. Jones, H.C. Connolly, M.M. Grady, J.N. Grossman, The Meteoritical Bulletin, No. 89, 2005 September. Meteorit. Planet. Sci. **40**(S9), A201–A263 (2005). https://doi.org/10.1111/j.1945-5100.2005.tb00425.x
A. Ruzicka, D.A. Kring, D.H. Hill, W.V. Boynton, R.N. Clayton, T.K. Mayeda, Silica-rich orthopyroxenite in the Bovedy chondrite. Meteoritics **30**(1), 57–70 (1995). https://doi.org/10.1111/j.1945-5100.1995.tb01212.x
J. Salmon, R.M. Canup, Lunar accretion from a Roche-interior fluid disk. Astrophys. J. **760**(1), 83 (2012). https://doi.org/10.1088/0004-637X/760/1/83
J. Salmon, R.M. Canup, Accretion of the Moon from non-canonical discs. Philos. Trans. R. Soc. A, Math. Phys. Eng. Sci. **372**(2024), 20130,256 (2014). https://doi.org/10.1098/rsta.2013.0256
P. Saxena, L. Elkins-Tanton, N. Petro, A. Mandell, A model of the primordial lunar atmosphere. Earth Planet. Sci. Lett. **474**, 198–205 (2017). https://doi.org/10.1016/j.epsl.2017.06.031
M. Schiller, E. Van Kooten, J.C. Holst, M.B. Olsen, M. Bizzarro, Precise measurement of chromium isotopes by MC-ICPMS. J. Anal. At. Spectrom. **29**(8), 1406–1416 (2014). https://doi.org/10.1039/c4ja00018h
R. Schoenberg, S. Zink, M. Staubwasser, F. von Blanckenburg, The stable Cr isotope inventory of solid Earth reservoirs determined by double spike MC-ICP-MS. Chem. Geol. **249**(3–4), 294–306 (2008). https://doi.org/10.1016/J.CHEMGEO.2008.01.009
H. Schulze, A. Bischoff, H. Palme, B. Spettel, G. Dreibus, J. Otto, Mineralogy and chemistry of Rumuruti: the first meteorite fall of the new R chondrite group. Meteoritics **29**(2), 275–286 (1994). https://doi.org/10.1111/j.1945-5100.1994.tb00681.x
D.W.G. Sears, J.D. Batchelor, B. Mason, E.R.D. Scott, R.N. Clayton, T.K. Mayeda, South Australian type 3 chondrites. Meteoritics **25**(4), 407–408 (1990). https://doi.org/10.1111/j.1945-5100.1990.tb00722.x
S. Seifert, A. E. Ringwood, The lunar geochemistry of chromium and vanadium. Earth Moon Planets **40**(1), 45–70 (1988). https://doi.org/10.1007/BF00057946
K. Shariff, Fluid mechanics in disks around young stars. Annu. Rev. Fluid Mech. **41**(1), 283–315 (2009). https://doi.org/10.1146/annurev.fluid.010908.165144
Z.D. Sharp, C.K. Shearer, K.D. McKeegan, J.D. Barnes, Y.Q. Wang, The chlorine isotope composition of the Moon and implications for an anhydrous mantle. Science **329**(5995), 1050–1053 (2010). https://doi.org/10.1126/science.1192606
Z.D. Sharp, J.A. Mercer, R.H. Jones, A.J. Brearley, J. Selverstone, A. Bekker, T. Stachel, The chlorine isotope composition of chondrites and Earth. Geochim. Cosmochim. Acta **107**, 189–204 (2013). https://doi.org/10.1016/j.gca.2013.01.003

C.K. Shearer, K. Righter, Behavior of tungsten and hafnium in silicates: a crystal chemical basis for understanding the early evolution of the terrestrial planets. Geophys. Res. Lett. **30**(1), 7-1–7-4 (2003). https://doi.org/10.1029/2002gl015523
A. Shukolyukov, G. Lugmair, Manganese–chromium isotope systematics of carbonaceous chondrites. Earth Planet. Sci. Lett. **250**(1–2), 200–213 (2006a). https://doi.org/10.1016/J.EPSL.2006.07.036
A. Shukolyukov, G.W. Lugmair, The Mn-Cr isotope systematics in the ureilites Kenna and LEW85440, in *Lunar and Planetary Science Conference Abstracts*, vol. 37 (2006b), p. 1478
A. Shukolyukov, G.W. Lugmair, A.J. Irving, Mn-Cr isotope systematics of angrite Northwest Africa 4801, in *Lunar and Planetary Science Conference Abstracts*, vol. 40 (2009), p. 1381
S.B. Simon, L. Grossman, I. Casanova, S. Symes, P. Benoit, D.W.G. Sears, J.F. Wacker, Axtell, a new CV3 chondrite find from Texas. Meteoritics **30**(1), 42–46 (1995). https://doi.org/10.1111/j.1945-5100.1995.tb01210.x
C.K. Sio, L.E. Borg, W.S. Cassata, The timing of lunar solidification and mantle overturn recorded in ferroan anorthosite 62237. Earth Planet. Sci. Lett. **538**, 116,219 (2020). https://doi.org/10.1016/j.epsl.2020.116219
G.A. Snyder, C.R. Neal, L.A. Taylor, A.N. Halliday, Processes involved in the formation of magnesian-suite plutonic rocks from the highlands of the Earth's Moon. J. Geophys. Res. **100**(E5), 9365–9388 (1995). https://doi.org/10.1029/95JE00575
V.S. Solomatov, D.J. Stevenson, Kinetics of crystal growth in a terrestrial magma ocean. J. Geophys. Res. **98**(E3), 5407 (1993a). https://doi.org/10.1029/92JE02839
V.S. Solomatov, D.J. Stevenson, Nonfractional crystallization of a terrestrial magma ocean. J. Geophys. Res. **98**(E3), 5391–5406 (1993b). https://doi.org/10.1029/92JE02579
V.S. Solomatov, D.J. Stevenson, Suspension in convective layers and style of differentiation of a terrestrial magma ocean. J. Geophys. Res. **98**(E3), 5375 (1993c). https://doi.org/10.1029/92JE02948
P.A. Sossi, F. Moynier, Chemical and isotopic kinship of iron in the Earth and Moon deduced from the lunar Mg-Suite. Earth Planet. Sci. Lett. **471**, 125–135 (2017). https://doi.org/10.1016/j.epsl.2017.04.029
P.A. Sossi, F. Moynier, K. van Zuilen, Volatile loss following cooling and accretion of the Moon revealed by chromium isotopes. Proc. Natl. Acad. Sci. **115**(43), 10,920–10,925 (2018). https://doi.org/10.1073/pnas.1809060115
P.A. Sossi, S. Klemme, H.S.C. O'Neill, J. Berndt, F. Moynier, Evaporation of moderately volatile elements from silicate melts: experiments and theory. Geochim. Cosmochim. Acta **260**, 204–231 (2019). https://doi.org/10.1016/j.gca.2019.06.021
P. Sprung, T. Kleine, E.E. Scherer, Isotopic evidence for chondritic Lu/Hf and Sm/Nd of the Moon. Earth Planet. Sci. Lett. **380**, 77–87 (2013). https://doi.org/10.1016/j.epsl.2013.08.018
E.S. Steenstra, N. Rai, J.S. Knibbe, Y.H. Lin, W. van Westrenen, New geochemical models of core formation in the Moon from metal-silicate partitioning of 15 siderophile elements. Earth Planet. Sci. Lett. **441**, 1–9 (2016). https://doi.org/10.1016/j.epsl.2016.02.028
E.S. Steenstra, Y. Lin, D. Dankers, N. Rai, J. Berndt, S. Matveev, W. van Westrenen, The lunar core can be a major reservoir for volatile elements S, Se, Te and Sb. Sci. Rep. **14**, 552 (2017). https://doi.org/10.1038/s41598-017-15203-0
E.S. Steenstra, J. Berndt, S. Klemme, Y. Fei, W. van Westrenen, A possible high-temperature origin of the Moon and its geochemical consequences. Earth Planet. Sci. Lett. **538**, 116,222 (2020). https://doi.org/10.1016/j.epsl.2020.116222
A. Stephant, M. Anand, X. Zhao, Q.H. Chan, M. Bonifacie, I.A. Franchi, The chlorine isotopic composition of the Moon: insights from melt inclusions. Earth Planet. Sci. Lett. **523**, 115,715 (2019). https://doi.org/10.1016/j.epsl.2019.115715
M. Stepniewski, J. Borucki, J. Siemiatkowski, New data on the L5(S1) chondrite Baszkowka (Poland). Meteorit. Planet. Sci. **33**, A150–A151 (1998)
R. Tartèse, M. Anand, K.H. Joy, I.A. Franchi, H and Cl isotope systematics of apatite in brecciated lunar meteorites Northwest Africa 4472, Northwest Africa 773, Sayh al Uhaymir 169, and Kalahari 009. Meteorit. Planet. Sci. **49**(12), 2266–2289 (2014). https://doi.org/10.1111/maps.12398
S.R. Taylor, *Planetary Science: A Lunar Perspective* (Lunar and Planetary Institute, Houston, 1982)
S.R. Taylor, The Moon re-examined. Geochim. Cosmochim. Acta **141**, 670–676 (2014). https://doi.org/10.1016/j.gca.2014.06.031
S.R. Taylor, S. McLennan, *Planetary Crusts: Their Composition, Origin and Evolution* (Cambridge University Press, Cambridge, 2009)
G.J. Taylor, M.A. Wieczorek, Lunar bulk chemical composition: a post-Gravity Recovery and Interior Laboratory reassessment. Philos. Trans. R. Soc. A, Math. Phys. Eng. Sci. **372**(2024), 20130,242 (2014). https://doi.org/10.1098/rsta.2013.0242
D.J. Taylor, K.D. McKeegan, T.M. Harrison, Lu-Hf zircon evidence for rapid lunar differentiation. Earth Planet. Sci. Lett. **279**(3–4), 157–164 (2009). https://doi.org/10.1016/j.epsl.2008.12.030

M.M. Thiemens, P. Sprung, R.O. Fonseca, F.P. Leitzke, C. Münker, Early Moon formation inferred from hafnium–tungsten systematics. Nat. Geosci. **12**(9), 696–700 (2019). https://doi.org/10.1038/s41561-019-0398-3
C. Thompson, D.J. Stevenson, Gravitational instability in two-phase disks and the origin of the Moon. Astrophys. J. **333**, 452 (1988). https://doi.org/10.1086/166760
Z. Tian, J. Wisdom, L. Elkins-Tanton, Coupled orbital-thermal evolution of the early Earth-Moon system with a fast-spinning Earth. Icarus **281**, 90–102 (2017). https://doi.org/10.1016/j.icarus.2016.08.030
M. Touboul, I.S. Puchtel, R.J. Walker, ^{182}W evidence for long-term preservation of early mantle differentiation products. Science **335**(6072), 1065–1069 (2012). https://doi.org/10.1126/science.1216351
M. Touboul, J. Liu, J. O'Neil, I.S. Puchtel, R.J. Walker, New insights into the Hadean mantle revealed by ^{182}W and highly siderophile element abundances of supracrustal rocks from the Nuvvuagittuq Greenstone Belt, Quebec, Canada. Chem. Geol. **383**, 63–75 (2014). https://doi.org/10.1016/j.chemgeo.2014.05.030
M. Touboul, I.S. Puchtel, R.J. Walker, Tungsten isotopic evidence for disproportional late accretion to the Earth and Moon. Nature **520**, 530–533 (2015). https://doi.org/10.1038/nature14355
A.H. Treiman, J.W. Boyce, J. Gross, Y. Guan, J.M. Eiler, E.M. Stolper, Phosphate-halogen metasomatism of lunar granulite 79215: impact-induced fractionation of volatiles and incompatible elements. Am. Mineral. **99**(10), 1860–1870 (2014). https://doi.org/10.2138/am-2014-4822
A. Trinquier, J. Birck, C.J. Allegre, Widespread ^{54}Cr heterogeneity in the inner solar system. Astrophys. J. **655**(2), 1179–1185 (2007). https://doi.org/10.1086/510360
A. Trinquier, T. Elliott, D. Ulfbeck, C. Coath, A.N. Krot, M. Bizzarro, Origin of nucleosynthetic isotope heterogeneity in the solar protoplanetary disk. Science **324**(5925), 374–376 (2009). https://doi.org/10.1126/science.1168221
J.M. Tucker, S. Mukhopadhyay, Evidence for multiple magma ocean outgassing and atmospheric loss episodes from mantle noble gases. Earth Planet. Sci. Lett. **393**, 254–265 (2014). https://doi.org/10.1016/j.epsl.2014.02.050
J.M. Tucker, S. Mukhopadhyay, J.G. Schilling, The heavy noble gas composition of the depleted MORB mantle (DMM) and its implications for the preservation of heterogeneities in the mantle. Earth Planet. Sci. Lett. **355–356**, 244–254 (2012). https://doi.org/10.1016/j.epsl.2012.08.025
J.W. Valley, A.J. Cavosie, T. Ushikubo, D.A. Reinhard, D.F. Lawrence, D.J. Larson, P.H. Clifton, T.F. Kelly, S.A. Wilde, D.E. Moser, M.J. Spicuzza, Hadean age for a post-magma-ocean zircon confirmed by atom-probe tomography. Nat. Geosci. **7**(3), 219–223 (2014). https://doi.org/10.1038/ngeo2075
C. Vockenhuber, F. Oberli, M. Bichler, I. Ahmad, G. Quitté, M. Meier, A.N. Halliday, D.C. Lee, W. Kutschera, P. Steier, R.J. Gehrke, R.G. Helmer, New half-life measurement of ^{182}Hf: improved chronometer for the early solar system. Phys. Rev. Lett. **172**, 501 (2004). https://doi.org/10.1103/PhysRevLett.93.172501
R.J. Walker, Highly siderophile elements in the Earth, Moon and Mars: update and implications for planetary accretion and differentiation. Chem. Erde **69**(2), 101–125 (2009). https://doi.org/10.1016/j.chemer.2008.10.001
K. Wang, S.B. Jacobsen, Potassium isotopic evidence for a high-energy giant impact origin of the Moon. Nature **538**(7626), 487–490 (2016). https://doi.org/10.1038/nature19341
X. Wang, Q. Amet, C. Fitoussi, B. Bourdon, Tin isotope fractionation during magmatic processes and the isotope composition of the bulk silicate Earth. Geochim. Cosmochim. Acta **228**, 320–335 (2018). https://doi.org/10.1016/j.gca.2018.02.014
X. Wang, C. Fitoussi, B. Bourdon, B. Fegley, S. Charnoz, Tin isotopes indicative of liquid–vapour equilibration and separation in the Moon-forming disk. Nat. Geosci. **12**(9), 707–711 (2019a). https://doi.org/10.1038/s41561-019-0433-4
Y. Wang, W. Hsu, Y. Guan, An extremely heavy chlorine reservoir in the Moon: insights from the apatite in lunar meteorites. Sci. Rep. **9**(1), 5727 (2019b). https://doi.org/10.1038/s41598-019-42224-8
H. Wänke, G. Dreibus, Chemical composition and isotopic evidence for the early history of the Earth-Moon system, in *Tidal Friction and the Earth's Rotation II*, ed. by P. Brosche, J. Sündermann (Springer, Berlin, 1982), pp. 322–344
H. Wänke, H. Baddenhausen, K. Blum, M. Cendales, G. Dreibus, H. Hofmeister, H. Kruse, E. Jagoutz, C. Palme, B. Spettel, On the chemistry of lunar samples and achondrites - primary matter in the lunar highlands: a re-evaluation, in *Lunar and Planetary Science Conference Abstracts*, vol. 2 (1977a), pp. 2191–2213
H. Wanke, H. Palme, H. Baddenhausen, G. Dreibus, H. Kruse, B. Spettel, Element correlations and the bulk composition of the Moon. Philos. Trans. R. Soc. A, Math. Phys. Eng. Sci. **285**(1327), 41–48 (1977b). https://doi.org/10.1098/rsta.1977.0041
W.R. Ward, Evolution of a protolunar disk in vapor/melt equilibrium. J. Geophys. Res., Planets **122**(2), 342–357 (2017). https://doi.org/10.1002/2016JE005198

P.H. Warren, The magma ocean concept and lunar evolution. Annu. Rev. Earth Planet. Sci. **13**(1), 201–240 (1985). https://doi.org/10.1146/annurev.ea.13.050185.001221
P.H. Warren, "New" lunar meteorites: implications for composition of the global lunar surface, lunar crust, and the bulk Moon. Meteorit. Planet. Sci. **40**(3), 477–506 (2005). https://doi.org/10.1111/j.1945-5100.2005.tb00395.x
D. Weber, R.N. Clayton, T.K. Mayeda, A. Bischoff, Unusual equilibrated carbonaceous chondrites and CO3 meteorites from the Sahara, in *Lunar and Planetary Science Conference Abstracts*, vol. 27 (1996), p. 1395
D. Weber, L. Schultz, H.W. Weber, R.N. Clayton, T.K. Mayeda, A. Bischoff, Hammadah AL Hamra 119 - a new, unbrecciated Saharan Rumuruti chondrite, in *Lunar and Planetary Science Conference Abstracts*, vol. 28 (1997), p. 1511
R.C. Weber, P.Y. Lin, E.J. Garnero, Q. Williams, P. Lognonné, Seismic detection of the lunar core. Science **331**(6015), 309–312 (2011). https://doi.org/10.1126/science.1199375
M.K. Weisberg, M. Prinz, H. Kojima, K. Yanai, R.N. Clayton, T.K. Mayeda, The Carlisle Lakes-type chondrites: a new grouplet with high $\Delta^{17}O$ and evidence for nebular oxidation. Geochim. Cosmochim. Acta **55**(9), 2657–2669 (1991). https://doi.org/10.1016/0016-7037(91)90380-N
M.K. Weisberg, M. Prinz, R.N. Clayton, T.K. Mayeda, The CR (Renazzo-type) carbonaceous chondrite group and its implications. Geochim. Cosmochim. Acta **57**(7), 1567–1586 (1993). https://doi.org/10.1016/0016-7037(93)90013-M
M. Weisberg, M. Prinz, R. Clayton, T. Mayeda, M. Grady, C. Pillinger, The CR chondrite clan, in *Proceedings of the NIPR Symposium on Antarctic Meteorites*, vol. 8 (1995), pp. 11–32
M.K. Weisberg, M. Prinz, R.N. Clayton, T.K. Mayeda, M.M. Grady, I. Franchi, C.T. Pillinger, G.W. Kallemeyn, The K (Kakangari) chondrite grouplet. Geochim. Cosmochim. Acta **60**(21), 4253–4263 (1996). https://doi.org/10.1016/S0016-7037(96)00233-5
M.K. Weisberg, M. Prinz, R.N. Clayton, T.K. Mayeda, CV3 chondrites: three subgroups, not two. Meteorit. Planet. Sci. **32**, A138–A139 (1997)
G.W. Wetherill, Radiometric chronology of early Solar-System. Annu. Rev. Nucl. Sci. **25**(1), 283–328 (1975). https://doi.org/10.1146/annurev.ns.25.120175.001435
U. Wiechert, A.N. Halliday, D.C. Lee, G.A. Snyder, L.A. Taylor, D. Rumble, Oxygen isotopes and the Moon-forming giant impact. Science **294**(5541), 345–348 (2001). https://doi.org/10.1126/science.1063037
S.A. Wilde, J.W. Valley, W.H. Peck, C.M. Graham, Evidence from detrital zircons for the existence of continental crust and oceans on the Earth 4.4 Gyr ago. Nature **409**(6817), 175–178 (2001). https://doi.org/10.1038/35051550
M. Willbold, T. Elliott, S. Moorbath, The tungsten isotopic composition of the Earth's mantle before the terminal bombardment. Nature **477**(7363), 195–198 (2011). https://doi.org/10.1038/nature10399
M. Willbold, S.J. Mojzsis, H.W. Chen, T. Elliott, Tungsten isotope composition of the Acasta Gneiss Complex. Earth Planet. Sci. Lett. **419**, 168–177 (2015). https://doi.org/10.1016/j.epsl.2015.02.040
J.G. Williams, A.S. Konopliv, D.H. Boggs, R.S. Park, D.N. Yuan, F.G. Lemoine, S. Goossens, E. Mazarico, F. Nimmo, R.C. Weber, S.W. Asmar, H.J. Melosh, G.A. Neumann, R.J. Phillips, D.E. Smith, S.C. Solomon, M.M. Watkins, M.A. Wieczorek, J.C. Andrews-Hanna, J.W. Head, W.S. Kiefer, I. Matsuyama, P.J. McGovern, G.J. Taylor, M.T. Zuber, Lunar interior properties from the GRAIL mission. J. Geophys. Res., Planets **119**(7), 1546–1578 (2014). https://doi.org/10.1002/2013JE004559
J. Wimpenny, N. Marks, K. Knight, J.M. Rolison, L. Borg, G. Eppich, J. Badro, F.J. Ryerson, M. Sanborn, M.H. Huyskens, Y. Qz, Experimental determination of Zn isotope fractionation during evaporative loss at extreme temperatures. Geochim. Cosmochim. Acta **259**, 391–411 (2019). https://doi.org/10.1016/j.gca.2019.06.016
J. Wisdom, Z. Tian, Early evolution of the Earth-Moon system with a fast-spinning Earth. Icarus **256**, 138–146 (2015). https://doi.org/10.1016/j.icarus.2015.02.025
A. Yamakawa, K. Yamashita, A. Makishima, E. Nakamura, Chromium isotope systematics of achondrites: chronology and isotopic heterogeneity of the inner solar system bodies. Astrophys. J. **720**(1), 150–154 (2010). https://doi.org/10.1088/0004-637X/720/1/150
K. Yamashita, T. Ueda, N. Najamura, N. Kita, L.M. Heaman, Chromium isotopic study of Mesosiderite and Ureilite: evidence for $\epsilon^{54}Cr$ deficit in differentiated meteorites, in *NIPR Symposium on Antarctic Meteorites*, vol. 29 (2005), pp. 100–101
K. Yamashita, S. Maruyama, A. Yamakawa, E. Nakamura, ^{53}Mn-^{53}Cr chronometry of CB chondrite: evidence for uniform distribution of ^{53}Mn in the early solar system. Astrophys. J. **723**(1), 20–24 (2010). https://doi.org/10.1088/0004-637X/723/1/20
Q. Yin, S.B. Jacobsen, K. Yamashita, Diverse supernova sources of pre-solar material inferred from molybdenum isotopes in meteorites. Nature **415**(6874), 881–883 (2002). https://doi.org/10.1038/415881a

R. Yokochi, B. Marty, Geochemical constraints on mantle dynamics in the Hadean. Earth Planet. Sci. Lett. **238**(1–2), 17–30 (2005). https://doi.org/10.1016/j.epsl.2005.07.020

T. Yoshino, Y. Makino, T. Suzuki, T. Hirata, Grain boundary diffusion of W in lower mantle phase with implications for isotopic heterogeneity in oceanic island basalts by core-mantle interactions. Earth Planet. Sci. Lett. **530**, 115,887 (2020). https://doi.org/10.1016/j.epsl.2019.115887

E.D. Young, H. Tang, Isotopic fractionation of moderately volatile elements during Moon formation, in *Lunar and Planetary Science Conference Abstracts*, vol. 50 (2019), p. 1941

E.D. Young, I.E. Kohl, P.H. Warren, D.C. Rubie, S.A. Jacobson, A. Morbidelli, Oxygen isotopic evidence for vigorous mixing during the Moon-forming giant impact. Science **351**(6272), 493–496 (2016). https://doi.org/10.1126/science.aad0525

J. Zhang, N. Dauphas, A.M. Davis, A. Pourmand, A new method for MC-ICPMS measurement of titanium isotopic composition: identification of correlated isotope anomalies in meteorites. J. Anal. At. Spectrom. **26**(11), 2197–2205 (2011). https://doi.org/10.1039/c1ja10181a

J. Zhang, N. Dauphas, A.M. Davis, I. Leya, A. Fedkin, The proto-Earth as a significant source of lunar material. Nat. Geosci. **5**(4), 251–255 (2012). https://doi.org/10.1038/ngeo1429

M.E. Zolensky, D.W. Mittlefehldt, M.E. Lipschutz, X. Xiao, R.N. Clayton, T.K. Mayeda, R.A. Barrett, M.M. Grady, The composition and mineralogy of EET 83334. Meteoritics **24**, 345 (1989)

M.E. Zolensky, D.W. Mittlefehldt, M.E. Lipschutz, M.S. Wang, R.N. Clayton, T.K. Mayeda, M.M. Grady, C. Pillinger, D. Barber, CM chondrites exhibit the complete petrologic range from type 2 to 1. Geochim. Cosmochim. Acta **61**(23), 5099–5115 (1997). https://doi.org/10.1016/S0016-7037(97)00357-8

Space Sci Rev (2020) 216:37
https://doi.org/10.1007/s11214-020-00656-z

On the Distribution and Variation of Radioactive Heat Producing Elements Within Meteorites, the Earth, and Planets

C. O'Neill[1] · H.S.C. O'Neill[1] · A.M. Jellinek[1]

Received: 1 October 2019 / Accepted: 5 March 2020 / Published online: 23 March 2020

Abstract The heat production budget of a planet exerts a first order control on its thermal evolution, tectonics, and likelihood for habitability. However, our knowledge of heat producing element concentrations for silicate-metal bodies in the solar system—including Earth—is limited. Here we review the chronicle of heat producing elements (HPEs) in the solar system, from the interstellar medium, to their incorporation in the protoplanetary disk and accreting planetesimals, to later collisional or atmospheric-erosion modifications. We summarise the state of knowledge of the HPEs in terrestrial planets and meteorites, and current Earth models from emerging constraints, and assess the effect variations may have on the thermal and tectonic history of terrestrial planets.

Keywords Heat production · Heat producing elements

1 Introduction

Earth is losing heat, and the mechanisms by which it does so largely power the surface geological cycle. High internal temperatures enable convection within the solid Earth, and thus plate tectonics. However, it has not historically been clear how much of this heat is produced contemporaneously by heat producing elements (HPEs), or whether there exists a large thermal repository of fossil-heat (Lenardic et al. 2011; Dye 2012; Korenaga 2003). This knowledge gap is critical, as the heating mode, and thus HPEs, exert a fundamental control over the tectonic regime of a terrestrial planet (e.g. O'Neill et al. 2007; Stein et al. 2004). The tectonic regime determines the rate and style of cooling, the extent and character of resurfacing, and the mode of volatile cycling that ultimately governs atmospheric production and loss, climate, and thus characteristics such as the occurrence

---

Reading Terrestrial Planet Evolution in Isotopes and Element Measurements
Edited by Helmut Lammer, Bernard Marty, Aubrey L. Zerkle, Michel Blanc, Hugh O'Neill and Thorsten Kleine

✉ C. O'Neill
craig.oneill@mq.edu.au

1 Planetary Research Centre, Macquarie University, 12 Wally's Walk, Sydney, NSW 2109, Australia

and longevity of habitable (i.e. surficial liquid water) conditions (e.g. Jellinek and Jackson 2015).

The sources of primordial heat within the Earth include heat from accretion (e.g. Birch 1965; Canup 2008) and core formation (Stevenson 1981; Shaw 1979), and heating due to ancient radioactive isotopes in the past. Early workers (e.g. Tozer 1965) suggested that the temperature dependence of mantle viscosity would result in faster convection for hot planets, leading to rapid heat loss, and equilibration of planetary cooling with the decay of radioactive isotopes.

The balance between contemporaneous heat production, and total heat loss, has previously been expressed in terms of the Urey ratio, which is defined as the ratio of internal heat generation, to total system heat loss (Christensen 1985; Lenardic et al. 2011). The most recent estimate for Earth's heat loss is 47 ± 1 TW (Davies and Davies 2010), and for its radiogenic heat production, about 21.5 TW, assuming a geochemical model for the Bulk Silicate Earth (BSE) based on chondrite meteorites (O'Neill 2016). This value gives a planetary Urey ratio of 0.45, implying that less than half of Earth's heat comes from contemporaneous heat production. Non-chondritic BSE models for the concentrations of the HPEs predict even lower values for the Urey ratio (O'Neill and Palme 2008; Jackson and Jellinek 2013).

A major challenge in interpreting these estimates comes from thermal history modelling, which can project the temperature of the mantle back through geological time using a coupled set of evolution equations for energy conservation and mantle convection scalings. For low Urey ratios, most of the heat flow through time comes from fossil and core heat, with lower contributions from radioactive heat production. This implies temperatures in the past were greater—often much greater than the upper mantle solidus for most of the Archaean. This result has been called the "Archaean thermal catastrophe", and it has historically presented a paradox as there is little evidence for an entirely molten Earth at this time (Christensen 1985; Lenardic 2006; Korenaga 2006). Resolutions to it include alternative tectonic regimes (Christensen 1985; Labrosse and Jaupart 2007; Höink et al. 2013; Moore and Webb 2013; O'Neill and Debaille 2014; Jellinek and Jackson 2015), modifications to subduction efficiency (Conrad and Hager 2001; Korenaga 2003), or deep layering (e.g. Kellogg et al. 1999; Butler and Peltier 2002). The latter class of solutions often involve partitioning of heat-producing elements into deep mantle reservoirs (Labrosse et al. 2007).

There are three critical issues surrounding planetary heat production: i) How many HPEs a planet starts with, and in what proportion—this involves some requisite knowledge of the birth environment of the solar system, and the local composition of the interstellar medium from which the solar system was sourced, as well as the sampling, and mixing efficiency during planetary formation; ii) The extent to which a planet retains its initial inventory of HPEs following accretion. This initial concentration may be modified by collisional erosion (O'Neill and Palme 2008) or atmospheric loss on growing planetesimals (Johnstone et al. 2019; Hin et al. 2017) or redistributed as a result of mantle differentiation and continental crust production; and iii) The interplay and character of thermal exchange among key heat-producing/thermal reservoirs within the Earth. A planet's heating mode through time modulates its tectonic state and, thus, initial thermal configurations (temperature and HPE concentrations and distribution) may determine the evolutionary path of a planet (O'Neill et al. 2016): The key for projecting a planet's future is knowing its formation history.

The purpose of this contribution is to bring new constraints on the heat-production budget of the Earth, from meteoritic, geochemical, and geophysical perspectives, to assess the distribution of heat production within the Earth, and understand its impact on planetary evolution.

2 History and Systematics

The heat producing elements within the Earth currently governing heating via radioactive decay are Uranium, which has two primary contributing isotopes, ^{235}U and ^{238}U, as well as ^{232}Th and ^{40}K. In addition to these four isotopes, ^{26}Al and ^{60}Fe may have contributed to internal heating early in the solar system. The Earth inherited these isotopes from the interstellar medium during the formation of the protoplanetary disk, with some components (notably ^{26}Al) being mixed in heterogeneously during its subsequent evolution (Bizzarro et al. 2007; Makide et al. 2011; Schiller et al. 2015).

These radioisotopes have distinct half-lives and decay energies, and contribute different amounts of power to a planet's heat budget through time. The short half-life ^{26}Al (it decays to ^{26}Mg either via a β^+ decay or electron capture, and had a half-life of ~0.717 Myr, Ruedas 2017) was, for example, probably critical in promoting melting and differentiation within developing planetesimals (e.g. Schiller et al. 2015). In contrast, the concentration of ^{60}Fe in the early disk has been questioned (e.g. Trappitsch et al. 2018), and may not have been present at high enough concentrations to be geodynamically relevant. ^{60}Fe's half-life was ~2.62 Myr (Ruedas 2017), and it decayed via beta (β^-) decays to ^{60}Co and then ^{60}Ni (the half-life applies to the whole chain).

The three elements contributing to present-day heat production (U, Th and K) share the geochemical characteristic of being incompatible during igneous differentiation of the Earth, and as a result have become concentrated in the Earth's continental crust over geological time, particular into the uppermost continental crust, and perhaps also in deep reservoirs (Labrosse et al. 2007).

^{238}U is the most abundant extant isotope of uranium, and has a half life of 4468 Myr (Ruedas 2017, and references therein). It decays via a very complicated decay chain (e.g. Ruedas 2017) to eventually ^{206}Pb (and eight ^{4}He atoms). ^{235}U is much less abundant than ^{238}U (currently 0.72% vs. 99.27% of natural Uranium), and has a much shorter half-life of 704 Myr. It also has a complicated decay chain to eventually form ^{207}Pb, and seven ^{4}He atoms. Lastly, ^{232}Th, essentially the only natural form of Thorium, has a long 14.05 Gyr (O'Neill 2016, and references therein), decaying eventually to ^{208}Pb (and six ^{4}He).

Both U and Th are "Refractory Lithophile Elements" (RLEs), which comprise a group of 28 elements that are calculated to condense from the canonical solar nebula at higher temperatures than the main constituents of the rocky planets (i.e. Mg, Si (and associated O) and Fe, the latter initially condensing as metal). The RLEs occur in approximately the same ratios to each other in most undifferentiated meteorites ("chondrites") and, within uncertainty, the solar composition. This observation is assumed to also apply to the Bulk Silicate Earth (known as the "chondritic model" of the Earth's composition), enabling its concentrations of U and Th to be estimated. The proportion of Th to U is much more tightly constrained than their individual concentrations. That both Th and U decay to isotopes of the same element, Pb, has long fascinated geochemists for its potential in deciphering the evolution of solar-system materials (especially the Earth), hence the ratio Th/U has been intensively studied. The isotopic ratio $^{232}Th/^{238}U$ is known in geochemistry as κ (Greek kappa). The current estimate of solar-system Th/U from κ via Th-U-Pb systematics is 3.88 ± 0.02 (Blichert-Toft et al. 2010). From the point of view of heat production, it seems reasonable on current knowledge to assume that all the terrestrial planets and meteorite parent bodies of the solar system have similar Th/U (see Sect. 3).

By contrast K is not a RLE, but behaves cosmochemically as a moderately volatile element. Therefore, its abundance in the Bulk Silicate Earth is not constrained by the chondritic model, but must be estimated empirically. This exercise has been traditionally carried out

Table 1 Summary of key physical data for important heat producing elements (HPEs) throughout Earth's history

Isotope	$\tau_{1/2}$: Half-life (Myr)[a,b]	A: Natural abundance[b] (%)	H: Heating rate[a] (W/kg)	C_0: Present mean BSE concentration[c] (kg/kg)	Final decay products[b]
^{60}Fe	2.62	~0	3.6579×10^{-2}	~0	^{60}Ni
^{26}Al	0.717	~0	0.3583	~0	^{26}Mg
^{40}K	1250	0.0117	28.761×10^{-6}	30.4×10^{-9}	^{40}Ar and ^{40}Ca
^{238}U	4468	99.27	94.946×10^{-6}	22.7×10^{-9}	^{206}Pb (+8 ^{4}He)
^{235}U	703.8	0.72	568.402×10^{-6}	0.16×10^{-9}	^{207}Pb (+7 ^{4}He)
^{232}Th	14050	100	26.368×10^{-6}	85×10^{-9}	^{208}Pb (+6 ^{4}He)

[a]Ruedas (2017), [b]O'Neill (2016), [c]Palme and O'Neill (2014)

by using the observation that the ratio K/U is reasonably constant in the major geochemical reservoirs of the Earth, insofar as these are accessible (Sun and McDonough 1989; see Table 2). More detailed investigation takes into account that K is somewhat less compatible than U in some of the most significant processes of planetary differentiation, such that K abundances are obtained from combining information from K/U and K/La ratios (e.g., Jackson and Jellinek 2013; Palme and O'Neill 2014). It is also the most abundant of the extant radioactive heat-producing nuclides. ^{40}K is present in natural potassium at a concentration of 1.668×10^{-4}, and decays into two daughter products: ^{40}Ca (via a β^- decay), which constitutes 89.28% of the decay product, and ^{40}Ar via electron capture, which constitutes the other 10.72%.

Due to their differing half-lives and decay energies, these isotopes have contributed variably to Earth's internal heat production over time. This total heat production can be quantified using the relationship (Turcotte and Schubert 2002):

$$H = \sum_i C_{0i} H_i \exp\left(\frac{t ln2}{\tau_{\frac{1}{2}i}}\right) \tag{1}$$

Here the summation is over each extant radioactive nuclide ($i = {}^{40}$K, ^{238}U, ^{235}U, and ^{232}Th), t is the time (in the same units as half-life), and the other values for each species are listed in Table 1 (note the natural abundance A has been factored into C_0 in Table 1). An example of the relative contribution of heat production through time can be seen in Fig. 1. The model here assumes a homogenous primordial mantle, that has not segregated into separate reservoirs—a discussion of Earth's internal reservoirs follows in Sect. 4.

^{26}Al and ^{60}Fe are long extinct, and are excluded from the calculation. ^{235}U and ^{40}K dominated Earth's early planetary heat budget, but due to their shorter half-lives, have waned in influence in the latter half of Earth's history. ^{238}U and ^{232}Th currently dominate Earth's internal heat production, due to their relative abundances, heating rates, and long half-lives, and ^{232}Th is expected to be the predominant heat source for ancient rocky exoplanets.

The heat production of the Earth was essentially set by the initial concentrations of HPEs inherited from the interstellar medium, which collapsed to form the protoplanetary disk. The production and seeding of the interstellar medium with elements other than H and He are

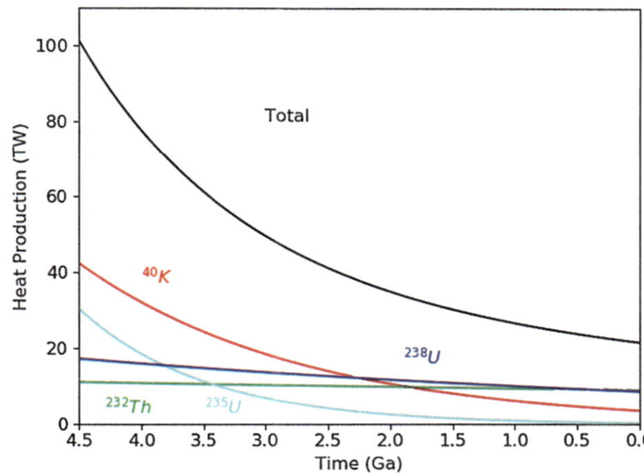

Fig. 1 Relative contribution of radioactive nuclides to Earth's internal heating rate, based on Table 1 and Equation (1)

largely governed by the life cycle of stars including main sequence evolution and supernovae (e.g. Lugaro et al. 2014), white dwarfs (via Type Ia supernovas, e.g. Tinsley 1979), and neutron-star mergers (Kasen et al. 2017; Bartos and Marka 2019). The radionuclides of uranium and thorium are the result of rapid neutron capture, or r-process, thought to occur either during specific supernova conditions, or alternatively via neutron-star mergers (Kasen et al. 2017; Bartos and Marka 2019). ^{40}K is more complicated, with contributions to its production rate from oxygen-burning stellar nucleosynthesis and s (slow neutron capture) processes (e.g. Frank et al. 2014), the latter event occurring largely in asymptotic giant branch (AGB) stars. On the basis of meteoritic concentrations of r-process (^{129}I) and s-process (^{182}Hf) radionuclides, Lugaro et al. (2014) suggested that the last r-process contributing to solar system r-process elements was 100 Myr prior to the formation of the first solar system solids. Brennecka et al. (2010) obtained similar intervals (110-140 Myr) from uranium isotopes. The last s-process addition (constrained by ^{182}Hf), via AGB star winds, was $\sim$30 Myr prior (Lugaro et al. 2014). Extremely short-lived isotopes (^{26}Al and ^{60}Fe, for example) are ascribed to 'self-pollution' of the nebula itself, which is supported by variably mixed ^{26}Al in the earliest solar system material (e.g. Holst et al. 2013).

On a galactic scale, the broad evolution of radionuclides can be considered from the perspective of galactic chemical evolution models. Frank et al. (2014) presented the results from an empirical galactic chemical evolution model that considered the production, mixing, and decay of ^{40}K, ^{232}Th, ^{235}U and ^{238}U throughout the evolution of the galaxy over 12.5 Gyr (Fig. 2). The critical dynamics of this model are that whilst these radionuclides are continually produced over time, they decay, whereas stable species do not. The net result is that the heat-producing radionuclides become diluted over time. Planets that formed early in galactic history would, other factors being constant, have higher heat production than rocky planets forming today. This observation has important implications for planetary evolution, tectonics, and predictions for the likelihood of habitability (Noack et al. 2014).

3 The Solar System Variations in HPEs

Abundances of heat producing elements in terrestrial planets inherit their concentrations from the protosolar nebula and, hence, the Sun's composition is the baseline for primitive disk material, meteorites and planets (e.g., Asplund et al. 2009). A difficulty in using the

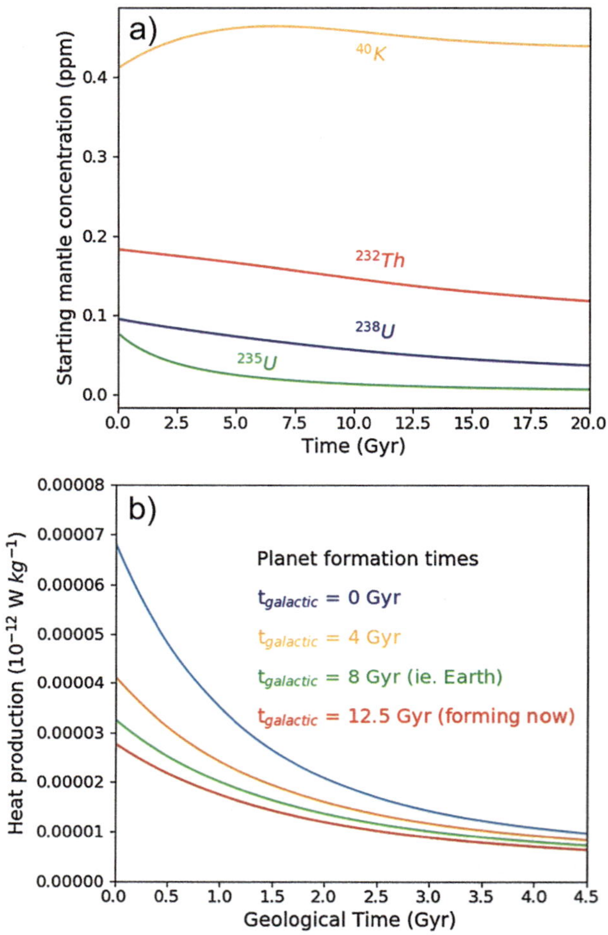

Fig. 2 (**a**) Starting mantle concentration at the time of planetary formation of the heat-producing radionuclides ^{40}K, ^{232}Th, ^{235}U and ^{238}U. (**b**) Evolution in mantle heat production for planets forming 0, 4, 8 and 12.5 Gyr after Galactic formation. Heat production curves are for the geological history of planet, starting from its formation age (set at 0 Gyr). Earth's heat production is equivalent to the example forming 8 Gyr after galaxy formation. Data from Frank et al. (2014)

composition of the Sun, though, is the lack of high-quality spectral lines for U and Th (Grevesse 1984; Grevesse et al. 2015)—a common problem in average or high metallicity stars (e.g. Cayrel et al. 2001). For our solar system, a small group of chondrites, the CI chondrites, are assumed to be reasonable proxies for stellar composition for all elements apart from the most volatile. This general constraint is on the basis of the good match between the concentrations of elements in CIs, irrespective of their distinct chemical properties, with relative solar abundances of those elements best determined in the solar composition, except for H, C, O and N (ice-forming elements), and the noble gases because of their volatility, and also Li, which is consumed by nucleosynthesis in the Sun. Note that there are only three CI meteorites that have been chemically analysed with modern methods, namely Ivuna (which is the meteorite giving the group its designation CI), Alais and Orgueil, and it is the latter meteorite and not the holotype, Ivuna, that overwhelmingly dominates the CI analytical database. However, inferences on solar abundances using the CI-proxy assumption are often supplemented with information from the more abundant CM group and its relations.

A number of syn- and post-accretion factors impact the final composition of planets formed from disk material, but first it is necessary to establish those factors that operated in the disk. It seems plausible that the chemical consequences of the earlier stages of the planet-building process are recorded in the compositional variability of the chondritic meteorites, which sample small parent bodies that have not differentiated by partial melting. Although most of these chondrite meteorites have been metamorphosed in their parent bodies (shock metamorphism associated with the break-up of the parent body is a different phenomenon), there are examples that are pristine, which record directly the agglomeration of the most primitive solar system material. These unmetamorphosed chondrites even host pre-solar grains. Confusingly for non-meteoricists, these most primitive meterorites are designated as having a metamorphic grade of 3. Grades 2 and 1 refer to aqueous alteration, 4 to 6 to thermal metamorphism, grade 7 is sometimes invoked to refer to incipient melting. CIs are all grade 1, so while they are the closest chondrites in composition to the solar composition, they are not texturally primitive. CMs are nearly all grade 2.

Chondrite meteorites in general record several major fractionations:

1. Fractionation of RLEs from the major rocky-planet-forming elements Mg and Si. Compared to the solar composition, both higher and lower RLE/Mg chondrites are known. Carbonaceous chondrites have similar or slightly higher RLE/Mg than the solar composition, but the ordinary chondrites and even more so the enstatite chondrites have lower. This factor affects Th and U contents directly because Th and U are RLEs.
2. Fractionation of Mg from Si, again with both higher and lower Mg/Si than solar. As these are major elements, varying Mg/Si necessarily affects the abundances of other elements.
3. Fractionation of metal (that is, Fe and other siderophile elements that condense into Fe) from silicates, also leads to higher and lower Fe/Mg. This fractionation should not be confused with the metal-silicate segregation that forms planetary cores. It is a nebular process that predates core formation, and is responsible for the division of the ordinary chondrites into their H, L and LL groups (high Fe, low Fe and very low Fe). That the extent of this fractionation among chondrites considered as a whole is usually not significant, is a remarkable feature that is consilient with the similar calculated condensation temperatures of Mg, Si and Fe (1348 K, 1314 K and 1333 K respectively according to Wood et al. 2019) under the canonical conditions used for the calculation, but a difference between the condensation temperatures of the Mg-silicates and Fe metal would open out at different H-C-O ratios or nebular pressures. Much greater metal-silicate fractionation might be expected in other solar systems.
4. Depletion in moderately volatile elements (MVEs), more-or-less according to their calculated condensation temperatures. The MVEs are those elements calculated to condense after Mg, Si and Fe, but before planetary ices (composed of H, C, N, O and noble gases). Chondrites enriched in the MVEs are unknown, except for some enstatite chondrites that show a slight enrichment in the least volatile of the MVEs.
5. Oxidation state – how much Fe is metallic (Fe^0), Fe^{2+} and Fe^{3+}. This is a matter of oxygen content, reflecting oxygen fugacity (fO_2) hence H-C-O ratios.
6. Depletions of the other ice-forming elements (H, C, N), and associated O (as in H_2O) and noble gases. The H content of texturally primitive (hence unmetamorphosed) chondrites varies from virtually nothing, to that in the relatively H_2O-rich CIs (which are not particularly primitive in texture, showing much hydrothermal alteration).

The processes behind these chemical fractionations remain incompletely understood, but likely reflect variations in local pressures and temperatures within the disk, H-C-O ratios, which control fO_2, and other factors that affect the condensation of different phases, generating gradients in major and minor elements (Lodders and Fegley 1999).

Geochemical assessments of the Earth's composition (e.g., Palme and O'Neill 2014 and references therein) show that it is not like any known chondrite group. Early attempts to reconcile this issue postulated complete mixing of different varieties of known chondrites, corresponding to accretion of the Earth from material sourced from different regions of the disk. Paradoxically, although the Earth appears isotopically most similar to the enstatite chondrites (e.g., Javoy 1995; Dauphas 2017), these meteorites have the least similarity to the Earth in terms of the abundances of major elements, which is closer to CV carbonaceous chondrites (Palme and O'Neill 2014). Recent work on the isotopic composition of Earth materials compared to known chondrites has indicated that the Earth may contain material not represented in our chondrite collections. The traditional view of chondrites as the "building blocks of the planets" may be misleadingly naïve (e.g., Burkhardt et al. 2019; Schiller et al. 2018).

Subsequent to the effective isolation of the chondrite parent bodies, further processes have modulated the compositions of the terrestrial planets or proto-planets. These include: i) thorough mixing driven by disk transport processes (e.g. Schiller et al. 2015); ii) impact erosion (O'Neill and Palme 2008) which may preferentially strip off incompatible elements enriched in planetary crusts; iii) the loss of moderately volatile elements after giant impacts, as occurred on the Moon (O'Neill 1991a, 1991b); and iv) hydrodynamic escape of atmospheres (e.g., Johnstone et al. 2019; Hin et al. 2017) which may preferentially blow off moderately volatile elements whilst stellar outflow is active (up to $\sim< 7$ Myr). We will consider the effects of these processes, in turn below.

Condensation Temperatures and Variations Within the Disk Following Urey (1952), many authors, notably Lord (1965), Larimer (1967) and Grossman (1972), see also Grossman and Larimer (1974), have calculated the equilibrium condensation phase relations of a gas of solar composition. Lodders (2003) updated the calculation, which extended it to all of the naturally occurring elements. Results have traditionally been summarized by reporting the temperature at which 50% of an element has condensed. Recently Wood et al. (2019) presented a significant revision of the calculation. To facilitate comparison with Lodders (2003) they used her solar abundances rather than those of Asplund et al. (2009), except for chlorine, which was recently been revised downward by a factor of 6 (Clay et al. 2017), although not uncontroversially (Ebihara and Sekimoto 2019). Among the new findings, one of relevance to the HPEs is that K is now calculated to condense at slightly lower temperatures than its cosmochemically more abundant fellow alkali metal, Na, in agreement with experimental observations (e.g., Kreutzberger et al. 1986).

The nebular condensation calculation should be viewed as a guide to an element's relative volatility under typical—but arguably not all—conditions relevant to the initial processing of material in the earliest history of the solar system. Specifically, the calculation only applies to the period before H_2, which makes up most of the solar composition (together with He), was dispersed. This process occurs over a relatively short timescale of one to a few million years or so (see below), much shorter that the estimated time needed for accretion of Earth-sized bodies. Furthermore, nebular condensation calculations have typically assumed a constant gas pressure (e.g., 10^{-4} bars), which is approximately the sum of the partial pressures of H_2 and He, since these are by far the most abundant gas species in the nebula. The calculations are, however, very sensitive to the selected H-C-O mixing ratios, which determine the important variable of oxygen fugacity. The exact values of H-C-O in the solar composition continue to be a matter of discussion, and have been periodically reassessed over the years (Asplund et al. 2009). Wood et al. (2019) have emphasised the importance of the chlorine abundance to the calculation, which was recently revised by Clay et al. (2017).

This condensation calculation should not be mistaken as modelling a physically plausible process, even if the nebular pressure distribution and the solar composition were well known.

The assumption of chemical equilibrium has long seemed unlikely, especially towards lower temperatures, by analogy with rates of reaction in various geological processes, but has never been investigated. To illustrate the problem, K itself provides a useful example. The canonical condensation calculation has K hosted as the component $KAlSi_3O_8$ in plagioclase feldspar, a solid solution between the components $CaAl_2Si_2O_8$ – $NaAlSi_3O_8$ (e.g., Wood et al. 2019). But the equilibrium calculation does not address the pathway by which this state is reached. Given the constraints of mass balance, condensed phase stoichiometry, and that the vapour pressures of gas-phase components with Si, and especially Al, a refractory element, are extremely low at the calculated condensation of K (that is, 993 K according to Wood et al. 2019), a hypothetical pathway to the state assumed by equilibrium condensation may reflect a multi-phase reaction like:

$$\underset{\text{Gas}}{K} + \underset{\text{Gas}}{H_2O} + \underset{\text{Corundum}}{1/2Al_2O_3} + \underset{\text{Pyroxene}}{3Mg_2Si_2O_6} = \underset{\text{Gas}}{H_2} + \underset{\text{Olivine}}{3Mg_2SiO_4} + \underset{\text{Feldspar}}{KAlSi_3O_8}$$

This reaction involves four condensed phases, uncompacted by pressure, and having to communicate in a tenuous gas; an additional kinetic constraint is the need for solid-state diffusion in those condensed phases in which the above components have solid solution. Note also that this reaction is not the only possibility: for example, we could equally well write a similar one involving $CaAl_2Si_2O_8$ in feldspar and $CaMgSi_2O_6$ in pyroxene in place of Al_2O_3 in corundum. This is a simple consequence of equilibrium thermodynamics not addressing reaction pathways. The point is, whether any such reaction pathway would proceed on the timescale of the cooling of the solar nebula ($\sim 10^5$ - 10^6 years). Geological experience prejudices us to suppose not – in the absence of hydrothermal fluids (whose existence requires high pressures), metamorphic rocks often preserve equilibrium mineral assemblages and compositions on geologically paced cooling ($10^6 - 10^7$ years) from not dissimilar temperatures. The actual condensation of K most likely takes place at lower temperatures via a metastable but kinetically feasible reaction pathway. In this regard, K exemplifies a typical consequence of non-equilibrium condensation: the actual condensation of those trace elements that are calculated to condense in pre-existing phases will be at lower temperatures than calculated assuming thermodynamic equilibrium.

Another, separate, problem is that the necessary thermodynamic data are not always available, let alone accurate. Especially difficult is selecting the appropriate thermodynamic components in the condensing phases for many of the less abundant elements, and then finding relevant thermodynamic data for these components, plus their activity-composition relations for their solid solution into the phases determined by major-element condensation. For example, the revision of Wood et al. (2019) places the calculated condensation temperature of Pb some 230 K lower than Lodders (2003) by assuming that it condenses as PbS and not into Fe metal. For these reasons the canonical nebular condensation calculation should still be seen as work in progress. We also emphasise that the sensitivity of an element's volatile tendencies to oxygen fugacities means that the volatility sequence enshrined in the nebular calculation may differ considerably for post-nebular processes (Norris and Wood 2017; Sossi et al. 2019). This is demonstrated by comparing the nearly solar ratio of Mn/Na seen in chondrite meteorites, with the much greater Mn/Na ratios characteristic of all known differentiated solar system bodies, except the Earth (O'Neill and Palme 2008; Siebert et al. 2018).

For our purposes it is interesting to note in which condensing phases U, Th and K first appear in. Figure 3 illustrates an abbreviated version of Lodder's oxidised condensation sequence. U and Th both strongly partition into one of the earliest high-temperature oxide

Fig. 3 Plot of the 50% condensation temperatures for the main hosts of the HPEs U, Th and K, based on the results of Lodders (2003) and Wood et al. (2019) for a solar system abundance gas. U and Th are largely taken up by hibonite (a calcium-aluminium oxide), K is not condensed out significantly until feldspars start to form at around ~1000 K. Bar size denotes nominal uncertainty, and major rock-forming elements Si and Fe are also shown for comparison

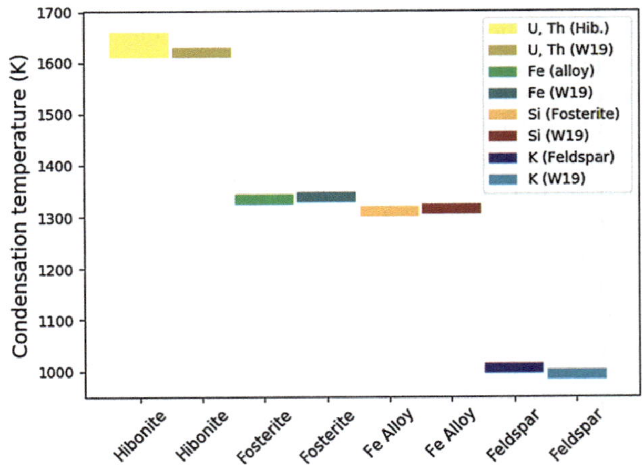

phases, hibonite (a calcium aluminium oxide, $CaAl_{12}O_{19}$). This is one of the highest temperature condensates (after corundum), and is expected to be well represented in the hot inner disk. The 50% condensation temperature for U (into hibonite) is 1610 K, and for Th 1659 K. As a result, U and Th should be well-represented in most rocky meteoritic material. K, in contrast, is mostly taken up by feldspars during the condensation sequence, and is 50% condensed at temperatures of ~1006 K—significantly less than U and Th. There exists the possibility that strong gradients in these heat producing elements may have been generated in the early solar system as a result of local disk temperature evolution, and variations in fO_2, if they existed, may also have had an effect (Lodders 2003).

Variations in Meteoritic Abundances Chondritic meteorites contain significant variations in their HPE concentrations, which can be conceptualized as arising from two factors: 1) variation in the concentrations of Refractory Lithophile Elements (thus Th and U) due to different amounts of the three other main compositional components in the meteorites, namely Mg-silicates, Fe-rich metal and/or sulfide, and H-C-O-N volatile-containing material; and 2) the further depletion of K relative to the RLEs, as expected from its status as a moderately volatile (but still lithophile) element. This factor is conveniently quantified as the K/U ratio. The resulting concentrations of U, Th and K, and the K/U ratio are listed in Table 2, and shown in Fig. 4. Note that even the CIs are greatly depleted in H, C and N (and noble gases) relative to their solar abundances, hence the amounts in meteorites should be regarded as a small difference between large numbers – that is, the solar H,C and N are lost. From this perspective, it is difficult to accord the precise amounts of H, C and N remaining in the volatile-rich chondrites (the CI and CM groups) any particular significance.

With this caveat in mind, primitive chondritic meteorites do demonstrate some significant variations in the HPE budget (Table 2 and Fig. 4). Much of this is related to bulk composition and formation environment. For instance, carbonaceous chondrites (in particular CI chondrites) contain more organic material and ice relative to silicate. Since HPEs are preferentially incorporated into silicates, these chondrite types as a whole tend to show lower HPE abundance (particularly U and Th) than ordinary or enstatite chondrites.

The exception to this is K, which, as a moderately volatile element, seems to show an increased concentration in the least-processed samples (CI, and to a lesser extent CM chondrites), compared to, for instance, CV chondrites, which may reflect a volatility trend in their disk formation environment (Morbidelli et al. 2016). U and Th also increase within the carbonaceous group from low abundances in the most solar-like, the CIs, to the RLE-enriched

Table 2 Concentration of K, Th and U in a variety of chondrites (in ppm). WK88 refers to Wasson and Kallemeyn (1988), CI chondrites and CV (Allende) are from Barrat et al. (2012); CM from Hewins et al. (2014); LL(6), EH(3), and EL(3) are from Barrat et al. (2012) for U and Th, Rubin et al. (2009) for EL K, Lehner et al. (2014) for EH K, and Wasson and Kallemeyn (1988) for K in LL ordinary chondrites

	CI	CM	CO	CV	LL	L	H	EL	EH
K (ppm)	550	363.5	–	277	790	–	–	713.42	735
SD*	77	88.03	–	2.77	158.0	–	–	64.50	30
K (WK88)	560	400	345	310	790	825	780	735	800
Th (ppm)	0.0283	0.0387	–	0.0594	0.04265			0.00233345	0.0296
SD	0.000566	0.0387	–	0.00119	0.00233345			0.00702258	0.00600571
Th(WK88)	0.029	0.040	0.045	0.060	0.043	0.043	0.042	0.035	0.030
U (ppm)	0.0077	0.009866		0.0152	0.0121			0.0083	0.008
SD	0.000385	0.000385		0.000304	0.00212132			0.0009737	0.00184481
U (WK88)	0.0089	0.011	0.013	0.017	0.013	0.013	0.012	0.010	0.009
K/U	62921	36363	26538	18235	60769	63462	65000	73500	88889

*SD is standard deviation

types (CV and CK), reflecting a decreasing amount of H-C-O-N volatile abundances, and an increased amount of a refractory component consisting of phases like hibonite (the main initial condensation carrier of U and Th), which formed in high-temperature inner disk environments.

The dichotomy between the trends observed in the carbonaceous, and non-carbonaceous chondrites, are reflected in nucleosynthetic anomalies in these meteorite groups (e.g. Carlson et al. 2018; Nielsen et al. 2019), which suggest meteorites at greater heliocentric distances are progressively depleted in s-process isotopes. This fundamental distinction is mirrored in the HPE trends observed in Fig. 4.

Impact Erosion and Atmospheric Loss The growth of planetary embryos occurred within a H_2-gas rich disk, for at least the first ~4 Myr of solar system evolution (e.g. Wang and Goodman 2017). After that time, the stellar activity (in particular, strong EUV radiation) rapidly blows the disk material away. However, if a growing planetary embryo reached 0.5 M_{Earth} within the lifetime of the disk, Johnstone et al. (2019) argue that it may have retained a tenuous H_2 gas envelope, which is subsequently rapidly lost due to initial H_2 boil-off, and later EUV insolation from the early Sun.

The process of early atmospheric loss may preferentially lead to loss of volatile and moderately volatile elements present in the atmosphere. K is a moderately volatile element that would be released from a high-temperature (1500–2500 K) magma ocean (Schaefer and Fegley 2010), to the hot early atmosphere. Having a moderate (~1000 K) condensation temperature it would have been sensitive to atmospheric loss via hydrodynamic escape (Scherff et al. 2019; Lammer et al. 2019), and this may explain the low K/U ratios of Earth and Venus (e.g. Albarede 2009; Kaula 1995), compared to primitive chondrites.

An argument against late K loss was presented by Humayun and Clayton (1995) who found no evidence of K isotopic variation between the Earth and Moon and suggested this constrained potassium loss due to partial volatilisation to < 2%. However, it was rapidly pointed out that lack of isotopic fractionation does not preclude loss by volatilization, only loss by simple Rayleigh fractionation into high vacuum (e.g., Esat 1996; Young 2000). For example, total loss of a volatile element (like K) from that part of an impacted body that was

Fig. 4 Distribution of heat producing elements within different chondrite types, in order of approximate decreasing heliocentric distance, from the data listed in Table 2. Black bars represent the data of Wasson and Kallemeyn (1988; note uncertainties were not reported, so the bar height is nominal). Blue bars represented the recompiled data from a number of chondrite groups (see Table 2), and height represents their uncertainties. The red line is a cubic interpolation of the trend. Sequence is ordered in approximate decreasing heliocentric distance from left to right

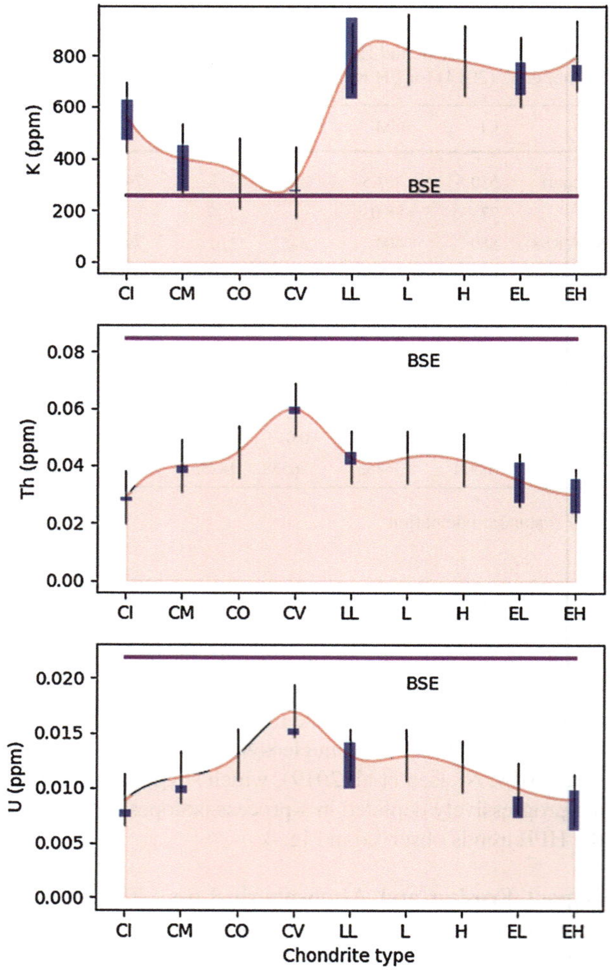

vaporized, with negligible volatile loss from its other part, would, after rehomogenization of the two parts, result in depletion with no isotopic fractionation whatsoever (e.g. Hin et al. 2017). Subsequently, a small difference in δ^{41}K has been found between the Earth and Moon (Wang and Jacobsen 2016), at a level of 0.4 ‰, smaller than that originally anticipated by Humayun and Clayton (1995), but consistent with the condensation of the Moon from a BSE vapour disk.

The ratio of K/U is far more sensitive to volatile loss than K isotopes, and depletion of K on developing planets is best tracked by considering the K/U ratio. For CI chondrites, this value is 6.3×10^4 (Table 2, Wasson and Kallemeyn (1988), Lodders et al. 2009). For the Earth, K/U ~13,800 (Lodders et al. 2009) and for Venus it has been constrained to > 7000 (Davis 2005). The K/U ratio has been argued to be sensitive to EUV flux on developing planets, whilst they were immersed in the gas disk, and possessed a more massive H_2-dominated early atmosphere, and blow-off of this gaseous envelope may have driven K/U to lower values. This ratio is perturbed further by the impact history of the planet; depleted (e.g. ureilite) vs. enriched (e.g. CI chondrites) will bring variable amounts of K to the late stages of accretion.

Table 3 Existing data on HPE concentrations within the mantles of terrestrial planets

	U (ppb)	Th (ppb)	K (ppm)	Reference
Mars—Primitive mantle	16	56	315	1
Mars—Primitive mantle	16	56	305	2
Mars—Primitive mantle	16	56	920	3
Earth—BSE—chondritic-RLE	22	85	260	4
Earth—BSE—non-chondritic	11	43	130	4
Venus—Crust (Venera 9)	600	3650	4700	5
Venus—Crust (Venera 10)	460	700	3000	5
Venus—Mean Crust	530	2175	3850	
Venus—Primitive mantle (Kaula 1999)	21	86	153	
Mercury—Northern Hemisphere, MESSENGER	90 ± 20	220 ± 60	1150 ± 220	6
Mercury—primitive mantle (25% melt fraction scaling)	22.5 ± 9.5	55 ± 26	287.5 ± 113	

1. McLennan (2012). 2. Wänke and Dreibus (1988, 1994). 3. Lodders and Fegley (1997). 4. O'Neill (2016). 5. Surkov et al. (1987). 6. Peplowski et al. (2011)

In addition to atmospheric loss, volatiles may also have been removed by impact erosion of planetary crusts. O'Neill and Palme (2008) suggested that Earth may have been shifted from its chondritic original composition, to a non-chondritic bulk-silicate Earth. In this scenario, early crusts on differentiated planetesimals or planetary embryos, or on the proto-Earth itself, enriched in incompatible elements including the HPEs (K, U and Th), would have been preferentially stripped off due to the shallow depth of impact excavation for most early impacts. This process left behind a mantle with a sub-chondritic bulk-silicate composition, and possibly with less than half its original HPE contingent. O'Neill and Palme (2008) noted that the Fe/Mg ratio of the Earth is larger than the solar ratio (2.1 ± 0.1, vs. 1.9 ± 0.1), and they suggested this implies 10% of silicates relative to metal were lost due to collisional erosion.

Variations Among Terrestrial Planets Variations among terrestrial planets have important consequences for the thermal evolution of each body, but are notoriously hard to constrain given the lack of samples from Venus and Mercury. Whilst we have direct samples of the mantle (as xenoliths, etc), these are representative of the depleted upper mantle, and not generally the primitive lower mantle or "bulk silicate Earth" (BSE). Elemental compositions of Earth's mantle can then be estimated from geochemically modelling, or comparisons with chondrites. On Mars, we have both direct measurements of the surface, and Martian meteorites. Together these also allow estimates of Mars' primitive mantle to be made (Wänke and Dreibus 1988, 1994). In addition, modelling Mars as a mixture of meteorite types also allows interior composition to be estimates (Lodders and Fegley 1997), and the results of these approaches are shown in Table 3.

Estimation of the surface composition of Venus and Mercury is more difficult. Venus has been visited by the Venera (8, 9 and 1) and Vega (1 and 2) landers, which analysed the concentrations of heat producing elements using a gamma-ray spectrometer. All the landing sites had compositions consistent with terrestrial basalts, with the exception of Venera 8, which appeared to be a differentiated crustal rock (and thus had elevated U, Th and K). In addition, the Vega missions targeted rocks from Aphrodite Terra (Vega 1 from Mermaid Valley to the north, and Vega 2 from the slopes of Aphrodite Terra). Both returned spectra, despite landing difficulties for Vega 1. However, Aphrodite Terra is a complicated crustal structure, and we have listed the results of Venera 9 and 10, which are probably the closest to mantle-derived volcanism, and presented a mean crustal value. The similarity in the uncompressed density of Venus to Earth, together with the broad similarity in the sampled Venera basalts to terrestrial basalts, is consistent with the composition of Venus's primitive mantle being similar to that of the BSE for the major elements, among which are included the refractory lithophile elements Ca and Al. Therefore, we expect the Th and U abundances of Venus's primitive mantle also to be Earth-like. There may, however, be differences at the level of a few tens of percent related to plausible differences in Venus's RLE/Mg and Mg/Si ratios, which are unconstrained by empirical evidence. The oft-quoted value of 21 ppb U for Venus's mantle given by Kaula (1999) is simply the BSE value from Wänke et al. (1984), under the chondritic assumption.

The K content of Venus's primitive mantle is estimated from this U content and the K/U ratio of 7220 ± 1220 of the Venera basalts, omitting one high K outlier. With only a few samples and such data selection, this estimate should viewed skeptically. The amount of ^{40}Ar, the decay product of ^{40}K, in Venus's atmosphere has been obtained from the ^{40}Ar/^{38}Ar ratio and the estimated abundance of ^{38}Ar (Hoffman et al. 1980), but with considerable uncertainty. This ^{40}Ar corresponds to much less K than estimated from K/U, which has been taken to mean that the mantle of Venus is much less efficiently degassed than the Earth's (Kaula 1999; O'Neill et al. 2014).

For Mercury, our information on its surface comes from gamma ray spectroscopy from the Messenger orbiter. Due to the orbit of Messenger, only results from the northern hemisphere were presented by Peplowski et al. (2011). They identified not only a crust enriched in U and Th, but also significant K, which was unexpected from formation models (e.g. Solomon et al. 2001). We have adopted their approach assuming the crust represents a 25% melt extraction from the mantle to reconstruct Mercury's primitive mantle composition listed in Table 3.

Significant average heat production variation may exist among the different terrestrial planets, and this variation may have been inherited from their disk environment. For instance, McCubbin et al. (2012) plotted the variations in K/U and K/Th for inner solar system bodies. Those results show that both Mars and Mercury are, comparatively, volatile rich bodies, with high K/Th and K/U ratios. Earth and Venus are both lower, and the Moon is extremely low—primarily as a result of it being largely in a rock-vapour state during its formation (e.g. Canup 2008). However, the HED meteorites, representative, of the 4 Vesta parent body, also showed extremely low K/U and K/Th. However, Vesta is not representative of the asteroid belt, and the largest asteroid belt body, constituting most of its mass, is the dwarf planet Ceres, has not been unequivocally sampled in the meteoritic collection. To calculate an additional 'asteroid belt' representative values of K/U and K/Th, we adopt values for CM chondrites in lieu of Ceres, and calculate mass weighted averages of Vesta and Ceres for K/U, K/Th and heliocentric distance. Our results are shown in Fig. 5.

We have left out the Moon in Fig. 5 due to its known formation artefacts. If we consider the "average" composition of the asteroid belt, there is an observed trend of increasing volatility, shown by increasing K/U and K/Th ratios, moving out of the inner solar system. This trend does not really hold for Vesta, and probably reflects the importance of stochastic processes, such as impacting, in the erosion and late delivery of volatiles. Mercury, as well, bucks any simple trend, and its S-composition also suggests additional complexity. Whether this is due to uncertainties in its bulk-silicate estimates, oddities in its accretion of K, or an unusual formation mechanism, remain to be seen. Further generalisations are limited due to our incomplete knowledge of their primitive mantle compositions—a problem even for Earth, as outlined next.

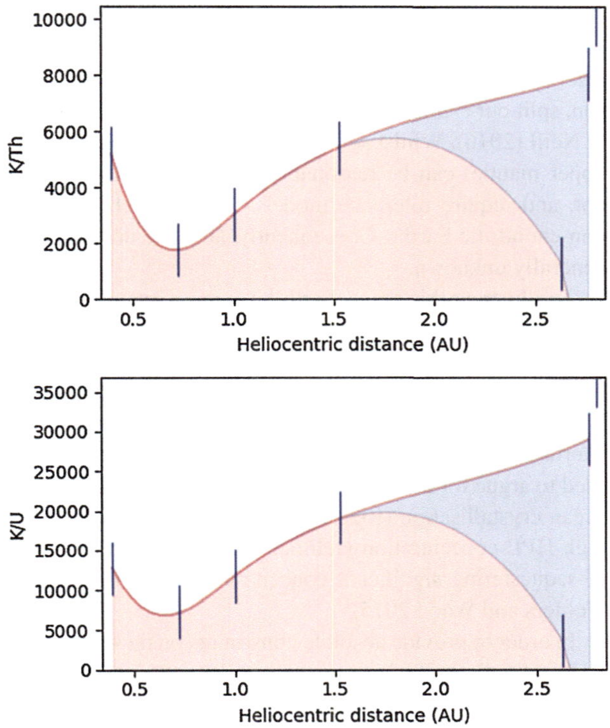

Fig. 5 Variations in K/Th (top) and K/U (bottom) for the terrestrial planets (Table 3), as well as an 'asteroid average' based on a weighted average of Vesta (McCubbin et al. 2012), and a 'CM' Ceres (CM values from Table 2). Individual points (including Vesta and 'CM' Ceres) shown as vertical blue lines at their respective heliocentric distance, and a cubic interpolation through the terrestrial planets and either the asteroid-belt average (red line) or Vesta (purple) is shown

4 Earth's HPE abundance

Earth's internal heat production budget can be estimated in a number of ways. There is a geodynamic approach, using Earth's total heat loss and assuming an energy budget to constrain heat production, and element ratios to constrain HPE concentrations (Turcotte and Schubert 2002). The geochemical approach utilises either direct measurements (including chondrites), and considers partitioning of HPEs into different reservoirs to constrain element concentrations (O'Neill 2016). And more recently, emerging geoneutrino measurements have been able to provide a direct measurement of the concentration (and eventually, the distribution) of HPEs in the Earth's interior.

Each of these approaches has significant limitations, discussed below. Ultimately, we wish to constrain the primary HPE budget of the planet. This is complicated by their enrichment into different reservoirs (e.g. crust, basal magma ocean) and depletion in others (e.g. upper mantle). The most primitive silicate reservoir, the lower mantle, is largely inaccessible (although, see Campbell and O'Neill 2012), and estimates for the bulk silicate Earth (BSE) are largely approached using mixing models.

Uncertainties in these values largely arise from our incomplete knowledge of the composition of the deep interior, and contribution of secular cooling to current heat flow. Geodynamic analysis has suggested that secular cooling contributes 20-25% of Earth's current heat flow (Turcotte and Schubert 2002)—however, this assumes a relationship between Rayleigh number and heat flow which may not be valid for non plate-tectonic regimes (Silver and Behn 2008) or mixed-heating mantle (Moore 2008). In addition, there exists significant time-lag between surface heat flux and the thermal state of the interior (Lenardic et al. 2019), and thus these geodynamically-constrained estimates should be treated with caution.

Cosmochemical models suggest that 40-50% of Earth's surface heat flow is from secular cooling (e.g. Schubert et al. 2001; Sun and McDonough 1989), however, observed reservoirs do not reflect cosmochemical (chondritic) abundances. We have, for comparison, split our estimates for the mantle into chondritic and non-chondritic families, following O'Neill (2016). Whilst some reservoirs (crustal, and to some degree mantle lithosphere and upper mantle) can be reasonably well constrained by observations (Table 2), others cannot, and require reference models—typically chondritic compositions, or, more recently, non-chondritic Earths. Consequently, absolute uncertainties for the quantities in Table 2 are generally unknown.

To illustrate this point, Earth has approximately a factor of 2 less K than CI chondrites (Fig. 3), and is enriched in U and Th by a factor of ~2-3. As a moderately volatile element, K is depleted in the Earth, and Earth has only around 50% of the atmospheric ^{40}Ar expected from ^{40}K decay throughout geological time for BSE models (Kaula 1999). This may be a factor of degassing efficiency under different tectonic regimes (O'Neill et al. 2014), or alternatively suggest lower total K abundances. Likewise, isotope systematics have been used to argue for a deep enriched reservoir within the Earth, segregated during early magma ocean crystallisation (Rizo et al. 2013; Rizo et al. 2016), which may potentially contain a high HPE concentration (Jellinek and Jackson 2015). There has also been the suggestion of sequestering significant concentrations of HPEs into the core (Blanchard et al. 2017; Wohlers and Wood 2015).

In order to provide absolute constraints on these models, we turn to geoneutrino measurements of radioactive decay in the mantle, and high-pressure experiments of HPE partitioning between silicate and metals.

Geoneutrinos A potential additional constraint on radiogenic heat production in the Earth can be made from the geoneutrino flux. Geoneutrinos are generally electron antineutrinos, produced during beta-minus decay, where a neutron within a nucleus decays to a proton, electron and electron-antineutrino. Present detectors use proton collisions, where an incoming antineutrino collides with a proton to form a neutron and positron, a process known as inverse beta-decay. The reaction is captured in large underground scintillation tanks, and consists of sequential light flashes associated with the initial positron destruction, and a later neutron capture. This proton collision, however, requires > 1.806 MeV to proceed, meaning these tanks are only sensitive to high-energy antineutrinos. Only isotopes on the ^{232}Th and ^{238}U decay scheme produce antineutrinos of sufficient energy (e.g. ^{228}Ac, ^{212}Bi for the ^{232}Th chain, and ^{234}Pa and ^{214}Bi for the ^{238}U chain (McDonough 2016)). As a result ^{40}K (and ^{87}Rb) are not currently detectable, despite making up most of Earth's geoneutrino flux, but may be in future electron collision detectors (e.g. Dye 2012).

Gando et al. (2011) combined results from Borexino and KamLAND to estimate that mantle ^{232}Th and ^{238}U contribute 20.0 +8.8/-8.6 TW to Earth's heat flux. Results from Borexino alone have tended towards higher values (Agostini et al. 2019), and recently, the Borexino team (Agostini et al. 2019) estimated the total radiogenic heat of the Earth at 38.2 +13:6/-12:7 TW (confidence intervals are $\pm$34%), and the total mantle heat contribution of 24.6 TW +11.1/-10.4 from ^{238}U and ^{232}Th. Therefore, even at the 95% confidence interval this estimate would only just be compatible with the largest radiogenic heat production values deduced from geochemistry. We can convert these contributions to mantle concentrations as follows. If we assume the mass of the mantle is 4×10^{24} kg, then a flux of 24.6 TW gives us a mantle heat production from these isotopes of 6.15×10^{-12} W/kg (range of

Table 4 Compilation of existing data on Earth's major geochemical reservoirs, and selected rock types. Drawn from existing compilations in O'Neill (2016), Turcotte and Schubert (2002) and Schubert et al. (2001)

Geochemical reservoir	Mass (kg)	U (ppb)	Th (ppb)	K (ppm)	Heat production (TW)
Continental sediments[1]	0.7	1730	8100	18300	0.3
Upper crust[1]	6.7	2700	10500	23200	4.2
Middle Crust[1]	6.9	970	4860	15200	1.9
Lower crust[1]	6.3	160	960	6500	0.4
Continental crust	20.6	1311	5612	15247	6.8
Oceanic sediments[1]	0.3	1730	8100	18300	0.1
N-MORB[2]	–	47	120	600	–
E-MORB[2]	–	180	600	2100	–
OIB[2]	–	1020	4200	12000	–
Oceanic crust[3]	6.3	20	52	280	0.03
Mantle (ocean crust source)[4]	–	4	10	60	–
Convecting mantle (chondritic)[4]	3942	15	55	181	13.9
Convecting mantle (non-chondritic)[4]	3942	4	11	47	3.2
Convecting fertile mantle (geodynamic)[5]	3942	31	124	310	–
BSE (chondritic)[6]	4060	22	85	260	21.4
BSE (non-chondritic)[7]	4060	11	43	130	10.8

[1]Huang et al. (2013), [2]Sun and McDonough (1989), [3]O'Neill and Jenner (2012), [4]O'Neill (2016), [5]Turcotte and Schubert (2002), [6]Palme and O'Neill (2014), [7]O'Neill and Palme (2008). Note that "Oceanic crust" is parental ocean floor basalt of O'Neill (2016), incorporating all groups of oceanic basalts, and mantle (ocean crust source) refers to the residue mantle after 20% melt extraction. The convecting mantle refers to the BSE minus continental crust/lithosphere contributions

3.55–8.93 $\times$ 10^{-12} W/kg). We can then use the relation (Turcotte and Schubert 2002):

$$H_0 = C_0^U \left(H^U + \frac{C_0^{Th}}{C_0^U} H^{Th} \right) \tag{2}$$

Here H denotes the heating rate of an isotope from Table 1 (H_0 is the mantle heating inferred above), and the ratio of Th to U is assumed to be ~3.9 (Gando et al. 2011; note from Table 1 the ratio of ^{232}Th:^{238}U is 4.0, and the uncertainty in this ratio is in excess of a correction for ^{235}U). This gives our current concentration of ^{238}U in the mantle to be 33.05 $\times$ 10^{-9} kg/kg (range 19.08–47.96 $\times$ 10^{-9} kg/kg), and the concentration of ^{232}Th to be 128.90 $\times$ 10^{-9} kg/kg (range 74.40–187.06 $\times$ 10^{-9} kg/kg).

These data may also be used to constrain nuclear activity within the core. An assessment of the viability of a natural nuclear reactor in the core limited core heat production from U and Th to 2.4 TW (KamLAND neutrino observatory; Agostini et al. 2019) and 3 TW (Borexino, which is further from any man-made reactors; Gando et al. 2011). Assuming a core mass of 1.883 $\times$ 10^{24} kg, and using Eq. (2) again gives a limit to the concentration of ^{238}U of 6.85 $\times$ 10^{-9} kg/kg, and a ^{232}Th limit in the core of 26.71 $\times$ 10^{-9} kg/kg.

Although argued by Agostini et al. (2019) to be marginally consistent with a geodynamic model for mantle heat generation, these values are not consistent with published chondritic or non-chondritic BSE compositions (Table 4). In particular, these results exclude the enstatite-chondrite or non-chondritic Earth models at a high degree of confidence and, thus, more work is required before geoneutrino constraints can be regarded as reliable.

High-Pressure Partitioning Experiments and the Core HPE Budget A recurrent trope in planetary geochemistry is whether K and sometimes also U but seldom Th are not completely lithophile during core formation in the Earth, with non-negligible abundances of K (±U) in the core (Corgne et al. 2007). This discussion has largely been motivated by the challenge of facilitating a geodynamo through time, and preventing over-growth of the inner core due to core cooling. As an example, U and Th may under reducing conditions with a high-S content partition into the metal phase (Wohlers and Wood 2015). Such conditions might occur in a Mercury-type or enstatite-chondrite type body, with high S content. Wohlers and Wood (2015) explored this behaviour and suggested that the impact of a Mercury type parent body 20% the mass of the Earth, with the proto-Earth, might have introduced up to 4-5 ppm of U into the Earth's core (and similar for Th). This value would be larger for bigger impactors, or assuming more generous Th/K ratios for the Earth, and Wohlers and Wood (2015) suggest these isotopes alone may have produced up to 2-2.4 TW of heat in the core, sufficient for a geodynamo.

All the metallic elements in the periodic table can be reduced to the metallic state given sufficiently reducing conditions (that is why they are classified as metals), and most can then form liquid alloys with liquid Fe, and those that are immiscible in liquid Fe, like Ag or Pb, are soluble in liquid FeS, a possible core constituent. But demonstrating this chemically commonplace in an experiment does not mean that it occurred in nature. To show that requires comparing the siderophile tendencies of K or U with those of other Refractory Lithophile Elements at the same conditions, and balancing these tendencies against whether these other elements are also depleted in the Bulk Silicate Earth by partitioning into the core (O'Neill 2016). In the case of U, its abundance in the BSE is constrained relative to Th by U-Th-Pb isotopic systematics. Th is one of the more difficult elements to reduce to the metallic state. A recent appraisal of the evidence concluded trenchantly that "Negligible Th/U fractionation accompanied accretion, core formation, and crust–mantle differentiation, and trivial amounts of these elements (<0.2 ng/g U) were added to the core and do not significantly power (~0.03 TW) the geodynamo" (Wipperfurth et al. 2018). In addition, Blanchard et al. (2017) argued that U metal partitioning depends strongly on temperature and oxygen in the metal. Under core-forming conditions, they found an upper limit to the U content in the core of 3.5 pbb.

The case for K in the core is not so easily dismissed because of the uncertainties in estimating the BSE concentration of this element. It has been argued that the incorporation of over 100 ppm K into the core early in its history may have generated sufficient heat to facilitate a geodynamo, in the absence of an inner core (Nimmo et al. 2004). K in the core has also been suggested as a partial resolution to its depletion in the BSE (e.g. Wohlers and Wood 2015). However, current estimates for the depletion of K in the BSE relative to the RLEs are broadly consistent with those of other moderately volatile elements without siderophile affiliations, such as boron and fluorine, and the cosmochemically abundant alkali metal sodium (Palme and O'Neill 2014). Significant Na in the core is not expected from considerations of BSE Mn/Na ratios (Siebert et al. 2018). Furthermore, as Na is a very light metal, significant Na in the core would have implications for the density deficit of the Earth's core compared to Fe-Ni alloy, although this is a complicated matter given the poor constraints on the amounts of the more plausible candidates: S, O, Si, H and C have all been suggested. Nevertheless, the margin of uncertainty in these considerations are such that minor amounts of K in the core cannot be excluded. Recent experiments (Blanchard et al. 2017), however, have suggested a limit to amount of K in the core of 26 ppm, which, together with their U constraints, seem to limit heat production as a viable driver for the early geodynamo.

5 Implications for the Long-Term Evolution of Planets

The variations in heat-producing elements documented here has implications for the thermal evolution of planetary bodies (Jaupart et al. 2007). We explore these potential consequences by employing thermal history calculations, and comparing the evolution of a bulk-silicate Earth thermal model, to alternative Earth-sized models with variable heat production.

The models use a parameterised convection approach outlined in Schubert et al. (2001) and Korenaga (2006), and details can be found there. Parameterised models utilise a well known relationship between convective vigour—reflected in the non-dimensional Rayleigh number (Ra), and the normalised heat flux, the Nusselt number (Nu). Boundary layer theory, laboratory, and numerical experiments have shown that the relationship between these terms follows a form

$$Ra \sim Nu^{\beta}$$

The exponent β is around $1/3$ for simple isoviscous systems (Turcotte and Schubert 2002), but has been argued to be slightly lower for mobile lid systems (β ~0.26, Moresi and Solomatov 1998), and significantly lower for stagnant-lid systems which lose heat inefficiently (β ~0.12, Christensen 1985). To avoid complexities around Earth's tectonic evolution (e.g. O'Neill et al. 2018), we have used these tectonic endmembers to bracket possible temperature evolution. The results are shown in Fig. 6.

Typically BSE heat production is greater than chondrites; this is largely due to the apparent enrichment of Earth in U and Th compared to many chondrite groups (Table 2 and 4). Whilst Earth is depleted in K compared to chondrites, this has less of an effect. The carbonaceous chondrites are somewhat 'diluted' in silicates due to the presence of other carbon and ice compounds, and all have not experienced core extraction, which has the effect of volumetrically enriching HPEs in the BSE. Nonetheless, the variation in HPEs between the Earth and chondrite group extremes could, for given end-members, result in internal temperature differences of up to 100K. This internal temperature divergence could also have important consequences for timing of tectonic transitions within planets, and thus have a non-linear effect on the final thermal state.

Whilst Earth and Mars have reasonable HPE constraints due to sample availability, Venus and Mercury do not, and their internal abundance must be estimated from gamma ray spectroscopy constraints. Our thermal history calculations for Earth-sized planets, with HPE abundances for Mercury, Mars and Venus (Table 3), are shown in Fig. 5c. Fortuitously our estimates for Venus and Mercury result in thermal histories that are bracketed by those of BSE and Mars. The variation in temperatures possible for a given tectonic regime are slightly less than 100 K (e.g. for the stagnant-lid end-member), but again, this divergent evolution could have larger effects if it affects the timing of tectonic transitions.

Finally, we have also plotted a chondritic BSE model against a non-chondritic BSE, and 'geodynamically constrained' BSE, in Fig. 6d. The latter uses present day heat flux, and argues (Turcotte and Schubert 2002) that secular cooling contributes around 20% to the present-day flux. However, this argument utilised convective relationships appropriate for simple convecting systems, heated solely from within, and does not consider convective lag-times between Ra and Nu, which may overestimate the efficacy of secular cooling (and thus overestimate HPE concentrations). The non-chondritic model, derived from consideration of impact-collision crustal loss on early planetary bodies, has much lower HPE abundances that the chondritic BSE. Unsurprisingly, its internal temperatures are much colder that the other more enriched models.

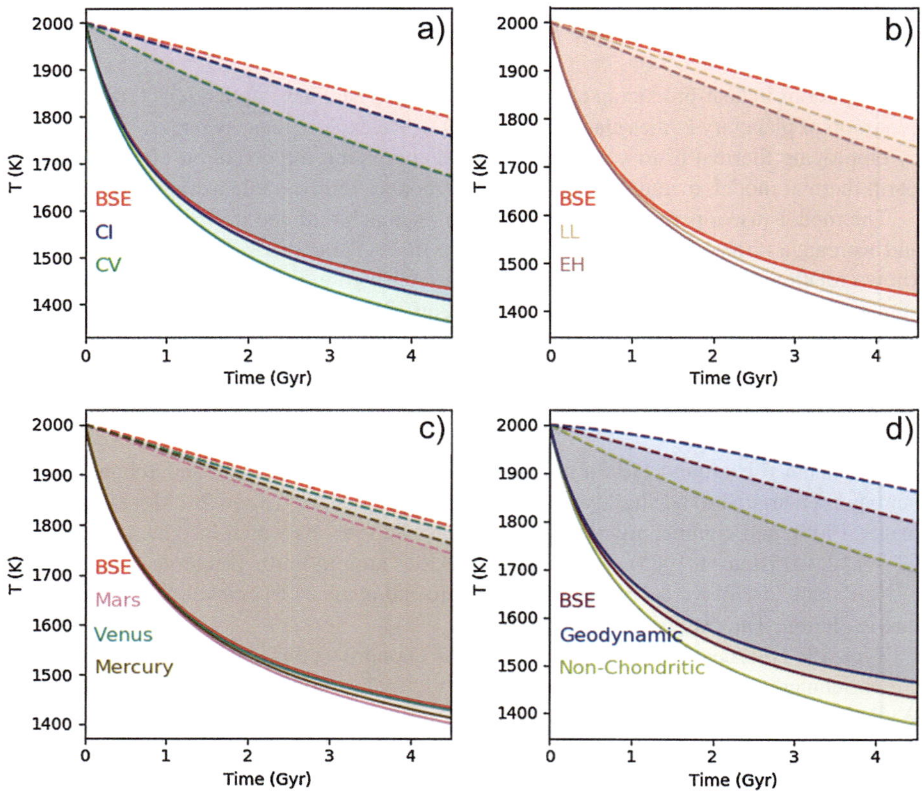

Fig. 6 Thermal evolution of Earth-sized planets for different abundances of heat-producing elements. The colour envelopes show the range of possible internal potential temperatures between a strict stagnant-lid regime (dashed lines), and a mobile lid regime (solid lines). (**a**) Difference in internal evolution between a bulk-silicate Earth model (see Table 4), and for an Earth with CI or CV chondrite HPE abundances (Table 2). (**b**) Comparison of BSE with LL ordinary chondrite and EH enstatite chondrite abundances (Table 2 and 4). (**c**) Comparison of BSE with estimated HPE abundances of Mars, Mercury and Venus (from Table 3). (**d**) Comparison of BSE based on chondritic abundances (red), non-chondritic estimates (brown) and geodynamic estimates (blue), from Table 4. Relevant physical parameters for the models are shown in Table 5

In order to assess the degree to which heat production may affect planetary tectonic regimes, and thus long-term evolution and habitability, we present a regime diagram (Fig. 7), showing the statistical likelihood for different tectonic regimes, depending on the magnitude of driving convective stresses compared to the retarding yield strength of the lithosphere.

Convective stresses are set by the mantle interior temperature through the exponential temperature-dependence of the mantle viscosity. In Fig. 7, the convective stress level for a mobile lid (plate tectonic) mode corresponds to a steady-state mantle interior temperature (yellow-filled black circles) that is given by the intersection (yellow-filled red circles) of a prescribed radiogenic heating power (dashed red lines) and mantle convective cooling power driven by three endmember mantle convective regimes (heavy red lines). In a probabilistic sense, the likelihood for a mobile lid- or episodic-style mantle convective regime is enhanced where the convective stress levels are within the lithospheric yield strength field. The yield strength is a wide field because this property evolves depending on the history of lithospheric damage and healing (e.g. Lenardic and Crowley 2012; Mulyukova and Bercovici 2018), as

Table 5 Parameters used to in the thermal history calculations of Fig. 6. For other details, see McNamara and Van Keken (2000), and Korenaga (2006)

Parameter	Value
Activation energy	300 kJ/mol
Density	4500 kg/m^3
Gravitational acceleration	9.81 m/s^2
Depth of convecting mantle	2890 km
Thermal conductivity	5.6 W/(m K)
Thermal expansivity	3×10^{-5} K^{-1}
Specific heat	1250 J/(kg K)
Reference viscosity	1×10^{21} Pa s

well as the thermal and volatile cycling history of a planet (Weller and Lenardic 2018; Lenardic et al. 2016). Lithospheric overturning in the presence of a water-rich atmosphere, for example, can cause the yield strength to decline over time through mineral hydration. Consequently, whether HPE concentrations permit a planet to enter into a mobile lid or intermittently mobile lid mode early in its evolution in the presence of a thick atmosphere can increase the likelihood that the planet will remain in a mobile lid regime. A similar early event in the presence of no atmosphere will induce a different evolutionary path.

In more detail, Fig. 7 enables a heuristic way to explore the sensitivity of Earth's (or any terrestrial planet's) current mode of plate tectonics to the HPE inventory. For example, if the present-day rate of internal heat production is increased by around a factor of 2, the mantle potential temperature increases to ~ 1600 °C (red arrows). Critically, the corresponding convective stress level falls well below the yield stress field (black arrows) and the initial mode of plate tectonics is unstable to natural perturbations to this temperature (Lenardic et al. 2011) through their effects on the convective stress (green-filled circle). There is, in turn, a strong likelihood for a transition to a stagnant lid mode of tectonics and a different thermal evolutionary path (lower red curve and red-filled circle). Although the specifics of these tectonic transitions are subject to ongoing debate in the geodynamics community, it is abundantly clear that the internal heating rates, and the HPE budget of a planet, exert an absolutely fundamental control on planetary evolutionary paths.

6 Conclusions and Directions

Heat producing elements exert a basic control over planetary evolution, and yet their present concentration and distribution within planets are not well understood. Significant variations in HPEs exist between primitive chondrites, and terrestrial planets. Although progress is being made on the processes involved in their early re-distribution, efforts are hampered by a lack of evidence at key stages, such as disk condensation, and early planetesimal evolution. New technologies (e.g. geoneutrino detection) are providing potential insights into the internal distribution of HPEs within the Earth, but rather fundamental issues regarding even Earth's total budget remain, including tighter constraints on deep mantle and core HPE concentrations, and ^{40}K concentrations, which are currently unconstrained by current generation proton-collision geoneutrino detectors, but may be amenable to future electron-collision detectors (Dye 2012).

Our current best estimates of the mantle, lithosphere, and crust's radiogenic heat contribution from geochemical and cosmochemical estimates place the total BSE heat production at 21.4 TW for a chondritic Earth, or 10.8 TW for non-chondritic models. Using the most

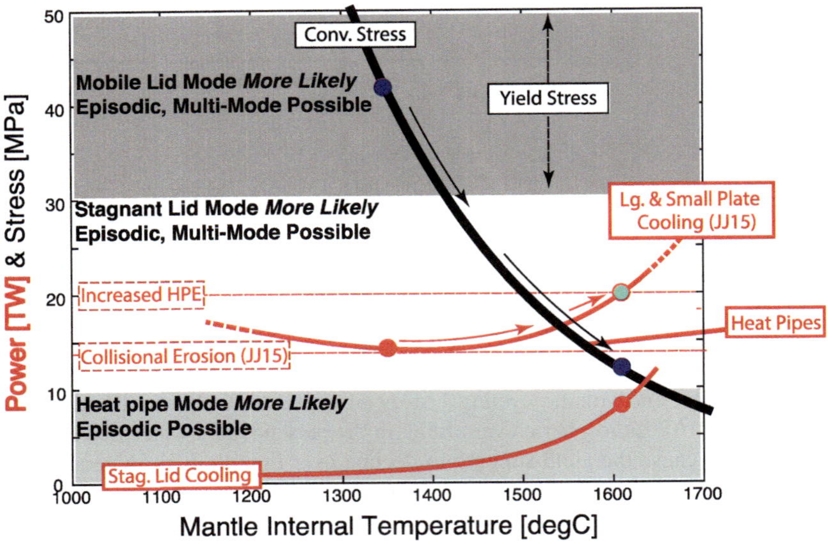

Fig. 7 Possible planetary evolutions (solid red lines) vs. tectonic endmember styles (shaded regions), as a function of the mantle convective stress imparted to the lithosphere (heavy black line) and HPE inventory expressed through present day radioactive heating power (dashed red lines) and mantle convective cooling regime (heavy red lines). This convective stress curve corresponds to the current "mobile lid" mode of plate tectonics and is governed by a strongly temperature-dependent mantle viscosity. Where stress levels are within the yield stress field, the likelihood for lithospheric failure and modern plate tectonics or episodic mantle convective regimes is increased in comparison to stagnant lid modes. For very low convective stress levels (below 10 MPa on this figure), lithospheric failure is impossible and the likelihood for cooling predominantly by mantle melting and magma transport to the surface in "heat pipes" is enhanced. The present-day convective stress level is well within the yield strength field (blue-filled circle)and corresponds to a mean mantle interior potential temperature (red-filled circle) given by the intersection of the non-chondritic "collisional erosion" radioactive heating power (dashed red line) and the cooling law for the "large and small plate regime" from Jellinek and Jackson (2015). This cooling law, which is consistent with Earth's current mode of plate tectonics and thermal history (Höink et al. 2013; van Summeren et al. 2012; Höink et al. 2013), reflects contributions to mantle cooling from large Pacific-sized (23%) plates, the motions of which are predominantly driven by plate buoyancy, and small Atlantic-sized plates (77%), the motions of which are mostly driven drag related to flow within an underlying asthenosphere. If the present day radiogenic heating power is increased to 20 TW, the interior temperature climbs to around 1625 °C (red arrows) and the corresponding convective stress drops well below the yield strength field (black arrows to blue-filled circle). This small- and large-plate solution for Earth's thermal history is consequently unstable (indicated by the green-filled circle). Natural fluctuations in the internal temperature and corresponding effects on the mantle convective stress will cause a transition to potentially a thermal evolution governed by the stagnant lid evolutionary path (filled red circle). See text

recent estimate for Earth's heat flux, 47 ± 1 TW (Davies and Davies 2010), then the implied Urey ratio is 0.45 or 0.23 respectively. The most recent heat production estimates from geoneutrinos is 38.2 +13.6/-12.7 TW, which would imply a Urey ratio of 0.81—though the uncertainties vary from 0.53 - 1.0. Uncertainties in geoneutrino estimates are large, as are variances between geochemical models, and the convergence of these approaches is critical to furthering models of Earth's internal HPE distribution.

The question of terrestrial exoplanet habitability—the long-term stability of surface conditions in the pressure-temperature window of liquid water—is closely tied to tectonic evolution and, thus heat production. These factors affect volcanic outgassing, and tectonically-associated in-gassing, and the major atmosphere gases associated with these processes (CO_2,

H_2O, SO_2, and noble gases) may be amenable to future spectroscopic measurements. With the next generation exoplanetary detection capability being able to detect Earth-sized planets, there is significant scope for refinement of our understanding of heat production variability within and beyond the solar system.

Publisher's Note Springer Nature remains neutral with regard to jurisdictional claims in published maps and institutional affiliations.

References

M. Agostini, K. Altenmüller, S. Appel, V. Atroshchenko, Z. Bagdasarian, D. Basilico, G. Bellini, J. Benziger, D. Bick, G. Bonfini, D. Bravo et al., Comprehensive geoneutrino analysis with Borexino. arXiv preprint (2019). arXiv:1909.02257

F. Albarede, Volatile accretion history of the terrestrial planets and dynamic implications. Nature **461**(7268), 1227 (2009)

M. Asplund, N. Grevesse, A.J. Sauval, P. Scott, The chemical composition of the Sun. Annu. Rev. Astron. Astrophys. **47**, 481–522 (2009)

J.A. Barrat, B. Zanda, F. Moynier, C. Bollinger, C. Liorzou, G. Bayon, Geochemistry of CI chondrites: major and trace elements, and Cu and Zn isotopes. Geochim. Cosmochim. Acta **83**, 79–92 (2012)

I. Bartos, S. Marka, A nearby neutron-star merger explains the actinide abundances in the early Solar System. Nature **569**(7754), 85–88 (2019)

F. Birch, Speculations on the Earth's thermal history. Geol. Soc. Am. Bull. **76**(2), 133–154 (1965)

M. Bizzarro, D. Ulfbeck, A. Trinquier, K. Thrane, J.N. Connelly, B.S. Meyer, Evidence for a late supernova injection of 60Fe into the protoplanetary disk. Science **316**(5828), 1178–1181 (2007)

I. Blanchard, J. Siebert, S. Borensztajn, J. Badro, The solubility of heat-producing elements in Earth's core. Geochem. Perspect. Lett. **5**, 1–5 (2017). https://doi.org/10.7185/geochemlet.1737.

J. Blichert-Toft, B. Zanda, D.S. Ebel, F. Albarède, The solar system primordial lead. Earth Planet. Sci. Lett. **300**(1–2), 152–163 (2010)

M. Boyet, R.W. Carlson, 142Nd evidence for early (> 4.53 Ga) global differentiation of the silicate Earth. Science **309**(5734), 576–581 (2005)

G.A. Brennecka, S. Weyer, M. Wadhwa, P.E. Janney, J. Zipfel, A.D. Anbar, 238U/235U variations in meteorites: extant 247Cm and implications for Pb-Pb dating. Science **327**(5964), 449–451 (2010)

C. Burkhardt, N. Dauphas, U. Hans, B. Bourdon, T. Kleine, Elemental and isotopic variability in solar system materials by mixing and processing of primordial disk reservoirs. Geochim. Cosmochim. Acta **261**, 145–170 (2019)

S.L. Butler, W.R. Peltier, Thermal evolution of Earth: models with time-dependent layering of mantle convection which satisfy the Urey ratio constraint. J. Geophys. Res., Solid Earth **107**(B6), ESE–3 (2002)

I.H. Campbell, H.S.C. O'Neill, Evidence against a chondritic Earth. Nature **483**(7391), 553 (2012)

R.M. Canup, Accretion of the Earth. Philos. Trans. R. Soc. Lond. Ser. A, Math. Phys. Sci. **366**(1883), 4061–4075 (2008)

R.W. Carlson, R. Brasser, Q.Z. Yin, M. Fischer-Gödde, L. Qin, Feedstocks of the terrestrial planets. Space Sci. Rev. **214**(8), 121 (2018)

R. Cayrel, V. Hill, T.C. Beers, B. Barbuy, M. Spite, F. Spite, B. Plez, J. Andersen, P. Bonifacio, P. Francois, P. Molaro, Measurement of stellar age from uranium decay. Nature **409**(6821), 691 (2001)

U.R. Christensen, Thermal evolution models for the Earth. J. Geophys. Res., Solid Earth **90**(B4), 2995–3007 (1985)

P.L. Clay, R. Burgess, H. Busemann, L. Ruzié-Hamilton, B. Joachim, J.M. Day, C.J. Ballentine, Halogens in chondritic meteorites and terrestrial accretion. Nature **551**(7682), 614 (2017)

C.P. Conrad, B.H. Hager, Mantle convection with strong subduction zones. Geophys. J. Int. **144**(2), 271–288 (2001)

A. Corgne, S. Keshav, Y. Fei, W.F. McDonough, How much potassium is in the Earth's core? New insights from partitioning experiments. Earth Planet. Sci. Lett. **256**(3–4), 567–576 (2007)

N. Dauphas, The isotopic nature of the Earth's accreting material through time. Nature **541**(7638), 521 (2017)

J.H. Davies, D.R. Davies, Earth's surface heat flux. Solid Earth **1**, 5–24 (2010)

A.M. Davis (ed.), *Meteorites, Comets, and Planets: Treatise on Geochemistry (Vol. 1)* (Elsevier, Amsterdam, 2005)

S.T. Dye, Geoneutrinos and the radioactive power of the Earth. Rev. Geophys. **50**(3), 3007 (2012)

M. Ebihara, S. Sekimoto, Halogen contents in meteorites (1) carbonaceous chondrites, in *50th Lunar and Planetary Science Conference*. LPI Contribution, vol. 2132 (2019), p. 2338

T.M. Esat, Comment on "Potassium isotope cosmochemistry: genetic implications of volatile element depletion" by Munir Humayun and RN Clayton. Geochim. Cosmochim. Acta **60**(19), 3755–3758 (1996)

E.A. Frank, B.S. Meyer, S.J. Mojzsis, A radiogenic heating evolution model for cosmochemically Earth-like exoplanets. Icarus **243**, 274–286 (2014)

A. Gando, Y. Gando, K. Ichimura, H. Ikeda, K. Inoue, Y. Kibe, Y. Kishimoto, M. Koga, Y. Minekawa, T. Mitsui, T. Morikawa, Partial radiogenic heat model for Earth revealed by geoneutrino measurements. Nat. Geosci. **4**(9), 647 (2011)

N. Grevesse, Abundances of the elements in the Sun, in *Frontiers of Astronomy and Astrophysics*, ed. by R. Pallavicini (Ital. Astron. Soc., Florence, 1984), pp. 71–82

N. Grevesse, P. Scott, M. Asplund, A.J. Sauval, The elemental composition of the Sun-III. The heavy elements Cu to Th. Astron. Astrophys. **573**, A27 (2015)

L. Grossman, Condensation in the primitive solar nebula. Geochim. Cosmochim. Acta **36**(5), 597–619 (1972)

L. Grossman, J.W. Larimer, Early chemical history of the solar system. Rev. Geophys. **12**(1), 71–101 (1974)

R.H. Hewins, M. Bourot-Denise, B. Zanda, H. Leroux, J.A. Barrat, M. Humayun, C. Göpel, R.C. Greenwood, I.A. Franchi, S. Pont, J.P. Lorand, The Paris meteorite, the least altered CM chondrite so far. Geochim. Cosmochim. Acta **124**, 190–222 (2014)

R.C. Hin, C.D. Coath, P.J. Carter, F. Nimmo, Y.J. Lai, P.A.P. von Strandmann, M. Willbold, Z.M. Leinhardt, M.J. Walter, T. Elliott, Magnesium isotope evidence that accretional vapour loss shapes planetary compositions. Nature **549**(7673), 511–515 (2017)

J.H. Hoffman, R.R. Hodges, T.M. Donahue, M.B. McElroy, Composition of the Venus lower atmosphere from the Pioneer Venus mass spectrometer. J. Geophys. Res. **85**, 7882–7890 (1980)

T. Höink, A. Lenardic, A.M. Jellinek, Earth's thermal evolution with multiple convection modes: a Monte-Carlo approach. Phys. Earth Planet. Inter. **221**, 22–26 (2013)

J.C. Holst, M.B. Olsen, C. Paton, K. Nagashima, M. Schiller, D. Wielandt, K.K. Larsen, J.N. Connelly, J.K. Jørgensen, A.N. Krot, Å. Nordlund, 182Hf–182W age dating of a 26Al-poor inclusion and implications for the origin of short-lived radioisotopes in the early Solar System. Proc. Natl. Acad. Sci. **110**(22), 8819–8823 (2013)

Y. Huang, V. Chubakov, F. Mantovani, R.L. Rudnick, W.F. McDonough, A reference Earth model for the heat-producing elements and associated geoneutrino flux. Geochem. Geophys. Geosyst. **14**(6), 2003–2029 (2013)

M. Humayun, R.N. Clayton, Potassium isotope cosmochemistry: genetic implications of volatile element depletion. Geochim. Cosmochim. Acta **59**(10), 2131–2148 (1995)

M.G. Jackson, A.M. Jellinek, Major and trace element composition of the high 3He/4He mantle: implications for the composition of a nonchonditic Earth. Geochem. Geophys. Geosyst. **14**(8), 2954–2976 (2013)

C. Jaupart, S. Labrosse, J.C. Mareschal, 7.06 - Temperatures, heat and energy in the mantle of the Earth, in *Treatise on Geophysics*, vol. 7 (2007), pp. 223–270

M. Javoy, The integral enstatite chondrite model of the Earth. Geophys. Res. Lett. **22**(16), 2219–2222 (1995)

A.M. Jellinek, M.G. Jackson, Connections between the bulk composition, geodynamics and habitability of Earth. Nat. Geosci. **8**(8), 587 (2015)

C.P. Johnstone, M.L. Khodachenko, T. Lüftinger, K.G. Kislyakova, H. Lammer, M. Güdel, Extreme hydrodynamic losses of Earth-like atmospheres in the habitable zones of very active stars. Astron. Astrophys. **624**, L10 (2019)

D. Kasen, B. Metzger, J. Barnes, E. Quataert, E. Ramirez-Ruiz, Origin of the heavy elements in binary neutron-star mergers from a gravitational-wave event. Nature **551**(7678), 80 (2017)

W.M. Kaula, Venus reconsidered. Science **270**(5241), 1460–1464 (1995)

W.M. Kaula, Constraints on Venus evolution from radiogenic argon. Icarus **139**(1), 32–39 (1999)

L.H. Kellogg, B.H. Hager, R.D. Van Der Hilst, Compositional stratification in the deep mantle. Science **283**(5409), 1881–1884 (1999)

J. Korenaga, Energetics of mantle convection and the fate of fossil heat. Geophys. Res. Lett. **30**(8), 1437 (2003)

J. Korenaga, *Archean Geodynamics and the Thermal Evolution of Earth*. Geophysical Monograph, vol. 164 (Am. Geophys. Union, Washington, 2006), p. 7

M.E. Kreutzberger, M.J. Drake, J.H. Jones, Origin of the Earth's Moon: constraints from alkali volatile trace elements. Geochim. Cosmochim. Acta **50**(1), 91–98 (1986)

S. Labrosse, C. Jaupart, Thermal evolution of the Earth: secular changes and fluctuations of plate characteristics. Earth Planet. Sci. Lett. **260**(3–4), 465–481 (2007)

S. Labrosse, J.W. Hernlund, N. Coltice, A crystallizing dense magma ocean at the base of the Earth's mantle. Nature **450**(7171), 866–869 (2007)

H. Lammer, M. Leitzinger, M. Scherf, P. Odert, C. Burger, D. Kubyshkina, C. Johnstone, T. Maindl, C.M. Schäfer, M. Güdel, N. Tosi, Measured atmospheric 36Ar/38Ar, 20Ne/22Ne, 36Ar/22Ne noble gas isotope and bulk K/U ratios constrain the early evolution of Venus and Earth. Icarus **339**, 113551 (2019)

J.W. Larimer, Chemical fractionations in meteorites—I. Condensation of the elements. Geochim. Cosmochim. Acta **31**(8), 1215–1238 (1967)

S.W. Lehner, W.F. McDonough, P. Németh, EH3 matrix mineralogy with major and trace element composition compared to chondrules. Meteorit. Planet. Sci. **49**(12), 2219–2240 (2014)

A. Lenardic, *Continental Growth and the Archean Paradox*. Geophysical Monograph, vol. 164 (Am. Geophys. Union, Washington, 2006), p. 33

A. Lenardic, J.W. Crowley, On the notion of well-defined tectonic regimes for terrestrial planets in this solar system and others. Astrophys. J. **755**(2), 132 (2012)

A. Lenardic, C.M. Cooper, L. Moresi, A note on continents and the Earth's Urey ratio. Phys. Earth Planet. Inter. **188**(1–2), 127–130 (2011)

A. Lenardic, A.M. Jellinek, B. Foley, C. O'Neill, W.B. Moore, Climate-tectonic coupling: variations in the mean, variations about the mean, and variations in mode. J. Geophys. Res., Planets **121**(10), 1831–1864 (2016)

A. Lenardic, M. Weller, T. Höink, J. Seales, Toward a boot strap hypothesis of plate tectonics: feedbacks between plates, the asthenosphere, and the wavelength of mantle convection. Phys. Earth Planet. Inter. **296**, 106299 (2019)

K. Lodders, Solar system abundances and condensation temperatures of the elements. Astrophys. J. **591**(2), 1220 (2003)

K. Lodders, B. Fegley Jr., An oxygen isotope model for the composition of Mars. Icarus **126**(2), 373–394 (1997)

K. Lodders, B. Fegley, Condensation chemistry of circumstellar grains, in *Symposium-International Astronomical Union*, vol. 191 (Cambridge University Press, Cambridge, 1999), pp. 279–290

K. Lodders, H. Palme, H.P. Gail, 4.4 Abundances of the elements in the Solar System, in *Solar System* (Springer, Berlin, 2009), pp. 712–770

H.C. Lord III., Molecular equilibria and condensation in a solar nebula and cool stellar atmospheres. Icarus **4**(3), 279–288 (1965)

M. Lugaro, A. Heger, D. Osrin, S. Goriely, K. Zuber, A.I. Karakas, B.K. Gibson, C.L. Doherty, J.C. Lattanzio, U. Ott, Stellar origin of the 182Hf cosmochronometer and the presolar history of solar system matter. Science **345**(6197), 650–653 (2014)

K. Makide, K. Nagashima, A.N. Krot, G.R. Huss, F.J. Ciesla, E. Hellebrand, E. Gaidos, L. Yang, Heterogeneous distribution of 26Al at the birth of the solar system. Astrophys. J. Lett. **733**, L31 (2011)

F.M. McCubbin, M.A. Riner, K.E. Vander Kaaden, L.K. Burkemper, Is Mercury a volatile-rich planet? Geophys. Res. Lett. **39**(9), L09202 (2012)

W.F. McDonough, Geoneutrino detection, in *Encyclopedia of Geochemistry*, ed. by W.M. White (Springer, Berlin, 2016). https://doi.org/10.1007/978-3-319-39193-9_213-1

S.M. McLennan, Geochemistry of sedimentary processes on Mars. Sediment. Geol. Mars **102**, 119–138 (2012)

A.K. McNamara, P.E. Van Keken, Cooling of the Earth: a parameterized convection study of whole versus layered models. Geochem. Geophys. Geosyst. **1**(11), 1027 (2000)

W.B. Moore, Heat transport in a convecting layer heated from within and below. J. Geophys. Res. Solid Earth **113**(B11), B11407 (2008)

W.B. Moore, A.A.G. Webb, Heat-pipe Earth. Nature **501**(7468), 501 (2013)

A. Morbidelli, B. Bitsch, A. Crida, M. Gounelle, T. Guillot, S. Jacobson, A. Johansen, M. Lambrechts, E. Lega, Fossilized condensation lines in the Solar System protoplanetary disk. Icarus **267**, 368–376 (2016)

L. Moresi, V. Solomatov, Mantle convection with a brittle lithosphere: thoughts on the global tectonic styles of the Earth and Venus. Geophys. J. Int. **133**(3), 669–682 (1998)

E. Mulyukova, D. Bercovici, Collapse of passive margins by lithospheric damage and plunging grain size. Earth Planet. Sci. Lett. **484**, 341–352 (2018)

S.G. Nielsen, M. Auro, K. Righter, D. Davis, J. Prytulak, F. Wu, J.D. Owens, Nucleosynthetic vanadium isotope heterogeneity of the early solar system recorded in chondritic meteorites. Earth Planet. Sci. Lett. **505**, 131–140 (2019)

F. Nimmo, G.D. Price, J. Brodholt, D. Gubbins, The influence of potassium on core and geodynamo evolution. Geophys. J. Int. **156**(2), 363–376 (2004)

L. Noack, M. Godolt, P. von Paris, A.C. Plesa, B. Stracke, D. Breuer, H. Rauer, Can the interior structure influence the habitability of a rocky planet? Planet. Space Sci. **98**, 14–29 (2014)

C.A. Norris, B.J. Wood, Earth's volatile contents established by melting and vaporization. Nature **549**(7673), 507 (2017)

H.S.C. O'Neill, The origin of the Moon and the early history of the Earth—A chemical model. Part 1: the Moon. Geochim. Cosmochim. Acta **55**(4), 1135–1157 (1991a)

H.S.C. O'Neill, The origin of the Moon and the early history of the Earth—A chemical model. Part 2: the Earth. Geochim. Cosmochim. Acta **55**(4), 1159–1172 (1991b)

H.S.C. O'Neill, Heat producing elements (HPEs), in *Encyclopedia of Geochemistry*, ed. by W.M. White (Springer, Berlin, 2016). https://doi.org/10.1007/978-3-319-39193-9_265-1

C. O'Neill, V. Debaille, The evolution of Hadean–Eoarchaean geodynamics. Earth Planet. Sci. Lett. **406**, 49–58 (2014)

H.S.C. O'Neill, F.E. Jenner, The global pattern of trace-element distributions in ocean floor basalts. Nature **491**(7426), 698 (2012)

H.S.C. O'Neill, H. Palme, Collisional erosion and the non-chondritic composition of the terrestrial planets. Philos. Trans. R. Soc. Lond. Ser. A, Math. Phys. Sci. **366**(1883), 4205–4238 (2008)

C. O'Neill, A. Lenardic, L. Moresi, T.H. Torsvik, C.T. Lee, Episodic precambrian subduction. Earth Planet. Sci. Lett. **262**(3–4), 552–562 (2007)

C. O'Neill, A. Lenardic, T. Höink, N. Coltice, Mantle convection and outgassing on terrestrial planets, in *Comparative Climatology of Terrestrial Planets* (2014), pp. 473–486

C. O'Neill, A. Lenardic, M. Weller, L. Moresi, S. Quenette, S. Zhang, A window for plate tectonics in terrestrial planet evolution? Phys. Earth Planet. Inter. **255**, 80–92 (2016)

C. O'Neill, S. Turner, T. Rushmer, The inception of plate tectonics: a record of failure. Philos. Trans. R. Soc. Lond. Ser. A, Math. Phys. Sci. **376**(2132), 20170414 (2018)

H. Palme, H.S.C. O'Neill, Cosmochemical estimates of mantle composition, in *Treatise on Geochemistry, Vol. 2: The Mantle and Core*, ed. by H.D. Holland, K.K. Turekian (2014)

P.N. Peplowski, L.G. Evans, S.A. Hauck, T.J. McCoy, W.V. Boynton, J.J. Gillis-Davis, D.S. Ebel, J.O. Goldsten, D.K. Hamara, D.J. Lawrence, R.L. McNutt, Radioactive elements on Mercury's surface from MESSENGER: Implications for the planet's formation and evolution. Science **333**(6051), 1850–1852 (2011)

H. Rizo et al., The elusive Hadean enriched reservoir revealed by 142Nd deficits in Isua Archaean rocks. Nature **491**, 96–100 (2013)

H. Rizo, R.J. Walker, R.W. Carlson, M.F. Horan, S. Mukhopadhyay, V. Manthos, D. Francis, M.G. Jackson, Preservation of Earth-forming events in the tungsten isotopic composition of modern flood basalts. Science **352**(6287), 809–812 (2016)

A.E. Rubin, H. Huber, J.T. Wasson, Possible impact-induced refractory-lithophile fractionations in EL chondrites. Geochim. Cosmochim. Acta **73**(5), 1523–1537 (2009)

T. Ruedas, Radioactive heat production of six geologically important nuclides. Geochem. Geophys. Geosyst. **18**(9), 3530–3541 (2017)

L. Schaefer, B. Fegley Jr., Chemistry of atmospheres formed during accretion of the Earth and other terrestrial planets. Icarus **208**(1), 438–448 (2010)

M. Scherff, H. Lammer, M. Leitzinger, P. Odert, C. Burger, D. Kubyshkina, C. Johnstone, T. Maindl, M. Güdel, N. Tosie, E. Marcq, Atmospheric noble gas isotope and bulk K/U ratios as a constraint on the origin and early evolution of Venus and Earth (2019)

M. Schiller, J.N. Connelly, A.C. Glad, T. Mikouchi, M. Bizzarro, Early accretion of protoplanets inferred from a reduced inner solar system 26Al inventory. Earth Planet. Sci. Lett. **420**, 45–54 (2015)

M. Schiller, M. Bizzarro, V.A. Fernandes, Isotopic evolution of the protoplanetary disk and the building blocks of Earth and the Moon. Nature **555**(7697), 507 (2018)

G. Schubert, D.L. Turcotte, P. Olson, *Mantle Convection in the Earth and Planets* (Cambridge University Press, Cambridge, 2001)

G.H. Shaw, Core formation in terrestrial planets. Phys. Earth Planet. Inter. **20**(1), 42–47 (1979)

J. Siebert, P.A. Sossi, I. Blanchard, B. Mahan, J. Badro, F. Moynier, Chondritic Mn/Na ratio and limited post-nebular volatile loss of the Earth. Earth Planet. Sci. Lett. **485**, 130–139 (2018)

P.G. Silver, M.D. Behn, Intermittent plate tectonics? Science **319**(5859), 85–88 (2008)

S.C. Solomon, R.L. McNutt Jr., R.E. Gold, M.H. Acuña, D.N. Baker, W.V. Boynton, C.R. Chapman, A.F. Cheng, G. Gloeckler, J.W. Head Iii, S.M. Krimigis, The MESSENGER mission to Mercury: scientific objectives and implementation. Planet. Space Sci. **49**(14–15), 1445–1465 (2001)

P.A. Sossi, S. Klemme, H.S.C. O'Neill, J. Berndt, F. Moynier, Evaporation of moderately volatile elements from silicate melts: experiments and theory. Geochim. Cosmochim. Acta **260**, 204–231 (2019)

C. Stein, J. Schmalzl, U. Hansen, The effect of rheological parameters on plate behaviour in a self-consistent model of mantle convection. Phys. Earth Planet. Inter. **142**(3–4), 225–255 (2004)

D.J. Stevenson, Models of the Earth's core. Science **214**(4521), 611–619 (1981)

S.S. Sun, W.F. McDonough, Chemical and isotopic systematics of oceanic basalts: implications for mantle composition and processes. Geol. Soc. (Lond.) Spec. Publ. **42**(1), 313–345 (1989)

Y.A. Surkov, F.F. Kirnozov, V.N. Glazov, A.G. Dunchenko, L.P. Tatsy, O.P. Sobornov, Uranium, thorium, and potassium in the Venusian rocks at the landing sites of Vega 1 and 2. J. Geophys. Res., Solid Earth **92**(B4), E537–E540 (1987)

B.M. Tinsley, Stellar lifetimes and abundance ratios in chemical evolution. Astrophys. J. **229**, 1046–1056 (1979). https://doi.org/10.1086/157039

D.C. Tozer, Heat transfer and convection currents. Philos. Trans. R. Soc. Lond. Ser. A, Math. Phys. Sci. **258**(1088), 252–271 (1965)

R. Trappitsch, P. Boehnke, T. Stephan, M. Telus, M.R. Savina, O. Pardo, A.M. Davis, N. Dauphas, M.J. Pellin, G.R. Huss, New constraints on the abundance of 60Fe in the Early Solar System. Astrophys. J. Lett. **857**(2), L15 (2018)

D.L. Turcotte, G. Schubert, *Geodynamics* (Cambridge University Press, Cambridge, 2002), pp. 23–48

H.C. Urey, The abundances of the elements. Phys. Rev. **88**(2), 248 (1952)

J. van Summeren, C.P. Conrad, C. Lithgow-Bertelloni, The importance of slab pull and a global asthenosphere to plate motions. Geochem. Geophys. Geosyst. **13**(2), Q0AK03 (2012)

L. Wang, J.J. Goodman, Wind-driven accretion in transitional protostellar disks. Astrophys. J. **835**(1), 59 (2017)

K. Wang, S.B. Jacobsen, Potassium isotopic evidence for a high-energy giant impact origin of the Moon. Nature **538**(7626), 487 (2016)

H. Wänke, G. Dreibus, Chemical composition and accretion history of terrestrial planets. Philos. Trans. R. Soc. Lond. Ser. A, Math. Phys. Sci. **325**(1587), 545–557 (1988)

H. Wänke, G. Dreibus, Chemistry and accretion history of Mars. Philos. Trans. R. Soc. Lond. Ser. A, Math. Phys. Sci. **349**(1690), 285–293 (1994)

H. Wänke, G. Dreibus, E. Jagoutz, Mantle chemistry and accretion history of the Earth, in *Archaean Geochemistry* (Springer, Berlin, 1984), pp. 1–24

J.T. Wasson, G.W. Kallemeyn, Compositions of chondrites. Philos. Trans. R. Soc. Lond. Ser. A, Math. Phys. Sci. **325**(1587), 535–544 (1988)

M.B. Weller, A. Lenardic, On the evolution of terrestrial planets: bi-stability, stochastic effects, and the non-uniqueness of tectonic states. Geosci. Front. **9**(1), 91–102 (2018)

S.A. Wipperfurth, M. Guo, O. Šrámek, W.F. McDonough, Earth's chondritic Th/U: negligible fractionation during accretion, core formation, and crust–mantle differentiation. Earth Planet. Sci. Lett. **498**, 196–202 (2018)

A. Wohlers, B.J. Wood, A Mercury-like component of early Earth yields uranium in the core and high mantle 142 Nd. Nature **520**(7547), 337 (2015)

B.J. Wood, D.J. Smythe, T. Harrison, The condensation temperatures of the elements: a reappraisal. Am. Mineral. **104**(6), 844–856 (2019)

E.D. Young, Assessing the implications of K isotope cosmochemistry for evaporation in the preplanetary solar nebula. Earth Planet. Sci. Lett. **183**, 321–333 (2000)

Space Sci Rev (2020) 216:74
https://doi.org/10.1007/s11214-020-00701-x

Loss and Fractionation of Noble Gas Isotopes and Moderately Volatile Elements from Planetary Embryos and Early Venus, Earth and Mars

Helmut Lammer[1] · Manuel Scherf[1] · Hiroyuki Kurokawa[2] · Yuichiro Ueno[3] · Christoph Burger[4] · Thomas Maindl[4] · Colin P. Johnstone[4] · Martin Leizinger[5] · Markus Benedikt[5,1] · Luca Fossati[1] · Kristina G. Kislyakova[4,1] · Bernard Marty[6] · Guillaume Avice[7] · Bruce Fegley[8] · Petra Odert[5]

Received: 4 October 2019 / Accepted: 3 June 2020 / Published online: 16 June 2020

Abstract Here we discuss the current state of knowledge on how atmospheric escape processes can fractionate noble gas isotopes and moderately volatile rock-forming elements that populate primordial atmospheres, magma ocean related environments, and catastrophically outgassed steam atmospheres. Variations of isotopes and volatile elements in different planetary reservoirs keep information about atmospheric escape, composition and even the source of accreting material. We summarize our knowledge on atmospheric isotope ratios and discuss the latest evidence that proto-Venus and Earth captured small H_2-dominated primordial atmospheres that were lost by hydrodynamic escape during and after the disk dispersed. All relevant thermal and non-thermal atmospheric escape processes that can fractionate various isotopes and volatile elements are discussed. Erosion of early atmospheres, crust and mantle by large planetary impactors are also addressed. Further, we discuss how moderately volatile elements such as the radioactive heat producing element ^{40}K and other

Reading Terrestrial Planet Evolution in Isotopes and Element Measurements
Edited by Helmut Lammer, Bernard Marty, Aubrey L. Zerkle, Michel Blanc, Hugh O'Neill and Thorsten Kleine

✉ H. Lammer

[1] Space Research Institute, Austrian Academy of Sciences, Schmiedlstr. 6, 8042 Graz, Austria

[2] Earth-Life Science Institute, Tokyo Institute of Technology, 2-12-1 Ookayama, Meguro-ku, Tokyo 152-8550, Japan

[3] Department of Earth and Planetary Sciences, Tokyo Institute of Technology, 2-12-1 Ookayama, Meguro-ku, Tokyo 152-8551, Japan

[4] Department of Astrophysics, University of Vienna, Türkenschanzstrasse 17, 1180 Vienna, Austria

[5] Institute of Physcis/IGAM, University of Graz, Universitätsplatz 5/II, 8010 Graz, Austria

[6] Centre de Recherches Pétrographiques et Geéochemiques, UMR CNRS - Université de Lorraine, 15 Rue Notre Dame des Pauvres BP 20, 54501 Vandoeuvre-lés-Nacy, France

[7] Université de Paris, Institut de Physique du Globe de Paris, CNRS, 75005 Paris, France

[8] Department of Earth & Planetary Sciences and McDonnell Center for the Space Sciences, Washington University, St. Louis, MO 63130, USA

rock-forming elements such as Na can also be outgassed and lost from magma oceans that originate on large planetary embryos and accreting planets. Outgassed elements escape from planetary embryos with masses that are $\leq M_{\rm Moon}$ directly, or due to hydrodynamic drag of escaping H atoms originating from primordial- or steam atmospheres at more massive embryos. We discuss how these processes affect the final elemental composition and ratios such as K/U, Fe/Mg of early planets and their building blocks. Finally, we review modeling efforts that constrain the early evolution of Venus, Earth and Mars by reproducing their measured present day atmospheric ^{36}Ar/^{38}Ar, ^{20}Ne/^{22}Ne, noble gas isotope ratios and the role of isotopes on the loss of water and its connection to the redox state on early Mars.

Keywords Protoplanetary disk · Primordial atmospheres · Steam atmospheres · Atmospheric escape · Noble gases · Isotopes · Magma oceans · Rock-forming elements · Planetary evolution

1 Introduction

The earliest atmospheric records of Venus, Earth and Mars are stored in the noble gases and their isotopes. The reason is that these elements do not react with other atmospheric species or the surface. Noble gas elements such as Ne, Ar, Kr and Xe and their isotopes contain historical records related to the early evolutionary processes starting during the accretion phase of planetesimals to planetary embryos before and after disk dispersal (Ozima and Podosek 2002; Porcelli et al. 2002).

Depending on how fast planetary embryos grow to larger masses within the disk and at which orbital locations they originate or migrate, embryos will evolve to objects with different volatile and elemental abundances, which results in different elemental ratios compared to their initial ones (Lammer et al. 2020a; this issue; Benedikt et al. 2020). Heating by short-lived ^{26}Al and ^{60}Fe radioisotopes shape the thermal history and interior structure of growing planetesimals and planetary embryos during the early stages of planetary formation (e.g., Lichtenberg et al. 2016; Young et al. 2019; Benedikt et al. 2020). The subsequent thermo-mechanical evolution, such as magma ocean-related internal differentiation, rapid volatile degassing, and the possible accumulation of H_2/He-envelopes has important implications for the final structure, composition and evolution of terrestrial protoplanets (Ikoma and Genda 2006; Elkins-Tanton 2012; Massol et al. 2016; Stökl et al. 2016; Lichtenberg et al. 2016; Benedikt et al. 2020). Under these extreme conditions, noble gases with low condensation temperatures such as Ar and Ne, significant amounts of major rock-forming elements (e.g., K, Si, Mg, Fe, Ca, Al, Na, S, P, Cl), and other moderately volatile elements like Rb or Zn with condensation temperatures of ≈ 1000 K and lower (Lodders et al. 2009) populate the hot accretionary atmospheres to significant amounts (Albarède and Blichert-Toft 2007; Schaefer and Fegley 2007; Fegley et al. 2016).

If one considers for instance a mixture of ordinary chondrites with enstatite-like material and/or carbonaceous chondrites with ureilites as recently suggested by Schiller et al. (2018) and as indicated by specific isotopic abundances to be the parent bodies of the Earth, then the terrestrial K/U ratio would be higher than the present-day ratio. This indicates that most likely some loss process during proto-Earth's accretion was necessary to deplete moderately volatile elements and fractionate K from U. Since radioactive decay of the heat producing element ^{40}K contributes to drive a long-lived magnetic dynamo and influences the long-term temperature evolution of a planet's interior (e.g. Turcotte and Schubert 2002; Murthy et al. 2003; Nimmo 2007; O'Neill et al. 2020; this issue), it is very important to investigate how atmospheric escape can modify the amount of ^{40}K and other moderately volatile elements such as Si, Mg, Fe, Ca, Al, Na, S, P, Cl, Rb, Zn until an Earth-like planet finishes its accretion.

It is currently believed that magma oceans played a large role in the early stages of terrestrial planet formation. Evidence that supports this hypothesis falls into four different categories:

(i) Samples found on the Moon, Mars and Vesta imply widespread silicate melting followed by fractional crystallization (e.g. Wakita and Schmitt 1970; Foley et al. 2005).
(ii) Ancient zircons have crystallization ages that fall into the time of planetary formation, which can be interpreted as accretion happening in a compressed timescale. As a result, heat production is more likely to cause melting instead of being radiated away into space (Harrison 2009).
(iii) Short accretionary timescales are also supported by ^{182}W and ^{142}Nd contents entrapped in crustal and mantle rocks, since those require early core formation (e.g. Foley et al. 2005; Caro et al. 2006).
(iv) Metallic core formation requires some kind of magma ocean (Terasaki et al. 2005).

According to magma ocean models, the slope of the adiabat in the pressure-temperature space is steeper than the solidus, they intersect at depth and the magma ocean starts to solidify upwards from the bottom (e.g., Elkins-Tanton 2012; Salvador et al. 2017). Cold plumes descend from the surface, resulting in a hotter return flow from the interior. In this way the solid region at the bottom grows as the magma ocean cools. The volatiles and trace gases contained in the magma ocean have low solid to liquid partition coefficients and as such do not crystallize (Ikoma et al. 2018). Hence, the liquid phase becomes more and more enriched in these elements as the magma ocean solidifies. Once the saturation limit for a given species is reached, it is outgassed catastrophically from the magma ocean into the atmosphere.

Although planetary embryos grow within the protoplanetary disk (Wade and Wood 2005), these bodies are differentiated and depleted in moderate volatile elements (Hin et al. 2017; Young et al. 2019) due to the loss of a magma ocean-related catastrophically outgassed steam atmospheres (Odert et al. 2018; Benedikt et al. 2020) as soon as they are not surrounded by the nebula gas anymore.

Hydrodynamic atmospheric escape, which is controlled by planetary parameters (e.g. Lammer et al. 2016; Owen and Wu 2016; Guo 2019) and the EUV flux of their young host stars (e.g. Koskinen et al. 2013; Erkaev et al., 2014, 2015, 2016; Lammer et al. 2014; Odert et al. 2018; Kubyshkina et al. 2018a; Guo 2019; Zahnle et al. 2019; Benedikt et al. 2020), can be very efficient if the mass of the protoplanetary body is small and the main upper atmospheric species is hydrogen. If hydrogen, which either originates from gas accretion in the disk or from dissociation of H_2O vapour of a magma ocean-related outgassed steam atmosphere, escapes efficiently, H atoms can drag along heavier species to space (Zahnle and Kasting 1986; Hunten et al. 1987; Zahnle et al. 1990, 2019; Chassefière 1996a, 1996b; Odert et al. 2018; Guo 2019; Benedikt et al. 2020). The conditions under which this so-called post-nebula atmospheric escape and volatilization processes take place, further lead to the modification of volatile composition and elemental abundances, as well as to the odd elemental and isotopic ratios that are observed in the Solar System planets (e.g. Gillmann et al. 2009; Marty 2012; Jellinek and Jackson 2015; Bonsor et al. 2015; Carter et al. 2015; Benedikt et al. 2020; Lammer et al. 2020b).

The main aim of this work is to review the latest knowledge on various atmospheric evolution and escape processes that fractionate isotopes and elements during Venus', Earth's and Mars' atmosphere evolution since their accretion phase in the disk. Section 2 summarizes our knowledge on the main atmospheric isotopic species and the composition of terrestrial planets obtained from various space missions. In Sect. 3 we discuss the evidence that accreting proto-Venus and Earth captured primordial atmospheres during their growth within the disk. Here we also discuss briefly the new knowledge obtained from exoplanet research and the discovery of H_2-dominated low mass exoplanets. In Sect. 4 we briefly review various thermal and non-thermal atmospheric loss processes and discuss their role in isotopic fractionation of different atmospheric species. In Sect. 5 the role of impact erosion of primordial atmospheres and crustal material as well as the role of this process in the modification of volatile elements of accreting protoplanets are discussed. In Sect. 6 we address the loss and isotopic and elemental fractionation of rock-forming volatile elements and noble gases from solidifying magma oceans from Moon- to Mars-like planetary embryos. In Sects. 7 and 8 we address the latest attempts for constraining the early evolution of Venus and Earth and Mars by reproducing their present noble gas and elemental ratios until these planets finished accretion. Section 9 concludes the review.

2 Atmospheric Isotopic Compositions of the Terrestrial Planets

Noble gases can be seen as a key to a planet's past because their abundances and isotope ratios represent the earliest record of atmospheres of terrestrial planets. Their evolved isotope ratios contain records of cataclysmic events that the evolving atmospheres experienced. These could be collisions with planetary embryos, comets, mass-fractionating atmospheric escape processes, outgassing from the planetary interior and geologic upheavals (Becker et al. 2003; Baines et al. 2013; Lammer et al. 2018).

Table 1 summarizes and compares the observed atmospheric isotope ratios in the atmospheres of Venus, Earth and Mars. The large Martian atmospheric $^{14}N/^{15}N$ disequilibrium (Fox and Hać 1997; Füri and Marty 2015) between the atmosphere and mantle and the similar but smaller disequilibrium on Earth (Cartigny and Marty 2013) combined with the fractionation of the significantly heavier Xe isotopes (e.g., Pepin 1991, 2000; Cassata 2017; Avice et al. 2018a, 2018b; Zahnle et al. 2019; Avice and Marty 2020; this issue) are a strong indication for different mass- and/or ionization-related atmospheric escape processes and atmosphere-magnetosphere environments of early Earth and Mars. These processes were shaped by the young Sun's radiation and plasma conditions (Lammer et al. 2018; and references therein). A detailed discussion on the nitrogen and Xe disequilibria in Earth's and Mars' atmosphere can be found in Avice and Marty (2020; this issue).

Additionally to Earth, icy moons such as Triton or the dwarf planet Pluto also possess N_2-dominated atmospheres. For example, Titan has a 1.45 bar atmosphere with a N_2-dominated atmosphere (Strobel and Shemansky 1982; Coustenis and Taylor 2008). However, the origin of such atmospheres is most likely related to early photolysis of accreted and outgassed NH_3 from subsurface H_2O-NH_3 oceans with contributions from refractory organics (Miller et al. 2019), leading to a very different evolution compared to Earth's N_2 atmosphere (Coustenis and Taylor 2008; Mandt et al. 2009, 2014; Miller et al. 2019). The origin and evolution of atmospheres that are constrained by $^{14}N/^{15}N$ isotope ratios at frozen worlds like Titan, Triton, and Pluto are discussed in Scherf et al. 2020; this issue). Here we focus mainly on noble gases and moderate volatile rock-forming element evolution on planetary embryos, early Venus, Earth and Mars that originated in an inner planetary system.

The present atmospheric $^{36}Ar/^{38}Ar$ and $^{20}Ne/^{22}Ne$ noble gas isotope ratios can provide historical records on the growth rate of proto-Venus and Earth in the presence of the nebula (Lammer et al. 2020b) and hence their early atmospheric evolution, which is also the case for

Table 1 Atmospheric isotope ratios of the terrestrial planets in the Solar System

Atmospheric isotope ratios	Venus	Earth	Mars
D/H	$(160 \pm 20) \times 10^{-4}$[a] 0.064 – 0.08 (above 70 km altitude[b])	$(1.49 \pm 0.03) \times 10^{-4}$[c] 1.56×10^{-4} (sea water[d])	$(7.58\text{-}10.9) \times 10^{-4}$[e] $< 1.99 \times 10^{-4}$ (interior[f])
$^3He/^4He$	Not measured; expected: $< 3 \times 10^{-4}$[g]	1.37×10^{-6}[h]	Not measured
$^{12}C/^{13}C$	88^{+2}_{-1}[i]	89 (inorganic[h])	85.1 ± 0.3[e,j]
$^{14}N/^{15}N$	273^{+70}_{-46}[k]	272[h]	173 ± 11[l]
$^{16}O/^{17}O$	Not measured	2520[h]	2577 ± 12[e,j]
$^{16}O/^{18}O$	500 ± 25[k]	489[h]	462 ± 2.5[e,j]
$^{20}Ne/^{22}Ne$	11.8 ± 0.7[a]	9.78[m]	10.1 ± 0.7[n]
$^{21}Ne/^{22}Ne$	Not measured; expected: < 0.067[a]	0.029[m]	Not measured[h]
$^{33}S/^{32}S$	Not measured; expected: $\approx 8 \times 10^{-3}$[h]	8.01×10^{-3}[h]	Not measured[h]
$^{34}S/^{32}S$	Not measured; expected: ≈ 0.04[h]	0.045[h]	Not measured[h]
$^{36}Ar/^{38}Ar$	5.6 ± 0.6[o]	5.32 ± 0.33[p]	4.2 ± 0.1[n]
$^{40}Ar/^{36}Ar$	1.03 ± 0.04[k] 1.19 ± 0.04[i]	298.56 ± 0.31[q]	$(1.714 \pm 0.17) \times 10^3$[e,j]
$^{136}Xe/^{132}Xe$	Not measured; expected: ≈ 1[h]	0.3294 ± 0.0004[p]	0.3451 ± 0.0023[p] ALH84001: 0.3450 ± 0.002[r]
$^{129}Xe/^{132}Xe$	Not measured; expected: ≈ 3[h]	0.983 ± 0.001[p]	2.522 ± 0.006[p] ALH84001: 2.151 ± 0.032[r]

[a]Donahue et al. (1982), [b]Bertaux et al. (2007), [c]Lécuyer et al. (1998), [d]Michael (1988), [e]Webster et al. (2013a, 2013b), [f]Hallis et al. (2012), [g]von Zahn et al. (1983), [h]Baines et al. (2013), [i]Istomin et al. (1980a), [j]Avice et al. (2018a, 2018b), [k]Hoffman et al. (1980), [l]Wong et al. (2013), [m]Marty (2012), [n]Pepin (1991), Atreya et al. (2013), [o]Istomin et al. (1980b), [p]Ozima and Podosek (2002), [q]Lee et al. (2006), [r]Cassata (2017)

Mars (e.g., Jakosky et al. 1994, 2017; Kurokawa et al. 2018). However, there are yet some relevant uncertainties on some of the noble gas isotope ratios at these planets. For Venus, one should also note that due to large uncertainties, the measured value of $^{20}Ne/^{22}Ne$-ratio is not precise enough for accurate scientific statements related to Venus' atmospheric evolution (von Zahn et al. 1983; Johnson and Oliveira 2019; Lammer et al. 2020b). New precise measurements by future spacecraft are necessary (Dandouras et al. 2020; this issue). It is also interesting to note that the $^{40}Ar/^{36}Ar$ ratio measured in the 4.1 Ga old Martian meteorite ALH84001 is ≤ 1400 (Murty and Mohapatra 1997; Grady et al. 1998) and therefore lower compared to the modern atmospheric value that has been measured both in meteorites and in-situ to be ≈ 1700 (Webster et al. 2013a, 2013b; Avice et al. 2018a, 2018b). One can expect that those ratios rise over time because of atmospheric escape and/or outgassing (Swindle 2002).

A reproduction of the primordial and steam atmosphere related D/H ratios yields information that may lead to a better understanding of the oxidation states and initial H_2O inventories of terrestrial planets. The terrestrial water (oceans) and meteorites are enriched by a factor of about 6 in D relative to the protosolar nebula value (Robert et al. 2000; Marty 2012; Pahlevan et al. 2019). Besides heterogeneities in the disk, mass fractionation during atmospheric escape processes from early terrestrial planets can enrich water in oceans in D, since the lighter isotopes are preferentially concentrated in the escaping upper atmosphere relative to ocean water (Genda and Ikoma 2008).

The preservation of a carbonaceous chondritic D/H signature with a mean statistical D/H ratio of $(150 \pm 10) \times 10^{-4}$ (e.g., Robert et al. 2000) in Earth's seawater (Marty 2012) provides evidence that early Earth's ocean underwent minimal D-enrichment via equilibration with H_2 followed by hydrodynamic escape. According to the recent results of Pahlevan et al. (2019) the close match of $\approx$ 10 - 20% between Earth's seawater and carbonaceous chondrites further constrains the prior existence of an atmospheric H_2-inventory of any origin on post-giant-impact Earth to < 20 bars. Moreover, the observed oxidation of silicate Earth occurred before or during the crystallization of the final magma ocean (Pahlevan et al. 2019). Unless D escape is very efficient on Venus, the present H escape flux (averaged over a solar cycle) cannot be larger than about 10^7 cm^{-2} s^{-1}, if the present water vapour in Venus' atmosphere is a remnant of water deposited billion years ago (Donahue 1999). On the other hand, these authors argued that, if the escape flux remains constant and is as large as about 3×10^7 cm^{-2} s^{-1}, water would be the remnant of H_2O outgassed only about 500 million years ago. If the hydrogen escape fluxes were $> 3 \times 10^7$ cm^{-2} s^{-1} during the past 100 Myrs then the present day D/H ratio in Venus' atmosphere could have been created since a massive resurfacing event a few 100 Myrs ago (Grinspoon 1993).

Hydrogen escape rates based on the loss of suprathermal H atoms and ions on present Venus were studied by Lammer et al. (2006). These authors modelled photochemically produced suprathermal H atom escape rates in the order of $\approx 3.8 \times 10^{25}$ s^{-1} which is in agreement with the estimates of Donahue and Hartle (1992).

Ion escape rates were uncertain before the arrival of Venus Express, and model-based H^+ escape rates in the order of $\approx 7.0 \times 10^{25}$ s^{-1} largely overestimated the escape from Venus over the planets nightside by assuming ion acceleration by an outward electric polarization force related to ionospheric holes (Hartle and Grebowsky 1993). A recent study that investigated the H^+ escape from Venus during 8 years of data from the Venus Express ion mass analyser instrument obtained much lower H^+ ion loss rates in the order of $\approx 7.6 \times 10^{24}$ s^{-1} at solar minimum and $\approx 2.1 \times 10^{24}$ s^{-1} during solar maximum (Persson et al. 2018). These authors found that the decrease of the H^+ ion escape from solar minimum to maximum is caused mainly by the change in the flow direction in the magnetotail. During solar maximum a significant Venus-ward ion flow close to the planet and its magnetotail was observed (Persson et al. 2018). Thus, the present day Venus total hydrogen escape is in the order of $\approx 4.5 \times 10^{25}$ s^{-1} ($\leq 10^7$ cm^{-2} s^{-1}) with suprathermal H atom escape as the dominant process.

The findings of lower H^+ ion escape rates by Persson et al. (2018) compared to previous estimates indicates that Venus' present D/H ratio is most likely a remnant of the ancient evaporated water inventory and subsequently outgassed water vapour that were modified by various non-thermal atmospheric escape processes during time periods that lasted longer than the last resurfacing event which occurred about 200 - 1000 Myr ago (e.g., Strom et al. 1994; Basilevsky et al. 1997; Kreslavsky et al. 2015).

Other isotopes such as $^{16}O/^{17}O$, $^{16}O/^{18}O$ may reveal information related to a common kinship of the planets (e.g., Clayton 1993; Yurimoto et al. 2006), $^{33}S/^{32}S$ and $^{34}S/^{32}S$ isotopes contain evidence on the past and current planetary volcanic activity and magmatic composition (e.g., Mahaffy et al. 2012), outgassing from the interior is recorded by ^{4}He, ^{40}Ar and $^{129,131\text{-}136}Xe$ (e.g., Allegre et al. 1987; Marty and Allé 1994; Watson et al. 2007), and $^{12}C/^{13}C$ can be related to a possible biological activity (e.g., Galimov 1985, 2003). As one can see in Table 1, several isotope ratios have large error bars when their measurements were analysed in laboratory settings, or were not measured/detected by spacecraft in the past. Therefore, new future in-situ measurements of these isotopes as discussed in Dandouras et al. (2020; this issue) in planetary atmospheres are necessary for constraining and understanding the state and evolution of terrestrial planets in the Solar System and beyond.

3 The Disk Phase: H_2-Dominated Envelopes on Low Mass Planets

In the Solar System, there are two well-separated classes of planets: rocky planets (Mercury, Venus, Earth, and Mars with average densities between 3.7 and 5.5 $g\,cm^{-3}$) and gas/ice giants (Jupiter, Saturn, Uranus, and Neptune with average densities between 0.7 and 1.6 $g\,cm^{-3}$). The rocky planets have a mass of about one Earth-mass (M_{Earth}), or less, while gas giants have masses of tens to hundreds of M_{Earth}. However, the largest population of exoplanets discovered to date has a mass/radius falling between those of the Earth and Neptune (e.g., Mullally et al. 2015), demonstrating that the variety of planets existing in nature is much larger than previously thought.

3.1 Low Mass Exoplanets with Primordial Atmospheres

The mass-radius, or equivalently mass-density, diagram is the first diagnostic enabling the basic classification of a planet, e.g. identify whether a planet hosts an H_2-dominated atmosphere or not. The rapidly increasing sample of Neptune-like planets and smaller for which we have a measurement of both mass and radius indicates that these planets present a surprisingly large variety of densities (see Fig. 1).

This can be clearly seen in the left panel of Fig. 1 showing the position of the known exoplanets with a mass smaller than Neptune in the mass vs. radius diagram, color-coded by the planetary equilibrium temperature (T_{eq}). This plot clearly indicates the presence of a large spread in planetary radius, and thus density, at a given mass, particularly for planets with masses larger than 1.5-2.0 M_{Earth}. The same can be seen also when considering only systems for which masses and radii are known to better than 25% (Fig. 1, right panel), indicating that this large spread in bulk density is not the result of large uncertainties on the planetary parameters.

Although planetary cores with various densities/compositions certainly exist, this spread is most likely tightly related to the existence of a large variety of planetary atmospheres that can range from being H_2-dominated primordial atmospheres to CO_2/H_2O-dominated steam atmospheres, or secondary atmospheres, thus more compact. Furthermore, this spread does not seem to be directly correlated with the amount of irradiation they receive from the host star. The origin of this phenomenon is unclear and several factors are likely to play a role, such as protoplanetary nebula lifetime (e.g., Stökl et al. 2015, 2016), mass accretion time (e.g., Schiller et al. 2018), migration history, collisions (e.g., Bonomo et al. 2019), received amount of high-energy irradiation (e.g., Fossati et al. 2017; Owen and Wu 2016; Zahnle and Catling 2017; Jin and Mordasini 2018; Kubyshkina et al. 2018b, 2018c), possible biases in the measurement of planetary radii due to the presence of high-altitude aerosols (e.g., Lammer et al. 2016; Cubillos et al. 2017a, 2017b).

It has been shown that these planets are likely to have Earth-like density cores (Owen and Wu 2016) and the observations conducted to date clearly show that a number of them still host an H_2-dominated atmosphere, most likely accreted from the protoplanetary nebula, despite their close distance to the host star. It is therefore possible that also the inner planets of the Solar System began their evolution with a rocky core surrounded by an H_2-dominated envelope.

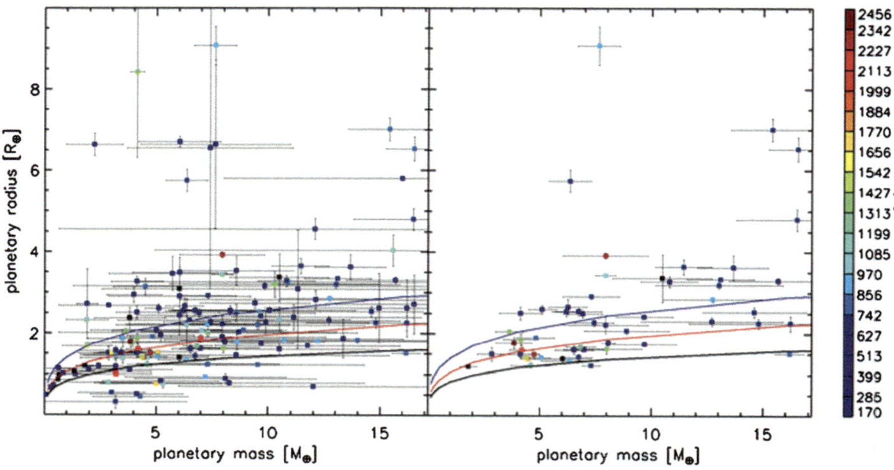

Fig. 1 Left: known exoplanets with masses smaller than Neptune in the mass vs radius plane, in Earth units. The color indicates the planetary equilibrium temperature T_{eq}. The black, red, and blue lines indicate compositions corresponding to 100% Fe, 100% rock, and 100% water, respectively (from Zeng et al. 2016). Right: same as left panel, but considering only planets with a mass and radius known to better than 25%. Data from the NASA exoplanet archive (https://exoplanetarchive.ipac.caltech.edu/)

3.2 Evidence of Captured Nebula Gas During Early Earth's and Venus' Accretion

Besides disk lifetime (e.g. Montmerle et al. 2006; Bollard et al. 2017; Wang et al. 2017) and the young Sun's X-ray, EUV and FUV fluxes and plasma winds (Tu et al. 2015; see also Güdel 2020; this issue), the accretion time is the relevant factor on whether a protoplanet captured enough mass to be able to accumulate and sustain an H_2-dominated envelope after the disk disappeared (e.g., Ikoma and Genda 2006; Lammer et al. 2016, 2020b; Stökl et al. 2015, 2016). In the case of the early terrestrial planets in the Solar System, scientists of various research areas debate since decades whether it was possible for proto-Venus and Earth to accrete enough mass during the disk lifetime for accumulating nebula gas before the disk disappeared. If a protoplanet reached a mass $> 0.5\ M_{\text{Earth}}$ before the disk evaporated it could have captured a small H_2-dominated envelope that remains after the nebula dissipated (e.g., Ikoma and Genda 2006; Stökl et al. 2015, 2016; see also Lammer et al. 2020a; this issue).

Captured nebula gas was used in past attempts to reproduce the present atmospheric $^{36}Ar/^{38}Ar$ and $^{20}Ne/^{22}Ne$ noble gas ratios of Venus and Earth by various researchers (e.g., Sekiya et al. 1980a, 1980b; Sasaki and Nakazawa 1989; Harper and Jacobsen 1996; Pepin 1991, 2000, Porcelli and Pepin 2000; Porcelli et al. 2001; Gillmann et al. 2009; Odert et al. 2018; Lammer et al. 2020b). In the case of the Earth, there is data from $^{20}Ne/^{22}Ne$ isotope ratios in mantle material indicating that a fraction of noble gases has been trapped in Earth's interior in solar composition from the protoplanetary disk (Mizuno et al. 1980; Porcelli et al. 2001; Yokochi and Marty 2004; Mukhopadhyay 2012). Some researchers argued that these $^{20}Ne/^{22}Ne$ isotope ratios between 12.52-12.75 could have also been reproduced through implantation of solar wind onto accreted material (Trieloff et al. 2000; Podosek 2003; Feigelson et al. 2002; Raquin and Moreira 2009; Moreira and Charnoz 2016; Péron et al. 2016, 2017). Later on, such solar wind related noble gas isotopes could then be outgassed during the magma ocean phase. A recent study, however, measured $^{20}Ne/^{22}Ne$ values up to 13.03 ± 0.04 from deep mantle plumes and determined a ratio for the primordial plume mantle of 13.23 ± 0.22 (Williams and Mukhopadhyay 2019). These values are only slightly fractionated but comparable with the nebular (solar) ratio. Even if the debate carries on, this can be seen as a robust evidence for a reservoir of nebular gas that is preserved in the deep mantle. If the accreting proto-Venus and Earth captured little H_2-dominated envelopes from the disk then these primordial atmospheres were lost during some time after disk dispersal.

Below we discuss various atmospheric escape processes, thier efficiencies and capabilities for the fractionation of various elements (i.e., isotopes and volatile elements, including rock-forming ones).

4 Atmospheric Escape Processes and Their Relevance for Elemental Fractionation

Losses of atmospheric gases to space are caused by a large number of physical processes taking place in the thermosphere, exosphere, and in the case of the Earth in the magnetosphere. These processes are mostly driven by inputs from the Sun in the forms of its high-energy radiative spectrum and the solar wind. For this reason, loss processes were more important during earlier phases of the Solar System when the Sun was more active.

4.1 Upper Atmospheres

The upper atmospheres of planets are of primarily importance for determining how rapidly atmospheric gases are lost to space and for determining the relative losses of different elements and isotopes. Understanding how Earth's upper atmosphere has evolved is important for understanding the effects of losses on isotopic fractionation. Various atmospheric layers of interest are outlined in Fig. 2. The main layers in a planetary atmosphere can be defined by the temperature structure. On Earth, above the troposphere and stratosphere, starting at approximately 50 km altitude, is a region called the mesosphere; in this region, the temperature decreases with altitude due to cooling by CO_2 molecules emitting infrared radiation (IR) to space. Above this is the thermosphere, starting at approximately 100 km altitude, which is heated by the Sun's X-ray, extreme ultraviolet and ultraviolet radiation, and therefore has a positive temperature gradient. On Earth, only negligible amounts of CO_2 reach the thermosphere, hence no efficient IR-coolant as on the CO_2-dominated atmospheres of Venus and Mars that more or less prevents the formation of an Earth-like thermosphere on these planets.

The top of the thermosphere is the exobase which is ≈ 500 km for the modern Earth but only about 210 km at present Venus and Mars (Lammer et al. 2018; and references therein). The exobase is the point above which the gas density is so low that the mean free paths of particles are long enough compared to the scale height of the atmosphere making the gas non-collisional. This non-collisional region, for the neutral gas above the exobase, is the exosphere. An additional important layer is the thermosphere-ionosphere region, which starts in the upper mesosphere and extends to the exobase. This is the domain in which most of the Sun's X-ray and extreme ultraviolet radiation is absorbed, which causes photoionization of the atmospheric constituents.

In the thermosphere, the composition of the gas not only changes due to molecular diffusion, but also due to photodissociation and photoionization, driven by the absorption of

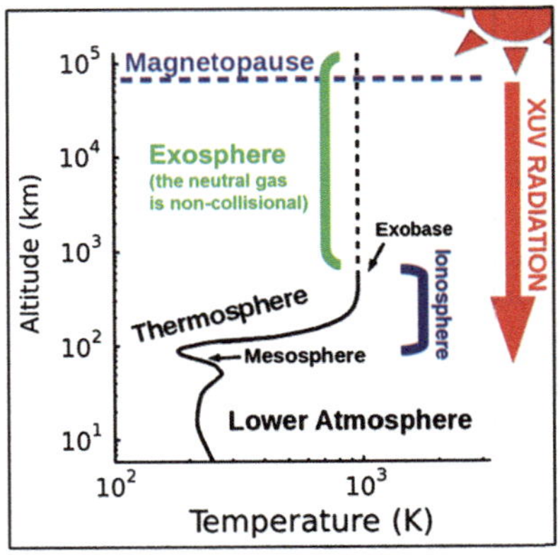

Fig. 2 Earth's atmospheric temperature structure and regions of interest

solar radiation at wavelengths below approximately 200 nm. For the modern Earth, the dissociation of O_2 and N_2 molecules, combined with molecular diffusion, mean that the gas in the exobase is primarily composed of atomic O and N.

The absorption of this radiation has several additional effects on the gas, with the most important being heating. When the atmosphere is in a steady state, this heating of the upper atmosphere is balanced by cooling from infrared emission to space, particularly by CO_2. Locally within the gas however, heating and cooling do not balance, and downward thermal conduction is in fact the main process responsible for cooling the upper thermosphere.

For Venus and Mars, the same basic set of physical processes are important in the upper atmospheres, with the only additional important process being heating by the absorption of solar infrared radiation by CO_2 (Fox and Bougher 1991). The high CO_2 concentration in these atmospheres also means that infrared cooling is much stronger and the thermospheres are therefore much cooler (e.g., de Pater and Lissauer 2015; Bauer and Lammer 2004). The higher concentrations of CO_2 expected in the Earth's atmosphere at earlier times (Walker 1985; Tian et al. 2005) would also likely have this effect, cooling the atmosphere and reducing the effects of atmospheric loss (Kulikov et al. 2007; Johnstone et al. 2018).

As illustrated in Fig. 3, it is common to break down the vertical structure of an atmosphere into the homosphere and heterosphere. In the homosphere, which for the Earth extends from the surface to the homopause at approximately 120 km altitude, turbulent mixing of the gas causes the mixing ratios of all long lived species (those that are not rapidly created and destroyed by chemical processes) to be uniform with altitude regardless of the particle masses. This process is often known as eddy diffusion. In the homosphere, diffusion is slower than mixing, allowing heavier particles such as CO_2 molecules and noble gases such as Ne, Ar, and even Xe atoms to reach the lower thermosphere. Without eddy diffusion, the abundances of heavier particles would rapidly drop with increasing altitude as molecular diffusion separates the particles by mass. This is indeed what happens above the homopause, with the abundances of heavier particles becoming extremely low in the mid and upper thermospheres. Here, diffusion is faster than mixing, and heavy species (e.g., N_2, Xe, etc.) decrease in density with altitude more rapidly than lighter species. This process contributes to causing lighter particles such as H atoms to be lost from the atmosphere more efficiently than heavier particles.

An important concept related to the escape of H atoms is diffusion-limited escape. This is itself not a loss process but an upper limit on how rapidly H atoms can be lost from the upper atmosphere of a planet that is dominated by an other species (N, O, etc.). This limit

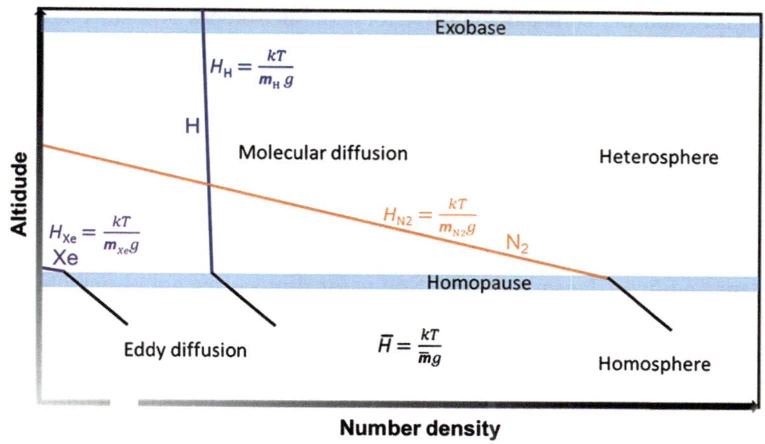

Fig. 3 Illustration of density-altitude structures of H atoms, N_2 molecules and Xe isotopes in the homosphere-heterosphere transition region. In the homosphere the atmosphere is mixed and all species are distributed by the same average scale height (homosphere), while above the homopause all species are distributed by their own scale heights as a function of mass. The density of heavy elements decreases very fast compared to light elements

exists because a minor species (i.e. H atoms) cannot be lost from the upper atmosphere more rapidly than they can be transported upwards by molecular diffusion at the homopause.

For instance, H atom escape on present Earth, Venus and Mars is limited by upward diffusion through the upper atmosphere (e.g. Hunten 1973; Kasting and Pollack 1983). On Earth, H_2O, H_2, and CH_4 move upward in the atmosphere. At $\approx$17 km, the cool temperatures at the tropopause freeze out most of the water vapour, preventing the upward transport of some hydrogen. Near the homopause hydrogen bearing molecules (i.e. CH_4) are dissociated by UV photons so that atomic and molecular hydrogen particles are produced. These H and H_2 particles diffuse upward through the upper atmosphere to the exobase level from where they can escape by thermal Jeans escape. On present Earth, the "bottleneck" for hydrogen escape is therefore diffusion-limited. This diffusion-limited flux is proportional to the hydrogen-mixing ratio at the homopause, which is $\approx 5.0 \times 10^{-7}$ at present Venus (Kasting and Pollack 1983) and $\approx 1.2 \times 10^{-5}$ and $\approx 3.0 \times 10^{-5}$ at present Earth and Mars, respectively (Krasnopolsky and Feldman 2001). For such low water vapour mixing ratios at the homopause the escape of hydrogen at the exobase level at Mars and Venus, but also at Earth are thermal Jeans escape.

Higher homopause water mixing ratios lead to higher escape rates that are not controlled by diffusion through the upper atmosphere anymore. Kasting and Pollack (1983) found that for H_2O mixing ratios at the homopause that are $\geq 5.0 \times 10^{-4}$ the stellar/solar EUV flux that is absorbed above the homopause level in a hydrostatic atmosphere becomes larger than the energy that is carried away by the thermally escaping particles. At such high H_2O mixing ratios at the homopause level the hydrogen-dominated upper atmosphere becomes unstable against expansion and Jeans escape transitions to hydrodynamic escape. According to Kasting and Pollack (1983), for H_2O mixing ratios at the homopause that are $\geq 10^{-3}$ the hydrogen escape flux reaches values that are comparable to energy limited escape. High H_2O-mixing ratios at the homopause level in the order of 0.1 and higher can originate in steam atmospheres (e.g., Lebrun et al. 2013; Massol et al. 2016; Salvador et al. 2017; Benedikt et al. 2020). Under such conditions thermal escape of H atoms from the upper atmosphere and the corresponding fractionation of isotopes and other heavier minor and/or trace species is then controlled by thermal hydrodynamic escape (e.g., Hunten 1973; Zahnle and Kasting 1986; Hunten et al. 1987; Zahnle et al. 1990; Odert et al. 2018; Lammer et al. 2020a, 2020b).

4.2 Atmospheric Loss Processes and Their Influence on Isotope Fractionation

Loss processes are often broken down into thermal and non-thermal processes. Thermal processes are related to the thermal energy of the gas and can be primarily broken down into Jeans escape and hydrodynamic escape. Non-thermal processes are essentially all others. Most escape processes preferentially remove lighter isotopes, though they differ in how efficiently they fractionate by mass. The ability of an escape process to fractionate isotopes is often described by a fractionation factor f, with $f > 1$ meaning that the heavier isotope is preferentially removed, $f < 1$ meaning that the escape process preferentially removes the lighter isotope, and $f = 1$ describing the case of no fractionation. The Rayleigh distillation relationship is commonly used to describe fractionation of an isotope ratio due to a separate process. This relationship connects the initial, n_0, and current, n, inventory of the constituent and a fractionation factor f of a particular escape process, "x" (e.g., Lunine et al. 1999; Donahue 1999; Mandt et al. 2009, 2014, 2015):

$$\frac{n_0}{n} = \left(\frac{R}{R_0}\right)^{\frac{1}{1-f_x}}, \tag{1}$$

where R is the current ratio of the heavy to light isotope abundance, and R_0 is the initial ratio. The ratio R/R_0 defines the degree of enrichment in the heavier isotope in comparison to the primordial value. The upper limit of the enrichment of the heavier isotope can be found as

$$\frac{R}{R_0} \leq \left(\frac{\varphi t}{n+1}\right)^{1-f_x}, \tag{2}$$

where φ is the maximum escape flux and t is time. Different escape processes have various possible maximum escape fluxes and different fractionation coefficients that can depend on the structure and the composition of the upper atmosphere. Below, we briefly summarize the main atmospheric escape processes and their influence on the atmospheric fractionation.

4.2.1 Fractionation by Thermal Atmospheric Escape

Jeans escape: when an upper atmosphere is hydrostatic and is not able to flow away from the planet hydrodynamically, it is still possible for particles in the high-speed tail of the thermal distribution of particles at the exobase to have speeds exceeding the escape velocity. The loss of such particles is known as Jeans escape. Jeans escape is highly mass fractionating only if the escape is not diffusion-limited. Only if Jeans escape is the bottleneck, this process leads to highly mass-dependent fractionation. For the present solar system planets, Jeans escape rates are quite low, so that it takes a long time to fractionate a large atmospheric reservoir assuming present escape rates (Mandt et al. 2014) though the process would have been much more rapid at earlier times when the Sun was much more active. The fractionation factor f_{Jeans} for Jeans escape can be written as (e.g., Mandt et al. 2009)

$$f_{\text{Jeans}} = \sqrt{\frac{m_1}{m_2}} \left[e^{(\lambda_1 - \lambda_2)} \frac{(1+\lambda_2)}{(1+\lambda_1)} \right], \tag{3}$$

where m_1 and m_2 are the masses of the respective isotopes and $\lambda_{1,2} = (GM_{pl}m_{1,2})/(k_B T_{ex} r_{ex})$ are the Jeans escape parameters of the lighter and heavier isotope at the exobase level, respectively. Here G is the Newton gravitational constant, M_{pl} is the planetary mass, k_B is the Bolzmann constant, T_{ex} the exobase temperature and r_{ex} the exobase distance.

Hydrodynamic escape: if the upper atmosphere is heated by the EUV flux of the planet's host star to very high temperatures, it is possible that the gas in the upper atmosphere flows away from the planet with a speed at the exobase exceeding the escape speed. Such a flow has the form of a transonic Parker wind (e.g. Parker 1958; Tian et al. 2005; Johnstone et al. 2019). Whether an atmosphere reaches this state depends on the gas temperature in the upper atmosphere and the mass of the planet, with hydrodynamic escape taking place significantly more readily for lower mass planets. The source of the energy that heats the gas is typically the X-ray and ultraviolet field of the star. Very early in a planet's history, immediately after the dissipation of the primordial nebula, it is also possible for hydrodynamic escape to be powered by the initial accretion energy of the envelope gas (Stökl et al. 2015) and the bolometric radiative emission of the star (Owen and Wu 2016), though the latter is unlikely to have been significant for the young solar system planets due to their large distances from the Sun. Previous works have found that hydrodynamic escape removes all isotopes very efficiently, which leads to small fractionations (e.g., Sekiya et al. 1980a; Mandt et al. 2009). For hydrodynamic escape, the Jeans escape parameter λ decreases to very small values (Chamberlain 1963; Volkov et al. 2011; Erkaev et al. 2015). In such a case, f_{Jeans} converges towards f_{hy}, i.e.

$$f_{hy} = \sqrt{\frac{m_1}{m_2}}, \tag{4}$$

due to the second part in Eq. (3) converging towards one for very small lambdas. When hydrogen atoms escape hydrodynamically, however, it will drag heavier elements with it. This effect can subsequently lead to a fractionation of lighter isotopes from heavier isotopes (e.g., Zahnle and Kasting 1986; Hunten et al. 1987; Zahnle et al. 1990; Odert et al. 2018). If, besides hydrogen, one assumes a second major species, e.g. oxygen, in the hydrodynamically escaping atmosphere, which is also dragged by hydrogen, then the fractionation factors f^i_{drag} for the second major species and f^j_{drag} for an arbitrary amount of additional minor species can be written as (e.g., Odert et al. 2018)

$$f^{i}_{drag} = 1 - \frac{g\,(m_i - m_H)\,b_{H,i}}{F_H k_B T\,(1+\xi_j)}, \tag{5}$$

and

$$f^{j}_{drag} = \frac{1 - \frac{g(m_j - m_H)b_{H,j}}{F_H k_B T} + \frac{b_{H,j}}{b_{H,i}}\xi_i\,(1+\xi_i) + \frac{b_{H,j}}{b_{i,j}}\xi_i f^{i}_{drag}}{1 + \frac{b_{H,j}}{b_{i,j}}\xi_i}, \tag{6}$$

where g is the gravitational acceleration of the planet, T the upper atmosphere temperature, m_i, m_H, and m_j the masses of the masses of the respective atoms, $b_{H,i}$, $b_{O,j}$, $b_{H,j}$ the binary diffusion parameters, F_H the hydrodynamic escape flux of hydrogen atoms, and ξ_i the mixing ratio of the second major species to hydrogen.

To summarize thermal escape and fractionation processes, one finds that hydrodynamic outflow is relatively inefficient at fractionating elements and isotopes with a small mass difference (i.e., D/H...), while it can be efficient if a light species such as atomic hydrogen drags heavier ones, whereas Jeans escape fractionates efficiently but total atmospheric escape is much less efficient. As a rule of thumb, high-rate loss processes, such as hydrodynamic escape, are very efficient in removing the atmospheric gas, but are inefficient fractionators in case of light elements where it takes a long time to change the isotopic ratio. On the other hand, a low-rate loss process such as Jeans escape can be an efficient fractionator, but take a very long time to influence a large inventory.

4.2.2 Fractionation by Non-thermal Atmospheric Escape

Ion pick-up: above the exobase, neutral atmospheric particles follow ballistic trajectories. If these particles reach the region above the magnetopause or ionopause, they can directly interact with the solar wind, where they can be ionized by solar EUV radiation and interactions with the solar wind. These ions are picked up by the solar wind and are lost from the planet. If a planet does not have an intrinsic magnetic field, as is the case for modern Mars and Venus, then the solar wind can directly interact with the ionosphere of the planet, thus making the ion pick-up escape more efficient. For a magnetized planet, atmospheric particles have first to reach the location of the magnetopause to be picked up by the solar wind, which can reduce the ion pick-up escape rates in comparison to the non-magnetized case. Due to diffusive separation of atmospheric particles by mass above the homopause (see also Fig. 2), isotopes with lighter masses reach higher altitudes and will therefore be preferentially removed from the top of the atmosphere. For non-thermal escape such as ion pick-up, the diffusive separation fractionation factor f_{ds} can be defined as follows (Lunine et al. 1999; Mandt et al. 2014), i.e.

$$f_{\mathrm{ds}} = \exp\left[-\frac{g(\Delta z)(m_2 - m_1)}{k_{\mathrm{B}}T}\right], \tag{7}$$

where $\Delta z = r_{\mathrm{ex}} - r_{\mathrm{h}}$ is the distance between the homopause r_{h} and the exobase r_{ex} levels (see also Fig. 2) if the planet has no intrinsic magnetic field, otherwise the exobase can be substituted by the magnetopause.

Sputtering: in this process, atmospheric atoms or molecules are ejected due to collisions between solar wind protons and backscattered pick-up ions with the planetary atmosphere (e.g., Johnson 1990). This process efficiently fractionates the isotopes through diffusion as a function of the temperature in the upper atmosphere and the distance between the exobase and the homopause. Like with other loss processes, the efficiency of the fractionation by sputtering depends on the exact structure of the upper atmosphere, with expanded atmospheres fractionating more easily. Sputtering preferentially removes the lighter atom or particle. On low mass planetary bodies such as Mars, ion pick-up and sputtering are likely the two fractionation processes that led to the observed $^{38}\mathrm{Ar}/^{36}\mathrm{Ar}$ fractionation (Jakosky et al. 2017). For atmospheric sputtering the diffusive separation, fractionation is relevant and can be calculated in the same way as for ion pick-up escape in Eq. (7).

Plasma instabilities: plasma instabilities, such as Kelvin-Helmholz instability, can drive an escape of plasma vortexes from the ionospheric boundary from planets which have no intrinsic magnetic dynamo (i.e., Venus, Mars, ...). A wave-like structure of initially small amplitudes can grow and form vortexes. These vortexes detach from the ionized upper atmosphere and escape from the planet in a plasma cloud (Terada et al. 2002; Amerstorfer et al. 2007). Since plasma instabilities remove a large number of ions at once, they are not very efficient isotope fractionators.

Cold ion-outflows: ions and free electrons produced by the absorption of solar extreme ultraviolet and X-ray photons are able to escape by a separate process. The expansion of the lighter electrons produces an ambipolar electric field that accelerates the ions away from the planet causing an outflow of ions and electrons (Ganguli 1996). For magnetised planets like the Earth, this flow takes place at the magnetic poles. This process can accelerate the planetary ions and, if they reach a sufficiently high velocity, lead to them escaping the planet's gravity (e.g., Hartle and Grebowsky 1993; Lundin 2011). Heavier ions can also be dragged by other atmospheric constituents through this process, as was probably the case for the easily ionized Xe^+ isotopes in the Earth's and Mars' atmosphere (e.g., Zahnle et al. 2019; Avice and Marty 2020; this issue).

Photochemical escape: this type of escape is related to chemical reactions that produce hot atoms with high energies, so called suprathermal or "hot" atoms. If the energy of an atom is high enough, it can exceed the energy of the gravitational binding to the planet, and then the atom is lost (e.g., Amerstorfer et al. 2017). The reactions can include various chemical reactions, including dissociative recombination or charge exchange. Photochemically produced suprathermal atom escape can preferentially remove lighter isotopes. On Mars, photochemical processes lead to removal of nitrogen and oxygen atoms, and similar to the fractionation by ion pick-up, the fractionation is driven by the diffusive separation of different isotopes by mass above the homopause (e.g., Jakosky et al. 1994).

On Earth, He escapes due to exothermic charge exchange reactions between He^+ ions and the major atmospheric constituents N_2, O_2, and O (e.g., Lie-Svendsen and Rees 1996) and on Venus photochemical processes are relevant for the D/H fractionation (Lammer et al. 2006).

Impact erosion: early in the planet's history, atmospheric gas can be removed by impacts due to a large mass of gas removed simultaneously (e.g., Schlichting et al. 2015; Schlichting and Mukhopadhyay 2018); this process is not an efficient fractionator of isotopes related elements in atmospheres and is discussed in more detail in Sect. 5.3.

5 Impact Erosion of Primordial Atmospheres, Early Crust, and Mantle

Understanding impact and collision processes and their consequences during accreting planetary embryos and protoplanets is a vital component in the overall picture, since those events happened during all stages of terrestrial planet formation. The currently established picture distinguishes several main stages, where theoretical knowledge about the early phases from dust to planetary embryos is still limited and various competing theories are discussed (e.g. Raymond et al. 2014; Johansen and Lambrechts 2017; Izidoro and Raymond 2018; Lammer et al. 2020a; this issue). However, it is relatively well established that the final phase of accretion, lasting for millions to several 10s of Myr is marked by giant collisions among planetary embryos and remaining planetesimals, which shape the bulk properties of eventually formed planets.

5.1 Impact Related Erosion of the Primordial Atmosphere

As discussed above, growing bodies accrete nebular gas once they become sufficiently massive, where cores larger than one Earth-mass accrete an H_2-dominated envelope that most likely remain during their whole lifetime (Johnstone et al. 2015; Owen and Wu 2016), while protoplanets below 0.5 M_{Earth} probably loose most of their primordial atmosphere, which is largely boiled off after the surrounding gas dissipates and potentially replaced by an outgassed one later (Stökl et al. 2016). Protoplanets with higher masses in orbits inside the habitable zone keep primordial atmospheres that can be lost afterwards by EUV-driven hydrodynamic escape and erosion by large impactors (e.g., Schlichting et al. 2015;

Schlichting and Mukhopadhyay 2018; Lammer et al. 2020b). It is important to mention that existing work on the interplay of internal and external (impact) heating, magma ocean and (proto-)atmosphere evolution of growing terrestrial planets has focused on roughly Earth-mass objects, even though smaller bodies – the size of planetary embryos and below – are the stepping stones on the way to fully grown planets.

Studies on impact efficiencies are controversial since all the work on that subject is theoretical. The first impact studies onto atmosphere and ocean-covered protoplanets were often based on 1-D hydrodynamic codes (Genda and Abe 2003, 2005; Inamdar and Schlichting 2016), analytical models in combination with such codes (Schlichting et al. 2015), and also 3-D hydrocode simulations of impacts of small bodies (Shuvalov 2009) up to head-on collisions in the Earth- to super-Earth mass range (Liu et al. 2015). An important result is that smaller impactors are - per unit mass - much more efficient in eroding atmosphere than larger ones (e.g. Schlichting et al. 2015; Schlichting and Mukhopadhyay 2018, for an overview). The mass loss efficiency (the ratio of lost atmospheric to impactor mass) is largest for small bodies, e.g. in the kilometer range for current Earth, and decreases with mass by orders of magnitude up to the giant impact regime ($\sim$ 1000 km bodies) This is mainly because for small impactors the majority of the energy is available for disturbing the atmosphere locally; above the tangent plane of the impact location, the gas can be entirely lost. It should also be noted that smaller impactors can also deliver material to the planet.

For impactors with energy deposition larger than what is required for ejecting all material above the tangent plane, considerable parts of their kinetic energy have to be used to drive a strong shock through the target body to eject atmosphere globally, once the shock-induced ground velocity is sufficient to accelerate atmosphere components to escape velocity. The presence of oceans on the surface has been found to aid the escape of overlying atmosphere, due to their lower shock impedance compared to a solid surface (Genda and Abe 2005). In recent 3-D smoothed-particle hydrodynamics (SPH) giant impact simulations, atmosphere losses of $\approx 5\%$ per collision for Moon-mass/size, and $\approx$ 10-20% for Mars-mass/size impactors in collisions with a $\approx$ 0.5-0.75 M_{Earth} protoplanet with primordial atmospheric mass fractions of ≥ 0.001 have been obtained (Lammer et al. 2020b).

The impact scenarios in this study cover the full range of impact angles and common impact velocities from 1 to $2 \times v_{\text{esc}}$. An SPH collision snapshot of a Moon-mass impactor that crashes into an H_2-covered protoplanet with 0.5 M_{Earth} is shown in panel (a) in Fig. 4 (Burger et al. 2018; Lammer et al. 2020b). Panel (b) in Fig. 4 shows eroded primordial H_2-dominated atmospheres in percent from protoplanets with masses $\approx$ 0.5 - 0.75 M_{Earth} caused by Moon-mars and Mars-mass impactors as a function of impact angle and escape velocity (Lammer et al. 2020b). The different results shown in Lammer et al. (2020b) indicate that studying atmospheric losses caused by giant impacts only by assuming head-on scenarios (e.g. Liu et al. 2015) may be insufficient to understand the full diversity and efficiency of atmospheric losses by such events. These results confirm that - like for small bodies - even in the giant impact regime lower-mass bodies are more efficient (normalized to impacting mass) than larger ones in eroding atmosphere (Schlichting et al. 2015; Schlichting and Mukhopadhyay 2018).

5.2 Collisional Erosion of Crust/Mantle Material by Large Impactors

During its accretion phase, in particular during the so-called giant impact phase, a growing protoplanet is expected to experience energetic collisions with other planetary embryos (e.g., Quintana et al. 2016). Such a collision can then remove part of the newly formed crust

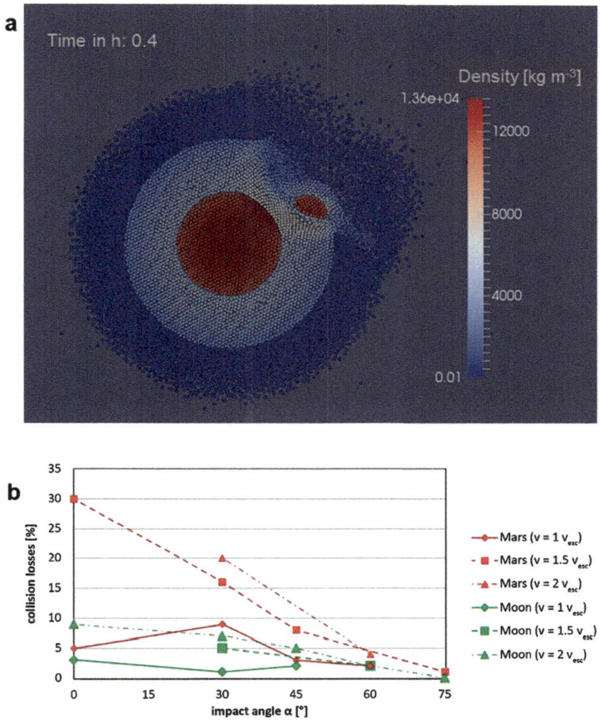

Fig. 4 Panel **a**: Snapshot of a SPH collision simulation between a Moon-mass embryo and a 0.55 M_{Earth} protoplanet that is surrounded by a primordial H_2-envelope, with an impact angle of ≈ 30 degrees and an impact velocity of $2v_{\text{esc}}$. The colour coding shows the density and thus the core-mantle-atmosphere structure. Panel **b**: Atmospheric collision losses in percent of H_2-envelope covered protoplanets caused by Moon- and Mars-mass impactors as a function of impact angle and velocity in units of v_{esc}

from the planet which will either re-accrete or be lost to space (Carter et al. 2018). This process will therefore preferentially remove incompatible lithophile elements from the growing planet leading to a fractionation between siderophile and lithophile elements (O'Neill and Palme 2008; O'Neill et al. 2020; this issue). While Fe/Mg or Fe/Si, of which Fe is siderophile and Mg and Si are lithophile, will be fractionated by collisional erosion, other ratios such as K/U, which are both lithophile elements, will not. This process could therefore account for the high terrestrial Fe/Mg ratio of 2.1 ± 0.1 compared to the solar ratio of 1.9 ± 0.1 (O'Neill and Palme 2008). Another ratio that was suggested to be fractionated by collisional erosion is Earth's ^{142}Nd/^{144}Nd ratio (Jellinek and Jackson 2015). Here, however, a more recent study found that ^{142}Nd/^{144}Nd might not have been fractionated at all (Bouvier and Boyet 2016).

Whether collisional erosion can account for the high terrestrial Fe/Mg ratio was numerically studied by Carter et al. (2015) and Bonsor et al. (2015). Even though they both conclude that collisional erosion can fractionate Earth's Fe/Mg from a chondritic to the present-day terrestrial value, they also found that this fractionation can only be reached if Earth was an extreme case. The growing planetary embryos that exhibit a different range of core mass fractions through giant impacts must have had a significant dynamical excitation to finally create the present-day Fe/Mg ratio of Earth (Bonsor et al. 2015; Carter et al. 2015).

However, the escape of Mg from a planetary embryo through hydrodynamic drag of hydrogen could additionally alter the Fe/Mg ratio, which might make the Earth a less extreme case than expected in these studies (Benedikt et al. 2020). A more recent study by Carter et al. (2018) found that a growing protoplanet on average can lose about one third of its crust through collisional erosion which in turn would lead to the loss of up to 20% of its heat producing elements if the crust does not re-accrete onto the planet. Such an alteration of the planetary heat budget might further also have implications on the habitability of a planet since this could strongly affect the evolution of tectonic activity (Jellinek and Jackson 2015; O'Neill et al. 2020; this issue).

Collisional erosion could further also have been important in view of Mercury's high iron abundance, as was recently simulated by Chau et al. (2018). These authors found that an energetic impactor could alter the iron distribution within a planet by up to 25%. According to their study, Mercury's abnormally high abundance of iron could therefore be the result of multiple collisions. Another recent study by Clement et al. (2019), however, concludes that, even though planets with elevated core fractions were produced in 90% of their simulations through the fragmentation of collided bodies, there is only a probability of $\leq 1\%$ to simultaneously reproduce Mercury and Venus after such an event.

5.3 Elemental Fractionation Caused by Large Impact Events

To summarize the impact erosion processes, one finds two scenarios where large impactors will change the composition of a growing proto-planet. Both cases are illustrated in Fig. 4. If a proto-planet accreted $> 0.5\ M_{\text{Earth}}$ inside the disk, elements in solar abundance and magma ocean-related outgassed moderately volatile rock-forming elements and noble gases will be embedded as trace elements in a captured H_2-dominated primordial atmosphere.

As illustrated in Fig. 5a, large impactors that mainly affect the atmosphere can then contribute to the fractionation of moderately volatile elements such as K and/or Mg compared to U or Fe, which remain in the crust/mantle. If a large impactor crashes into the body, crustal /mantle material including refractory elements (e.g., U, Th) might also be lost to space if not re-accreted by the protoplanet (see Fig. 5b), (e.g. Palme and O'Neill 2003; see also O'Neill et al. 2020; this issue). Depending on the impactor and its parameters, impact erosion can be a strong fractionating process.

Impacts, however, will not fractionate isotopes of atmospheric noble gases (e.g. ^{36}Ar, ^{38}Ar, ^{20}Ne, ^{22}Ne, etc.) if they only affect the atmosphere (Fig. 5a) but contribute to a depletion of volatile elements from a planetary body. If impactors with different isotopic compositions and ratios are involved in the accretion of a protoplanet (Fig. 5b), noble gas isotope ratios, however, can be modified by their mixing with compositions, abundances and ratios of the impacting bodies as discussed in Sect. 6. Below we discuss atmospheric escape and fractionation of such elements from planetary embryos, which experience magma oceans and the therewith related outgassing of volatile elements.

6 Escape of Rock-Forming Elements and Volatiles from Planetary Embryos

Terrestrial planets accrete mass by numerous collisions between small objects that accumulate to larger ones, including Moon and Mars-mass planetary embryos that accrete onto full planets (e.g., Johansen and Lambrechts 2017; Izidoro and Raymond 2018; Lammer et al. 2020a, this issue). Growing planetesimals, which are embedded in the circumstellar disk, with a size between hundred to thousand kilometers develop global magma oceans, caused by the heating of short-lived radioactive elements, like ^{26}Al (e.g., Lichtenberg et al. 2016; 2017; Young et al. 2019), and mutual collisions (e.g., Safronov and Zvjagina 1969; Wetherill 1980; Tonks and Melosh 1993). Schaefer and Fegley (2007) and Fegley et al. (2016) performed chemical kinetic and equilibrium calculations for modeling the chemistry of elements (e.g., K, Si, Mg, Fe, Ca, Al, Na, S, P, Cl, F) released by heating of different types of carbonaceous, ordinary and enstatite chondrites as a function of temperature and pressure

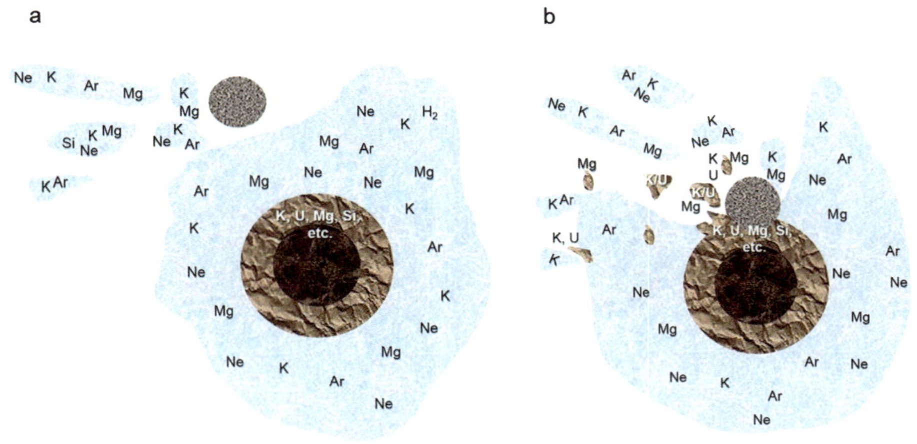

Fig. 5 Illustration of the two impact related erosion scenarios. (**a**) Impact erosion of atmospheres, and (**b**) direct impacts and related erosion of crustal and mantle material. Impacts that mainly affect a primordial atmosphere can contribute to the modification of ratios between moderately volatile rock-forming elements that can also be in a primordial atmosphere and refractory elements that remain in the surface, such as U or Th (panel **a**), while direct impacts can also erode crustal and mantle material (panel **b**)

of early atmospheres. These studies yield significant amounts of major rock-forming and other volatile elements, such as Rb and Zn, and noble gases (Ar, Ne, etc.) in primordial and/or catastrophically outgassed steam atmospheres at high temperatures (1000-2500 K).

When the disk dissipates, after a short but efficient boil-off phase (Owen and Wu 2016; Lammer et al. 2018) the magma oceans of the former disk-embedded bodies solidify and steam atmospheres are catastrophically outgassed (see Fig. 6). Depending on the redox state these catastrophically outgassed atmospheres contain H_2, CO (reduced interior) and/or H_2O and CO_2 (oxidized interior; Nikolaou et al. 2019). The outgassed amount per element depends on the bulk composition of a particular planetary embryo and the solidification process of the mentioned magma ocean (e.g., Elkins-Tanton 2008, 2012; Salvador et al. 2017; Nikolaou et al. 2019).

Recently, Benedikt et al. (2020) studied the losses of magma ocean-related catastrophically outgassed steam atmospheres and outgassed rock-forming elements as well as Ar and Ne noble gases from planetary embryos with masses between 1.0 M_{Moon} and 1.5 M_{Mars} along the EUV-flux evolution tracks of slow and moderate/fast rotating young G-type stars (Tu et al. 2015; Güdel 2020; this issue) at orbital distances between 0.5-1.5 AU.

Based on studies from Wakita and Schmitt (1970), Righter and Chabot (2011), and Elkins-Tanton (2012), one can assume magma ocean depths for planetary embryos within the mass ranges mentioned above between 500 - 1000 km for surface temperatures of 1500 - 3000 K. Because of the low mass of planetary embryos and these high surface temperatures, according to 1-D radiative-convective atmosphere models (Marcq 2012; Marcq et al. 2017; Pluriel et al. 2019), it was found that the steam atmosphere heights are increasing up to a few thousands of kilometers (Benedikt et al. 2020). The main process, which is dragging heavier atoms into space, is related to the dissociation of H_2O molecules around the 1 μbar (Guo 2019) level, so that atomic hydrogen dominates the upper atmosphere and is lost hydrodynamically, thereby potentially dragging heavier elements.

For planetary embryos with masses ≤ 1 M_{Moon}, no dense silicate atmosphere or steam atmosphere can build up (Hin et al. 2017; Benedikt et al. 2020; Young et al. 2019), the embryo will continuously lose outgassed elements very fast to space. Therefore, one can

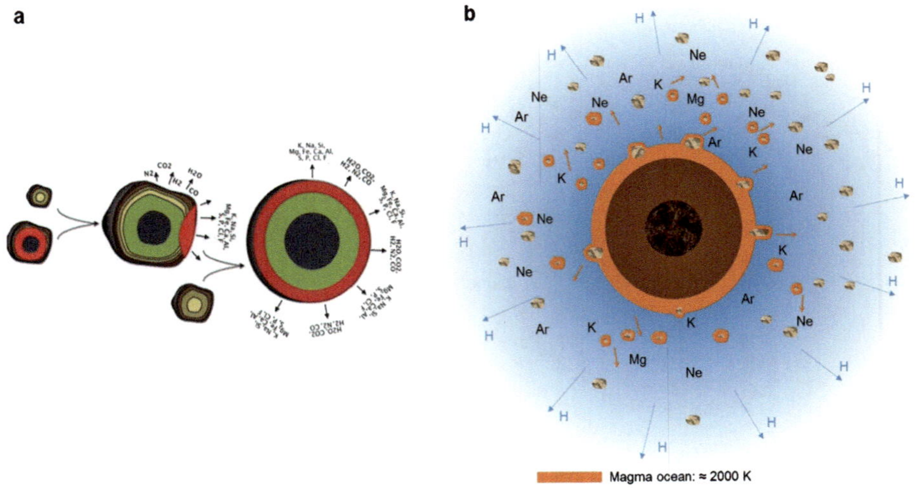

Fig. 6 (**a**) Illustration of planetary embryo growth after Elkins-Tanton (2012) due to collisions between differentiated planetesimals/planetary embryos. The red areas correspond to magma from which various elements evaporate and can be lost directly at bodies with masses $\leq M_{\mathrm{Moon}}$ (Benedikt et al. 2020; Young et al. 2019) and by dragging of hydrodynamically escaping H atoms dissociated either from H_2O molecules originated from catastrophically outgassed steam atmospheres during magma ocean solidification (Benedikt et al. 2020) or (**b**) from remnants of captured H_2-dominated nebula-based envelopes (Lammer et al. 2020b)

expect that these bodies will be strongly depleted in volatiles, noble gases and outgassed rock-forming elements.

One should note that in the recent study by Young et al. (2019) it is assumed that hydrostatic conditions for an outgassed rock vapour atmosphere above a magma ocean with a surface temperature of 1830 K on a planetary embryo with half of the mass of Pluto build up. As discussed in Benedikt et al. (2020), and shown in Fig. 7, under such extreme conditions where the ratio between the gravitational energy and the thermal energy, i.e. the Jeans parameter λ is < 6 (in the case of Young et al. 2019 the λ at the surface is 1.38) it is impossible that a stable atmosphere can build up at such low mass and hot bodies. In such cases, direct escape of outgassed rock-forming elements occurs unless there is a steady supply from the interior to compensate for these fast losses. For silicate atmospheres, this means that the entire mantle will eventually depleted in the outgassed elements. The assumption of a hydrostatic equilibrium as assumed in the study of Young et al. (2019) is thus only justified for cases in which the Jeans parameter of the outgassed elements at the surface of an embryo exceed ≈ 6 (Erkaev et al. 2015; Volkov et al. 2011).

Figure 7 shows the evolution of the atmospheric temperature (red curves) and Jeans parameter (green curves) by using the same embryo and silicate atmosphere parameters with 34 amu as given in Young et al. (2019) for an assumed hydrostatic (dotted lines) and realistic non-hydrostatic (solid lines) atmosphere.

In the realistic non-hydrostatic case, the atmospheric temperature is decreasing much faster due to adiabatic cooling of the outward flowing bulk gas. At about 9 embryo radii the atmosphere is heated again because of absorption of the EUV radiation that is scaled in Benedikt et al. (2020) to the value expected for the early Solar System at the time when the disk dissipated (Tu et al. 2015; Güdel 2020; this issue). Because of the low λ value of ≈ 1.4, the hydrodynamic escaping bulk atmosphere is supersonic just form the surface.

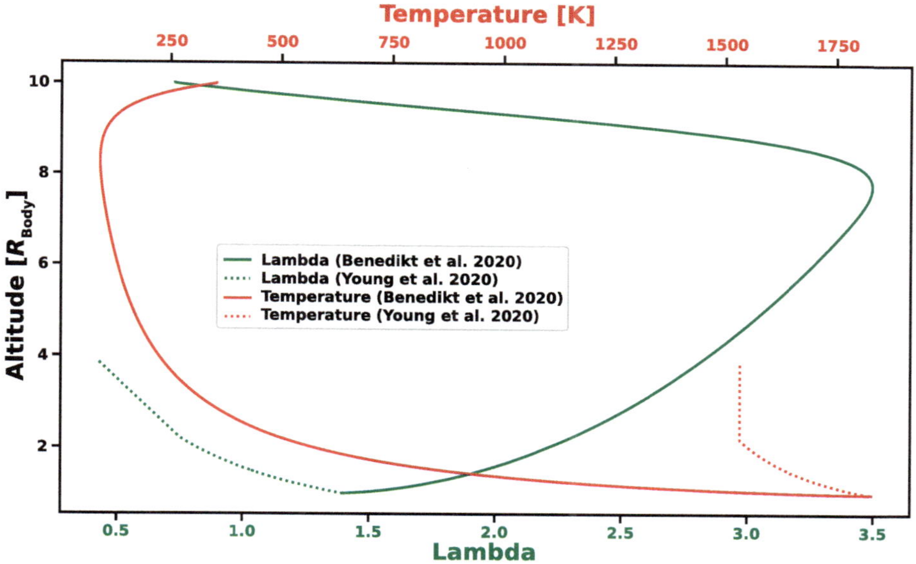

Fig. 7 Change of temperature (red lines) and Jeans parameter lambda (green lines) for the assumed but not realistic hydrostatic (dotted lines) outgassed silicate atmosphere (Young et al. 2019) and a non-hydrostatic (solid lines) silicate atmosphere (Benedikt et al. 2020) around a planetary embryo with 0.5 $M_{\rm Pluto}$ (after Benedikt et al. 2020)

According to Benedikt et al. (2020), more massive embryos with masses $> 0.5\ M_{\rm Mars}$ also lose outgassed heavy trace elements when they are incorporated into solidified magma ocean-related steam atmospheres. As long as steam atmospheres do not condense and outgassed rock-forming elements reach the upper atmosphere, they can also be lost by the dragging of escaping H atoms.

Figure 8 shows an example of escape times of magma ocean-related catastrophically outgassed steam atmospheres and embedded trace elements for a planetary embryo with a mass of 0.5 $M_{\rm Mars}$ at various orbital distances, partial surface pressures of 50/5 (green), 100/10 (red), 150/30 (blue) bar H_2O/CO_2 and surface temperatures of 1500 K. The bold blue line corresponds to the condensation temperature of a steam atmosphere based on the results of Lebrun et al. (2013). As one can see in Fig. 7, depending on the partial pressure and surface temperature, embryos with a mass of 0.5 $M_{\rm Mars}$ can lose steam atmospheres at an orbital distance of 1 AU within the condensation time. For heavier planetary embryos of up to $\approx 1.5\ M_{\rm Mars}$ almost all outgassed trace elements can be lost together with the escaping steam atmosphere within ≈ 10 Myr (Benedikt et al. 2020), which lies within the time frame of the formation of the first protocrust on Mars (Bouvier et al. 2018; Kruijer et al. 2020).

This indicates that frequent smaller impactors (Maindl et al. 2015) keep shallow magma oceans molten most likely over longer periods so that the steam atmosphere can longer remain in vapour form compared to the modelled time scales given in Lebrun et al. (2013). A more active young Sun-like star leads to a shorter escape time for the whole steam atmosphere as for the slow rotator in Fig. 8.

Figure 9 shows an example of the remaining amount of various elements on a planetary embryo with 0.5 $M_{\rm Mars}$ at 1 AU, with a magma ocean-related steam atmosphere partial surface pressure of 50 bar H_2O and 5 bar CO_2. If all outgassed elements escape, the particular volatiles are depleted by $\approx 75\%$ within the planetary embryo. The remaining amount of Ne

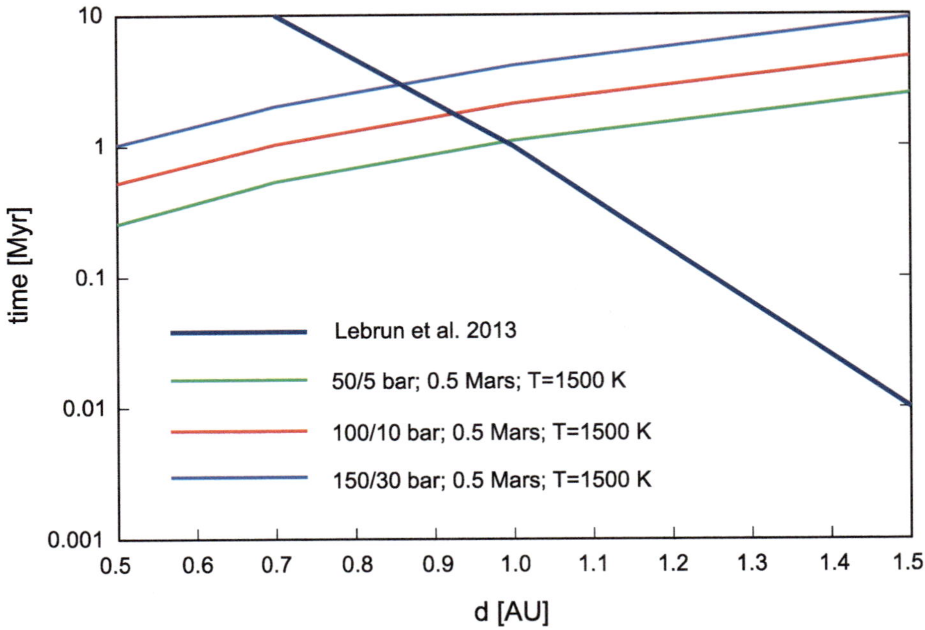

Fig. 8 Escape time t of steam atmospheres for a planetary embryo with a mass of 0.5 M_{Mars}, various partial surface pressures of H_2O/CO_2, for a surface temperature of 1500 K and for a weakly active slowly rotating young G-type star in dependence of the orbital distance d. The solid dark blue line in each panel shows the steam atmosphere condensation time based on the study of Lebrun et al. (2013) (after Benedikt et al. 2020)

and Ar isotopes in mol/g and $^{20}Ne/^{22}Ne$ and $^{36}Ar/^{38}Ar$ ratios are also shown. One can see that on such low mass bodies the Ne and Ar isotopes only escape to space but do not fractionate due to the high loss rates. Isotope fractionation, however, between the vapour and melt of small planetary embryos could at least fractionate the isotopic ratios of Mg and Si at near-equilibrium (Young et al. 2019).

It should also be noted, that the depletions and ratios of the volatile rock-forming elements of the larger planetary embryos with masses $\geq 0.5\ M_{\text{Mars}}$ most likely represent upper values. First, it is unclear whether 100% of the outgassed elements will indeed be incorporated into the steam atmosphere or if a fraction remains inside the magma ocean. Second, the outgassed rock-forming elements have to overcome a bottleneck from the lower atmosphere to the upper atmosphere, the so-called cold-trap in the mesosphere; otherwise, they will reach the condensation temperature. For noble gases such as Ne or Ar, however, the atmospheric cold-trap of larger bodies does not exist due to their low condensation temperatures and rain back to the surface.

One should also note that for reduced atmospheres with high contents of H_2 and CO molecules, the atmosphere would only marginally bound on such a low mass body. For surface temperatures of ≥ 2000 K, reduced H_2 and CO dominated atmospheres would be unbound. Therefore, outgassed atmospheres that are highly reduced will escape from such low mass bodies even faster than the oxidized steam atmospheres shown in Fig. 9.

Because of this, it seems reasonable that large planetary embryos should be more depleted in noble gases (i.e., Ne and Ar) than in rock-forming elements such as K, Si, Mg, Fe, Ca, Al, Na, S, P, Cl, F due to their significantly higher condensation temperatures.

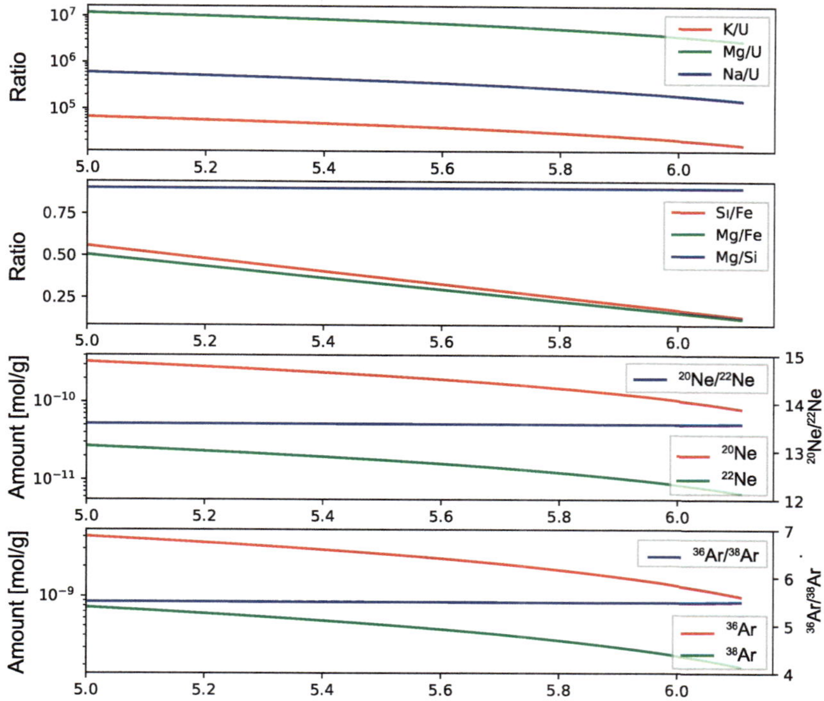

Fig. 9 Escape related evolution of the K/U, Mg/U, Na/U, Si/Fe, Mg/Fe, Mg/Si, ^{20}Ne/^{22}Ne, ^{36}Ar/^{38}Ar ratios and losses of Ar and Ne as a function of time for a 50 bar H_2O and 5 bar CO_2 outgassed steam atmosphere from a planetary embryo with 0.5 M_{Mars} at 1 AU. The embryo's steam atmosphere is exposed to the EUV flux of a slowly rotating young G-type star after the magma ocean solidification (assumed to be at $\approx$ 5 Myr). About 75% of the total initial amounts of the elements are lost more or less within 1 Myr with surface temperatures corresponding to 1500 K. The lines end when the steam atmospheres are lost from the embryo. The ratios were calculated for mol/g and U and Fe assumed not to escape but remain at the body with their initial amounts (after Benedikt et al. 2020)

In general, these studies also agree with the recent experimental and theoretical based results of Sossi and Fegley (2018) and Sossi et al. (2019), that atmospheric escape caused by one large evaporation event at Earth is ineffective in fractionating moderately volatile element abundances to an extend that match the observations. Instead, the present moderately volatile element abundance in terrestrial planets most likely reflects accretion from several lower-mass embryos where each of them experienced different degrees of volatilization. One can conclude from these recent studies that a late accretion of chondritic material via planetary embryos that underwent melting processes might therefore not necessarily lead to a volatile rich planet.

7 Evolution of Earth's and Venus' Ar and Ne Noble Gas Isotope and K/U Bulk Elemental Ratios

Chondritic meteorites have long been linked to the bulk composition of the silicate part of Solar System terrestrial planets (e.g. Taylor 1964; Anders and Grevesse 1989; Javoy et al. 2010; Dauphas 2017; Schiller et al. 2018). Dauphas (2017) analyzed isotopes of lithophile (O, Ca, Ti, Nd), highly siderophile (Ru) and moderately siderophile (Cr, Ni and Mo) elements from Earth's mantle and found that they record different accretion stages. According to analysed isotope compositions by Dauphas (2017), the first 60% of proto-Earth consisted of about 50% enstatite-like and 5% chondrite-like material, whereas all of the remaining accreted material consisted of enstatite-like impactors. According to Budde et al. (2019), most isotopic compositions can also be explained if mixtures of carbonaceous chondritic and non-carbonaceous chondritic material composed the building blocks of the accreting Earth. On the other hand, the oxygen isotopic composition of the Earth is nearly identical to that of enstatite chondrites. Oxygen is the most abundant element in rocks and one of the most abundant element in Earth depending on the estimated FeO content in the mantle and the Fe content in the core. Unless the Earth has a large enstatite chondrite component, the oxygen-isotopic composition and the iron content of the Earth cannot be matched (Dauphas 2017). The Dauphas (2017) O-isotope model and composition hypothesis agrees also with the two-component models of Ringwood (1979), Wänke and Dreibus (1988) and Lodders (2000), which assume almost the same percentage of O-isotope related enstatite chondritic material.

Recently, Schiller et al. (2018) investigated the $^{48}Ca/^{44}Ca$ isotope evolution of the composition of different Solar System objects such as Lunar and Martian meteorites, ordinary and carbonaceous chondrites, angrites, eucrites, ureilites, as well as very young chondrules from ordinary and carbonaceous chondrites and its role as a precursor element to rocky planets. The results of this study indicate that the $^{48}Ca/^{44}Ca$ composition of the earliest material in the inner disk originated from a ureilite-like composition and evolved to a terrestrial within the first million years after the formation of the Sun. Calculated accretion rates of the inner protoplanetary disk inferred from the Ca isotope composition of the involved objects reveal that the bulk disk composition at $\approx$ 5-6 Myr after the formation of the Solar System consisted of a mixture of about 60% ureilite-like and 40% outer Solar System CI chondritic material (Schiller et al. 2018).

In contrast to this, $^{36}Ar/^{38}Ar$ and $^{20}Ne/^{22}Ne$ noble gases of the present Earth atmosphere and isotopes of water (D/H), and nitrogen ($^{14}N/^{15}N$) can also be reproduced by assuming a volatile content equivalent to a mixture of about 98% non-chondritic dry material with a contribution of about 2% undepleted carbonaceous CI-CM-like material (Marty 2012). Although, the results for the proto-Earth obtained by Dauphas (2017) and Schiller et al. (2018) might seem to contradict the findings of Marty (2012) representing present Earth, it is not necessarily the case. As shown and discussed in Odert et al. (2018), Benedikt et al. (2020) and in Sect. 6, the building blocks of terrestrial planets were most likely depleted in noble gases and other moderately volatile rock-forming elements. Additionally, one should not forget that proto-Earth and Venus underwent different evolutionary processes since the evaporation of the disk, including the possible escape of a primordial atmosphere, magma oceans and impactors with a depletion-modified abundance of their composition. These processes certainly played a crucial role in shaping the noble gas isotope, elemental and volatile evolution of these planets over their history.

Since decades, various researchers tried to model the compositional evolution of early Venus and Earth by reproducing the observed isotope ratios of the non-radiogenic noble gases Ar and Ne in their atmospheres (e.g., Cameron 1983; Pepin 1991, 1997; Becker et al. 2003; Gillmann et al. 2009). For instance, Gillmann et al. (2009) studied the escape and fractionation of $^{36}Ar/^{38}Ar$, and $^{20}Ne/^{22}Ne$ noble gas isotope ratios of Venus' early atmosphere with a simple hydrodynamic escape model that is based on approaches and inputs provided by earlier studies of Hunten (1987), Kasting and Pollack (1983), Zahnle and Kasting (1986) and Chassefière (1996a, 1996b, 1997).

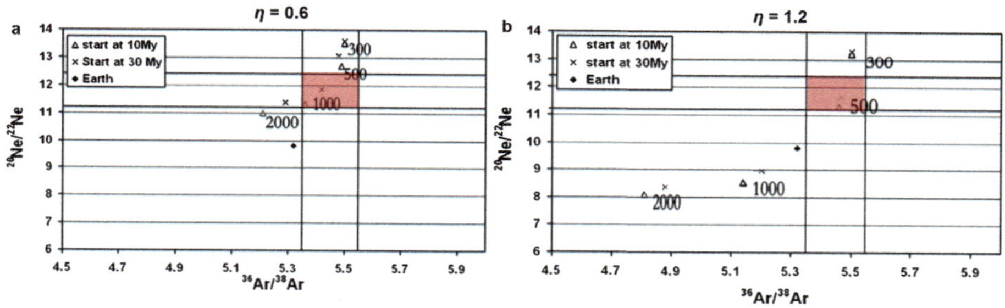

Fig. 10 Reproduction attempts of $^{36}Ar/^{38}Ar$ and $^{20}Ne/^{22}Ne$ isotope ratios according to different evolution scenarios related to atmospheric escape of early Venus with assumed average thermospheric temperatures between 300 - 2000 K. The two horizontal and vertical lines limit the ranges of the measured values of the isotopic ratios of Ne and Ar. They are delineated with the shaded area, which marks the scenarios that are in agreement with the measured isotopic ratios. a) Assumed initial atmospheric hydrogen content based on H_2O-amounts of 4 - 14 terrestrial oceans (TOs) for an η value of 0.6. b) Assumed initial atmospheric hydrogen content based on H_2O-amounts of 20 - 40 TOs for an η value of 1.2. Atomic oxygen does not escape in these scenarios (Gillmann et al. 2009)

In this approach Gillmann et al. (2009) introduced a factor $\eta = \varepsilon\zeta + \Phi_{sw}/\Phi_{EUV}$, that depends on the heating efficiency ε of the solar EUV flux Φ_{EUV}, generally assumed to be $\approx 15\%$ (Chassefière 1996b; Shematovich et al. 2014; Lammer et al. 2014), a geometrical amplification factor ζ related to the EUV heated and extended upper atmosphere and the relative contribution Φ_{sw} of the solar wind energy deposition to the total energy deposition (solar wind + EUV) near the exobase level. Since at the time of this study only very rough estimates existed on the young Sun's activity evolution, Gillmann et al. (2009) assumed η values between 0.3 - 20. The introduced geometrical amplification factor ζ was also based on arbitrary assumptions within a range of 4 - 8, since no upper atmosphere structures have been modelled in detail in these pioneering but simplified models.

Figure 10 shows resulting $^{36}Ar/^{38}Ar$ and $^{20}Ne/^{22}Ne$ isotope fractionation patterns in Venus atmosphere, obtained by the simple model approaches presented in Gillmann et al. (2009). One can see that their obtained evolutionary scenarios are consistent within the present error bar ranges. As mentioned in Sect. 2, one should note that the currently available $^{20}Ne/^{22}Ne$ ratio in Venus atmosphere and especially the quoted errors should be considered as very unprecise.

Since no upper atmosphere structure and temperature profiles were modelled in Gillmann et al. (2009), these authors assumed various average thermospheric temperatures between 300 and 2000 K. For high thermospheric temperatures > 1000 K, they obtained too large isotopic fractionations, because the densities of Ne and Ar and the escape fluxes are too large in their model approach. One can also see that for average thermosphere temperatures < 400 K, Ne is not fractionated enough to fit the present fractionation value due to its low abundance at the escape level. One should also note that the amount of hydrogen necessary to obtain the fractionation results for assumed η values of 0.6 (Fig. 10a) and 1.2 (Fig. 10b) contained 6 - 14 terrestrial oceans (TO) and 30 - 40 TOs, respectively (Gillmann et al. 2009).

More recently, Lichtenberg et al. (2016) studied the expected relative contribution of the solar wind energy deposition Φ_{sw} on early Venus, by assuming that the young Sun was either a weak or a moderately active young G star. These authors assumed a hydrogen-dominated upper atmosphere and modelled the precipitation of energetic hydrogen atoms (ENAs) that originate due to charge exchange between the planet's extended exosphere and solar wind protons (Chassefière 1996b, 1997). Chassefière (1997) and Gillmann et al. (2009) expected that these ENAs deposit a huge amount of energy into the upper atmosphere and as a result enhance the H escape rate that influenced the noble gas fractionation. According to Lichtenberg et al. (2016), however, ENAs deposit their energy and heat mainly above the main EUV energy deposition layer. Although these particles modify the thermal structure of the upper atmosphere, the enhancement of the thermal escape rates caused by ENAs remains negligible (Lichtenberg et al. 2016). Since several pioneering studies including that of Cameron (1983), Pepin (1991) and Gillmann et al. (2009) lacked knowledge on the X-ray and EUV evolution of the young Sun/stars, magma ocean physics or impact studies, assumed arbitrarily reservoirs of H_2-molecules from primordial atmospheres and/or catastrophically outgassed H_2O inventories, one can consider that these arbitrarily chosen assumptions are insufficiently accurate and hence outdated.

Recently, Odert et al. (2018) tried to reproduce the $^{36}Ar/^{38}Ar$ and $^{20}Ne/^{22}Ne$ isotope ratios of Venus' atmosphere within a more realistic parameter space and found that EUV-driven hydrodynamic escape of an H_2-envelope that remained around proto-Venus after the disk evaporated can reproduce the observed atmospheric Ne and Ar isotope ratios within their large error bars. In the study of these authors, the EUV activity evolution of the young Sun corresponded to a slowly to moderately rotating young Sun (Tu et al. 2015; Lammer et al. 2020b). These findings stimulated Lammer et al. (2020b) to apply more complex models to figure out whether one can reproduce the measured atmospheric $^{36}Ar/^{38}Ar$, $^{20}Ne/^{22}Ne$, and, in case of the Earth, also $^{36}Ar/^{22}Ne$ noble gas isotope ratios and bulk planet K/U elemental ratios of Earth and Venus to constrain their early evolution.

While the atmospheric $^{36}Ar/^{22}Ne$ ratio of $\approx$18.8 in Earth's present day atmosphere cannot be explained by a pure nebula origin (Marty 2012; Marty and Allé 1994), the $^{36}Ar/^{38}Ar$, $^{20}Ne/^{22}Ne$ ratios have been proposed to originate from the nebula only (e.g., Sasaki et al. 1989; Pepin 1991). Lammer et al. (2020b) assumed a reasonable set of initial protoplanet and impactor compositions based on the hypotheses of Dauphas (2017) and Schiller et al. (2018) and figured out under which protoplanetary maximum masses, at the time when the disk dissipated, Earth's present day atmospheric $^{36}Ar/^{22}Ne$ ratio can be reproduced. Additionally to these noble gas ratios, these authors also investigated whether the measured K/U ratios of both planets can be reproduced through the same process. As discussed in Sect. 6, elements like K are released at high temperatures ($\approx$ 1500 – 2500 K) from magma oceans that form below primordial atmospheres (Schaefer and Fegley 2010; Fegley et al. 2016; see also O'Neill et al. 2020; this issue). That a fraction of K compared to its initial composition was lost during accretion can be seen in the K/U ratio of each planet (e.g. $\approx$ 13800 for the bulk silicate Earth (BSE) (Arevalo et al. 2009; Lodders et al. 2009), and $\approx$ 7000 on Venus' surface (e.g., Surkov et al. 1976; Abdrakhimov and Basilevsky 2002), compared to a K/U abundance ratio of $\approx$ 64650 in the solar photosphere (Lodders et al. 2009) and $\approx$ 68930 in unprocessed chondrites (Lodders et al. 2009).

The model efforts carried out in Lammer et al. (2020b) are illustrated in Fig. 11 and summarized below:

- Protoplanetary masses between 0.5 - 0.8 M_{Earth} are placed within the disk at Venus and Earth orbit locations. In case a primordial H_2-envelope survives hydrodynamic boil-off during disk dispersal, the atmosphere mass and upper atmosphere structure including its photospheric radius and effective EUV absorption radius are modelled by a hydrodynamic upper atmosphere model and embedded trace elements (^{36}Ar, ^{38}Ar, ^{20}Ne, ^{22}Ne, K) in solar composition, and outgassed from an assumed 1000 km deep magma ocean in abundances according to different assumed compositions - i.e., ureilite (UR), enstatite (EN), carbonaceous chondrites (CC); as illustrated in Fig. 10, are included into the remaining H_2-dominated primordial atmosphere.

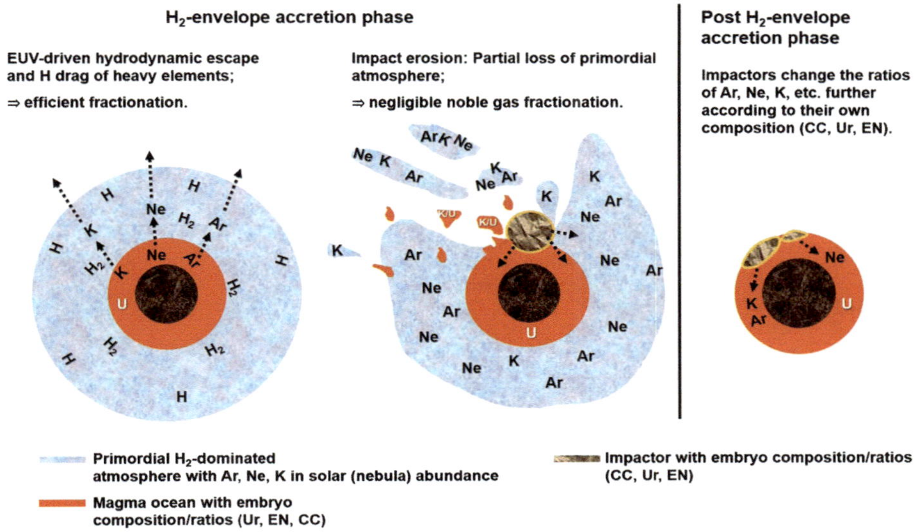

Fig. 11 Illustration of main processes and scenarios involved in the reproduction attempts of present day atmospheric ^{36}Ar, ^{38}Ar, ^{20}Ne, ^{22}Ne, noble gas isotope ratios and bulk K/U ratios on Venus and Earth by Lammer et al. (2020b)

- Hydrodynamic upper atmosphere models (Lammer et al. 2014; Kubyshkina et al. 2018b, 2018c) that can calculate the EUV-driven escape of H atoms of the primordial atmosphere and an evolution code that drag the heavier trace elements (^{36}Ar, ^{38}Ar, ^{20}Ne, ^{22}Ne, etc.) (Zahnle and Kasting 1986; Odert et al. 2018) out of the gravity field of accreting protoplanets are applied for various expected EUV-flux evolution scenarios from a slowly to a moderately and a fast rotating young Sun.
- Atmospheric mass loss of remaining H_2-envelopes caused by Moon- to Mars-mass impactors (Sect. 5) is considered additionally to the EUV-driven thermal escape and modelled with a multi-material SPH code (Schäfer et al. 2016), allowing simulation of gases, non-viscous fluids (Burger et al. 2018), granular media (Schäfer et al. 2017), solid bodies (Maindl et al. 2015; Burger and Schäfer 2014), and porous continua (Haghighipour et al. 2018). The masses of these impactors are assumed in Lammer et al. (2020b) to be admixtures of CC, UR, EN, and are added to the protoplanetary mass where they are also altering the ^{36}Ar, ^{38}Ar, ^{20}Ne, ^{22}Ne, and K content of proto-Earth and Venus.
- Evolution of the photospheric radii within the escaping H_2-envelope according to mass additions of accreting bodies, gravity change and atmospheric escape as well as its effects on the isotope/element ratio evolution is also included in the study of Lammer et al. (2020b).

The illustrations in Fig. 11 show the main phases and processes that are included in the recent model simulation attempts of Lammer et al. (2020b) for reproducing proto-Earth's and Venus' atmospheric Ar and Ne noble gas isotope and bulk K/U ratios. By reproducing present day measurements within their error bars these authors were able to constrain the growth rate of both planets in the presence of the solar nebular as well as a certain range for the EUV flux evolution of the young Sun.

Lammer et al. (2020b) present several model runs that cannot reproduce the above-mentioned elemental ratios and some that could. Here we present three cases for which

Planet	Earth	**Impactor depletion**	
Atmospheric mass	3.6 x 10^{-5} M_{Earth}	Dauphas:	EC: Ne 99%, Ar 99%, K 85%
Nebula life time	4 million years		CC: Ne 33%, Ar 5%, K 1%
Post H_2-env. impactor	95% EC, 5% CC (Dauphas)	Schiller:	CC: Ne 70%, Ar 32.5%, K 10%
	100% CC (Schiller)	Marty:	EC: Ne 99%, Ar 99%, K 60%
	98%EC, 2% CC (Marty)		CC: Ne 99%, Ar 0%, K 0%
Protoplanetary mass	0.55 M_{Earth} after disk dispersal (~4 Myr)		
EUV flux	slow to moderate		

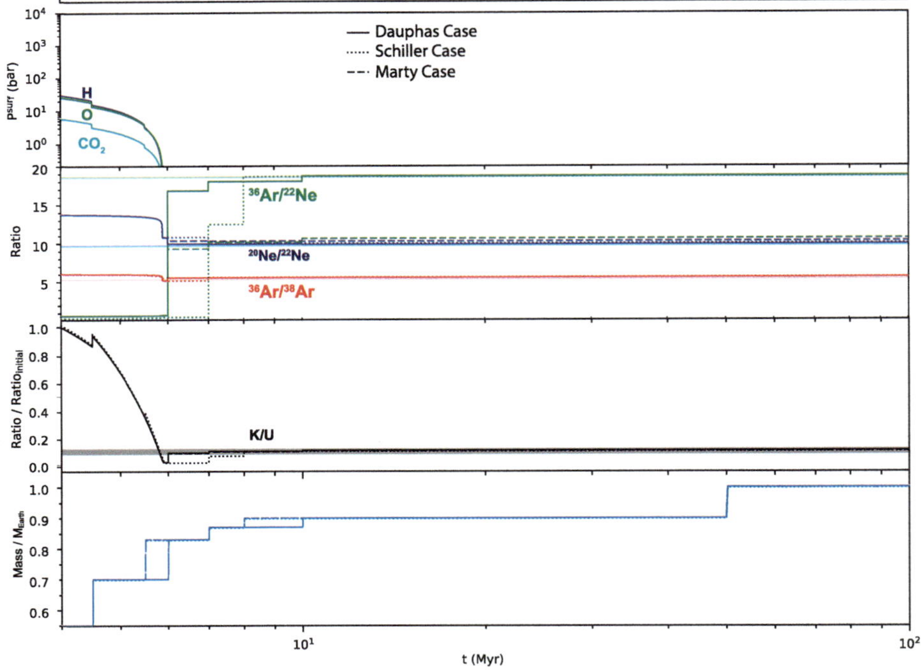

Fig. 12 Reproduction attempts of Earth's present-day atmospheric $^{36}Ar/^{38}Ar$, $^{20}Ne/^{22}Ne$, $^{36}Ar/^{22}Ne$ and bulk K/U ratios by assuming different post-H_2-envelope impactor compositions (after Lammer et al. 2020b). The rise of Earth's final mass at $\approx$ 50 Myr is related to the Moon-forming impact event for which it is assumed that the final impactor does not significantly alter the composition of proto-Earth

the atmospheric Ar and Ne isotope ratios and Earth's K/U bulk ratio can be reproduced if one assumes that proto-Earth accreted enough mass within the disk to accrete a small H_2-dominated primordial atmosphere which was subsequently lost by EUV-driven hydrodynamic escape.

Figure 12 shows evolution scenarios for early Earth with an initial protoplanetary mass of 0.55 M_{Earth} at the end of the disk at $\approx$ 4 Myr (Bollard et al. 2017; Wang et al. 2017) orbiting a young Sun that evolves along an EUV activity that corresponds to an intermediate rotating young G-type star (Tu et al. 2015; Güdel 2020; this issue). In such a case proto-Earth would have a post-H_2-envelope impactor composition of $\approx$ 95% EN and $\approx$ 5% CC (Dauphas 2017; solid lines), 100% CC (Schiller et al. 2018; dotted lines) and 98% EC and 2% CC (Marty 2012; dashed lines). Earth's atmospheric $^{36}Ar/^{22}Ne$ ratio of $\approx$ 18.8 could be reproduced by the model methods applied in Lammer et al. (2020b) for the Dauphas and Schiller scenarios but with different assumed growth rates and depletions of the post-H_2-envelope impactors. A significant difference in these production attempts is the amount of volatile rich CCs. The more CCs are delivered to the growing protoplanet the faster it has to grow so that the volatile elements can escape together with the captured primordial H_2-dominated atmosphere. If one uses Ar and Ne isotope ratios corresponding to CI material instead of CCs the outcomes only change marginally yielding more or less similar results. If proto-Earth captured such a tiny primordial atmosphere, the initial composition suggested by Marty (2012) cannot reproduce Earth's present atmospheric ^{36}Ar/^{22}Ne ratio. However, if proto-Earth accreted less mass so that no primordial atmosphere remained after disk dispersal the composition derived in Marty (2012) cannot be excluded. Due to this, Lammer et al. (2020b) concluded that the reproduction of the present day atmospheric Ar and Ne isotope ratios, by including a nebula component that escaped during a few million years after the disk dissipated, yields an upper limit for the mass of the proto-Earth of $\leq 0.58\ M_{\rm Earth}$ at the time when the disk evaporated.

As mentioned before, an important constraint for the composition of the post-nebula accretion is concealed in the atmospheric ratio of ^{36}Ar/^{22}Ne. Lammer et al. (2020b) found that different admixtures of UR, EC and CC-like material, for the impactors that were delivered after the primordial atmospheres escaped, could reproduce the atmospheric Ar and Ne ratios only if the impactors are depleted in the studied elements. As discussed in Sect. 6 and shown in Benedikt et al. (2020), one can expect that lower mass planetary embryos that developed their own magma ocean phases should be highly depleted in noble gases and outgassed volatiles.

By assuming an initial proto-Earth composition (95% EC, 5% CC) as suggested by Dauphas (2017), one can see in Fig. 12 that Earth's atmospheric Ar and Ne isotope ratios can only be reproduced if the accreting EC material is strongly depleted in noble gases ($\approx 99\%$) and volatile elements, while the remaining CCs are less depleted (Ne: $\approx 33\%$; Ar $\approx 5\%$). In such a case, most of the delivered CCs would be unfractionated material instead of highly depleted larger planetary embryos. For reproducing, the atmospheric Ar and Ne isotope ratios, if one assumes that $\approx 100\%$ of CCs were involved in early Earth's post-nebula accretion as suggested by Schiller et al. (2018), one needs impactors that are more efficiently depleted in these volatile elements (Ne: $\approx 70\%$; Ar $\approx 32.5\%$). In such a case, a larger fraction of the CC-related building blocks should have been bigger bodies that underwent magma ocean related outgassing events thereby losing much of its volatiles. If one assumes an initial proto-Earth composition of $\approx 98\%$ volatile poor material and $\approx 2\%$ CCs, as suggested by Marty (2012), then Earth's atmospheric ^{36}Ar/^{22}Ne ratio can only be reproduced if the Earth accreted slower with most of its mass being accreted after the disk dissipated. In such a case, no primordial atmosphere was captured by proto-Earth. One can see from Fig. 12 that such a scenario cannot reproduce the ^{36}Ar/^{22}Ne ratio if only even a tiny primordial atmosphere was present for a few million years after the disk evaporated. One should also note that the assumed depletions in the model runs of Lammer et al. (2020b) for the post-H_2-envelope impactor material are not chosen completely arbitrarily. For instance, planetary building blocks where only one noble gas element is strongly depleted such as Ne, but Ar and/or moderately volatile elements such as K are not depleted at all, are physically not realistic (Benedikt et al. 2020).

Figure 13 shows a best-case evolution scenario obtained by Lammer et al. (2020b) that can reproduce the present known atmospheric Ar and Ne isotope and measured surface K/U ratios on Venus by assuming a similar initial composition and the same young Sun EUV flux evolution track as assumed for the successful reproduction of the elements on Earth in Fig. 13.

In this particular scenario, proto-Venus accreted its whole mass during the disk lifetime and was released from the disk at ≈ 4 Myr with an H_2-dominated primordial atmosphere with a hydrogen partial pressure of ≈ 1000 bar. The escaping H atoms dragged

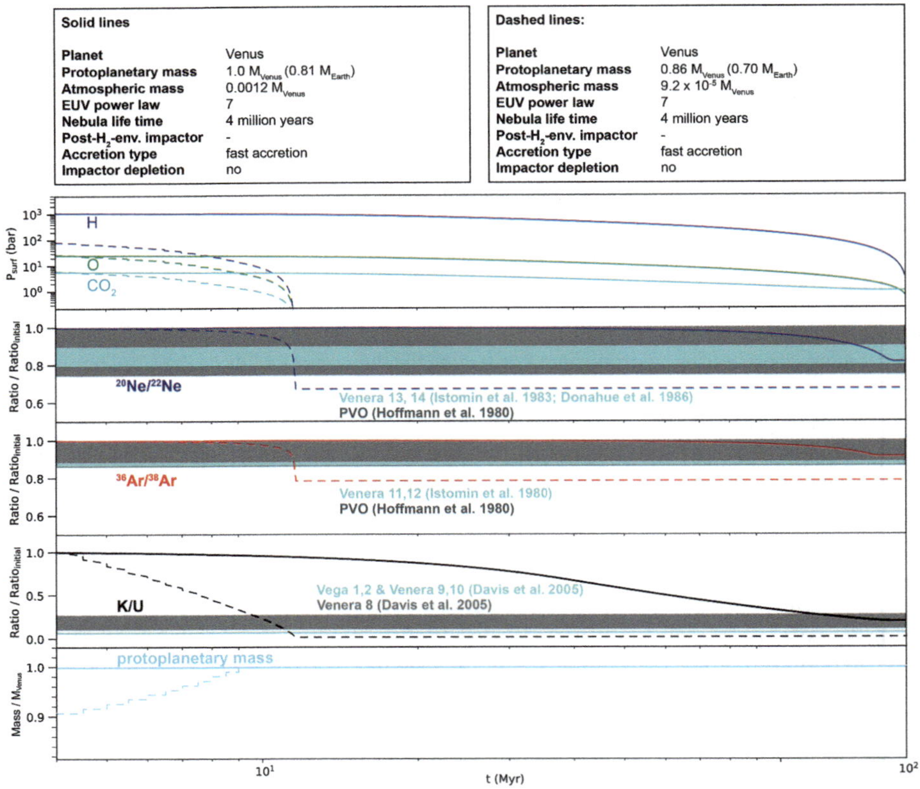

Fig. 13 Successful (solid lines) and failed (dashed-lines) reproduction attempts of the atmospheric $^{36}Ar/^{38}Ar$ and $^{20}Ne/^{22}Ne$ isotope and K/U surface ratios at Venus. A totally accreted Venus at the time when the disk evaporated can reproduce the present observations if the captured primordial H_2-dominated atmosphere is exposed to a slowly to less than moderately rotating young Sun (solid lines). All ratios are normalized to the initial ratio. A lower mass proto-Venus with a mass of about 0.86 M_{Venus} over-fractionates these noble gas isotopes and yields lower K/U ratios as measured by Venera 8, 9, 10, and Vega 1 and 2 (after Lammer et al. 2020b)

away the heavier trace species and fractionated $^{36}Ar/^{38}Ar$, $^{20}Ne/^{22}Ne$ from the initially solar-dominated ratios to the ones in today's Venus' atmosphere (Lammer et al. 2020b). Besides the noble gas isotope ratios, the corresponding K/U ratio lies slightly above Vega 1, 2 and Venera 9, 10 data but within the range of the Venera 8 landing site (Basilevsky et al. 1997; Abdrakhimov and Basilevsky 2002).

The dashed lines in Fig. 13 show an example modelled by Lammer et al. (2020b) that failed to reproduce the present-day ratios for Venus, even within the large uncertainties of the analysed measurements. In this failed attempt a less massive proto-Venus with about 0.86 M_{Venus} (0.69 M_{Earth}) leads to an under fractionation of the Ar, Ne isotope and K/U ratios. It should also be noted that for Venus' case Lammer et al. (2020b) did not model the $^{36}Ar/^{22}Ne$ ratio due to the large error bars. New precise measurements in future Venus missions are necessary so that one can test whether the planets building blocks can be linked to material related to EC or to a mixture of EC and CC with remnants from a primordial atmosphere.

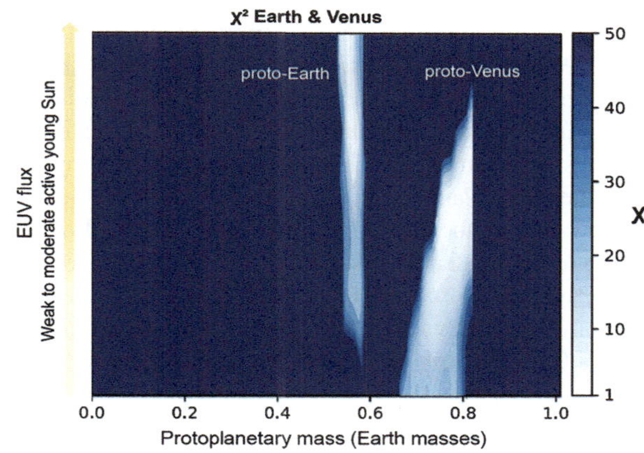

Fig. 14 Statistical χ^2 parameter study for constraining proto-planetary masses of early Earth and Venus at the time when the disk evaporated at about 3-5 Myr after the origin of the Solar System by successful reproduction (bright areas) of today's atmospheric ^{36}Ar/^{38}Ar, ^{20}Ne/^{22}Ne and bulk K/U ratios. χ^2 values that are larger than 5.0 indicate that at least one element is not within the error bars of today's measurements and scenarios with values larger than 30 show distinct deviations (after Lammer et al. 2020b)

Figure 14 shows the statistical outcome of many modelled proto-Earth and proto-Venus masses at the end of the disk lifetime (3 - 5 Myr) by Lammer et al. (2020b). The white areas illustrate the parameter space of EUV flux and protoplanetary mass in which the ^{36}Ar/^{38}Ar and ^{20}Ne/^{22}Ne noble gas isotope ratios and bulk planet K/U elemental ratios were successfully reproduced if the protoplanets accreted enough mass during the disk life time so that small primordial atmospheres were captured and lost afterwards, for particular compositions on both planets within their error bars.

One can see that the present day ratios can be reproduced if proto-Earth and Venus captured primordial atmospheres before the disk evaporated between 3 - 5 Myr with masses of about 0.53 - 0.58 M_{Earth} and 0.68 - 0.81 M_{Earth}, respectively, if the young Sun followed an EUV flux evolution path between a slowly and moderately rotating young G-type star (Tu et al. 2015; Lammer et al. 2020b; Güdel 2020, this issue). One should also note that in Earth's case the atmospheric ^{36}Ar/^{22}Ne ratio can also be reproduced as in Fig. 14 for the initial compositions as suggested by Dauphas (2017) and/or Schiller et al. (2018) if one adjusts the assumed depletion of the studied elements in the post-H_2-envelope impactors. For proto-Earth-masses larger than $\approx 60\%$ of its present mass at the time when the disk evaporated, Lammer et al. (2020b) could not reproduce the atmospheric Ar and Ne isotope ratios. Furthermore, for proto-Earth masses $\geq 80\%$ the accreting Earth would not have been able to lose its captured primordial atmosphere. These findings are also in agreement with the study of Pahlevan et al. (2019) that the accordance within $\approx$ 10–20 % between Earth's seawater and carbonaceous chondrites in D/H indicate that the prior existence of early Earth's atmospheric H_2 inventory on post-giant-impact Earth was $<$ 20 bars (see also Sect. 2).

According to the resent work of Burkhardt et al. (2019) an analysis of measured Ti and Sr isotopic compositions of Ca, Al-rich inclusions (CAIs) from the Allende CV3 chondrite indicate that CAI-like material was the carrier of the nucleosynthetic isotopic anomalies in most minor and trace elements. These authors found that a variable admixing of CAI-like refractory material to an average inner solar nebula component could explain the planetaryscale Ti and Sr isotope anomalies and the elemental and isotopic difference between noncarbonaceous (NC) and carbonaceous (CC) nebular reservoirs for these elements. If this is the case, then the differences between meteorites and Earth may simply reflect the timing or spatial location for adding this CAI-like material to the solar nebula. The findings by Lammer et al. (2020a) that the post-H_{2-}envelope impactors should have also be at least partially related to the CC-reservoir agrees also with the dating of terrestrial planet building blocks from CAIs, from which it is inferred that CCs finished their accretion at times around $\approx 3-4$ Myr after the formation of the Solar System (Kruijer et al. 2017).

Compared to both large terrestrial planets in the Solar System, the evolution for the lower mass planet Mars was different and one cannot constrain it in a similar way as in the case of Venus and Earth. Mars' mass is too low to capture a relevant H_2-envelope during the phase the small planet grew within the gas disk. Noble gases could have also been lost and dragged away by the escaping H atoms as long as hydrogen, which originated from solidified magma ocean-related outgassed steam atmospheres, dominated the upper atmosphere (Pepin 1994; Pepin and Porcelli 2002). Afterwards ionization related fractionation through thermal and non-thermal escape (i.e., ion pick up, sputtering, dissociative recombination) of the planet's CO_2 atmosphere resulted in today's observed Ar, Ne and N isotopic compositions (e.g., Jakosky et al. 1994; Hutchins and Jakosky 1996). In the next section, the isotopic fractionation and escape of the Martian atmosphere and initial water inventory are discussed.

8 Isotopic Fractionation as a Constraint on the Atmospheric Evolution and Water Inventory on Mars

The small gravity and absence of an intrinsic magnetic field allow efficient atmospheric escape from Mars induced by the solar EUV radiation and the solar wind. Its isotopic compositions show heavy-isotope enrichment in hydrogen (D/H), carbon ($^{13}C/^{12}C$), nitrogen ($^{15}N/^{14}N$), and noble gases ($^{22}Ne/^{20}Ne$, $^{36}Ar/^{38}Ar$, and non-radiogenic Xe isotopes) except Kr compared to their possible origins (nebula gas, chondrites, and comets), as an outcome of atmospheric escape. As mentioned above, noble gases are particularly useful to constrain the atmospheric evolution since they are mostly partitioned into the atmosphere and do not form molecules so that the escape processes are limited to hydrodynamic escape and to atmospheric sputtering for lower mass bodies without magnetospheres (e.g., Pepin 1991, 1994; Jakosky et al. 1994). Nitrogen behaves similarly to noble gases as it is chiefly partitioned into the atmosphere but ion-molecule photochemical reactions also induce atmospheric escape (e.g., Fox and Dalgarno 1983; Fox 1993a, 1993b; Fox and Hać 1999; see also Avice and Marty 2020; this issue). Carbon is partitioned into multiple reservoirs - the atmosphere, carbonate, organic compounds, CO_2/CH_4-clathrate, and CO_2 ice (e.g., Kurahashi-Nakamura and Tajika 2006; Longhi 2006; Hu et al. 2015; Wray et al. 2016) so that the isotopic fractionation is more complicated.

As briefly discussed in Sect. 4, fractionation mechanisms include diffusive separation by mass between the homopause and exobase (e.g., McElroy and Yung 1976; Jakosky et al. 1994, 2017). The number densities of photochemically long-lived atmospheric species follow their own scale height above the homopause. As the escape processes remove atoms, molecules, and ions from the vicinity of the exobase, the diffusive separation causes mass-dependent fractionation. Since atmospheric sputtering is so energetic that mass-dependent fractionation in the removal process itself is limited, the isotopic fractionation through sputtering is chiefly determined by the diffusive separation (Jakosky et al. 1994).

The diffusive separation between the homopause and exobase was recently probed with the help of the MAVEN spacecraft (Jakosky et al. 2017; Slipski et al. 2018). Measurements of the $N_2/^{40}Ar$ ratio indicated substantial variations at the homopause ($\approx 60-140$ km) and exobase ($\approx 140-200$ km) altitudes. This represents an unresolved combination of real geophysical variations with time and location but confirms the concept of elemental and isotopic fractionation. Photochemical processes fractionate isotopic ratios due to the difference in reaction rates between isotopic species. Both photodissociation of CO and

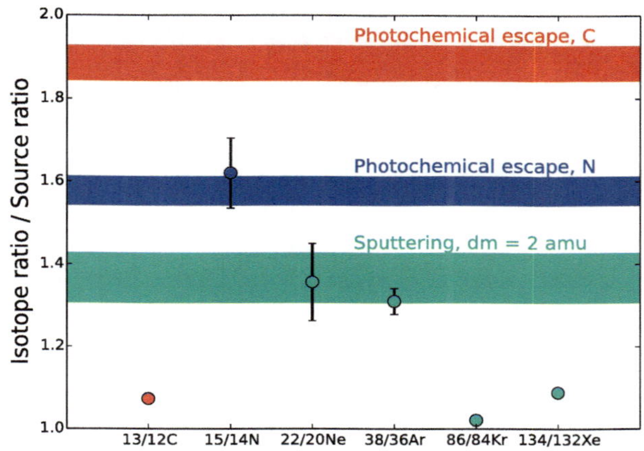

Fig. 15 Isotope ratios of volatile elements in the present-day Martian atmosphere with respect to their source isotope ratio (data points). We assumed the source values of Martian mantle informed from Martian meteorites for C and N (Wright et al. 1992; Mathew and Marti 2001), and solar abundances for noble gases (Pepin 1991; Pepin et al. 2012). Data for the Martian atmosphere are from Mahaffy et al. (2013a), Pepin (1991), Wong et al. (2013), Atreya et al. (2013), Conrad et al. (2016). The expected steady state values are shown as horizontal ranges. We assumed the upper atmospheric temperature to be 200 K and the homopause-exobase separation 60-80 km to calculate the fractionation factor for the diffusive separation and adopted additional fractionation for photochemical escape of suprathermal C and N atoms that originate mainly from dissociative recombination of N_2^+ and CO^+ molecules, $f = 0.62$ and 0.74, respectively (Manning et al. 2008; Hu et al. 2015)

dissociative recombination of CO^+ and CO_2^+ could induce preferential loss of ^{12}C rather than ^{13}C (Fox and Hać 1999; Hu et al. 2015). Dissociative recombination of N_2^+ causes isotopic fractionation between ^{14}N and ^{15}N (Wallis 1989; Fox and Hać 1997; Manning et al. 2008).

8.1 Isotopic Compositions of the Present-Day Martian Atmosphere

The observed isotopic compositions of N_2, Ne, and of the non-radiogenic Ar ($\delta^{15}N = 572 \pm 82‰$, $^{20}Ne/^{22}Ne = 10.1 \pm 0.7$, and $^{36}Ar/^{38}Ar = 4.2 \pm 0.1$, Pepin 1991; Wong et al. 2013; Atreya et al. 2013) are well described by the steady state where the loss via atmospheric escape balances the supply via, for example, volcanic outgassing (Fig. 15), and is given by $I/I_0 = 1/f$, where I, I_0, and f are the isotope ratio in the atmosphere, the isotope ratio of the source, and the net fractionation factor.

The estimated lifetimes of N (via sputtering and photochemical escape) and Ne (via sputtering) are shorter than one billion years, supporting the hypothesis of steady state (Jakosky et al. 1994; Manning et al. 2008; Kurokawa et al. 2018). The lifetime of Ar is comparable to the age of Mars (Jakosky et al. 1994), but its isotopic ratio being close to the value of steady state suggests that the lifetime was shorter on earlier time (Kurokawa et al. 2018). Being close to the steady state suggests

- the model for their isotopic fractionation is valid, and
- the present-day isotopic compositions do not reflect the earlier period (i.e. before ≈ 4 billion years ago (Ga).

The observed $^{13}C/^{12}C$ ratio being smaller than the ratio expected from the steady state with sputtering and/or photochemical escape (Fig. 14) suggests that atmospheric

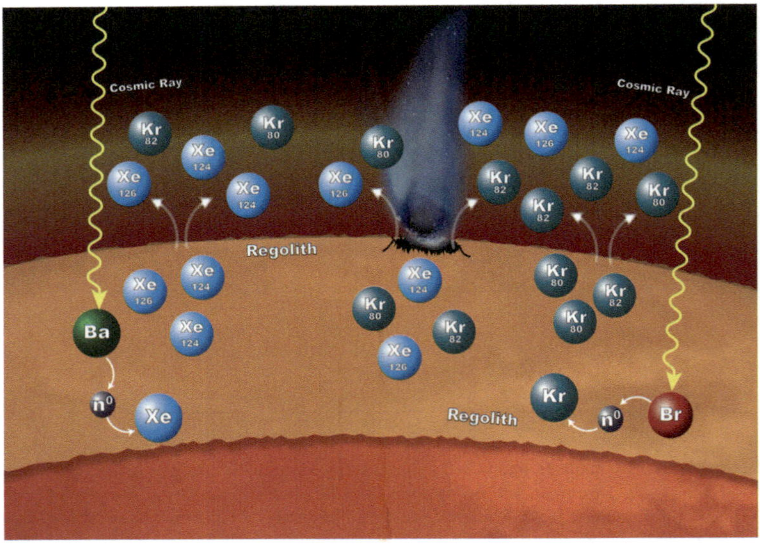

Fig. 16 Illustration of the cosmic ray-related surface chemistry that takes place in the Martian regolith, as an explanation why particular Xe and Kr isotopes are more abundant in the atmosphere than expected (Conrad et al. 2016). (NASA/GSFC/JPL-Caltech)

escape is not the only sink of an ancient dense atmosphere. Other sinks include carbonate precipitation (Hu et al. 2015; Wray et al. 2016), organic compounds deposition, and CO_2-clathrate formation following the basal melting of CO_2-ice caps (Longhi 2006; Kurahashi-Nakamura and Tajika 2006). Hu et al. (2015) considered atmospheric escape, carbonate precipitation, and concluded that Mars at ≈ 3.8 Ga had an atmospheric pressure ≤ 1 bar. Based on an experimental study (Eiler et al. 2000), part of the atmospheric CO_2 may convert into CO_2 ice without significant isotopic fractionation ($\delta^{13}C_{\text{ice-vapor}} < -0.4‰$ at 135 K). Considering additional sinks may allow a denser atmosphere.

The isotopic composition of Xe is also less fractionated compared to the value expected from the steady state with atmospheric sputtering. It suggests that the present-day Xe is a remnant of early hydrodynamic loss (Pepin 1994) or a result of ionization-induced escape during the first 500 Myr (Zahnle et al. 2019; see also Avice and Marty 2020; this issue). Ionization-induced escape can lead to high Xe^+ escape rates, while the escape of other heavy noble gases such as the hardly ionizable Kr is not significant. Due to the ionization of Xe, its cross-section becomes significantly larger, which makes it possible to escape and fractionate, whereas Kr stays neutral and, therefore, cannot escape due to its lower cross-section (Avice and Marty 2020; this issue). Furthermore, as illustrated in Fig. 16, atmosphere-surface interaction processes in combination with cosmic ray exposure and its related chemistry in the Martian surface can also contribute to an enhancement of some of the Xe and Kr isotopes in the Martian atmosphere (Conrad et al. 2016).

Because of the present thin Martian atmosphere, galactic cosmic rays penetrate into the surface material. According to these authors, if cosmic rays hit a barium (Ba) atom, Ba can lose neutrons. Xe atoms can catch some of those neutrons to form the isotopes ^{124}Xe and ^{126}Xe. In a similar way, Br can lose neutrons to Kr, leading to the formation of ^{80}Kr and ^{82}Kr isotopes. Impacts and abrasion allow that these isotopes enter the atmosphere, where they underwent escape by impact erosion.

8.2 Hydrodynamic Escape from Early Mars

Models of the upper atmosphere have shown that hydrodynamic escape prohibits early Mars from developing a stable atmosphere (Tian et al. 2009; Erkaev et al. 2014; Odert et al. 2018). The primordial H_2/He atmosphere that originated from the solar nebula (Pepin 1991, 1994; Stökl et al. 2015, 2016; Saito and Kuramoto 2018) was lost due to a short and efficient boil-off phase during disk dispersal (Lammer et al. 2018). This was followed by a catastrophically outgassed steam atmosphere during the solidification of a magma ocean (Elkins-Tanton 2008; Lebrun et al. 2013; Erkaev et al. 2014) that has been lost early on, and a secondary atmosphere later from volcanic outgassing (Lammer et al. 2013).

Hydrodynamic escape involves mass-dependent elemental and isotopic fractionation: heavier species are bound by the gravity more strongly and enrich in the remaining atmosphere (e.g., Hunten et al. 1987; Pepin 1991, 1994; Odert et al. 2018). The isotopic compositions of trapped non-radiogenic Xe in Martian meteorites Allan Hills 84001 (ALH84001) formed 4.1 Ga and North West Africa 7034 (NWA7034) formed 4.4 Ga are indistinguishable from current atmospheric Xe (Cassata 2017; see also Avice and Marty 2020; this issue), suggesting that the fractionation has completed before 4.4 Ga. If atmospheric Xe fractionated before 4.4 Ga and was later influenced by impact erosion and replenishment through time (de Niem et al. 2012; Kurokawa et al. 2018; Sakuraba et al. 2019), atmospheric escape of Xe in the later period via a photo-ionized hydrogen wind (Zahnle et al. 2019) mentioned before might compensate the replenishment.

8.3 Atmospheric Evolution After the Build-up of a Secondary Atmosphere

Since except Xe the present-day isotopic compositions do not probe the early atmospheric history, Kurokawa et al. (2018) utilized the isotopic compositions in the atmosphere 4.1 Ga recorded in ALH84001 (Mathew and Marti 2001). While a dense atmosphere preserves primitive isotopic compositions in their model, a thin atmosphere on early Mars is severely influenced by stochastic impact events and following escape-induced fractionation (Fig. 17).

Though the timing when Mars started to build up a stable atmosphere depends on the initial spin rate and related activity of the Sun (Tu et al. 2015; Güdel 2020; this issue) as well as on the volcanic outgassing activity on early Mars (Amerstorfer et al. 2017; Lammer et al. 2018), the less fractionated N ($\delta^{15}N = 7‰$) and Ar ($^{36}Ar/^{38}Ar > 5.0$) recorded in the ALH84001 meteorite (Mathew and Marti 2001) which formed at 4.1 Ga (Lapen et al. 2010) suggest that a denser atmosphere whose pressure was > 0.5 bar has already existed at the time (Kurokawa et al. 2018).

Other geochemical and geological estimates of paleo-pressures also point to a moderately dense atmosphere whose pressure is comparable to or slightly ≤ 1 bar (Kite 2019 and references therein). Elevated radiogenic Ar ($^{40}Ar/^{36}Ar$) and Xe ($^{129}Xe/^{132}Xe$) ratios recorded in ALH84001 (Cassata et al. 2010; Cassata 2017) are consistent with the Martian atmosphere originating at ≈ 4.1 Ga from late outgassing. The atmospheric ^{40}Ar on Mars is mainly the product of the past cumulative outgassing of ^{40}K from Mars' mantle and crust. This Ar isotope can therefore be used as a tracer of the past evolution of volatiles in Mars' crust, mantle and atmosphere. Similar as ^{36}Ar and ^{38}Ar, ^{40}Ar can also escape from the atmosphere into space via atmospheric sputtering. Leblanc et al. (2012) showed that atmospheric escape, which is more efficient in the early evolution, does not significantly affect the present abundance of the atmospheric ^{40}Ar, which indicates that the outgassing of this isotope from the mantle occurred relatively late in the Martian evolution, making ^{40}Ar to a tracer of outgassing but not for atmospheric escape.

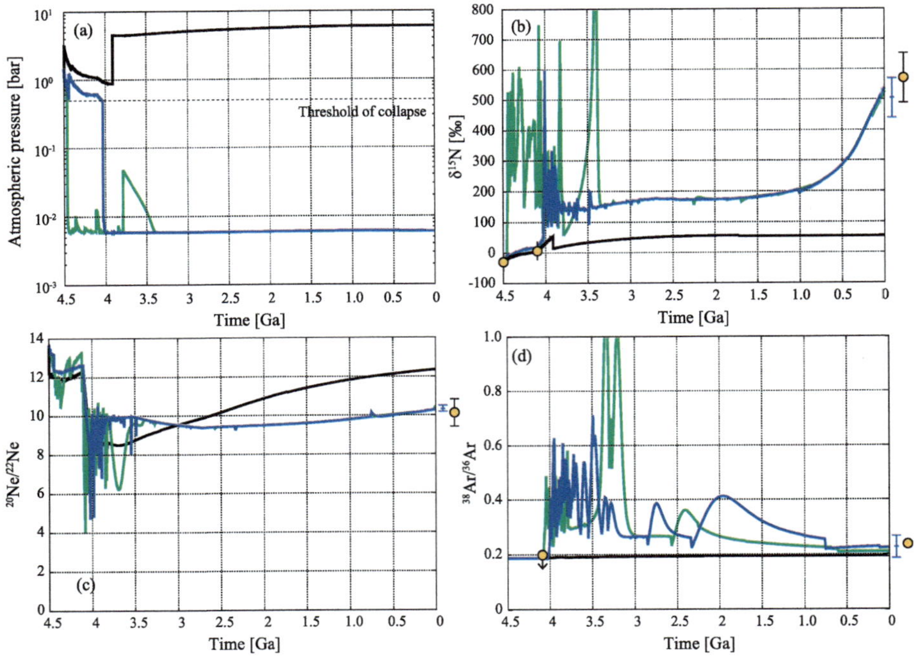

Fig. 17 Results of Monte Carlo simulations for the evolution of (**a**) atmospheric pressure, (**b**) N, (**c**) Ne, and (**d**) Ar isotope ratios (modified after Kurokawa et al. 2018). Three cases are shown: the atmosphere has been dense for 4.5 Gas (black), the atmosphere becomes thin after 4.1 Ga (blue), and the atmosphere becomes thin before 4.1 Ga (green). Data points in blue show the average value in the cases where the atmosphere collapsed at some point. Data points in yellow are the value of Martian atmosphere (Wong et al. 2013; Pepin 1991; Atreya et al. 2013; Mathew and Marti 2001)

8.4 Loss of Water on Mars

The D/H ratio recorded in ALH84001 (Boctor et al. 2003; Greenwood et al. 2008; Usui et al. 2017) is in the range of $\delta D = 500 - 3000‰$, which means that more than 1/3 of the surface water could have been lost at 4.1 Ga (Fig. 18). The surface water has experienced further loss to reach $\delta D = 5000 - 6000‰$ (Usui et al. 2012; Kurokawa et al. 2014; Mahaffy et al. 2013b; Villanueva et al. 2015). An intermediate D/H ratio ($\delta D = 1000 - 2000‰$) recorded in impact glasses in Martian meteorites suggests that the majority of near-surface water might have been stored as subsurface ice isolated from the atmosphere (Usui et al. 2015; Kurokawa et al. 2016; Grimm et al. 2017).

On the other hand, if one assume for a Noachean ocean volume of about 500 m global equivalent layer (GEL) or 0.05 times the terrestrial ocean mass (M_{TO}) (Di Ahille and Hynek 2010), the water loss before 4.1 Ga is > 160 m GEL, which might exceed the integrated oxygen escape (Jakosky et al. 2018). This imbalance between H and O escape drives the oxidation of the Martian surface environment (Lammer et al. 2003b; Zahnle et al. 2008).

That Mars lost large amounts of water was also suggested by Lammer et al. (2003a), who estimated the present and past H_2O-ice reservoirs, which are and were in exchange with the atmosphere, by using the observed D/H ratio in the atmospheric water vapour, measured D/H ratios in Martian SNC meteorites and D/H isotope ratios based on a study regarding asteroid and cometary water delivery to early Mars by Lunine et al. (2003).

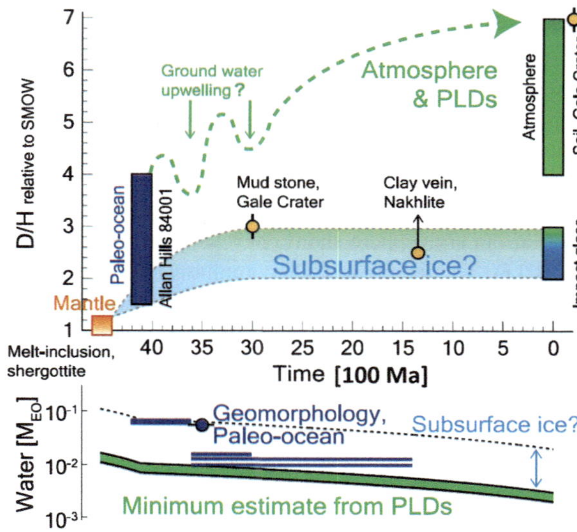

Fig. 18 Evolution of D/H (upper panel) and water volume (lower panel) on Mars (modified after Kurokawa et al. 2014, 2016; Usui et al. 2017). Data for D/H are from Usui et al. (2012, 2015, 2017), Boctor et al. (2003), Greenwood et al. (2008), Webster et al. (2013a, 2013b), Mahaffy et al. (2015), Villanueva et al. (2015), Krasnopolsky (2015), and Aoki et al. (2015). Water volume was computed by assuming Rayleigh fractionation with $f = 0$

These authors estimated the present H_2O-ice reservoir to be equivalent to about 11 – 27 m GEL. Here, the upper value more or less agrees with a more recent study by Carr and Head (2015) who found that a 34 m GEL resides in the Martian polar-layered deposits and shallow ground ice. By considering such a GEL value, a reservoir exchangeable with the atmosphere on Mars 3.5 Ga with a thickness of between ≈ 60 m is estimated (Lammer et al. 2003a; Carr and Head 2015).

By using these reservoirs from 3.5 Ga, Lammer et al. (2003a) found a global ocean at ≈ 4.5 Ga with a depth of ≈ 115 m GEL, if the initial isotopic ratio of Mars' water was 1.2 times the terrestrial sea water ratio (Lunine et al. 2003). A GEL of ≈ 115 m at ≈ 4.5 Ga is also in agreement with the initial water inventory in Mars mantle at the end of its accretion as estimated by Dreibus and Wänke (1987) by comparing the solubility of H_2O and HCl in molten silicates. In this way, Dreibus and Wänke (1987) obtained an H_2O content in the Martian mantle of ≈ 36 ppm corresponding to a global Martian ocean with a depth of about 130 m. The difference with respect to the initial value (di Achille and Hynek 2010) may have escaped unfractionated to space between 3.5 - 4.5 Ga and/or partly stored in the subsurface (Orosei et al. 2018) (see also Fig. 18).

8.5 Redox State of the Martian Atmosphere in Relation to the Atmospheric Escape

The preferential loss of hydrogen into space causes the oxidation of the Martian surface (e.g. Lammer et al. 2003b). Additionally, carbon escape may have dominated against oxygen escape on early Mars (Tian et al. 2009; Amerstorfer et al. 2017) which may have also oxidized Mars' surface. If this is the case, the redox state of the early Martian atmosphere could be more reducing than modern Mars. Thus, it is worth considering the possibility of an early reducing Martian atmosphere.

First, the outgassing composition of presumably early Martian volcanos could have been more reducing than Earth's since the oxygen fugacity of the Martian mantle could be up to 3 log-units lower than of the terrestrial mantle and possibly close to the IW-buffer especially during its early period (Wadhwa 2001; Herd et al. 2002; Righter et al. 2008;

McSween et al. 2009). When assuming an oxygen fugacity of log f_{O2} = IW+2 and a temperature of ≈ 1500 K, with ≈ 100 m GEL and ≈ 1 bar CO_2, the outgassing composition should be roughly H_2: H_2O: CO: CO_2 ≈ 10: 1: 20: 1, which is much more reducing than Earth's present volcanic flux (e.g. Zolotov 2003). Although the oxygen fugacity of the Martian crust may vary and could be higher than in the mantle (Wadhwa 2001), the Martian basaltic volatiles could be dominated by H_2 and CO relative to CO_2 based on a thermodynamic calculation of Gaillard and Scaillet (2014). Given the H_2-rich volcanic flux, the early Martian atmosphere should have been very reducing and possibly rich in CO (Sholes et al. 2017). This is due to CO_2 continuously being photolyzed into CO, whereas the oxidation of CO back into CO_2 is not efficient if the volcanic flux provides sufficient reducing agents into the atmosphere. This might then lead to a CO-runaway (Yung and DeMore 1998; Zahnle et al. 2008).

The reducing early Martian atmosphere may provide a solution to explain why the atmospheric CO_2 of Mars ($\delta^{13}C_{PDB} = +46 \pm 4‰$; Webster et al. 2013a, 2013b) is so enriched in ^{13}C compared to that of the Martian mantle ($\delta^{13}C_{PDB}$ = -25 ± 5‰; Wright et al. 1992). Galimov (2000) originally proposed that CO_2 could have been a minor component in the early Martian atmosphere as compared with CH_4 or CO from a thermodynamic point of view. CH_4 and CO are depleted in ^{13}C and preferentially escaped into space, whereas the ^{13}C-enriched CO_2 is ultimately deposited as carbonate. Also, as mentioned above, Hu et al. (2015) argue that photodissociation of CO may significantly induce preferential loss of ^{13}C. These authors estimate that the current atmospheric $^{13}C/^{12}C$ ratio and rock and soil carbonate measurements indicate an early Martian atmosphere with a surface pressure < 1 bar at ≈ 3.8 Ga ago. Hu et al. (2015) points out that only scenarios with large amounts of carbonate formation in open lakes would permit higher values of up to 1.8 bar.

These isotopic evidence and studies on atmospheric escape (Lammer et al. 2018) as well as a recent investigation on the formation of Martian geologic features, including degraded impact craters, lakes and valley networks by Palumbo et al. (2020) indicate that Mars most likely did not need a warm and wet Noachian climate, with surface temperatures above-freezing that allowed rainfall.

In these scenarios, the atmospheric ^{13}C-enrichment is caused by preferential escape of ^{13}C into space though time. Based on the isotopic studies of carbonate globules in ALH84001 (Eiler et al. 2002; Halevy et al. 2011), the enrichment in ^{13}C may have possibly occurred before 3.9 Ga, when the geomagnetic field may still have existed Lillis et al. (2013) and prohibited carbon escape into space via pick-up-ion sputtering induced by the solar wind. On the other hand, C atoms most likely escaped efficiently during this period by thermal escape (Tian et al. 2009; Amerstorfer et al. 2017). Moreover, due to the higher EUV flux of the young Sun, early Mars upper atmosphere expanded to larger distances so that the intrinsic magnetic field may not have shielded exospheric particles against escape. A reducing atmosphere (a smaller CO_2 mixing ratio) leads to a higher temperature in the upper atmosphere due to inefficient cooling and therefore allows efficient escape (Kulikov et al. 2007; Tian et al. 2009; Johnstone et al. 2018). Further experimental and model studies are required to test the reducing early Martian atmosphere scenario.

An additional alternative mechanism for early ^{13}C-enrichment recorded in ALH84001 than atmospheric escape is isotopic fractionation by solar UV photodissociation of CO_2, which may cause significantly large carbon isotopic fractionation of about 200‰ based on theoretical calculations of absorption spectra of the CO_2 isotopologues (Schmidt et al. 2013). Such a process is also relevant if Mars' early atmosphere would have been lost during younger ages compared to that of ALH84001. Although the effect has not yet been tested by laboratory experiment, solar UV could dissociate $^{12}CO_2$ preferentially, and thus provide ^{13}C-enriched carbonate precipitating from the remaining CO_2. In this case, highly ^{13}C-depleted CO is produced in the atmosphere. Under this reducing condition and the presence of water vapor, CO could be converted mainly into formaldehyde and removed from the atmosphere (Bar-Nun and Hartman 1978; Pinto et al. 1980; Bar-Nun and Chang 1983). If this scenario is correct, then ^{12}C preferentially deposited as organics into sediment, but did not escape into space.

9 Conclusion

We discussed the relevance of various loss processes that led most likely to today's atmospheric Ar, and Ne noble gas isotope anomalies and the depletion of moderately volatile rock-forming elements on the terrestrial planets. A complex interplay between the young Sun's EUV flux, the accreted protoplanetary mass within the disk lifetime, a possible accumulated primordial H_2-dominated atmosphere, the impacting bodies that were involved in the accretion before and after disk dispersal, as well as the depletion of volatile elements from planetary embryos and the final protoplanets led eventually to today's measured ratios. The discovery of many low mass exoplanets that have captured a primordial H_2-dominated atmosphere that never escaped to space during billions of years indicates that the accretion of smaller planets is also a fast process. One finds from early atmosphere evolution models that measured atmospheric Ar and Ne isotope ratios can be reproduced if proto-Venus and Earth were released from the solar nebula at ≈ 4 Myr with a small captured H_2-dominated envelope and maximum protoplanetary masses at that time of $\approx 0.85 - 1.0\ M_{\text{Venus}}$ and $0.53 - 0.58\ M_{\text{Earth}}$, respectively.

If proto-Earth accreted such a mass from enstatite-like material during the disk lifetime, a tiny primordial atmosphere would have been lost soon after the disk dissipated and the presentday atmospheric Ar and Ne isotope ratios can be reproduced if the post-nebula impactors contained $\leq 5 \approx 95\ \%$ strongly depleted enstatite chondrites. If higher amounts of carbonaceous chondrites were involved in early Earth's accretion then the present atmospheric Ar and Ne ratios can only be reproduced if the delivered carbonaceous chondritic material taht was delivered after the disk dissipated was also highly depleted in these noble gases and/or had to be mainly delivered as long as the primordial atmosphere was yet escaping. If proto-Earth accreted less mass during the disk lifetime atmospheric escape cannot explain the present-day ratios, then these ratios have to be explained differently (Marty, 2012).

Impact erosion and outgassing of volatiles from magma pools, oceans, and subsequent escape to space should have depleted the planetary embryos that acted in the final accretion of the early terrestrial planets. Latest reproduction attempts of atmospheric Ar and Ne isotope ratios and bulk K/U ratios on Venus agree with the findings related to early Earth's evolution that the young Sun was less active than a moderate rotator and EUV fluxes corresponding to a faster rotating young Sun were very unlikely.

Thus, new precise measurements by future Venus' space missions of atmospheric Ar, Ne, Kr, and Xe noble gas isotope ratios and more data of moderately volatile elements distributed over a wide surface area would give a precise evolution scenario of Venus and would further help to constrain the Sun's history. For terrestrial exoplanets one can expect that different accretion scenarios and different radiation fields of their host stars will result in different heat producing elements and hence in different tectonic regimes (see also O'Neill et al. 2020; this issue), which can have tremendous implication for the planet's evolution and habitability.

From D/H analysis, one finds that Mars lost a huge fraction of its initial water inventory during the first 500 Myr after its origin. The low gravity and high EUV fluxes of the young Sun led also to high escape rates of its bulk atmosphere. After the EUV flux decreased at $\approx$ 4–4.1 Ga, volcanic outgassing and impacts could have resulted in the build-up of a denser atmosphere of ≤ 1 bar, which is also constrained by analysis of $^{36}Ar/^{38}Ar$ and $^{13}C/^{12}C$ data. Analysis of the atmosphere enriched in ^{13}C isotopes compared to Mars' mantle may be an indication that early Mars was more reduced than today. CO_2 could have been a minor component of the early Martian atmosphere as compared with CO or CH_4 from a thermodynamic point of view. Both molecules are depleted in ^{13}C and preferentially escaped into space, whereas ^{13}C-enriched CO_2 is ultimately deposited as carbonate. CO_2 could have been released from surface reservoirs during periods of climate change or impacts and may have modified the atmospheric surface pressure several times during the planet's history. Since $\approx$ 3.5 Ga, when the young Sun's EUV flux decreased to lower values, Mars' atmospheric evolution until today was determined by a complex interplay of ion escape, atmospheric sputtering and suprathermal hot atom escape, carbonate precipitation, and serpentinization that led to today's low surface pressure.

Acknowledgements We acknowledge support by the Austrian Fonds zur Förderung der Wissenschaftlichen Forschung, Nationales Forschungs Netzwerk (FWF NFN) project S116-N16 and the subprojects S11603-N16, S11604-N16, S11606-N16, S11607-N16 and S11608-N16. H. Lammer, M. Leitzinger and P. Odert acknowledge support of the FWF projects P27256-N27 and P30949-N36. M. Benedikt and H. Lammer acknowledge support from the Austrian Forschungsförderungsgesellschaft (FFG) project RASEN. L. Fossati acknowledge also the FFG project "TAPAS4CHEOPS" P853993. H. Kurokawa acknowledges support from JSPS KAKENHI Grant 17H01175, 17H06457, 18K13602, 19H01960, and 19H05072. Finally, we thank an anonymous referee for very valuable and important suggestions that helped to improve the manuscript.

Publisher's Note Springer Nature remains neutral with regard to jurisdictional claims in published maps and institutional affiliations.

References

A.M. Abdrakhimov, A.T. Basilevsky, Geology of the Venera and Vega landing-site regions. Sol. Syst. Res. **36**, 136–159 (2002)

F. Albarède, J. Blichert-Toft, The split fate of the early Earth, Mars, Venus, and Moon. C. R. Géosci. **339**, 917–927 (2007)

C.J. Allegre, T. Staudacher, P. Sarda, Rare gas systematics – formation of the atmosphere, evolution and structure of the Earth's mantle. Earth Planet. Sci. Lett. **81**, 127–150 (1987)

U.V. Amerstorfer, N.V. Erkaev, D. Langmayr, H.K. Biernat, On Kelvin Helmholtz instability due to the solar wind interaction with unmagnetized planets. Planet. Space Sci. **55**, 1811–1816 (2007)

U.V. Amerstorfer, H. Gröller, H.I.M. Lichtenegger, H. Lammer, F. Tian, L. Noack, M. Scherf, C. Johnstone, L. Tu, M. Güdel, Escape and evolution of Mars's CO_2 atmosphere: influence of suprathermal atoms. J. Geophys. Res., Planets **122**, 1321–1337 (2017)

E. Anders, N. Grevesse, Abundances of the elements - Meteoritic and solar. Geochim. Cosmochim. Acta **53**, 197–214 (1989)

S. Aoki, H. Nakagawa, H. Sagawa, M. Giuranna, G. Sindoni, A. Aronica, Y. Kasaba, Seasonal variation of the HDO/H_2O ratio in the atmosphere of Mars at the middle of northern spring and beginning of northern summer. Icarus **260**, 7–22 (2015)

R. Arevalo, W.F. McDonough, M. Luong, The K/U ratio of the silicate Earth: insights into mantle composition, structure and thermal evolution. Icarus **278**, 361–369 (2009)

S.K. Atreya, M.G. Trainer, H.B. Franz, M.H. Wong (the SAM Team), Primordial argon isotope fractionation in the atmosphere of Mars measured by the SAM instrument on Curiosity and implications for atmospheric loss. Geophys. Res. Lett. **40**, 5605–5609 (2013)

G. Avice, B. Marty, Perspectives on atmospheric evolution from Xe and nitrogen isotopes on Earth, Mars & Venus. Space Sci. Rev. (2020, this issue)

G. Avice, D.V. Bekaert, H. Chennaoui Aoudjehane, B. Marty, Noble gases and nitrogen in Tissint reveal the composition of the Mars atmosphere. Geochem. Perspect. Lett. **6**, 11–16 (2018a)

G. Avice, B. Marty, R. Burgess, A. Hofmann, P. Philippot, K.J. Zahnle, D. Zakharov, Evolution of atmospheric xenon and other noble gases inferred from Archean to Paleoproterozoic rocks. Geochim. Cosmochim. Acta **232**, 82–100 (2018b)

K.H. Baines, S.K. Atreya, M.A. Bullock, D.H. Grinspoon, P. Mahaffy, C.T. Russell, G. Schubert, K. Zahnle, The atmospheres of the terrestrial planets: Clues to the origins and early evolution of Venus, Earth, and Mars, in *Comparative Climatology of Terrestrial Planets*, ed. by S.J. Mackwell, A.A. Simon-Miller, J.W. Harder, M.A. Bullock (University of Arizona Press, Tucson, 2013), pp. 137–160, 610 pp.

A. Bar-Nun, S. Chang, Photochemical reactions of water and carbon monoxide in Earth's primitive atmosphere. J. Geophys. Res., Oceans **88**(C11), 6662–6672 (1983)

A. Bar-Nun, H. Hartman, Synthesis of organic compounds from carbon monoxide and water by UV photolysis. Orig. Life **9**(2), 93–101 (1978)

A.T. Basilevsky, J.W. Head, G.G. Schaber, R.G. Strom, The resurfacing history of Venus, in *Venus II*, ed. by W. Bougher, D.M. Hunten, R.J. Phillips (University of Arizona Press, Tucson, 1997), p. 1047

S.J. Bauer, H. Lammer, *Planetary Aeronomy: Atmosphere Environments in Planetary Systems* (Springer, New York, 2004), p. 207

R.H. Becker, R.N. Clayton, E.M. Galimov, H. Lammer, B. Marty, R.O. Pepin, R. Wieler, Isotopic signatures of volatiles in terrestrial planets. Space Sci. Rev. **106**, 377–410 (2003)

M.R. Benedikt, M. Scherf, H. Lammer, E. Marcq, P. Odert, M. Leitzinger, N.V. Erkaev, Escape of rock-forming volatile elements and noble gases from planetary embryos. Icarus **347**, 113772 (2020). https://doi.org/10.1016/j.icarus.2020.113772

J.-L. Bertaux, A.-C. Vandaele, O. Korablev (the SPICAV/SOIR Team), A warm layer in Venus' cryosphere and high-altitude measurements of HF, HCl, H_2O and HDO. Nature **450**, 646–649 (2007)

N.Z. Boctor, C.M.O. Alexander, J. Wang, E. Hauri, The sources of water in Martian meteorites: clues from hydrogen isotopes. Geochim. Cosmochim. Acta **67**, 3971–3989 (2003)

J. Bollard, J.N. Connelly, M.J. Whitehouse, E.A. Pringle, L. Bonal, J.K. Jørgensen, Å. Nordlund, F. Moynier, M. Bizzarro, Early formation of planetary building blocks inferred from Pb isotopic ages of chondrules. Sci. Adv. **3**, 1–9 (2017)

A.S. Bonomo, L. Zeng, M. Damasso, Z.M. Leinhardt, A.B. Justesen (the Kepler Team), A giant impact as the likely origin of different twins in the Kepler-107 exoplanet system. Nat. Astron. **3**, 416–423 (2019)

A. Bonsor, Z.M. Leinhardt, P.J. Carter, T. Elliott, M.J. Walter, S.T. Stewart, A collisional origin to Earth's non-chondritic composition? Icarus **247**, 291–300 (2015)

A. Bouvier, M. Boyet, Primitive Solar System materials and Earth share a common initial ^{142}Nd abundance. Nature **537**, 399–402 (2016)

L. Bouvier, M. Costa, J. Connelly, N. Jensen, D. Wielandt, M. Storey, A. Nemchin, M. Whitehouse, J. Snape, J. Bellucci, F. Moynier, A. Agranier, B. Gueguen, M. Schönbächler, M. Bizzarro, Evidence for extremely rapid magma ocean crystallization and crust formation on Mars. Nature **558**, 586–589 (2018)

G. Budde, C. Burkhardt, T. Kleine, Molybdenum isotopic evidence for the late accretion of outer Solar System material to Earth. Nat. Astron. **3**, 736–741 (2019). https://doi.org/10.1038/s41550-019-0779-y

C. Burger, C.M. Schäfer, Applicability and limits of simple hydrodynamic scaling for collisions of water-rich bodies in different mass regimes, in *Proceedings of the First Greek-Austrian Workshop on Extrasolar Planetary Systems* (2014), pp. 63–81. arXiv:1704.06075

C. Burger, T.I. Maindl, C. Schäfer, Transfer, loss and physical processing of water in hit-and run collisions of planetary embryos. Celest. Mech. Dyn. Astron. **130**, 2 (2018)

C. Burkhardt, N. Dauphas, U. Hans, B. Bourdon, T. Kleine, Elemental and isotopic variability in solar system materials by mixing and processing of primordial disk reservoirs. Geochim. Cosmochim. Acta **261**, 145–170 (2019)

A.G.W. Cameron, Origin of the atmospheres of the terrestrial planets. Icarus **56**, 195–201 (1983)

G. Caro, B. Bourdon, J.-L. Birck, S. Moorbath, High-precision ^{142}Nd/^{144}Nd measurements in terrestrial rocks: constraints on the early differentiation of the Earth's mantle. Geochim. Cosmochim. Acta **70**, 164–191 (2006)

M.H. Carr, J.W. Head Jr., Martian surface/near – surface water inventory: Sources, sinks, and changes with time. Geophys. Res. Lett. **42**, 726–732 (2015)

P.J. Carter, Z.M. Leinhardt, T. Elliott, M.J. Walter, S.T. Stewart, Compositional evolution during rocky protoplanet accretion. Astrophys. J. **460**, 276–286 (2015)

P.J. Carter, Z.M. Leinhardt, T. Elliot, S.T. Stewart, M.J. Walter, Collisional stripping of planetary crusts. Earth Planet. Sci. Lett. **484**, 276–286 (2018)

P. Cartigny, B. Marty, Nitrogen isotopes and mantle geodynamics: the emergence of life and the atmosphere-crust-mantle connection. Elements **9**, 359–366 (2013)

W.S. Cassata, Meteorite constraints on Martian atmospheric loss and paleoclimate. Earth Planet. Sci. Lett. **479**, 322–329 (2017)

W.S. Cassata, D.L. Shuster, P.R. Renne, B.P. Weiss, Evidence for shock heating and constraints on Martian surface temperatures revealed by $^{40}Ar/^{39}Ar$ thermochronometry of Martian meteorites. Geochim. Cosmochim. Acta **74**, 6900–6920 (2010)

J.W. Chamberlain, Planetary coronae and atmospheric evaporation. Planet. Space Sci. **11**, 901–996 (1963)

E. Chassefière, Hydrodynamic escape of hydrogen from a hot water-rich atmosphere: the case of Venus. J. Geophys. Res. **101**, 26039–26056 (1996a)

E. Chassefière, Hydrodynamic escape of oxygen from primitive atmospheres: applications to the cases of Venus and Mars. Icarus **124**, 537–552 (1996b)

E. Chassefière, Note: loss of water on the young Venus: the effect of a strong primitive solar wind. Icarus **126**, 229–232 (1997)

A. Chau, C. Reinhardt, R. Helled, J. Stadel, Forming mercury by giant impacts. Astrophys. J. **865**, 35 (2018)

R.N. Clayton, Oxygen isotopes in meteorites. Annu. Rev. Earth Planet. Sci. **21**, 115–149 (1993)

M.S. Clement, N.A. Kaib, J.E. Chambers, Dynamical constraints in Mercury's collisional origin. Astrophys. J. **157**, 208 (2019)

P.G. Conrad, C.A. Malespin, H.B. Franz, R.O. Pepin, M.G. Trainer, S.P. Schwenzer, S.K. Atreya, C. Freissinet, J.H. Jones, H. Manning, T. Owen, A.A. Pavlova, R.C. Wiens, M.H. Wong, P.R. Mahaffy, In situ measurement of atmospheric krypton and xenon on Mars with Mars Science Laboratory. Earth Planet. Sci. Lett. **454**, 1–9 (2016)

A. Coustenis, F.W. Taylor, Titan: exploring an Earth-like world, in *Series on Atmospheric, Oceanic and Planetary Physics*, vol. 4, ed. by A. Coustenis, F.W. Taylor 2nd edn. (World Scientific Publishing, Singapore, 2008), pp. 154–155

P.E. Cubillos, N.V. Erkaev, I. Juvan, L. Fossati, C.P. Johnstone, H. Lammer, M. Lendl, P. Odert, K.G. Kislyakova, An overabundance of low-density Neptune-like planets. Mon. Not. R. Astron. Soc. **466**, 1868–1879 (2017a)

P.E. Cubillos, L. Fossati, N.V. Erkaev, M. Malik, T. Tokano, M. Lendl, C.P. Johnstone, H. Lammer, A. Wyttenbach, Aerosol constraints on the atmosphere of the hot Saturn-mass planet WASP-49b. Astrophys. J. **849**, 145–154 (2017b)

I. Dandouras, M. Blanc, L. fossati, M. Gerasimov, E.W. Guenther, K.G. Kislyakova, H. Lammer, Y. Lin, B. Marty, S. Rugheimer, C. Sotin, S. Tachibana, P. Wurz, M. Yamauchi, Future missions related to isotope and element measurements. Space Sci. Rev. (2020, this issue)

N. Dauphas, The isotopic nature of the Earth's accreting material through time. Nature **541**, 521–524 (2017)

D. de Niem, E. Kührt, A. Morbidelli, U. Motschmann, Atmospheric erosion and replenishment induced by impacts upon the Earth and Mars during a heavy bombardment. Icarus **221**, 495–507 (2012)

I. de Pater, J.J. Lissauer, *Planetary Sciences* (Cambridge University Press, Cambridge, 2015), ISBN: 9781316165270. https://doi.org/10.1017/CBO9781316165270

G. di Achille, B.M. Hynek, Ancient ocean on Mars supported by global distribution of deltas and valleys. Nat. Geosci. **3**, 459–463 (2010)

T.M. Donahue, New analysis of hydrogen and deuterium escape from Venus. Icarus **141**, 226–235 (1999)

T.M. Donahue, R. Hartle, Solar cycle variations in H^+ and D^+ densities in the Venus ionosphere - implications for escape. Geophys. Res. Lett. **19**, 2449–2452 (1992)

T.M. Donahue, J.H. Hoffman, R.R. Hodges Jr., A.J. Watson, Venus was wet: A measurement of the ratio of D/H. Science **216**, 630–633 (1982)

G. Dreibus, H. Wänke, Volatiles on Earth and Mars: A comparison. Icarus **71**, 225–240 (1987)

J.M. Eiler, N. Kitchen, T.A. Rahn, Experimental constraints on the stable-isotope systematics of CO2 ice/vapor systems and relevance to the study of Mars. Geochim. Cosmochim. Acta **64**(4), 733–746 (2000)

J.M. Eiler, J.W. Valley, C.M. Graham, J. Fournelle, Two populations of carbonate in ALH84001: Geochemical evidence for discrimination and genesis. Geochim. Cosmochim. Acta **66**(7), 1285–1303 (2002)

L.T. Elkins-Tanton, Linked magma ocean solidication and atmospheric growth for Earth and Mars. Earth Planet. Sci. Lett. **271**, 181–191 (2008)

L.T. Elkins-Tanton, Magma oceans in the inner solar system. Annu. Rev. Earth Planet. Sci. **40**, 113–139 (2012)

N.V. Erkaev, H. Lammer, L.T. Elkins-Tanton, A. Stökl, P. Odert, E. Marcq, E.A. Dorfi, K.G. Kislyakova, Yu.N. Kulikov, M. Leitzinger, M. Güdel, Escape of the Martian protoatmosphere and initial water inventory. Planet. Space Sci. **98**, 106–119 (2014)

N.V. Erkaev, H. Lammer, P. Odert, Yu.N. Kulikov, K.G. Kislyakova, Extreme hydrodynamic atmospheric loss near the critical thermal escape regime. Mon. Not. R. Astron. Soc. **448**, 1916–1921 (2015)

N.V. Erkaev, H. Lammer, P. Odert, K.G. Kislyakova, C.P. Johnstone, M. Güdel, M.L. Khodachenko, EUV-driven mass-loss of protoplanetary cores with hydrogen-dominated atmospheres: the influences of ionization and orbital distance. Mon. Not. R. Astron. Soc. **460**, 1300–1309 (2016)

B. Fegley Jr., N.S. Jacobson, K.B. Williams, J.M.C. Plane, L. Schaefer, K. Lodders, Solubility of rock in steam atmospheres of planets. Astrophys. J. **824**, 103 (2016)
E.D. Feigelson, G.P. Garmire, S.H. Pravdo, Magnetic flaring in the pre-main-sequence Sun and implications for the early solar system. Astrophys. J. **572**, 335–349 (2002)
C.N. Foley, M. Wadhwa, L.E. Borg, P.E. Janney, R. Hines, T.L. Grove, The early differentiation history of Mars from ^{182}W-^{142}Nd isotope systematics in the SNC meteorites. Geochim. Cosmochim. Acta **69**, 4557–4571 (2005)
L. Fossati, N.V. Erkaev, H. Lammer, P.E. Cubillos, P. Odert, I. Juvan, K.G. Kislyakova, M. Lendl, D. Kubyshkina, S.J. Bauer, Aeronomical constraints to the minimum mass and maximum radius of hot low-mass planets. Astron. Astrophys. **598**, A90–A99 (2017)
J.L. Fox, On the escape of oxygen and hydrogen from Mars. Geophys. Res. Lett. **20**, 1847–1850 (1993a)
J.L. Fox, The production and escape of nitrogen atoms on Mars. J. Geophys. Res. **98**, 3297–3310 (1993b)
J.L. Fox, S.W. Bougher, Structure, luminosity, and dynamics of the Venus thermosphere. Space Sci. Rev. **55**, 357 (1991)
J.L. Fox, A. Dalgarno, Nitrogen escape from Mars. J. Geophys. Res. **88**, 9027–9032 (1983)
J.L. Fox, A. Hać, The ^{15}N/^{14}N isotope fractionation in dissociative recombination of N_2^+. J. Geophys. Res., Planets **102**, 9191–9204 (1997)
J.L. Fox, A. Hać, Velocity distribution of C atoms in CO^+ dissociative recombination: implications for photochemical escape of C from Mars. J. Geophys. Res. **104**, 24729–24737 (1999)
E. Füri, B. Marty, Nitrogen isotope variations in the Solar System. Nat. Geosci. **8**, 515–522 (2015)
F. Gaillard, B. Scaillet, A theoretical framework for volcanic degassing chemistry in a comparative planetology perspective and implications for planetary atmospheres. Earth Planet. Sci. Lett. **403**, 307–316 (2014)
E.M. Galimov, *The Biological Fractionation of Isotopes* (Academ. Press, Toronto, 1985), p. 262
E.M. Galimov, On the phenomenon of enrichment of Mars in ^{13}C: a suggestion on the reduced initial atmosphere. Icarus **147**(2), 472–476 (2000)
E.M. Galimov, Isotopic criteria for identification of organic carbon on Earth and meteorites. Space Sci. Rev. **106**, 249–262 (2003)
S.B. Ganguli, The polar wind. Rev. Geophys. **34**, 311 (1996)
H. Genda, Y. Abe, Survival of a proto-atmosphere through the stage of giant impacts: The mechanical aspects. Icarus **164**, 149–162 (2003)
H. Genda, Y. Abe, Enhanced atmospheric loss on protoplanets at the giant impact phase in the presence of oceans. Nature **433**, 842–844 (2005)
H. Genda, I. Masahiro, Origin of the ocean on the Earth: early evolution of water D/H in a hydrogen-rich atmosphere. Icarus **194**, 42–52 (2008)
C. Gillmann, E. Chassefière, P. Lognonné, A consistent picture of early hydrodynamic escape of Venus atmosphere explaining present Ne and Ar isotopic ratios and low oxygen atmospheric content. Earth Planet. Sci. Lett. **286**, 503–513 (2009)
M.M. Grady, I.P. Wright, C.T. Pillinger, A nitrogen and argon stable isotope study of Allan Hills 84001: implications for the evolution of the Martian atmosphere. Meteorit. Planet. Sci. **33**, 795–802 (1998)
J.P. Greenwood, S. Itoh, N. Sakamoto, E.P. Vicenzi, H. Yurimoto, Hydrogen isotope evidence for loss of water from Mars through time. Geophys. Res. Lett. **35**, L05203 (2008)
R.E. Grimm, K.P. Harrison, D.E. Stillman, M.R. Kirchoff, On the secular retention of ground water and ice on Mars. J. Geophys. Res., Planets **122**, 94–109 (2017)
D.H. Grinspoon, Implications of the high D/H ratio for the sources of water in Venus' atmosphere. Nature **363**, 428–431 (1993)
M. Güdel, The young Sun and relevance for planet evolution. Space Sci. Rev. (2020, this issue)
J.H. Guo, The effect of photoionization on the loss of water of the planet. Astrophys. J. **872**, 1 (2019)
N. Haghighipour, T.I. Maindl, C.M. Schaefer, O.J. Wandel, Triggering the activation of main belt comets: the effect of porosity. Astrophys. J. **855**, 60 (2018). arXiv:1801.08247
I. Halevy, W.W. Fischer, J.M. Eiler, Carbonates in the Martian meteorite Allan Hills 84001 formed at 18±4 C in a near-surface aqueous environment. Proc. Natl. Acad. Sci. **108**(41), 16895–16899 (2011)
L.J. Hallis, G.J. Taylor, K. Nagashima, G.R. Huss, Hydrogen isotope analyses of alteration phases in the nakhlite Martian meteorites. Geochim. Cosmochim. Acta **97**, 105–119 (2012)
C.L. Harper, S.B. Jacobsen, Noble gases and Earth's accretion. Science **273**, 1814–1818 (1996)
T.M. Harrison, The Hadean crust: Evidence from > 4 Ga zircons. Annu. Rev. Earth Planet. Sci. **37**, 479–505 (2009)
E. Hartle, J.M. Grebowsky, Light ion flow in the nightside ionosphere of Venus. J. Geophys. Res. **98**, 7437–7445 (1993)

C.D. Herd, L.E. Borg, J.H. Jones, J.J. Papike, Oxygen fugacity and geochemical variations in the martian basalts: Implications for martian basalt petrogenesis and the oxidation state of the upper mantle of Mars. Geochim. Cosmochim. Acta **66**(11), 2025–2036 (2002)
R.C. Hin, C.D. Coath, P.J. Carter, F. Nimmo, Y.J. Lai, P.A.E. Pogge von Strandmann, M. Willbold, Z.M. Leinhardt, M.J. Walter, T. Elliott, Magnesium isotope evidence that accretional vapour loss shapes planetary compositions. Nature **549**, 511–515 (2017)
J.H. Hoffman, R.R. Hodges Jr., T.M. Donahue, M.B. McElroy, Composition of the Venus lower atmosphere from the Pioneer Venus spectrometer. J. Geophys. Res. **85**, 7882–7890 (1980)
R. Hu, D.M. Kass, B.L. Ehlmann, Y.L. Yung, Tracing the fate of carbon and the atmospheric evolution of Mars. Nat. Commun. **6**, 10003 (2015)
E.M. Hunten, The escape of light gases from planetary atmospheres. J. Atmos. Sci. **30**, 1481–1494 (1973)
E.M. Hunten, R.O. Pepin, J.C.G. Walker, Mass fractionation in hydrodynamic escape. Icarus **69**, 532–549 (1987)
K.S. Hutchins, B.M. Jakosky, Evolution of Martian atmospheric argon: implications for sources of volatiles. J. Geophys. Res. **101**, 14933–14950 (1996)
M. Ikoma, H. Genda, Constraints on the mass of a habitable planet with water of nebular origin. Astrophys. J. **648**, 696–706 (2006)
M. Ikoma, L.T. Elkins-Tanton, K. Hamano, J. Suckale, Water portioning in planetary embryos and protoplanets with magma oceans. Space Sci. Rev. **214**(4), 76 (2018). https://doi.org/10.1007/s11214-018-0508-3
N.K. Inamdar, H.E. Schlichting, Stealing the gas: Giant impacts and the large diversity in exoplanet densities. Astrophys. J. **817**, L13 (2016)
V.G. Istomin, K.V. Grechnev, C.A. Kochnev, Mass spectrometry of the lower atmosphere of Venus: Krypton isotopes and other recent results of the Venera-11 and -12 data processing, in *23rd COSPAR Meeting*, Budapest, Hungary (1980a)
V.G. Istomin, K.V. Grechnev, C.A. Kochnev, Mass spectrometer measurements of the lower atmosphere of Venus. Space Res. **20**, 215–218 (1980b)
A. Izidoro, S.N. Raymond, Formation of terrestrial planets, in *Handbook of Exoplanets*, ed. by H. Deeg, J. Belmonte (Springer, Heidelberg, 2018), pp. 2365–2423
B.M. Jakosky, R.O. Pepin, R.E. Johnson, J.L. Fox, Mars atmospheric loss and isotopic fractionation by solar-wind-induced sputtering and photochemical escape. Icarus **111**, 271–288 (1994)
B.M. Jakosky, M. Slipski, M. Benna (the MAVEN Team), Mars' atmospheric history derived from upper-atmosphere measurements of $^{38}Ar/^{36}Ar$. Science **355**, 1408–1410 (2017)
B.M. Jakosky, D. Brain, M. Chaffin, S. Curry, J. Deighan (the MAVEN Team), Loss of the Martian atmosphere to space: present-day loss rates determined from MAVEN observations and integrated loss through time. Icarus **315**, 146–157 (2018)
M. Javoy, E. Kaminski, F. Guyot, D. Ansrault, C. Sanloup, M. Moreira, S. Labrosse, A. Jambon, P. Agrinier, A. Davaille, C. Jaupart, The chemical composition of the Earth: Enstatite chondrite models. Earth Planet. Sci. Lett. **293**, 259–268 (2010)
A.M. Jellinek, M.G. Jackson, Connections between the bulk composition, geodynamics and habitability of Earth. Nat. Geosci. **8**, 587–593 (2015)
S. Jin, C. Mordasini, Compositional imprints in density-distance-time: A rocky composition for close-in low-mass exoplanets from the location of the valley of evaporation. Astrophys. J. **853**, 163–186 (2018)
A. Johansen, M. Lambrechts, Forming planets via pebble accretion. Annu. Rev. Earth Planet. Sci. **45**, 359–387 (2017)
R.E. Johnson, *Energetic Charged Particle Interactions with Atmospheres and Surfaces* (Springer, Berlin, 1990)
N.M. Johnson, Venus atmospheric composition in situ data: a compilation. Earth Space Sci. **6**, 1299–1318 (2019)
C.P. Johnstone, M. Güdel, A. Stökl, H. Lammer, L. Tu, K.G. Kislyakova, T. Lüftinger, P. Odert, N.V. Erkaev, A.E. Dorfi, The evolution of stellar rotation and the hydrogen atmospheres of habitable-one terrestrial planets. Astrophys. J. **815**, L12 (2015)
C.P. Johnstone, M. Güdel, H. Lammer, K.G. Kislyakova, Upper atmospheres of terrestrial planets: Carbon dioxide cooling and the Earth's thermospheric evolution. Astron. Astrophys. **617**, A107 (2018)
C.P. Johnstone, M.L. Khodachenko, T. Lüftinger, K.G. Kislyakova, H. Lammer, M. Güdel, Extreme hydrodynamic losses of Earth-like atmospheres in the habitable zones of very active stars. Astron. Astrophys. **624**, L10 (2019)
J.F. Kasting, J.B. Pollack, Loss of water from Venus. I - hydrodynamic escape of hydrogen. Icarus **53**, 479–508 (1983)
E.S. Kite, Geologic constraints on early Mars climate. Space Sci. Rev. **215**, 10 (2019)
T.T. Koskinen, M.J. Harris, R.V. Yelle, P. Lavvas, The escape of heavy atoms from the ionosphere of HD209458b. I. A photochemical dynamical model of the thermosphere. Icarus **226**, 1678–1694 (2013)

V.A. Krasnopolsky, Variations of the HDO/H_2O ratio in the martian atmosphere and loss of water from Mars. Icarus **257**, 377–386 (2015)
V.A. Krasnopolsky, P.D. Feldman, Detection of molecular hydrogen in the atmosphere of Mars. Science **294**, 1914–1917 (2001)
M.A. Kreslavsky, M.A. Ivanov, J.W. Head, The resurfacing history of Venus: Constraints from buffered crater densities. Icarus **250**, 438–450 (2015)
T.S. Kruijer, C. Burkhardt, G. Budde, T. Kleine, Age of Jupiter inferred from the distinct genetics and formation times of meteorites. Proc. Natl. Acad. Sci. **114**, 6712–6716 (2017)
T.S. Kruijer, L.E. Borg, J. Wimpenny, S. Sio, K. Corliss, Onset of magma ocean solidification on Mars inferred from Mn-Cr chronometry. Earth Planet. Sci. Lett. **542**, 116315 (2020)
D. Kubyshkina, L. Fossati, N.V. Erkaev, P.E. Cubillos, C.P. Johnstone, K.G. Kislyakova, H. Lammer, M. Lendl, P. Odert, Overcoming the limitations of the energy-limited approximation for planet atmospheric escape. Astrophys. J. Lett. **866**, L18 (2018a)
D. Kubyshkina, L. Fossati, N.V. Erkaev, C.P. Johnstone, P.E. Cubillos, K.G. Kislyakova, H. Lammer, M. Lendl, P. Odert, Grid of upper atmosphere models for 1-40$M_\oplus$ planets: application to CoRoT-7 b and HD 219134 b,c. Astron. Astrophys. **619**, A151, 21 pp. (2018b)
D. Kubyshkina, M. Lendl, L. Fossati, P. Cubillos, H. Lammer, N.V. Erkaev, C.P. Johnstone, Young planets under extreme UV irradiation. I. Upper atmosphere modelling of the young exoplanet k2-33b. Astron. Astrophys. **612**, A25, 10 pp. (2018c)
Y.N. Kulikov, H. Lammer, H.I.M. Lichtenegger, T. Penz, D. Breuer, T. Spohn, R. Lundin, H.K. Biernat, A comparative study of the influence of the active young Sun on the early atmospheres of Earth, Venus, and Mars. Space Sci. Rev. **129**(1–3), 207–243 (2007)
T. Kurahashi-Nakamura, E. Tajika, Atmospheric collapse and transport of carbon dioxide into the subsurface on early Mars. Geophys. Res. Lett. **33**, L18205 (2006)
H. Kurokawa, M. Sato, M. Ushioda, T. Matsuyama, R. Moriwaki, J.M. Dohm, T. Usui, Evolution of water reservoirs on Mars: constraints from hydrogen isotopes in martian meteorites. Earth Planet. Sci. Lett. **394**, 179–185 (2014)
H. Kurokawa, T. Usui, M. Sato, Interactive evolution of multiple water-ice reservoirs on Mars: insights from hydrogen isotope compositions. Geochem. J. **50**, 67–79 (2016)
H. Kurokawa, K. Kurisawa, T. Usil, A lower limit of atmospheric pressure on early Mars inferred from nitrogen and argon isotopic compositions. Icarus **299**, 443–459 (2018)
H. Lammer, C. Kolb, T. Penz, U.V. Amerstorfer, H.K. Biernat, B. Bodiselitsch, Estimation of the past and present Martian water-ice reservoirs by isotopic constraints on exchange between the atmosphere and the surface. Int. J. Astrobiol. **2**, 195–202 (2003a)
H. Lammer, H.I.M. Lichtenegger, C. Kolb, I. Ribas, S.J. Bauer, Loss of water from Mars: Implications for the oxidation of the soil. Icarus **165**, 9–25 (2003b)
H. Lammer, H.I.M. Lichtenegger, H.K. Biernat, N.V. Erkaev, I.L. Arshukova, C. Kolb, H. Gunell, A. Lukyanov, M. Holmström, S. Barabash, T.L. Zhang, W. Baumjohann, Loss of hydrogen and oxygen from the upper atmosphere of Venus. Planet. Space Sci. **54**, 1445–1456 (2006)
H. Lammer, E. Chassefière, Ö. Karatekin, A. Morschhauser, P.B. Niles, O. Mousis, P. Odert, U.V. Möstl, D. Breuer, V. Dehant, M. Grott, H. Gröller, E. Hauber, L.B. Pham, Outgassing history and escape of the Martian atmosphere and water inventory. Space Sci. Rev. **174**, 113–154 (2013)
H. Lammer, A. Stökl, N.V. Erkaev, E.A. Dorfi, P. Odert, M. Güdel, K.N. Kulikov, K.G. Kislyakova, M. Leitzinger, Origin and loss of nebula-captured hydrogen envelopes from 'sub'- to 'super-Earths' in the habitable zone of Sun-like stars. Mon. Not. R. Astron. Soc. **439**, 3225–3238 (2014)
H. Lammer, N.V. Erkaev, L. Fossati, I. Juvan, P. Odert, P.E. Cubillos, E. Guenther, K.G. Kislyakova, C.P. Johnstone, T. Lüftinger, M. Güdel, Identifying the 'true' radius of the hot sub-Neptune CoRoT-24b by mass-loss modelling. Mon. Not. R. Astron. Soc. **461**, L62–L66 (2016)
H. Lammer, A.L. Zerkle, S. Gebauer, N. Tosi, L. Noack, M. Scherf, E. Pilat-Lohinger, M. Güdel, J.L. Grenfell, M. Godolt, A. Nikolaou, Origin and evolution of the atmospheres of early Venus, Earth and Mars. Astron. Astrophys. Rev. **26**(2), 72 (2018)
H. Lammer, M. Leitzinger, M. Scherf, P. Odert, C. Burger, D. Kubyshkina, C.P. Johnstone, T. Maindl, C.M. Schäfer, M. Güdel, N. Tosi, A. Nikolaou, E. Marcq, N.V. Erkaev, L. Noak, K.G. Kisylakova, L. Fossati, E. Pilat-Lohinger, F. Ragossnig, E.A. Dorfi, Constraining the early evolution of Venus and Earth through atmospheric Ar, Ne isotope and bulk K/U ratios. Icarus **339**, 1–25 (2020b)
H. Lammer, R. Brasser, A. Johansen, M. Scherf, M. Leitzinger, Formation of Venus, Earth and Mars: Constrained by isotopes. Space Sci. Rev. (2020a, this issue)
T.J. Lapen, M. Righter, A.D. Brandon, V. Debaille, B.L. Beard, J.T. Shafer, A.H. Peslier, A younger age for ALH 84001 and its geochemical link to Shergottite sources in Mars. Science **328**, 347–351 (2010)
F. Leblanc, E. Chassefière, C. Gillmann, D. Breuer, Mars' atmospheric ^{40}Ar: a tracer for past crustal erosion. Icarus **218**, 561–570 (2012)

T. Lebrun, H. Massol, E. Chassefière, A. Davaille, E. Marcq, P. Sarda, F. Leblanc, G. Brandeis, Thermal evolution of an early magma ocean in interaction with the atmosphere. J. Geophys. Res., Planets **118**, 1155–1176 (2013)

C. Lécuyer, P. Gillet, F. Robert, The hydrogen isotope composition of seawater and the global water cycle. Chem. Geol. **145**, 249–261 (1998)

J.-Y. Lee, M. Kurt, J.P. Severenghaus, K. Kawamura, H.-S. Yoo, J.B. Lee, J.S. Kim, A redetermination of the isotopic abundances of atmospheric Ar. Geochim. Cosmochim. Acta **70**, 4507–4512 (2006)

T. Lichtenberg, G.J. Golabek, T.V. Gerya, M.R. Meyer, The effects of short-lived radionuclides and porosity on the early thermo-mechanical evolution of planetesimals. Icarus **274**, 350–365 (2016)

T. Lichtenberg, G.J. Golabek, C.P. Dullemond, M. Schönbächler, T.V. Gerya, M.R. Meyer, Impact splash chondrule formation during planetesimal recycling. Icarus **302**, 27–43 (2017)

Ø. Lie-Svendsen, M.H. Rees, Helium escape from the terrestrial atmosphere: The ion outflow mechanism. J. Geophys. Res. **101**, 2435–2443 (1996)

R.J. Lillis, S. Robbins, M. Manga, J.S. Halekas, H.V. Frey, Time history of the Martian dynamo from crater magnetic field analysis. J. Geophys. Res., Planets **118**(7), 1488–1511 (2013)

S.-F. Liu, Y. Hori, D.N.C. Lin, E. Asphaug, Giant impact: An efficient mechanism for the devolatilization of super-Earths. Astrophys. J. **812**, 164, 9 pp. (2015)

K. Lodders, An oxygen isotope mixing model for the accretion and composition of rocky planets. Space Sci. Rev. **92**, 341–354 (2000)

K. Lodders, H. Palme, H.-P. Gail, *Abundances of the Elements in the Solar System. Solar System, Landolt-Börnstein - Group VI.* Astron. Astrophys., vol. 4B (Springer, Berlin, 2009), p. 712

J. Longhi, Phase equilibrium in the system CO_2-H_2O: application to Mars. J. Geophys. Res. **111**, E06011 (2006)

R. Lundin, Ion acceleration and outflow from Mars and Venus: An overview. Space Sci. Rev. **162**, 309–334 (2011)

J.I. Lunine, Y.L. Yung, R.D. Lorenz, On the volatile inventory of Titan from isotopic abundances in nitrogen and methane. Planet. Space Sci. **47**, 1291–1303 (1999)

J.I. Lunine, J. Chambers, A. Morbidelli, L.A. Leshin, The origin of water on Mars. Icarus **165**, 1–8 (2003)

P.R. Mahaffy, R.C. Webster, M. Cabane (the SAM Team), The Sample Analysis at Mars investigation and instrument suite. Space Sci. Rev. **70**, 401–478 (2012)

P.R. Mahaffy, M. Cabane, C.R. Webster (the SAM Team), Sample Analysis at Mars (SAM) investigation: Overview of results from the first 120 sols on Mars (abstract), in *Lunar Planet. Sci. Conf. 44th, Abstract #1395* (Lunar and Planetary Institute, Houston, 2013a)

P.R. Mahaffy, C.R. Webster, S.K. Atreya, H. Franz, M. Wong (the MSL Science Team), Abundance and isotopic composition of gases in the Martian atmosphere from the Curiosity rover. Science **341**, 263–266 (2013b)

P.R. Mahaffy, C.R. Webster, J.C. Stern, A.E. Brunner, S.K. Atreya (the MSL Science Team), The imprint of atmospheric evolution in the D/H of Hesperian clay minerals on Mars. Science **347**, 412–414 (2015)

T.I. Maindl, R. Dvorak, H. Lammer, M. Güdel, C. Schäfer, R. Speith, P. Odert, N.V. Erkaev, K.G. Kislyakova, E. Pilat-Lohinger, Impact induced surface heating by planetesimals on early Mars. Astron. Astrophys. **574**, A22 (2015)

K. Mandt, J.H. Waite, W. Lewis, B. Magee, J. Bell, J. Lunine, O. Mousis, D. Cordier, Isotopic evolution of the major constituents of Titan's atmosphere based on Cassini data. Planet. Space Sci. **57**, 1917–1930 (2009)

K. Mandt, O. Mousis, J. Lunine, D. Gautier, Protosolar ammonia as the unique source of Titan's nitrogen. Astrophys. J. Lett. **788**, L24 (2014)

K. Mandt, O. Mousis, E. Chassefière, Isotopic evolution of the major constituents of Titan's atmosphere based on Cassini data. Icarus **254**, 259–261 (2015)

C.V. Manning, C.P. McKay, K.J. Zahnle, The nitrogen cycle on Mars: impact decomposition of near-surface nitrates as a source for a nitrogen steady state. Icarus **197**, 60–64 (2008)

E. Marcq, A simple 1-D radiative-convective atmospheric model designed for integration into coupled models of magma ocean planets. J. Geophys. Res., Planets **117**, 1–9 (2012)

E. Marcq, A. Salvador, H. Massol, A. Davaille, Thermal radiation of magma ocean planets using a 1-D radiative-convective model of H_2O-CO_2 atmospheres. J. Geophys. Res., Planets **122**, 1539–1553 (2017)

A. Marty, The origins and concentrations of water, carbon, nitrogen and noble gases on Earth. Earth Planet. Sci. Lett. **313**, 56–66 (2012)

B. Marty, P. Allé, Neon and Argon isotopic constraints on Earth-atmosphere evolution, in *Noble Gas Geochem. Cosmochem.*, ed. by J. Matsuda (Terra Sci. Pub., Tokyo, 1994), pp. 191–204

H. Massol, K. Hamano, F. Tian, M. Ikoma, Y. Abe, E. Chassefíere, A. Davaille, H. Genda, M. Güdel, Y. Hori, F. Leblanc, E. Marcq, P. Sarda, V.I. Shematovich, A. Stökl, H. Lammer, Formation and evolution of protoatmospheres. Space Sci. Rev. **205**, 153–211 (2016)

K.J. Mathew, K. Marti, Early evolution of Martian volatiles: nitrogen and noble gas components in ALH84001 and Chassigny. J. Geophys. Res. **106**, 1401–1422 (2001)

M.B. McElroy, Y.L. Yung, Isotopic composition of nitrogen: implications for the past history of Mars' atmosphere. Science **194**, 70–72 (1976)

H.Y. McSween, G.J. Taylor, M.B. Wyatt, Elemental composition of the Martian crust. Science **324**(5928), 736–739 (2009)

P.J. Michael, The concentration, behavior and storage of H_2O in the suboceanic upper mantle: implications for mantle metasomatism. Geochim. Cosmochim. Acta **52**, 555–566 (1988)

K.E. Miller, C.R. Glein, J.H. Waite Jr., Contributions form accreted organics to Titan's atmosphere: New insights from cometary and chondritic data. Astrophys. J. **871**(59), 1–13 (2019)

H. Mizuno, K. Nakazawa, C. Hayashi, Dissolution of the primordial rare gases into the molten Earth's material. Earth Planet. Sci. Lett. **50**, 202–210 (1980)

T. Montmerle, J.-C. Augereau, M. Chaussidon, M. Gounelle, B. Marty, A. Morbidelli, Solar System formation and early evolution: the first 100 million years. Earth Moon Planets **98**, 39–95 (2006)

M. Moreira, S. Charnoz, The origin of the neon isotopes in chondrites and on Earth. Earth Planet. Sci. Lett. **433**, 249–256 (2016)

S. Mukhopadhyay, Early differentiation and volatile accretion recorded in neon and xenon. Nature **486**, 101–106 (2012)

F. Mullally, J.L. Coughlin, S.E. Thompson, J. Rowe, C. Burke (the Kepler team), Planetary candidates observed by Kepler. VI. Planet sample from Q1-Q16 (47 months). Astrophys. J. Suppl. Ser. **217**, 31–47 (2015)

V.R. Murthy, W. van Westrenen, Y. Fei, Experimental evidence that potassium is a substantial radioactive heat source in planetary cores. Nature **423**, 163–165 (2003)

S.V.S. Murty, R.K. Mohapatra, Nitrogen and heavy noble gases in ALH 84001: signatures of ancient Martian atmosphere. Geochim. Cosmochim. Acta **61**, 5417–5428 (1997)

A. Nikolaou, N. Katyal, N. Tosi, M. Godolt, J.L. Grenfell, H. Rauer, What factors affect the duration and outgassing of the terrestrial magma ocean? Astrophys. J. **875**, 1, 24 pp. (2019)

F. Nimmo, Thermal and compositional evolution of the core. Treatise Geophys. **9**, 201–209 (2007)

P. Odert, H. Lammer, N.V. Erkaev, A. Nikolaou, H.I.M. Lichtenegger, C.P. Johnstone, K.G. Kislyakova, M. Leitzinger, N. Tosi, Escape and fractionation of volatiles and noble gases from Mars-sized planetary embryos and growing protoplanets. Icarus **307**, 327–346 (2018)

H.S.C. O'Neill, H. Palme, Collisional erosion and the non-chondritic composition of the terrestrial planets. Philos. Trans. R. Soc. A **366**, 4205–4238 (2008)

C. O'Neill, A.M. Jellinek, H.S.C. O'Neill, On the distribution of radioactive heat producing elements within meteorites, the Earth, and planets. Space Sci. Rev. (2020, this issue)

R. Orosei, S.E. Lauro, E. Pettinelli, A. Cicchetti, M. Coradini, B. Cosciotti, F. Di Paolo, E. Flamini, E. Mattei, M. Pajola, F. Soldovieri, M. Cartacci, F. Cassenti, A. Frigeri, S. Giuppi, R. Martufi, A. Masdea, G. Mitri, C. Nenna, R. Noschese, M. Restano, R. Seu, Radar evidence of subglacial liquid water on Mars. Science **361**, 490–493 (2018)

J.E. Owen, Y. Wu, Atmospheres of low-mass planets: the "boil-off". Astrophys. J. **817**, 107, 14 pp. (2016)

M. Ozima, F.A. Podosek, *Noble Gas Geochemistry*, 2nd edn. (Cambridge Univ., New York, 2002), 300 pp.

K. Pahlevan, L. Schaefer, M.M. Hirschmann, Magma ocean outgassing and hydrogen isotopic constraints on the Hadean Earth. Earth Planet. Sci. Lett. **526**, 115770 (2019)

H. Palme, H.St.C. O'Neill, Cosmochemical estimates of mantle composition, in *Treatise on Geochemistry*, vol. 2 (2003), pp. 664–676

A.M. Palumbo, J.W. Head, L. Wilson, Rainfall on Noachian Mars: nature, timing, and influence on geologic processes and climate history. Icarus **347**, 664–676 (2020)

E.N. Parker, Dynamics of the interplanetary gas and magnetic fields. Astrophys. J. **128**, 664–676 (1958)

R.O. Pepin, On the origin and early evolution of terrestrial planet atmospheres and meteoritic volatiles. Icarus **92**, 2–79 (1991)

R.O. Pepin, Evolution of the Martian atmosphere. Icarus **111**, 289–304 (1994)

R.O. Pepin, Evolution of Earth's noble gases: consequences of assuming hydrodynamic loss driven by giant impact. Icarus **126**, 148–156 (1997)

R.O. Pepin, On the isotopic composition of primordial xenon in terrestrial planet atmospheres. Space Sci. Rev. **92**, 371–395 (2000)

R.O. Pepin, D. Porcelli, Origin of noble gases in the terrestrial planets. Rev. Mineral. Geochem. **47**, 191–246 (2002)

R.O. Pepin, D. Schulter, R. Becker, D. Reisenfeld, Helium, neon, and argon composition of the solar wind as recorded in gold and other Genesis collector materials. Geochim. Cosmochim. Acta **89**, 62–80 (2012)

S. Péron, M. Moreira, A. Colin, L. Arbaret, B. Putlitz, M.D. Kurz, Neon isotopic composition of the mantle constrained by single vesicle analyses. Earth Planet. Sci. Lett. **449**, 145–154 (2016)

S. Péron, M. Moreira, B. Putlitz, M.D. Kurz, Solar wind implantation supplied light volatiles during the first stage of Earth accretion. Geochem. Perspect. Lett. **3**, 151–159 (2017)
M. Persson, Y. Futaana, A. Fedorov, H. Nilsson, M. Hamrin, S. Barabash, H^+/O^+ escape rate ratio in the Venus magnetotail and its dependence on the solar cycle. Geophys. Res. Lett. **45**, 10805–10811 (2018)
J.P. Pinto, G.R. Gladstone, Y.L. Yung, Photochemical production of formaldehyde in Earth's primitive atmosphere. Science **210**, 183–185 (1980)
W. Pluriel, E. Marcq, M. Turbet, Modeling the albedo of Earth-like magma ocean planets with H_2O-CO_2 atmospheres. Icarus **317**, 583–590 (2019)
F.A. Podosek, Noble gases, in *Treatise on Geochemistry, vol. 1*, ed. by A.M. Davis, H.D. Holland, K.K. Turekian (Elsevier, Amsterdam, 2003), pp. 381–405, 711 pp.
D. Porcelli, R.O. Pepin, Rare gas constraints on early Earth history, in *Origin of the Earth and Moon*, ed. by R.M. Canup, K. Righter (University of Arizona Press, Tucson, 2000), pp. 435–458
D. Porcelli, P. Cassen, W.D. Deep, Earth rare gases: initial inventories, capture from the solar nebula and losses during Moon formation. Earth Planet. Sci. Lett. **193**, 237–251 (2001)
D. Porcelli, C.J. Ballentine, R. Wieler, An overview of noble gas geochemistry and cosmochemistry. Rev. Mineral. Geochem. **47**(1), 1–19 (2002)
E.V. Quintana, T. Barclay, W.J. Borucki, J.F. Rowe, J.E. Chambers, The frequency of giant impacts on Earth-like worlds. Astrophys. J. **821**, 126 (2016)
A. Raquin, M. Moreira, Atmospheric $^{38}Ar/^{36}Ar$ in the mantle: Implications for the nature of the terrestrial parent bodies. Earth Planet. Sci. Lett. **287**, 551–558 (2009)
S.N. Raymond, E. Kokubo, A. Morbidelli, R. Morishima, K.J. Walsh, Terrestrial planet formation at home and abroad, in *Protostars and Planets VI* (2014), pp. 595–618
K. Righter, N.L. Chabot, Moderately and slightly siderophile element constraints on the depth and extent of melting in early Mars. Meteorit. Planet. Sci. **46**, 157–176 (2011)
K. Righter, H. Yang, G. Costin, R.T. Downs, Oxygen fugacity in the Martian mantle controlled by carbon: New constraints from the nakhlite MIL 03346. Meteorit. Planet. Sci. **43**(10), 1709–1723 (2008)
A.E. Ringwood, *Origin of the Earth and Moon* (Springer, New York, 1979)
F. Robert, D. Gautier, B. Dubrulle, The Solar System D/H ratio: observations and theories. Space Sci. Rev. **92**, 201–224 (2000)
V.S. Safronov, E.V. Zvjagina, Relative sizes of the largest bodies during the accumulation of planets. Icarus **10**, 109–115 (1969)
H. Saito, K. Kuramoto, Formation of hybrid-type proto-atmosphere on Mars accreting in the solar nebula. Mon. Not. R. Astron. Soc. **475**, 1274–1287 (2018)
H. Sakuraba, H. Kurokawa, H. Genda, Impact degassing and atmospheric erosion on Venus, Earth, and Mars during the late accretion. Icarus **317**, 48–58 (2019)
A. Salvador, H. Massol, A. Davaille, E. Marcq, P. Sarda, E. Chassefière, The relative influence of H_2O and CO_2 on the primitive surface conditions and evolution of rocky planets. J. Geophys. Res., Planets **122**, 1458–1486 (2017)
S. Sasaki, K. Nakazawa, Did a primary solar-type atmosphere exist around the proto-Earth? Icarus **85**, 21–42 (1989)
L. Schaefer, B. Fegley, Outgassing of ordinary chondritic material and some of its implications for the chemistry of asteroids, planets, and satellites. Icarus **186**, 462–483 (2007)
L. Schaefer, B. Fegley, Chemistry of atmospheres formed during accretion of the Earth and other terrestrial planets. Icarus **208**, 438–448 (2010)
C.M. Schäfer, S. Riecker, T.I. Maindl, R. Speith, W. Kley, A smooth particle hydrodynamics code to model collisions between solid, self-gravitating objects. Astrophys. J. **590**, A19 (2016)
C.M. Schäfer, S. Scherrer, R. Buchwald, T.I. Maindl, R. Speith, W. Kley, Numerical simulations of regolith sampling processes. Planet. Space Sci. **141**, 35–44 (2017)
M. Scherf, H. Lammer, N.V. Erkaev, K.E. Mandt, S. Thaller, B. Marty, Nitrogen atmospheres of the icy bodies in the solar system. Space Sci. Rev. (2020, this issue)
M. Schiller, M. Bizzarro, V.A. Fernandes, Isotopic evolution of the protoplanetary disk and the building blocks of Earth and the Moon. Nature **555**, 507–510 (2018)
H.E. Schlichting, S. Mukhopadhyay, Atmosphere impact losses. Space Sci. Rev. **214**(34), 1–31 (2018)
H.E. Schlichting, R. Sari, A. Yalinewich, Atmospheric mass loss during planet formation: The importance of planetesimal impacts. Icarus **247**, 81–94 (2015)
J.A. Schmidt, M.S. Johnson, R. Schinke, Carbon dioxide photolysis from 150 to 210 nm: Singlet and triplet channel dynamics, UV-spectrum, and isotope effects. Proc. Natl. Acad. Sci. USA **110**(44), 17691–17696 (2013)
M. Sekiya, K. Nakazawa, C. Hayashi, Dissipation of the rare gases contained in the primordial Earth's atmosphere. Earth Planet. Sci. Lett. **50**, 197–201 (1980a)

M. Sekiya, K. Nakazawa, C. Hayashi, Dissipation of the primordial terrestrial atmosphere due to irradiation of the solar EUV. Prog. Theor. Phys. **64**, 1968–1985 (1980b)
V.I. Shematovich, D.E. Ionov, H. Lammer, Heating efficiency in hydrogen-dominated upper atmospheres. Astron. Astrophys. **571**, A94 (2014)
S.F. Sholes, M.L. Smith, M.W. Claire, K.J. Zahnle, D.C. Catling, Anoxic atmospheres on Mars driven by volcanism: Implications for past environments and life. Icarus **290**, 46–62 (2017)
V. Shuvalov, Atmospheric erosion induced by oblique impacts. Meteorit. Planet. Sci. **44**, 1095–1105 (2009)
M. Slipski, B.M. Jakosky, M. Benna, M. Elrod, P. Mahaffy, D. Kass, S. Stone, R. Yelle, Variability of Martian turbopause altitudes. J. Geophys. Res. **123**, 2939–2957 (2018)
P.A. Sossi, B. Fegley Jr., Thermodynamics of element volatility and its application to planetaryprocesses. Rev. Mineral. Geochem. **84**, 393–459 (2018)
P.A. Sossi, S. Klemme, H.St.C. O'Neill, J. Berndt, F. Moynier, Evaporation of moderately volatile elements from silicate melts: experiments and theory. Geochim. Cosmochim. Acta **260**, 204–231 (2019)
A. Stökl, E.A. Dorfi, C.P. Johnstone, H. Lammer, Hydrodynamic simulations of captured protoatmospheres around Earth-like planets. Astron. Astrophys. **576**, 87–97 (2015)
A. Stökl, E.A. Dorfi, C.P. Johnstone, H. Lammer, Dynamical accretion of primordial atmospheres around planets with masses between 0.1 and 5 $M_\oplus$ in the habitable zone. Astrophys. J. **825**, 86 (2016)
D.F. Strobel, D.E. Shemansky, EUV emission from Titan's upper atmosphere: Voyager 1 encounter. J. Geophys. Res. **87**, 1361–1368 (1982)
R.G. Strom, G.G. Schaber, D.D. Dawson, The global resurfacing of Venus. J. Geophys. Res. **99**(E5), 10899–10926 (1994)
I.A. Surkov, F.F. Kirnozov, V.N. Glazov, A.G. Dunchenko, L.P. Tatsii, Content of natural radioactive elements in Venusian soils from Venera 9 and 10 data. Kosm. Issled. **14**, 704–709 (1976)
T.D. Swindle, Martian noble gases. Rev. Mineral. Geochem. **47**, 171–187 (2002)
S.R. Taylor, Chondritic Earth model. Nature **202**, 281–282 (1964)
N. Terada, S. Machida, H. Shinagawa, Global hybrid simulation of the Kelvin-Helmholtz instability at the Venus ionopause. J. Geophys. Res. Space Phys. **107**, 1471 (2002)
H. Terasaki, D.C. Rubie, U. Mann, D.J. Frost, F. Langenhorst, The effects of oxygen, sulfur, and silicon on the dihedral angles between Fe-rich liquid metal and olivine, ringwoodite, and silicate perovskite: implications for planetary core formation, in *Proc. Lunar Planet. Sci. Conf., 36th*, League City, Tex., March 14–18, Abst. 1129 (Lunar Planet. Inst., Houston, 2005)
F. Tian, O.B. Toon, A.A. Pavlov, H. de Sterck, A hydrogen-rich early Earth atmosphere. Science **13, 308(5724)**, 1014–1017 (2005)
F. Tian, J.F. Kasting, C.S. Solomon, Thermal escape of carbon from the early Martian atmosphere. Geophys. Res. Lett. **36**, L02205 (2009)
W.B. Tonks, H.J. Melosh, Magma ocean formation due to giant impacts. J. Geophys. Res. **98**, 5319–5333 (1993)
M. Trieloff, J. Kunz, D.A. Clague, D. Harrison, C.J. Allègre, The nature of pristine noble gases in mantle plumes. Science **288**, 1036–1038 (2000)
L. Tu, C.P. Johnstone, M. Güdel, H. Lammer, The extreme ultraviolet and X-ray Sun in Time: High-energy evolutionary tracks of a solar-like star. Astron. Astrophys. **577**, L3 (2015)
D.L. Turcotte, G. Schubert, *Geodynamics* (Cambridge University Press, Cambridge, 2002), p. 456
T. Usui, C.M.O.'D. Alexander, J. Wang, J.I. Simon, J.H. Jones, Origin of water and mantle–crust interactions on Mars inferred from hydrogen isotopes and volatile element abundances of olivine-hosted melt inclusions of primitive shergottites. Earth Planet. Sci. Lett. **357**, 119–129 (2012)
T. Usui, C.M.O.'D. Alexander, J. Wang, J.I. Simon, J.H. Jones, Meteoritic evidence for a previously unrecognized hydrogen reservoir on Mars. Earth Planet. Sci. Lett. **410**, 140–151 (2015)
T. Usui, H. Kurokawa, J. Wang, C.M.O.'D. Alexander, J.I. Simon, J.H. Jones, Hydrogen isotopic constraints on the evolution of surface and subsurface water on Mars, in *Lunar Planet. Science XLVIII* (2017), p. 1278
G.L. Villanueva, M.J. Mumma, R.E. Novak, H.U. Käufl, P. Hartogh, T. Encrenaz, A. Tokunaga, A. Khayat, M.D. Smith, Strong water isotopic anomalies in the Martian atmosphere: Probing current and ancient reservoirs. Science **348**, 218–221 (2015)
A.N. Volkov, R.E. Johnson, O.J. Tucker, J.T. Erwin, Thermally driven atmospheric escape: transition from hydrodynamic to jeans escape. Astrophys. J. **729**, L24 (2011)
U. von Zahn, S. Komer, H. Wieman, R. Prinn, Composition of the Venus atmosphere, in *Venus*, ed. by D.M. Hunten, L. Colin, T.M. Donahue, V.I. Moroz (University of Arizona Press, Tucson, 1983), pp. 297–430
J. Wade, B.J. Wood, Core formation and the oxidation state of the Earth. Earth Planet. Sci. Lett. **236**, 78–95 (2005)
M. Wadhwa, Redox state of Mars' upper mantle and crust from Eu anomalies in shergottite pyroxenes. Science **291**, 1527–1530 (2001)

H. Wakita, R.A. Schmitt, Lunar anorthosites: rare-Earth and other elemental abundances. Science **170**, 969–974 (1970)

J.C.G. Walker, Carbon dioxide on the early Earth. Orig. Life **16**, 117–127 (1985)

M.K. Wallis, C, N, O isotope fractionation on Mars: implications for crustal H2O and SNC meteorites. Earth Planet. Sci. Lett. **93**, 321–324 (1989)

H. Wang, B.P. Weiss, X.N. Bai, B.G. Downey, J. Wang, C. Suavet, R.R. Fu, M.E. Zucolotto, Lifetime of the solar nebula constrained by meteorite paleomagnetism. Science **355**, 623–627 (2017)

H. Wänke, G. Dreibus, Chemical composition and accretion history of terrestrial planets. Philos. Trans. R. Soc. A **325**, 545–557 (1988)

A.B. Watson, J.B. Thomas, D.J. Cherniak, ^{40}Ar retention in the terrestrial planets. Nature **449**, 299–304 (2007)

B.R. Webster, P.R. Mahaffy, G.J. Flesch, P.B. Niles, J.H. Jones (the MSL Science Team), Isotope ratios of H, C, and O in CO_2 and H_2O of the martian atmosphere. Science **341**, 260–263 (2013a)

C.R. Webster, P.R. Mahaffy, L.A. Leshin (the SAM Team), Mars atmospheric escape recorded by H, C and O isotope ratios in carbon dioxide and water measured by the SAM Tunable Laser Spectrometer on the Curiosity Rover (abstract), in *Lunar Planet. Sci. Conf. 44th, Abstract #1365* (Lunar and Planetary Institute, Houston, 2013b)

F.W. Wetherill, Formation of the terrestrial planets. Annu. Rev. Astron. Astrophys. **18**, 77–113 (1980)

D.C.D. Williams, S. Mukhopadhyay, Capture of nebular gases during Earth's accretion is preserved in deep-mantle neon. Nature **565**, 78–81 (2019)

M.H. Wong, S.K. Atreya, P.N. Mahaffy, H.B. Franz, C. Malespin, M.G. Trainer, J.C. Stern, P.G. Conrad, H.L.K. Manning, R.O. Pepin, R.H. Becker, C.P. McKay, T.C. Owen, R. Navarro-González, J.H. Jones, B.M. Jakosky, A. Steele, Isotopes of nitrogen on Mars: atmospheric measurements by Curiosity's mass spectrometer. Geophys. Res. Lett. **40**, 6033–6037 (2013)

J.J. Wray, S.L. Murchie, J.L. Bishop, B.L. Ehlmann, R.E. Milliken, M.B. Wilhelm, K.D. Seelos, M. Chojnacki, Orbital evidence for more widespread carbonate-bearing rocks on Mars. J. Geophys. Res., Planets **121**, 652–677 (2016)

I.P. Wright, M.M. Grady, C.T. Pillinger, Chassigny and the Nakhlites: carbon-bearing components and their relationship to Martian environmental conditions. Geochim. Cosmochim. Acta **56**, 817–826 (1992)

R. Yokochi, B. Marty, A determination of the neon isotopic composition of the deep mantle. Earth Planet. Sci. Lett. **225**, 77–88 (2004)

E.D. Young, A. Shahar, F. Nimmo, H.E. Schlichting, E.A. Schauble, H. Tang, J. Labidi, Near-equilibrium isotope fractionation during planetesimal evaporation. Icarus **323**, 1–15 (2019)

Y.L. Yung, W.B. DeMore, *Photochemistry of Planetary Atmospheres* (Oxford University Press, London, 1998)

H. Yurimoto, K. Kuramoto, A.N. Krot, E.R.D. Scott, J.N. Cuzzi, M.H. Thiemans, J.R. Lyons, Origin and evolution of oxygen isotopic compositions of the solar system, in *Protostars and Planets*, ed. by V.B. Reipurth et al. (Univ. of Arizona, Tucson, 2006), pp. 849–862

K.J. Zahnle, D.C. Catling, The cosmic shoreline: The evidence that escape determines which planets have atmospheres, and what this may mean for Proxima Centauri B. Astrophys. J. **843**, 122, 23 pp. (2017)

K.J. Zahnle, J.F. Kasting, Mass fractionation during transonic escape and implications for loss of water from Mars and Venus. Icarus **68**, 462–480 (1986)

K.J. Zahnle, J.B. Pollack, J.F. Kasting, Mass fractionation of noble gases in diffusion-limited hydrodynamic hydrogen escape. Icarus **84**, 502–527 (1990)

K.J. Zahnle, R.M. Haberle, D.C. Catling, J.F. Kasting, Photochemical instability of the ancient Martian atmosphere. J. Geophys. Res. **113**, E11004 (2008)

K.J. Zahnle, M. Gacesa, D.C. Catling, Strange messenger: A new history of hydrogen on Earth, as told by Xenon. Geochim. Cosmochim. Acta **244**, 56–85 (2019)

L. Zeng, D.D. Sasselov, S.B. Jacobsen, Mass-radius relation for rocky planets based on PREM. Astrophys. J. **819**, 127–132 (2016)

M.Y. Zolotov, Martian volcanic gases: are they terrestrial-like? in *Lunar and Planetary Science Conference*, vol. 34 (2003), abs. 1795

Space Sci Rev (2020) 216:123
https://doi.org/10.1007/s11214-020-00752-0

Nitrogen Atmospheres of the Icy Bodies in the Solar System

M. Scherf[1] · H. Lammer[1] · N.V. Erkaev[2,3] · K.E. Mandt[4] · S.E. Thaller[5] · B. Marty[6]

Received: 15 October 2019 / Accepted: 13 October 2020 / Published online: 27 October 2020

Abstract This brief review will discuss the current knowledge on the origin and evolution of the nitrogen atmospheres of the icy bodies in the solar system, particularly of Titan, Triton and Pluto. An important tool to analyse and understand the origin and evolution of these atmospheres can be found in the different isotopic signatures of their atmospheric constituents. The $^{14}N/^{15}N$ ratio of the N_2-dominated atmospheres of these bodies serve as a footprint of the building blocks from which Titan, Triton and Pluto originated and of the diverse fractionation processes that shaped these atmospheres over their entire evolution. Together with other measured isotopic and elemental ratios such as $^{12}C/^{13}C$ or $^{36}Ar/N_2$ these atmospheres can give important insights into the history of the icy bodies in the solar system, the diverse processes that affect their N_2-dominated atmospheres, and the therewith connected solar activity evolution. Titan's gaseous envelope most likely originated from ammonia ices with possible contributions from refractory organics. Its isotopic signatures can yet be seen in the – compared to Earth – comparatively heavy $^{14}N/^{15}N$ ratio of 167.7, even though this value slightly evolved over its history due to atmospheric escape and photodissociation of N_2. The origin and evolution of Pluto's and Triton's tenuous nitrogen atmospheres remain unclear, even though it might be likely that their atmospheres originated from the protosolar nebula

Reading Terrestrial Planet Evolution in Isotopes and Element Measurements
Edited by Helmut Lammer, Bernard Marty, Aubrey L. Zerkle, Michel Blanc, Hugh O'Neill and Thorsten Kleine

✉ M. Scherf
manuel.scherf@oeaw.ac.at; helmut.lammer@oeaw.ac.at

N.V. Erkaev
erkaev@icm.krasn.ru

1 Space Research Institute, Austrian Academy of Sciences, Schmiedlstr. 6, 8042 Graz, Austria

2 Institute of Computational Modeling, SB-RAS, 660036 Krasnoyarsk, Russian Federation

3 Siberian Federal University, 660036 Krasnoyarsk, Russian Federation

4 Johns Hopkins University Applied Physics Laboratory, Laurel MD, USA

5 Institute of Physics/IGAM, University of Graz, Universitätsplatz 5, 8010 Graz, Austria

6 Centre de Recherches Pétrographiques et Géochimiques, UMR CNRS & Université de Lorraine, Vandoeuvre-les-Nancy Cedex, France

or from comets. An in-situ space mission to Triton such as the recently proposed Trident mission, and/or to the ice giants would be a crucial cornerstone for a better understanding of the origin and evolution of the icy bodies in the outer solar system and their atmospheres in general. Due to the importance of the isotopic measurements for understanding the origin and evolution of the icy bodies in the solar system, this review will also give a brief discussion on the diverse isotope measurement techniques with a focus on nitrogen.

Keywords Titan · Nitrogen atmospheres · Isotopes · Icy satellites · Atmospheric evolution · Isotope measurement techniques

1 Introduction

Only a small fraction of the more than 190 satellites in the solar system are currently known to have tenuous atmospheres (see e.g., Coustenis et al. 2010). These atmospheres are dominated by different species such as oxygen at Europa (Hall et al. 1995, 1998) or Ganymede (Noll et al. 1996; Hall et al. 1998; Feldman et al. 2000), CO_2 (Carlson 1999), CO, O_2, O and C in Callisto's exosphere (Vorburger et al. 2019), water vapor at Enceladus (Dougherty et al. 2006), SO_2 at Io (Pearl et al. 1979; Lellouch et al. 2007) or tenuous nitrogen atmospheres at Triton and Pluto (Hart 1974; Smith et al. 1989; Gladstone et al. 2016; Gladstone and Young 2019). Among all known satellites in the solar system only Titan has a thick atmosphere with nitrogen as its main constituent. This makes Titan, Triton and Pluto the only icy bodies in the solar system with an N_2-dominated atmosphere.

Besides these icy bodies, only Earth's atmosphere is in addition dominated by nitrogen which likely originated from chondritic meteorites (Marty 2012; Harries et al. 2015). The origin – but also the evolution – of N_2 within the atmospheres of the icy bodies might be a different one. An indication for this can be found in the isotopic signature of N_2 which gives clues on the origin and evolution of the respective atmospheres. While Earth's atmospheric nitrogen has an isotopic ratio of $^{14}N/^{15}N \sim 272$ (Marty and Zimmermann 1999; Füri et al. 2015; see also Table 1), which is close to chondritic, Titan's value is 167.7 ± 0.6 (Niemann et al. 2010); for Pluto and Triton this is currently unknown. The difference between the terrestrial and Titan's $^{14}N/^{15}N$ ratio can be explained either by different building blocks or by fractionation processes that changed the respective ratios over time – or by both.

There are several primordial nitrogen reservoirs in the solar system with distinct $^{14}N/^{15}N$ ratios from which the building blocks of Titan's – and of Triton's and Pluto's – N_2-atmosphere might have originated (see also Table 1). These reservoirs include

i) chondrites (e.g., Kerridge 1985; Alexander et al. 2012), the likely reservoir of the terrestrial nitrogen (Marty and Zimmermann 1999; Marty 2012),
ii) the solar wind and Jupiter, which are believed to represent N_2 within the protosolar nebula (e.g., Meibom et al. 2007; Marty et al. 2010),
iii) ammonia ices (Rousselot et al. 2014; Shinnaka et al. 2014, 2016; Shinnaka and Kawakita 2016), cyanide (CN) and hydrogen cyanide (HCN; e.g., Bockelée-Morvan et al. 2008; Manfroid et al. 2009) in comets, and
iv) complex refractory organics which are observed in cometary grains (e.g., McKeegan et al. 2006; De Gregorio et al. 2010), interplanetary dust particles (IDPs; e.g., Messenger 2000; Aléon et al. 2003; McKeegan et al. 2006; Busemann et al. 2009), and as insoluble organic matter (IOM) in primitive chondrites (e.g., Aléon 2010; Alexander et al. 2017).

Table 1 $^{14}N/^{15}N$ throughout the solar system

Solar system object	$^{14}N/^{15}N$	Reference
Sun/Protosolar Nebula	440.5 ± 5.8	Marty et al. 2011
Solar wind	458.7 ± 4.1	Marty et al. 2011
Venus' atmosphere	273 ± 56	Hoffman et al. 1979
Earth's atmosphere	272 ± 0.3	Marty and Zimmermann 1999
Earth's mantle	273.4 ± 0.8	Cartigny 2005
Earth's primordial mantle	~283	Cartigny and Marty 2013
Mars' atmosphere	168 ± 17	Wong et al. 2013
Mars' interior	~280	Mathew and Marti 2001
Jupiter	435 ± 65	Owen et al. 2001
Saturn	>357	Fletcher et al. 2014
Titan (present-day)	167.7 ± 0.6	Niemann et al. 2010
Titan (past)	≤190;	Mandt et al. 2014;
	≤129;	Krasnopolsky 2016;
	≤166–172	Erkaev et al. 2020
Comet C/2012 S1 (ISON)	139 ± 38	Shinnaka et al. 2014
NH_2 of Jupiter family and Oort cloud comets	127 ± 32	Rousselot et al. 2014
NH_2 of additional 18 comets from various dynamical groups	135.7 ± 5.9	Shinnaka et al. 2016
Refractory organics	160-320 (mean: 231)	Miller et al. 2019
Chondrites	259 ± 15	Alexander et al. 2012
Triton	unknown	
Pluto	unknown	

Since organic matter, however, is the main carrier of nitrogen within chondrites (e.g., Aléon 2010), these may not be classified as an independent reservoir but as an admixture of organics with another nitrogen-bearing reservoir.

By reconstructing the initial fractionation of Titan's building blocks one can determine the specific reservoir(s) out of which Titan's N_2 originated. Various processes, however, can fractionate isotopic ratios and must be considered to determine the initial $^{14}N/^{15}N$ ratio of a satellite or a planet. While in the specific case of Earth biological, geological, and anthropogenic processes are also fractionating nitrogen, it is expected that the main drivers for fractionation on the icy bodies are

i) atmospheric escape (Lammer et al. 2002; Lammer and Bauer 2003; Mandt et al. 2009, 2014) and
ii) photochemical processes such as photodissociation of N_2 (Liang et al. 2007; Krasnopolsky 2016), even though this process might be more important on Titan than on Pluto (Mandt et al. 2017).

While atmospheric escape preferentially removes the lighter isotope from the atmosphere (see also Lammer et al. (2020b), this issue, and Lammer et al. (2008) for a review on fractionation by atmospheric escape), photolysis of N_2 preferentially removes the heavier isotope (Liang et al. 2007). All these processes must be taken into account to gain insights into the origin and evolution of the nitrogen dominated atmospheres of Titan, Triton and

Pluto. This brief review will mainly focus on these aspects while discussing the current knowledge of the nitrogen atmospheres around these icy bodies. In Sect. 2, we will further review different measurement techniques for obtaining the isotopic composition of nitrogen throughout the solar system, since its measurement accuracy – and our therewith derived scientific knowledge from these icy bodies – strongly relies on the involved technology. Section 3 will discuss Titan's nitrogen atmosphere and Sect. 4 will give an overview on the tenuous atmospheres of Pluto and Triton. A discussion and conclusion finalize the review.

2 Nitrogen Isotopes Measurement Techniques

The isotopic ratio of $^{14}N/^{15}N$ on Titan was long believed to be around 60–70 as determined through remote measurements of HCN in its atmosphere (Marten et al. 1997, 2002; Gurwell 2004). However, as the space mission Cassini-Huygens has subsequently illustrated, the isotopic ratio of Titan's N_2 cannot be derived adequately by just measuring nitrogen-bearing molecules such as HCN. This further exemplifies the difficulties in determining isotopic ratios of N_2 throughout the Solar System. It is, therefore, crucial to comprehend how these isotopic ratios were obtained, what are its measurement methods and the difficulties connected with these techniques. These will be addressed in this chapter.

Determining the abundance and ratio of the stable isotopes of nitrogen, ^{14}N and ^{15}N, throughout the solar system can principally be performed either through mass spectrometry or high-resolution spectroscopy ranging from the optical up to submillimetre wavelengths. Measurements of these ratios for different N-bearing molecules, including the involved measurement techniques for different solar system bodies, are illustrated in Fig. 1.

In case of mass spectrometry, the instrument has either to be transported to the target or the target to a laboratory on Earth. Besides a few sample return missions (e.g., Kallenbach et al. 1998; McKeegan et al. 2006; Marty et al. 2011) or investigations of primitive meteorites (e.g., Kerridge 1985; Busemann et al. 2006; Alexander et al. 2007; Alexander 2011) and IDPs (e.g., Messenger 2000; Floss et al. 2006; Busemann et al. 2009) that, among other research questions, also investigated the nitrogen isotopic composition of its target, mass spectrometers will have to be adopted to be suitable for transportation to the sample, such as the nitrogen-dominated atmosphere of Titan. The respective instrument has to be miniaturized from its complex laboratory setup to the small-sized, low-weight requirements of a planetary mission, thereby inevitably reducing its sensitivity and resolution (for reviews on mass spectrometry in planetary sciences see e.g., Palmer and Limero 2001; Ireland 2013; Ren et al. 2018; Yokota 2018; McIntyre et al. 2019; Arevalo et al. 2020).

Consequently, mass interferences between different isotopologues of similar mass significantly complicate the determination of isotopic ratios and increase its error bars. The Galileo Probe Mass Spectrometer (GPMS; Niemann et al. 1992), a quadrupole mass spectrometer (see e.g., Arevalo et al. 2020, for a thorough description of different types of mass spectrometers adopted for space missions), measured the Jovian $^{14}N/^{15}N$ ratio from the doubly charged $^{14}NH_3^{++}$ and $^{15}NH_3^{++}$ signals (see also Fig. 1) at 9 and 8.5 m/z, since $H_2^{16}O$ at 18 m/z interfered with $^{15}NH_3^+$, making a measurement through $^{14}NH_3^+$ and $^{15}NH_3^+$ not possible (Wong et al. 2004). Similarly, the Mars Science Laboratory (MSL) instrument Sample Analysis at Mars/Quadrupole Mass Spectrometer (SAM/QMS; Mahaffy et al. 2012) could not infer the Martian $^{14}N/^{15}N$ through direct atmospheric experiments from N_2^+ due to interferences of CO^+ ions from CO_2 but only from N^+ ions or through enrichment experiments where CO_2 and H_2O were removed by chemical scrubbers (Wong et al. 2013).

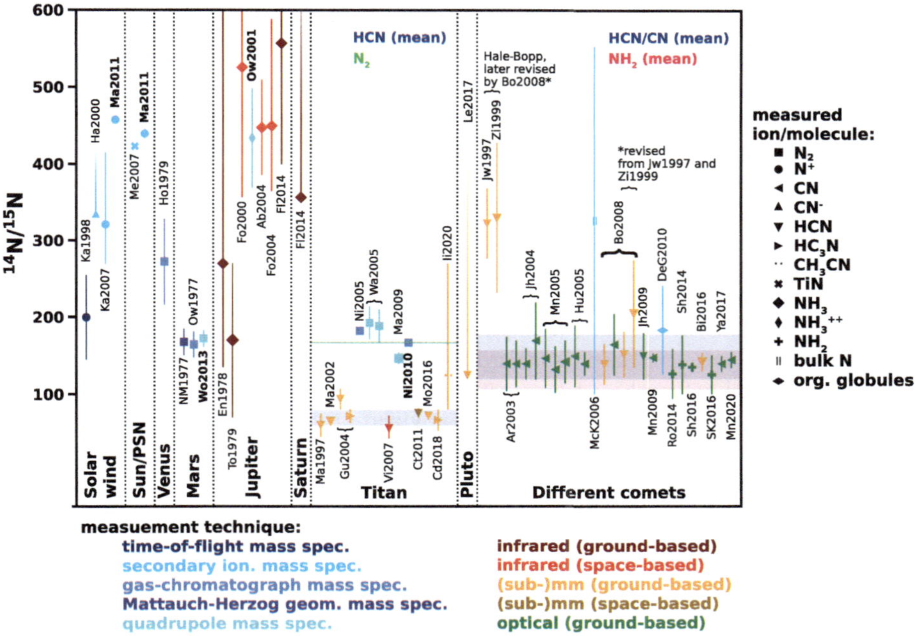

Fig. 1 Measurements of $^{14}N/^{15}N$ throughout the solar system (except for the Earth, interplanetary dust particles, and chondrites which are generally measured with high precision in sophisticated Earth-based laboratories) for different molecules and measurement techniques including sector-type mass spectrometry and secondary ionization mass spectrometry for sample return missions. The bluish and reddish areas illustrate the mean values for HCN(/CN) and NH_2 for Titan and comets, respectively, while the greenish line illustrates the value of N_2 at Titan as a reference. The bold references are the current standards for these respective ratios. References are Ka1998 (Kallenbach et al. 2003), Ha2000 (Hashizume et al. 2000), Ka2007 (Kallenbach et al. 2007; re-evalutation of Ka1998), Ma2011 (Marty et al. 2011), Me2007 (Meibom et al. 2007), Ho1979 (Hoffman et al. 1979), NM1977 (Nier and McElroy 1977), Ow1977 (Owen et al. 1977), Wo2013 (Wong et al. 2013), En1978 (Encrenaz et al. 1978), To1979 (Tokunaga et al. 1979), Fo2000 (Fouchet et al. 2000), Ow2001 (Owen et al. 2001), Ab2004 (Abbas et al. 2004), Fo2004 (Fouchet et al. 2004), Fl2014 (Fletcher et al. 2014), Ma1997 (Marten et al. 1997), Ma2002 (Marten et al. 2002), Gu2004 (Gurwell 2004), Ni2005 (Niemann et al. 2005), Wa2005 (Waite et al. 2005), Ma2009 (Mandt et al. 2009), Ni2010 (Niemann et al. 2010), Ct2010 (Courtin et al. 2011), Mo2016 (Molter et al. 2016), Cd2018 (Cordiner et al. 2018), Ii2020 (Iino et al. 2020), Le2017 (Lellouch et al. 2017), Jw1997 (Jewitt et al. 1997), Zi1999 (Ziurys et al. 1999), Ar2003 (Arpigny et al. 2003), Jh2004 (Jehin et al. 2004), Mn2005 (Manfroid et al. 2005), Hu2005 (Hutsemékers et al. 2005), McK2006 (McKeegan et al. 2006), Bo2008 (Bockelée-Morvan et al. 2008), Jh2009 (Jehin et al. 2009), Mn2009 (Manfroid et al. 2009), DeG2010 (De Gregorio et al. 2010), Ro2014 (Rousselot et al. 2014), Sh2014 (Shinnaka et al. 2014), Sh2016 (Shinnaka et al. 2016), Bi2016 (Biver et al. 2016), SK2016 (Shinnaka and Kawakita 2016), Ya2017 (Yang et al. 2018), Mn2020 (Moulane et al. 2020)

However, in-situ mass spectrometry, besides sample return, nevertheless provides the most accurate means for direct measuring of nitrogen isotopes on distant solar system bodies, if the instrument's resolution allows to bypass interferences, and its sensitivity for a detection of the less abundant isotopologue. Another inaccuracy introduced by mass spectrometry might be coming from the spatial area of the measurement, i.e. where the measurements were performed. The Ion and Neutral Mass Spectrometer (INMS) of Cassini (Waite et al. 2004) determined the $^{14}N/^{15}N$ ratio at Titan by crossings through its tenuous upper atmosphere above the homopause. Due to diffusive separation between the lighter and heavier isotopes above the well-mixed lower atmosphere, the measured values (Waite et al. 2005) from Cassini/INMS of $^{14}N/^{15}N = 188^{+14}_{-16}$, as determined through the integrated sums of the measured abundances below 1230 km, and $^{14}N/^{15}N = 214 \pm 1$, as retrieved through mass deconvolution of the average spectra between the height of 1230 and 1180 km, must be extrapolated to the surface, thereby necessarily introducing further inaccuracies. Waite et al. (2005) extrapolated these values to be $^{14}N/^{15}N = 168$–211 on Titan's surface, while Mandt et al. (2009) estimated the surface value based on Cassini/INMS data to be $^{14}N/^{15}N = 148 \pm 7.5$.

Another instrument on-board Cassini-Huygens, the Gas-Chromatograph Mass Spectrometer (GCMS; Niemann et al. 2002) measured the value directly at the surface to preliminarily be $^{14}N/^{15}N = 183 \pm 5$ directly from molecular nitrogen (Niemann et al. 2005). This was later re-evaluated to 167.7 ± 0.6 due to a refined instrument calibration (Niemann et al. 2010) and is now considered as the standard-value for Titan's atmosphere.

Another possibility for in-situ determination of isotopic ratios is the highly-sensitive Tunable Laser Spectroscopy (TLS) which is currently part of the SAM instrument-suite (Mahaffy et al. 2012) on-board of MSL. TLS can provide mass resolution sensitivities of up to $\approx$1‰ for planetary low-mass instruments (e.g., Webster et al. 2014) and works through spectroscopy in the IR region with a tunable diode laser light source that can be tuned over the specific molecular absorption line wavelengths of interest, thereby being absorbed by the respective molecule. While this method is well suited for the characterization of H_2O, CO, CO_2, CH_4, or O_3, and indeed recorded high-precision stable isotope ratios on Mars for hydrogen, oxygen and carbon isotopes (e.g., Webster et al. 2013), it is less appropriate for the characterization of N_2. Molecular nitrogen does not show significant absorption lines in the IR region (e.g., Lofthus and Krupenie 1977). However, TLS is principally also suited to observe NO, NO_2, HNO_3, and NH_3 (Webster et al. 2014) and was already proposed for a mission to Titan (Webster et al. 1990), and, more recently, to Venus (Webster et al. 2015).

That molecular nitrogen only shows very weak absorption lines is also a problem for spectroscopic observations from ground and space, since spectroscopic techniques that are normally used to measure other isotopic ratios cannot be applied to N_2. This also prohibits a remote determination of the $^{14}N/^{15}N$ ratio in Titan's atmosphere through spectroscopic observations of N_2. However, other nitrogen-bearing molecules show spectral features and can be observed from Earth and from space, which has been performed in the past for several different molecular species in Titan's atmosphere, as can be seen in Table 2.

Depending on the molecule, fingerprints of nitrogen can be observed in the optical, such as the molecules NH_2, a primer for NH_3 (Rousselot et al. 2014; Shinnaka et al. 2014, 2016; Shinnaka and Kawakita 2016) and CN (Arpigny et al. 2003; Jehin et al. 2004, 2009; Hutsemékers et al. 2005; Manfroid et al. 2005, 2009; Bockelée-Morvan et al. 2008; Yang et al. 2018; Moulane et al. 2020) as both observed for different comets, HCN (Vinatier et al. 2007, for Titan) and NH_3 (Encrenaz et al. 1978; Tokunaga et al. 1979; Fouchet et al. 2000, 2004; Abbas et al. 2004; Fletcher et al. 2014, all for Jupiter) in the infrared, and HCN, HC_3N, and CHC_3N for Titan (Marten et al. 1997, 2002; Gurwell 2004; Courtin et al. 2011; Molter et al. 2016; Cordiner et al. 2018; Iino et al. 2020), Pluto (Lellouch et al. 2017), and for comets (Jewitt et al. 1997; Ziurys et al. 1999; Bockelée-Morvan et al. 2008; Biver et al. 2016) in the sub- to millimetre wavelength range.

A few space-bourne measurements of some of these molecules have been performed from afar and through flybys. During the latter, Cassini's Composite Infrared Spectrometer (CIRS; Flasar et al. 2004) performed measurements on HCN at Titan (Vinatier et al. 2007) and NH_3 at Jupiter (Abbas et al. 2004; Fouchet et al. 2004). These were the only successful attempts so far to determine nitrogen isotopic ratios in-situ with spacecraft instrumentation.

Table 2 Measurements and techniques of $^{14}N/^{15}N$ at Titan

Species	$^{14}N/^{15}N$ ratio	Reference	Technique	Instrument	Remarks
HCN	60 ± 15	Marten et al. 1997	(sub-)mm (ground)	IRAM 30-m telescope (Baars et al. 1987)	
HCN	65 ± 5	Marten et al. 2002	(sub-)mm (ground)	IRAM 30-m telescope (Baars et al. 1987)	renewal of telescope & receiver in 1998; better position of Saturn
HCN, HC_3N	94 ± 13 (a) 72 ± 9 (b)	Gurwell 2004	(sub-)mm (ground)	Submillimetre Array	temperature profiles of (a) Coustenis and Bézard 1995, and (b) Lellouch 1990
N_2	183 ± 5	Niemann et al. 2005	gcms[a]	Huygens/GCSM (Niemann et al. 2002)	preliminary results
N_2	172-215 (a) 168-211 (b)	Waite et al. 2005	qms[b]	Cassini/INMS (Waite et al. 2004)	(a) measured above homopause (b) extrapolated to surface
HCN	56^{+16}_{-13}	Vinatier et al. 2007	IR (space)	Cassini/CIRS (Flasar et al. 2004)	
N_2	147.5 ± 7.5	Mandt et al. 2009	qms[b]	Cassini/INMS (Waite et al. 2004)	extrapolated to surface
N_2	167.7 ± 0.6	Niemann et al. 2010	gcms[a]	Huygens/GCSM (Niemann et al. 2002)	refined results based on better laboratory calibration
HCN	76 ± 6	Courtin et al. 2011	(sub-)mm (space)	Herschel-SPIRE (Griffin et al. 2010)	
HCN	72.3 ± 2.2	Molter et al. 2016	(sub-)mm (ground)	ALMA (see Molter et al. 2016)	
HC_3N	67 ± 14	Cordiner et al. 2018	(sub-)mm (ground)	ALMA (see Cordiner et al. 2018)	
CH_3CN	125^{+145}_{-44}	Iino et al. 2020	(sub-)mm (ground)	ALMA (see Iino et al. 2020)	

[a]Gas-chromatograph spectrometer

[b]Quadrupole mass spectrometer

For Saturn, Cassini/CIRS was unable to determine $^{14}N/^{15}N$ through NH_3. Due to the giant planet's lower atmospheric temperature and the dominance of phosphine in its atmosphere in the mid-infrared, the sensitivity requirements were too high for Cassini/CIRS (e.g., Fletcher et al. 2014).

Other space-based spectroscopic determinations of $^{14}N/^{15}N$ were performed from the distance; infrared observations of NH_3 at Jupiter by Fouchet et al. (2000) with ISO-SWS (De Graauw et al. 1996) and submillimetre observations of HCN with Herschel-SPIRE (Griffin et al. 2010) by Courtin et al. (2011). Several different measurements of $^{14}N/^{15}N$ ratios on a diverse set of solar system bodies were additionally performed from the ground. While ground-based instrumentation has the advantage of not having the need to be miniaturized as for a space-based detector, interferences and low transmission rates due to e.g. H_2O and O_3 in the Earth's atmosphere, as well as local weather conditions, complicate observations from the ground. However, the first determination of the $^{14}N/^{15}N$ ratio in Titan's atmosphere was performed with the IRAM 30-m telescope (Baars et al. 1987) by Marten et al. (1997), who measured the ratio in HCN to be $^{14}N/^{15}N = 65 \pm 15$. This was later refined by Marten et al. (2002) to $^{14}N/^{15}N = 65 \pm 5$, which was made possible due to higher sensitivity observations with upgraded receivers and a renewed IRAM 30-m telescope, but also due to a better position of Saturn in the sky as for their previous observations.

At the time of these measurements, the nitrogen isotopic ratio of HCN was expected to represent Titan's atmospheric N_2 as well. While other subsequent observations of HCN by Gurwell (2004) more or less confirmed the results of Marten et al. (2002), preliminary measurements of Cassini-Huygens/GCSM found an $^{14}N/^{15}N$ ratio for N_2 that was much higher, i.e. $^{14}N/^{15}N = 183 \pm 5$ (Niemann et al. 2005). This discrepancy was later to be resolved by photolytic fractionation of $^{14}N^{14}N$ and $^{14}N^{15}N$ (Liang et al. 2007) from N_2, which leads to a preferential incorporation of ^{15}N into HCN, thereby explaining the different isotopic ratios of N_2 and HCN, one of its daughter products.

This, however, also illustrates the difficulty to determine the ratio of ^{14}N and ^{15}N in N_2 through measurements of other molecules such as HCN, CN, or NH_2, which all show lower $^{14}N/^{15}N$ ratios than molecular nitrogen. Further problems in determining isotopic ratios from spectral observations come from missing experimental determinations of the spectral lines of particularly the less abundant isotopes (e.g., Rousselot et al. 2014), but also from theoretical models that are needed to translate emission lines into abundances and isotopic ratios. Derived atmospheric temperature profiles, for instance, significantly influence emission lines and the subsequent determination of isotopic abundances. Gurwell (2004) retrieved two slightly diverging values for the $^{14}N/^{15}N$ ratio in Titan's HCN due to the usage of different temperature profiles. They determined $^{14}N/^{15}N$ to be 94 ± 13 for a profile provided by Coustenis and Bézard (1995), and $^{14}N/^{15}N = 72 \pm 9$ for another one by Lellouch (1990).

Furthermore, observations of the $^{14}N/^{15}N$ ratio in ammonia ices in comets only became possible due to experimental determination of the $^{15}NH_2$ emission lines by Rousselot et al. (2014), which serves as proxy for the very faint lines of $^{15}NH_3$. Likewise, the fractionation between HCN and N_2 could have only been determined due to quantum-mechanical advances in the computation of isotopic photoabsorption and photodissociation cross sections of N_2 (Liang et al. 2007). This fractionation process further turns out to be dependent on the incident EUV flux (Mandt et al. 2017; Hily-Blant et al. 2019).

These examples demonstrate the difficulties for determining accurate $^{14}N/^{15}N$ ratios for any N-bearing molecules but in particular for N_2 through the extrapolation from CN, HCN, or NH_2. It also provides good arguments for in-situ measurements with mass spectrometers or for sample return missions, such as the Stardust mission (McKeegan et al. 2006; De Gregorio et al. 2010), or Genesis (Marty et al. 2011). The latter returned a sample of solar wind N^+ ions to be analysed via a high-precision technique called secondary ionization mass spectrometry (SIMS; see e.g., Schaepe et al. 2019, for an introduction into this technique) at the Cameca 1280HR2 instrument at CRPG Nancy (Marty et al. 2011), and confirmed the highly ^{14}N-enriched (proto-)solar reservoir out of which Jupiter accreted its nitrogen. The sample return of the Stardust mission from comet 81P/Wild 2 confirmed the existence of nitrogen-bearing insoluble organic matter in cometary dust (McKeegan et al. 2006). This was as well investigated with the SIMS technique, but by the nanoSIMS instrument of the

Washington University in St. Louis. Another high precision measurement technique, the so-called sector-type mass spectrometry, which either uses a static electric or a magnetic sector, or a combination of both, to retrieve high resolution and reproducibility measurements, was applied on the ground for analyzing noble gases in the Genesis targets (e.g., Meshik et al. 2014, 2020), and for analyzing Martian atmospheric nitrogen and noble gases trapped in Martian meteorites with high precision, supplementing the in-situ measurements done by the Viking and the MSL mass spectrometers (e.g., Becker and Pepin 1984; Cassata 2017). This technique was further also applied for determining the evolution of atmospheric xenon and other noble gas isotopes over the Earth's Archean eon (see, Avice et al. 2018; Avice and Marty 2020, this issue).

As this overview of different measurement techniques for the determination of the $^{14}N/^{15}N$ ratios of diverse solar system bodies illustrates, one must be cautious in extrapolating a specific measurement to another species or even solar system body. While there are already several determinations of $^{14}N/^{15}N$ existing for different molecular species at Titan, these results cannot simply be generalized to isotopic nitrogen ratios at Triton and Pluto. As the following chapters will show, determining their isotopic ratios independently in the future will be crucial for better understanding the origin and evolution of their nitrogen atmospheres.

3 Titan

Saturn's large satellite Titan (Fig. 2; Tables 1 and 2) is the only moon in the solar system with a substantial gaseous envelope, and besides Earth the only body with a thick N_2-dominated atmosphere. It has a surface pressure of 1.5 bar and consists of 98.4% N_2 and 1.5% CH_4 in the lower stratosphere (Niemann et al. 2010); at the surface the methane fraction even rises up to 5.7% (Samuelson et al. 1997; Niemann et al. 2010). While the existence of its atmosphere was first suggested already in 1908 (Solá 1908) and proven in 1944 via the discovery of methane (Kuiper 1944), N_2 being its main constituent was not confirmed until the flybys of Voyager 1 and 2 in the early 1980s (Broadfoot et al. 1981; Lindal et al. 1983; Hunten et al. 1984; Vervack et al. 2004) due to the challenging detection and confirmation of N_2 (see Sect. 2).

The origin of nitrogen in Titan's atmosphere, however, is still a matter of debate (Niemann et al. 2010) with its most likely sources being proposed to be either N_2 or NH_3 (e.g. Strobel 1982). Lewis (1971), who first suggested that Titan possesses a thick N_2-atmosphere, proposed that it accreted from NH_3, i.e. ammonia, during the formation of the moon. This ammonia was then either decomposed to N_2 via photolysis (Lewis 1971; Atreya et al. 1978) or impact-induced chemistry (McKay et al. 1988; Ishimaru et al. 2011; Sekine et al. 2011), or outgassed as N_2 through thermal decomposition from Titan's interior (Glein et al. 2009; Glein 2015). If NH_3 might have been efficiently altered to N_2 via impacts into an early ammonium hydrate rich crust (Sekine et al. 2011), such impacts would have also led to significant atmospheric erosion (Marounina et al. 2015).

As described in Sect. 2, first measurements of Titan's atmospheric $^{14}N/^{15}N$-ratio (see Table 2) suggested a strong enrichment of ^{15}N by about 4.5 times (Marten et al. 1997; Lunine et al. 1999) compared to the terrestrial atmospheric value of $\sim$272 (Marty and Zimmermann 1999). Since the conversion processes of NH_3 to N_2 could not have significantly altered the $^{14}N/^{15}N$-ratio (Berezhnoi 2010), it was proposed that, in case Titan's atmosphere would have the same building blocks than the Earth's atmospheric nitrogen, this enrichment might indicate strong atmospheric loss of N_2, most likely through atmospheric escape early

Fig. 2 Cassini image of Saturn's moon Titan (Courtesy NASA/JPL/University of Arizona)

in Titan's history (e.g. Lunine et al. 1999; Lammer et al. 2000, 2008; Penz et al. 2005). These results also suggested that Titan's volatile inventory was accreted at least partially from the Saturnian subnebula which was more reducing than the surrounding solar nebula (Lunine et al. 1999). The $^{14}N/^{15}N$-ratio, however, was more precisely measured by the Cassini-Huygens/GCMS to be 167.7 (see Tables 1 and 2; Niemann et al. 2010) which was found to be close to the $^{14}N/^{15}N$-ratios of cometary ammonia (Rousselot et al. 2014; Shinnaka et al. 2014, 2016).

As has been shown by Mandt et al. (2009, 2014), atmospheric loss through hydrodynamic escape, sputtering and Jeans escape cannot explain the strong enrichment of ^{15}N in Titan's atmosphere compared to the terrestrial value, since all of these processes would have required a longer duration than the existence of the solar system for fractionating Titan's value from 272 to 167.7. Mandt et al. (2014), therefore, concluded that the most likely source of Titan's N_2-atmosphere must have been ammonia ices with an initial ratio of <190 (see Fig. 3). This result was recently confirmed by a new study by Erkaev et al. (2020) who applied a hydrodynamic upper atmosphere model (Erkaev et al. 2015) that solves the equations of mass, momentum and energy for Titan's N_2-atmosphere under different EUV-fluxes from the young Sun. These authors found that depending on the solar EUV evolution, the $^{14}N/^{15}N$-ratio of Titan's atmosphere could have only been fractionated by thermal and non-thermal atmospheric escape, and photochemical processes at maximum from an initial value of 167 for a weakly active (i.e. a so-called "slow rotator") to 172 for a moderately active, and 185 for a highly active (fast rotating) young Sun according to Tu et al. (2015). The total loss of nitrogen to space, if one assumes that Titan's atmosphere originated ~4.5 Gyr ago, varies significantly in their study. While for a slow and moderate rotator the total loss would only have been ~0.6 bar, and ~2 bar, respectively, it would have increased to ~16 bar for a fast rotating Sun (Erkaev et al. 2020). For high EUV fluxes, a nitrogen-dominated atmosphere with negligible abundances of CO_2 will start to rapidly expand and

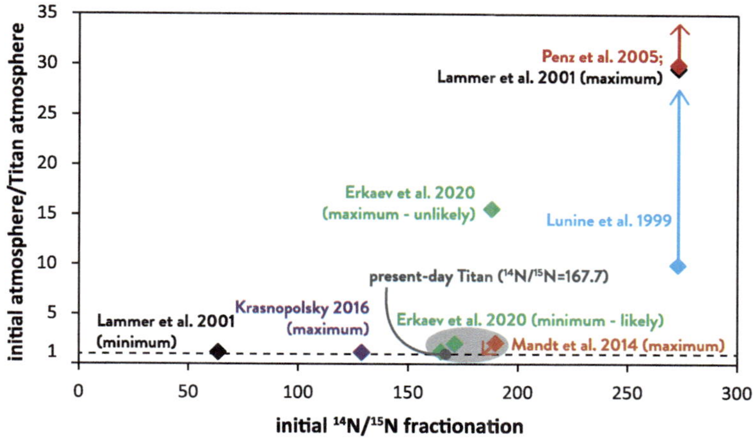

Fig. 3 Different attempts to reproduce the initial $^{14}N/^{15}N$-fractionation of Titan. While older studies like Lunine et al. (1999) or Penz et al. (2005) attempted to reproduce the terrestrial fractionation (i.e. ~273) and therefore yielded unrealistically high losses, newer studies calculated the loss through time to yield a maximum initial fractionation. The grey shaded area most likely represents the initial ratios

hydrodynamically escape to space (see e.g., Erkaev et al. 2020; Johnstone et al. 2019; Tian et al. 2009). To experience such a behaviour, the irradiation onto Titan's atmosphere must have reached values of above 100 times the present-day EUV flux (Erkaev et al. 2020). Even though the young Sun most probably reached such high values very early in its history for a few million years, only a fast rotating Sun could have maintained these high fluxes for a significant period of time (Tu et al. 2015; Güdel 2020, this issue).

Figure 4a shows the exobase altitudes of Titan's nitrogen atmosphere for different solar EUV fluxes for a slow, moderate and fast rotator, respectively, as simulated with the model of Erkaev et al. (2020); Fig. 4b shows the related cumulative loss in Titan's nitrogen atmospheres over time. Contrary to the present-day, for which sputtering is the dominant loss process, thermal escape would have clearly dominated early in Titan's history (with escape rates up to 10^{30} s^{-1} for 400 EUV), in case that its nitrogen-dominated atmosphere was already present at that time.

If, however, the proposed endogenic origin of Titan's atmosphere through decomposition of NH_3 and subsequent outgassing as N_2, as proposed by Glein (2015), is substantially true, then the study of Erkaev et al. (2020) further suggests that either

i) for a moderate to fast rotating young Sun the outgassing of Titan's atmosphere did not start before 20-67 million years to 425-495 million years after the formation of the solar system, respectively, or

ii) that the young Sun was a slow rotating young G-type star.

The reason for this is the following. Besides the outgassing efficiency ε, an important variable in the model of Glein (2015) is the so-called reaction progress variable ξ_{N2}, which describes the yield of N_2 from its precursor NH_3. If $\xi_{N2} = 1$ then all endogenic NH_3 would have needed to be decomposed to N_2 via $2NH_3 \rightarrow N_2 + 3H_2$. Values higher than $\xi_{N2} = 1$ are therefore clearly not possible (see Fig. 5). However, if Titan's atmosphere already started outgassing ~4.5 Gyr ago, the thermal escape produced by the strong EUV flux of a fast rotating Sun would have been so strong that ξ_{N2} would have to be as high as up to $\xi_{N2} \sim 11$. By considering the model of Glein (2015) and therefore assuming an endogenic origin of Titan's atmosphere, Erkaev et al. (2020) restricted the initial fractionation of $^{14}N/^{15}N$

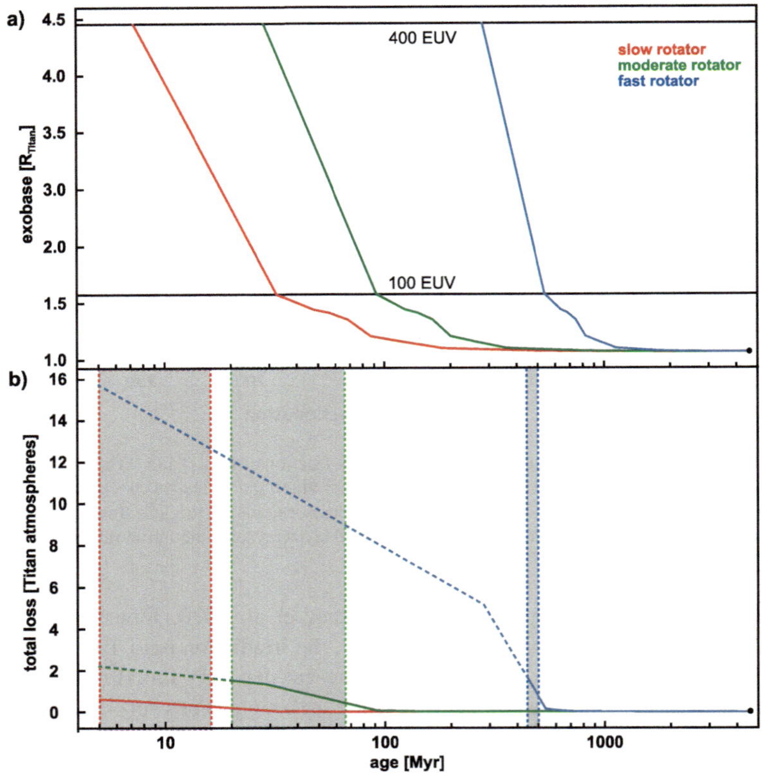

Fig. 4 Upper panel: Exobase levels of Titan's nitrogen atmosphere for different EUV fluxes for a slow, moderate, and fast rotating Sun according to Erkaev et al. (2020). Lower panel: Total cumulative loss of Titan's atmosphere over time in units of Titan atmospheres. The grey areas surrounded by the respective dotted lines correspond to the earliest possible outgassing of Titan's atmosphere for a slow, moderate and fast rotator, respectively, based on the model of Glein (2015) but with the escape of Erkaev et al. (2020) included. (After Erkaev et al. 2020)

to be <172, which is close to the present-day value and in agreement with an origin from ammonia ices (e.g. Mandt et al. 2014). The red, green, and blue encased areas in Fig. 4b illustrate the earliest possible times for the outgassing of Titans nitrogen atmosphere for a slowly, moderately, and fast rotating young Sun according to Erkaev et al. (2020).

It further has to be noted that some recent studies found that the Sun originated most likely as a slow to moderate rotating star (e.g. Odert et al. 2018; Saxena et al. 2019; Lammer et al. 2020a; Johnstone et al. 2021). Lammer et al. (2020a) found that the present-day $^{36}Ar/^{38}Ar$ and $^{20}Ne/^{22}Ne$ ratios in Venus and Earth's atmosphere cannot be reproduced if the Sun originated as a moderate to fast rotator. In addition, another study by Johnstone et al. (2021) found that the terrestrial atmosphere would not have been stable during the Hadean and early Archean eons under the strong EUV flux of a fast rotating Sun, which is also in agreement with the work of Saxena et al. (2019), who found that the Sun likely has been a slow rotator based on sodium and potassium constraints from the Lunar regolith. These findings indicate that the nitrogen atmosphere of Titan could theoretically have been outgassed already very early after the formation of the solar system and not only after about

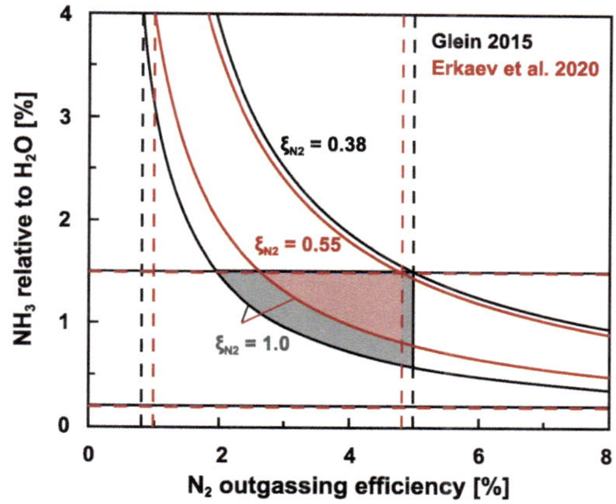

Fig. 5 The parameter space for making Titan's nitrogen atmosphere according to Glein (2015) and, for a slow rotator, to Erkaev et al. (2020). Here it is assumed that the atmosphere originated endogenically from geochemical synthesis of NH_3 into N_2 within the deep interior of Titan; $\xi_{N2} = 1$ corresponds to 100% synthesis. The dashed lines illustrate limiting values according to Glein (2015), and slightly adopted by Erkaev et al. (2020), on the abundance of NH_3 and the outgassing efficiency of N_2. The greyish and reddish areas give the parameter space within which the present-day atmosphere would be consistent with all constraints. Glein (2015) did not include atmospheric escape into his model, while Erkaev et al. (2020) did. This reduces the parameter space for an initially outgassed atmosphere

500 million years as it would have been the case for a fast rotating young Sun (Erkaev et al. 2020). A later outgassing, however, is not excluded by this study.

Another clue to the origin of Titan's atmosphere can be drawn from the atmospheric abundances of the noble gases (see e.g. Hörst 2017). Since the ratio of ^{36}Ar to ^{14}N is orders of magnitudes below solar (Niemann et al. 2005, 2010), this is an indication that only an unsubstantial amount of N_2 could have been accreted during the formation of the Saturnian moon, since ^{36}Ar is trapped in ices at much colder temperatures than N_2 (Bar-Nun et al. 1988; Owen and Bar-Nun 1995; Niemann et al. 2005). However, aerosols in Titan's atmosphere can trap ^{36}Ar, which might remove enough ^{36}Ar from its atmosphere over its lifetime to explain the present-day $^{36}Ar/N_2$ ratio (Jacovi and Bar-Nun 2008). But the existence of the radiogenic ^{40}Ar in the atmosphere of Titan further suggests that outgassing from the interior played a major role in Titan's history (Niemann et al. 2005) which might be another indication for a later origin of its atmosphere through outgassing from the interior. An origin from the protosolar nebula seems to also be confirmed by the CO-chemistry in Titan's atmosphere (Hörst et al. 2008; Hörst 2017, and references therein). Concerning its age, interior models further indicate that major outgassing mainly took place during the last $\sim$1 Gyr (Tobie et al. 2012; Castillo-Rogez and Lunine 2010), which is supported by the $^{12}C/^{13}C$ ratio in CH_4 (Mandt et al. 2012; Nixon et al. 2012).

Determining the noble gas abundances and its isotopic ratios should give further clues on whether Titan's nitrogen originated from NH_3 in its interior or from another source. The endogenic model by Glein (2015) predicts quite specific ratios of $^{20}Ne/^{22}Ne \approx 8.9$ and $^{36}Ar/^{38}Ar \approx 5.3$ based on their outgassing efficiencies and on their potential origin. The latter was tentatively determined by the Cassini-Huygens/GCMS instrument (Niemann et al. 2010) to be $\approx$5, (Niemann et al. 2010), but this will have to be confirmed by future missions. In a more recent article, Glein (2017) also speculates on a nebula origin of Titan's Ne, which might be another explanation of its potentially close-to solar $^{22}Ne/^{36}Ar$ ratio of about (0.05–

2.5) times the Sun's value (Niemann et al. 2010), and would not be in contradiction with a non-solar origin of Titan's nitrogen (Glein 2017). Other noble gas ratio could so far, however, not have been measured at Titan. Abundances or ratios of noble gases in comets, which could give another clue on the origin of Titan's atmosphere, were by now not measured at all.

Further isotopic measurements at Titan itself, can be performed by the upcoming Dragonfly mission, which arrives at the satellite in the mid-2030s to investigate its building blocks (Lorenz et al. 2018; Turtle et al. 2018). It consists of a lander that uses rotors for flying to different sites, and will carry on-board the Dragonfly Mass Spectrometer DraMS (Lorenz et al. 2018). This will perform Laser Desorption/ionization Mass Spectrometry (LDMS) and Gas Chromatography Mass Spectrometry (GCMS) and uses elements of Curiosity's SAM-suite (Mahaffy et al. 2012) and the Exomars/MOMA instrument (Goesmann et al. 2017). Since this mission will also study Titan's surface, it might be able to attain the $^{14}NH_3/^{15}NH_3$ ratio in its ices, a crucial measurement for determining the origin of Titan's atmospheric N_2, an issue that is still not entirely settled.

A recent study by Miller et al. (2019) proposes that accreted refractory organics may contribute up to 50% to Titan's present-day atmosphere. They argue that the $^{14}N/^{15}N$-ratio of cometary ammonia ($\sim$136; Shinnaka et al. (2016)) is lower than in Titan's atmosphere whereas cometary $^{36}Ar/N_2$-ratios seem to be too high (Miller et al. 2019). Refractory organics on the other hand show higher mean $^{14}N/^{15}N$-ratios (McKeegan et al. 2006) and lower mean $^{36}Ar/N_2$-ratios (Lodders 2010) than Titan's atmosphere. Miller et al. (2019) therefore conclude that the isotopic and noble gas constraints of Titan suggest NH_3 together with refractory organics to be the dominant sources for its atmospheric N_2. They further conclude that this mixture may have been the result of an accretion of material from the warm Saturnian subnebula followed by the heating of organics in Titan's interior and a subsequently outgassing into the atmosphere. It also has to be noted that photochemical fractionation, which tends to remove the heavier isotope from the atmosphere (Krasnopolsky 2016), likely was not the reason for the fractionation from the cometary value to the present-day value according to Miller et al. (2019), since this process can only be relevant with methane being abundant in the atmosphere. But methane, however, might have been depleted in Titan's atmosphere early on, most likely as long as 1 Gyr ago or even longer (Yung et al. 1984; Mandt et al. 2012). Interior models (Tobie et al. 2006), on the other hand, also suggest earlier episodic outgassing of methane, so there could have been extended periods of methane in the atmosphere even before 1 Gyr ago which could have been a driver of photochemical fractionation.

A further insight into the origin of Titan's N_2 might be revealed by a measurement of the $^{14}N/^{15}N$-ratio of N_2 in Enceladus' plumes (Glein et al. 2009) which is expected to originate from the oxidation of NH_3 into N_2 (Matson et al. 2007; Glein and Zolotov 2008). Since N_2 at Enceladus is not affected by fractionation through atmospheric escape this might also lead to a better understanding of the initial $^{14}N/^{15}N$-ratio of Titan and hence of its origin (Glein et al. 2009).

4 Pluto

That Pluto (Fig. 6) possesses an atmosphere was suggested as early as 1974 (Hart 1974; Golitsyn 1975; Cruikshank and Silvaggio 1980; Fink et al. 1980; Trafton and Stern 1983; Brosch 1995) and confirmed via ground-based observations through stellar occultation's in 1988 (Hubbard et al. 1988; Elliot et al. 1989). These and further studies (e.g. Young et al. 1997; Lellouch et al. 2011; Olkin et al. 2014; Bosh et al. 2015; Dias-Oliveira et al. 2015) showed that the main constituent of its atmosphere was N_2 which was recently confirmed by the New Horizons space mission (Stern et al. 2015; Gladstone et al. 2016). Contrary to Titan, however, Pluto's tenuous atmosphere only has a surface pressure of about 6-24 μbar depending on its orbital location (Young et al. 1997; Lellouch et al. 2011). But as for Titan, a determination of Pluto's atmospheric $^{14}N/^{15}N$-ratio might reveal important hints on its origin (Jessup et al. 2013; Mandt et al. 2016). Observations with the Atacama Large Millimeter Array (ALMA) of $HC^{14}N$ and a non-detection of $HC^{15}N$ in Pluto's atmosphere provide a lower limit of 125 for $^{14}N/^{15}N$ in HCN (Lellouch et al. 2017), which would be significantly higher than the value in Titan's HCN (see Fig. 1 and Table 1). Based on this lower limit, Krasnopolsky (2020) recently estimated isotopic ratios of Pluto's N_2 and found lower limits of $^{14}N/^{15}N>253$ for the lower and $^{14}N/^{15}N>228$ for the upper atmosphere. A pre-New Horizons study by Mandt et al. (2016) showed that if Pluto's nitrogen originated as N_2 from the protosolar nebula its ratio could not be lower than 324, and it would be less than 154 if its origin would be ammonia. The lower limit by Krasnopolsky (2020), therefore, potentially indicates a significant protosolar contribution to Pluto's N_2 reservoir. Updated work, however, based on the observations made by New Horizons (Gladstone et al. 2016; Young et al. 2018) showed that the uncertainties in how condensation of HCN fractionates the isotopes create large uncertainties on how to interpret a current ratio in N_2 (Mandt et al. 2017).

Due to the low gravity of Pluto it was expected that its atmospheric N_2 would easily escape. While it was debated whether this escape might be hydrodynamic (Krasnopolsky 1999; Tian and Toon 2005; Strobel 2008) with loss rates of up to 10^{27}–10^{28} molecules s^{-1} (Tian and Toon 2005; Zhu et al. 2014), or enhanced Jeans escape (Tucker et al. 2011), New Horizons showed that its atmosphere is currently lost via Jeans escape with the loss rate being in the range of $\sim 10^{23}$ molecules s^{-1}. Based on the pre-New Horizons escape models, Singer and Stern (2015) suggested that Pluto's N_2 should be of endogenic origin, if the estimated escape rates would have been so high that cometary impacts could not have resupplied its atmospheric N_2. In this case internal processes would be needed to outgas N_2 from the interior into the atmosphere (Singer and Stern 2015). A post-New Horizons model by Glein and Waite (2018) considered whether the reservoir of Pluto's N_2 might be of primordial origin for which they developed two different model approaches, i.e. i) a "cometary model" which assumes a composition similar to comets with trapped N_2 from the solar nebula, and ii) a "solar model" which assumes solar abundances of N_2. They found that, while their cometary model can account for the amount of N_2 stored in Sputnik Planitia, their solar model can provide such a large inventory of N_2 that strong atmospheric escape could have been possible in the past. Glein and Waite (2018) therefore conclude that the origin of Pluto's nitrogen inventory is likely primordial, i.e. accreted early as N_2 and might not originate from a secondary source. However, a solar as well as a cometary origin would result in a CO/N_2 ratio being significantly higher than the atmospheric ratio as measured by New Horizons and ALMA; Glein and Waite (2018) suggest a burial mechanism that depletes CO from Pluto's atmosphere to solve this paradox.

Even though New Horizons showed that the present-day escape rates are in the Jeans escape regime, it is important to determine whether the dominant escape process over Pluto's history was hydrodynamic or Jeans escape, as the studies of Singer and Stern (2015) and Glein and Waite (2018) illustrate. Since both processes fractionate very differently (e.g. Volkov et al. 2011; Mandt et al. 2014), measuring Pluto's atmospheric $^{14}N/^{15}N$-ratio would be crucial to not only gain better insights into its origin but also into its escape mechanisms. In addition, constraining the ratio of $^{36}Ar/N_2$ at Pluto's surface can give an important constraint on whether its N_2 really is primordial. A high amount of ^{36}Ar would be consistent

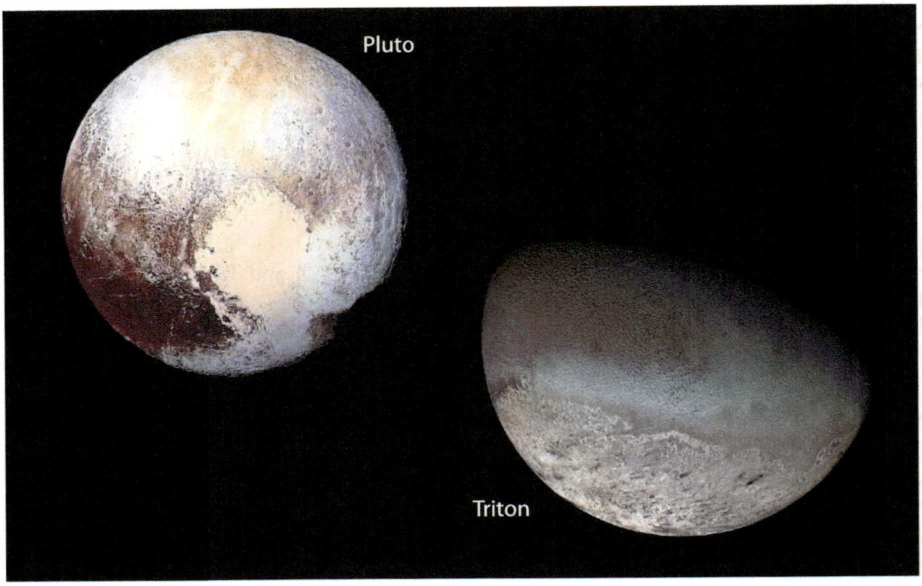

Fig. 6 Left: New Horizons image of Pluto (Courtesy NASA/JHUAPL/SwRI); right: Voyager 2 image of Neptune's moon Triton (Courtesy NASA/Jet Propulsion Lab/U.S. Geological Survey)

with the primordial hypothesis, while low amounts would imply an origin from a secondary source (Glein and Waite 2018).

Further interesting questions in view of the evolution of Pluto's nitrogen atmosphere are how its escape and pressure vary over its elongated orbit. While the escape rate seems to change only by a factor of 2 (Tian and Toon 2005), its atmosphere might even collapse due to cooling on the outer parts of its orbit (e.g. Hansen et al. (2015)). From 1988 to 2002, however, its atmospheric pressure increased by a factor of about two from 2.33 ± 0.24 μbar at $r = 1215$ km in 1988 (Yelle and Elliot 1997) to 5.0 ± 0.6 μbar in 2002 (Sicardy et al. 2003) even though Pluto was receding from the Sun (Elliot et al. 2003; Sicardy et al. 2003). This effect might be due to Pluto's volatile cycle which is not only determined by its distance from the Sun but also by its high obliquity and therefore its subsolar latitude as was predicted by Hansen and Paige (1996). Meza et al. (2019), who measured the surface pressure of Pluto through eleven stellar occultation events between 2002 and 2016, further reported a continuous increase in pressure of up to 12.04 ± 0.41 μbar at the surface in 2015 (compared to 4.28 ± 0.44 μbar in 1988) and 6.92 ± 0.076 μbar at $r = 1215$ km. This rise is due to i) the heating of nitrogen ice in Sputnik Planitia and in the northern mid latitudes because of the direct insolation from the Sun, and ii) the orbital position of Pluto close to the Sun (Bertrand and Forget 2016).

5 Triton

Neptune's moon Triton (Fig. 6) also has a tenuous nitrogen-dominated atmosphere similar to Pluto's. While the Voyager flyby of 1989 measured an N_2-surface pressure of about 14 μbar (Gurrola 1995), it has risen to about 40 μbar in 2009 (Lellouch et al. 2010). First models of the present-day thermal nitrogen escape rates at Triton estimated values in the range of 10^{24}–10^{25} s^{-1} (Yung and Lyons 1990; Summers and Strobel 1991; Lyons et al. 1992; Krasnopolsky et al. 1993; Krasnopolsky and Cruikshank 1995). Lammer (1995) found that non-thermal escape rates via sputtering through particles within the magnetosphere of Neptune could also lead to comparable values in the order of 5×10^{24} s^{-1}. These escape rates are about one to two orders higher than the thermal escape rates that New Horizons found at Pluto. If, for thermal escape, one considers the closer location to the Sun during the Voyager 2 flyby, on which most of the Triton models are based, then such higher escape rates might indeed be likely. Also the value for sputtering at Triton seems reasonable if one compares it with Titan, for which the sputtering escape rate is estimated to be around $3-4 \times 10^{25}$ s^{-1} (Shematovich et al. 2003; Michael et al. 2005). Within a more recent study, Strobel and Zhu (2017), however, modelled thermal escape rates from Triton's atmosphere that would be significantly higher, i.e. about 2.6×10^{27} molecules s^{-1} which would also be orders of magnitudes higher than the escape from present-day Pluto's atmosphere. Whether this is likely or not could be resolved by a new mission to the ice giants. The same model, however, that was used by Strobel and Zhu (2017) for Triton, significantly overestimated nitrogen escape from Pluto's atmosphere as already described in Sect. 4.

As for Pluto, also in the case of Triton the atmospheric $^{14}N/^{15}N$-ratio is yet unknown and its origin unclear, but a similar source as Pluto's nitrogen-dominated atmosphere might be likely. Glein and Waite (2018) speculate that if Triton formed from the same reservoir than Pluto, then the non-detection of CO_2 at Pluto but its detection at Triton might be an indicator on some geochemical endowment or geophysical processes at Triton. The "solar model" by Glein and Waite (2018) interestingly might also provide sufficient N_2 for the idea of a massive early N_2 atmosphere around Triton that could have been accreted from the protosolar or proto-Neptunian nebula, as suggested by Lunine and Nolan (1992). They proposed that the lifetime of such a thick atmosphere could have been more than a billion years, depending on the escape rates. However, the current knowledge on the origin and evolution of Triton's atmosphere is poorly known and a new mission to Neptune or particularly to Triton, such as the recently proposed NASA flyby-mission Trident (Mitchell et al. 2019; Prockter et al. 2019), which carries a plasma spectrometer that will probe Triton's tenuous atmosphere by passing through its upper layers at about 500 km (Mitchell et al. 2019), would lead to new insights not only into the origin and evolution of Triton's atmosphere but also into Pluto's.

6 Conclusion

Within the last 15 years our knowledge of the nitrogen atmospheres of the icy bodies in the solar system significantly advanced particularly due to Cassini around Saturn and the flyby of New Horizons at Pluto. However, other space missions such as Rosetta and ground-based observations of the icy bodies and particularly of comets also helped to trace the origin of their atmospheric N_2. The different isotopic signatures of nitrogen in cometary ammonia ices, refractory organics, chondrites, and in the proto-solar nebula define different reservoirs of which the icy bodies could have accreted their nitrogen. In contrast to Earth – its nitrogen atmosphere most probably originated from the chondritic reservoir – it currently seems likely that their building blocks might have originated from NH_3 in comets and potentially from refractory organics in case of Titan, and from comets and/or the protosolar nebula in case of Pluto and potentially Triton. For understanding their origin, however, it is crucial to study the various fractionation processes that take place in the atmospheres of these bodies as several studies successfully did in the recent past. Here, it is important to note that the

Table 3 Key measurements for a better understanding of the nitrogen atmospheres of icy satellites

Key measurement	Solar system body	Scientific question
$^{20}Ne/^{22}Ne$, $^{36}Ar/^{38}Ar$, $^{36}Ar/^{38}Ar$ and its abundances; Kr and Xe abundances	Titan, Pluto, Triton, comets	testing the endogenic model of Glein (2015) for Titan; clues on the building blocks of the atmospheres of Titan, Triton, and Pluto; origin of the noble gases on these bodies; history of atmospheric escape due to the fractionation of the lighter from the heavier isotope
$^{14}N/^{15}N$ in N_2	Pluto, Triton	testing the origin of Pluto's and Titan's N_2 (protosolar vs. cometary origin); clues on the history of atmospheric escape at Triton and Pluto
$^{14}N/^{15}N$ in NH_3	Titan's surface ices	confirming the origin of Titan's N_2 from ammonia
$^{14}N/^{15}N$	Saturn	determining the isotopic ratio of Saturn and its subnebula as clues to Titan's origin
$^{14}N/^{15}N$	Enceladus' plumes	comparing Enceladus hydrothermal plumes with Titan's composition
radiogenic $^{12}9Xe$ and ^{4}He	Titan	constraining the outgassing history of Titan
atmospheric escape of nitrogen	Triton	retrieving escape rates for Triton's atmosphere for better understanding its stability

evolution of Titan, Triton and Pluto is strongly dependent on the evolution of the Sun which has profound effects on their atmospheres, promoting escape and chemical reactions, and therefore on isotopic fractionation. Only if we understand the solar evolution, will it be possible to trace back the origin of the nitrogen atmospheres of the icy bodies, and the history of the Sun can best be reconstructed via comparative planetology.

Even though comparative planetology is helpful to understand the origin and evolution of nitrogen atmospheres in general, future in-situ missions to the icy bodies will be required to answer currently unknown questions and to better understand their origin (see Table 3 for future key measurements for a better understanding of the nitrogen atmospheres of the icy satellites in the solar system, and, particularly, Glein (2015) for a thorough analysis on future measurements for a better understanding of Titan). In 2019 NASA announced that the space mission Dragonfly (Lorenz et al. 2018; Turtle et al. 2018) is going to be funded for flying to Titan. The mission will start in 2026 and arrive in 2034. Its main goal is to study Titan's origin and habitability. It will explore the building blocks of the moon in unprecedented detail and will also tell us a lot about the origin of the other icy bodies in the solar system such as Triton and Pluto. Our knowledge about Triton, however, is mainly based on the flyby of Voyager 2 and on comparative planetology. Here, a mission to Neptune and its moon such as the recently proposed NASA Trident mission (Prockter et al. 2019)—or even a twin ice giants mission (see e.g. Hofstadter et al. (2017), ESA's CDF Study Report on Ice Giants (2019) and Fletcher et al. (2019) for the scientific potential of such missions)—would be crucial to significantly enhance our scientific knowledge about this satellite, its tenuous nitrogen atmosphere and about icy bodies in general. Finally, another mission that was recently selected by ESA, will help to better understand the potential building blocks of Titan, Pluto, and Triton. Comet Interceptor (Snodgrass and Jones 2019), which will launch in 2028, will be the first spacecraft to visit a pristine comet.

Acknowledgements MS acknowledges the support of Europlanet 2020 RI. Europlanet 2020 RI has received funding from the European Union's Horizon 2020 research and innovation programme under grant agreement No 654208. NVE and HL acknowledge the FWF NFN Project S11607-N27. NVE acknowledges RFBR grant No 18-05-00195-a. KM acknowledges NASA grants 80NSSC18K1233 and 80NSSC19K1306. BM acknowledges the European Research Council grant 695618. We finally thank an anonymous referee who helped to significantly enhance the value of our review.

References

M.M. Abbas, A. LeClair, T. Owen, B.J. Conrath, F.M. Flasar, V.G. Kunde, C.A. Nixon, R.K. Achterberg, G. Bjoraker, D.J. Jennings, G. Orton, P.N. Romani, The nitrogen isotopic ratio in Jupiter's atmosphere from observations by the composite infrared spectrometer on the Cassini spacecraft. Astrophys. J. **602**, 1063–1074 (2004). https://doi.org/10.1086/381084

J. Aléon, Multiple origins of nitrogen isotopic anomalies in meteorites and comets. Astrophys. J. **722**, 1342–1351 (2010). https://doi.org/10.1088/0004-637X/722/2/1342

J. Aléon, F. Robert, M. Chaussidon, B. Marty, Nitrogen isotopic composition of macromolecular organic matter in interplanetary dust particles. Geochim. Cosmochim. Acta **67**, 3773–3783 (2003). https://doi.org/10.1016/S0016-7037(03)00170-4

C.M.O.D. Alexander, A common origin for organics in meteorites and comets: was it interstellar? Proc. Int. Astron. Union **7**, 288–301 (2011). https://doi.org/10.1017/S1743921311025051

C.M.O.D. Alexander, M. Fogel, H. Yabuta, G.D. Cody, The origin and evolution of chondrites recorded in the elemental and isotopic compositions of their macromolecular organic matter. Geochim. Cosmochim. Acta **71**, 4380–4403 (2007). https://doi.org/10.1016/j.gca.2007.06.052

C.M.O. Alexander, R. Bowden, M.L. Fogel, K.T. Howard, C.D.K. Herd, L.R. Nittler, The provenances of asteroids, and their contributions to the volatile inventories of the terrestrial planets. Science **80**(337), 721–723 (2012). https://doi.org/10.1126/science.1223474

C.M.O.D. Alexander, G.D. Cody, B.T. De Gregorio, L.R. Nittler, R.M. Stroud, The nature, origin and modification of insoluble organic matter in chondrites, the major source of Earth's C and N. Chem. Erde **77**, 227–256 (2017). https://doi.org/10.1016/j.chemer.2017.01.007

R. Arevalo, Z. Ni, R.M. Danell, Mass spectrometry and planetary exploration: a brief review and future projection. J. Mass Spectrom. **55**, e4454 (2020). https://doi.org/10.1002/jms.4454

C. Arpigny, E. Jehin, J. Manfroid, D. Hutsemékers, R. Schulz, J.A. Stüwe, J.M. Zucconi, I. Ilyin, Anomalous nitrogen isotope ratio in comets. Science **80**(301), 1522–1524 (2003). https://doi.org/10.1126/science.1086711

S.K. Atreya, T.M. Donahue, W.R. Kuhn, Evolution of a nitrogen atmosphere on Titan. Science **80**(201), 611–613 (1978). https://doi.org/10.1126/science.201.4356.611

G. Avice, B. Marty, Perspectives on atmospheric evolution from noble gas and nitrogen isotopes on Earth, Mars & vEnus. Space Sci. Rev. (2020). https://doi.org/10.1007/s11214-020-00655-0

G. Avice, B. Marty, R. Burgess, A. Hofmann, P. Philippot, K.J. Zahnle, D. Zakharov, Evolution of atmospheric xenon and other noble gases inferred from Archean to Paleoproterozoic rocks. Geochim. Cosmochim. Acta **232**, 82–100 (2018). https://doi.org/10.1016/j.gca.2018.04.018

J.W.M. Baars, B.G. Hooghoudt, P.G. Mezger, M.J. de Jonge, The IRAM 30-m millimeter radio telescope on Pico Veleta, Spain. Astron. Astrophys. **175**(1–2), 319–326 (1987)

A. Bar-Nun, I. Kleinfeld, E. Kochavi, Trapping of gas mixtures by amorphous water ice. Phys. Rev. B **38**, 7749–7754 (1988). https://doi.org/10.1103/PhysRevB.38.7749

R.H. Becker, R.O. Pepin, The case for a martian origin of the shergottites: nitrogen and noble gases in EETA 79001. Earth Planet. Sci. Lett. **69**, 225–242 (1984). https://doi.org/10.1016/0012-821X(84)90183-3

A.A. Berezhnoi, The role of photochemical processes in evolution of the isotopic composition of the atmosphere of Titan. Sol. Syst. Res. **44**, 498–506 (2010). https://doi.org/10.1134/s0038094610060031

T. Bertrand, F. Forget, Observed glacier and volatile distribution on Pluto from atmosphere–topography processes. Nature **540**, 86–89 (2016). https://doi.org/10.1038/nature19337

N. Biver, R. Moreno, D. Bockelée-Morvan, A. Sandqvist, P. Colom, J. Crovisier, D.C. Lis, J. Boissier, V. Debout, G. Paubert, S. Milam, A. Hjalmarson, S. Lundin, T. Karlsson, M. Battelino, U. Frisk, D. Murtagh, t.O. team, Isotopic ratios of H, C, N, O, and S in comets C/2012 F6 (Lemmon) and C/2014 Q2 (Lovejoy). Astron. Astrophys. **589**, A78 (2016). https://doi.org/10.1051/0004-6361/201528041

D. Bockelée-Morvan, N. Biver, E. Jehin, A.L. Cochran, H. Wiesemeyer, J. Manfroid, D. Hutsemékers, C. Arpigny, J. Boissier, W. Cochran, P. Colom, J. Crovisier, N. Milutinovic, R. Moreno, J.X. Prochaska, I. Ramirez, R. Schulz, J.-M. Zucconi, Large excess of heavy nitrogen in both hydrogen cyanide and cyanogen from Comet 17P/Holmes. Astrophys. J. **679**, L49–L52 (2008). https://doi.org/10.1086/588781

A.S. Bosh, M.J. Person, S.E. Levine, C.A. Zuluaga, A.M. Zangari, A.A.S. Gulbis, G.H. Schaefer, E.W. Dunham, B.A. Babcock, A.B. Davis, J.M. Pasachoff, P. Rojo, E. Servajean, F. Förster, T. Oswalt, D. Batcheldor, D. Bell, P. Bird, D. Fey, T. Fulwider, E. Geisert, D. Hastings, C. Keuhler, T. Mizusawa, P. Solenski, B. Watson, The state of Pluto's atmosphere in 2012–2013. Icarus **246**, 237–246 (2015). https://doi.org/10.1016/j.icarus.2014.03.048

A.L. Broadfoot, B.R. Sandel, D.E. Shemansky, J.B. Holberg, G.R. Smith, D.F. Strobel, J.C. McConnell, S. Kumar, D.M. Hunten, S.K. Atreya, T.M. Donahue, H.W. Moos, J.L. Bertaux, J.E. Blamont, R.B. Pomphrey, S. Linick, Extreme ultraviolet observations from voyager 1 encounter with Saturn. Science **80**(212), 206–211 (1981). https://doi.org/10.1126/science.212.4491.206

N. Brosch, The 1985 stellar occultation by Pluto. Mon. Not. R. Astron. Soc. **276**, 571–578 (1995). https://doi.org/10.1093/mnras/276.2.571

H. Busemann, A.F. Young, C.M.O.D. Alexander, P. Hoppe, S. Mukhopadhyay, L.R. Nittler, Interstellar chemistry recorded in organic matter from primitive meteorites. Science **80**(312), 727–730 (2006). https://doi.org/10.1126/science.1123878

H. Busemann, A.N. Nguyen, G.D. Cody, P. Hoppe, A.L.D. Kilcoyne, R.M. Stroud, T.J. Zega, L.R. Nittler, Ultra-primitive interplanetary dust particles from the comet 26P/Grigg-Skjellerup dust stream collection. Earth Planet. Sci. Lett. **288**, 44–57 (2009). https://doi.org/10.1016/j.epsl.2009.09.007

R.W. Carlson, A tenuous carbon dioxide atmosphere on Jupiter's Moon Callisto. Science **(80–.)**. **283**, 820–821 (1999). https://doi.org/10.1126/science.283.5403.820

P. Cartigny, Stable isotopes and the origin of diamond. Elements **1**, 79–84 (2005). https://doi.org/10.2113/gselements.1.2.79

P. Cartigny, B. Marty, Nitrogen isotopes and mantle geodynamics: the emergence of life and the atmosphere-crust-mantle connection. Elements **9**, 359–366 (2013). https://doi.org/10.2113/gselements.9.5.359

W.S. Cassata, Meteorite constraints on Martian atmospheric loss and paleoclimate. Earth Planet. Sci. Lett. **479**, 322–329 (2017). https://doi.org/10.1016/j.epsl.2017.09.034

J.C. Castillo-Rogez, J.I. Lunine, Evolution of Titan's rocky core constrained by Cassini observations. Geophys. Res. Lett. **37**, 1–5 (2010). https://doi.org/10.1029/2010GL044398

M.A. Cordiner, C.A. Nixon, S.B. Charnley, N.A. Teanby, E.M. Molter, Z. Kisiel, V. Vuitton, Interferometric imaging of Titan's HC 3 N, H 13 CCCN, and HCCC 15 N. Astrophys. J. **859**, L15 (2018). https://doi.org/10.3847/2041-8213/aac38d

R. Courtin, B.M. Swinyard, R. Moreno, T. Fulton, E. Lellouch, M. Rengel, P. Hartogh, First results of Herschel-SPIRE observations of Titan. Astron. Astrophys. **536**, 2–5 (2011). https://doi.org/10.1051/0004-6361/201118304

A. Coustenis, B. Bézard, Titan's atmosphere from voyager infrared observations. IV. Latitudinal variations of temperature and composition. Icarus **115**, 126–140 (1995). https://doi.org/10.1006/icar.1995.1084

A. Coustenis, T. Tokano, M.H. Burger, T.A. Cassidy, R.M. Lopes, R.D. Lorenz, K.D. Retherford, G. Schubert, Atmospheric/exospheric characteristics of icy satellites. Space Sci. Rev. **153**, 155–184 (2010). https://doi.org/10.1007/s11214-009-9615-5

D.P. Cruikshank, P.M. Silvaggio, The surface and atmosphere of Pluto. Icarus **41**, 96–102 (1980). https://doi.org/10.1016/0019-1035(80)90162-1

T. De Graauw, L.N. Haser, D.A. Beintema, P.R. Roelfsema, H. Van Agthoven, L. Barl, O.H. Bauer, H.E.G. Bekenkamp, A.-J. Boonstra, D.R. Boxhoorn, J. Coté, P. De Groene, C. Van Dijkhuizen, S. Drapatz, J. Evers, H. Feuchtgruber, M. Frericks, R. Genzel, G. Haerendel, A.M. Heras, K.A. Van Der Hucht, T. Van Der Hulst, R. Huygen, H. Jacobs, G. Jakob, T. Kamperman, R.O. Katterloher, D.J.M. Kester, D. Kunze, D. Kussendrager, F. Lahuis, H.J.G.L.M. Lamers, K. Leech, S. Vanderlei, R. Van Der Linden, W. Luinge, D. Lutz, F. Melzner, P.W. Morris, D. Vannguyen, G. Ploeger, S. Price, A. Salama, S.G. Schaeidt, N. Sijm, C. Smoorenburg, J. Spakman, H. Spoon, J. Stoecker, B. Vandenbussche, H. Visser, L.B.F.M. Waters, J. Wensink, P.R. Wesselius, E. Wiezorrek, E. Wieprecht, J.J. Wijnbergen, K.J. Wildeman, E. Young, Observing with the ISO short-wavelength spectrometer? Astron. Astrophys. **315**, L49–L54 (1996)

B.T. De Gregorio, R.M. Stroud, L.R. Nittler, C.M.O.D. Alexander, A.L.D. Kilcoyne, T.J. Zega, Isotopic anomalies in organic nanoglobules from Comet 81P/Wild 2: comparison to Murchison nanoglobules and isotopic anomalies induced in terrestrial organics by electron irradiation. Geochim. Cosmochim. Acta **74**, 4454–4470 (2010). https://doi.org/10.1016/j.gca.2010.05.010

A. Dias-Oliveira, B. Sicardy, E. Lellouch, R. Vieira-Martins, M. Assafin, J.I.B. Camargo, F. Braga-Ribas, A.R. Gomes-Júnior, G. Benedetti-Rossi, F. Colas, A. Decock, A. Doressoundiram, C. Dumas, M. Emilio, J.F. Polleri, R. Gil-Hutton, M. Gillon, J. Girard, G. Hau, V.D. Ivanov, E. Jehin, J. Lecacheux,

R. Leiva, C. Lopez-Sisterna, L. Mancini, A. Maury, E. Meza, N. Morales, L. Nagy, C. Opitom, J.L. Ortiz, J. Pollock, F. Roques, C. Snodgrass, J.F. Soulier, A. Thirouin, L. Vanzi, T. Widemann, D.E. Reichart, A.P. LaCluyze, J.B. Haislip, K.M. Ivarsen, M. Dominik, U. Jørgensen, J. Skottfelt, J. Skottfelt, Pluto's atmosphere from stellar occultations in 2012 and 2013. Astrophys. J. **811**(1), 53 (2015). https://doi.org/10.1088/0004-637X/811/1/53.

M.K. Dougherty, K.K. Khurana, F.M. Neubauer, C.T. Russell, J. Saur, J.S. Leisner, M.E. Burton, Identification of a dynamic atmosphere at Enceladus with the Cassini magnetometer. Science **80**(311), 1406–1409 (2006). https://doi.org/10.1126/science.1120985

J.L. Elliot, E.W. Dunham, A.S. Bosh, S.M. Slivan, L.A. Young, L.H. Wasserman, R.L. Millis, Pluto's atmosphere. Icarus **77**, 148–170 (1989). https://doi.org/10.1016/0019-1035(89)90014-6

J.L. Elliot, A. Ates, B.A. Babcock, A.S. Bosh, M.W. Buie, K.B. Clancy, E.W. Dunham, S.S. Eikenberry, D.T. Hall, S.D. Kern, S.K. Leggett, S.E. Levine, D.-S. Moon, C.B. Olkin, D.J. Osip, J.M. Pasachoff, B.E. Penprase, M.J. Person, S. Qu, J.T. Rayner, L.C. Roberts, C.V. Salyk, S.P. Souza, R.C. Stone, B.W. Taylor, D.J. Tholen, J.E. Thomas-Osip, D.R. Ticehurst, L.H. Wasserman, The recent expansion of Pluto's atmosphere. Nature **424**, 165–168 (2003). https://doi.org/10.1038/nature01762

T. Encrenaz, M. Combes, Y. Zeau, The spectrum of Jupiter between 10 and 13 ~i. Astron. Astrophys. **70**, 29 (1978). http://adsabs.harvard.edu/abs/1978A&A....70...29E/abstract

N.V. Erkaev, H. Lammer, P. Odert, Y.N. Kulikov, K.G. Kislyakova, Extreme hydrodynamic atmospheric loss near the critical thermal escape regime. Mon. Not. R. Astron. Soc. **448**(2), 1916–1921 (2015). https://doi.org/10.1093/mnras/stv130

N.V. Erkaev, M. Scherf, S.E. Thaller, H. Lammer, A.V. Mezentsev, V.A. Ivanov, K.E. Mandt, Escape and evolution of Titan's N_2 atmosphere constrained by $^{14}N/^{15}N$ isotope ratios. Mon. Not. R. Astron. Soc. (2020). https://doi.org/10.1093/mnras/staa3151

ESA, *CDF Study Report Ice Giants a Mission to the Ice Giants – Neptune and Uranus*, vol. 187 (2019), pp. 1–431

P.D. Feldman, M.A. McGrath, D.F. Strobel, H.W. Moos, K.D. Retherford, B.C. Wolven, HST/STIS ultraviolet imaging of polar aurora on Ganymede. Astrophys. J. **535**(2), 1085–1090 (2000). https://doi.org/10.1086/308889

U. Fink, B.A. Smith, D. Chris Benner, J.R. Johnson, H.J. Reitsema, J.A. Westphal, Detection of a CH4 atmosphere on Pluto. Icarus **44**, 62–71 (1980). https://doi.org/10.1016/0019-1035(80)90055-X

F.M. Flasar, V.G. Kunde, M.M. Abbas, R.K. Achterberg, P. Ade, Exploring the Saturn system in the thermal infrared: the composite infrared spectrometer, in *The Cassini-Huygens Mission*, ed. by C.T. Russell (Springer, Dordrecht, 2004). https://doi.org/10.1007/1-4020-3874-7_4

L.N. Fletcher, T.K. Greathouse, G.S. Orton, P.G.J. Irwin, O. Mousis, J.A. Sinclair, R.S. Giles, The origin of nitrogen on Jupiter and Saturn from the 15N/14N ratio. Icarus **238**, 170–190 (2014). https://doi.org/10.1016/j.icarus.2014.05.007

L.N. Fletcher, N. André, D. Andrews, M. Bannister, E. Bunce, T. Cavalié, S. Charnoz, F. Ferri, J. Fortney, D. Grassi, L. Griton, P. Hartogh, R. Helled, R. Hueso, G. Jones, Y. Kaspi, L. Lamy, A. Masters, H. Melin, J. Moses, O. Mousis, N. Nettleman, C. Plainaki, E. Roussos, J. Schmidt, A. Simon, G. Tobie, P. Tortora, F. Tosi, D. Turrini, *Ice Giant Systems: The Scientific Potential of Missions to Uranus and Neptune (ESA Voyage 2050 White Paper)* (2019). arXiv:1907.02963

C. Floss, F.J. Stadermann, J.P. Bradley, Z.R. Dai, S. Bajt, G. Graham, A.S. Lea, Identification of isotopically primitive interplanetary dust particles: a NanoSIMS isotopic imaging study. Geochim. Cosmochim. Acta **70**, 2371–2399 (2006). https://doi.org/10.1016/j.gca.2006.01.023

T. Fouchet, E. Lellouch, B. Bézard, T. Encrenaz, P. Drossart, H. Feuchtgruber, T. De Graauw, ISO-SWS observations of Jupiter: measurement of the ammonia tropospheric profile and of the 15N/14N isotopic ratio. Icarus **143**, 223–243 (2000). https://doi.org/10.1006/icar.1999.6255

T. Fouchet, P.G.J. Irwin, P. Parrish, S.B. Calcutt, F.W. Taylor, C.A. Nixon, T. Owen, Search for spatial variation in the jovian 15N/14N ratio from Cassini/CIRS observations. Icarus **172**, 50–58 (2004). https://doi.org/10.1016/j.icarus.2003.11.011

E. Füri, B. Marty, Nitrogen isotope variations in the Solar System. Nat. Geosci. **8**, 515–522 (2015). https://doi.org/10.1038/ngeo2451

G.R. Gladstone, L.A. Young, New Horizons Observations of the Atmosphere of Pluto. Annu. Rev. Earth Planet. Sci. **47**, 119–140 (2019). https://doi.org/10.1146/annurev-earth-053018-060128

G.R. Gladstone, S.A. Stern, K. Ennico, C.B. Olkin, H.A. Weaver, L.A. Young, M.E. Summers, D.F. Strobel, D.P. Hinson, J.A. Kammer, A.H. Parker, A.J. Steffl, I.R. Linscott, J.W. Parker, A.F. Cheng, D.C. Slater, M.H. Versteeg, T.K. Greathouse, K.D. Retherford, H. Throop, N.J. Cunningham, W.W. Woods, K.N. Singer, C.C.C. Tsang, E. Schindhelm, C.M. Lisse, M.L. Wong, Y.L. Yung, X. Zhu, W. Curdt, P. Lavvas, E.F. Young, G.L. Tyler, F. Bagenal, W.M. Grundy, W.B. McKinnon, J.M. Moore, J.R. Spencer, T. Andert, J. Andrews, M. Banks, B. Bauer, J. Bauman, O.S. Barnouin, P. Bedini, K. Beisser, R.A. Beyer, S. Bhaskaran, R.P. Binzel, E. Birath, M. Bird, D.J. Bogan, A. Bowman, V.J. Bray, M. Brozovic, C. Bryan,

M.R. Buckley, M.W. Buie, B.J. Buratti, S.S. Bushman, A. Calloway, B. Carcich, S. Conard, C.A. Conrad, J.C. Cook, D.P. Cruikshank, O.S. Custodio, C.M.D. Ore, C. Deboy, Z.J.B. Dischner, P. Dumont, A.M. Earle, H.A. Elliott, J. Ercol, C.M. Ernst, T. Finley, S.H. Flanigan, G. Fountain, M.J. Freeze, J.L. Green, Y. Guo, M. Hahn, D.P. Hamilton, S.A. Hamilton, J. Hanley, A. Harch, H.M. Hart, C.B. Hersman, A. Hill, M.E. Hill, M.E. Holdridge, M. Horanyi, A.D. Howard, C.J.A. Howett, C. Jackman, R.A. Jacobson, D.E. Jennings, H.K. Kang, D.E. Kaufmann, P. Kollmann, S.M. Krimigis, D. Kusnierkiewicz, T.R. Lauer, J.E. Lee, K.L. Lindstrom, A.W. Lunsford, V.A. Mallder, N. Martin, D.J. McComas, R.L. McNutt, D. Mehoke, T. Mehoke, E.D. Melin, M. Mutchler, D. Nelson, F. Nimmo, J.I. Nunez, A. Ocampo, W.M. Owen, M. Paetzold, B. Page, F. Pelletier, J. Peterson, N. Pinkine, M. Piquette, S.B. Porter, S. Protopapa, J. Redfern, H.J. Reitsema, D.C. Reuter, J.H. Roberts, S.J. Robbins, G. Rogers, D. Rose, K. Runyon, M.G. Ryschkewitsch, P. Schenk, B. Sepan, M.R. Showalter, M. Soluri, D. Stanbridge, T. Stryk, J.R. Szalay, M. Tapley, A. Taylor, H. Taylor, O.M. Umurhan, A.J. Verbiscer, M.H. Versteeg, M. Vincent, R. Webbert, S. Weidner, G.E. Weigle, O.L. White, K. Whittenburg, B.G. Williams, K. Williams, S. Williams, A.M. Zangari, E. Zirnstein, The atmosphere of Pluto as observed by New Horizons. Science **351**, aad8866 (2016). https://doi.org/10.1126/science.aad8866

C.R. Glein, Noble gases, nitrogen, and methane from the deep interior to the atmosphere of Titan. Icarus **250**, 570–586 (2015). https://doi.org/10.1016/j.icarus.2015.01.001

C.R. Glein, A whiff of nebular gas in Titan's atmosphere – potential implications for the conditions and timing of Titan's formation. Icarus **293**, 231–242 (2017). https://doi.org/10.1016/j.icarus.2017.02.026

C.R. Glein, J.H. Waite, Primordial N2 provides a cosmochemical explanation for the existence of Sputnik Planitia, Pluto. Icarus **313**, 79–92 (2018). https://doi.org/10.1016/j.icarus.2018.05.007

C. Glein, M. Zolotov, The oxidation state of hydrothermal systems on early Enceladus. Icarus **197**, 157–163 (2008). https://doi.org/10.1016/j.icarus.2008.03.021

C.R. Glein, S.J. Desch, E.L. Shock, The absence of endogenic methane on Titan and its implications for the origin of atmospheric nitrogen. Icarus **204**, 637–644 (2009). https://doi.org/10.1016/j.icarus.2009.06.020

F. Goesmann, W.B. Brinckerhoff, F. Raulin, W. Goetz, R.M. Danell, S.A. Getty, S. Siljeström, H. Mißbach, H. Steininger, R.D. Arevalo, A. Buch, C. Freissinet, A. Grubisic, U.J. Meierhenrich, V.T. Pinnick, F. Stalport, C. Szopa, J.L. Vago, R. Lindner, M.D. Schulte, J.R. Brucato, D.P. Glavin, N. Grand, X. Li, F.H.W. Van Amerom, The Mars Organic Molecule Analyzer (MOMA) instrument: characterization of organic material in Martian sediments. Astrobiology (2017). https://doi.org/10.1089/ast.2016.1551

G.S. Golitsyn, A possible atmosphere on Pluto. Sov. Astron. Lett. **1**, 19, 20 (1975). Transl.: Pisma Astron. Zh. **1**, 38–42. http://adsabs.harvard.edu/abs/1975SvAL....1...19G

M.J. Griffin, A. Abergel, A. Abreu, P.A.R. Ade, P. André, J.L. Augueres, T. Babbedge, Y. Bae, T. Baillie, J.P. Baluteau, M.J. Barlow, G. Bendo, D. Benielli, J.J. Bock, P. Bonhomme, D. Brisbin, C. Brockley-Blatt, M. Caldwell, C. Cara, N. Castro-Rodriguez, R. Cerulli, P. Chanial, S. Chen, E. Clark, D.L. Clements, L. Clerc, J. Coker, D. Communal, L. Conversi, P. Cox, D. Crumb, C. Cunningham, F. Daly, G.R. Davis, P. DeAntoni, J. Delderfield, N. Devin, A. Di Giorgio, I. Didschuns, K. Dohlen, M. Donati, A. Dowell, C.D. Dowell, L. Duband, L. Dumaye, R.J. Emery, M. Ferlet, D. Ferrand, J. Fontignie, M. Fox, A. Franceschini, M. Frerking, T. Fulton, J. Garcia, R. Gastaud, K. Gear W, J. Glenn, A. Goizel, D.K. Griffin, T. Grundy, S. Guest, L. Guillemet, P.C. Hargrave, M. Harwit, P. Hastings, E. Hatziminaoglou, M. Herman, B. Hinde, V. Hristov, M. Huang, P. Imhof, K.J. Isaak, U. Israelsson, R.J. Ivison, D. Jennings, B. Kiernan, K.J. King, A.E. Lange, W. Latter, G. Laurent, P. Laurent, S.J. Leeks, E. Lellouch, L. Levenson, B. Li, J. Li, J. Lilienthal, T. Lim, S.J. Liu, N. Lu, S. Madden, G. Mainetti, P. Marliani, D. McKay, K. Mercier, S. Molinari, H. Morris, H. Moseley, J. Mulder, M. Mur, D.A. Naylor, H. Nguyen, B. O'Halloran, S. Oliver, G. Olofsson, H.G. Olofsson, R. Orfei, J. Page M, I. Pain, P. Panuzzo, A. Papageorgiou, G. Parks, P. Parr-Burman, A. Pearce, C. Pearson, I. Pérez-Fournon, F. Pinsard, G. Pisano, J. Podosek, M. Pohlen, E.T. Polehampton, D. Pouliquen, D. Rigopoulou, D. Rizzo, G. Roseboom I, H. Roussel, M. Rowan-Robinson, B. Rownd, P. Saraceno, M. Sauvage, R. Savage, G. Savini, E. Sawyer, C. Scharmberg, D. Schmitt, N. Schneider, B. Schulz, A. Schwartz, R. Shafer, D.L. Shupe, B. Sibthorpe, S. Sidher, A. Smith, A.J. Smith, D. Smith, L. Spencer, B. Stobie, R. Sudiwala, K. Sukhatme, C. Surace, J.A. Stevens, B.M. Swinyard, M. Trichas, T. Tourette, H. Triou, S. Tseng, C. Tucker, A. Turner, M. Vaccari, I. Valtchanov, L. Vigroux, E. Virique, G. Voellmer, H. Walker, R. Ward, T. Waskett, M. Weilert, R. Wesson, G.J. White, N. Whitehouse, C.D. Wilson, B. Winter, A.L. Woodcraft, G.S. Wright, C.K. Xu, A. Zavagno, M. Zemcov, L. Zhang, E. Zonca , The Herschel-SPIRE instrument and its in-flight performance. Astron. Astrophys. **518** (2010). https://doi.org/10.1051/0004-6361/201014519

M. Güdel, The Sun through time. Space Sci. Rev. (2020, accepted)

E.M. Gurrola, Interpretation of Radar Data from the Icy Galilean Satellites and Triton. Thesis (Ph.D.), Stanford Univ., 1995. Source Diss. Abstr. Int., vol. 56-06, Sect. B (1995), p. 3353. http://adsabs.harvard.edu/abs/1995PhDT........12G

M.A. Gurwell, Submillimeter observations of Titan: global measures of stratospheric temperature, CO, HCN, HC 3 N, and the isotopic ratios 12 C/ 13 C and 14 N/ 15 N. Astrophys. J. **616**, L7–L10 (2004). https://doi.org/10.1086/423954
D.T. Hall, D.F. Strobel, P.D. Feldman, M.A. McGrath, H.A. Weaver, Detection of an oxygen atmosphere on Jupiter's moon Europa. Nature **373**, 677–679 (1995). https://doi.org/10.1038/373677a0
D.T. Hall, P.D. Feldman, M.A. McGrath, D.F. Strobel, The far-ultraviolet oxygen airglow of Europa and Ganymede. Astrophys. J. **499**, 475–481 (1998). https://doi.org/10.1086/305604
C.J. Hansen, D.A. Paige, Seasonal nitrogen cycles on Pluto. Icarus **120**, 247–265 (1996). https://doi.org/10.1006/icar.1996.0049
C.J. Hansen, D.A. Paige, L.A. Young, Pluto's climate modeled with new observational constraints. Icarus **246**, 183–191 (2015). https://doi.org/10.1016/j.icarus.2014.03.014
D. Harries, P. Hoppe, F. Langenhorst, Reactive ammonia in the solar protoplanetary disk and the origin of Earth's nitrogen. Nat. Geosci. **8**, 97–101 (2015). https://doi.org/10.1038/ngeo2339
M.H. Hart, A possible atmosphere for Pluto. Icarus **21**, 242–247 (1974). https://doi.org/10.1016/0019-1035(74)90039-6
K. Hashizume, M. Chaussidon, B. Marty, F. Robert, Solar wind record on the moon: deciphering presolar from planetary nitrogen. Science **80**(290), 1142–1145 (2000). https://doi.org/10.1126/science.290.5494.1142
P. Hily-Blant, V. Magalhaes de Souza, J. Kastner, T. ForveilleDonahue, Multiple nitrogen reservoirs in a protoplanetary disk at the epoch of comet and giant planet formation. Astron. Astrophys. **632**, L12 (2019). https://doi.org/10.1051/0004-6361/201936750
J.H. Hoffman, R.R. Hodges, M.B. McElroy, T.M. Donahue, M. Kolpin, Composition and structure of the Venus atmosphere: results from pioneer Venus. Science **80**(205), 49–52 (1979). https://doi.org/10.1126/science.205.4401.49
M.D. Hofstadter, A. Simon, K. Reh, J. Elliot, Ice Giants Pre-Decadal Study Final. Report 473–488 (2017). https://www.lpi.usra.edu/icegiants/mission_study/
S. Hörst, Titan's atmosphere and climate. J. Geophys. Res., Planets **122**(3), 432–482 (2017). https://doi.org/10.1002/2016JE005240
S.M. Hörst, V. Vuitton, R.V. Yelle, Origin of oxygen species in Titan's atmosphere. J. Geophys. Res. **113**, E10006 (2008). https://doi.org/10.1029/2008JE003135
W.B. Hubbard, D.M. Hunten, S.W. Dieters, K.M. Hill, R.D. Watson, Occultation evidence for an atmosphere on Pluto. Nature **336**, 452–454 (1988). https://doi.org/10.1038/336452a0
D.M. Hunten, M.G. Tomasko, F.M. Flasar, R.E. Samuelson, D.F. Strobel, D.J. Stevenson, Titan, in *Saturn (A85-33976 15-91)* (Univ. Arizona Press, Tucson, 1984), pp. 671–759. http://adsabs.harvard.edu/abs/1984satn.book..671H
D. Hutsemékers, J. Manfroid, E. Jehin, C. Arpigny, A. Cochran, R. Schulz, J.A. Stüwe, J.M. Zucconi, Isotopic abundances of carbon and nitrogen in Jupiter-family and Oort Cloud comets. Astron. Astrophys. **440**, 21–24 (2005). https://doi.org/10.1051/0004-6361:200500160
T. Iino, H. Sagawa, T. Tsukagoshi, 14 N/ 15 N isotopic ratio in CH 3 CN of Titan's atmosphere measured with ALMA. Astrophys. J. **890**, 95 (2020). https://doi.org/10.3847/1538-4357/ab66b0
T.R. Ireland, Invited review article: recent developments in isotope-ratio mass spectrometry for geochemistry and cosmochemistry. Rev. Sci. Instrum. (2013). https://doi.org/10.1063/1.4765055
R. Ishimaru, Y. Sekine, T. Matsui, O. Mousis, Oxidizing proto-atmosphere on Titan: constraint from N_2 formation by impact shock. Astrophys. J. **741**, L10 (2011). https://doi.org/10.1088/2041-8205/741/1/L10
R. Jacovi, A. Bar-Nun, Removal of Titan's noble gases by their trapping in its haze. Icarus **196**, 302–304 (2008). https://doi.org/10.1016/j.icarus.2008.02.014
E. Jehin, J. Manfroid, A.L. Cochran, C. Arpigny, J.-M. Zucconi, D. Hutsemékers, W.D. Cochran, M. Endl, R. Schulz, The anomalous 14 N/ 15 N ratio in Comets 122P/1995 S1 (de Vico) and 153P/2002 C1 (Ikeya-Zhang). Astrophys. J. **613**, L161–L164 (2004). https://doi.org/10.1086/425254
E. Jehin, D. Bockelée-Morvan, N. Dello Russo, J. Manfroid, D. Hutsemékers, H. Kawakita, H. Kobayashi, R. Schulz, A. Smette, J. Stüwe, M. Weiler, C. Arpigny, N. Biver, A. Cochran, J. Crovisier, P. Magain, H. Rauer, H. Sana, R.J. Vervack, H. Weaver, J.M. Zucconi, A multi-wavelength simultaneous study of the composition of the Halley Family comet 8P/Tuttle. Earth Moon Planets **105**, 343–349 (2009). https://doi.org/10.1007/s11038-009-9317-8
K.L. Jessup, G.R. Gladstone, A.N. Heays, S.T. Gibson, B.R. Lewis, G. Stark, 14N15N detectability in Pluto's atmosphere. Icarus **226**, 1514–1526 (2013). https://doi.org/10.1016/j.icarus.2013.08.012
D.C. Jewitt, H.E. Matthews, T. Owen, R. Meier, Measurements of 12C/13C, 14N/15N, and 32S/34S ratios in comet Hale-Bopp (C/1995 O1). Science **80**(278), 90–93 (1997). https://doi.org/10.1126/science.278.5335.90
C.P. Johnstone, M.L. Khodachenko, T. Lüftinger, K.G. Kislyakova, H. Lammer, M. Güdel, Extreme hydrodynamic losses of Earth-like atmospheres in the habitable zones of very active stars. Astron. Astrophys. **624**, L10 (2019). https://doi.org/10.1051/0004-6361/201935279

C.P. Johnstone, H. Lammer, K.G. Kislyakova, M. Scherf, M. Güdel, High atmospheric carbon dioxide levels and low solar activity during the Earth's Archean. Earth Planet. Sci. Lett. (2021, under revision)

R. Kallenbach, J. Geiss, F.M. Ipavich, G. Gloeckler, P. Bochsler, F. Gliem, S. Hefti, M. Hilchenbach, D. Hovestadt, Isotopic composition of solar wind nitrogen: first in situ determination with the CELIAS/MTOF spectrometer on board [ITAL]SOHO[/ITAL]. Astrophys. J. **507**, L185–L188 (1998). https://doi.org/10.1086/311702

R. Kallenbach, T. Encrenaz, J. Geiss, K. Mauersberger, T.C. Owen, F. Robert, Solar System History from Isotopic Signatures of Volatile Elements: Volume Resulting from an ISSI Workshop, 14-18 January 2002, Bern, Switzerland. Springer, Bern (2003)

R. Kallenbach, K. Bamert, M. Hilchenbach, Isotopic composition of the solar wind inferred from in-situ spacecraft measurements. Space Sci. Rev. **130**, 173–182 (2007). https://doi.org/10.1007/s11214-007-9216-0

J.F. Kerridge, Carbon, hydrogen and nitrogen in carbonaceous chondrites: abundances and isotopic compositions in bulk samples. Geochim. Cosmochim. Acta **49**, 1707–1714 (1985). https://doi.org/10.1016/0016-7037(85)90141-3

V.A. Krasnopolsky, Hydrodynamic flow of N_2 from Pluto. J. Geophys. Res., Planets **104**, 5955–5962 (1999). https://doi.org/10.1029/1998JE900052

V.A. Krasnopolsky, Isotopic ratio of nitrogen on Titan: photochemical interpretation. Planet. Space Sci. **134**, 61–63 (2016). https://doi.org/10.1016/j.pss.2016.10.008

V.A. Krasnopolsky, On the methylacetylene abundance and nitrogen isotope ratio in Pluto's atmosphere. Planet. Space Sci. **192**, 105044 (2020). https://doi.org/10.1016/j.pss.2020.105044

V.A. Krasnopolsky, D.P. Cruikshank, Photochemistry of Triton's atmosphere and ionosphere. J. Geophys. Res. **100**, 21271 (1995). https://doi.org/10.1029/95JE01904

V.A. Krasnopolsky, B.R. Sandel, F. Herbert, R.J. Vervack, Temperature, N_2, and N density profiles of Triton's atmosphere: observations and model. J. Geophys. Res., Planets **98**, 3065–3078 (1993). https://doi.org/10.1029/92JE02680

G.P. Kuiper, Titan: a satellite with an atmosphere. Astrophys. J. **100**, 378 (1944). https://doi.org/10.1086/144679

H. Lammer, Mass loss of N2 molecules from Triton by magnetospheric plasma interaction. Planet. Space Sci. **43**, 845–850 (1995). https://doi.org/10.1016/0032-0633(94)00214-C

H. Lammer, W. Stumptner, G.J. Molina-Cuberos, S.J. Bauer, T. Owen, Nitrogen isotope fractionation and its consequence for Titan's atmospheric evolution. Planet. Space Sci. **48**, 529–543 (2000). https://doi.org/10.1016/S0032-0633(00)00043-X

H. Lammer, S.J. Bauer, Isotopic fractionation by gravitational escape. Space Sci. Rev. **106**, 281–291 (2003). https://doi.org/10.1023/A:1024602124097

H. Lammer, W. Stumptner, G. Molina-Cuberos, S. Bauer, T. Owen, Nitrogen isotope fractionation and its consequence for Titan's atmospheric evolution. Planet. Space Sci. **48**, 529–543 (2002). https://doi.org/10.1016/s0032-0633(00)00043-x

H. Lammer, J.F. Kasting, E. Chassefière, R.E. Johnson, Y.N. Kulikov, F. Tian, Atmospheric escape and evolution of terrestrial planets and satellites. Space Sci. Rev. **139**, 399–436 (2008). https://doi.org/10.1007/s11214-008-9413-5

H. Lammer, M. Leitzinger, M. Scherf, P. Odert, C. Burger, D. Kubyshkina, C. Johnstone, T. Maindl, C.M. Schäfer, M. Güdel, N. Tosi, A. Nikolaou, E. Marcq, N.V. Erkaev, L. Noack, K.G. Kislyakova, L. Fossati, E. Pilat-Lohinger, F. Ragossnig, E.A. Dorfi, Constraining the early evolution of Venus and Earth through atmospheric Ar, Ne isotope and bulk K/U ratios. Icarus **339**, 113551 (2020a). https://doi.org/10.1016/j.icarus.2019.113551

H. Lammer, M. Scherf, H. Kurokawa, Y. Ueno, C. Burger, Z. Leinhart, T. Maindl, C. Johnstone, M. Leizinger, M. Benedikt, L. Fossati, K.G. Kislyakova, B. Marty, G. Avice, B. Fegley, P. Odert, Loss and fractionation of noble gas isotopes and moderately volatile elements from planetary embryos and early. Space Sci. Rev. **216**, 74 (2020b). https://doi.org/10.1007/s11214-020-00701-x

E. Lellouch, Atmospheric models of Titan and Triton. Ann. Geophys. **8**, 653 (1990). https://ui.adsabs.harvard.edu/abs/1990AnGeo...8..653L/abstract

E. Lellouch, M.A. McGrath, K.L. Jessup, Io's atmosphere, in *Io After Galileo* (Springer, Berlin, 2007), pp. 231–264. https://doi.org/10.1007/978-3-540-48841-5_10

E. Lellouch, C. de Bergh, B. Sicardy, S. Ferron, H.-U. Käufl, Detection of CO in Triton's atmosphere and the nature of surface-atmosphere interactions. Astron. Astrophys. **512**, L8 (2010). https://doi.org/10.1051/0004-6361/201014339

E. Lellouch, C. de Bergh, B. Sicardy, H.U. Käufl, A. Smette, High resolution spectroscopy of Pluto's atmosphere: detection of the 2.3 μm CH_4 bands and evidence for carbon monoxide. Astron. Astrophys. **530**, L4 (2011). https://doi.org/10.1051/0004-6361/201116954

E. Lellouch, M. Gurwell, B. Butler, T. Fouchet, P. Lavvas, D.F. Strobel, B. Sicardy, A. Moullet, R. Moreno, D. Bockelée-Morvan, N. Biver, L. Young, D. Lis, J. Stansberry, A. Stern, H. Weaver, E. Young, X. Zhu, J. Boissier, Detection of CO and HCN in Pluto's atmosphere with ALMA. Icarus **286**, 289–307 (2017). https://doi.org/10.1016/j.icarus.2016.10.013

J.S. Lewis, Satellites of the outer planets: their physical and chemical nature. Icarus **15**, 174–185 (1971). https://doi.org/10.1016/0019-1035(71)90072-8

M.-C. Liang, A.N. Heays, B.R. Lewis, S.T. Gibson, Y.L. Yung, Source of nitrogen isotope anomaly in HCN in the atmosphere of Titan. Astrophys. J. **664**, L115–L118 (2007). https://doi.org/10.1086/520881

G.F. Lindal, G.E. Wood, H.B. Hotz, D.N. Sweetnam, V.R. Eshleman, G.L. Tyler, The atmosphere of Titan: an analysis of the Voyager 1 radio occultation measurements. Icarus **53**, 348–363 (1983). https://doi.org/10.1016/0019-1035(83)90155-0

K. Lodders, Solar system abundances of the elements, in *Princ. Perspect. Cosmochem. Astrophys. Sp. Sci. Proceedings* (Springer, Berlin, 2010), pp. 379–417. https://doi.org/10.1007/978-3-642-10352-0_8. ISBN 978-3-642-10351-3

A. Lofthus, P.H. Krupenie, The spectrum of molecular nitrogen. J. Phys. Chem. Ref. Data **6**, 113–307 (1977). https://doi.org/10.1063/1.555546

R.D. Lorenz, E.P. Turtle, J.W. Barnes, M.G. Trainer, D.S. Adams, K.E. Hibbard, C.Z. Sheldon, K. Zacny, P.N. Peplowski, D.J. Lawrence, M.A. Ravine, T.G. McGee, K.S. Sotzen, S.M. MacKenzie, J.W. Langelaan, S. Schmitz, L.S. Wolfarth, P.D. Bedini, Dragonfly: a rotorcraft lander concept for scientific exploration at Titan. Johns Hopkins APL Tech. Dig. **34**, 374–387 (2018)

J.I. Lunine, M.C. Nolan, A massive early atmosphere on Triton. Icarus **100**, 221–234 (1992). https://doi.org/10.1016/0019-1035(92)90031-2

J.I. Lunine, Y.L. Yung, R.D. Lorenz, On the volatile inventory of Titan from isotopic abundances in nitrogen and methane. Planet. Space Sci. **47**, 1291–1303 (1999)

J.R. Lyons, Y. Yung, M. Allen, Solar control of the upper atmosphere of Triton. Science **80**(256), 204–206 (1992). https://doi.org/10.1126/science.11540928

P.R. Mahaffy, C.R. Webster, M. Cabane, P.G. Conrad, P. Coll, S.K. Atreya, R. Arvey, M. Barciniak, M. Benna, L. Bleacher, W.B. Brinckerhoff, J.L. Eigenbrode, D. Carignan, M. Cascia, R.A. Chalmers, J.P. Dworkin, T. Errigo, P. Everson, H. Franz, R. Farley, S. Feng, G. Frazier, C. Freissinet, D.P. Glavin, D.N. Harpold, D. Hawk, V. Holmes, C.S. Johnson, A. Jones, P. Jordan, J. Kellogg, J. Lewis, E. Lyness, C.A. Malespin, D.K. Martin, J. Maurer, A.C. McAdam, D. McLennan, T.J. Nolan, M. Noriega, A.A. Pavlov, B. Prats, E. Raaen, O. Sheinman, D. Sheppard, J. Smith, J.C. Stern, F. Tan, M. Trainer, D.W. Ming, R.V. Morris, J. Jones, C. Gundersen, A. Steele, J. Wray, O. Botta, L.A. Leshin, T. Owen, S. Battel, B.M. Jakosky, H. Manning, S. Squyres, R. Navarro-González, C.P. McKay, F. Raulin, R. Sternberg, A. Buch, P. Sorensen, R. Kline-Schoder, D. Coscia, C. Szopa, S. Teinturier, C. Baffes, J. Feldman, G. Flesch, S. Forouhar, R. Garcia, D. Keymeulen, S. Woodward, B.P. Block, K. Arnett, R. Miller, C. Edmonson, S. Gorevan, E. Mumm, The sample analysis at Mars investigation and instrument suite. Space Sci. Rev. **170**, 401–478 (2012). https://doi.org/10.1007/s11214-012-9879-z

K.E. Mandt, J.H. Waite, W. Lewis, B. Magee, J. Bell, J. Lunine, O. Mousis, D. Cordier, Isotopic evolution of the major constituents of Titan's atmosphere based on Cassini data. Planet. Space Sci. **57**, 1917–1930 (2009). https://doi.org/10.1016/j.pss.2009.06.005

K.E. Mandt, J.H. Waite, B. Teolis, B.A. Magee, J. Bell, J.H. Westlake, C.A. Nixon, O. Mousis, J.I. Lunine, The $^{12}C/^{13}C$ ratio on Titan from Cassini INMS measurements and implications for the evolution of methane. Astrophys. J. **749**, 160 (2012). https://doi.org/10.1088/0004-637X/749/2/160

K.E. Mandt, O. Mousis, J. Lunine, D. Gautier, Protosolar ammonia as the unique source of Titan's nitrogen. Astrophys. J. Lett. **788**, L24 (2014). https://doi.org/10.1088/2041-8205/788/2/L24

K.E. Mandt, O. Mousis, A. Luspay-Kuti, Isotopic constraints on the source of Pluto's nitrogen and the history of atmospheric escape. Planet. Space Sci. **130**, 104–109 (2016). https://doi.org/10.1016/j.pss.2016.02.011

K. Mandt, A. Luspay-kuti, M. Hamel, K. Jessup, in *Photochemistry on Pluto: Part II HCN and Nitrogen Isotope Fractionation* (2017), pp. 1–46

J. Manfroid, E. Jehin, D. Hutsemékers, A. Cochran, J.M. Zucconi, C. Arpigny, R. Schulz, J.A. Stüwe, Isotopic abundance of nitrogen and carbon in distant comets. Astron. Astrophys. **432**, 5–8 (2005). https://doi.org/10.1051/0004-6361:200500009

J. Manfroid, E. Jehin, D. Hutsemékers, A. Cochran, J.M. Zucconi, C. Arpigny, R. Schulz, J.A. Stüwe, I. Ilyin, The CN isotopic ratios in comets. Astron. Astrophys. **503**, 613–624 (2009). https://doi.org/10.1051/0004-6361/200911859

N. Marounina, G. Tobie, S. Carpy, J. Monteux, B. Charnay, O. Grasset, Evolution of Titan's atmosphere during the Late Heavy Bombardment. Icarus **257**, 324–335 (2015). https://doi.org/10.1016/j.icarus.2015.05.011

A. Marten, T. Hidayat, R. Moreno, G. Paubert, B. Bezard, D. Gautier, H. Matthews, T. Owen, Saturn VI (Titan), in *IAU Circ.*, vol. 6702 (1997), p. 1. http://adsabs.harvard.edu/abs/1997IAUC.6702....1M/abstract

A. Marten, T. Hidayat, Y. Biraud, R. Moreno, New millimeter heterodyne observations of Titan: vertical distributions of nitriles HCN, HC3N, CH3CN, and the isotopic ratio 15N/14N in its atmosphere. Icarus **158**, 532–544 (2002). https://doi.org/10.1006/icar.2002.6897

B. Marty, The origins and concentrations of water, carbon, nitrogen and noble gases on Earth. Earth Planet. Sci. Lett. **313–314**, 56–66 (2012). https://doi.org/10.1016/j.epsl.2011.10.040

B. Marty, L. Zimmermann, Volatiles (He, C, N, Ar) in mid-ocean ridge basalts: assesment of shallow-level fractionation and characterization of source composition. Geochim. Cosmochim. Acta **63**, 3619–3633 (1999). https://doi.org/10.1016/S0016-7037(99)00169-6

B. Marty, L. Zimmermann, P.G. Burnard, R. Wieler, V.S. Heber, D.L. Burnett, R.C. Wiens, P. Bochsler, Nitrogen isotopes in the recent solar wind from the analysis of Genesis targets: evidence for large scale isotope heterogeneity in the early solar system. Geochim. Cosmochim. Acta **74**, 340–355 (2010). https://doi.org/10.1016/j.gca.2009.09.007

B. Marty, M. Chaussidon, R.C. Wiens, A.J.G. Jurewicz, D.S. Burnett, A 15N-poor isotopic composition for the solar system as shown by genesis solar wind samples. Science **80**(332), 1533–1536 (2011). https://doi.org/10.1126/science.1204656

K.J. Mathew, K. Marti, Early evolution of martian volatiles: nitrogen and noble gas components in ALH84001 and Chassigny. J. Geophys. Res., Planets **106**, 1401–1422 (2001). https://doi.org/10.1029/2000JE001255

D.L. Matson, J.C. Castillo, J. Lunine, T.V. Johnson, Enceladus' plume: compositional evidence for a hot interior. Icarus **187**, 569–573 (2007). https://doi.org/10.1016/j.icarus.2006.10.016

S.R.N. McIntyre, C.H. Lineweaver, M.J. Ireland, Planetary magnetism as a parameter in exoplanet habitability. Mon. Not. R. Astron. Soc. **485**, 3999–4012 (2019). https://doi.org/10.1093/mnras/stz667

C.P. McKay, T.W. Scattergood, J.B. Pollack, W.J. Borucki, H.T. Van Ghyseghem, High-temperature shock formation of N2 and organics on primordial Titan. Nature **332**, 520–522 (1988). https://doi.org/10.1038/332520a0

K.D. McKeegan, J. Aleon, J. Bradley, D. Brownlee, H. Busemann, A. Butterworth, M. Chaussidon, S. Fallon, C. Floss, J. Gilmour, M. Gounelle, G. Graham, Y. Guan, P.R. Heck, P. Hoppe, I.D. Hutcheon, J. Huth, H. Ishii, M. Ito, S.B. Jacobsen, A. Kearsley, L.A. Leshin, M.-C. Liu, I. Lyon, K. Marhas, B. Marty, G. Matrajt, A. Meibom, S. Messenger, S. Mostefaoui, S. Mukhopadhyay, K. Nakamura-Messenger, L. Nittler, R. Palma, R.O. Pepin, D.A. Papanastassiou, F. Robert, D. Schlutter, C.J. Snead, F.J. Stadermann, R. Stroud, P. Tsou, A. Westphal, E.D. Young, K. Ziegler, L. Zimmermann, E. Zinner, Isotopic compositions of cometary matter returned by stardust. Science **80**(314), 1724–1728 (2006). https://doi.org/10.1126/science.1135992

A. Meibom, A.N. Krot, F. Robert, S. Mostefaoui, S.S. Russell, M.I. Petaev, M. Gounelle, Nitrogen and carbon isotopic composition of the Sun inferred from a high-temperature solar nebular condensate. Astrophys. J. **656**, L33–L36 (2007). https://doi.org/10.1086/512052

A. Meshik, C. Hohenberg, O. Pravdivtseva, D. Burnett, ScienceDirect heavy noble gases in solar wind delivered by Genesis mission. Geochim. Cosmochim. Acta **127**, 326–347 (2014). https://doi.org/10.1016/j.gca.2013.11.030

A. Meshik, O. Pravdivtseva, D. Burnett, Refined composition of solar wind xenon delivered by Genesis NASA mission: comparison with xenon captured by extraterrestrial regolith soils. Geochim. Cosmochim. Acta **276**, 289–298 (2020). https://doi.org/10.1016/j.gca.2020.03.001

S. Messenger, Identification of molecular-cloud material in interplanetary dust particles. Nature **404**, 968–971 (2000). https://doi.org/10.1038/35010053

E. Meza, B. Sicardy, M. Assafin, J.L. Ortiz, T. Bertrand, E. Lellouch, J. Desmars, F. Forget, D. Bérard, Lower atmosphere and pressure evolution on Pluto from ground-based stellar occultations, 1988–2016. Astron. Astrophys. **625**, A42 (2019)

M. Michael, R.E. Johnson, F. Leblanc, M. Liu, J.G. Luhmann, V.I. Shematovich, Ejection of nitrogen from Titan's atmosphere by magnetospheric ions and pick-up ions. Icarus **175**, 263–267 (2005). https://doi.org/10.1016/j.icarus.2004.11.004

K.E. Miller, C.R. Glein, J.H. Waite, Contributions from accreted organics to Titan's atmosphere: new insights from cometary and chondritic data. Astrophys. J. **871**, 59 (2019). https://doi.org/10.3847/1538-4357/aaf561

K.L. Mitchell, L.M. Prockter, W.E. Frazier, W.D. Smythe, B.M. Sutin, D.A. Bearden, T. Team, Implementation of trident: a discovery-class mission to Triton, in *50th Lunar and Planetary Science Conference, LPI Contribution No. 2132, id. 3200* (2019)

E.M. Molter, C.A. Nixon, M.A. Cordiner, J. Serigano, P.G.J. Irwin, N.A. Teanby, S.B. Charnley, J.E. Lindberg, ALMA observations of HCN and its isotopologues on Titan. Astron. J. **152**, 42 (2016). https://doi.org/10.3847/0004-6256/152/2/42

Y. Moulane, E. Jehin, P. Rousselot, J. Manfroid, Y. Shinnaka, F.J. Pozuelos, D. Hutsemékers, C. Opitom, B. Yang, Z. Benkhaldoun, Photometry and high-resolution spectroscopy of comet 21P/Giacobini-Zinner during its 2018 apparition 1–14 (2020). arXiv:2006.05017

H.B. Niemann, D.N. Harpold, S.K. Atreya, G.R. Carignan, D.M. Hunten, T.C. Owen, Galileo Probe Mass Spectrometer experiment. Space Sci. Rev. **60**, 111–142 (1992). https://doi.org/10.1007/BF00216852

H.B. Niemann, S.K. Atreya, S.J. Bauer, K. Biemann, B. Block, G.R. Carignan, T.M. Donahue, R.L. Frost, D. Gautier, J.A. Haberman, D. Harpold, D.M. Hunten, G. Israel, J.I. Lunine, K. Mauersberger, T.C. Owen, F. Raulin, J.E. Richards, S.H. Way, The gas chromatograph mass spectrometer for the Huygens probe. Space Sci. Rev. **104**, 553–591 (2002). https://doi.org/10.1023/A:1023680305259

H.B. Niemann, S.K. Atreya, S.J. Bauer, G.R. Carignan, J.E. Demick, R.L. Frost, D. Gautier, J.A. Haberman, D.N. Harpold, D.M. Hunten, G. Israel, J.I. Lunine, W.T. Kasprzak, T.C. Owen, M. Paulkovich, F. Raulin, E. Raaen, S.H. Way, The abundances of constituents of Titan's atmosphere from the GCMS instrument on the Huygens probe. Nature **438**, 779–784 (2005). https://doi.org/10.1038/nature04122

H.B. Niemann, S.K. Atreya, J.E. Demick, D. Gautier, J.A. Haberman, D.N. Harpold, W.T. Kasprzak, J.I. Lunine, T.C. Owen, F. Raulin, Composition of Titan's lower atmosphere and simple surface volatiles as measured by the Cassini-Huygens probe gas chromatograph mass spectrometer experiment. J. Geophys. Res., Planets **115**, 1–22 (2010). https://doi.org/10.1029/2010JE003659

A.O. Nier, M.B. McElroy, Composition and structure of Mars' Upper atmosphere: results from the neutral mass spectrometers on Viking 1 and 2. J. Geophys. Res. **82**, 4341–4349 (1977). https://doi.org/10.1029/js082i028p04341

C.A. Nixon, B. Temelso, S. Vinatier, N.A. Teanby, B. Bézard, R.K. Achterberg, K.E. Mandt, C.D. Sherrill, P.G.J. Irwin, D.E. Jennings, P.N. Romani, A. Coustenis, F.M. Flasar, Isotopic ratios in titan's methane: measurements and modeling. Astrophys. J. **749**, 159 (2012). https://doi.org/10.1088/0004-637X/749/2/159

K.S. Noll, R.E. Johnson, A.L. Lane, D.L. Domingue, H.A. Weaver, Detection of ozone on Ganymede. Science **80**(273), 341–343 (1996). https://doi.org/10.1126/science.273.5273.341

P. Odert, H. Lammer, N.V. Erkaev, A. Nikolaou, H.I.M. Lichtenegger, C.P. Johnstone, K.G. Kislyakova, M. Leitzinger, N. Tosi, Escape and fractionation of volatiles and noble gases from Mars-sized planetary embryos and growing protoplanets. Icarus **307**, 327–346 (2018). https://doi.org/10.1016/j.icarus.2017.10.031

C.B. Olkin, L.A. Young, R.G. French, E.F. Young, M.W. Buie, R.R. Howell, J. Regester, C.R. Ruhland, T. Natusch, D.J. Ramm, Pluto's atmospheric structure from the July 2007 stellar occultation. Icarus **239**, 15–22 (2014). https://doi.org/10.1016/j.icarus.2014.05.018

T. Owen, A. Bar-Nun, Comets, impacts, and atmospheres. Icarus **116**, 215–226 (1995). https://doi.org/10.1006/icar.1995.1122

T. Owen, K. Biemann, D.R. Rushneck, J.E. Biller, D.W. Howarth, A.L. Lafleur, The composition of the atmosphere at the surface of Mars. J. Geophys. Res. **82**, 4635–4639 (1977). https://doi.org/10.1029/js082i028p04635

T. Owen, P.R. Mahaffy, H.B. Niemann, S. Atreya, M. Wong, Protosolar nitrogen. Astrophys. J. **553**, L77–L79 (2001). https://doi.org/10.1086/320501

P.T. Palmer, T.F. Limero, Mass spectrometry in the U.S. space program: past, present, and future. J. Am. Soc. Mass Spectrom. **12**, 656–675 (2001). https://doi.org/10.1016/S1044-0305(01)00249-5

J. Pearl, R. Hanel, V. Kunde, W. Maguire, K. Fox, S. Gupta, C. Ponnamperuma, F. Raulin, Identification of gaseous SO2 and new upper limits for other gases on Io. Nature **280**, 755–758 (1979). https://doi.org/10.1038/280755a0

T. Penz, H. Lammer, Y.N. Kulikov, H.K. Biernat, The influence of the solar particle and radiation environment on Titan's atmosphere evolution. Adv. Space Res. **36**, 241–250 (2005). https://doi.org/10.1016/j.asr.2005.03.043

L.M. Prockter, K.L. Mitchell, C.J.A. Howett, W.D. Smythe, B.M. Sutin, D.A. Bearden, W.E. Frazier, Exploring triton with trident: a discovery class mission, in *50th Lunar Planet. Sci. Conf.*, Woodlands, Texas, 18-22 March, 2019 (2019). LPI Contrib. No. 2132, id. 3188. http://adsabs.harvard.edu/abs/2019LPI....50.3188P

Z. Ren, M. Guo, Y. Cheng, Y. Wang, W. Sun, H. Zhang, M. Dong, G. Li, A review of the development and application of space miniature mass spectrometers. Vacuum **155**, 108–117 (2018). https://doi.org/10.1016/j.vacuum.2018.05.048

P. Rousselot, O. Pirali, E. Jehin, M. Vervloet, D. Hutsemékers, J. Manfroid, D. Cordier, M.A. Martin-Drumel, S. Gruet, C. Arpigny, A. Decock, O. Mousis, Toward a unique nitrogen isotopic ratio in cometary ICES. Astrophys. J. Lett. **780**, 2–6 (2014). https://doi.org/10.1088/2041-8205/780/2/L17

R.E. Samuelson, N.R. Nath, A. Borysow, Gaseous abundances and methane supersaturation in Titan's troposphere. Planet. Space Sci. **45**, 959–980 (1997)

P. Saxena, R.M. Killen, V. Airapetian, N.E. Petro, N.M. Curran, A.M. Mandell, Was the Sun a slow rotator? Sodium and potassium constraints from the lunar regolith. Astrophys. J. **876**, L16 (2019). https://doi.org/10.3847/2041-8213/ab18fb

K. Schaepe, H. Jungnickel, T. Heinrich, J. Tentschert, A. Luch, W.E.S. Unger, Secondary ion mass spectrometry, in *Characterization of Nanoparticles: Measurement Processes for Nanoparticles* (Elsevier, Amsterdam, 2019), pp. 481–509. https://doi.org/10.1016/B978-0-12-814182-3.00025-0

Y. Sekine, H. Genda, S. Sugita, T. Kadono, T. Matsui, Replacement and late formation of atmospheric N2 on undifferentiated Titan by impacts. Nat. Geosci. **4**, 359–362 (2011). https://doi.org/10.1038/ngeo1147

V.I. Shematovich, R.E. Johnson, M. Michael, J.G. Luhmann, Nitrogen loss from Titan. J. Geophys. Res. **108**, 5087 (2003). https://doi.org/10.1029/2003JE002094

Y. Shinnaka, H. Kawakita, Nitrogen isotopic ratio of cometary ammonia from high-resolution optical spectroscopic observations of C/2014 Q2 (Lovejoy). Astron. J. **152**, 145 (2016). https://doi.org/10.3847/0004-6256/152/5/145

Y. Shinnaka, H. Kawakita, H. Kobayashi, M. Nagashima, D.C. Boice, $^{14}NH_2/^{15}NH_2$ ratio in comet C/2012 S1 (ISON) observed during its outburst in 2013 November. Astrophys. J. **782**, L16 (2014a). https://doi.org/10.1088/2041-8205/782/2/L16

Y. Shinnaka, H. Kawakita, E. Jehin, A. Decock, D. Hutsemékers, J. Manfroid, A. Arai, Nitrogen isotopic ratios of NH_2 in comets: implication for ^{15}N-fractionation in cometary ammonia. Mon. Not. R. Astron. Soc. **462**, S195–S209 (2016). https://doi.org/10.1093/mnras/stw2410

B. Sicardy, T. Widemann, E. Lellouch, C. Veillet, J.-C. Cuillandre, F. Colas, F. Roques, W. Beisker, M. Kretlow, A.-M. Lagrange, E. Gendron, F. Lacombe, J. Lecacheux, C. Birnbaum, A. Fienga, C. Leyrat, A. Maury, E. Raynaud, S. Renner, M. Schultheis, K. Brooks, A. Delsanti, O.R. Hainaut, R. Gilmozzi, C. Lidman, J. Spyromilio, M. Rapaport, P. Rosenzweig, O. Naranjo, L. Porras, F. Díaz, H. Calderón, S. Carrillo, A. Carvajal, E. Recalde, L.G. Cavero, C. Montalvo, D. Barría, R. Campos, R. Duffard, H. Levato, Large changes in Pluto's atmosphere as revealed by recent stellar occultations. Nature **424**, 168–170 (2003). https://doi.org/10.1038/nature01766

K.N. Singer, S.A. Stern, On the provenance of Pluto's nitrogen (N2). Astrophys. J. Lett. **808**(2), L50 (2015). https://doi.org/10.1088/2041-8205/808/2/l50.

B.A. Smith, L.A. Soderblom, D. Banfield, c. Barnet, A.T. Basilevsky, R.F. Beebe, K. Bollinger, J.M. Boyce, A. Brahic, G.A. Briggs, R.H. Brown, c. Chyba, s.A. Collins, T. Colvin, A.F. Cook, D. Crisp, S.K. Croft, D. Cruikshank, J.N. Cuzzi, G.E. Danielson, M.E. Davies, E. De Jong, L. Dones, D. Godfrey, J. Goguen, I. Grenier, V.R. Haemmerle, H. Hammel, c.J. Hansen, c.P. Helfenstein, C. Howell, G.E. Hunt, A.P. Ingersoll, T.V. Johnson, J. Kargel, R. Kirk, D.I. Kuehn, S. Limaye, H. Masursky, A. McEwen, D. Morrison, T. Owen, W. Owen, J.B. Pollack, C.C. Porco, K. Rages, P. Rogers, D. Rudy, C. Sagan, J. Schwartz, E.M. Shoemaker, M. Showalter, B. Sicardy, D. Simonelli, J. Spencer, L.A. Sromovsky, C. Stoker, R.G. Strom, V.E. Suomi, S.P. Synott, R.J. Terrile, P. Thomas, W.R. Thompson, A. Verbiscer, J. Veverka, Imaging science results. Science **80**(246), 1422–1449 (1989). https://doi.org/10.1126/science.246.4936.1422. Voyager 2 at Neptune

C. Snodgrass, G.H. Jones, The European Space Agency's comet interceptor lies in wait. Nat. Commun. (2019). https://doi.org/10.1038/s41467-019-13470-1

J.C. Solá, Observations des satellites principaux de Jupiter et de Titan. Astron. Nachr. **179**, 289–290 (1908). https://doi.org/10.1002/asna.19081791807

S.A. Stern, F. Bagenal, K. Ennico, G.R. Gladstone, W.M. Grundy, W.B. McKinnon, J.M. Moore, C.B. Olkin, J.R. Spencer, H.A. Weaver, L.A. Young, T. Andert, J. Andrews, M. Banks, B. Bauer, J. Bauman, O.S. Barnouin, P. Bedini, K. Beisser, R.A. Beyer, S. Bhaskaran, R.P. Binzel, E. Birath, M. Bird, D.J. Bogan, A. Bowman, V.J. Bray, M. Brozovic, C. Bryan, M.R. Buckley, M.W. Buie, B.J. Buratti, S.S. Bushman, A. Calloway, B. Carcich, A.F. Cheng, S. Conard, C.A. Conrad, J.C. Cook, D.P. Cruikshank, O.S. Custodio, C.M.D. Ore, C. Deboy, Z.J.B. Dischner, P. Dumont, A.M. Earle, H.A. Elliott, J. Ercol, C.M. Ernst, T. Finley, S.H. Flanigan, G. Fountain, M.J. Freeze, T. Greathouse, J.L. Green, Y. Guo, M. Hahn, D.P. Hamilton, S.A. Hamilton, J. Hanley, A. Harch, H.M. Hart, C.B. Hersman, A. Hill, M.E. Hill, D.P. Hinson, M.E. Holdridge, M. Horanyi, A.D. Howard, C.J.A. Howett, C. Jackman, R.A. Jacobson, D.E. Jennings, J.A. Kammer, H.K. Kang, D.E. Kaufmann, P. Kollmann, S.M. Krimigis, D. Kusnierkiewicz, T.R. Lauer, J.E. Lee, K.L. Lindstrom, I.R. Linscott, C.M. Lisse, A.W. Lunsford, V.A. Mallder, N. Martin, D.J. McComas, R.L. McNutt, D. Mehoke, T. Mehoke, E.D. Melin, M. Mutchler, D. Nelson, F. Nimmo, J.I. Nunez, A. Ocampo, W.M. Owen, M. Paetzold, B. Page, A.H. Parker, J.W. Parker, F. Pelletier, J. Peterson, N. Pinkine, M. Piquette, S.B. Porter, S. Protopapa, J. Redfern, H.J. Reitsema, D.C. Reuter, J.H. Roberts, S.J. Robbins, G. Rogers, D. Rose, K. Runyon, K.D. Retherford, M.G. Ryschkewitsch, P. Schenk, E. Schindhelm, B. Sepan, M.R. Showalter, K.N. Singer, M. Soluri, D. Stanbridge, A.J. Steffl, D.F. Strobel, T. Stryk, M.E. Summers, J.R. Szalay, M. Tapley, A. Taylor, H. Taylor, H.B. Throop, C.C.C. Tsang, G.L. Tyler, O.M. Umurhan, A.J. Verbiscer, M.H. Versteeg, M. Vincent, R. Webbert, S. Weidner, G.E. Weigle, O.L. White, K. Whittenburg, B.G. Williams, K. Williams, S. Williams, W.W. Woods, A.M.

Zangari, E. Zirnstein, The Pluto system: initial results from its exploration by New Horizons. Science **350**(6258), aad1815 (2015). https://doi.org/10.1126/science.aad1815.
D.F. Strobel, Chemistry and evolution of Titan's atmosphere. Planet. Space Sci. **30**, 839–848 (1982). https://doi.org/10.1016/0032-0633(82)90116-7
D.F. Strobel, N2 escape rates from Pluto's atmosphere. Icarus **193**, 612–619 (2008). https://doi.org/10.1016/j.icarus.2007.08.021
D.F. Strobel, X. Zhu, Comparative planetary nitrogen atmospheres: density and thermal structures of Pluto and Triton. Icarus **291**, 55–64 (2017). https://doi.org/10.1016/j.icarus.2017.03.013
M.E. Summers, D.F. Strobel, Triton's atmosphere: a source of N and H for Neptune's magnetosphere. Geophys. Res. Lett. **18**, 2309–2312 (1991). https://doi.org/10.1029/91GL01334
F. Tian, O.B. Toon, Hydrodynamic escape of nitrogen from Pluto. Geophys. Res. Lett. **32**, L18201 (2005). https://doi.org/10.1029/2005GL023510
F. Tian, J.F. Kasting, S.C. Solomon, Thermal escape of carbon from the early Martian atmosphere. Geophys. Res. Lett. **36**, 1–5 (2009). https://doi.org/10.1029/2008GL036513
G. Tobie, J.I. Lunine, C. Sotin, Episodic outgassing as the origin of atmospheric methane on Titan. Nature **440**, 61–64 (2006). https://doi.org/10.1038/nature04497
G. Tobie, D. Gautier, F. Hersant, Titan's bulk composition constrained by Cassini-Huygens: implication for internal outgassing. Astrophys. J. **752**, 125 (2012). https://doi.org/10.1088/0004-637X/752/2/125
A.T. Tokunaga, S.T. Ridgway, L. Wallace, R.F. Knacke, High-resolution spectra of Jupiter in the 744-980 inverse centimeter spectral range. Astrophys. J. **232**, 603 (1979). https://doi.org/10.1086/157319
L. Trafton, S.A. Stern, On the global distribution of Pluto's atmosphere. Astrophys. J. **267**, 872 (1983). https://doi.org/10.1086/160921
L. Tu, C.P. Johnstone, M. Güdel, H. Lammer, The extreme ultraviolet and X-ray Sun in time: high-energy evolutionary tracks of a solar-like star. Astron. Astrophys. **577**, L3 (2015). https://doi.org/10.1051/0004-6361/201526146
O.J. Tucker, J.T. Erwin, J.I. Deighan, A.N. Volkov, R.E. Johnson, Thermally driven escape from Pluto's atmosphere: a combined fluid/kinetic model. Icarus **217**(1), 408–415 (2011). https://doi.org/10.1016/j.icarus.2011.11.017
E.P. Turtle, J.W. Barnes, M.G. Trainer, R.D. Lorenz, K.E. Hibbard, D.S. Adams, P. Bedini, W.B. Brinckerhoff, M.L. Cable, C. Ernst, C. Freissinet, K. Hand, A.G. Hayes, S.M. Horst, J.R. Johnson, E. Karkoschka, J.W. Langelaan, D.J. Lawrence, A. Le Gall, J.M. Lora, S.M. MacKenzie, C.P. McKay, C.D. Neish, C.E. Newman, J. Palacios, M.P. Panning, A.M. Parsons, P.N. Peplowski, J. Radebaugh, S.C.R. Rafkin, M.A. Ravine, S. Schmitz, J.M. Soderblom, K.S. Sotzen, A.M. Stickle, E.R. Stofan, T. Tokano, C. Wilson, R.A. Yingst, K. Zacny, Dragonfly: in situ exploration of Titan's organic chemistry and habitability, in *49th Lunar Planet. Sci. Conf.*, Woodlands, Texas, 19-23 March, 2018 (2018). LPI Contrib. No. 2083, id. 1641. http://adsabs.harvard.edu/abs/2018LPI....49.1641T
R.J. Vervack, B.R. Sandel, D.F. Strobel, New perspectives on Titan's upper atmosphere from a reanalysis of the Voyager 1 UVS solar occultations. Icarus **170**, 91–112 (2004). https://doi.org/10.1016/j.icarus.2004.03.005
S. Vinatier, B. Bézard, C.A. Nixon, The Titan 14N/15N and 12C/13C isotopic ratios in HCN from Cassini/CIRS. Icarus **191**, 712–721 (2007). https://doi.org/10.1016/j.icarus.2007.06.001
A.N. Volkov, R.E. Johnson, O.J. Tucker, J.T. Erwin, Thermally driven atmospheric escape: transition from hydrodynamic to Jeans escape. Astrophys. J. Lett. **729**, L24 (2011). https://doi.org/10.1088/2041-8205/729/2/L24
A. Vorburger, M. Pfleger, J. Lindkvist, M. Holmström, H. Lammer, H.I.M. Lichtenegger, A. Galli, M. Rubin, P. Wurz, Three-Dimensional Modeling of Callisto's Surface Sputtered Exosphere Environment. J. Geophys. Res. Space Phys. (2019). https://doi.org/10.1029/2019JA026610
J.H. Waite, W.S. Lewis, W.T. Kasprzak, V.G. Anicich, B.P. Block, T.E. Cravens, G.G. Fletcher, J.G. Luhmann, R.L. Mcnutt, H.B. Niemann, J.K. Parejko, J.E. Richards, R.L. Thorpe, E.M. Walter, R.V. Yelle, The Cassini Ion and Neutral Mass Spectrometer (INMS) investigation. Space Sci. Rev. **114**, 113–231 (2004). https://doi.org/10.1007/s11214-004-1408-2
J.H. Waite, H. Niemann, R.V. Yelle, W.T. Kasprzak, T.E. Cravens, J.G. Luhmann, R.L. McNutt, W.-H. Ip, D. Gell, V. De La Haye, I. Müller-Wordag, B. Magee, N. Borggren, S. Ledvina, G. Fletcher, E. Walter, R. Miller, S. Scherer, R. Thorpe, J. Xu, B. Block, K. Arnett, Ion Neutral Mass Spectrometer results from the first flyby of Titan. Science **80**(308), 982–986 (2005). https://doi.org/10.1126/science.1110652
C.R. Webster, S.P. Sander, R. Beer, R.D. May, R.G. Knollenberg, D.M. Hunten, J. Ballard, Tunable diode laser IR spectrometer for in situ measurements of the gas phase composition and particle size distribution of Titan's atmosphere. Appl. Opt. **29**, 907 (1990). https://doi.org/10.1364/ao.29.000907
C.R. Webster, P.R. Mahaffy, G.J. Flesch, P.B. Niles, J.H. Jones, L.A. Leshin, S.K. Atreya, J.C. Stern, L.E. Christensen, T. Owen, H. Franz, R.O. Pepin, A. Steele, Isotope ratios of H, C, and O in CO2 and H2O of the Martian atmosphere. Science **80**(341), 260–263 (2013). https://doi.org/10.1126/science.1237961

C.R. Webster, L.E. Christensen, G.J. Flesch, S. Forouhar, R. Briggs, D. Keymeulen, J. Blacksberg, P.R. Mahaffy, Tunable laser spectrometers for space science, in *Proceedings of the International Workshop on Instrumentation for Planetary Missions*, IPM-2014 (2014)

C.R. Webster, J. Blacksberg, L.E. Christensen, G.J. Flesch, S. Forouhar, R. Briggs, D. Keymeulen, P.R. Mahaffy, P.R.G. Mahaffy@nasa, Digital Tunable Laser Spectrometer for Venus atmospheric isotope ratios, in *LPICo*, vol. 1838, (2015), p. 4012. https://doi.org/10.1088/2041-8205/713/1/L59.[5

M.H. Wong, P.R. Mahaffy, S.K. Atreya, H.B. Niemann, T.C. Owen, Updated Galileo probe mass spectrometer measurements of carbon, oxygen, nitrogen, and sulfur on Jupiter. Icarus **171**, 153–170 (2004). https://doi.org/10.1016/j.icarus.2004.04.010

M.H. Wong, S.K. Atreya, P.N. Mahaffy, H.B. Franz, C. Malespin, M.G. Trainer, J.C. Stern, P.G. Conrad, H.L.K. Manning, R.O. Pepin, R.H. Becker, C.P. McKay, T.C. Owen, R. Navarro-González, J.H. Jones, B.M. Jakosky, A. Steele, Isotopes of nitrogen on Mars: atmospheric measurements by Curiosity's mass spectrometer. Geophys. Res. Lett. **40**, 6033–6037 (2013). https://doi.org/10.1002/2013GL057840

B. Yang, D. Hutsemékers, Y. Shinnaka, C. Opitom, J. Manfroid, E. Jehin, K.J. Meech, O.R. Hainaut, J.V. Keane, M. Gillon, Isotopic ratios in outbursting comet C/2015 ER61. Astron. Astrophys. **609**, 1–4 (2018). https://doi.org/10.1051/0004-6361/201732100

R.V. Yelle, J.L. Elliot, Atmospheric Structure and Composition: Pluto and Charon, in *Pluto Charon*, ed. by S.A. Stern, D.J. Tholen, A.S. Rus, M.L. Guerr, M.S. Matthews (University of Arizona Press, Tucson, 1997), p. 347. http://adsabs.harvard.edu/abs/1997plch.book..347Y

S. Yokota, Isotope mass spectrometry in the solar system exploration. J. Mass Spectrom. **7**, S0076–S0076 (2018). https://doi.org/10.5702/massspectrometry.s0076

L.A. Young, J.L. Elliot, A. Tokunaga, C. de Bergh, T. Owen, Detection of gaseous methane on Pluto. Icarus **127**, 258–262 (1997). https://doi.org/10.1006/icar.1997.5709

L.A. Young, J.A. Kammer, A.J. Steffl, G.R. Gladstone, M.E. Summers, D.F. Strobel, D.P. Hinson, S.A. Stern, H.A. Weaver, C.B. Olkin, K. Ennico, D.J. McComas, A.F. Cheng, P. Gao, P. Lavvas, I.R. Linscott, M.L. Wong, Y.L. Yung, N. Cunningham, M. Davis, J.W. Parker, E. Schindhelm, O.H.W. Siegmund, J. Stone, K. Retherford, M. Versteeg, Structure and composition of Pluto's atmosphere from the New Horizons solar ultraviolet occultation. Icarus **300**, 174–199 (2018). https://doi.org/10.1016/j.icarus.2017.09.006

Y.L. Yung, J.R. Lyons, Triton: topside ionosphere and nitrogen escape. Geophys. Res. Lett. **17**, 1717–1720 (1990). https://doi.org/10.1029/GL017i010p01717

Y.L. Yung, M. Allen, J.P. Pinto, Photochemistry of the atmosphere of Titan—comparison between model and observations. Astrophys. J. Suppl. Ser. **55**, 465 (1984). https://doi.org/10.1086/190963

X. Zhu, D.F. Strobel, J.T. Erwin, The density and thermal structure of Pluto's atmosphere and associated escape processes and rates. Icarus **228**, 301–314 (2014). https://doi.org/10.1016/j.icarus.2013.10.011

L.M. Ziurys, C. Savage, M.A. Brewster, A.J. Apponi, T.C. Pesch, S. Wyckoff, Cyanide chemistry in Comet Hale-Bopp (C/1995 O1). Astrophys. J. **527**, L67–L71 (1999). https://doi.org/10.1086/312388

Space Sci Rev (2020) 216:36
https://doi.org/10.1007/s11214-020-00655-0

Perspectives on Atmospheric Evolution from Noble Gas and Nitrogen Isotopes on Earth, Mars & Venus

Guillaume Avice[1] · Bernard Marty[2]

Received: 7 August 2019 / Accepted: 5 March 2020 / Published online: 23 March 2020

Abstract The composition of an atmosphere has integrated the geological history of the entire planetary body. However, the long-term evolutions of the atmospheres of the terrestrial planets are not well documented. For Earth, there were until recently only few direct records of atmosphere's composition in the distant past, and insights came mainly from geochemical or physical proxies and/or from atmospheric models pushed back in time. Here we review innovative approaches on new terrestrial samples that led to the determination of the elemental and isotopic compositions of key geochemical tracers, namely noble gases and nitrogen. Such approaches allowed one to investigate the atmosphere's evolution through geological period of time, and to set stringent constraints on the past atmospheric pressure and on the salinity of the Archean oceans. For Mars, we review the current state of knowledge obtained from analyses of Martian meteorites, and from the direct measurements of the composition of the present-day atmosphere by rovers and spacecrafts. Based on these measurements, we explore divergent models of the Martian and Terrestrial atmospheric evolutions. For Venus, only little is known, evidencing the critical need for dedicated missions.

Keywords Atmosphere · Noble gases · Nitrogen · Geochemistry

1 Introduction

Planetary atmospheres have accumulated some of their volatile elements either by early impact delivery or by degassing of the planetary body over long periods of time. They also

Reading Terrestrial Planet Evolution in Isotopes and Element Measurements
Edited by Helmut Lammer, Bernard Marty, Aubrey L. Zerkle, Michel Blanc, Hugh O'Neill and Thorsten Kleine

✉ G. Avice
avice@ipgp.fr

1 Université de Paris, Institut de physique du globe de Paris, CNRS, 75005 Paris, France

2 Centre de Recherches Pétrographiques et Géochimiques, UMR 7358 CNRS & Université de Lorraine, 15 Rue Notre Dame des Pauvres, 54501 Vandoeuvre-les-Nancy, France

have lost some of these elements by atmospheric escape or subduction into the silicate portions of the planet. For these reasons, they constitute an integrated archive of the evolution of the entire planet. Studying the relative elemental ratios and isotopic compositions of key tracers like noble gases and nitrogen provides crucial information on the origin and evolution of planetary atmospheres. Here, we review results from recent studies on the elemental and isotopic composition of volatile elements in ancient rocks and their implications for the evolution of the Earth's atmosphere (Sect. 1). Some of these results are then compared with the case of Mars (Sect. 2). Finally, the state of (lack of) knowledge for the Venus atmosphere is briefly reviewed (Sect. 3).

2 Earth's Atmosphere

2.1 Noble Gases

Noble gases are excellent tracers of the geodynamics of a planetary system (see reviews by Moreira 2013; Ozima and Podosek 2002; Porcelli et al. 2002)). Due to their inertness, their low abundances and the production of some of their isotopes by radioactive decay of extinct (e.g. ^{129}I, $t_{1/2} = 15.7$ Myr) and extant (e.g. ^{40}K, $t_{1/2} = 1.25$ Gyr) radionuclides, their measurement can track physical processes such as volatile delivery during Earth's accretion, magmatic degassing, atmospheric escape etc.

2.1.1 The Long-Standing Xenon Paradox

Xenon (Xe), the least abundant stable noble gas, has nine stable isotopes which relative proportions reflect the origin of this element as well as processes that changed its abundance through time. Atmospheric xenon on Earth and Mars presents two intriguing features. Firstly, its abundance is lower than what would be predicted if noble gases in planetary atmospheres were following the same abundance pattern as in chondritic meteorites (Fig. 1a). In other words, the $^{132}Xe/^{84}Kr$ ratio is one order of magnitude lower than that of chondritic gases (Pepin 1991). This depletion suggests that xenon is "missing" in the atmospheres of Earth and Mars. The only data for Xe in the Venus atmosphere gives an upper limit of 10 ppb (Wieler (2002) and refs. therein) and does not allow one to evaluate if xenon is also missing in the atmosphere of this planet. Secondly, the isotopic spectra of Earth and Mars atmospheric xenon show enrichments in heavy isotopes relative to the light ones compared to other cosmochemical Xe components (Fig. 1b) such as Xe in the Q phase, the major carrier of noble gases in chondrites (Busemann et al. 2000), Xe in the solar gas (SW-Xe) from which the solar wind (SW) originates (Meshik et al. 2014) or U-Xe the progenitor of atmospheric xenon (Pepin 1991). The strong mass-dependent fractionation of Xe isotopes (3-4% per u, i.e., atomic mass unit) is the second feature of the so-called "xenon paradox": atmospheric xenon, the heaviest noble gas, and thus the least prone among noble gases to escape by mass-dependent thermal processes, is depleted in abundance and isotopically fractionated in favor of heavy isotopes in both Earth and Mars atmospheres.

Previous theoretical studies (Pepin 1991; Shcheka and Keppler 2012; Tolstikhin and O'Nions 1996) explained this paradox by early processing of noble gases during Earth's forming events (see also the review by Dauphas and Morbidelli (2014)). Overall, these models assume separation of noble gases by magmatic/petrological processes, storage of these but xenon in solid Earth reservoirs (e.g., core, mantle or magma ocean), escape of Xe from the early atmosphere and fractionation of its isotopes, and subsequent degassing of the other

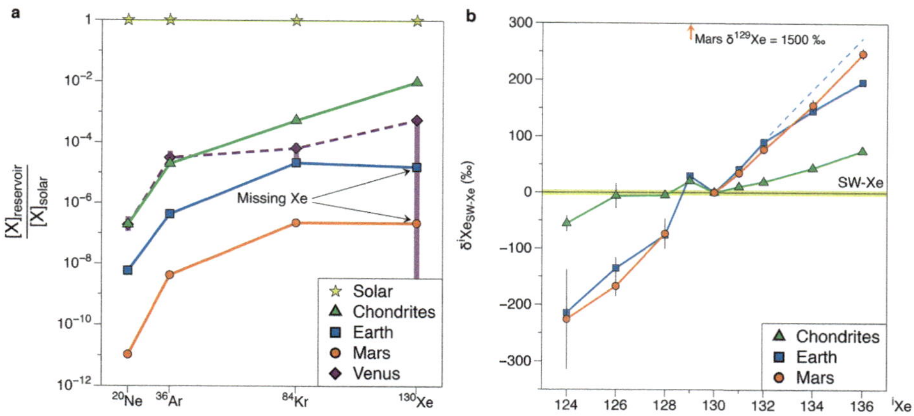

Fig. 1 Illustration of the xenon paradox. (**a**) Abundances of noble gases in planetary atmospheres and in chondritic meteorites normalized to the solar composition. Modified from Dauphas and Morbidelli (2014). Data for Venus are from Wieler (2002). (**b**) Isotopic composition of Xe in chondrites (Busemann et al. 2000) and Earth (Ozima and Podosek 2002) and Mars (Mathew et al. 1998) atmospheres. Isotopic compositions are given using the delta notation normalized to ^{130}Xe and to the isotopic composition of Solar Wind Xe (SW-Xe, $\delta^i Xe_{SW\text{-}Xe} = 0‰$): $\delta^i Xe_{SW\text{-}Xe} = 1000 \times (\frac{(^i Xe/^{130}Xe)_{component}}{(^i Xe/^{130}Xe)_{SW\text{-}Xe}} - 1)$. The dashed line illustrates the underabundance of ^{134}Xe and ^{136}Xe relative to mass-dependent fractionation of SW-Xe

noble gases. These models were somehow *ad hoc* and require a suite of complex, fine-tuned events. Recent experimental studies propose that missing Xe might be stored inside silicate structures in the deep Earth (Crépisson et al. 2018, 2019; Sanloup 2005). However, a selective retention of Xe inside the Earth could only explain the elemental depletion of atmospheric Xe, but not the large isotopic fractionation relative to cosmochemical end-members. Additionally, storage of atmospheric Xe, which is ultimately derived from U-Xe (see next paragraph), in the deep Earth is not consistent with a chondritic origin of mantle Xe (Caracausi et al. 2016; Péron and Moreira 2018).

There is an additional remarkable feature of atmospheric Xe aside of its elemental depletion and isotopic fractionation. Isotopic fractionation of atmospheric Xe follows a mass-dependent law in which the extent of fractionation between two isotopes relates to the mass difference between these two isotopes. This law applies to almost all Xe isotopes (except for a mono-isotopic excess of ^{129}Xe due to a nuclear effect, the decay of extinct ^{129}I). However, the two heaviest Xe isotopes, ^{134}Xe and ^{136}Xe, depart from this law and are under-abundant with respect to other Xe isotopes (Pepin 1991). This Xe component has been nicknamed U-Xe and has resisted decades of investigation (Pepin 1994). Recently, the analysis of noble gases released by Comet 67P/Churyumov-Gerasimenko by the ROSINA instrument onboard of the Rosetta spacecraft (Balsiger et al. 2015; Marty et al. 2017; Rubin et al. 2018) revealed that cometary Xe is depleted in these two ^{134}Xe and ^{136}Xe isotopes, corresponding to a depletion in xenon nuclides created by the r-process of stellar nucleosynthesis. This feature, shared with U-Xe, suggests the persistence of large-scale nucleosynthetic heterogeneities across the Solar System (Avice et al. 2020). Mass balance calculations show that the addition of ~20% cometary Xe to meteoritic Xe accounts for the peculiar composition of U-Xe for $^{128,130\text{-}136}$Xe isotopes (Marty et al. 2017). The interior of the Earth contains meteoritic Kr and Xe (Holland et al. 2009; Péron and Moreira 2018), implying that the addition of cometary xenon to the Earth's atmosphere took place when the terrestrial mantle, or at least some of its regions, were already isolated from contribution of atmospheric volatiles. Thus,

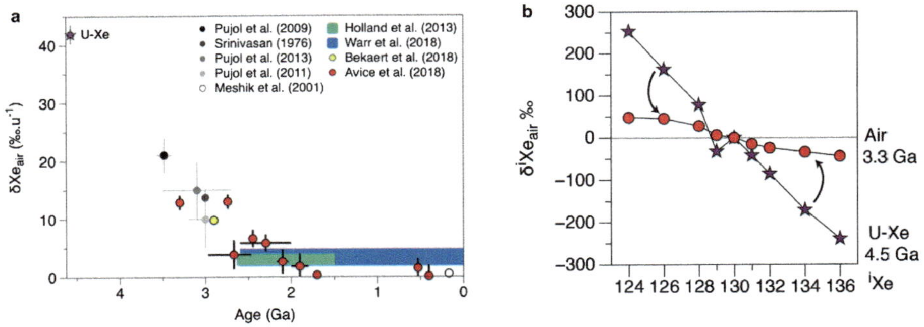

Fig. 2 Evolution of atmospheric Xe. (**a**) Evolution of the isotopic composition of atmospheric Xe with time. Most data are from analyses of gases trapped in mineral samples (Avice et al. 2017, 2018a; Bekaert et al. 2018; Holland et al. 2013; Meshik et al. 2001; Pujol et al. 2013, 2009; Srinivasan 1976). References to Holland et al. (2013) and Warr et al. 2018) correspond to measurements of gases contained in ancient fluids collected underground. Error bars correspond to 1σ. (**b**) Isotopic composition of atmospheric Xe at 4.5 Ga (U-Xe) and 3.3 Ga (Barberton, Avice et al. 2017)). Isotopic compositions are given using the delta notation normalized to ^{130}Xe and to the isotopic composition of Earth atmospheric Xe ($\delta^i \mathrm{Xe_{air}} = 0‰$): $\delta^i \mathrm{Xe_{air}} = 1000 \times (\frac{(^i\mathrm{Xe}/^{130}\mathrm{Xe})_{\mathrm{sample}}}{(^i\mathrm{Xe}/^{130}\mathrm{Xe})_{\mathrm{air}}} - 1)$. Curved arrows depict the progressive evolution of the isotopic composition of atmospheric Xe

cometary contributions appear to have been late with respect to the main phases of terrestrial accretion. In summary, xenon in the terrestrial atmosphere originates from a peculiar component, U-Xe, that may consist in a mixture of cometary and meteoritic gases. This mixture is likely to have occurred during "late" contribution of material from the outer solar system, by the end of the Earth's building stage (O'Brien et al. 2018). Once formed, this component has been isotopically fractionated by losing light Xe isotopes with respect to heavy ones. The analysis of ancient atmospheric xenon, reported in the next section gives clues about this process.

2.1.2 Evolution of the Isotopic Composition of Earth's Atmospheric Xe: Data and Models

Recent measurements of paleo-atmospheric gases trapped in fluid inclusions, in lattices of ancient minerals, in carbon-rich lithologies or in free fluids collected underground reveal that the isotopic composition of atmospheric xenon evolved through time (Fig. 2a). Analyses of Archean samples show isotopic compositions of past-atmospheric Xe intermediate between U-Xe and modern Xe (Fig. 2b), demonstrating that Xe isotopic fractionation did not take place during the Earth's building stages but was progressive through time (Avice et al. 2017, 2018a; Bekaert et al. 2018; Bekaert et al. 2020; Pujol et al. 2011). This evolution corresponds to a progressive enrichment of atmospheric Xe in heavy isotopes relative to the light ones (Fig. 2b), presumably initiating from U-Xe at 4.5 Ga and reaching the modern value around 2 Ga ago. In contrast, there is no evidence for a change during the last 3.5 Gyr of the isotopic compositions of neon, argon (for its non-radiogenic isotopes) and krypton (Avice et al. 2018a, 2017 and refs. therein).

Even if the temporal resolution of the dataset remains low due to the difficulties to date paleo-atmospheric signals (Avice et al. 2017; Pujol et al. 2013; Warr et al. 2018; Bekaert et al. 2020), some general features of the evolution of Xe are observed. Firstly, the evolution of the isotopic composition of atmospheric Xe is a long-term process covering several

Gyr. At 3.5 Ga, only half of the isotopic fractionation was achieved. This suggests that the isotopic composition of atmospheric Xe cannot be solely explained by early episodes of vigorous hydrodynamic escape (Pepin 1991; Tolstikhin and O'Nions 1996). Secondly, the evolution was progressive without retrograde evolution. These observations are consistent with a transfer of atmospheric xenon, concomitant with its isotopic fractionation, from the atmosphere to another reservoir. Because mantle Xe appears to be a mixture of modern Xe and meteoritic Xe, it cannot constitute the complementary reservoir, and the most likely solution is a time-dependent escape of atmospheric xenon from the atmosphere to space. Such escape had to be specific to Xe. Since it did not affect the other noble gases, it cannot have resulted from standard thermal (mass-dependent) escape processes. This supports a process where part of Xe escaped from the atmosphere and left the remaining fraction isotopically fractionated. Some pause in the isotopic fractionation of atmospheric Xe might have existed (see for example data of Avice et al. (2018a) between 3.3 and 2.7 Ga) but remains to be confirmed (Bekaert et al. 2018). Finally, the evolution appears to have ended between 2.5 and 2.0 Ga. Even if the exact timing could be determined by future studies with higher temporal and isotopic resolutions, this temporal sequence overlaps with a major change in the chemistry of the Earth's atmosphere, the Great Oxidation Event. The facts that (i) the modern isotopic composition was attained only ~2 Ga ago and (ii) mantle-derived Xe does not show evidence of an Archean fractionated Xe component in the mantle demonstrate that Xe was not efficiently subducted and retained in the Earth's mantle before 2.5 Ga (Avice et al. 2017; Parai and Mukhopadhyay 2018; Péron and Moreira 2018).

2.1.3 Escape Models and Research Avenues

Geochemical data obtained on paleo-atmospheric samples show a progressive evolution of the isotopic composition of atmospheric Xe. However, until recently, the physical mechanism behind this trend remained elusive. Some studies suggested a selective long-term escape and isotopic fractionation of atmospheric Xe. A model was first proposed by Hébrard and Marty (2014) for which, during the Archean, a fraction of atmospheric Xe was ionized and trapped in organic haze in the Archean atmosphere. Xenon was preferentially ionized and trapped in organic haze relative to other noble gases (e.g., Kr), for two reasons. First, the ionization potential of xenon (12.13 eV) is the lowest among noble gases (>14 eV) and other atmospheric species, e.g., CO_2 (13.8 eV), CO (14 eV), and N_2 (15.6 eV). Second, the 1D model for the Archean atmosphere developed by these authors showed that the atmospheric height at which haze production from CH_4 took place coincided with that of maximum Xe ionization. Trapping of xenon ions in organics results in isotopic fractionation of Xe in favor of the heavy isotopes in the trapped phase relative to the starting gas (Kuga et al. 2017; Marrocchi et al. 2011), with a fractionation factor around 1%/u. The escape of the non-trapped fraction and the subsequent release of the trapped and isotopically fractionated Xe would lead to progressive depletion and isotopic fractionation of atmospheric Xe. One caveat of this model was the absence of a physical explanation for Xe escape. More recently Zahnle et al. (2019) proposed a physical framework for selective Xe escape and isotopic fractionation. The basis of this model is that Xe ions are dragged by escaping hydrogen (H) ions from the Archean atmosphere. Because of the low ionization potential of this element, Xe ions are thus likely to exist in the presence of hydrogen ions and can persist in the atmosphere due to their low recombination coefficient with oxygen, especially during the absence of oxygen in the atmosphere before the G.O.E. Coulomb interactions between Xe and H ions permitted the escaping H^+ to lift Xe^+. Even if a magnetic field was probably present and protected the atmosphere (Tarduno et al. 2014), Xe ions could have still escaped along open magnetic field lines at the magnetic poles. Hydrogen escape could have been operant on Earth until 2.0 Ga as a driving mechanism for the progressive oxidation of the Earth's surface reservoirs (Zahnle et al. 2013). If the model of Zahnle et al. (2019) is correct, the evolution of the isotopic composition of atmospheric Xe would thus be an indirect but powerful tracer of the evolution of the composition of the Earth's atmosphere. See also Catling and Zahnle (2020) for a review on the Archean atmosphere and on the potential meanings of the isotopic evolution of atmospheric xenon. The model of Zahnle et al. (2019), through escape of Xe ions together with hydrogen ions, requires a source of hydrogen. For Xe to escape, the total H_2 mixing ratio had to be at least 1%. If hydrogen was only in H_2 molecules, this value seems to be an upper limit for the Archean atmosphere (Kadoya and Catling 2019). If hydrogen originated from the photo-dissociation of H_2O vapor, its escape could have resulted in change on the volume and composition of the hydrosphere. Zahnle et al. (2019) suggested that a volume of water corresponding to about 30% of the present oceans could have escaped although this amount would be overestimated if Xe escape happened in short and intense episodes. Alternatively, hydrogen could have originated from CH_4 dissociation (Catling et al. 2001). Hydrogen escape has the potential to increase the D/H ratio of hydrogen remaining in the atmosphere. This enrichment would be of the order of 25% (Zahnle et al. 1990, 2019), which is of the same order of magnitude as the difference between bulk Earth's D/H and some D/H values for chondritic meteorites (Marty et al. 2016).

2.2 Other Noble Gases

There is no other evidence for a secular evolution of other noble gases except for a gradual increase of the abundance of ^{40}Ar in the Earth's atmosphere (Cadogan 1977; Pujol et al. 2013; Stuart et al. 2016). ^{40}Ar is continuously produced in Earth's silicate reservoirs by radioactive decay of ^{40}K ($T_{1/2} = 1.25$ Gyr) and, as an atmophile element, subsequently degassed into the atmosphere. Potassium is an incompatible element and concentrates in the continental crust produced by differentiation of mafic lithologies yielding felsic compositions. This behavior implies a link between the abundance of ^{40}Ar in the atmosphere and the rate of growth of continental crust (e.g. Allègre et al. 1987). Paleo-atmospheric data on the ^{40}Ar/^{36}Ar ratios coupled with box models have been used to put constraints on the volume of continental crust with time (Cadogan 1977; Pujol et al. 2013; Stuart et al. 2016). Overall, around 75% of the volume of the continental crust was already formed at 2.5 Ga. The rate of crustal growth was lower for the last 2.5 Gyr. These results are in broad agreement with other recent crustal growth models based on hafnium isotopes in zircons (e.g. Dhuime et al. 2012).

2.3 Paleo-Atmospheric Dinitrogen

The N_2 molecule, because of its triple covalent bond, is particularly stable and N_2 is sometimes called the "6th noble gas" due to its relative inertness. Under oxidizing conditions, with an oxygen fugacity higher than the iron-wüstite buffer, N_2 is incompatible during partial melting (Mikhail and Sverjensky 2014). It readily degasses from basaltic magmas due to its low solubility, comparable to that of argon (Libourel et al. 2003). Today, N_2 is the most abundant molecule in the Earth's atmosphere, and was certainly an important player in the distant past (e.g. Catling and Zahnle 2020). Nitrogen is also a key chemical element in biomolecules (DNA, RNA, proteins) and its bio-mediated speciation allows exchanges between the biosphere and the geosphere. Atmospheric dinitrogen is reduced to NH_3 by bacteria carrying the nitrogenase enzyme and nitrogen is further processed to become incorporated into organic compounds. Organic N is converted to ammonium by bacteria or fungi. NH_4^+ ions can substitute to K^+ in soils and sediments. This is one major pathway to incorporate nitrogen into the rock cycle. Sedimentary N can go back in the atmosphere as N_2 during metamorphism induced by continental collision and in subduction zones. Oxidation of reduced N can also take place when specific bacteria convert ammonia to nitrites or nitrates. Nitrates are dissolved in groundwater or oceanic water, where denitrification can take place converting nitrates back to N_2. Several other reactions such as anamox, direct conversion of nitrite and ammonia to N_2 permits the liberation of reduced and oxidized N back to the atmosphere. Further details about the global bio-geochemical cycle of nitrogen are given by Stüeken et al. (2020).

Overall, nitrogen is exchanged between the interior of the Earth and its surface as the result of mantle degassing at ridges and plumes, the development of the continental crust storing ammonium, and subduction permitting the return of nitrogen trapped in oceanic crust back to the mantle. The cycle of nitrogen probably varied a lot throughout the course of Earth's history due to major changes such as the colonization of the oceans by bacteria, the onset and production of continental crust, the onset of subduction, the Great Oxidation Event etc. Current views on the evolution of the nitrogen cycle are debated (e.g. Johnson and Goldblatt 2018; Zerkle 2018) but a record of these variations might have been kept in ancient sedimentary rocks.

Atmospheric N_2 has also the potential to have recorded past atmospheric escape processes (Lichtenegger et al. 2010). Contrary to the case of Mars (McElroy et al. 1976) dinitrogen is presently not escaping from the terrestrial atmosphere. The terrestrial exobase (the height at which molecules do not interact each other and can escape to space if they have sufficient velocity) extends at altitudes a few hundreds of km, whereas the magnetosphere expands up to several tens of thousands km. Hence, atmospheric species are shielded from interaction with extraterrestrial charged particles like those of the solar wind, which is a major process to escape nitrogen in the case of Mars. In the distant past, the Earth's atmosphere might not have been well shielded when the terrestrial magnetic field was weaker. Note that the shielding efficiency of planetary magnetic fields against atmospheric escape remains debated (e.g. Gunell et al. 2018). The altitude of the exobase also strongly depends on the composition of the atmosphere (Tian et al. 2008; Johnstone et al. 2018), which is likely to have varied. The flux of charged particles from space might also have been much higher, than today, like during exceptional solar flare events which are observed on stars younger than the Sun. Without a magnetic field, the ancient atmosphere might have lost N_2 in a few million years, unless some CO_2 in excess of the present concentration by at least two orders of magnitude would have been present to allow cooling and shrinking of the thermosphere (Lichtenegger et al. 2010).

On Mars, atmospheric nitrogen is enriched in its rare isotope ^{15}N relative to abundant ^{14}N by 60% compared to the Martian mantle, meteoritic, and terrestrial (atmosphere, mantle) nitrogen (McElroy et al. 1976). Although ^{15}N-rich cometary contributions cannot be dismissed (Marty et al. 2016) such enrichment is generally attributed to isotope fractionation during non-thermal (charge dependent) escape to space (McElroy et al. 1976). On Earth, atmospheric nitrogen is enriched in the heavy isotope ^{15}N by $\sim\geq 0.5\%$, possibly up to 3%, compared to mantle N, an enrichment that could have resulted from past nitrogen escape processes (Tolstikhin and Marty 1998). From a modeling point of view, it is difficult to decipher if atmospheric escape could have affected nitrogen in the distant past, because the interplay between potentially responsible processes such as a reduced magnetosphere, and/or an expanded exobase, are model-dependent (Lichtenegger et al. 2010) and not constrained by external observations. For these reasons, it is utterly important to determine the partial pressure and the isotope composition of atmospheric nitrogen in the distant past. Several approaches have been attempted, which converge towards coherent results. They are described in the following paragraphs.

Constraints on the total paleo-atmospheric pressure have been obtained from the morphological analysis of fossil raindrops imprinted in the 2.7 Ga volcanic tuffs of the Ventersdorp Supergroup, South Africa (Som et al. 2012). The rationale is that the maximum size a falling raindrop can reach before fragmentation is a direct function of the ambient atmospheric pressure. Assuming that the studied rocks were emplaced at the sea level, the distribution of fossil raindrops, together with analogic experiments aimed at reproducing this distribution in the laboratory allowed Som et al. (2012) to propose that the total atmospheric pressure could not have been larger than two times the modern atmospheric pressure, and more probably comparable to, or lower than, the latter. However, the approach by Som et al. (2012) was subsequently criticized by Kavanagh and Goldblatt (2015) who claimed that the pressure determined by the "fossil raindrop method" is likely to be uncertain within one order of magnitude. This criticism was later refuted by Goosmann et al. (2018). More recently, Som et al. (2016) attempted to estimate the absolute Archean barometric pressure using the size distribution of gas bubbles in lava flows that solidified at sea level 2.7 Ga ago in the Pilbara craton, Australia. They proposed that the atmospheric pressure was low, about 0.23 bar.

A different approach was proposed by Marty et al. (2013) and Avice et al. (2018a) (see also Nishizawa et al. (2007)). They analyzed the elemental and isotopic compositions of noble gases and nitrogen trapped in fluid inclusions in hydrothermal quartz by vacuum crushing and static mass spectrometry. The samples were from 3.5 Ga-old Apex and Dresser formations, Pilbara Craton, Australia, and from the meso-Archean Barberton belt, South Africa. Ar-Ar dating constrained ages of the trapped fluids to be $\geq$3.0 Ga (Marty et al. 2013) and 3.3 Ga (Avice et al. 2018a, 2017), respectively. By doing sequential crushing, these authors found that data define correlations in a $N_2/^{36}Ar$ vs. $^{40}Ar/^{36}Ar$ space interpreted as a mixing between a hydrothermal end-member rich in crustal N and radiogenic ^{40}Ar, and a low $^{40}Ar/^{36}Ar$, low $N_2/^{36}Ar$ end-member. The latter had $^{40}Ar/^{36}Ar$ ratios comparable to that of the Archean atmosphere, and was considered to represent atmospheric noble gases dissolved in Archean seawater (Fig. 3). The $N_2/^{36}Ar$ ratio of this end-member corresponds to dissolved atmospheric N_2 and ^{36}Ar. Measurements of different samples led to distinct correlations with variable slopes reflecting different compositions for the hydrothermal end member. However, the correlations converged toward a common low $^{40}Ar/^{36}Ar$, $N_2/^{36}Ar$ composition. Atmospheric N_2 and Ar dissolve in water with known solubilities. The dissolved $N_2/^{36}Ar$ ratio is thus directly representative of the atmospheric ratio. One assumption is that the abundance of atmospheric ^{36}Ar did not vary with time, which is comforted by the fact that the atmospheric $^{36}Ar/^{38}Ar$ ratio is similar to that of primitive meteorites (Ozima and Podosek 2002); and by the fact that there is no known process allowing the escape of neutral Ar atoms. The partial pressure of N_2 can thus be deduced from the $N_2/^{36}Ar$ ratio of the Archean water end-member. Marty et al. (2013) proposed an Archean P_{N_2} in the range of 0.5-1 bar at $\geq$3 Ga, and Avice et al. (2018a) confirmed that at 3.3 Ga, the Archean P_{N_2} was lower than the modern one.

Altogether, these studies suggest that the partial pressure of paleo-atmospheric N_2 was comparable to, or lower than, the modern one (see also a compilation provided by Silverman et al. (2018)). These results do not support the suggestion by Goldblatt et al. (2009) that the faint young Sun paradox (see Chaps. 1 & 9) could be resolved if the paleo-atmospheric P_{N_2} was 2-3 times more than the modern one leading to an enhancement of the greenhouse effect by pressure broadening of N_2. Similarly, warming induced by H_2-N_2 collision-induced

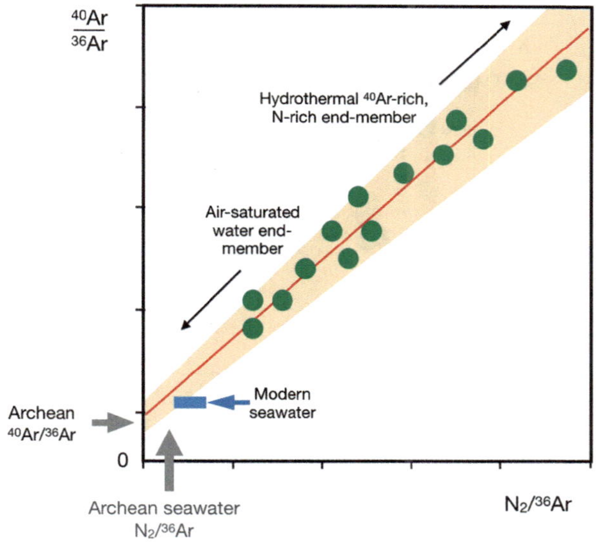

Fig. 3 Schematic illustration of the principle of the geochemical determination of the partial pressure of N_2 of the Archean atmosphere. The samples are Archean hydrothermal quartz hosting fluid inclusions. Ages of the fluids are constrained with the Ar-Ar method. Sequential heating and crushing under static high vacuum, followed by mass spectrometry analysis of Ar and N isotopes and abundances permit to obtain a series of data points for a given samples (green dots). Data define a linear relationship in a $^{40}Ar/^{36}Ar$ vs. $N_2/^{36}Ar$ diagram regarded as representing fluid mixing between a crustal component rich in radiogenic ^{40}Ar and in nitrogen, and a low $^{40}Ar/^{36}Ar$, $N_2/^{36}Ar$ component representing Archean sea- or fresh-water component, which has the Ar isotopic composition of dissolved Archean air, and a $N_2/^{36}Ar$ ratio representative of Archean air, after correction for Bunsen (solution) coefficients for both gases. As the abundance of ^{36}Ar is likely to have been conservative since at least the meso-Archean, the $N_2/^{36}Ar$ ratio of the Archean water can be compared directly to the modern seawater ratio. Values lower than the latter indicate P_{N_2} lower than today. Correlations (Marty et al. 2013; Avice et al. 2018a, 2017) indicate that the Archean P_{N_2} was equal to, and probably lower than, the modern P_{N_2}, thus dismissing a P_{N_2} three times higher than the modern one proposed to counterbalance a fainter Sun than today (Goldblatt et al. 2009)

absorption (Wordsworth and Pierrehumbert 2013) seems excluded since it requires a paleo-atmospheric P_{N_2} three times higher than the modern one. Finally, a relatively low P_{N_2} in the past supports the view that, in the deep past, a nitrogen-dominated atmosphere could not have been stable against thermal escape fueled by high EUV fluxes from the young Sun (Johnstone et al. 2019).

The nitrogen isotope composition was also investigated in the studies described above (Marty et al. 2013; Avice et al. 2018a) and also in the study by Pinti et al. (2001). The $\delta^{15}N$ variations (where $\delta^{15}N = [(^{15}N/^{14}N_{sample}/^{15}N/^{14}N_{atm}) - 1] \times 1000$, in permil; ‰) can also be accounted for by mixing between a paleoseawater component having a N isotope composition similar to that of modern seawater and a crustal component having values in the range of modern crust and sediments (Fig. 4). Thus, the N isotope composition of Archean atmosphere seems to have been similar within a few permil to that of the modern atmosphere.

Taken together, these different studies converge towards an atmospheric composition still dominated by N_2. The Archean P_{N_2} was at best equal to, and probably lower than, the modern total and N partial pressures. A low Archean atmospheric N_2 is unlikely to be the result of atmospheric N escape, since the N isotope composition did not vary. These results are somewhat unexpected for a period of time when subduction and the possibility of N recy-

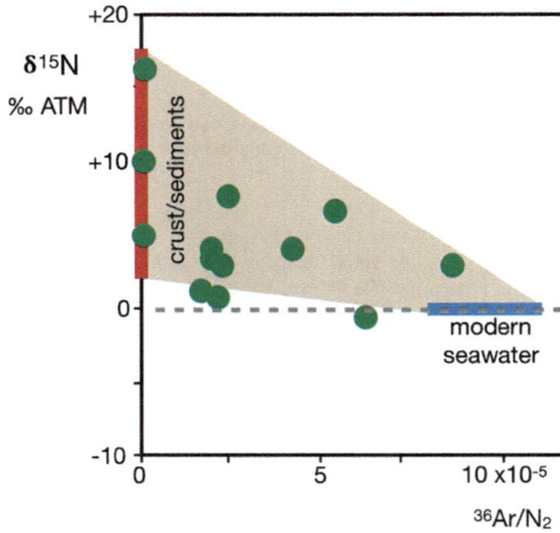

Fig. 4 Principle of the determination of the nitrogen isotope composition of Archean atmospheric N_2 ($\delta^{15}N$ is the $^{15}N/^{14}N$ ratio of the sample normalized to the isotopic ratio of nitrogen in the modern atmosphere, in parts per mil—‰). In this space, mixing between two component results in a linear relationship. Here the data (green dots) obtained from the sequential analysis of fluid inclusions trapped in Archean hydrothermal quartz rather define an area corresponding to mixing between modern seawater and the range of values typical of oceanic and continental crust as well as sediments. The Archean seawater end-member appears to have had a N isotope composition not drastically different from the modern one, within a few permil (modified from Marty et al. 2013, Avice et al. 2018a and Pinti et al. 2001)

cling to the mantle might not have been operative. The Archean atmospheric N "depletion" might have been the result of nitrogen sequestration in organic matter and in the crust at a time when environment reducing conditions did not permit fixed nitrogen to be returned to the atmosphere by denitrification and oxidative weathering (Stüeken et al. 2016, 2020). Combining estimates of the total atmospheric pressure with those of the P_{N_2} suggests that there might have been some significant amount of other atmospheric species up to 0.7 bar when uncertainties are taken into account, which leaves space to have enough greenhouse gas like CO_2 to counterbalance a low energy delivery by a fainter Sun at that time (Catling and Zahnle (2020) and refs. therein). The most extreme terrestrial N isotope values are about $\delta^{15}N = -40‰$ as found in some diamonds, and the convective mantle is characterized by $\delta^{15}N$ values around $-5‰$ (Cartigny and Marty 2013). The maximum isotope fractionation of nitrogen in the terrestrial system was thus at most 40‰, and since 3.5 Ga it may have not varied by more than a few permil. Contrary to the case of xenon, these results suggest low nitrogen atmospheric escape since three billion years, attesting for efficient atmospheric shielding against interactions with charged particles in the upper atmosphere.

Interestingly, the argon correlation approach used to determine the Archean P_{N_2} permitted also to constrain the salinity of the Archean oceans (Marty et al. 2018). Some of the above samples were neutron-irradiated, which synthetized Ar isotopes from Cl and K isotopes dissolved in fluid inclusions. Two end-member Cl-K-Ar isotope correlations were obtained, regarded as representing mixing between an Archean seawater component and a hydrothermal fluid end-member. Different hydrothermal end-members were identified yielding correlations with different slopes which converged towards a unique seawater end-member (Marty et al. 2018). An example of such correlation is given in (Fig. 5). The seawater end-member appears to have a $Cl/^{36}Ar$-$K/^{36}Ar$ composition comparable to that of the modern oceans. Assuming again that ^{36}Ar has been conservative in the atmosphere since 3.5 Ga, it

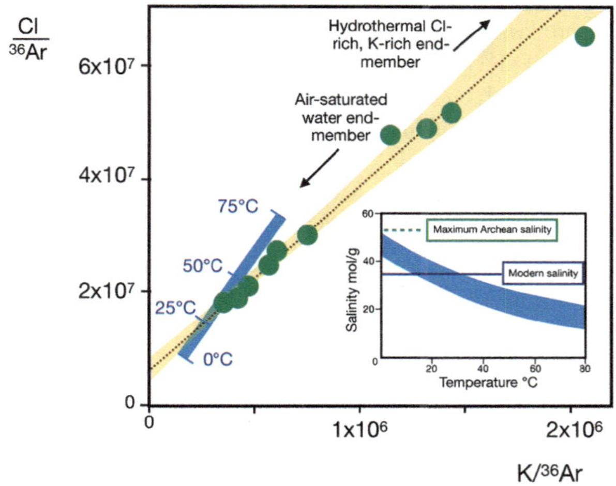

Fig. 5 $Cl/^{36}Ar$ vs. $K/^{36}Ar$ data (green dots) for stepwise crushing of 3.4 Ga-old Barberton sample PI02-39-2 (adapted from Marty et al. (2018)). Data define a correlation representing mixing between an old seawater component and a hydrothermal end-member rich in crustal chlorine and potassium. Error bars are not shown here but are given in the original work. The yellow area represents the 95% confidence interval of the error-weighted correlation. The blue bar represents the composition of modern seawater for different temperatures between 0 °C and 75 °C. The correlation intercepts the modern composition for temperatures between 25 °C and 50 °C, suggesting an oceanic $Cl/^{36}Ar$ ratio similar to the modern value and, for constant atmospheric ^{36}Ar, a salinity (Cl) of the Archean oceans similar to the modern one. The insert shows the link between salinity and temperature (Marty et al. 2018). The constraint for the maximum Archean salinity is a temperature of water above the freezing point

follows that the salinity (Cl) of the oceans did not vary significantly through time. In detail, the abundance of potassium in the oceans might have been ~30% lower than today, in possible relation with a lower volume of continental crust 3.5 Ga ago. Interestingly, the equilibrium temperature for atmospheric gas dissolution appears to have been in the range 25-50 °C, suggesting that the temperature of the oceans might have been higher than today, as argued (and disputed) by studies reporting oxygen/silicon isotope measurements on cherts and carbonaceous matter (e.g. Knauth 2005; Robert and Chaussidon 2006; Tartèse et al. 2016)). Alternatively, such temperature range could reflect local surface conditions where atmospheric gases were in equilibrium with surface water, and not global oceans' temperature. The corollary is that the halogen/seawater ratio remained constant within ~20-30% from 3.5 Ga to present, implying that the potential escape of hydrogen from photodissociation of atmospheric H_2O might have been limited. A ~30% water loss is consistent with modeling of the escape of atmospheric xenon which requires approximately this amount of oceanic water to be dissociated and resulting hydrogen to be lost into space in order to lift up Xe ions (Zahnle et al. 2019).

3 Martian Atmosphere

The Martian atmosphere is much less dense (P_{atm} = 600 Pa) than that of the Earth (P_{atm} = 101,325 Pa). The major atmospheric species (CO_2, N_2) do not escape by thermal (mass-dependent) processes, at least at present. However, the isotopic compositions of hydrogen

(D/H) and nitrogen are largely fractionated by several hundreds to thousands per mil, respectively (see Bogard et al. (2001) and refs therein). The two stable isotopes of argon are also isotopically fractionated, but krypton is not, and xenon is as fractionated as terrestrial xenon. Martian volatiles present an intriguing puzzle which is not resolved yet. The diversity and extent of isotope fractionation call for several types of escape processes having acted on the atmosphere of the red planet during extended periods of time. In this part we focus on the isotopic effects which affected xenon and nitrogen.

3.1 The Strange Case of Martian Atmospheric Xenon

If the abundance and the isotopic composition of Earth atmospheric Xe result from a specific escape mechanism, it is interesting to turn ourselves toward Mars and evaluate if such a mechanism was operating on the red planet. The isotopic fractionation of terrestrial Xe is 3.6%/u (computed with the atmospheric $^{128}Xe/^{130}Xe$ ratio) relative to its precursor, U-Xe (Pepin and Porcelli 2002). For Mars the precursor seems to be SW-Xe and not U-Xe, as observed in the Chassigny meteorite thought to have sampled noble gases from the Martian mantle (Ott 1988) on one hand, and from the isotopic composition of Martian atmospheric Xe after correction for mass-dependent fractionation, on another hand. The isotope fractionation of Xe in the Martian atmosphere is 3.3%/u when the Xe isotope composition of the Martian atmospheric gases trapped in Martian meteorites (Swindle 2002) is considered (relative to SW-Xe (Meshik et al. 2014)). It is only 2.4%/u when the direct measurement of the atmosphere by the Mars Sample Laboratory's mass spectrometer (Conrad et al. 2016) is considered. The difference between these two values could be due to potential spallation excesses on light Xe isotopes for the latter, but the debate remains open (Avice et al. 2018b; Conrad et al. 2016). In any case, the extent of fractionation is comparable for Mars and Earth. Having a similar isotopic fractionation for atmospheric Xe on Mars and Earth is striking. Indeed, the escape process proposed by Zahnle et al. (2019) depends on parameters such as the gravity of the body, the ionization rate of Xe in the atmosphere and the Solar irradiation activity, and probably the strength and shape of the magnetic field since Xe escapes as ions. All these parameters have different values on Earth and Mars and some of them have also varied with time. It is thus surprising to end up with a similar isotopic fractionation and depletion factor for Mars and Earth atmospheric Xe. If the extent of isotopic fractionation of atmospheric Xe on Earth and Mars is similar, the timing of events might be different. Recently, Cassata (2017) argued that the >4 Ga ALH84001 and NWA7084 ("Black Beauty") Martian meteorites contain ancient Martian atmospheric gases with little if any Xe isotopic fractionation relative to modern Martian atmospheric Xe. One conclusion of this study is that the isotopic fractionation and escape of Mars atmospheric Xe occurred very early in the planet's history and ceased around 4.4-4.0 Ga (Fig. 6). This observation is in line with studies arguing for an early stage of strong atmospheric escape on Mars starting from the time when a steam atmosphere was outgassed from the magma ocean and lasting until 4 Ga (e.g. Lammer et al. (2013)). It is thus tempting to link the escape of xenon with a major transition in the Martian environment and the disappearance of permanent liquid water on Mars by the end of the Noachian (Bibring et al. 2006). As in the case of terrestrial Xe (Zahnle et al. 2019), hydrogen from dissociation of atmospheric H_2O could have played a role in the Xe escape and isotopic fractionation, but at different periods for Earth and Mars. It remains to be understood why the extent of Xe isotopic fractionations are comparable for both planetary atmospheres.

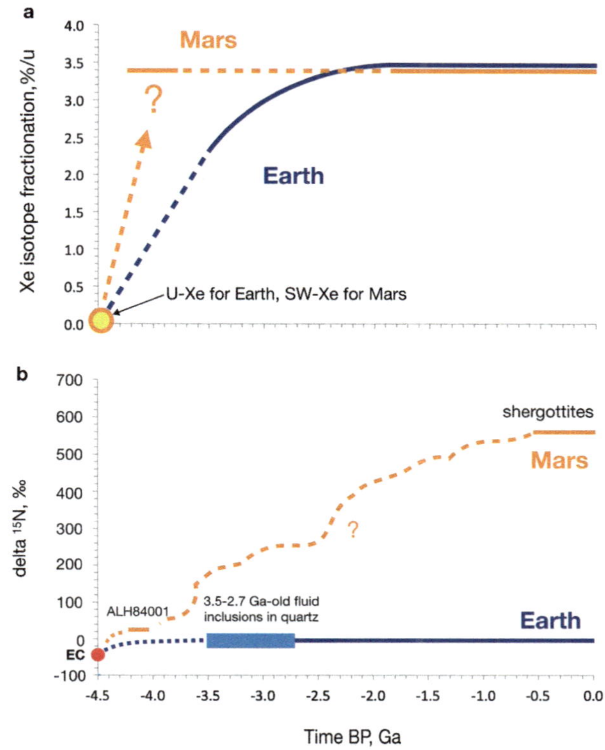

Fig. 6 Comparative evolutions of xenon and nitrogen isotopes as constrained by available data. a: Xe isotope fractionation through time. The terrestrial ancestor of Xe is thought to be U-Xe, a composite composition resulting from mixing cometary and solar, or chondritic Xe (Marty et al. 2017). For Mars, a solar Xe composition has been advocated (Conrad et al. 2016) and this component is identified in mantle-derived Xe (Mathew et al. 1998). At this scale, considering U-Xe or solar Xe does not make a difference. Xenon in >4 Ga Martian meteorites appears to be already isotopically fractionated (Cassata 2017), suggesting that Xe isotope fractionation took place early. On Earth, this has been a continuous process still in progress during the Archean (Avice et al. 2018a, 2017). In the case of nitrogen, the situation appears reverse. The "Old" Martian meteoritic component did not show ^{15}N enrichments (Mathew and Marti 2001) whereas "recent" (<1.3 Ga) Martian meteorites show Xe isotope fractionation similar to that of the modern atmosphere (Mathew et al. 1998). Martian samples (meteorites, returned samples) with intermediate ages would allow investigating more precisely the long-term evolution of the Martian atmosphere

3.2 Martian Atmospheric Nitrogen

The present-day isotopic composition of Martian atmospheric N_2 measured by the Mars Science Laboratory (Curiosity rover) is $\delta^{15}N = 572 \pm 90‰$ (Wong et al. 2013). The initial Martian value could have been $\delta^{15}N = -30‰$ as measured in the Chassigny meteorite (Mathew et al. 1998). Thus, either the Martian atmosphere was contributed by a component rich in ^{15}N such as a comet (see Marty et al. 2016), or N isotopes were fractionated during non-thermal nitrogen escape to space (McElroy et al. 1976). The modern nitrogen isotopic composition is consistent with values measured in Martian meteorites, most of them being younger than 1.3 Ga. However, the 4.1 Ga ALH84001 Martian meteorite contains different N components including a presumed Martian atmospheric end-member, identified with Xe isotopes, which is less rich in ^{15}N. This suggests that the evolution of the N isotope composition by atmospheric escape processes is younger than ~4 Ga. Figure 6 compares the different evolutions of N and Xe isotope fractionation through time as constrained by the present data. Clearly, the history of the Martian atmosphere is far to be unveiled, and more data for various geological ages are needed.

4 A Desperate Lack of Data for Venus

The striking similarities between atmospheric Xe on Earth and Mars call for a new point of comparison and the Venusian atmosphere is an ideal candidate here. Comparative planetology was already successful at deciphering the origin and evolution of Earth and Mars. Venus is often depicted as the "sister" planet of Earth. Yet, Venus took a very different evolutionary path that led to an unhabitable world and reasons for this different history are matter of debate. Contrary to the case of Mars, noble gas data for the Venusian atmosphere are lacking and further investigations are mandatory (Chassefière et al. 2012). The current state of knowledge was already reviewed by Wieler (2002), and data obtained *in-situ* have also been recently reviewed by Johnson and de Oliveira (2019). Only some major points are highlighted below.

The abundances of Ne and Ar in the Venusian atmosphere are higher than those of the Earth by about one order of magnitude. Even if the Ne/Ar ratio is close to chondritic values, these high abundances of Ne and Ar and a ^{20}Ne/^{22}Ne ratio close to 12, higher than the chondritic end-member (about 10.7, Busemann et al. (2000)) suggest a contribution from a solar-derived component to the Venusian atmosphere (Solar Wind Ne implanted on grains (Bogard 1988) or solar neon from the nebula). The Venusian ^{40}Ar/^{36}Ar ratio of 1.11 ± 0.02 is lower than the terrestrial ratio of 298.6 (Lee et al. 2006), and this value has been used to argue for incomplete degassing of Venus of its radiogenic ^{40}Ar produced by the radioactive decay of ^{40}K ($T_{1/2} = 1.25$ Ga) (Kaula 1999). Based on the atmospheric ^{36}Ar abundance, the ^{40}Ar/^{36}Ar ratio, the K/U ratios measured by the Venera landers and assuming a terrestrial-like uranium abundance, Kaula (1999) computed that on Venus only 25% of ^{40}Ar produced in silicate reservoirs is degassed into the atmosphere. This is half the fraction of ^{40}Ar degassed on Earth (Allègre et al. 1987). The abundance of ^{84}Kr in the Venusian atmosphere is 25 (+13/−18) ppb vol. The ^{84}Kr/^{36}Ar elemental ratio suggests a solar-like abundance pattern for heavy noble gases (Fig. 1a). However, there is no data on the isotope composition of Kr to test this hypothesis. The situation is even worse for Xe with only an upper limit of 10 ppb for ^{132}Xe having been determined. The absence of precise data on the abundance of Xe precludes from evaluating if the abundance of Xe is close to solar values or if there is a "missing Xe" also on Venus. The absence of isotope data also prevents one from checking if the precursor of Venus atmospheric Xe is Chondritic, Solar or U-Xe and if this starting isotopic composition has been mass-dependently fractionated similarly to the case of Earth and Mars.

The current abundance of N_2 in the Venus atmosphere is about 3.5% vol., the remaining of the atmosphere being 96.5% vol. of CO_2 (see the review by Fegley (2003) on the composition of the Venusian atmosphere). Interestingly, Venus has a higher abundance of nitrogen than the Earth in terms of column mass (surface pressure divided by gravitational acceleration) with 23.7×10^3 kg m^{-2} of N_2 for Venus and 7.8×103 kg m^{-2} of N_2 for the Earth (Catling and Kasting 2017). This difference suggests that, relatively to their respective masses, Venus could have been endowed with more nitrogen than the Earth. Alternatively, some studies suggest that a large fraction of terrestrial nitrogen also resides in silicate reservoirs of the planet (Cartigny and Marty 2013 and refs. therein). Unfortunately, the isotopic composition of nitrogen in the atmosphere of Venus remains largely unknown with estimates of terrestrial-like value but with an uncertainty of about 20% covering a wide range of reference values for reservoirs of nitrogen in the solar system (Chassefière et al. 2012; Hoffman et al. 1979). This prevents to compare the cycle of nitrogen on both planets, to estimate which cosmochemical reservoirs contributed this element to Venus and to evaluate if the budget of nitrogen has been altered by atmospheric escape processes.

New space mission efforts toward a measurement of the elemental and isotopic composition of noble gases and nitrogen in the Venus atmosphere are required and proposed to space agencies worldwide (cf. Chap. 7). The paucity of noble gases and their inertness prevent remote observations. To our knowledge, three types of investigations could be carried out: i) in situ measurement in the Venusian atmosphere at low velocity, e.g., on board a balloon or a vehicle having a slow descent through the atmosphere: ii) sample collection at high velocity, e.g., by a probe skimming into the Venusian atmosphere followed by measurement in space; iii) an atmosphere's sample return. Collecting a sample below the homopause and measuring in-situ its noble gas composition in space with a compact probe look promising, especially with the development of a new generation of miniature mass spectrometers (Avice et al. 2019).

5 Concluding Remarks

Recent studies allowed one to access the composition of the Earth's atmosphere and of the ancient seawater, up to 3.3 Ga. A major result is that the isotopic composition of atmospheric xenon evolved through time, at least until 2 Ga. This evolution was probably linked to a specific Xe escape process taking place in the early anoxic atmosphere under irradiation, and thus ionization, by the young Sun. Nitrogen-argon systematics demonstrates that the partial pressure of nitrogen in the Archean was similar to, or lower than, the modern value. This implies that a higher abundance of nitrogen in the atmosphere than today cannot explain the faint young Sun paradox. The isotopic composition of atmospheric nitrogen was also close to modern values implying little isotopic evolution since the original delivery of nitrogen to Earth. Additionally, analyses of Archean fluid inclusions point toward oceanic waters having salinities in the same range as today if oceans were warmer than present (Robert and Chaussidon 2006: Tartèse et al. 2016). The atmospheres of Earth and Mars followed distinct paths as evidenced by geochemical studies of Xe and N. The isotope fractionation and potential loss of Xe happened early on Mars, contrary to a protracted evolution for the Earth, but the interplay between the parameters controlling xenon escape and its isotopic evolution remains to be understood. For the Earth, atmospheric nitrogen seems to have suffered little isotopic fractionation while on Mars nitrogen is heavily isotopically fractionated.

The exercise of comparative planetology for the Solar system cannot be completed without considering Venus, the Earth's sister planet. However, too little is known about this planet, especially about the elemental and isotopic composition of noble gases in its atmosphere. Future space missions will certainly unveil some of the remaining mysteries and help to understand the origin and evolution of planetary atmospheres and, ultimately, what makes a planetary body habitable.

Acknowledgements G.A. acknowledges the DIM ACAV+ program (Région Ile-de-France) for its financial support. B.M. acknowledges the European Research Council (grant PHOTONIS 695618 to B.M.).

References

C.J. Allègre, T. Staudacher, P. Sarda, Earth Planet. Sci. Lett. **81**, 127 (1987)
G. Avice, B. Marty, R. Burgess, Nat. Commun. **8**, 15455 (2017)
G. Avice, B. Marty, R. Burgess, A. Hofmann, P. Philippot, K. Zahnle, D. Zakharov, Geochim. Cosmochim. Acta **232**, 82 (2018a)
G. Avice, D.V. Bekaert, H. Chennaoui Aoudjehane, B. Marty, Geochem. Perspect. Lett. **6**, 11 (2018b)
G. Avice, A. Belousov, K.A. Farley, S.M. Madzunkov, J. Simcic, D. Nikolić, M.R. Darrach, C. Sotin, J. Anal. At. Spectrom. **34**, 104 (2019)
G. Avice, M. Moreira, J.D. Gilmour, Astrophys. J. **889**, 68 (2020)
H. Balsiger, K. Altwegg, A. Bar-Nun, J.J. Berthelier, A. Bieler, P. Bochsler, C. Briois, U. Calmonte, M. Combi, J. De Keyser, P. Eberhardt, B. Fiethe, S.A. Fuselier, S. Gasc, T.I. Gombosi, K.C. Hansen, M. Hässig, A. Jackel, E. Kopp, A. Korth, L. Le Roy, U. Mall, B. Marty, O. Mousis, T. Owen, H. Reme, M. Rubin, T. Semon, C.Y. Tzou, J.H. Waite, P. Wurz, Sci. Adv. **1**, 1 (2015)
D.V. Bekaert, M.W. Broadley, F. Delarue, G. Avice, F. Robert, B. Marty, Sci. Adv. **4**, (2018). https://doi.org/10.1126/sciadv.aar2091
D.V. Bekaert, M.W. Broadley, F. Delarue, Z. Druzhinina, G. Paris, F. Robert, K. Sugitani, B. Marty, Precambrian Res. **336**, 105505 (2020)
J.P. Bibring, Y. Langevin, J.F. Mustard, R. Arvidson, A. Gendrin, B. Gondet, F. Forget, the OMEGA team, M. Berthe, J.P. Bibring, A. Gendrin, C. Gomez, D. Jouglet, F. Poulet, A. Soufflot, M. Vincendon, M. Combes, P. Drossart, T. Encrenaz, T. Fouchet, R. Merchiorri, G. Belluci, F. Altieri, V. Formisano, F. Capaccioni, P. Cerroni, A. Coradini, S. Fonti, O. Korablev, V. Kottsov, N. Ignatiev, V. Moroz, D. Titov, L. Zasova, D. Loiseau, N. Mangold, P. Pinet, S. Doute, B. Schmitt, C. Sotin, E. Hauber, H. Hoffmann, R. Jaumann, U. Keller, T. Duxbury, G. Neukum, Science **312**, 400 (2006)
D.D. Bogard, Icarus **74**, 3 (1988)
D.D. Bogard, R.N. Clayton, K. Marti, T. Owen, G. Turner, in *Chronology and Evolution of Mars*, Chronology and Evolution of Mars, Netherlands, (2001), pp. 425–458
H. Busemann, H. Baur, R. Wieler, Meteorit. Planet. Sci. **35**, 949 (2000)
P.H. Cadogan, Nature **268**, 38 (1977)
A. Caracausi, G. Avice, P.G. Burnard, E. Füri, B. Marty, Nature **533**, 82 (2016)
P. Cartigny, B. Marty, Elements **9**, 359 (2013)
W.S. Cassata, Earth Planet. Sci. Lett. **479**, 322 (2017)
D.C. Catling, J.F. Kasting, *Atmospheric Evolution on Inhabited and Lifeless Worlds* (Cambridge University Press, Cambridge, 2017)
D.C. Catling, K.J. Zahnle, Sci. Adv. **6**, eaax1429 (2020)
D.C. Catling, K.J. Zahnle, C.P. McKay, Science **293**, 839 (2001)
E. Chassefière, R. Wieler, B. Marty, F. Leblanc, Planet. Space Sci. **63–64**, 15 (2012)
P.G. Conrad, C.A. Malespin, H.B. Franz, R.O. Pepin, M.G. Trainer, S.P. Schwenzer, S.K. Atreya, C. Freissinet, J.H. Jones, H. Manning, T. Owen, A.A. Pavlov, R.C. Wiens, M.H. Wong, P.R. Mahaffy, Earth Planet. Sci. Lett. **454**, 1 (2016)
C. Crépisson, M. Blanchard, M. Lazzeri, E. Balan, C. Sanloup, Geochim. Cosmochim. Acta **222**, 146 (2018)
C. Crépisson, C. Sanloup, M. Blanchard, J. Hudspeth, K. Glazyrin, F. Capitani, Geochem. Geophys. Geosyst. **20**, 992 (2019)
N. Dauphas, A. Morbidelli, in *Treatise on Geochemistry* 2nd edn. (Elsevier, Oxford, 2014), pp. 1–35
B. Dhuime, C.J. Hawkesworth, P.A. Cawood, C.D. Storey, Science **335**, 1334 (2012)
B. Fegley, in *Treatise on Geochemistry* (Elsevier, Amsterdam, 2003), pp. 487–507
C. Goldblatt, M.W. Claire, T.M. Lenton, A.J. Matthews, A.J. Watson, K.J. Zahnle, Nat. Geosci. **2**, 891 (2009)
E.A. Goosmann, D.C. Catling, S.M. Som, W. Altermann, R. Buick, J. Geophys. Res., Planets **123**, 2506 (2018)
H. Gunell, R. Maggiolo, H. Nilsson, G. STenberg Wieser, R. Slapak, J. Lindkvist, M. Hamrin, J. De Keyser, Astron. Astrophys. **614**, L3 (2018)
E. Hébrard, B. Marty, Earth Planet. Sci. Lett. **385**, 40 (2014)
J.H. Hoffman, R.R. Hodges, M.B. McElroy, T.M. Donahue, M. Kolpin, Science **205**, 49 (1979)
G. Holland, M. Cassidy, C.J. Ballentine, Science **326**, 1522 (2009)
G. Holland, B.S. Lollar, L. Li, G. Lacrampe-Couloume, G.F. Slater, C.J. Ballentine, Nature **497**, 357 (2013)
B.W. Johnson, C. Goldblatt, Geochem. Geophys. Geosyst. **19**, 2516 (2018)
N.M. Johnson, M.R.R. de Oliveira, Earth Space Sci. **6**, 1299 (2019)
C. Johnstone, M. Güdel, H. Lammer, K.G. Kislyakova, Astron. Astrophys. **617**, A107 (2018)
C. Johnstone, M.L. Khodachenko, T. Lüftinger, K.G. Kislyakova, H. Lammer, M. Güdel, Astron. Astrophys. **624**, L10 (2019)
S. Kadoya, D.C. Catling, Geochim. Cosmochim. Acta **262**, 207 (2019)

W.M. Kaula, Icarus **139**, 32 (1999)
L. Kavanagh, C. Goldblatt, Earth Planet. Sci. Lett. **413**, 51 (2015)
L.P. Knauth, Palaeogeogr. Palaeoclimatol. Palaeoecol. **219**, 53 (2005)
M.A. Kuga, G. Cernogora, Y. Marrocchi, L. Tissandier, B. Marty, Geochim. Cosmochim. Acta **217**, 219 (2017)
H. Lammer, E. Chassefière, Ö. Karatekin, A. Morschhauser, P.B. Niles, O. Mousis, P. Odert, U.V. Möstl, D. Breuer, V. Dehant, M. Grott, H. Groller, E. Hauber, L.B.S. Pham, Space Sci. Rev. **174**, 113 (2013)
J.-Y. Lee, K. Marti, J.P. Severinghaus, K. Kawamura, H.-S. Yoo, J.B. Lee, J.S. Kim, Geochim. Cosmochim. Acta **70**, 4507 (2006)
G. Libourel, B. Marty, F. Humbert, Geochim. Cosmochim. Acta **67**, 4123 (2003)
H.I.M. Lichtenegger, H. Lammer, J.M. Grießmeier, Y.N. Kulikov, P. von Paris, W. Hausleitner, S. Krauss, H. Rauer, Icarus **210**, 1 (2010)
Y. Marrocchi, B. Marty, P. Reinhardt, F. Robert, Geochim. Cosmochim. Acta **75**, 6255 (2011)
B. Marty, L. Zimmermann, M. Pujol, R. Burgess, P. Philippot, Science **342**, 101 (2013)
B. Marty, G. Avice, Y. Sano, K. Altwegg, H. Balsiger, M. Hässig, A. Morbidelli, O. Mousis, M. Rubin, Earth Planet. Sci. Lett. **441**, 91 (2016)
B. Marty, K. Altwegg, H. Balsiger, A. Bar-Nun, D.V. Bekaert, J.J. Berthelier, A. Bieler, C. Briois, U. Calmonte, M. Combi, J. De Keyser, B. Fiethe, S.A. Fuselier, S. Gasc, T.I. Gombosi, K.C. Hansen, M. Hässig, A. Jackel, E. Kopp, A. Korth, L. Le Roy, U. Mall, O. Mousis, T. Owen, H. Reme, M. Rubin, T. Semon, C.Y. Tzou, J.H. Waite, P. Wurz, Science **356**, 1069 (2017)
B. Marty, G. Avice, D.V. Bekaert, M.W. Broadley, C. R. Géosci. **350**, 154 (2018)
K.J. Mathew, K. Marti, J. Geophys. Res. **106**, 1401 (2001)
K.J. Mathew, J.S. Kim, K. Marti, Meteorit. Planet. Sci. **33**, 655 (1998)
M.B. McElroy, Y.L. Yung, A.O. Nier, Science **194**, 70 (1976)
A.P. Meshik, C.M. Hohenberg, O.V. Pravdivtseva, Y.S. Kapusta, Phys. Rev. C **64**, 035205 (2001)
A. Meshik, C. Hohenberg, O. Pravdivtseva, D. Burnett, Geochim. Cosmochim. Acta **127**, 326 (2014)
S. Mikhail, D.A. Sverjensky, Nat. Geosci. **7**, 816 (2014)
M. Moreira, Geochem. Perspect. **2**, 229 (2013)
M. Nishizawa, Y. Sano, Y. Ueno, S. Maruyama, Earth Planet. Sci. Lett. **254**, 332 (2007)
D.P. O'Brien, A. Izidoro, S.A. Jacobson, S.N. Raymond, D.C. Rubie, Space Sci. Rev. **214**, 47 (2018)
U. Ott, Geochim. Cosmochim. Acta **52**, 1937 (1988)
M. Ozima, F.A. Podosek, *Noble Gas Geochemistry*, 2nd edn. (Cambridge University Press, Cambridge, 2002)
R. Parai, S. Mukhopadhyay, Nature **560**, 223 (2018)
R.O. Pepin, Icarus **92**, 2 (1991)
R.O. Pepin, Meteoritics **29**, 568 (1994)
R.O. Pepin, D. Porcelli, Rev. Mineral. Geochem. **47**, 191 (2002)
S. Péron, M. Moreira, Geochem. Perspect. Lett. **9**, 21 (2018)
D.L. Pinti, K. Hashizume, J.-I. Matsuda, Geochim. Cosmochim. Acta **65**, 2301 (2001)
D. Porcelli, C.J. Ballentine, R. Wieler, Rev. Mineral. Geochem. **47**, 1 (2002)
M. Pujol, B. Marty, P. Burnard, P. Philippot, Geochim. Cosmochim. Acta **73**, 6834 (2009)
M. Pujol, B. Marty, R. Burgess, Earth Planet. Sci. Lett. **308**, 298 (2011)
M. Pujol, B. Marty, R. Burgess, G. Turner, P. Philippot, Nature **498**, 87 (2013)
F. Robert, M. Chaussidon, Nature **443**, 969 (2006)
M. Rubin, K. Altwegg, H. Balsiger, A. Bar-Nun, J.-J. Berthelier, C. Briois, U. Calmonte, M. Combi De Keyser Johan, B. Fiethe, S.A. Fuselier, S. Gasc, T.I. Gombosi, K.C. Hansen, E. Kopp, A. Korth, D. Laufer, L. Le Roy, U. Mall, B. Marty, O. Mousis, T. Owen, H. Rème, T. Sémon, C.-Y. Tzou, J.H. Waite, P. Wurz, Sci. Adv. **4**, 1 (2018)
C. Sanloup, Science **310**, 1174 (2005)
S.S. Shcheka, H. Keppler, Nature **490**, 531 (2012)
S.N. Silverman, S.H. Kopf, B.M. Bebout, R. Gordon, S.M. Som, Geobiology **135**, 839 (2018)
S.M. Som, D.C. Catling, J.P. Harnmeijer, P.M. Polivka, R. Buick, Nature **484**, 359 (2012)
S.M. Som, R. Buick, J.W. Hagadorn, T.S. Blake, J.M. Perreault, J.P. Harnmeijer, D.C. Catling, Nat. Geosci. **9**, 448 (2016)
B. Srinivasan, Earth Planet. Sci. Lett. **31**, 129 (1976)
F.M. Stuart, D.F. Mark, P. Gandanger, P. McConville, Earth Planet. Sci. Lett. **446**, 21 (2016)
E.E. Stüeken, M.A. Kipp, M.C. Koehler, E.W. Schwieterman, B. Johnson, R. Buick, Modeling pN_2 through geological time: implications for planetary climates and atmospheric biosignatures. Astrobiology **16**, 12 (2016). https://doi.org/10.1089/ast.2016.1537
E.E. Stüeken, S.M. Som, M. Claire, S. Rugheimer, M. Scherf, L. Sproß, N. Tosi, Y. Ueno, H. Lammer, Space Sci. Rev. **216**, 1–37 (2020)

T.D. Swindle, Rev. Mineral. Geochem. **47**, 171 (2002)
J.A. Tarduno, E.G. Blackman, E.E. Mamajek, Phys. Earth Planet. Inter. **233**, 68 (2014)
R. Tartèse, M. Chaussidon, A. Gurenko, F. Delarue, F. Robert, Geochem. Perspect. Lett. **3**, 55 (2016)
F. Tian, J.F. Kasting, H.-L. Liu, R.G. Roble, J. Geophys. Res. **113**, E05008 (2008)
I. Tolstikhin, B. Marty, Chem. Geol. **147**, 27–52 (1998)
I.N. Tolstikhin, R.K. O'Nions, Chem. Geol. **129**, 185 (1996)
O. Warr, B.S. Lollar, J. Fellowes, C.N. Sutcliffe, J.M. McDermott, G. Holland, J.C. Mabry, C.J. Ballentine, Geochim. Cosmochim. Acta **222**, 340 (2018)
R. Wieler, Rev. Mineral. Geochem. **47**, 21 (2002)
M.H. Wong, S.K. Atreya, P.N. Mahaffy, H.B. Franz, C. Malespin, M.G. Trainer, J.C. Stern, P.G. Conrad, H.L.K. Manning, R.O. Pepin, R.H. Becker, C.P. McKay, T.C. Owen, R. Navarro-González, J.H. Jones, B.M. Jakosky, A. Steele, Geophys. Res. Lett. **40**, 6033 (2013)
R. Wordsworth, R. Pierrehumbert, Science **339**, 64 (2013)
K. Zahnle, J.F. Kasting, J.P. Icarus, Icarus **84**, 502 (1990). 1990
K.J. Zahnle, D.C. Catling, M.W. Claire, Chem. Geol. **362**, 26 (2013)
K.J. Zahnle, M. Gacesa, D.C. Catling, Geochim. Cosmochim. Acta **244**, 56 (2019)
A.L. Zerkle, Philos. Trans. R. Soc. A **376**, 20170401 (2018)

Space Sci Rev (2020) 216:31
https://doi.org/10.1007/s11214-020-00652-3

Mission to Planet Earth: The First Two Billion Years

E.E. Stüeken[1,2] · S.M. Som[3] · M. Claire[1,2,3] ·
S. Rugheimer[4] · M. Scherf[5] · L. Sproß[5,6] · N. Tosi[7,8] ·
Y. Ueno[9,10] · H. Lammer[5]

Received: 14 July 2019 / Accepted: 25 February 2020 / Published online: 16 March 2020

Abstract Solar radiation and geological processes over the first few million years of Earth's history, followed soon thereafter by the origin of life, steered our planet towards an evolutionary trajectory of long-lived habitability that ultimately enabled the emergence of complex life. We review the most important conditions and feedbacks over the first 2 billion years of this trajectory, which perhaps represent the best analogue for other habitable worlds in the galaxy. Crucial aspects included: (1) the redox state and volatile content of Earth's building blocks, which determined the longevity of the magma ocean and its ability to degas H_2O and other greenhouse gases, in particular CO_2, allowing the condensation of a water ocean; (2) the chemical properties of the resulting degassed mantle, including oxygen fugacity, which would have not only affected its physical properties and thus its ability to recycle volatiles and nutrients via plate tectonics, but also contributed to the timescale of atmospheric oxygenation; (3) the emergence of life, in particular the origin of autotrophy, biological N_2 fixation, and oxygenic photosynthesis, which accelerated sluggish abiotic pro-

Reading Terrestrial Planet Evolution in Isotopes and Element Measurements
Edited by Helmut Lammer, Bernard Marty, Aubrey L. Zerkle, Michel Blanc, Hugh O'Neill and Thorsten Kleine

✉ E.E. Stüeken
ees4@st-andrews.ac.uk

1 University of St Andrews, St Andrews, UK

2 Virtual Planetary Laboratory, Seattle, USA

3 Blue Marble Space Institute of Science, Seattle, USA

4 Oxford University, Oxford, UK

5 Space Research Institute, Austrian Academy of Sciences, Graz, Austria

6 Institute of Physics, University of Graz, Graz, Austria

7 Technische Universität Berlin, Berlin, Germany

8 German Aerospace Center (DLR), Berlin, Germany

9 Department of Earth and Planetary Sciences, Tokyo Institute of Technology, Tokyo, Japan

10 Earth-Life Science Institute, Tokyo Institute of Technology, Tokyo, Japan

cesses of transferring some volatiles back into the lithosphere; (4) strong stellar UV radiation on the early Earth, which may have eroded significant amounts of atmospheric volatiles, depending on atmospheric CO_2/N_2 ratios and thus impacted the redox state of the mantle as well as the timing of life's origin; and (5) evidence of strong photochemical effects on Earth's sulfur cycle, preserved in the form of mass-independent sulfur isotope fractionation, and potentially linked to fractionation in organic carbon isotopes. The early Earth presents itself as an exoplanet analogue that can be explored through the existing rock record, allowing us to identify atmospheric signatures diagnostic of biological metabolisms that may be detectable on other inhabited planets with next-generation telescopes. We conclude that investigating the development of habitable conditions on terrestrial planets, an inherently complex problem, requires multi-disciplinary collaboration and creative solutions.

Keywords Early Earth · Biosignatures · Atmospheric evolution

1 Introduction

Earth is the only inhabited planet that we know of. Our planet therefore provides a key anchor point for understanding, predicting, and modelling the formation of habitable worlds as well as for testing remote and *in situ* biosignatures. Importantly, Earth and its biosphere have undergone major evolutionary changes over the last 4.5 billion years, enabling the investigation of geo-biological feedbacks under a range of environmental conditions. Until about 2.4 billion years ago (Ga) Earth's atmosphere was essentially devoid of O_2 gas. Complex life only evolved roughly 600 million years ago (Ma). For its first two billion years (4.5–2.5 Ga), ecosystems were thus dominated by anaerobic microbial communities and perhaps more strongly tied to planetary processes than today. These processes include outgassing of volatiles, recycling of nutrients and greenhouse gases after the establishment of plate tectonics, and photochemical reactions that can erode atmospheric gases or convert them into soluble molecules. Life originated from this framework and eventually took control over many of the associated feedbacks. It is this interval of Earth's history that can perhaps be considered the best analogue for other worlds that may be inhabited by simple life forms, incapable of complex metabolisms such as oxygenic photosynthesis. This paper reviews what is known about the establishment of habitable conditions spanning geological (Sects. 2, 3), biological (Sect. 4), solar/photochemical constraints (Sects. 5, 6, 7) on the atmosphere; and how these feed into the frontier challenge of detecting biospheres remotely via atmospheric signatures (Sect. 8).

2 From Magma to a Water Ocean

An important transition towards habitability in Earth's early history was the establishment of a liquid water ocean under an atmosphere with a stabilizing degree of greenhouse warming. This transition required cooling and solidification of the primitive mantle following Moon formation (Lock et al. 2018; Canup and Asphaug 2001), as well as outgassing of H_2O and CO_2 (Elkins-Tanton 2008). These two volatiles were perhaps the major players of the so-called secondary atmosphere from which the ocean ultimately condensed. The formation period of a solid crust following Moon formation and its duration are crucial for understanding the initial secondary atmospheres on Earth-like planets. During the solidification period of the primitive mantle, volatiles are relatively easily exchanged between the planetary interior and the atmosphere. Once the surface solidifies to form the crust, this exchange is severely hampered for the rest of the planet's lifetime, causing volatile exchange rates to drop dramatically. In the classical scenario of a magma ocean that crystallizes from the bottom upwards (e.g. Elkins-Tanton 2012), some volatile components can become supersaturated in the magma and be released into the atmosphere. By this mechanism, most of the CO_2 tends to be outgassed in the earliest stages of the magma ocean because of its low solubility in silicate melts. H_2O, in contrast, is more soluble and is outgassed only when it is highly enriched in the liquid magma ocean—a condition that is generally reached towards the end of magma solidification. At that time, atmospheric CO_2 and H_2O may have reached partial pressures (p_{CO_2}, p_{H_2}) of approximately 100 bar and a few hundred bar, respectively (Zahnle et al. 1988; Schaefer and Fegley 2010; Sleep 2010; Salvador et al. 2017). All outgassed particles, in combination with possible remnants from the hydrogen envelope that formed during the planet's accretion phase (the "primary atmosphere"), form a dense steam atmosphere.

The behaviour of nitrogen in this period, the most abundant atmospheric component at present day, is not fully understood. On the one hand, several mass-balance and modelling studies have suggested N_2 partial pressures similar to today or slightly higher (e.g. Barry and Hilton 2016; Stüeken et al. 2016b; Johnson and Goldblatt 2018; Mallik et al. 2018). In fact, higher-than-modern atmospheric p_{N_2}, paired with relatively low p_{CO_2}, was proposed as a potential solution to the Faint Young Sun Paradox later in the Archean by pressure broadening (e.g. Goldblatt et al. 2009; Goldblatt and Zahnle 2011; Byrne and Goldblatt 2015). On the other hand, in an atmospheric steam N_2 could have been reduced to NH_3 which may have been soluble in the magma ocean. Through this mechanism, nitrogen could thus have been transferred efficiently from the atmosphere back into the lithosphere as long as the magma ocean was liquid (Solomatov 2000; Libourel et al. 2003; Zahnle et al. 2010; Kadik et al. 2011; Wordsworth 2016). In addition to this temperature-driven conversion of N_2 to NH_3 at the interface between the atmosphere and the crust, high EUV irradiation from the young Sun (e.g. Tu et al. 2015; Airapetian et al. 2016), cosmic rays (e.g. Grenfell et al. 2012; Tabataba-Vakili et al. 2016; and see also Sect. 5), and frequent meteoric impacts (e.g. Fegley et al. 1986; Parkos et al. 2018) are all capable of dissociating atmospheric N_2 and enabling N incorporation into surface materials. Although significant uncertainties remain, emerging views on ingassing and escape of N allow for depletion of N (and potentially other volatiles) in Earth's early secondary atmosphere.

If nitrogen was outgassed during the magma ocean phase with a similar C/N ratio as observed on present Earth and did not escape, then the partial pressure of N_2 (p_{N_2}) is expected to be two orders lower than that of CO_2. It would not exceed 500 mbar for realistic amounts of CO_2 (Lammer et al. 2019). Following the estimates of abiotic lightning-driven N_2 fixation rates proposed by Navarro-González et al. (1998), and assuming complete conversion of this fixed nitrogen to NH_3, this partial pressure would imply a drawdown of 100 mbar nitrogen in around 5 Myr assuming that the fixed nitrogen is trapped as NH_3 at reducing conditions in the mantle and is not outgassed until the mantle reaches a high enough oxidation state for the conversion of NH_3 to N_2 (discussed in Lammer et al. 2019). However, it remains uncertain if nitrogen outgassed at all during the magma ocean stage (Libourel et al. 2003). In summary, it is thus conceivable that nitrogen partial pressures were rather low towards the end of the Hadean, but outgassing rates during the early Earth remain a crucial unknown.

The H_2O and CO_2 that outgassed from the magma ocean would have acted as greenhouse gases. H_2O in particular absorbs infrared radiation over a wide range of wavelengths, which may have slowed down the cooling and solidification of the magma ocean (Elkins-Tanton

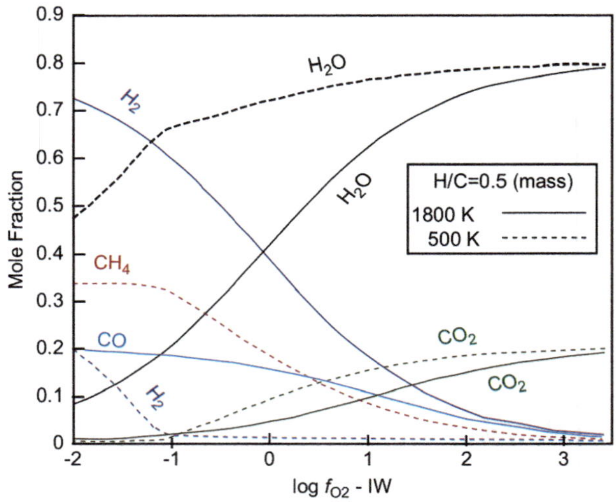

Fig. 1 Composition of outgassed gases from a magma ocean in dependence of oxidation state and temperature. The gas has a fix H/C ratio of 0.5. IW describes the oxygen fugacity through the buffer in coexisting pure iron and wüstite (figure by Hirschmann 2012)

2008, 2012). Although the solidification timescale depends on several poorly constrained parameters (Nikolaou et al. 2019), it is influenced to first order by the bulk abundance of volatiles and by the incoming stellar radiation (Hamano et al. 2013; Lebrun et al. 2013; Nikolaou et al. 2019). A volatile-depleted magma ocean that radiates its heat to space in the absence of an atmosphere is expected to solidify relatively quickly within a few thousand years (e.g. Monteux et al. 2016). In contrast, a magma ocean with a conservative initial H_2O budget equivalent to 300 bar outgasses about 220 bar H_2O into the atmosphere over its evolution, and at a distance from the Sun of 1 AU, such an atmosphere extends the solidification time to about 1 Myr (e.g. Lebrun et al. 2013; Nikolaou et al. 2019). This duration scales roughly linearly with the initial abundance of water in the mantle, which is poorly constrained (Korenaga 2013). Since these timescales are relatively short compared to the period over which large-scale, magma-ocean-forming collisions with other planetary bodies are expected, the occurrence of multiple magma ocean phases is possible over the early planetary history. Indeed, for the Earth, two episodes of magma ocean outgassing have been proposed to explain the isotopic difference of $^{3}He/^{22}Ne$ between the depleted mantle and a primitive deep reservoir (Tucker and Mukhopadhyay 2014). It has to be noted that the oxidation state of the melts has a crucial influence on the outgassed composition (see Fig. 1).

The end of the magma ocean is usually identified as the time when the planet enters a so-called "mush phase" (Lebrun et al. 2013), where the whole mantle up to its surface exhibits a solid-like behavior despite the presence of residual melt (below 30–40%) at shallow depths. The beginning of the mush phase coincides with a dramatic drop in the surface temperature and heat flux, which generally causes part of the outgassed steam atmosphere to condense, forming a water ocean. The possibility to achieve condensation depends on the amount of CO_2 present in the atmosphere at the end of solidification (Lebrun et al. 2013; Massol et al. 2016; Salvador et al. 2017). For a given amount of outgassed water, the higher the CO_2 partial pressure, the more difficult it is to achieve condensation, which occurs when the saturation vapor pressure of water at the surface is reached. At the Earth's orbit and for plausible amounts of H_2O and CO_2 (comparable to the present-day extent of the two reservoirs), the condition to achieve condensation is easily met (e.g. Salvador et al. 2017).

The formation of a water ocean is thus a natural outcome of the solidification of a water-rich magma ocean and is expected to occur within up to few million years after the last major impact event capable of causing large-scale mantle melting. The last such event was probably the Moon-forming impact taking place up to 130 Myr after the condensation of the first solids in the solar system (e.g. Jacobson et al. 2014; Bottke et al. 2015). The close-to-chondritic D/H ratio of sea water provides evidence that the oxidation of silicate Earth occurred during or before the crystallization of the final global-scale magma ocean (Pahlevan et al. 2019, see Fig. 1). If the mantle had been highly reducing at that time, then the outgassed atmosphere would have been dominated by H_2 and CO instead of H_2O and CO_2, and such a gas mixture would have facilitated hydrogen escape to space, including loss of H relative to D, leading to higher D/H ratios than observed. D/H ratios thus suggest that the silicate Earth was at least oxidized enough to outgas mainly H_2O and CO_2, leading to reduced hydrogen escape, because of CO_2 would have provided IR-cooling in the upper atmosphere. Similarly, if the steam atmosphere had survived for several Myr, one would expect to see a more significant increase in the D/H ratio through hydrodynamic escape over time. Hence the present-day D/H ratio implicates a short-lived steam atmosphere presence. This view is further supported by unfractionated Ne isotopes (Avice and Marty 2020, this issue).

CO_2 would have played a key role in protecting the secondary atmosphere from escaping to space, because it cools the upper atmosphere and therefore reduces its expansion. For sufficient atmospheric CO_2, an appropriate oxidation state of the final magma ocean is required (Fig. 1); for reduced conditions the atmosphere is therefore highly susceptible to strong escape rates (Kasting 1982; Lammer et al. 2018). After the magma-ocean phase and during ensuing tectonic conditions that may have been similar to a stagnant-lid regime (e.g. O'Neill and Debaille 2014), CO_2 would have dissolved in the ocean and reacted with oceanic crust as in classic weathering reactions. This process would have decreased the partial pressure CO_2 over time (e.g. Hessler et al. 2004; Kanzaki and Murakami 2015; Sleep and Zahnle 2001). For 1 bar of atmospheric CO_2, in the presence of a stagnant-lid and a water ocean, estimates of the amount of CO_2 that could have been removed by seafloor weathering range from 70 to 97 % (Walker 1985; Jacob 1999). With the onset of plate tectonics, which allowed replacement of weatherable surface material, CO_2 levels would likely have been reduced further. Hence overall, atmospheric p_{CO_2} likely decreased continuously throughout the Archean, especially if volcanic regimes did not change significantly during this time to counteract high weathering rates, but this remains unconstrained. As discussed in Birmingham et al. (2020, this issue), the young Sun's bolometric luminosity was only about 70% of today's value, which demands higher amounts of greenhouse gases like CO_2, CH_4, etc. to keep the ocean liquid (Feulner 2012; Wolf and Toon 2013; Kunze et al. 2014; Charnay et al. 2017).

A high CO_2 to N_2 ratio in the early Archean was proposed by Lammer et al. (2019): High EUV radiation likely eroded N_2 from the atmosphere, but since there is no detectable evidence of this process in the $^{14}N/^{15}N$ record (Füri and Marty 2015), the atmospheric N_2 reservoir was probably small at that time and only constituted a small fraction of Earth's total nitrogen inventory. However, sometime in Earth's history, maybe as late as the Great Oxidation Event at 2.4 Ga, nitrogen became the most abundant constituent in the atmosphere, possibly due to continuous volcanogenic or biogenic N_2 liberation (Zerkle et al. 2017; Lammer et al. 2019; Stüeken et al. 2016b), although see Sect. 3.2. By that time, solar EUV irradiation would have been sufficiently low and the magnetosphere sufficiently intense and stable to minimize strong atmospheric escape of nitrogen, as supported by stable isotopic compositions in atmospheric N_2 since that time (Marty et al. 2013).

3 Beyond the Magma Ocean: Geological Controls on Planetary Surface Evolution

The development of habitable conditions at the Earth's surface was controlled by geological parameters, i.e. the redox state and composition of the primitive mantle and the internal heat of the planet (Sect. 2). While the details of these processes provided the initial conditions that enabled the evolution of the Earth system, additional processes were pivotal in establishing the habitability of our planet and may prove fundamental to the search for life in the Universe. A clear impediment to this line of research is the scarcity of a preserved geologic record of Earth's most ancient history. As a result, the state of the early Earth is predominantly investigated through numerical simulations. Two types of models are in use: (a) understanding the controls on the modern Earth and seeking how these contributions have changed backwards in time ("backward models"), and (b) developing initial conditions bounded by fundamental physical and chemical laws (Sleep 2010) and evolving these simulations forward in time ("forward models").

A planetary state can be thought of as a snapshot of a planet at a certain instant of time. The planetary state of modern Earth is determined by the transfer of energy between the planet's interior and the planet's surface, coupled with the transfer of energy between the planet's surface and the space environment. As a result, the state of Earth's surface depends on energy delivered from the Earth's interior and from the Sun (Lenardic et al. 2016). The Earth's interior has a comparatively greater role than the Sun in determining Earth's state for two main reasons: The first is in the delivery and recycling of volatiles to and from the atmosphere by way of plate tectonics (mantle convection and plate subduction) and associated volcanism, and the second is the protection of the atmosphere from solar wind stripping through the magnetic field generated by convection of the liquid outer core.

3.1 Earth Formation, Ocean Evolution, and the Onset of Plate Tectonics

The proto-Earth formed by the accretion of smaller precursors analogues of which still exist today as remnants of the early solar system: chondritic meteorites ("chondrites"). The Earth-Moon system was likely formed following a high-energy high-angular-momentum impact between the proto-Earth and "Theia", another solar system body. This cataclysmic impact may have formed a synestia, a rapidly spinning mass of vaporized rock from which the Earth-Moon system emerged (Lock et al. 2018). This proposed mechanism differs from the canonical model of the "Moon-forming impact" based on smooth particle hydrodynamics (Canup and Asphaug 2001). In the latter mechanism, forward models predict a hydrated initial mantle (Zahnle et al. 2010). Backward models argue for a dry mantle (Korenaga 2011). Regardless of the initial state of the mantle, the resulting angular momentum at the end of the synestia or canonical Moon-forming impact is hypothesized to have been conserved since then.

As a result of the tidal torque on the Earth's ocean by the Moon and by conservation of angular momentum, the Earth's rotation rate has been slowing down and the Moon has receded. Orbital mechanics calculations suggest an Earth spinning twice the current rate 3.5 billion years ago (Webb 1982). Such predictions mostly agree with geological indicators of the diurnal cycle (Williams 2000). Any predictions older than 3.5 Ga must be used with caution as the uncertainties in the controlling parameters become high. For example, the origin and initial evolution of the oceans, a major player in tidal dissipation (Bills and Ray 1999), is poorly constrained. Based on geochemical arguments, the oceans' volume has been proposed to be 26% larger 3.8 billion years ago, slowly losing mass through H_2O

dissociation and hydrogen escape (Pope et al. 2012). Other studies suggests an ocean 50% larger 3.0 billion years ago, based on assumptions about the thermal evolution of the Earth and constancy of the continental freeboard, which states that the mean sea-level has always been close to the mean continental level (Korenaga 2013). Ocean salinity was likely similar or slightly (30%) higher than modern values (Marty et al. 2018; Sleep 2010). How a larger ocean would have affected tidal dissipation is unclear. A much faster spinning planet would have had dynamical implications for the atmosphere, for example thinning the latitudinal extent of the Hadley cell among other properties sensitive to the Coriolis force (Held and Hou 2002).

The atmospheric volatile evolution of modern Earth is strongly controlled by plate tectonics, mantle redox and outgassing rates. How outgassing rates scale with geothermal heat flux remains poorly constrained. Subduction carries sediments that contain atmospheric volatiles into the mantle, from where they can be speciated and either stored in the mantle wedge or released back into the atmosphere via volcanism. Mantle redox can control the oxidation state of released volatiles (Trail et al. 2011; Mikhail and Sverjensky 2014) coupled with hydrogen escape following methane photolysis on early Earth (Catling et al. 2001). The oxidation state of the atmosphere is thus coupled to that of the mantle (Nicklas et al. 2019), emphasizing the role of the planetary interior in affecting the planetary state. Oxygen fugacity (fO_2) investigations of Precambrian lavas reveal a secular oxidation of the mantle from 3.5 Ga to 1.9 Ga. No single process explains this oxidation, with recycling of altered oceanic crust, venting of oxygen from the core due to inner core crystallization, and convection-driven homogenization of an initially redox-heterogeneous primordial mantle as the proposed end-member contributors (Nicklas et al. 2019).

Predictions of past behavior of atmospheric volatiles, such as C and N have been performed predominantly via box modeling (Hart 1978). Such efforts necessarily require assumptions that become harder to defend in deep time, especially prior to $\sim$3 Ga. Common first-order assumptions include that the same mechanism of mantle recycling has existed over geological history and that arc and ridge fluxes scale with the heat production of the radioactive elements that fuel Earth's interior. Other important parametrizations include the known luminosity changes of our Sun as stars increase in luminosity as they age (Gough 1981), which contributes to surface temperature and thus weathering rates of C and N (Walker et al. 1981). However, the assumed coupled history between tectonic regime and planetary heat flux is far from settled (Lenardic et al. 2016), with suggestions regarding the initiation of modern plate tectonics regime spanning the geological record (Korenaga 2013; Sleep and Schubert 2007).

Observational insights on the thermal evolution of the Earth do exist. Herzberg et al. (2010) computationally predict the MgO and FeO content of the primary magmas responsible for basaltic rocks spanning Earth history. These chemical parameters are a function of mantle temperature. They find a maximum T of $\sim$1500–1600 °C (compared to a modern value of 1350 °C) at 2.5–3.0 Ga. Mantle recycling with subduction is inferred from 2.7 Ga rocks. Continental rifting has also been proposed as far back as 3.2 Ga (Van Kranendonk et al. 2007). Several plate tectonic models fit these constraints, and the correct one may ultimately depend on advances in knowledge regarding the interaction of the ocean and the mantle over Earth history and on additional geological constraints.

Taken together, a heuristic model can be presented if we assume that the ocean volume has indeed decreased over time to the present value. Based on evidence from ancient zircons, primordial oceans appear to have existed by 4.3 Ga (Ushikubo et al. 2008). Such primordial oceans, formed by the collapse of a steam atmosphere, could have been underlain by a dry mantle (Korenaga 2013). The water would act to weaken the previously stiff lithosphere (akin to present-day Venus), and a dry mantle would lead to high convective stress.

Modern-style subduction was initiated when convective stresses overcame the strength of the lithosphere (plates), but initiation may have been delayed. Distinct geodynamic regimes compared to modern plate tectonics are inferred at 4 Ga from zircon geochemistry (Reimink et al. 2016).

It is thus plausible that subduction processes began as late as the mid-Archean (3.8–2.5 Ga) rather than in the Hadean (4.5–3.8 Ga) (Avice et al. 2017; Parai and Mukhopadhyay 2018; Péron and Moreira 2018).The near-surface material would have then begun to participate in mantle overturn by being dragged down. Such subduction would have started hydrating the mantle, thus increasing the viscosity contrast ($\Delta\eta$) across the lithosphere. Because the mantle's internal Raleigh number scales with $\Delta\eta^2$, continued subduction would have further promoted convection (Korenaga 2013). Net water influx would eventually subside to reach a steady state as the ocean volume decreased and further ocean loss was limited to hydrogen escape. Initially, internal heating (H) would have exceeded the mantle convective heatflux Q ($H/Q > 1$), leading to an increase in mantle temperature until ~3 Ga where the ratio would have flipped and Earth would then witness a slow cooling to achieve the modern Urey ratio ($Ur = H/Q$) of ~0.3. This increase in Urey ratio from the Hadean to the mid-Archean may explain the possible delay in the onset of subduction.

3.2 Geological Sources and Sinks of Atmospheric Carbon and Nitrogen

Atmospheric volatiles like C and N cycle through the various planetary reservoirs. Common reservoirs for C include the ocean/atmosphere system, the seafloor, continents, and the mantle (Sleep and Zahnle 2001; Tajika and Matsui 1992). For geological carbon cycle investigations (i.e. "long-term" cycle), biology is often omitted as the corresponding reservoir size of biota is small (Tajika and Matsui 1990), even though fluxes can be substantial (see Sect. 4.2). From the mantle, C can return to the ocean/atmosphere via arc and ridge volcanism. From the ocean/atmosphere, C can return to the mantle via precipitation of carbonate and organic carbon on the seafloor, followed by subduction. Continental weathering, which pulls CO_2 out of the atmosphere through the formation of bicarbonate ions, is balanced as continents "scrape" sediments off the seafloor during subduction. This process is called accretion. The geological carbon cycle is important as a "planetary thermostat" in regulating planetary surface temperature via the greenhouse effect driven by p_{CO_2} (Walker et al. 1981). Potentially higher levels of atmospheric C on the early Earth would have been driven by a higher degassing flux, itself driven by hotter mantle temperatures. However, this higher flux was perhaps balanced to some extent by faster plate velocities and corresponding faster subduction rates resulting from the higher heat production.

The N cycle can be assessed using a similar framework, yet N cycle studies are in their infancy compared with the abundance of work on the C cycle. A major difference between the modern and the ancient N cycle is the oxidized nature of modern Earth. For modern Earth, a quantitative flux balance can be written (Som et al. 2016). The total N source flux (0.33 ± 0.08 Tmol/year), composed of volcanic outgassing (0.18 ± 0.07 Tmol/year) and oxidative weathering (0.15 ± 0.03) balances the total N sink flux composed entirely of N burial (0.4 ± 0.2 Tmol/year). The subducted flux can essentially be ignored as it is within the uncertainty (0.094 ± 0.015 Tmol/year).

On the ancient Earth, oxygen was essentially absent from the environment (Reinhard et al. 2016). Given the flux balance above, oxidative weathering can thus be neglected, dropping the N_2 source flux to 50% of its present value, if all other fluxes remained constant (Som et al. 2016). At steady state, the global N sink would balance the source flux and be likewise diminished. Assuming that the atmospheric N reservoir is proportional to source and sink fluxes, a 50% reduction in the N source would leave the atmospheric reservoir depleted by 50% compared to modern values. Given that the modern air pressure is composed of 0.8 bar N_2, and 0.2 bar O_2, the considerations above (p_{N_2} of 50% of 0.8 bar) predict an Archean pressure of 0.4 bar. While this framework is first-order and neglecting potential non-linearities in the flux-reservoir relationship, low air pressure has been corroborated by independent constraints from the Archean rock record (Avice et al. 2018; Endo et al. 2019; Marty et al. 2013; Som et al. 2012, 2016). Additional considerations for the Archean include the existence of ammonium in the ocean as a sink for N via ion replacement in clays that are eventually subducted (Stüeken et al. 2016b), and redox controls on N_2 volcanic outgassing (Mikhail and Sverjensky 2014).

Taken together, a heuristic model of N evolution can be presented. The rocky Earth started with a dense CO_2/H_2O atmosphere with perhaps lesser amounts of N_2 (Sleep 2010; Zahnle et al. 2010). H_2O condensed out to form the oceans, and life perhaps took a foot-hold early (Sect. 4). The growing biomass soon outpaced the ability of abiotically produced fixed nitrogen to provide the biologically available nitrogen needed for biochemistry (Navarro-González et al. 2001a, 2001b). Under this environmental pressure, and assuming no other limiting nutrients (Zerkle 2018), biology capable of undergoing nitrogen fixation would be selected for. Indeed, nitrogen fixation is known to be an ancient metabolic pathway (Raymond et al. 2004; Sect. 4.3). In the absence of oxygen, ammonium rather than nitrate would have been the dominant nitrogen species in the ocean (Zerkle et al. 2017). Ionic substitution with potassium in clay minerals ($NH_4^+ \leftrightarrow K^+$) within sedimentary pore waters would have led to burial and eventual subduction of N (Som et al. 2016; Stüeken et al. 2016b). At steady state, N burial may have balanced the volcanic flux. With the advent of oxygen in the environment, new processes enabled the atmospheric N reservoir to increase to modern values. Oxidative weathering would have introduced a new source of N into the atmosphere, and subducted oxidized sediments may have changed the redox state of the mantle wedge and further enabled N_2 outgassing (Mikhail and Sverjensky 2014). Advances in N-cycle modeling (Johnson and Goldblatt 2018) coupled with the development of additional geological markers of air density (Goosmann et al. 2018; Silverman et al. 2018) will test this model.

4 Biological Imprints on the Early Atmosphere

Geological and photochemical processes set the stage for habitable conditions (Sects. 2, 3, 5), but the emergence of life likely had a significant impact on volatile inventories of the atmosphere and solid Earth with implications for global climate and habitability (e.g., Lenton 1998). Apart from the noble gases, all atmospheric constituents, including N_2 and O_2, as well as the major greenhouse gases CO_2, CH_4 and N_2O, have biological sources and sinks. Determining the antiquity and evolution of these biological fluxes—against a backdrop of abiotic processes—is therefore important for reconstructing the habitability of the early Earth as an analogue for other worlds.

4.1 The Surficial Carbon Cycle and the Role of Prebiotic Reactions

To first order, the major fluxes even in the modern carbon cycle are not directly dependent on biology. Atmospheric carbon is sourced from volcanic outgassing of CO_2, and it is ultimately removed by the burial of carbonate rocks. Burial of organic carbon constitutes a relatively minor fraction of around 15–20% (Mason et al. 2017). Although many carbonate deposits today are biogenic in origin, it is conceivable that they would also form on an abiotic planet once carbonate minerals reach saturation in seawater. Indeed, aragonitic crystal fans from the Archean rock record have been taken as evidence for widespread abiotic carbonate precipitation in the early ocean under a CO_2-rich atmosphere (Grotzinger and Kasting 1993). However, biological activity still has a significant effect on the carbon cycle, because it changes the redox state of gaseous and sedimentary carbon. Autotrophic organisms convert CO_2 with a +IV oxidation state into organic matter or methane (CH_4) with oxidation states of 0 and −IV, respectively. Burial of organic matter in sediments thus removes a reductant from the Earth's surface, rendering surface environments overall more oxidized (Catling 2014b). This process may have contributed to the rise of atmospheric O_2 around the Archean-Proterozoic boundary, and it may have been the major driver of later oxygenation events in the Neoproterozoic and Paleozoic (Rothman et al. 2003; Husson and Peters 2017; Lenton et al. 2014; Butterfield 2018). In contrast, the production of biogenic methane injects a reductant into the atmosphere, which may have slowed down the accumulation of O_2 (Zahnle et al. 2006) while also acting as a strong greenhouse gas. On the Archean Earth, where the Sun was fainter, biogenic CH_4 production may have been important for maintaining a warm climate (Haqq-Misra et al. 2008; Arney et al. 2016). The onset of biological carbon burial (reductant removal) as well as the biological conversion of carbon into CH_4 (reductant liberation and greenhouse gas production) were thus probably two key processes that shaped the long-term evolution of our planet.

Prior to the origin of life, some abiotic process would have converted CO_2 into reduced organic carbon and methane. In fact, such abiotic sources of reduced carbon were probably essential for prebiotic chemical reactions that ultimately culminated in the rise of autotrophic cells. The major energy sources for the abiotic reduction of CO_2 were likely geothermal heat, UV radiation, and lightning reactions. Geothermal settings have been shown to favor the production of methane, longer-chain alkanes and more complex organic molecules in both experiments and natural environments through so-called Fischer-Tropsch-type reactions (Cody 2004; Russell and Martin 2004; McCollom and Seewald 2007; Proskurowski et al. 2008). Hydrogen gas produced by serpentinization (i.e., aqueous alteration of fresh olivine and pyroxene) acts as a reductant, while metallic minerals such as sulfides, magnetite or elemental iron and nickel act as reaction catalysts. In atmospheric reactions, such catalysts are not needed, as lightning and UV radiation provide sufficient energy to produce organic carbon molecules from CO_2 and CH_4. The organic carbon products of UV reactions are HCN and CH_2O (Zahnle 1986; Deamer 2007; Cleaves 2008), while lightning reactions have been reported to generate a variety of organic compounds, including amino acids and nucleobases (Miller 1953; Ferus et al. 2017). The magnitude of these abiotic sources of reduced organic carbon as well as its stability in the early ocean on the prebiotic Earth are not well constrained. However, it is conceivable that geothermally catalyzed reactions were relatively more abundant than today, because Earth's earliest crust was likely more mafic (Dhuime et al. 2015; Tang et al. 2016), leading to more widespread serpentinization (Smit and Mezger 2017) with the potential to produce large amounts of reduced carbon molecules.

4.2 The Emergence and Early Diversity of Biological CO_2 Fixation Pathways

Although some form of abiotic CO_2 reduction was undoubtedly present on the early Earth, the emergence of autotrophic life is evidence that these abiotic sources of reduced carbon were not large enough to sustain a significant biosphere. Hence the rise and radiation of autotrophic organisms probably caused a sharp increase in the burial flux of organic carbon. The most widely cited record of biological CO_2 fixation is archived in carbon isotope ratios of sedimentary rocks (Schidlowski 2001). Enzymes favor ^{12}C over ^{13}C in specific ratios that are characteristic of their respective carbon fixation pathway. The most widespread pathway today is the Calvin cycle, catalyzed by the enzyme Rubisco, and $^{12}C/^{13}C$ ratios of ancient organic matter suggest that this metabolism may date back to the early Archean (Nisbet et al. 2007). The oldest (though controversial, see e.g. Whitehouse et al. 2019) records of the Calvin cycle go back to more than 3.9 billion years ago (Bell et al. 2015; Tashiro et al. 2017). These ancient records are controversial because carbon isotope fractionations can also be produced by abiotic processes that generate organic matter, such as Fischer-Tropsch-type reactions (McCollom and Seewald 2006). More widely accepted evidence has been documented from around 3.5 Ga, where it coincides with occurrences of stromatolites and biogenic sulphur isotope fractionation (Buick 2007). Rocks of this age also preserve the oldest evidence of biological CH_4 production (Ueno et al. 2006; Flannery et al. 2018; Schopf et al. 2018). Methanogenesis is driven by the acetyl-CoA pathway, and it is characterized by large isotopic fractionations imparted on the resulting CH_4 and biomass (House et al. 2003). Isotopically depleted CH_4 in gas inclusions in cherts from >3.46 Ga (Ueno et al. 2006) and isotopically light kerogen in similarly aged rocks (3.43–3.47 Ga, Flannery et al. 2018; Schopf et al. 2018) thus indicate that the Archean biosphere also contributed to the production of a major greenhouse gas. Phylogenetic studies suggest that the acetyl-CoA pathway is the most ancient CO_2 fixation pathway (Fuchs 2011). Importantly, several enzymes involved in methanogenesis contain metallic catalysts similar to those that are considered to be important in driving abiotic Fischer-Tropsch-type reactions in hydrothermal settings (Russell and Martin 2004; Stüeken et al. 2013). This correlation may point to a hydrothermal origin of autotrophy on the early Earth.

By the Mesoarchean (3.2–2.8 Ga), evidence of life spans from the deep sea to continental lakes and soils (e.g. Rasmussen 2000; Nabhan et al. 2016; Homann et al. 2018; Stüeken and Buick 2018), suggesting that the biosphere was now diverse enough to colonize a range of environments and large enough to leave a recognizable imprint in the geologic record. Biological CH_4 production at this time has been documented in particular from non-marine and brackish environments (Stüeken and Buick 2018). The Neoarchean perhaps saw an expansion of biological methane production as suggested by widespread occurrences of isotopically reduced organic carbon (Thomazo et al. 2009a, 2009b; Zerkle et al. 2012; Sect. 7). The lightest values are found in shallow marine and non-marine environments (Eigenbrode and Freeman 2006; Flannery et al. 2016; Stüeken et al. 2017), which may reflect biological consumption and oxidation of CH_4 by methanotrophic organisms in a world that experienced increasing local (Lalonde and Konhauser 2015; Sumner et al. 2015) and/or temporal (Anbar et al. 2007; Koehler et al. 2018) oxygenation of surface environments by growing populations of oxygenic phototrophs. Biological and geological feedbacks associated with these processes ultimately led to the Great Oxidation Event (GOE) in the Paleoproterozoic (Lyons et al. 2014).

4.3 The Nitrogen Cycle and Its Abiotic Roots

Like carbon, nitrogen is a major constituent of all life on Earth. Its largest reservoir at the Earth's surface is atmospheric N_2 gas, which is relatively inert. Roughly 15% of cultured prokaryotic organisms are capable of breaking the N_2 triple bond with a so-called nitrogenase enzyme (Dos Santos et al. 2012). In this process, termed biological N_2 fixation, the end product is ammonium (NH_4^+), the most reduced form of nitrogen. Today, the estimated global marine N_2 fixation rate is on the order of 10^{11} kg/yr (Wang et al. 2019). About half as much is fixed by terrestrial ecosystems (Vitousek et al. 2013), some of which gets transported to the ocean by rivers. However, only about 1–2% of this biologically fixed nitrogen is buried in sediments (Stüeken et al. 2016b). The majority is recycled through remineralization of biomass to nitrate (NO_3^-), which is subsequently reduced back to N_2 in suboxic sedimentary pore waters and marine oxygen minimum zones (Kritee et al. 2012). Long-term removal of nitrogen from the atmosphere is therefore directly dependent on biological productivity and the efficiency of biomass burial. Within sediments, buried biomass slowly breaks down and releases ammonium into pore waters, where it can build up to millimolar concentrations (e.g., Boudreau and Canfield 1988). Ammonium has the same charge and roughly the same size as K^+ and can therefore partition into diagenetic potassium-rich clay minerals, such as illite (Schroeder and McLain 1998). During subsequent lithification, metamorphism and suduction, some of this silicate-bound ammonium can be stored in the continental crust and the Earth's mantle (Palya et al. 2011). Estimates suggest that the continental crust contains about half as much nitrogen as the modern atmosphere, and most of this crustal nitrogen is stored in (meta-)sedimentary rocks (Johnson and Goldblatt 2015). The contribution of biomass burial to the nitrogen reservoir in the mantle is difficult to estimate (Zerkle and Mikhail 2017), but it seems likely that biological activity has removed a significant amount of N_2 from the atmosphere over time.

Although biological N_2 fixation constitutes the major flux of nitrogen to seawater today, there are significant abiotic fixation fluxes, which would have dominated the nitrogen cycle prior to the emergence of life. The most important abiotic fluxes are hydrothermal reduction of N_2 to NH_4^+ (Brandes et al. 1998; Smirnov et al. 2008), photochemical HCN production by UV radiation (Zahnle 1986) and volcanically or lightning-driven conversion of N_2 to nitrogen oxides (mainly NO and N_2O) (Kasting and Walker 1981; Navarro-González et al. 1998; Mather et al. 2004), followed by reduction of nitrogen oxides to NH_4^+ with dissolved Fe^{2+} (Summers and Chang 1993). The magnitude of the various abiotic sources of fixed nitrogen on the early Earth has been appraised in computational models and laboratory experiments (reviewed by Stüeken et al. 2016a). By some estimates, these fluxes could have approached the modern biological N_2 fixation rate and may thus have been sufficient for driving the early biosphere and for modulating atmospheric N_2 pressures over long periods of time. However, the estimated global fluxes are necessarily based on parameters that are very poorly constrained, such as the rate of lightning, hydrothermal circulation, or oxidative volcanism on the early Earth. It is therefore equally likely that abiotic N_2 fixation flux was insufficient to sustain a significant biosphere, which would have favored an early origin of the nitrogenase enzyme.

4.4 The Emergence of Nitrogenase and Aerobic Nitrogen Metabolisms

Early phylogenetic reconstructions placed the origin of biological N_2 fixation into the Paleoproterozoic, around 2 Ga (Boyd et al. 2011). This conclusion was supported by the observation that most nitrogenases strictly require molybdenum as a catalyst (McGlynn et al. 2012), because geochemical data suggest that molybdenum was scarce in seawater until after the Paleoproterozoic Great Oxidation Event ($\sim$2.3 Ga, Anbar 2008; Scott et al. 2008). However, more recent phylogenetic analyses point to a much earlier origin of nitrogenase, possibly going back as far as the last universal common ancestor (LUCA), which probably lived in the Hadean or early Archean (Weiss et al. 2016). This inference can be reconciled with the strict requirement for molybdenum if hydrothermal source fluxes are taken into account. Off-axis hydrothermal vents contribute around 13% of the modern marine molybdenum budget (Wheat et al. 2002), and data from a hydrothermally altered Archean seafloor (Sulphur Springs Group, 3.3 Ga) show that hydrothermal fluids actively leached molybdenum from volcanic rocks at that time (Huston et al. 2001). Furthermore, culturing studies have demonstrated that biological N_2 fixation can be maintained down to around 5 nM of molybdenum (Zerkle et al. 2006; Glass et al. 2010), which is within the estimated range for the early ocean (1–10 nM; Reinhard et al. 2013). It is therefore likely that molybdenum was available in sufficient quantities, especially in the vicinity of hydrothermal settings. Importantly, the nitrogenase enzyme also contains an iron-sulfur cluster as well as the only known carbide atom in modern biology (Preiner et al. 2019). Both iron-sulfur clusters and carbides are produced in hydrothermal environments, which may indicate that this enzyme originated in a hydrothermal environment, possibly by modification of a geochemical process. In support of this model, biological N_2 fixation has been documented from thermophilic methanogenic Archaea that are deeply rooted in the tree of life (Mehta and Baross 2006). Enzymatic N_2 and CO_2 fixation may thus have originated in similar environmental settings early in Earth's history.

The oldest geochemical evidence of biological N_2 fixation in the form of nitrogen isotope ratios goes back to 3.2 Ga (Stüeken et al. 2015). The nitrogenase enzyme only imparts a small isotopic fractionation during the conversion of N_2 to NH_4^+ such that sedimentary $^{15}N/^{14}N$ ratios close to that of atmospheric N_2 are indirect evidence for burial of biomass that depended on this metabolism. Abiotic processes such as lightning and photochemical reactions seem to impart large negative fractionations (Kuga et al. 2014; Moore 1977) that are inconsistent with this record. Well-preserved sedimentary rocks throughout the Meso- and Neoarchean preserve signatures of biological N_2 fixation in a range of environmental settings (Stüeken et al. 2015; Stüeken and Buick 2018; Koehler et al. 2019; Ossa Ossa et al. 2019). Such geochemical records preserved in bulk rocks are unlikely to capture rare metabolisms that are only expressed by a small fraction of the ecosystem. Hence these data collectively indicate that a large proportion of the Archean biosphere had become independent from abiotic sources of fixed nitrogen. The drawdown of atmospheric N_2 (Som et al. 2016) would therefore have been dependent on biological productivity and biomass burial, which was in turn perhaps limited by low supplies of phosphorus (Kipp and Stüeken 2017; Reinhard et al. 2017).

The magnitude of the nitrogen flux from the atmosphere into geological archives via biomass burial also depends on the efficiency of remineralization, which ultimately returns nitrogen back to the atmosphere. Ferric iron can convert ammonium to N_2 (Yang et al. 2012), but in the Archean this metabolism would have been restricted to shelf environments where ferric iron may have been produced by photoferrotrophs (Crowe et al. 2008). Conversion of ammonium to nitrate (i.e. nitrification) above pH 6.5 requires O_2 (Stüeken et al. 2016a). Hence the efficient production of nitrate (which can subsequently be reduced to N_2 by denitrifying organisms) would have required the emergence of oxygenic photosynthesis. Isotopic evidence of at least locally aerobic biological nitrogen cycling dates back to around 2.66 Ga (Godfrey and Falkowski 2009; Koehler et al. 2018) and possibly earlier (Delarue et al. 2018; Homann et al. 2018). It has been proposed that the rise of coupled nitrification-denitrification would have induced a nitrogen crisis, because it leads to a net nitrogen loss from the ocean (Fennel et al. 2005). However, increasing supplies of molybdenum with the onset of oxidative weathering (Stüeken et al. 2012) would probably also have allowed for higher rates of biological N_2 fixation to offset the loss of fixed nitrogen from the water column.

Regarding atmospheric evolution, the rise of nitrification-denitrification would probably not have had a significant effect on net nitrogen burial in marine sediments, because, as noted above, the nitrogen burial flux is primarily controlled by biomass burial. The Archean ocean prior to the onset of nitrification was probably ammonium-poor, prohibiting efficient abiotic nitrogen burial through adsorption on clay minerals (Stüeken 2016). Biological N_2 fixation is an energetically costly process, and hence organisms are unlikely to have built up an excess reservoir of ammonium in Archean seawater. However, the expansion of ammonium oxidation around the GOE would probably have accelerated the rate of oxidative ammonium weathering on land (Stüeken et al. 2016b), which constitutes a significant flux of nitrogen today (Houlton et al. 2018). Nitrification-denitrification of this continental ammonium reservoir may have contributed to increasing atmospheric N_2 levels at that time (Stüeken et al. 2016b). It is therefore conceivable that Earth's biosphere has had a significant impact on not just atmospheric composition but also on total air pressure over geologic timescales (Som et al. 2016).

5 Stellar Evolution and Atmospheric Erosion

In addition to the geological and biological processes discussed above, the evolution of solar radiative properties would have had a significant effect on the evolution of Earth's atmosphere. While the evolution of the solar bolometric luminosity over time is well known (e.g. Sackmann and Boothroyd 2002), the solar XUV evolution and the magnetic activity of the Sun are less well understood. By the time the Sun had developed into a main sequence star, its bolometric luminosity was only about 70% of the present-day value but increased over time whereas the magnetic and XUV activity (i.e. the x-ray and EUV flux activity) steadily declined from the beginning to the present-day (e.g. Ribas et al. 2005; Guedel 2007; Füri et al. 2020, this issue). Recent studies of the magnetic and XUV evolution of solar-like stars (Johnstone et al. 2015a,b; Tu et al. 2015) found that these are strongly linked to the star's rotational evolution. As the rotation slows down, the XUV flux and the mass loss of a star also decline. After about a billion years, all stars converge towards one rotational track, but their initial rotational speeds can be very different, with the fastest rotators having the highest XUV fluxes and stellar mass losses and the slowest rotators the lowest, respectively (Johnstone et al. 2015a,b; Tu et al. 2015). For example, during the late Hadean eon, the XUV flux from a slow rotating Sun would have been about 15 times the present-day value, whereas for a fast rotating Sun it would have been up to 150 times as high as of today. Even though it is not yet entirely known whether the Sun was a slow, moderate or fast rotator, the latter seems unlikely (e.g. Lammer et al. 2020). Understanding the Sun's rotational evolution is crucial for reconstructing the evolution of the terrestrial atmosphere since higher XUV fluxes can have profound effects on atmospheric escape. In addition, atmospheres of different compositions respond very differently to an increase in solar XUV.

Several studies (Tian et al. 2008a,b; Johnstone et al. 2018) have shown that a nitrogen-dominated atmosphere (with a present-day N_2–CO_2 mixing ratio) adiabatically expands when the solar XUV flux exceeds a certain threshold (see Fig. 2), which is typically around 5–6 times the present-day value. This effect is accompanied by fast atmospheric bulk flow velocities that start to approach the escape velocity for values around 10 times the present-day XUV flux. At that point, the atmosphere transitions into hydrodynamic escape. Under such high XUV fluxes the exobase of a nitrogen-dominated atmosphere extends several Earth radii above the surface, thus reaching above the magnetopause standoff-distance (Lichtenegger et al. 2010). In addition to the strong thermal losses (Tian et al. 2008a,b), such an atmosphere would be strongly susceptible to non-thermal escape processes like ion pick-up (Lichtenegger et al. 2010; Lammer et al. 2018), facilitated by a much stronger early solar wind (Johnstone et al. 2015a). In fact, even without thermal escape, such a nitrogen-dominated atmosphere would have been lost within a few million years (Lichtenegger et al. 2010). A more recent study (Johnstone et al. 2019) has additionally shown that such atmospheres would also quickly be lost thermally around active G-type stars comparable to the young Sun during the Hadean and early Archean eons. A nitrogen-dominated atmosphere was thus unlikely to survive the early stages of the solar system. However, there is no sign

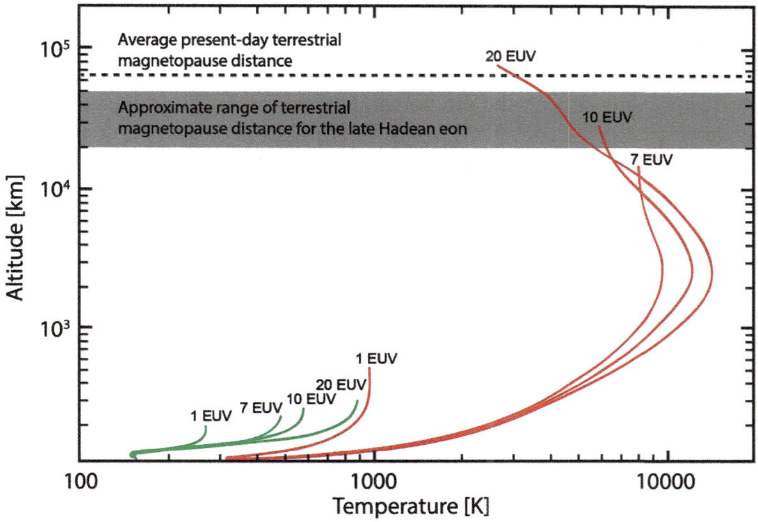

Fig. 2 Different thermospheric temperature profiles and corresponding exobase levels for 1, 7, 10 and 20 times the present-day XUV flux for a nitrogen-dominated atmosphere in red (as for the present-day Earth) (Tian et al. 2008a) and for a Venus-like atmosphere in green with 96% CO_2 (Kulikov et al. 2006). Due to the efficient infrared-cooling of CO_2 in the thermosphere of Venus exobase temperatures and levels are significantly lower than for the nitrogen-dominated atmosphere. The grey shaded area represents potential ranges for the terrestrial paleo-magnetosphere at the late Hadean eon (Scherf et al. 2018). Figure adopted from Lammer et al. (2008)

of such escape in the terrestrial $^{14}N/^{15}N$ isotopic record. The lighter isotope should have escaped more easily than the heavier one if a strong atmospheric escape took place in the past (Cartigny and Marty 2013; Füri and Marty 2015; Avice and Marty 2020, this issue). The small $^{14}N/^{15}N$-disequilibrium between the atmosphere and the terrestrial mantle compared to, for example, Mars indicates that only a very small fraction of nitrogen escaped over the history of the Earth (Lammer et al. 2018), and this escape process likely ended before 3.5 Ga, as isotopic data show nearly constant values since that time (Marty et al. 2013; Avice et al. 2018). So what happened to the atmosphere? A solution might be the following.

As briefly mentioned in Sect. 3, atmospheres with different compositions respond very differently to high XUV fluxes. A higher atmospheric partial pressure of CO_2, for instance, can prevent the exobase from expanding dramatically (Fig. 2), because CO_2 serves as an infrared cooler in the upper atmosphere (Kulikov et al. 2007; Lammer et al. 2008; Johnstone et al. 2018).

The N_2–CO_2-mixing ratio might therefore be crucial for the survival of the early terrestrial atmosphere. It has indeed long been suggested that the CO_2 partial pressure during the Hadean and early Archean was significantly higher than today (e.g. Holland 1984; Berner and Kothavala 2001; Dauphas and Kasting 2011; Reinhard and Planavsky 2011; Goldblatt and Zahnle 2011) with values of up to about 100–1000 times the Present Atmospheric Level (PAL) (Hessler et al. 2004). A more recent study (Kanzaki and Murakami 2015) has even shown that around 2.7 billion years ago the CO_2 partial pressure in the atmosphere was as high as 80–100 PAL. These numbers make it plausible that high CO_2 levels limited N_2 escape by the mechanism described above. It is, however, not yet entirely clear which mixing ratios could have prevented an atmosphere with a significant of N_2 from escaping to space. A later build-up of the Earth's nitrogen-dominated atmosphere might be an alternative solution to this paradox (Lammer et al. 2018, 2019).

There is one geological marker that indicates significant atmospheric loss of hydrogen from the terrestrial atmosphere: the fractionation of the heavy Xe-isotopes (e.g. Srinivasan 1976; Zahnle et al. 1990; Pepin 2000; Pujol et al. 2009, 2011; Avice et al. 2018; Zahnle et al. 2019; Avice and Marty 2020, this issue). The underlying fractionation process is expected to be mass-dependent dragging of Xe^+ by hydrodynamically escaping H^+ due to a strong repulsive Coulomb interaction between these ions, which preferentially happens via the polar cusps (Zahnle et al. 2019). Since Xe is easily ionized compared to other elements including hydrogen, this process can lead to strong Xe^+ escape rates while the escape of lighter species such as N and other noble gases (Ne, Ar, Kr, Avice et al. 2018) is not significant. For this process to be efficient, however, high XUV fluxes, probably >10 times the present-day value, and a significant hydrogen mixing ratio in the upper atmosphere might be needed (Zahnle et al. 2019). It is, however, not entirely clear how the atmospheric composition influences the escape of Xe; Zahnle et al. (2019) did not include atmospheric N_2 into their model, whose total pressure might influence the escape process of the heavy Xe^+.

Fractionation of Xe isotopes appears to have ended towards the end of the Archean (Pujol et al. 2011), which indicates a complex interplay of solar XUV flux, escape of hydrogen, atmospheric composition and the terrestrial magnetic field. Understanding the "missing Xenon" paradox (Ozima and Podosek 1983)—in particular together with the small terrestrial $^{14}N/^{15}N$-disequilibrium—is therefore a crucial component for the reproduction of not only the terrestrial atmospheric evolution but also of the XUV flux evolution and potential bursts and flares of the young Sun.

6 Evidence and Mechanisms of Photochemistry in Earth's Anoxic Atmosphere

Clues to paleo-environmental conditions are left behind in Earth's sedimentary rock record. The ancient sediments that have somehow survived the gauntlet of tectonic recycling can be investigated for traces they record about the oceanic environment they were deposited in. Various proxies have been developed to constrain the temperature and chemistry of the originating waters, along with quantifying the disequilibrium imparted by the metabolisms of ancient life (Sect. 4). Unfortunately, Earth's sedimentary record is effectively absent for the first billion years of Earth's history and somewhat sparse during the second billion. As a result, the paleo-environments that we have to interrogate do not constrain Earth's tumultuous early days but are limited to post- accretion/heavy bombardment/blow-off conditions known as the "secondary atmosphere" (see Sects. 2–5). This term refers to what most Earth Scientists consider as simply the atmosphere—the envelope of gas derived primarily from volcanic outgassing subject to intense biological meddling. Photochemistry—the interaction of solar light and absorption/scattering/transformation of input gases—sets the bulk composition of the atmosphere and the UV environment at the surface, both of which play strong roles in climate and planetary habitability. Understanding the bulk composition of the atmosphere from ancient ocean sediments is a frontier interdisciplinary challenge.

As atmospheric gases are generally not preserved directly by the rock record, various geochemical measurements have been proposed as proxy for atmospheric information. Proxies are developed from studying modern analog processes that generally involve quantitative involvement of atmospheric gases, but most of these processes are indirect (e.g., via chemical weathering) and can suffer from non-uniqueness (e.g., oxidation can occur in the water column or sediments without being atmospheric). The only direct proxies for atmospheric chemistry that exist are mass-independent isotope fractionations, or MIF, which originate primarily in gas-phase photochemical reactions (Thiemens 2006). MIF signals are known to arise in ozone formation (Thiemens et al. 2012), and via various photochemical reactions involving sulfur (Ono 2017) and mercury (Blum et al. 2014). MIF has additional strength as atmospheric proxy given that these anomalous isotopic signals persist through most of the transformations that occur during biogeochemical cycling prior to deposition into the sediments. These atmospheric signatures are therefore relatively easy to transfer and store in the permanent rock record, where they will remain unless subjected to significant metamorphic temperatures. While MIF in O, S, and Hg have many interesting roles in modern and paleoclimates, this section will focus on reviewing MIF-S, which has the most data relevant to the first two billion years of Earth history.

6.1 Discovery and Expansion of the MIF-S Dataset

Sulfur exists in multiple redox states in atmospheric and marine environments but is preserved in the rock record in two distinct forms—sulfide and sulfate. MIF-S (quantified simply as $\Delta^{33}S = \delta^{33}S–0.52\delta^{34}S$ and $\Delta^{36}S = \delta^{36}S–1.9\delta^{34}S$) measurements of opposing magnitudes were discovered in both sulfides and sulfates in the year 2000. In the subsequent 19 years, the global database of MIF-S measurements has grown substantially revealing both temporal and stratigraphic variations, with substantial reviews in 2011 (Johnston 2011) and 2017 (Ono 2017) (see also Fig. 3). Like nearly every other geochemical measurement, the ability to measure the signal has to date outpaced our ability to interpret it. The most profound interpretation that has been derived from MIF-S is that the atmosphere prior to about 2.3 billion years ago was devoid of oxygen gas, which effectively resolved a century-old debate. Critically, this interpretation requires only the mere presence of MIF-S. As the data also show substantial variations in magnitude, sign, facies, and derived slopes, there is significant hope that additional constraints about the specific gases in the reducing atmosphere can to be determined from this record.

6.2 Interpreting Sulfur MIF

The primary interpretation of a reducing atmosphere stemmed from three foundational papers which combined data, experiments, and theory and collectively demonstrated that MIF-S could not be exported from atmospheres with more than 1 ppm of O_2. Farquhar et al. (2000) presented the first data and hypothesized a role for reducing atmospheric chemistry. Follow-on experiments irradiated SO_2 gas and quantified MIF-S signals in the elemental sulfur and residual SO_2 (Farquhar et al. 2001). The magnitudes, signs, and $\Delta^{36}S/\Delta^{33}S$ slope of the measured signal changed with wavelength of light and were best explained by SO_2 photolysis in the absence of O_2. Pavlov and Kasting (2002) used a 1-D photochemical model and showed that MIF-S propagates through all atmospheric S species, highlighted the role that multiple exit channels play in signal preservation, and demonstrated that atmospheres with more than 1 ppm p_{O_2} will turn all input sulfur to sulfate, so that MIF-S preservation should only occur in reducing atmospheres. The most substantial challenge to the interpretation was raised by Ohmoto et al. (2006) who argued that the low magnitudes of $\Delta^{33}S$ seen between 3.2–2.7 Ga might reflect mass conservation effects in an oxic atmosphere. However, Farquhar et al. (2007) demonstrated using $\Delta^{36}S/\Delta^{33}S$ slopes that these signals unequivocally preserve MIF-S, even though the magnitudes are low.

6.2.1 Modern Updates to the Initial Interpretation

Follow-on work has confirmed the primary interpretation of a reducing early atmosphere, although the specific mechanism and theory behind it have changed, and are still undergoing revision (Ono 2017; Sect. 7) Atmospheric exit channels remain a key part of the story, but the mechanism is no longer thought to be solely SO_2 photolysis. While a single dominant source mechanism can produce variable magnitudes of MIF-S depending on the exit channels (Claire et al. 2014), subsequent processing in the oceans would likely reduce magnitudes of signatures that are preserved in various facies (Halevy 2013). A current consensus view is that multiple mechanisms involving sulfur photochemistry may be integrated into the rock record. This is also true of the experimental work done to constrain the relevant fractionation factors.

6.3 SO_2 Photolysis Experiments

Experiments following on from Farquhar et al. (2001), revealed that SO_2 photolysis with broadband lamps displayed varying sulfur MIF magnitudes with SO_2 and total pressure (Masterson et al. 2011; Whitehill and Ono 2012; Endo et al. 2016). Interpretation of experimental results to the natural atmosphere are complicated by self-shielding conditions in the chamber experiments, and multiple important gas mixtures remain to be tested under atmospherically relevant conditions. Hints towards understanding the mechanism were provided by measurements of SO_2 absorption cross-section (Danielache et al. 2008; Endo et al. 2015), although these measurements did not resolve the potential effect of hyper-fine structure (Stark et al. 1999; Lyons 2009). Recent work has untangled the role of background pressure from self-shielding (Endo et al. 2019), and implies that pressure broadening in the terrestrial atmosphere may remove the need to consider the SO_2 fine-structure. Overall, SO_2 photolysis has been unequivocally shown to produce MIF-S, but the magnitude of $\Delta^{33}S$ produced in low pressure atmospheres appears to be smaller than the observed data, and a consensus viewpoint is that another mechanism in addition to SO_2 photolysis is contributing to the observed signal.

6.3.1 Other Mechanisms Known to Produce MIF-S

MIF-S mechanisms have become quite a bit more complicated than the corresponding case of MIF-O, for which only one atmospheric mechanism has been revealed in 40 years of study. MIF-S has been demonstrated to arise from CS_2 photolysis (Zmolek et al. 1999), and SO_2 photodeexcitation (Whitehill and Ono 2012). Theory supports MIF-S formation in SO_2 photoexcitation (Danielache et al. 2012), SO photolysis (Danielache et al. 2014) and in the polymerization of elemental sulfur (Babikov 2017; Harman et al. 2018b). There are a number of complications involved with extrapolating from experimental and theoretical results to natural atmospheric conditions, as multiple mechanisms are likely to be involved during and experiment. Extrapolation of fractionation factors to natural atmospheric conditions also remains challenging. Recently theorized effects in sulfur polymerization can create large magnitude MIF-S signals but need experimental verification.

6.4 What Is Needed to Move Forward?

It is clear that the extensive MIF-S dataset contains far more information than a single on/off switch for O_2 (Ono 2017). There are changes in the sulfur MIF record between the Paleo-,

Meso-, and Neo-Archean that are best described as changes in the transparency to UV driven by varying gas compositions at these times. In addition, stratigraphic changes in MIF-S hint at time-dependent shifts in atmospheric chemistry on shorter timescales (Zerkle et al. 2012; Izon et al. 2015, 2017). Quantitative constraints on the composition of reducing atmosphere appears feasible given the quality of the observed record, but all hinges on identifying and quantifying all of the anomalous fractionation factors which emerge from sulfur gas phase chemistry. It is not clear that this has occurred yet—it may have, but there certainly exists the possibility that one or more major processes have yet to be uncovered.

Once all fractionation factors are identified and quantified, a number of challenges remain. This include a full calibration and validation of the modern sulfur MIF record generated by stratospheric volcanic eruptions as recorded in ice cores (Savarino et al. 2003). Recent methodological advances have enabled sampling of $\Delta^{33}S$ at $\sim$ monthly resolution (Burke et al. 2019) providing a robust dataset for exploration with a time-dependent photochemical plume model, although detection of $\Delta^{36}S$ at small sample size remains a frontier challenge. Once calibrated against the modern (and all experimental work), photochemical modelling could be undertaken with a full Monte-Carlo exploration over the unknown Archean parameter space, with the goal of identifying which background atmospheres reproduce, for example, the Neoarchean reference array of $\delta^{34}S/\Delta^{33}S \sim 1$ and $\Delta^{36}S/\Delta^{33}S \sim -0.9$. Monte-Carlo uncertainty propagation coupled to sensitivity analysis could be used to identify, measure, and thus reduce uncertainty in any reactions, cross-sections, or other internal model parameterizations which contribute significant uncertainty to MIF-S prediction (Hébrard et al. 2009). An ultimate goal would be an inverse model—one in which the data itself are allowed to constrain a Monte-Carlo parameter search—which could identify not only the portions of unknown parameter space which can match the data, but also constrain their likelihood. In summary, interpretation of atmospheric composition from sulfur MIF measurements and experiments remains a challenging frontier but is ultimately a surmountable problem.

7 Global Methanotrophy and the Possibility of a Neoarchean CO-World

7.1 The Negative $\delta^{13}C_{org}$ Anomaly in the Late Archean: "Global Methanotrophy"?

Stable isotopes of sedimentary rocks reflect biogeochemical cycling and its evolution through time (Fig. 3). Since the pioneering works of the Precambrian Palaeontology Research Group (Schidlowski et al. 1983; Hayes 1983), it has been known that carbon isotopes of sedimentary carbonate ($\delta^{13}C_{carb}$) and organic matter ($\delta^{13}C_{org}$) show an isotopic difference of about 25‰ on average since at least 3.5 billion years ago (Ga). This observed isotopic fractionation is primarily derived from biological CO_2 fixation: autotrophic organisms preferentially convert ^{12}C from mantle-derived CO_2 (~ -5‰) into organic carbon (-25‰), pushing the remaining inorganic carbon, which is deposited as carbonate, to heavier values (~ 0‰) relative to the starting composition.

This simplest explanation may not be applicable for the late Archean record from 2.7 to 2.5 Ga, when the organic carbon exhibit large ^{13}C-depletion below -40‰ (Fig. 1). Apart from localized occurrences, such a large depletion over extended temporal and spatial scales is not observed outside of the late Archean time window, suggesting a distinct biogeochemical carbon cycling during in the period. A leading interpretation of these data is the "*global methanotrophy" hypothesis*" (Hayes 1994).

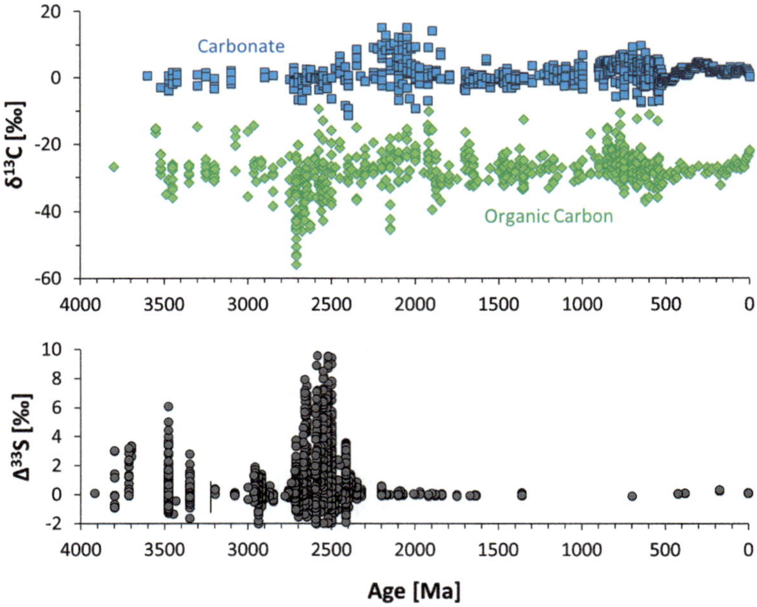

Fig. 3 Stable carbon and sulfur isotopes through time. In the upper panel, blue and orange symbols show carbonate carbon ($\delta^{13}C_{org}$) and organic carbon ($\delta^{13}C_{carb}$) isotopes in sedimentary rocks, respectively. Lower panel shows Sulfur Mass Independent Fractionation ($\Delta^{33}S$) of sedimentary sulfide and sulfate measured by bulk rock chemical extraction methods. In the late Archean from 2.7 to 2.5 Ga, both the $\delta^{3}C_{org}$ and $\Delta^{33}S$ values show larger variation including negative $\delta^{13}C$ below −40‰

The highly negative $\delta^{13}C_{org}$ are unlikely to have been produced solely by photosynthesis and other autotrophic carbon fixation pathways, which exhibits fractionations less than 40‰ against CO_2 (e.g., House et al. 2003). Instead, these values are consistent with methanotrophic metabolisms in which isotopically-light CH_4 is incorporated into biomass, leading to $\delta^{13}C_{org}$ well below −40‰. These observations are what led Hayes (1994) to propose the "*global methanotrophy hypothesis*" where methanotrophic organisms appeared at around 2.7 Ga and started to incorporate atmospheric CH_4 globally in the late Archean. Atmospheric model calculations indicate that Archean CH_4 levels were likely high in the anoxic Archean atmosphere (Pavlov et al. 2001). Importantly, microbial production of CH_4 (methanogenesis) dates back to at least 3.5 Ga (Ueno et al. 2006). Global methanotrophy may have been terminated by increasing atmospheric oxygen levels, which removed CH_4 from the atmosphere, and thus the negative $\delta^{13}C_{org}$ signature disappeared after about 2.4 Ga. The onset of the global methanotrophy episode is more difficult to explain, but it could potentially be related to localized appearances of oxygenic photosynthesis at around 2.7 Ga (Eigenbrode and Freeman 2006; Stüeken et al. 2017). Aerobic methanotrophs not only assimilate CH_4 but also couple the oxidation of CH_4 to O_2 for energy. Anaerobic methanotrophs, which use nitrate, ferric iron or sulfate as an oxidant instead of O_2, would benefit indirectly from increased O_2 levels, because O_2 generates more of these alternative oxidants through oxidation of ammonium, ferrous iron and sulfide, respectively. Originally, Hayes (1994) explained the expansion of methanotrophy by the concomitant appearance of oxygenic phototrophs based on the occurrence of stromatolites and cyanobacterial microfossils at about 2.7 Ga (Packer 1990). However, it remains conceivable that the isotopic anomaly was caused by anaerobic methanotrophs that received their oxidants without the need for O_2 (Hinrichs 2002).

Irrespective of the exact mechanism, the period of highly negative $\delta^{13}C_{org}$ value likely points to a redox transition of in the atmosphere and ocean. It is important to note that not only carbon, but also sedimentary nitrogen and iron isotopes are anomalous during this time period compared to the earlier Archean (Pinti and Hashizume 2010; Koehler et al. 2018), possibly indicating increasing abundances of oceanic nitrate and iron oxide. It is tempting to conclude that a gradual and perhaps localized oxidation of surface environments not only transformed the biogeochemical carbon cycle, but also sulfur, nitrogen, and iron cycling in the time period from 2.7 to 2.5 Ga.

7.2 Sulfur Isotopic Anomalies in the Neoarchean and Its Correlation with the $\delta^{13}C_{Org}$

In contrast to the anomalous $\delta^{13}C_{org}$, $\delta^{15}N$ and $\delta^{54}Fe$, the $\delta^{34}S$ record of sedimentary sulfide and sulfate do not show a distinctive signature in the Neoarchean. Instead, the $\Delta^{33}S$ value (Sect. 6) shows the largest scatter during this time (Fig. 1). Since large perturbations in MIF-S are thought have been produced in the atmosphere, the apparent correlation between $\Delta^{33}S$ and $\delta^{13}C_{org}$ may suggest a link between biogeochemical cycles and atmospheric chemistry. So far, the environmental factors controlling the magnitude and preservation of $\Delta^{33}S$ values is largely unknown because a clear understanding of the primary photochemical mechanism of S-MIF production and its transfer from the atmosphere to sediments is lacking. Based on experimental studies, large $\Delta^{33}S$ fractionations over 10‰ can be produced when the optical thickness of SO_2 is large enough (Lyons 2009; Ono et al. 2013; Endo et al. 2016). Such a high optical thickness, i.e. higher atmospheric p_{SO_2}, could have been produced by enhanced subaerial volcanism at that time. This suggestion is roughly consistent with the geological record of continental crust formation estimated from age populations of zircons (Condie 2005). However, although the link to "global methanotrophy" remains unclear in this scenario, it is also important to keep in mind that the atmospheric $\Delta^{33}S$ signal can be diluted by the addition of mass-dependently fractionated "normal" sulfur sources before preservation in sediment (e.g., sulfate supplied from oxidative weathering of continents, hydrothermal input, or volcanic H_2S; Halevy 2013). Thus, the magnitude of $\Delta^{33}S$ alone cannot be interpreted unambiguously.

An alternative interpretation may be obtained from the analysis of both $\Delta^{33}S$ and $\Delta^{36}S$. Like $\Delta^{33}S$, $\Delta^{36}S$ is an expression of the photochemically-induced mass independent fractionation that was produced in the Archean atmosphere. Even though the atmospheric signal of both $\Delta^{33}S$ and $\Delta^{36}S$ can be diluted by the mechanisms listed above, the $\Delta^{36}S/\Delta^{33}S$ ratio would remain unchanged because the "normal" sulfur sources should exhibit $\Delta^{33}S = \Delta^{36}S = 0$. Interestingly, in the late Archean, the $\Delta^{36}S/\Delta^{33}S$ ratio shows systematic fluctuations around a mean of -0.9, and these fluctuations seem to correlate with $\delta^{13}C_{org}$ anomalies in multiple depositional basins (Ono et al. 2003; Kaufman et al. 2007; Zerkle et al. 2012; Izon et al. 2015; Mishima et al. 2017). The apparent correlation may possibly imply the connection between the atmosphere and the biosphere through CH_4 metabolisms. Based on a photochemical model, Izon et al. (2017) and Claire et al. (2014) argued that the formation and thickness of a hydrocarbon haze in the atmosphere can change the S-MIF signal by altering the transmission spectrum of UV radiation. Changes in the UV wavelength affect SO_2 photolysis and thus may control the change in $\Delta^{36}S/\Delta^{33}S$ ratio. The effect is primarily derived from the difference in absorption cross sections of SO_2 isotopologues, although a later experimental study demonstrated that the isotopologue cross-section should be refined for predicting $\Delta^{33}S$ and $\Delta^{36}S$ more precisely with a proper consideration of total pressure and temperature (Endo et al. 2019). Furthermore, laboratory experiments of SO_2 photochemistry have demonstrated that the $\Delta^{36}S/\Delta^{33}S$ ratio is more sensitive to the availability of reducing gas species (potentially H_2, CH_4 or CO) rather than haze thickness (Whitehill et al. 2013; Endo et al. 2016). Thus, our understanding of the potential link with the atmospheric state and the late Archean $\delta^{13}C_{org}$ anomaly is still limited.

7.3 An Alternative Scenario: A Late Archean CO-World

Alternative to the "global methanotorophy" scenario, the anomalously low $\delta^{13}C_{org}$ in the late Archean may possibly be explained by non-biological processes. Pavlov et al. (2001) pointed out that organic haze produced from CH_4 by photochemical reactions in the atmosphere should be highly depleted in ^{13}C, and thus the deposition of the haze in surface environments may possibly contribute the observed negative $\delta^{13}C_{org}$ anomaly in sediment at that time. However, also in this case, the link between the carbon isotopes and S-MIF remains uncertain.

As noted above, high p_{SO_2} seems to be required to explain the observed large $\Delta^{33}S$ over +10‰ due to the significant SO_2-self-shielding effect during SO_2 photolysis. However, self-shielding alone should produce the S-MIF with a $\Delta^{36}S/\Delta^{33}S$ ratio of about −2.5 (Lyons 2009; Ono et al. 2013; Endo et al. 2016), which does not match the late Archean record ($\Delta^{36}S/\Delta^{33}S$ ranges from −1.52 to −0.85). To resolve this issue, Endo et al. (2016) demonstrated experimentally that the shallower $\Delta^{36}S/\Delta^{33}S$ slope about −1 can be produced when the photoproduct SO comes not only from SO_2 photolysis (destruction of the molecules, 185–220 nm) but also from the SO_2 photoexcitation channel (excitation of an electron to a higher energy level, 250–320 nm). The contribution from the SO_2 photoexcitation channel is expected in a reducing atmosphere, especially in the presence of carbon monoxide (CO roughly 0.01 $\sim$ 0.1 bar: Endo et al. 2016). The observed $\Delta^{36}S/\Delta^{33}S$ ratio around −1 may thus be evidence that the Neoarchean atmosphere was not only rich in CH_4 but also CO. A plausible source of CO would have been photochemical CO_2 reduction coupled to the presence of a strong reducing agents (most likely volcanic H_2) that was supplied directly to the atmosphere (Kharecha et al. 2005; Yung and DeMore 1982).

In such a hypothetical CO-rich atmosphere, low-^{13}C organic carbon in the form of haze particles may have rained out into surface environments. Based on theoretical calculation of absorption spectra of the CO_2 isotopologues (Schmidt et al. 2013), solar UV photolysis of CO_2 may cause significant carbon isotopic fractionation of about 200‰ because ^{13}C and ^{12}C have quite distinct UV absorption properties. Since the most probable source of atmospheric CO is from the photodissociation of CO_2, the atmospheric CO would have been highly depleted in ^{13}C. Furthermore, the atmospheric CO could have been converted mainly into formaldehyde and deposited on the surface (Bar-Nun and Hartman 1978; Pinto et al. 1980; Bar-Nun and Chang 1983), where it could have been assimilated into the biosphere.

Although the large isotope effect has not yet been tested quantitatively by laboratory experiment, and the chemical/biological processes to transfer the isotopically depleted CO into sediment are largely uncertain, the alternative CO-world hypothesis may explain both carbon and sulfur isotope anomalies in the late Archean, and it should be considered as a possible scenario on other exoplanetary worlds.

8 Atmospheric Biosignatures on the Early Earth

Using the early Earth as an analogue for exoplanets not only provides insights into the mechanisms that drive planetary habitability and evolution, but it also allows us to develop methods of detecting microbial biospheres remotely. Remote detection relies on absorption features of diagnostic gases in reflectance spectra of planets. Such diagnostic gases are those that are those that have an overwhelmingly biological source. These signatures are thus dependent on biological metabolisms but also require the characterization of potential abiotic sources that could create 'false positives'.

8.1 Methane and Other Metabolic Products

When considering the atmospheric biosignatures on the early Earth we first note that while life likely produces thousands of molecules in small quantities (see Seager et al. 2016), only a few have fluxes high enough to contribute substantially to the atmosphere. Even fewer would be detectable from the perspective of an exoplanet that is similar to early Earth (Rugheimer and Kaltenegger 2018). While it is possible that alternative biochemistries exist, to date we only know of life that uses a carbon-based biochemistry and liquid water. Since carbon dioxide and water are abundant in the Universe (Fortney 2012), it is conceivable that at least some life forms elsewhere evolved using similar chemical energy pathways and would thus potentially be detectable by methods developed based on Earth's biosphere.

While the first life form on Earth is not known, it is postulated that chemoheterotrophic organisms with metabolisms such as iron reduction and methanogenesis arose first due to the relative simplicity of their metabolic pathways. Methane is a gas with both biological and geological sources (Sects. 4, 7). On Earth, however, the biological flux is much higher than the abiotic flux, which is primarily focused around hydrothermal vents and subduction zones (Guzmán-Marmolejo et al. 2013). Given the antiquity of methanogens, the abundance of metabolic substrates for them (CO_2, H_2 and metal catalysts) and the absence of free oxygen—a major sink for methane—in the Archean, it is thought that atmospheric methane was a large component of Earth's early atmosphere (Claire et al. 2006; Zahnle et al. 2006). Estimates suggest that the residence time of methane would be 100 times greater than it is today. However, as noted above, methane is not unique to biology and 'false positives' could arise on planets around starts with a relatively low UV flux, which may allow abiotic methane to accumulate in the planet's atmosphere (Guzmán-Marmolejo et al. 2013; Domagal-Goldman et al. 2014; Rugheimer et al. 2015a, 2015b).

The Archean Earth may have experienced an episode where methane accumulation was so strong that a hydrocarbon haze formed in the atmosphere (Zerkle et al. 2012). Such a haze creates an additional biosignature, because it would indicate large amounts of methane production and a biosphere dominated by methanogenesis (Arney et al. 2016, 2018). Furthermore, a haze could protect surface environments on the planet from harmful UV-C radiation, which may result in an expansion of the biosphere and thus create a positive feedback on the production of biogenic gases, including hydrocarbons (Arney et al. 2017; Rugheimer et al. 2015b; Hörst et al. 2018; Ozaki et al. 2017). A haze-dominated planet would have detectable spectral features in the UV which could serve as a biosignature of an anoxic world (Arney et al. 2018).

Besides methane, biogenic sulfur gases have been proposed as a biosignature in reducing atmospheres (Domagal-Goldman et al. 2011), although their detection is challenging due to spectral overlap. Phosphine is another gas which has recently been proposed as a biosignature on an anoxic world (Sousa-Silva et al. 2019). Phosphine is produced by some anaerobic organisms on Earth (Bains et al. 2019), and it is the simplest, isolated gas without known false positives at detectable levels. Importantly, it has strong spectral features distinct from other biosignatures which if present, could indicate an anoxic biosphere on an extrasolar planet, although its fluxes on modern (and early) Earth are not well understood.

On the modern Earth, O_2 is the most notable biosignature gas, produced by oxygenic photosynthesis. However, models have identified many abiotic mechanisms that build up appreciable quantities of this gas and create a false positive. Several of those mechanisms would be present on a young planet (e.g. Hu et al. 2012; Domagal-Goldman et al. 2014; Tian et al. 2014; Rameriz and Kaltenegger 2014; Wordsworth and Pierrehumbert 2014; Luger and Barns 2015; Harman et al. 2015; Gao et al. 2015; or see recent review by Meadows et al. 2018). These abiotic mechanisms of O_2 production largely require UV photolysis of H_2O followed by hydrogen escape, or the photolysis of CO_2. The final amount of O_2 in the atmosphere, however, will ultimately depend on the redox potential of the surface, making it vital to measure other indicators to put constraints on the redox budget of the planet (see Schaefer et al. 2016; Wordsworth et al. 2017). The additional detection of CH_4 or H_2 or constraints on CO_2 or CO could eliminate the "false positive problem" (Domagal-Goldman et al. 2014), as does the inclusion of lightning which is thought to be ubiquitous in planetary atmospheres (Harman et al. 2018a). For most of Earth's early history, O_2 levels were low, around 10^{-7} to 10^{-5} bar as constrained by geological evidence from mass independent sulfur isotope fractionation (Sect. 6). It is debated when oxygenic photosynthesis first evolved (Brocks et al. 1999; Planavsky et al. (2014a); French et al. 2015; Lalonde and Konhauser 2015; Gold et al. 2017). However, a permanent and relatively quick rise in atmospheric O_2 seemingly occurred around 2.33 Ga, termed the Great Oxidation Event (Luo et al. 2016; Bekker et al. 2004).

Oxygenic photosynthesis requires a second photosystem to harvest electrons from water, suggesting that this evolutionary innovation postdates the origin of anoxygenic photosynthesis, which requires only one photosystem (Hohmann-Marriott and Blankenship 2011). However, once oxygenic photosynthesis evolved, it probably began to dominate ecosystems relatively quickly, because the required substrates (photons, water and CO_2) are abundant. Hence there may be a strong evolutionary selection pressure in favor of this metabolism, maintaining the possibility that it also evolved elsewhere. O_2 gas remains a primary observational goal on exoplanets (Meadows et al. 2018). However, oxygen only produces a significant absorption feature in the spectrum of a planet if its partial pressure is above 10^{-2} times present atmospheric levels (PAL) (Segura et al. 2003). Below this threshold also O_3 becomes difficult to detect, although it has stronger absorption features than O_2 (Kasting and Catling 2003). Hence the invention of oxygenic photosynthesis is not a sufficient condition for the detection of an aerobic biosphere. It also depends strongly on O_2 sinks, i.e. fluxes of reducing gases and minerals to a planet's surface. There is extensive debate about when in Earth's history detectable amounts of O_2 were present. Estimates of Proterozoic O_2 levels range from 0.1% PAL to a maximum of 40% PAL (Canfield 2005, Kump 2008, Planavsky et al. 2014b). The presence of charcoals after 420 Ma indicates O_2 levels of at least 60% PAL and up to 160% PAL (Scott and Glasspool 2006, Kump 2008). See Lyons et al. (2014) for a comprehensive review of oxygen through Earth's history and Meadows (2018) for a review of oxygen as a biosignature target on exoplanets.

8.2 N_2 and O_2 as Possible Geo-Biosignature

Nitrogen gas (N_2) is largely of geological origin, but it has a dynamic biogeochemical cycle with multiple oxidation states, which has led to the idea that a long-term stable N_2–O_2 atmosphere may be indicative of aerobic lifeforms and even point to an active tectonic regime

(Krissansen-Totton et al. 2016; Stüeken et al. 2016b; Lammer et al. 2019). These arguments are based on the premise that (a) N_2 and O_2 are in thermodynamic equilibrium and may ultimately be converted to nitrogen oxides by high energy processes, unless both of them are constantly replenished by life (Krissansen-Totton et al. 2016); (b) net burial of nitrogen proceeds mostly under anoxic conditions where oxidative weathering of crustal ammonium is suppressed, which implies that the rise of O_2 also leads to a decrease in N burial and therefore a build-up of N_2 in the atmosphere, provided that buried N becomes re-exposed to weathering by tectonic uplift (Stüeken et al. 2016b), and (c) N_2 may be easily removed from the atmosphere on young anoxic planets with a weak magnetic field where cosmic radiation is strong, whereas most return fluxes of N_2 to the atmosphere are biologically driven and proceed via nitrate or nitrite, which require the presence of O_2 (Lammer et al. 2019). Of course, these deliberations are only valid for terrestrial planets in their respective habitable zone, but these are the class of planets where the conditions for life as we know it are most promising. The detection of an atmosphere with N_2 and O_2 as bulk gases and only a considerably lower share of CO_2 most likely constitutes a geo-biosignature.

9 Conclusions and Outlook

The development and sustenance of habitable conditions on Earth depends on numerous geological, solar, and biological factors. Initially, the composition and volatile content of Earth's building blocks determined the amount of water and greenhouse gases that could be released and maintained after Moon formation. The release itself was in turn dependent on the longevity of the magma ocean phase (Sect. 2), which would have been a function of Earth's internal heat, i.e. its endowment of radioactive elements, planetary differentiation and gravitational contraction. Further, the accumulation of gases within the atmosphere may have been limited by stellar erosion, if the young Sun radiated more strongly in the XUV range than it does today and if the upper atmosphere was not sufficiently cooled by high levels of CO_2 (Sect. 5). It is also conceivable that some volatiles were converted back into magma-soluble molecules (e.g. N_2 to NH_3) by atmospheric processes (Sect. 2). Direct evidence for photochemistry on the early Earth is archived in the sulfur isotope record (Sects. 6, 7) and possibly in the record of organic carbon isotopes (Sect. 7). The return flux of volatiles into the mantle may have been accelerated by the onset of plate tectonics, and the timing of this onset was probably determined by the degree of volatile loss during the magma ocean phase (Sect. 3). The emergence of life likely enhanced the drawdown of both CO_2 and N_2 while also affecting the redox state of the atmosphere through the production of CH_4 and O_2 (Sect. 4). These biological imprints have probably had a substantial effect on the evolution of our planet and may serve as biosignatures on other terrestrial worlds (Sect. 8).

Our review highlights the complex interplay between physical and chemical processes many of which are so far only poorly constrained. The future of this field of research will greatly benefit from new cross-disciplinary approaches.

Acknowledgements We thank the organizers of the ISSI workshop on Planetary Evolution 2018 for making this volume possible. Bernard Marty provided detailed, constructive comments that improved the manuscript.

References

V.S. Airapetian, A. Glocer, G. Gronoff, E. Hébrard, W. Danchi, Prebiotic chemistry and atmospheric warming of early Earth by an active young Sun. Nat. Geosci. **9**(6), 452–455 (2016)

A. Anbar, Elements and evolution. Science **322**, 1481–1483 (2008)

A. Anbar, Y. Duan, T.W. Lyons, G.L. Arnold, B. Kendall, R.A. Creaser, A.J. Kaufman, G.W. Gordon, C.T. Scott, J. Garvin, R. Buick, A whiff of oxygen before the Great Oxidation Event? Science **317**(5846), 1903–1906 (2007)

G. Arney, S.D. Domagal-Goldman, V.S. Meadows, E.T. Wolf, E. Schwieterman, B. Charnay, M. Claire, E. Hébrard, M.G. Trainer, The pale orange dot: the spectrum and habitability of hazy Archean Earth. Astrobiology **16**, 873–899 (2016). https://doi.org/10.1089/ast.2015.1422

G.N. Arney, V.S. Meadows, S.D. Domagal-Goldman, D. Deming, T.D. Robinson, G. Tovar, E.T. Wolf, E. Schwieterman, Pale orange dots: the impact of organic haze on the habitability and detectability of earthlike exoplanets. Astrophys. J. **836**, 49 (2017). https://doi.org/10.3847/1538-4357/836/1/49

G. Arney, S.D. Domagal-Goldman, V.S. Meadows, Organic haze as a biosignature in anoxic Earth-like atmospheres. Astrobiology **18**(3), 311–329 (2018). https://doi.org/10.1089/ast.2017.1666

G. Avice, B. Marty, Perspectives on atmospheric evolution from noble gas and nitrogen isotopes on Earth, Mars Venus. Space Sci. Rev. (2020). https://doi.org/10.1007/s11214-020-00655-0. 2020

G. Avice, B. Marty, R. Burgess, The origin and degassing history of the Earth's atmosphere revealed by Archean xenon. Nat. Commun. **8**, 15455 (2017)

G. Avice et al., Evolution of atmospheric xenon and other noble gases inferred from Archean to Paleoproterozoic rocks. Geochim. Cosmochim. Acta **232**, 82–100 (2018)

D. Babikov, Recombination reactions as a possible mechanism of mass-independent fractionation of sulfur isotopes in the Archean atmosphere of Earth. Proc. Azerb. Natl. Acad. Sci. **114**(12), 3062–3067 (2017). https://doi.org/10.1073/pnas.1620977114

W. Bains, J.J. Petkowski, C. Sousa-Silva, S. Seager, Trivalent phosphorus and phosphines as components of biochemistry in anoxic environments. Astrobiology **19**, 885–902 (2019). https://doi.org/10.1089/ast.2018.1958

A. Bar-Nun, S. Chang, Photochemical reactions of water and carbon monoxide in Earth's primitive atmosphere. J. Geophys. Res., Oceans **88**(C11), 6662–6672 (1983)

A. Bar-Nun, H. Hartman, Synthesis of organic compounds from carbon monoxide and water by UV photolysis. Orig. Life **9**(2), 93–101 (1978)

P.H. Barry, D.R. Hilton, Release of subducted sedimentary nitrogen throughout Earth's mantle. Geochem. Perspect. Lett. **2**, 148–159 (2016)

A. Bekker, H.D. Holland, P.-L. Wang, D. Rumble III., H.J. Stein, J.L. Hannah, L.L. Coetzee, N.J. Beukes, Dating the rise of atmospheric oxygen. Nature **427**, 117–120 (2004)

E.A. Bell, P. Boehnke, T.M. Harrison, W.L. Mao, Potentially biogenic carbon preserved in a 4.1 billion-year-old zircon. Proc. Azerb. Natl. Acad. Sci. **112**(47), 14518–14521 (2015)

R.A. Berner, Z. Kothavala, GEOCARB III: a revised model of atmospheric CO_2 over Phanerozoic time. Am. J. Sci. **301**, 182–204 (2001)

B.G. Bills, R.D. Ray, Lunar orbital evolution: a synthesis of recent results. Geophys. Res. Lett. **26**(19), 3045–3048 (1999)

K.R. Birmingham, E. Füri, K. Lodders, B. Marty, Chemical and isotopic evolution of the early Solar System. Space Sci. Rev. **201**, 25–48 (2020)

J.D. Blum, L.S. Sherman, M.W. Johnson, Mercury isotopes in Earth and environmental sciences. Annu. Rev. Earth Planet. Sci. **42**(1), 249–269 (2014). https://doi.org/10.1146/annurev-earth-050212-124107

W.F. Bottke, D. Vokrouhlický, S. Marchi, T. Swindle, E.R.D. Scott, J.R. Weirich, H. Levison, Dating the Moon-forming impact event with asteroidal meteorites. Science **348**(6232), 321–323 (2015)

B.P. Boudreau, D.E. Canfield, A provisional diagenetic model for pH in anoxic porewaters: application to the FOAM site. J. Mar. Res. **46**(2), 429–455 (1988)

E.S. Boyd, A.D. Anbar, S. Miller, T.L. Hamilton, M. Lavin, J.W. Peters, A late methanogen origin for molybdenum-dependent nitrogenase. Geobiology **9**(3), 221–232 (2011)

J.A. Brandes, N.Z. Boctor, G.D. Cody, B.A. Cooper, R.M. Hazen, H.S. Yoder Jr., Abiotic nitrogen reduction on the early Earth. Nature **395**, 365–367 (1998)

J.J. Brocks, G.A. Logan, R. Buick, R.E. Summons, Archean molecular fossils and the early rise of eukaryotes. Science **285**, 1033–1036 (1999)

R. Buick, The earliest records of life on Earth, in *Planets and Life: The Emerging Science of Astrobiology*, ed. by W.T.I. Sullivan, J. Baross (Cambridge University Press, Cambridge, 2007), pp. 237–264

A. Burke, K.A. Moore, M. Sigl, D.C. Nita, J.R. McConnell, J.F. Adkins, Stratospheric eruptions from tropical and extra-tropical volcanoes constrained using high-resolution sulfur isotopes in ice cores. Earth Planet. Sci. Lett. **521**, 113–119 (2019). https://doi.org/10.1016/j.epsl.2019.06.006

N.J. Butterfield, Oxygen, animals and aquatic bioturbation: an updated account. Geobiology **16**(1), 3–16 (2018)

B. Byrne, C. Goldblatt, Diminished greenhouse warming from Archean methane due to solar absorption lines. Clim. Past **11**(3), 559–570 (2015)

D.E. Canfield, The early history of atmospheric oxygen: homage to Robert M. Garrels. Annu. Rev. Earth Planet. Sci. **33**, 1–36 (2005)

R.M. Canup, E. Asphaug, Origin of the Moon in a giant impact near the end of the Earth's formation. Nature **412**(8), 708–712 (2001)

P. Cartigny, B. Marty, Nitrogen isotopes and mantle geodynamics: the emergence of life and the atmosphere-crust-mantle connection. Elements **9**(5), 359–366 (2013)

D. Catling, The great oxidation event transition. Treatise Geochem. **6**, 177–195 (2014b)

D.C. Catling, K.J. Zahnle, C.P. McKay, Biogenic methane, hydrogen escape, and the irreversible oxidation of early Earth. Science **293**(5531), 839–843 (2001)

B.D. Charnay, G. Le Hir, F. Fluteau, F. Forget, D.C. Catling, A warm or a cold early Earth? New insights from a 3-D climate-carbon model. Earth Planet. Sci. Lett. **474**, 97–109 (2017)

M.W. Claire, D.C. Catling, K.J. Zahnle, Biogeochemical modelling of the rise in atmospheric oxygen. Geobiology **4**(4), 239–269 (2006). https://doi.org/10.1111/j.1472-4669.2006.00084.x

M.W. Claire, J.F. Kasting, S.D. Domagal-Goldman, E.E. Stüeken, R. Buick, V.S. Meadows, Modeling the signature of sulfur mass-independent fractionation produced in the Archean atmosphere. Geochim. Cosmochim. Acta **141**, 365–380 (2014)

H.J. Cleaves II., The prebiotic geochemistry of formaldehyde. Precambrian Res. **164**(3–4), 111–118 (2008)

G.D. Cody, Transition metal sulfides and the origins of metabolism. Annu. Rev. Earth Planet. Sci. **32**, 569–599 (2004)

K.C. Condie, E. Beyer, E. Belousova, W.L. Griffin, S.Y. O'Reilly, U–Pb isotopic ages and Hf isotopic composition of single zircons: the search for juvenile Precambrian continental crust. Precambrian Res. **139**(1–2), 42–100 (2005)

S.A. Crowe, C. Jones, S. Katsev, C. Magen, A.H. O'Neill, A. Sturm, D.E. Canfield, G.D. Haffner, A. Mucci, B. Sundby, D.A. Fowle, Photoferrotrophs thrive in an Archean ocean analogue. Proc. Natl. Acad. Sci. **105**(41), 15938–15943 (2008)

S.O. Danielache, C. Eskebjerg, M.S. Johnson, Y. Ueno, N. Yoshida, High-precision spectroscopy of 32S, 33S, and 34S sulfur dioxide: ultraviolet absorption cross sections and isotope effects. J. Geophys. Res. **113**(D17), 84–89 (2008). https://doi.org/10.1029/2007JD009695

S.O. Danielache, S. Hattori, M.S. Johnson, Y. Ueno, S. Nanbu, N. Yoshida, Photoabsorption cross-section measurements of 32S, 33S, 34S, and 36S sulfur dioxide for the B1B1-X1A1 absorption band. J. Geophys. Res. **117**(D24), n/a–n/a (2012). https://doi.org/10.1029/2012JD017464

S.O. Danielache, S. Tomoya, A. Kondorsky, I. Tokue, S. Nanbu, Nonadiabatic calculations of ultraviolet absorption cross section of sulfur monoxide: isotopic effects on the photodissociation reaction. J. Chem. Phys. **140**(4), 044319 (2014). https://doi.org/10.1063/1.4862429

N. Dauphas, J.F. Kasting, Low pCO_2 in the pore water, not in the Archean air. Nature **474**, E1–E1 (2011). https://doi.org/10.1038/nature09960

D.W. Deamer, The origin of cellular life, in *Planets and Life: The Emerging Science of Astrobiology*, ed. by W.T. Sullivan, J. Baross (Cambridge University Press, Cambridge, 2007), pp. 187–209

F. Delarue, F. Robert, K. Sugitani, R. Tartèse, R. Duhamel, S. Derenne, Nitrogen isotope signatures of microfossils suggest aerobic metabolism 3.0 Gyr ago. Geochem. Perspect. Lett. **7**, 32–36 (2018)

B. Dhuime, A. Wuestefeld, C.J. Hawkesworth, Emergence of modern continental crust about 3 billion years ago. Nat. Geosci. **8**(7), 552–555 (2015)

S.D. Domagal-Goldman, V.S. Meadows, M.W. Claire, J.F. Kasting, Using biogenic sulfur gases as remotely detectable biosignatures on anoxic planets. Astrobiology **11**(5), 419–441 (2011). https://doi.org/10.1089/ast.2010.0509

S. Domagal-Goldman, A. Segura, M. Claire, T. Robinson, V. Meadows, Abiotic ozone and oxygen in atmospheres similar to prebiotic Earth. Astrophys. J. **792**, 90 (2014)

P.C. Dos Santos, Z. Fang, S.W. Mason, J.C. Setubal, R. Dixon, Distribution of nitrogen fixation and nitrogenase-like sequences amongst microbial genomes. BMC Genomics **13**(1), 162 (2012)

J.L. Eigenbrode, K.H. Freeman, Late Archean rise of aerobic microbial ecosystems. Proc. Natl. Acad. Sci. **103**(43), 15759–15764 (2006)

L.T. Elkins-Tanton, Linked magma ocean solidification and atmospheric growth for Earth and Mars. Earth Planet. Sci. Lett. **271**(1–4), 181–191 (2008)

L.T. Elkins-Tanton, Magma oceans in the inner solar system. Annu. Rev. Earth Planet. Sci. **40**(1), 113–139 (2012)

Y. Endo, S. Danielache, Y. Ueno, S. Hattori, M.S. Johnson, N. Yoshida, H.G. Kjaergaard, Photoabsorption cross-section measurements of 32S, 33S, 34S and 36S sulfur dioxide from 190 to 220 nm. J. Geophys. Res., Atmos. **120**, 2546–2557 (2015). https://doi.org/10.1002/2014JD021671

Y. Endo, Y. Ueno, S. Aoyama, S.O. Danielache, Sulfur isotope fractionation by broadband UV radiation to optically thin SO_2 under reducing atmosphere. Earth Planet. Sci. Lett. **453**(C), 9–22 (2016). https://doi.org/10.1016/j.epsl.2016.07.057

Y. Endo, S.O. Danielache, Y. Ueno, Total pressure dependence of sulfur mass-independent fractionation by SO_2 photolysis. Geophys. Res. Lett. **46**(1), 483–491 (2019)

J. Farquhar, H. Bao, M. Thiemens, Atmospheric influence of Earth's earliest sulfur cycle. Science **289**, 756–759 (2000)

J. Farquhar, J. Savarino, S. Airieau, M.H. Thiemens, Observation of wavelength-sensitive mass-independent sulfur isotope effects during SO_2 photolysis: implications for the early atmosphere. J. Geophys. Res. **106**, 32829–32839 (2001)

J. Farquhar, M. Peters, D.T. Johnston, H. Strauss, A. Masterson, U. Wiechert, A.J. Kaufman, Isotopic evidence for Mesoarchaean anoxia and changing atmospheric sulphur chemistry. Nature **449**(7163), 706–709 (2007). https://doi.org/10.1038/nature06202

B. Fegley, R.G. Prinn, H. Hartman, G.H. Watkins, Chemical effects of large impacts on the Earth's primitive atmosphere. Nature **319**(6051), 305–308 (1986)

K. Fennel, M. Follows, P.G. Falkowski, The co-evolution of the nitrogen, carbon and oxygen cycles in the Proterozoic ocean. Am. J. Sci. **305**, 526–545 (2005)

M. Ferus, F. Pietrucci, A.M. Saitta, A. Knížek, P. Kubelík, O. Ivanek, V. Shestivska, S. Civiš, Formation of nucleobases in a Miller–Urey reducing atmosphere. Proc. Natl. Acad. Sci. **114**(17), 4306–4311 (2017)

G. Feulner, The faint young Sun problem. Rev. Geophys. **50**(2), RG2006 (2012). https://doi.org/10.1029/2011RG000375

D.T. Flannery, A.C. Allwood, M.J. van Kranendonk, Lacustrine facies dependence of highly 13C-depleted organic matter during the global age of methanotrophy. Precambrian Res. **285**, 216–241 (2016)

D.T. Flannery, A.C. Allwood, R.E. Summons, K.H. Williford, W. Abbey, E.D. Matys, N. Ferralis, Spatially-resolved isotopic study of carbon trapped in ~3.43 Ga Strelley Pool Formation stromatolites. Geochim. Cosmochim. Acta **223**, 21–35 (2018)

J.J. Fortney, On the carbon-to-oxygen ratio measurement in nearby Sun-like stars: implications for planet formation and the determination of stellar abundances. Astrophys. J. Lett. **747**(2), 2 (2012). https://doi.org/10.1088/2041-8205/747/2/L27

K.L. French, C. Hallmann, J.M. Hope, P.L. Schoon, J.A. Zumberge, Y. Hoshino, C.A. Peters, S.C. George, G.D. Love, J.J. Brocks, R. Buick, Reappraisal of hydrocarbon biomarkers in Archean rocks. Proc. Natl. Acad. Sci. **112**(19), 5915–5920 (2015)

G. Fuchs, Alternative pathways of carbon dioxide fixation: insights into the early evolution of life? Annu. Rev. Microbiol. **65**, 631–658 (2011)

E. Füri et al., Space Sci, Rev. (2020, this issue)

E. Füri, B. Marty, Nitrogen isotope variations in the Solar System. Nat. Geosci. **8**(7), 515–522 (2015)

P. Gao, R. Hu, T.D. Robinson, L. Cheng, Y.L. Yung, Stability of CO_2 atmospheres on desiccated M dwarf exoplanets. Astrophys. J. **806**, 249 (2015)

J.B. Glass, F. Wolfe-Simon, J.J. Elser, A.D. Anbar, Molybdenum-nitrogen co-limitation in freshwater and coastal heterocystous cyanobacteria. Limnol. Oceanogr. **55**(2), 667–676 (2010)

L.V. Godfrey, P.G. Falkowski, The cycling and redox state of nitrogen in the Archaean ocean. Nat. Geosci. **2**, 725–729 (2009)

D.A. Gold, A. Caron, G.P. Fournier, R.E. Summons, Paleoproterozoic sterol biosynthesis and the rise of oxygen. Nature **543**(7645), 420–423 (2017)

C. Goldblatt, K.J. Zahnle, Faint young Sun paradox remains. Nature **474**(7349), E1–E1 (2011). https://doi.org/10.1038/nature09961

C. Goldblatt, M.W. Claire, T.M. Lenton, A.J. Matthews, A.J. Watson, K.J. Zahnle, Nitrogen-enhanced greenhouse warming on early Earth. Nat. Geosci. **2**(12), 891–896 (2009)

E.A. Goosmann, D.C. Catling, S.M. Som, W. Altermann, R. Buick, Eolianite grain size distributions as a proxy for large changes in planetary atmospheric density. J. Geophys. Res., Planets **123**(10), 1–21 (2018)

D.O. Gough, Solar interior structure and luminosity variation. Phys. Sol. Var. **74**, 21–34 (1981)

J.L. Grenfell, J.-M. Grießmeier, P. von Paris, A.B.C. Patzer, H. Lammer, B. Stracke et al., Response of atmospheric biomarkers to NO_x-induced photochemistry generated by stellar cosmic rays for Earth-like planets in the habitable zone of M dwarf stars. Astrobiology **12**(12), 1109–1122 (2012)

J.P. Grotzinger, J.F. Kasting, New constraints on Precambrian ocean composition. J. Geol. **101**, 235–243 (1993)

M. Guedel, The Sun in time: activity and environment. Living Rev. Sol. Phys. **4**(1), 3 (2007). https://doi.org/10.12942/lrsp-2007-3

A. Guzmán-Marmolejo, A. Segura, E. Escobar-Briones, Abiotic production of methane in terrestrial planets. Astrobiology **13**(6), 550–559 (2013). https://doi.org/10.1089/ast.2012.0817

I. Halevy, Production, preservation, and biological processing of mass-independent sulfur isotope fractionation in the Archean surface environment. Proc. Natl. Acad. Sci. **110**(44), 17644–17649 (2013). https://doi.org/10.1073/pnas.1213148110

K. Hamano, Y. Abe, H. Genda, Emergence of two types of terrestrial planet on solidification of magma ocean. Nature **497**(7451), 607–610 (2013)

J.D. Haqq-Misra, S.D. Domagal-Goldman, P.J. Kasting, J.F. Kasting, A revised, hazy methane greenhouse for the Archean Earth. Astrobiology **8**(6), 1127–1137 (2008)

C. Harman, E. Schwieterman, J. Schottelkotte, J. Kasting, Abiotic O_2 levels on planets around F, G, K, and M stars: possible false positives for life? Astrophys. J. **812**, 137 (2015)

C.E. Harman, R. Felton, R. Hu, S.D. Domagal-Goldman, A. Segura, F. Tian, J.F. Kasting, Abiotic O_2 levels on planets around F, G, K, and M stars: effects of lightning-produced catalysts in eliminating oxygen false positives. Astrophys. J. **866**(1), 1 (2018a). https://doi.org/10.3847/1538-4357/aadd9b

C.E. Harman, A.A. Pavlov, D. Babikov, J.F. Kasting, Chain formation as a mechanism for mass-independent fractionation of sulfur isotopes in the Archean atmosphere. Earth Planet. Sci. Lett. **496**, 238–247 (2018b). https://doi.org/10.1016/j.epsl.2018.05.041

M.H. Hart, The evolution of the atmosphere of the Earth. Icarus **39**, 23–39 (1978)

J.M. Hayes, Geochemical evidence bearing on the origin of aerobiosis, a speculative hypothesis, in *Earth's Earliest Biosphere*, ed. by J.W. Schopf (Princeton University Press, Princeton, 1983), pp. 291–301

J.M. Hayes, Global methanotrophy at the Archean-Proterozoic transition, in *Early Life on Earth*, ed. by S. Bengtson (Columbia University Press, New York, 1994), pp. 220–236

E. Hébrard, M. Dobrijevic, P. Pernot, N. Carrasco, A. Bergeat, K.M. Hickson et al., How measurements of rate coefficients at low temperature increase the predictivity of photochemical models of Titan's atmosphere. J. Phys. Chem. A **113**(42), 11227–11237 (2009). https://doi.org/10.1021/jp905524e

I.M. Held, A.Y. Hou, Nonlinear axially symmetric circulations in a nearly inviscid atmosphere. J. Atmos. Sci. **37**(3), 515–533 (2002)

C. Herzberg, K. Condie, J. Korenaga, Thermal history of the Earth and its petrological expression. Earth Planet. Sci. Lett. **292**(1–2), 79–88 (2010)

A.M. Hessler, D.R. Lowe, R.L. Jones, D.K. Bird, A lower limit for atmospheric carbon dioxide levels 3.2 billion years ago. Nature **428**(6984), 736–738 (2004)

K. Hinrichs, Microbial fixation of methane carbon at 2.7 Ga: was an anaerobic mechanism possible? Geochem. Geophys. Geosyst. **3**, U1–U10 (2002)

M.M. Hirschmann, Magma ocean influence on early atmosphere mass and composition. Earth Planet. Sci. Lett. **341–344**, 48–57 (2012)

M.F. Hohmann-Marriott, R.E. Blankenship, Evolution of photosynthesis. Annu. Rev. Plant Biol. **62**(1), 515–548 (2011). https://doi.org/10.1146/annurev-arplant-042110-103811

H.D. Holland, *The Chemical Evolution of the Atmosphere and Oceans*. Chem. Evol. Atmos. Ocean (North-Holland/Princeton University Press, Princeton, 1984), 583 pp. Price US 24.50 Pap. US 75.00 cloth

M. Homann, P. Sansjofre, M. Van Zuilen, C. Heubeck, J. Gong, B. Killingsworth, I.S. Foster, A. Airo, M.J. Van Kranendonk, M. Ader, S.V. Lalonde, Microbial life and biogeochemical cycling on land 3,220 million years ago. Nat. Geosci. **11**(9), 665–671 (2018)

S.M. Hörst, C. He, M.S. Ugelow, A.M. Jellinek, R.T. Pierrehumbert, M.A. Tolbert, Exploring the atmosphere of neoproterozoic Earth: the effect of O_2 on haze formation and composition. Astrophys. J. **858**, 1–9 (2018). https://doi.org/10.3847/1538-4357/aabd7d

B.Z. Houlton, S.L. Morford, R.A. Dahlgren, Convergent evidence for widespread rock nitrogen sources in Earth's surface environment. Science **360**(6384), 58–62 (2018)

C.H. House, J.W. Schopf, K.O. Stetter, Carbon isotopic fractionation by Archaeans and other thermophilic prokaryotes. Org. Geochem. **34**, 345–356 (2003)

R. Hu, S. Seager, W. Bains, Photochemistry in terrestrial exoplanet atmospheres. I. Photochemistry model and benchmark cases. Astrophys. J. **761**, 166 (2012)

J.M. Husson, S.E. Peters, Atmospheric oxygenation driven by unsteady growth of the continental sedimentary reservoir. Earth Planet. Sci. Lett. **460**, 68–75 (2017)

D.L. Huston, C.W. Brauhart, S.L. Drieberg, G.J. Davidson, D.I. Groves, Metal leaching and inorganic sulfate reduction in volcanic-hosted massive sulfide mineral systems: evidence from the paleo-Archean Panorama district, Western Australia. Geology **29**(8), 687–690 (2001)

G. Izon, A.L. Zerkle, I. Zhelezinskaia, J. Farquhar, R.J. Newton, S.W. Poulton et al., Multiple oscillations in Neoarchaean atmospheric chemistry. Earth Planet. Sci. Lett. **431**, 264–273 (2015). https://doi.org/10.1016/j.epsl.2015.09.018

G. Izon, A.L. Zerkle, K.H. Williford, J. Farquhar, S.W. Poulton, M. Claire, Biological regulation of atmospheric chemistry en route to planetary oxygenation, in *Proceedings of the National Academy of Sciences of the United States of America* (2017)

D.J. Jacob, *Introduction to Atmospheric Chemistry* (Princeton University Press, Princeton, 1999). Retrieved from: https://press.princeton.edu/titles/6767.html

S.A. Jacobson, A. Morbidelli, S.N. Raymond, D.P. O'Brien, K.J. Walsh, D.C. Rubie, Highly siderophile elements in Earth's mantle as a clock for the Moon-forming impact. Nature **508**, 84–87 (2014)

B. Johnson, C. Goldblatt, The nitrogen budget of Earth. Earth-Sci. Rev. **148**, 150–173 (2015)

B.W. Johnson, C. Goldblatt, EarthN: a new Earth system nitrogen model. Geochem. Geophys. Geosyst. **19**(8), 2516–2542 (2018)

D.T. Johnston, Multiple sulfur isotopes and the evolution of Earth's surface sulfur cycle. Earth-Sci. Rev. **106**(1–2), 161–183 (2011). https://doi.org/10.1016/j.earscirev.2011.02.003

C.P. Johnstone, M. Güdel, I. Brott, T. Lüftinger, Stellar winds on the main-sequence II: the evolution of rotation and winds. Astron. Astrophys. **577**(25), A28 (2015a). https://doi.org/10.1051/0004-6361/201425301

C.P. Johnstone, M. Güdel, T. Lüftinger, G. Toth, I. Brott, Stellar winds on the main-sequence I: wind model. Astron. Astrophys. **577**, A27 (2015b). https://doi.org/10.1051/0004-6361/201425300

C.P. Johnstone, M. Güdel, H. Lammer, K.G. Kislyakova, Upper atmospheres of terrestrial planets: carbon dioxide cooling and the Earth's thermospheric evolution. Astron. Astrophys. **617**, A107 (2018). https://doi.org/10.1051/0004-6361/201832776

C.P. Johnstone, M.L. Khodachenko, T. Lüftinger, K.G. Kislyakova, H. Lammer, M. Güdel, Extreme hydrodynamic losses of Earth-like atmospheres in the habitable zones of very active stars. Astron. Astrophys. **624**, L10 (2019). https://doi.org/10.1051/0004-6361/201935279

A.A. Kadik, N.A. Kurovskaya, Y.A. Ignat'ev, N.N. Kononkova, V.V. Koltashev, V.G. Plotnichenko, Influence of oxygen fugacity on the solubility of nitrogen, carbon, and hydrogen in FeO-Na_2O-SiO_2-Al_2O_3 melts in equilibrium with metallic iron at 1.5 GPa and 1400 °C. Geochem. Int. **49**(5), 429–438 (2011)

Y. Kanzaki, T. Murakami, Estimates of atmospheric CO_2 in the Neoarchean-Paleoproterozoic from paleosols. Geochim. Cosmochim. Acta **159**, 190–219 (2015)

J.F. Kasting, Stability of ammonia in the primitive terrestrial atmosphere. J. Geophys. Res. **87**(C4), 3091 (1982)

J.F. Kasting, D. Catling, Evolution of a habitable planet. Annu. Rev. Astron. Astrophys. **41**(1), 429–463 (2003)

J.F. Kasting, J.C. Walker, Limits on oxygen concentration in the prebiological atmosphere and the rate of abiotic fixation of nitrogen. J. Geophys. Res., Oceans **86**(C2), 1147–1158 (1981)

A.J. Kaufman, D.T. Johnston, J. Farquhar, A.L. Masterson, T.W. Lyons, S. Bates, A.D. Anbar, G.L. Arnold, J. Garvin, R. Buick, Late Archean biospheric oxygenation and atmospheric evolution. Science **317**, 1900–1903 (2007)

P. Kharecha, J.F. Kasting, J. Siefert, A coupled atmosphere-ecosystem model of the early Archean Earth. Geobiology **3**, 53–76 (2005)

M.A. Kipp, E.E. Stüeken, Biomass recycling and Earth's early phosphorus cycle. Sci. Adv. **3**(11), eaao4795 (2017). https://doi.org/10.1126/sciadv.aao4795

M.C. Koehler, R. Buick, M.A. Kipp, E.E. Stüeken, J. Zaloumis, Transient surface ocean oxygenation recorded in the $\sim$ 2.66-Ga Jeerinah formation, Australia. Proc. Natl. Acad. Sci. **115**(30), 7711–7716 (2018)

M.C. Koehler, R. Buick, M.E. Barley, Nitrogen isotope evidence for anoxic deep marine environments from the Mesoarchean Mosquito Creek formation, Australia. Precambrian Res. **320**, 281–290 (2019)

J. Korenaga, Thermal evolution with a hydrating mantle and the initiation of plate tectonics in the early Earth. J. Geophys. Res., Solid Earth **116**(12), 1–20 (2011)

J. Korenaga, Initiation and evolution of plate tectonics on Earth: theories and observations. Annu. Rev. Earth Planet. Sci. **41**(1), 117–151 (2013)

J. Krissansen-Totton, D.S. Bergsman, D.C. Catling, On detecting biospheres from chemical thermodynamic disequilibrium in planetary atmospheres. Astrobiology **16**(1), 39–67 (2016). https://doi.org/10.1089/ast.2015.1327

K. Kritee, D.M. Sigman, J. Granger, B.B. Ward, A. Jayakumar, C. Deutsch, Reduced isotope fractionation by denitrification under conditions relevant to the ocean. Geochim. Cosmochim. Acta **92**, 243–259 (2012)

M. Kuga, N. Carrasco, B. Marty, Y. Marrocchi, S. Bernard, T. Rigaudier, B. Fleury, L. Tissandier, Nitrogen isotopic fractionation during abiotic synthesis of organic solid particles. Earth Planet. Sci. Lett. **393**, 2–13 (2014)

Y.N. Kulikov, E.F. Guinan et al., Atmospheric and water loss from early Venus atmospheric and water loss from early Venus. Planet. Space Sci. (2006). https://doi.org/10.1016/j.pss.2006.04.021

Y.N. Kulikov, H. Lammer, H.I.M. Lichtenegger, T. Penz, D. Breuer, T. Spohn, R. Lundin, H.K. Biernat, A comparative study of the influence of the active young Sun on the early atmospheres of Earth, Venus, and Mars. Space Sci. Rev. **129**, 207–243 (2007)

L.R. Kump, The rise of atmospheric oxygen. Nature **451**(7176), 277–278 (2008)

M. Kunze, M. Godolt, U. Langematz, J.L. Grenfell, A. Hamann-Reinus, H. Rauer, Investigating the early Earth faint young Sun problem with a general circulation model. Planet. Space Sci. **98**, 77–92 (2014)

S.V. Lalonde, K.O. Konhauser, Benthic perspective on Earth's oldest evidence for oxygenic photosynthesis. Proc. Natl. Acad. Sci. **112**(4), 995–1000 (2015)

H. Lammer, J.F. Kasting, E. Chassefière, R.E. Johnson, Y.N. Kulikov, F. Tian, Atmospheric escape and evolution of terrestrial planets and satellites. Space Sci. Rev. **139**, 399–436 (2008)

H. Lammer, A.L. Zerkle, S. Gebauer, N. Tosi, L. Noack, M. Scherf, E. Pilat-Lohinger, M. Güdel, J.L. Grenfell, M. Godolt, A. Nikolaou, Origin and evolution of the atmospheres of early Venus, Earth and Mars. Astron. Astrophys. Rev. **26**(1), 2 (2018). https://doi.org/10.1007/s00159-018-0108-y

H. Lammer, L. Sproß, J.L. Grenfell, M. Scherf, L. Fossati, M. Lendl, P.E. Cubillos, The role of N_2 as a geo-biosignature for the detection and characterization of Earth-like habitats. Astrobiology **19**(7), 927–950 (2019). https://doi.org/10.1089/ast.2018.1914

H. Lammer, M. Leitzinger, P. Odert, C. Burger, C.M. Schäfer, D. Kubyshkina et al., Constraining the early evolution of Venus and Earth through atmospheric Ar, Ne isotope and bulk K/U ratios. Icarus **339**, 113551 (2020). https://doi.org/10.1016/j.icarus.2019.113551

T. Lebrun, H. Massol, E. Chassefière, A. Davaille, E. Marcq, P. Sarda et al., Thermal evolution of an early magma ocean in interaction with the atmosphere. J. Geophys. Res. E, Planets **118**(6), 1155–1176 (2013)

A. Lenardic, B. Foley The Craig, O. Neill, W.B. Moore, in *Tectonic-Climate Coupling?: Variations in the Mean, Variations about the Mean, Variations in Mode*, vol. 121 (2016), pp. 1–34

T.M. Lenton, Gaia and natural selection. Nature **394**(6692), 439–447 (1998)

T.M. Lenton, R.A. Boyle, S.W. Poulton, G.A. Shields-Zhou, N.J. Butterfield, Co-evolution of eukaryotes and ocean oxygenation in the Neoproterozoic era. Nat. Geosci. **7**(4), 257–265 (2014)

G. Libourel, B. Marty, F. Humbert, Nitrogen solubility in basaltic melt. Part I. Effect of oxygen fugacity. Geochim. Cosmochim. Acta **67**(21), 4123–4135 (2003)

H.I.M. Lichtenegger, H. Lammer, J.-M. Grießmeier, Aeronomical evidence for higher CO_2 levels during Earth's Hadean epoch. Icarus **210**(1), 1–7 (2010)

S.J. Lock et al., The origin of the Moon within a terrestrial synestia. J. Geophys. Res., Planets **123**(4), 910–951 (2018)

R. Luger, R. Barns, Extreme water loss and abiotic O_2 buildup on planets throughout the habitable zones of M dwarfs. Astrobiology **15**, 119 (2015)

G. Luo, S. Ono, N.J. Beukes, D.T. Wang, S. Xie, R.E. Summons, Rapid oxygenation of Earth's atmosphere 2.33 billion years ago. Sci. Adv. **2**(5), e1600134 (2016)

J.R. Lyons, Atmospherically-derived mass-independent sulfur isotope signatures, and incorporation into sediments. Chem. Geol. **267**, 164–174 (2009)

T.W. Lyons, C.T. Reinhard, N.J. Planavsky, The rise of oxygen in Earth's early ocean and atmosphere. Nature **506**, 307–315 (2014)

A. Mallik, Y. Li, M. Wiedenbeck, Nitrogen evolution within the Earth's atmosphere–mantle system assessed by recycling in subduction zones. Earth Planet. Sci. Lett. **482**, 556–566 (2018)

B. Marty, L. Zimmermann, M. Pujol, R. Burgess, P. Philippot, Nitrogen isotopic composition and density of the Archean atmosphere. Science **342**(6154), 101–104 (2013)

B. Marty, G. Avice, D.V. Bekaert, M.W. Broadley, Salinity of the Archaean oceans from analysis of fluid inclusions in quartz. C. R. Géosci. **350**(4), 154–163 (2018)

E. Mason, M. Edmonds, A.V. Turchyn, Remobilization of crustal carbon may dominate volcanic arc emissions. Science **357**(6348), 290–294 (2017)

H. Massol, K. Hamano, F. Tian, M. Ikoma, Y. Abe, E. Chassefière et al., Formation and evolution of protoatmospheres. Space Sci. Rev. **205**(1–4), 153–211 (2016)

A.L. Masterson, J. Farquhar, B.A. Wing, Sulfur mass-independent fractionation patterns in the broadband UV photolysis of sulfur dioxide: pressure and third body effects. Earth Planet. Sci. Lett. **306**(3–4), 253–260 (2011). https://doi.org/10.1016/j.epsl.2011.04.004

T.A. Mather, D.M. Pyle, A.G. Allen, Volcanic source for fixed nitrogen in the early Earth's atmosphere. Geology **32**(10), 905–908 (2004)

T.M. McCollom, J.S. Seewald, Carbon isotope composition of organic compounds produced by abiotic synthesis under hydrothermal conditions. Earth Planet. Sci. Lett. **243**(1–2), 74–84 (2006)

T.M. McCollom, J.S. Seewald, Abiotic synthesis of organic compounds in deep-sea hydrothermal environments. Chem. Rev. **107**, 382–401 (2007)

S.E. McGlynn, E.S. Boyd, J.W. Peters, V.J. Orphan, Classifying the metal dependence of uncharacterized nitrogenases. Front. Microbiol. **30**(3), 419 (2012). https://doi.org/10.3389/fmicb.2012.00419

V.S. Meadows, C.T. Reinhard, G.N. Arney, M.N. Parenteau, E.W. Schwieterman, S.D. Domagal-Goldman, A.P. Lincowski, K.R. Stapelfeldt, H. Rauer, S. DasSarma, S. Hegde, Exoplanet biosignatures: understanding oxygen as a biosignature in the context of its environment. Astrobiology **18**(6), 630–662 (2018)

V.S. Meadows, C.T. Reinhard, G.N. Arney, M.N. Parenteau, E.W. Schwieterman, S.D. Domagal-Goldman, A.P. Lincowski, K.R. Stapelfeldt, H. Rauer, S. DasSarma, S. Hegde, N. Narita, R. Deitrick, T.W. Lyons, N. Siegler, J. Lustig-Yaeger, Exoplanet biosignatures: understanding oxygen as a biosignature in the context of its environment. Astrobiology **18**, 6 (2018)

M.P. Mehta, J.A. Baross, Nitrogen fixation at 92 C by a hydrothermal vent archaeon. Science **314**(5806), 1783–1786 (2006)

S. Mikhail, D.A. Sverjensky, Nitrogen speciation in upper mantle fluids and the origin of Earth's nitrogen-rich atmosphere. Nat. Geosci. **7**(11), 816–819 (2014)

S.L. Miller, A production of amino acids under possible primitive Earth conditions. Science **117**, 528–529 (1953)

K. Mishima, R. Yamazaki, M. Satish-Kumar, Y. Ueno, T. Hokada, T. Toyoshima, Multiple sulfur isotope geochemistry of Dharwar Supergroup, Southern India: late Archean record of changing atmospheric chemistry. Earth Planet. Sci. Lett. **464**, 69–83 (2017)

J. Monteux, D. Andrault, H. Samuel, On the cooling of a deep terrestrial magma ocean. Earth Planet. Sci. Lett. **448**, 140–149 (2016)

H. Moore, The isotopic composition of ammonia, nitrogen dioxide and nitrate in the atmosphere. Atmos. Environ. (1967) **11**(12), 1239–1243 (1977)

S. Nabhan, M. Wiedenbeck, R. Milke, C. Heubeck, Biogenic overgrowth on detrital pyrite in ca. 3.2 Ga Archean paleosols. Geology **44**(9), 763–766 (2016)

R. Navarro-González, M.J. Molina, L.T. Molina, Nitrogen fixation by volcanic lightning in the early Earth. Geophys. Res. Lett. **25**(16), 3123–3126 (1998)

R. Navarro-González, C.P. McKay, D.N. Mvondo, A possible nitrogen crisis for Achaean life due to reduced nitrogen fixation by lightning. Nature **412**(6842), 61–64 (2001a)

R. Navarro-González, C. Mckay, D. Nda Myondo, A possible nitrogen crisis for Archean life due to reduce nitrogen fixation by lighting. Nature **412**(412), 61–64 (2001b)

R.W. Nicklas, I.S. Puchtel, R.D. Ash, P.M. Piccoli, E. Hanski, E.G. Nisbet, P. Waterton, D.G. Pearson, A.D. Anbar, Secular mantle oxidation across the Archean-Proterozoic boundary: evidence from V partitioning in komatiites and picrites. Geochim. Cosmochim. Acta **250**, 49–75 (2019)

A. Nikolaou, N. Katyal, N. Tosi, M. Godolt, J.L. Grenfell, H. Rauer, What factors affect the duration and outgassing of the terrestrial magma ocean? Astrophys. J. **875**(1), 11 (2019). https://doi.org/10.3847/1538-4357/AB08ED

E.G. Nisbet, N.V. Grassineau, C.J. Howe, P.I. Abell, M. Regelous, R.E.R. Nisbet, The age of Rubisco: the evolution of oxygenic photosynthesis. Geobiology **5**, 311–335 (2007)

H. Ohmoto, Y. Watanabe, H. Ikemi, S.R. Poulson, B.E. Taylor, Sulphur isotope evidence for an oxic Archaean atmosphere. Nature **442**(7105), 908–911 (2006). https://doi.org/10.1038/nature05044

C. O'Neill, V. Debaille, The evolution of Hadean–Eoarchaean geodynamics. Earth Planet. Sci. Lett. **406**, 49–58 (2014)

S. Ono, Photochemistry of sulfur dioxide and the origin of mass-independent isotope fractionation in Earth's atmosphere. Annu. Rev. Earth Planet. Sci. **45**(1), 301–329 (2017). https://doi.org/10.1146/annurev-earth-060115-012324

S. Ono, J.L. Eigenbrode, A.A. Pavlov, P. Kharecha, D. Rumble III., J.F. Kasting, K.H. Freeman, New insights into Archean sulfur cycle from mass-independent sulfur isotope records from the Hamersley basin, Australia. Earth Planet. Sci. Lett. **213**, 15–30 (2003)

S. Ono, A.R. Whitehill, J.R. Lyons, Contribution of isotopologue self-shielding to sulfur mass-independent fractionation during sulfur dioxide photolysis. J. Geophys. Res., Atmos. **118**, 2444–2454 (2013)

F. Ossa Ossa, A. Hofmann, J.E. Spangenberg, S.W. Poulton, E.E. Stüeken, R. Schoenberg, B. Eickmann, M. Wille, M. Butler, A. Bekker, Limited oxygen production in the Mesoarchean ocean. Proc. Natl. Acad. Sci. (2019). https://doi.org/10.1073/pnas.1818762116

K. Ozaki, E. Tajika, P.K. Hong, Y. Nakagawa, C.T. Reinhard, Effects of primitive photosynthesis on Earth's early climate system. Nat. Geosci. **11**, 55–59 (2017). https://doi.org/10.1038/s41561-017-0031-2

M. Ozima, F.A. Podosek, *Noble Gas Geochemistry* (Cambridge University Press, Melbourne, 1983), https://doi.org/10.1017/S0016756800029587. x+367 pp. Price £40.00, U.S. $79.50. ISBN 0-521-23939-7. Geol. Mag. 121, 384

B.M. Packer, Sedimentology, paleontology, and stable-isotope geochemistry of selected formations in the 2.7-billion-year-old Fortescue Group, Western Australia. UCLA (1990), p. 187

K. Pahlevan, L. Schaefer, M.M. Hirschmann, Hydrogen isotopic evidence for early oxidation of silicate Earth. Earth and Planetary Science Letters **526**, 15770 (2019)

A.P. Palya, I.S. Buick, G.E. Bebout, Storage and mobility of nitrogen in the continental crust: Evidence from partially melted metasedimentary rocks, Mt. Stafford, Australia. Chem. Geol. **281**(3), 211–226 (2011)

R. Parai, S. Mukhopadhyay, Publisher correction: xenon isotopic constraints on the history of volatile recycling into the mantle. Nature **563**(7732), E28 (2018)

D. Parkos, A. Pikus, A. Alexeenko, H.J. Melosh, HCN production via impact ejecta reentry during the late heavy bombardment. J. Geophys. Res., Planets **123**(4), 892–909 (2018)

A.A. Pavlov, J.F. Kasting, Mass-independent fractionation of sulfur isotopes in Archean sediments: strong evidence for an anoxic Archean atmosphere. Astrobiology **2**, 27–41 (2002)

A.A. Pavlov, J.F. Kasting, J.L. Eigenbrode, K.H. Freeman, Organic haze in Earh's early atmosphere: source of low-C-13 late Archean kerogens? Geology **29**, 1003–1006 (2001)

R.O. Pepin, On the isotopic composition of primordial xenon in terrestrial planet atmospheres. Space Sci. Rev. **92**, 371–395 (2000)

S. Péron, M. Moreira, Onset of volatile recycling into the mantle determined by xenon anomalies. Geochem. Perspect. Lett. **9**, 21 (2018)

D.L. Pinti, K. Hashizume, Early life record from nitrogen isotopes, in *Earliest Life on Earth: Habitats, Environments and Methods of Detection*, ed. by S.D. Golding, M. Glikson (Springer, Berlin, 2010), pp. 183–205

J.P. Pinto, G.R. Gladstone, Y.L. Yung, Photochemical production of formaldehyde in Earth's primitive atmosphere. Science **210**(4466), 183–185 (1980)

N.J. Planavsky, D. Asael, A. Hofmann, C.T. Reinhard, S.V. Lalonde, A. Knudsen, X. Wang, F.O. Ossa, E. Pecoits, A.J. Smith, N.J. Beukes, Evidence for oxygenic photosynthesis half a billion years before the Great Oxidation Event. Nat. Geosci. **7**(4), 283–286 (2014a)

N.J. Planavsky, C.T. Reinhard, X. Wang, D. Thomson, P. McGoldrick, R.H. Rainbird, T. Johnson, W.W. Fischer, T.W. Lyons, Low Mid-Proterozoic atmospheric oxygen levels and the delayed rise of animals. Science **346**(6209), 635–638 (2014b)

E.C. Pope, D.K. Bird, M.T. Rosing, Isotope composition and volume of Earth's Early Oceans. Proc. Natl. Acad. Sci. **109**(12), 4371–4376 (2012)

M. Preiner, J. Xavier, F. Sousa, V. Zimorski, A. Neubeck, S. Lang, H. Greenwell, K. Kleinermanns, H. Tüysüz, T. McCollom, N. Holm, Serpentinization: connecting geochemistry, ancient metabolism and industrial hydrogenation. Life **8**(4), 41 (2019). https://doi.org/10.3390/life8040041

G. Proskurowski, M.D. Lilley, J.S. Seewald, G.L. Früh-Green, E.J. Olson, J.E. Lupton, S.P. Sylva, D.S. Kelley, Abiogenic hydrocarbon production at Lost City hydrothermal field. Science **319**(5863), 604–607 (2008)

M. Pujol, B. Marty, P. Burnard, P. Philippot, Xenon in Archean barite: weak decay of 130Ba, mass-dependent isotopic fractionation and implication for barite formation. Geochim. Cosmochim. Acta **73**, 6834–6846 (2009)

M. Pujol, B. Marty, R. Burgess, Chondritic-like xenon trapped in Archean rocks: a possible signature of the ancient atmosphere. Earth Planet. Sci. Lett. **308**, 298–306 (2011)

R.M. Ramirez, L. Kaltenegger, The habitable zones of pre-main-sequence stars. Astrophys. J. Lett. **797**(2), L25 (2014)

B. Rasmussen, Filamentous microfossils in a 3,235-million-year-old volcanogenic massive sulphide deposit. Nature **405**(6787), 676–679 (2000)

J. Raymond, J.L. Siefert, C.R. Staples, R.E. Blankenship, The natural history of nitrogen fixation. Mol. Biol. Evol. **21**(3), 541–554 (2004)

J.R. Reimink et al., No evidence for Hadean continental crust within Earth's oldest evolved rock unit. Nat. Geosci. **9**(10), 777–780 (2016)

C.T. Reinhard, N.J. Planavsky, Mineralogical constraints on Precambrian pCO_2. Nature **474**, E1–E1 (2011). https://doi.org/10.1038/nature09959

C.T. Reinhard, N.J. Planavsky, L.J. Robbins, C.A. Partin, B.C. Gill, S.V. Lalonde, A. Bekker, K.O. Konhauser, T.W. Lyons, Proterozoic ocean redox and biogeochemical stasis. Proc. Natl. Acad. Sci. **110**(14), 5357–5362 (2013)
C.T. Reinhard, N.J. Planavsky, S.L. Olson, T.W. Lyons, D.H. Erwin, Earth's oxygen cycle and the evolution of animal life. Proc. Natl. Acad. Sci. **113**(32), 8933–8938 (2016)
C.T. Reinhard, N.J. Planavsky, B.C. Gill, K. Ozaki, L.J. Robbins, T.W. Lyons, W.W. Fischer, C. Wang, D.B. Cole, K.O. Konhauser, Evolution of the global phosphorus cycle. Nature **541**(7637), 386–389 (2017)
Ribas et al., Evolution of the solar activity over time and effects on planetary atmospheres. I. High-energy irradiances (1–1700Å). Astrophys. J. **622**, 680 (2005). https://doi.org/10.1086/427977
D.H. Rothman, J.M. Hayes, R.E. Summons, Dynamics of the Neoproterozoic carbon cycle. Proc. Natl. Acad. Sci. **100**(14), 8124–8129 (2003)
S. Rugheimer, L. Kaltenegger, Spectra of Earth-like planets through geological evolution around FGKM stars. Astrophys. J. **854**, 19 (2018)
S. Rugheimer, L. Kaltenegger, A. Segura, J. Linsky, S. Mohanty, Effect of UV radiation on the spectral fingerprints of earth-like planets orbiting M dwarfs. Astrophys. J. **809**, 57 (2015a)
S. Rugheimer, A. Segura, L. Kaltenegger, D. Sasselov, UV surface environment of Earth-like planets orbiting FGKM stars through geological evolution. Astrophys. J. **806**, 137 (2015b)
M.J. Russell, W. Martin, The rocky roots of the acetyl-CoA pathway. Trends Biochem. Sci. **29**(7), 358–363 (2004)
I.-J. Sackmann, A.I. Boothroyd, Our Sun. V. A bright young Sun consistent with helioseismology and warm temperatures on ancient Earth and Mars. Astrophys. J. **583**(2), 1024–1039 (2002)
A. Salvador, H. Massol, A. Davaille, E. Marcq, P. Sarda, E. Chassefière, The relative influence of H_2O and CO_2 on the primitive surface conditions and evolution of rocky planets. J. Geophys. Res., Planets **122**(7), 1458–1486 (2017)
J. Savarino, A. Romero, J. Cole-Dai, S. Bekki, M.H. Thiemens, UV induced mass-independent sulfur isotope fractionation in stratospheric volcanic sulfate. Geophys. Res. Lett. **30**(21), 3645 (2003). https://doi.org/10.1029/2003GL018134
L. Schaefer, B. Fegley, Chemistry of atmospheres formed during accretion of the Earth and other terrestrial planets. Icarus **208**(1), 438–448 (2010)
L. Schaefer, R. Wordsworth, Z. Berta-Thompson, D. Sasselov, Predictions of the atmospheric composition of GJ 1123b. Astrophys. J. **829**(2), 63 (2016)
M. Scherf, M. Khodachenko, H. Lammer, I. Alexeev, C. Johnstone, M. Guedel, L. Tu, M. Blokhina, J. Tarduno, H. Lichtenegger, Y. Kulikov, The terrestrial paleo-magnetosphere and its implications on the origin and evolution of the nitrogen-dominated atmosphere, in *EGU General Assembly Conference Abstracts*, 20 (2018) p. 16210
M. Schidlowski, Carbon isotopes as biogeochemical recorders of life over 3.8 Ga of Earth history: evolution of a concept. Precambrian Res. **106**, 117–134 (2001)
M. Schidlowski, J.M. Hayes, I.R. Kaplan, Isotopic inferences of ancient biochemistries: carbon, sulfur, hydrogen, and nitrogen, in *Earth's Earliest Biosphere*, ed. by J.W. Schopf (Princeton University Press, Princeton, 1983), pp. 149–186
J.A. Schmidt, M.S. Johnson, R. Schinke, Carbon dioxide photolysis from 150 to 210 nm: Singlet and triplet channel dynamics, UVspectrum, and isotope effects. Proc. Natl. Acad. Sci. **110**(44), 17691–17696 (2013)
J.W. Schopf, K. Kitajima, M.J. Spicuzza, A.B. Kudryavtsev, J.W. Valley, SIMS analyses of the oldest known assemblage of microfossils document their taxon-correlated carbon isotope compositions. Proc. Natl. Acad. Sci. **115**(1), 53–58 (2018)
P.A. Schroeder, A.A. McLain, Illite-smectites and the influence of burial diagenesis on the geochemical cycling of nitrogen. Clay Miner. **33**(4), 539–546 (1998)
A.C. Scott, I.J. Glasspool, The diversification of Paleozoic fire systems and fluctuations in atmospheric oxygen concentration. Proc. Natl. Acad. Sci. USA **103**, 10861–10865 (2006)
C. Scott, T.W. Lyons, A. Bekker, Y. Shen, S.W. Poulton, X. Chu, A.D. Anbar, Tracing the stepwise oxygenation of the Proterozoic ocean. Nature **452**, 456–459 (2008)
S. Seager, W. Bains, J.J. Petkowski, Toward a list of molecules as potential biosignature gases for the search for life on exoplanets and applications to terrestrial biochemistry. Astrobiology **16**(6), 1–21 (2016)
A. Segura, K. Krelove, J.F. Kasting, D. Sommerlatt, V. Meadows, D. Crisp et al., Ozone concentrations and ultraviolet fluxes on Earth-like planets around other stars. Astrobiology **3**(4), 689–708 (2003). https://doi.org/10.1089/153110703322736024
S.N. Silverman, S.H. Kopf, B.M. Bebout, R. Gordon, S.M. Som, Morphological and isotopic changes of heterocystous cyanobacteria in response to N_2 partial pressure. Geobiology **17**, 60–75 (2018)
N.H. Sleep, The Hadean-Archaean environment. Cold Spring Harb. Perspect. Biol. **2**, a002527 (2010)
N.H. Sleep, G. Schubert, Plate tectonics through time. Evolution of the Earth **9**, 145–169 (2007)

N.H. Sleep, K. Zahnle, Carbon dioxide cycling and implications for climate on ancient Earth. J. Geophys. Res. **106**(E1), 1373–1399 (2001)

A. Smirnov, D. Hausner, R. Laffers, D.R. Strongin, M.A. Schoonen, Abiotic ammonium formation in the presence of Ni–Fe metals and alloys and its implications for the Hadean nitrogen cycle. Geochem. Trans. **9**, 5 (2008). https://doi.org/10.1186/1467-4866-9-5

M.A. Smit, K. Mezger, Earth's early O_2 cycle suppressed by primitive continents. Nat. Geosci. **10**(10), 788–792 (2017)

V.S. Solomatov, Fluid dynamics of a terrestrial magma ocean, in *Origin of the Earth and Moon, vol. 1*, ed. by R. Canup, K. Righter (University of Arizona Press, Tucson, 2000), pp. 323–338

S.M. Som, D.C. Catling, J.P. Harnmeijer, P.M. Polivka, R. Buick, Air density 2.7 billion years ago limited to less than twice modern levels by fossil raindrop imprints. Nature **484**(7394), 359–362 (2012)

S.M. Som, M. Sanjoy et al., Earth's air pressure 2.7 billion years ago constrained to less than half of modern levels. Nat. Geosci. **9**(5), 1–5 (2016)

C. Sousa-Silva, S. Seager, S. Ranjan, J.J. Petkowski, Z. Zhan, R. Hu, W. Bains, Phosphine as a biosignature gas in exoplanet atmospheres. Astrobiology **20**, 235–268 (2019)

B. Srinivasan, Barites: anomalous xenon from spallation and neutron-induced reactions. Earth Planet. Sci. Lett. **31**, 129–141 (1976)

G. Stark, P.L. Smith, J. Rufus, A.P. Thorne, J.C. Pickering, G. Cox, High-resolution photoabsorption cross-section measurements of SO_2 at 295 K between 198 and 220 nm. J. Geophys. Res., Planets **104**(E7), 16585–16590 (1999)

E.E. Stüeken, Nitrogen in ancient mud: a biosignature? Astrobiology **16**(9), 730–735 (2016)

E.E. Stüeken, R. Buick, Environmental control on microbial diversification and methane production in the Mesoarchean. Precambrian Res. **304**, 64–72 (2018)

E.E. Stüeken, D.C. Catling, R. Buick, Contributions to late Archaean sulphur cycling by life on land. Nat. Geosci. **5**, 722–725 (2012)

E.E. Stüeken, R.E. Anderson, J.S. Bowman, W.J. Brazelton, J. Colangelo-Lillis, A.D. Goldman, S.M. Som, J.A. Baross, Did life originate from a global chemical reactor? Geobiology **11**(2), 101–126 (2013)

E.E. Stüeken, R. Buick, B.M. Guy, M.C. Koehler, Isotopic evidence for biological nitrogen fixation by molybdenum-nitrogenase at 3.2 Gyr. Nature **520**, 666–669 (2015)

E.E. Stüeken, M.A. Kipp, M.C. Koehler, R. Buick, The evolution of Earth's biogeochemical nitrogen cycle. Earth-Sci. Rev. **160**, 220–239 (2016a)

E.E. Stüeken, M.A. Kipp, M.C. Koehler, E.W. Schwieterman, B. Johnson, R. Buick, Modeling pN_2 through geological time: implications for planetary climates and atmospheric biosignatures. Astrobiology **16**(12), 949–963 (2016b)

E.E. Stüeken, R. Buick, R.E. Anderson, J.A. Baross, N.J. Planavsky, T.W. Lyons, Environmental niches and metabolic diversity in Neoarchean lakes. Geobiology **15**(6), 767–783 (2017)

D.P. Summers, S. Chang, Prebiotic ammonia from reduction of nitrite by iron (II) on the early Earth. Nature **365**, 630–633 (1993)

D.Y. Sumner, I. Hawes, T.J. Mackey, A.D. Jungblut, P.T. Doran, Antarctic microbial mats: a modern analog for Archean lacustrine oxygen oases. Geology **43**(10), 887–890 (2015)

F. Tabataba-Vakili, J.L. Grenfell, J.-M. Grießmeier, H. Rauer, Atmospheric effects of stellar cosmic rays on Earth-like exoplanets orbiting M-dwarfs. Astron. Astrophys. **585**, A96 (2016). https://doi.org/10.1051/0004-6361/201425602

E. Tajika, T. Matsui, The evolution of the terrestrial environment. Orig. Earth 347–70 (1990)

E. Tajika, T. Matsui, Evolution of terrestrial proto-CO_2 atmosphere coupled with thermal history of the Earth. Earth Planet. Sci. Lett. **113**(1–2), 251–566 (1992)

M. Tang, K. Chen, R.L. Rudnick, Archean upper crust transition from mafic to felsic marks the onset of plate tectonics. Science **351**(6271), 372–375 (2016)

T. Tashiro, A. Ishida, M. Hori, M. Igisu, M. Koike, P. Méjean, N. Takahata, Y. Sano, T. Komiya, Early trace of life from 3.95 Ga sedimentary rocks in Labrador, Canada. Nature **549**(7673), 516–518 (2017)

M.H. Thiemens, History and applications of mass-independent isotope effects. Annu. Rev. Earth Planet. Sci. **34**(1), 217–262 (2006). https://doi.org/10.1146/annurev.earth.34.031405.125026

M.H. Thiemens, S. Chakraborty, G. Dominguez, The physical chemistry of mass-independent isotope effects and their observation in nature. Annu. Rev. Phys. Chem. **63**(1), 155–177 (2012). https://doi.org/10.1146/annurev-physchem-032511-143657

C. Thomazo, M. Ader, J. Farquhar, P. Philippot, Methanotrophs regulated atmospheric sulfur isotope anomalies during the Mesoarchean (Tumbiana formation, Western Australia). Earth Planet. Sci. Lett. **279**, 65–75 (2009a)

C. Thomazo, D.L. Pinti, V. Busigny, M. Ader, K. Hashizume, P. Phillippot, Biological activity and the Earth's surface evolution: insights from carbon, sulfur, nitrogen and iron stable isotopes in the rock record. C. R. Palevol **8**, 665–678 (2009b)

F. Tian, J.F. Kasting, H.L. Liu, R.G. Roble, Hydrodynamic planetary thermosphere model: 1. Response of the Earth's thermosphere to extreme solar EUV conditions and the significance of adiabatic cooling. J. Geophys. Res. E, Planets **113**, 1–19 (2008a)

F. Tian, S.C. Solomon, L. Qian, J. Lei, R.G. Roble, Hydrodynamic planetary thermosphere model: 2. Coupling of an electron transport/energy deposition model. J. Geophys. Res. E, Planets **113**, 1–13 (2008b)

F. Tian, K. France, J. Linsky, P. Mauas, M. Vieytes, High stellar FUV/NUV ratio and oxygen contents in the atmospheres of potentially habitable planets. Earth Planet. Sci. Lett. **385**, 22–27 (2014)

D. Trail, E.B. Watson, N.D. Tailby, The oxidation state of Hadean magmas and implications for early Earth's atmosphere. Nature **480**(7375), 79 (2011)

L. Tu, C.P. Johnstone, M. Güdel, H. Lammer, The extreme ultraviolet and X-ray Sun in time: high-energy evolutionary tracks of a solar-like star. Astron. Astrophys. **577**, L3 (2015). https://doi.org/10.1051/0004-6361/201526146

J.M. Tucker, S. Mukhopadhyay, Evidence for multiple magma ocean outgassing and atmospheric loss episodes from mantle noble gases. Earth Planet. Sci. Lett. **393**, 254–265 (2014)

Y. Ueno, K. Yamada, N. Yoshida, S. Maruyama, Y. Isozaki, Evidence from fluid inclusions for microbial methanogenesis in the early Archaean era. Nature **440**, 516–519 (2006)

T. Ushikubo et al., Lithium in Jack Hills Zircons: evidence for extensive weathering of Earth's earliest crust. Earth Planet. Sci. Lett. **272**(3–4), 666–676 (2008)

M.J. Van Kranendonk, R. Hugh Smithies, A.H. Hickman, D.C. Champion, Review: secular tectonic evolution of Archean continental crust: interplay between horizontal and vertical processes in the formation of the Pilbara Craton, Australia. Terra Nova **19**(1), 1–38 (2007)

P.M. Vitousek, D.N. Menge, S.C. Reed, C.C. Cleveland, Biological nitrogen fixation: rates, patterns and ecological controls in terrestrial ecosystems. Philos. Trans. R. Soc. B, Biol. Sci. **368**(1621), 20130119 (2013). https://doi.org/10.1098/rstb.2013.0119

J.C.G. Walker, Carbon dioxide on the early Earth, in *Origins of Life and Evolution of the Biosphere*, vol. 16 (Kluwer Academic, Dordrecht, 1985), pp. 117–127

J.C.G. Walker, P.B. Hays, J.F. Kasting, A negative feedback mechanism for the long-term stabilization of Earth's surface temperature. J. Geophys. Res. **86**(C10), 9776–9782 (1981)

W.L. Wang, J.K. Moore, A.C. Martiny, F.W. Primeau, Convergent estimates of marine nitrogen fixation. Nature **566**(7743), 205–211 (2019)

D.J. Webb, Tides and the evolution of the Earth—Moon system. Geophys. J. R. Astron. Soc. **70**(1), 261–271 (1982)

M.C. Weiss, F.L. Sousa, N. Mrnjavac, S. Neukirchen, M. Roettger, S. Nelson-Sathi, W.F. Martin, The physiology and habitat of the last universal common ancestor. Nat. Microbiol. (2016). https://doi.org/10.1038/nmicrobiol.2016.116

C.G. Wheat, M.J. Mottl, M. Rudnicki, Trace element and REE composition of a low-temperature ridge-flank hydrothermal spring. Geochim. Cosmochim. Acta **66**(21), 3693–3705 (2002)

A.R. Whitehill, S. Ono, Excitation band dependence of sulfur isotope mass-independent fractionation during photochemistry of sulfur dioxide using broadband light sources. Geochim. Cosmochim. Acta **94**, 238–253 (2012)

A.R. Whitehill, C. Xie, X. Hu, D. Xie, H. Guo, S. Ono, Vibronic origin of sulfur mass-independent isotope effect in photoexcitation of SO_2 and the implications to the early Earth's atmosphere. Proc. Natl. Acad. Sci. USA **110**, 17697–17702 (2013)

M.J. Whitehouse, D.J. Dunkley, M.A. Kusiak, S.A. Wilde, On the true antiquity of Eoarchean chemofossils–assessing the claim for Earth's oldest biogenic graphite in the Saglek Block of Labrador. Precambrian Res. **323**, 70–81 (2019)

G.E. Williams, Geological constraints on the Precambrian history of Earth's rotation and the Moon's orbit. Rev. Geophys. **38**(1), 37–59 (2000)

E.T. Wolf, O.B. Toon, Hospitable Archean climates simulated by a general circulation model. Astrobiology **13**(7), 656–673 (2013)

R.D. Wordsworth, Atmospheric nitrogen evolution on Earth and Venus. Earth Planet. Sci. Lett. **447**, 103–111 (2016)

R. Wordsworth, R. Pierrehumbert, Abiotic oxygen-dominated atmospheres on terrestrial habitable zone planets. Astrophys. J. Lett. **785**(2), L20 (2014)

R.D. Wordsworth, L. Schaefer, R.A. Fischer, Redox evolution via gravitational differentiation on low mass planets: implications for biosignatures, water loss and habitability. Astron. J. **155**, 5 (2017)

W.H. Yang, K.A. Weber, W.L. Silver, Nitrogen loss from soil through anaerobic ammonium oxidation coupled to iron reduction. Nat. Geosci. **5**(8), 538–541 (2012)

Y.L. Yung, W.B. DeMore, Photochemistry of the stratosphere of Venus: implications for atmospheric evolution. Icarus **51**, 199–247 (1982)

K.J. Zahnle, Photochemistry of methane and the formation of hydrocyanic acid (HCN) in the Earth's early atmosphere. J. Geophys. Res. **91**(D2), 2819–2834 (1986)

K.J. Zahnle, J.F. Kasting, J.B. Pollack, Evolution of a steam atmosphere during Earth's accretion. Icarus **74**(1), 62–97 (1988)

K. Zahnle, J.B. Pollack, J.F. Kasting, Xenon fractionation in porous planetesimals. Geochim. Cosmochim. Acta **54**, 2577–2586 (1990)

K.J. Zahnle, M.W. Claire, D.C. Catling, The loss of mass-independent fractionation in sulfur due to a Palaeoproterozoic collapse of atmospheric methane. Geobiology **4**(4), 271–283 (2006)

K.J. Zahnle, L. Schaefer, B. Fegley, Earth's earliest atmospheres. Cold Spring Harb. Perspect. Biol. **2**(10), a004895 (2010)

K.J. Zahnle, M. Gacesa, D.C. Catling, Strange messenger: a new history of hydrogen on Earth, as told by xenon. Geochim. Cosmochim. Acta **244**, 56–85 (2019)

A.L. Zerkle, Biogeodynamics: bridging the gap between surface and deep Earth processes. Philos. Trans. R. Soc., Math. Phys. Eng. Sci. **376**(2132), 20170401 (2018)

A.L. Zerkle, S. Mikhail, The geobiological nitrogen cycle: from microbes to the mantle. Geobiology **15**(3), 343–352 (2017)

A.L. Zerkle, C.H. House, R.P. Cox, D.E. Canfield, Metal limitation of cyanobacterial N_2 fixation and implications for the Precambrian nitrogen cycle. Geobiology **4**(4), 285–297 (2006)

A.L. Zerkle, M.W. Claire, S.D. Domagal-Goldman, J. Farquhar, S.W. Poulton, A bistable organic-rich atmosphere on the Neoarchaean Earth. Nat. Geosci. **5**, 359–363 (2012)

A.L. Zerkle, S.W. Poulton, R.J. Newton, C. Mettam, M.W. Claire, A. Bekker, C.K. Junium, Onset of the aerobic nitrogen cycle during the Great Oxidation Event. Nature **542**(7642), 465–467 (2017)

P. Zmolek, X. Xu, T. Jackson, M.H. Thiemens, W.C. Trogler, Large mass independent sulfur isotope fractionations during the photopolymerization of 12CS 2 and 13CS 2. J. Phys. Chem. A **103**(15), 2477–2480 (1999). https://doi.org/10.1021/jp9907140

Space Sci Rev (2020) 216:112
https://doi.org/10.1007/s11214-020-00730-6

The Isotopic Imprint of Life on an Evolving Planet

M.K. Lloyd[1,2] · H.L.O. McClelland[3,4] · G. Antler[5] ·
A.S. Bradley[6] · I. Halevy[3] · C.K. Junium[7] ·
S.D. Wankel[8] · A.L. Zerkle[9]

Received: 25 December 2019 / Accepted: 26 August 2020 / Published online: 6 October 2020

Abstract Stable isotope compositions of biologically cycled elements encode information about the interaction between life and environment. On Earth, geochemical biomarkers have been used to probe the extent, nature, and activity of modern and ancient organisms. However, extracting biological information from stable isotopic compositions requires untangling the interconnected nature of the Earth's biogeochemical system, and must be viewed through the lens of evolving metabolisms on an evolving planet. In this chapter, we provide an introduction to isotope geobiology and to the geobiological history of Earth. We discuss the isotope biogeochemistry of the biologically essential elements carbon, nitrogen and sul-

M.K. Lloyd and H.L.O. McClelland contributed equally to this work

Reading Terrestrial Planet Evolution in Isotopes and Element Measurements
Edited by Helmut Lammer, Bernard Marty, Aubrey L. Zerkle, Michel Blanc, Hugh O'Neill and Thorsten Kleine

✉ M.K. Lloyd
mlloyd@psu.edu

✉ H.L.O. McClelland
hmcclelland@unimelb.edu.au

1 Department of Earth and Planetary Science, University of California, Berkeley, CA, USA

2 Department of Geosciences, The Pennsylvania State University, University Park, PA, USA

3 Department of Earth and Planetary Sciences, Weizmann Institute of Science, Rehovot, Israel

4 School of Earth Sciences, University of Melbourne, Parkville, VIC, Australia

5 Department of Earth and Environmental Sciences, Faculty of Natural Sciences, Ben-Gurion University of the Negev, Beer Sheva, Israel

6 Department of Earth and Planetary Sciences, Washington University in St Louis, St Louis, MO, USA

7 Department of Earth Sciences, Syracuse University, Syracuse, NY, USA

8 Department of Marine Chemistry and Geochemistry, Woods Hole Oceanographic Institution, Woods Hole, MA, USA

9 School of Earth and Environmental Sciences and Centre for Exoplanet Science, University of St Andrews, St Andrews, UK

fur, and we summarize their distribution on the modern Earth as an interconnected network of isotopically fractionated reservoirs with contrasting residence times. We show how this framework can be used to explore the evolution of life and environments on the ancient Earth, which is our closest accessible analogue for an extraterrestrial planet.

Keywords Isotope geochemistry · Biogeochemical cycles · Geobiology · Evolution of life · Carbon · Nitrogen · Sulfur

1 Introduction

Living things exist in thermodynamic disequilibrium with their environment. This can be reflected in the stable isotope ratios of materials that are produced by organisms or otherwise impacted by metabolic processes. Isotopic analyses can therefore be used to study life where direct observations are not possible. Isotopic information has been used to understand metabolism on Earth in environments that are otherwise inaccessible in space or time, such as the deep subsurface (e.g., Kotelnikova 2002; Hinrichs et al. 2006) and in the geologic past (e.g., Leavitt et al. 2013). Soon, with the realization of a sample return mission to Mars, isotopic signatures of putative biogenic materials will be an important diagnostic tool in the search for extraterrestrial life (e.g., Franz et al. 2017).

Metabolic processes can broadly be grouped as either anabolic (involving biomass synthesis) or catabolic (purely energy yielding). Both of these are required by all living things on Earth. In life as we know it, all anabolic processes involve carbon. An individual organism can synthesize biomass from existing organic matter (heterotrophy) or from CO_2 (autotrophy). Catabolic process meanwhile include all reactions that can be harnessed to provide the energy required to grow, move, and propagate. The energy released—that is not lost as heat—is used to maintain cellular organization and to drive chemical reactions that would be otherwise unfavorable, such as the production of complex organic molecules. Most organisms obtain this energy from their environment by catalyzing systems of net exergonic chemical reactions, which are kinetically inhibited in the absence of life. The important exception to this is photosynthesis, where an additional source of chemical potential energy to the system is generated through the conversion of light energy and which subsequently feeds into a cascade of diverse metabolic activity (see Middelburg 2019, for review). On a planet-wide scale, the ensuing interactions are observable as global biogeochemical cycles. On a sub-cellular scale, enzymes—biology's catalysts—direct the flow of energy. Information about enzyme structure, localization and regulation within the cell is encoded in an organism's DNA. Through selective propagation of this information, the efficiency of these catalysts is subject to evolutionary adaptation. Metabolic processes can thus be viewed as sets of stoichiometrically tuned reactions that maximize energy release, biomass synthesis, and propagation over long timescales.

Thermodynamic disequilibrium between the intracellular (/near-cell) and ambient environment is maintained with the energy released from metabolic reactions, and results in spatial heterogeneity. Organisms create microenvironments that are conducive to the production of organic molecules and biominerals (such as calcite, opal, and bone). Organisms also create reactive chemical species that subsequently precipitate as minerals through abiotic processes. Much of the solid material produced at the Earth's surface at low temperature that makes it into the geologic record is precipitated in these biologically controlled/influenced environments. The isotopic composition of these phases in sediments and sedimentary rocks can thus record the biological processes and environments in which the organisms lived.

In this chapter we discuss the isotopic imprint of life on materials found on Earth. We first describe some general concepts for considering biological isotope fractionations. We then provide summaries of the isotope biogeochemistry of the biologically essential elements carbon, nitrogen and sulfur on the modern Earth and outline the geological evidence for the evolution of the Earth system over the course of its history. Finally, we synthesize the previous sections to discuss how the structures of Earth's biogeochemical cycles influence their sedimentary isotopic records and speculate on how these cycles and records would differ in the absence of life on Earth.

1.1 Isotope Notation

Carbon, nitrogen and sulfur stable isotope fractionations in natural environments are modest. It is therefore useful to express isotope ratios in terms of deviations from a reference ratio. In this chapter we use δ notation, defined as follows:

$$\delta^r \mathrm{E}_{\mathrm{sample}} = \frac{R_{\mathrm{sample}}}{R_{\mathrm{standard}}} - 1, \tag{1}$$

where

$$R = \frac{[^{\mathrm{r}}\mathrm{E}]}{[^{\mathrm{c}}\mathrm{E}]}, \tag{2}$$

and where $[^{\mathrm{r}}\mathrm{E}]$ = the concentration of the rare isotope of element, E, and of mass r, $[^{\mathrm{c}}\mathrm{E}]$ is the concentration of the common isotope of element, E, and of mass c. In the isotope systems considered here ($\frac{^{13}\mathrm{C}}{^{12}\mathrm{C}}$, $\frac{^{15}\mathrm{N}}{^{14}\mathrm{N}}$, and $\frac{^{34}\mathrm{S}}{^{32}\mathrm{S}}$), all rare isotopes are heavier than their common counterparts, so more positive isotopic compositions are more enriched in the heavy isotope. The standard for carbon is the Vienna Pee Dee Belemnite (VPDB), for nitrogen it is the modern atmosphere (AIR), and for sulfur it is the Vienna Canyon Diablo Troilite (VCDT). Here and elsewhere, δ values are expressed in permil (‰; i.e., $\delta \times 1000$).

Differences in stable isotopic ratios of substrates (S) and products (P) are given by the fractionation factor ($\alpha_{\mathrm{P-S}}$):

$$\alpha_{\mathrm{P-S}} = \frac{R_{\mathrm{P}}}{R_{\mathrm{S}}} = \frac{1 + \delta^r \mathrm{E}_{\mathrm{P}}}{1 + \delta^r \mathrm{E}_{\mathrm{S}}}. \tag{3}$$

The equivalent of $\alpha_{\mathrm{P-S}}$, in permil fractionation is $\varepsilon_{\mathrm{P-S}}$, defined as follows:

$$\varepsilon_{\mathrm{P-S}} \equiv (\alpha_{\mathrm{P-S}} - 1) \approx \Delta_{\mathrm{P-S}} \equiv \delta^r \mathrm{E}_{\mathrm{P}} - \delta^r \mathrm{E}_{\mathrm{S}}, \tag{4}$$

where Δ is an approximation for ε, valid when α is close to unity. According to this definition, reactions associated with a positive fractionation factor result in a product that is enriched in the rare isotope compared with the substrate; negative fractionation factors correspond to depleted product. As with δ values, ε values are expressed in permil (‰; i.e., $\varepsilon \times 1000$).

1.2 Factors Affecting Isotopic Fractionations

Enzymatically catalyzed chemical reactions are often associated with (relatively) large isotope effects, whereby a reaction product is enriched or depleted in the rare isotope of a given element compared with the substrate by tens of permil. Isotopic fractionations are typically categorized as: *i*) kinetic—where a unidirectional reaction occurs at different rates for compounds containing different isotopes; or *ii*) equilibrium—where a reversible reaction, occurring at equal rates in each direction, results in the partitioning of isotopes largely as a function of their zero point energy in the reactant vs. product (Criss 1999). Heavy isotopes tend to form stronger chemical bonds than their light counterparts. In chemical reactions, a kinetic isotope effect reflects the difference in energy required to break a bond when it involves a heavy vs. a light isotope of the same element. Kinetic isotope effects are usually negative in both the forward and reverse directions for a given chemical reaction, though there are apparent exceptions (e.g., Casciotti 2009). Equilibrium isotope effects reflect the difference between the forward and reverse kinetic effects, and manifest as an enrichment of the heavy isotope in the phase where the bonding environment is most stable. Owing to the relative strength of the bonding partners involved, for a redox reaction allowed to reach equilibrium, the heavy isotope tends to be enriched in the more oxidized species (Schauble 2004). Kinetic and equilibrium isotopic fractionation factors are idealized, end-member scenarios. Enzymatically catalyzed reactions that occur inside organisms are often characterized by an intermediate state, where the *net* flux in the forward direction is larger than zero (equilibrium) but smaller than the *gross* flux (kinetic). While chemical kinetics dictate the gross rate of the forward reaction, the thermodynamic driving force of the reaction determines the ratio of the reverse to forward reaction rates. The net reaction rate, therefore, is a combination of both kinetics and thermodynamics (Britton 1965; Beard and Qian 2007; Noor et al. 2013).

Irreversible reactions cause localized chemical and isotopic disequilibrium that impacts the composition of reaction products. The spatial scale over which these departures from equilibrium occur can range from a physical membrane-bound compartment such as a cell or organelle to, e.g., the entire ocean. In every case, for a reaction flux associated with a given fractionation factor (ε), changes in the isotopic composition of the compartmental reservoir result in changes in the net fractionation between the reactant entering the compartment and the product leaving it. Utilization within a compartment is the fraction of substrate entering the compartment that is redirected to the reaction flux (Hayes 2001).

Classically, systems are described as ‘open’ when the removal of substrate by reaction is continually replenished from an external source, or ‘closed’ when the system is isolated and there is no replenishment (Hayes 2001). A one-compartment system can only be at steady state when it is ‘open’. Microbial cells are often treated as open systems and characterized by steady state growth due to the relatively high rates of intracellular substrate turnover (or short residence time—see Sect. 1.3). Higher reaction rates inside the cell (perhaps corresponding to higher growth rates) result in greater utilization, while a higher flux of substrate to the cell (perhaps due to higher concentrations of substrate in the environment) results in lower utilization (Farquhar et al. 1982; Rau et al. 1996). It is therefore common for even simple systems to be characterized by a complex isotopic interplay between reaction kinetics, reversibility, and reservoir effects through substrate and product concentrations. Closed system reservoir effects can result in very large drifts in isotopic composition when not at steady state, due to continued reaction with no replenishment of substrate.

Due to these complexities, the same reaction can express different isotopic fractionations depending on the conditions under which it occurs. Such variations are exacerbated in biological systems, where physiological and environmental factors can influence the reversibility, openness, and utilization of a given reaction. In Table 1, we summarize the isotopic fractionations associated with key biological and abiotic reactions in the cycles of carbon, nitrogen, and sulfur on Earth. These reactions, and their roles in the relevant biogeochemical cycle, are detailed in the following sections. We emphasize that no single number characterizes the isotopic fractionation of any of the reactions listed; rather, reactions express characteristic *ranges* of isotopic fractionations, and variations within these ranges result from the

Table 1 Biological and abiotic reactions that govern the exogenic C, N, and S cycles. Bars indicate ranges of observed isotopic fractionations for every reaction (*in vivo* or in environmental materials). Vertical dashes indicate estimated mean isotopic fractionations in the environment, where appropriate. Note that ε scales are nonlinear. All fractionations reported as ε_{P-S}, as per definition in Eqn. (4) (i.e., negative numbers indicate 'normal' isotope effects). Ranges in expressed fractionations depend on various factors (general considerations described in Sect. 1.2; specific considerations listed in the table where known). Ranges in net isotopic fractionations manifest as variability in the isotope compositions of the reaction products

Element	Reaction	Sources of variability	Selected references	Text section
Carbon	C3 photosynthesis	Various environmental	Farquhar et al. (1989);	2.3.1
	C4 photosynthesis	& physiologic parameters	Still et al. (2003);	2.3.1
	rTCA pathway	that directly/indirectly influence openness/utilization, incl.	Sirevåg et al. (1977); Berg et al. (2010)	2.3.1
	Wood-Ljungdahl pathway	T, water availibility, soil pH	Summons et al. (1998); House et al. (2000); Blaser et al. (2013)	2.3.1
	Post-primary metabolite biosynthesis	Metabolic demands at branching points; reservoir effects	DeNiro and Epstein (1977); Monson and Hayes (1982); Van Der Meer et al. (1998); Hayes (2001)	2.3.2
	Heterotrophy		DeNiro and Epstein (1978)	2.3.3
	Selective organic C degradation	Highly variable, depends on specific compounds involved	Tissot and Welte (1984)	2.3.3
	Equil. $CO_{2(aq)}-CO_{2(g)}$	Temperature/pH/salinity	Zhang et al. (1995); Zeebe and Wolf-Gladrow (2001)	2.4
	Equil. $CO_3^{2-}-HCO_3^-$			2.4
	Equil. $CO_{2(aq)}-HCO_3^-$			2.4
	Vital effects during carbonate precipitation	Physiology, growth-conditions	Zeebe et al. (1999a); McClelland et al. (2017); Chen et al. (2018)	2.5
Nitrogen	Biological nitrogen fixation (Mo-nitrogenase)		Minagawa and Wada (1986); Macko et al. (1987); Bauersachs et al. (2009); Zhang et al. (2014)	3.3
	Biological nitrogen fixation (Alternative nitrogenases)		Zhang et al. (2014)	3.3
	Biological denitirfication		Delwiche and Steyn (1970); Granger et al. (2008)	3.3
	Annamox: $NH_4^+ \rightarrow N_2$	Enzyme reversibility	Brunner et al. (2013)	3.3
	Annamox: $NO_2^- \rightarrow N_2$			3.3
	Dissolved organic N uptake	Uptake pathway; dissolved N/C availability; utilization	Hoch et al. (1992)	3.3
	Sedimentary diagenesis		Freudenthal et al. (2001)	3.3
	Metamorphic rxns.		Boyd and Philippot (1998); Stüeken et al. (2017)	3.3
	Hydrothermal rxns.		Bebout and Fogel (1992); Stüeken et al. (2017)	3.3
	Equil. $NH_4^+-NH_3$ partitioning		Stüeken et al. (2015a)	3.4
Sulfur	Dissimilatory SO_4^{2-} reduction	Electron, sulfate availibility; cell-specific reduction rate	Canfield (2001); Leavitt et al. (2013); Zaarur et al. (2017)	4.2
	Phototrophic H_2S oxidation	Dependent on strain, sulfide availibility	Zerkle et al. (2009)	4.3
	Chemotrophic H_2S oxidation	Dependent on strain, sulfide availability	Zerkle et al. (2016); Pellerin et al. (2019)	4.3
	S^0 disproportionation		Habicht et al. (1998); Johnston et al. (2005)	4.3

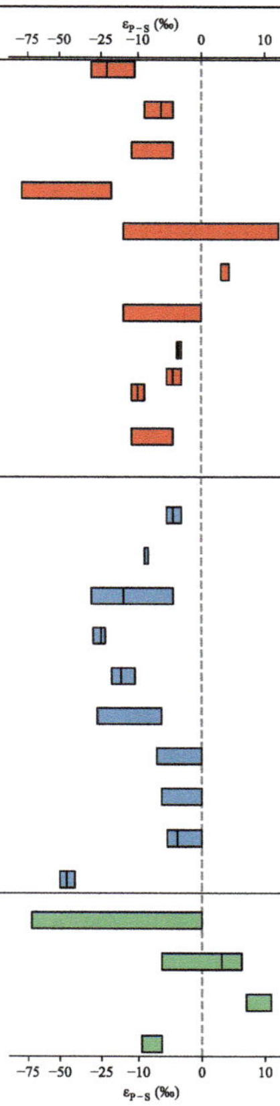

various factors described above. As expected, variability is greater for biologically catalyzed reactions.

1.3 Sedimentary Isotope Records in the Context of Reservoir Residence Times

An isotope record is a sequence of isotopic measurements made on the same sedimentary phase from samples that have been resolved into a geologic time series. Compilations of sedimentary carbon, nitrogen, and sulfur isotope records are shown in Fig. 1. These records are superpositions of multiple features: secular changes, transient events (excursions), and repeating cycles, each with different amplitudes/magnitudes and different characteristic timescales ranging from shorter than years to billions of years. Such variations are often expressions of the changes in the makeup of the respective elemental cycle at any one time. The carbon, nitrogen and sulfur cycles on Earth comprise pools of varying sizes, and are interconnected by fluxes associated with isotopic fractionations. While the isotopic frac-

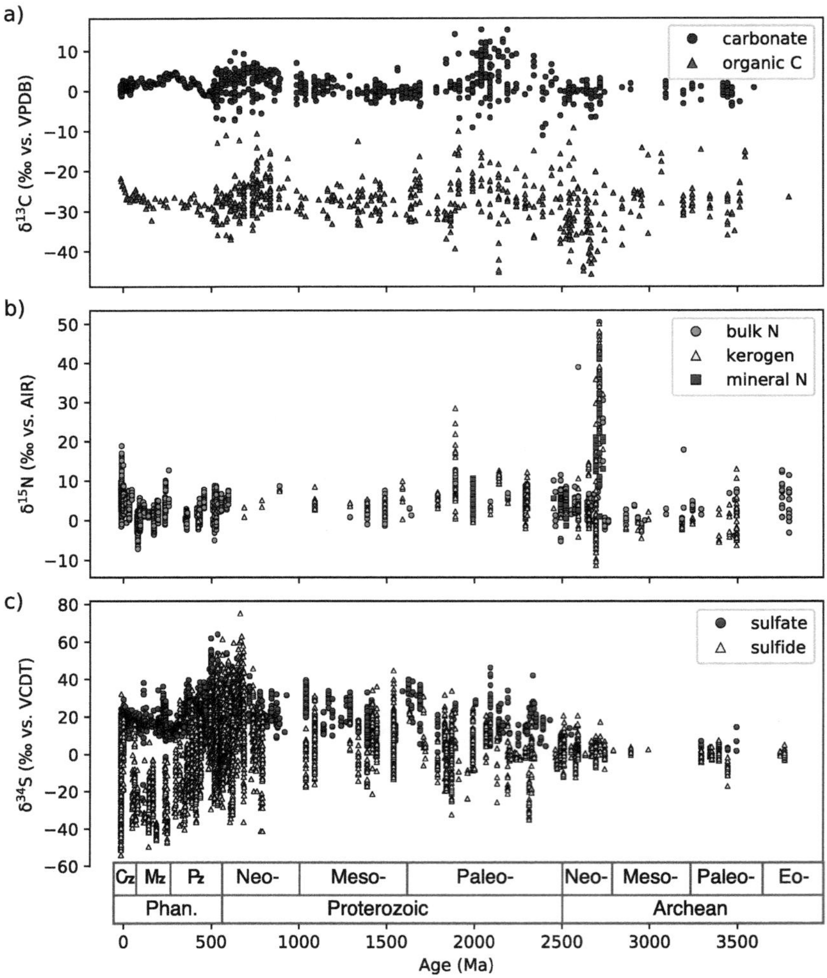

Fig. 1 The isotopic records of C, N, and S cycling on earth: Compilation of isotope compositions of biomolecules and minerals through time. a) Carbon data from Krissansen-Totton et al. (2015), Zachos et al. (2001) and references therein. b) Nitrogen data from Yang et al. (2019) and references therein. c) Sulfur data from Fike et al. (2015), Crockford et al. (2019), Canfield (2004), and references therein

tionations and magnitudes of these fluxes set the steady state isotopic composition of each pool, the time it takes for any given pool's isotopic composition to respond to a change in input or output is approximately the residence time of the element in that pool. Changes in the composition of any given pool are therefore dynamic on approximately the residence time of the pool, which is defined as the ratio of the size of pool (in moles) to the flux through the pool at steady state (in moles per year). Compared with a timescale of interest (e.g., the temporal duration of an isotope excursion), pools with longer residence times change little and can be assumed to be approximately constant, while pools with shorter residence times change rapidly and can be assumed to be approximately in steady state.

Most pools do not have geologic records. In the carbon cycle (Sect. 2, below) sedimentary carbonate carbon and organic carbon are the principal archives of past conditions

(Fig. 1). Attempts to reconstruct the makeup of a past state of the carbon cycle (i.e., the size and isotopic composition of every pool, plus the size and isotopic fractionation of the fluxes between them) are typically grounded in two observations: the $\delta^{13}C$ values of carbonate and organic carbon. Such an exercise is clearly underconstrained. Considering the Earth's biogeochemical system in this way—as separated into steady state, dynamic, and constant components at a given timescale—provides additional constraints on the system. The relevant timescale determines which pools need be treated dynamically (i.e., where the composition of the pool depends on its previous state) and which can be assumed to be in steady state or constant. Moreover, a clue as to the mechanism behind a specific feature in an isotopic time series lies in the timescale of that feature—e.g., the onset and decay time of an isotope excursion or the period of a repeating cycle. The characteristic timescale of a feature may be imposed by an external 'forcing' on the biogeochemical cycle (such as temperature), by an independent process (such as biological evolution) driving a secular change in the factors controlling fluxes or isotopic fractionations between these pools, or by dynamics inherited from a pool with the appropriate residence time. In the following sections, we develop graphical spaces relating isotopic compositions and residence times for the C, N and S pools on the modern Earth (Figs. 2, 3, and 4). We use these spaces as a framework for interpreting the evolving imprint of biology on the isotope chronologies shown above (Fig. 1) and, thus, on Earth's biogeochemical cycles though time.

2 Carbon

Carbon (C) is central to life on Earth. It forms the chemical backbone of all organic molecules. Transitions between its wide range of oxidation states is ultimately the means by which energy from the sun drives most metabolisms on Earth. Carbon from the mantle enters Earth's surface environment through volcanic activity mostly in its most oxidized form, CO_2 (Ciais et al. 2013). The reduction of CO_2 to organic carbon requires a strong reducing agent. On the modern surface Earth, oxygenic photosynthesis dominates primary production (CO_2 reduction to organic carbon; also called CO_2 fixation). This conversion is powered by sunlight, with water as the terminal electron donor, and results in the production of oxygen. CO_2 reduction can also occur via anoxygenic photosynthesis, which is also powered by sunlight but uses sulfide, molecular hydrogen or reduced iron as a terminal electron donor (Blankenship et al. 2004), or via chemoautotrophic processes like methanogenesis, which couples the oxidation of molecular hydrogen to the reduction of CO_2 (Demirel and Scherer 2008). These strongly reducing terminal electron donors are often hydrothermal in origin (McCollom and Shock 1997). Molecular hydrogen, in particular, can be also generated through radiolysis of water in sediments (Blair et al. 2007), through serpentenization reactions (Russell et al. 2010), or by organisms from fermentative and hydrolytic metabolisms. Once reduced, organic carbon can be used to generate energy by coupling its oxidation to the reduction of a wide range of oxidized compounds including molecular oxygen, sulfate, nitrate and oxidized metals (e.g., Jørgensen 1982; Thamdrup 2000; Achtnich et al. 1995; Straub et al. 1996; Vargas et al. 1998, and see Sects. 3 and 4 in this Chapter). All known life needs carbon to build biomass, and most life (including all eukaryotes) also uses carbon-based redox chemistry to generate energy.

Two stable isotopes of carbon exist on Earth—^{12}C (six protons, six neutrons) and ^{13}C (six protons, seven neutrons)—in a ratio of ~90:1. The extra mass of ^{13}C causes it to react more sluggishly and bond more stably. As carbon is cycled between various phases, these

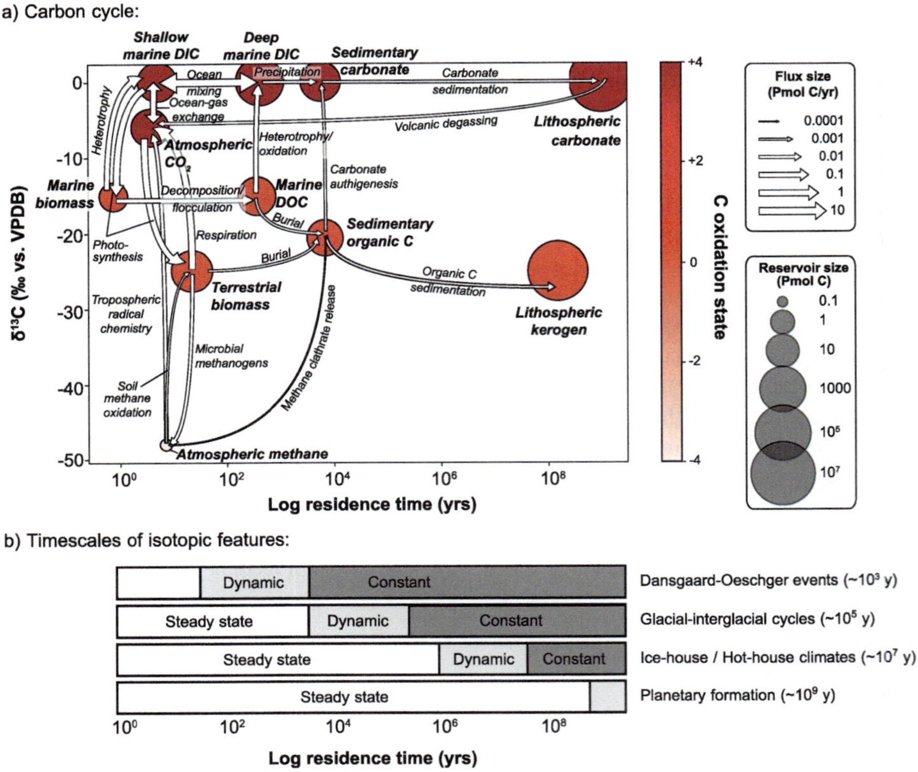

Fig. 2 a) Quantitative depiction of the exogenic pre-industrial carbon cycle: Carbon reservoirs (circles) are drawn proportional to their size and colored according to their weighted average carbon oxidation state. To highlight differences among surface and near-surface reservoirs, deep mantle and core carbon reservoirs are omitted. Reservoir locations are plotted according to their average $\delta^{13}C$ composition versus turnover time (Sec. 1.3). Widths of fluxes (arrows) are proportional to their magnitude. Estimates of flux sizes, reservoir sizes, and isotopic compositions from Falkowski et al. (2000), Ciais et al. (2013), Bar-On et al. (2018). $\delta^{13}C$ values shown are estimates of the mean values of these pools today; deviations from these mean $\delta^{13}C$ values can be large (especially among organic species; $\sim 10-20‰$), and are approximated by the ranges in the isotopic fractionations of the relevant reactions shown in Table 1. b) Timescales of isotopic features that may be recorded by various sets of reservoirs in the carbon cycle

differences fractionate carbon isotopes. Fractionations manifest as small differences in the isotopic ratio of different carbon pools, typically on the order of a few to tens of ‰. As the carbon isotope compositions of different reservoirs within the Earth-system depend on their relative sizes, the carbon isotopic compositions of ancient materials are sensitive to the sizes of these reservoirs in the past.

Earth's 350 million petamol (Pmol = 10^{15} mol) carbon inventory is distributed between the metallic core ($\sim 90\%$), the mantle ($\sim 8-9\%$) and the crust and surface pools ($\sim 1.5\%$) (McDonough and Sun 1995) (Fig. 2). The atmosphere, biosphere and oceans, where we can observe the interactions between life and the carbon cycle, together house only 0.001% of Earth's carbon. The atmospheric carbon pool ($\sim$ 50 Pmol C on the pre-industrial Holocene Earth) exchanges with the other surface reservoirs, in which we include the surface ocean ($\sim$ 80 Pmol C), biosphere ($\sim$ 50 Pmol C), and soils ($\sim$ 160 Pmol C), on a timescale of years to decades (Ciais et al. 2013). The pool of carbon in the deep ocean, which is an order of magnitude larger than the combined surface pools (at around 3100 Pmol C) has a residence time of around 1000 years (Ciais et al. 2013). Carbonate in marine surface sediments in contact with seawater can precipitate or dissolve in response to changes in deep ocean seawater chemistry. Due to these feedbacks the size of this pool is difficult to constrain (nominally $\sim$ 140 Pmol C; Ciais et al. 2013). Given the long residence time of the deep ocean, any change to the ocean–sediment system takes over 10,000 years to reach a new steady state (Sigman and Boyle 2000). Earth's exosphere comprises the atmosphere, terrestrial biosphere, the ocean, and the reactive surface sediments (note that this usage differs from the other common usage of 'exosphere' to describe the outermost layer of a planet's atmosphere). The combined exospheric pool has a residence time of less than 10^5 years (Fig. 2), and is thus approximately at steady state on million year timescales, but is dynamic on timescales of < 1 Myr (Sect. 1.3; Walker et al. 1981). The amount of carbon stored as carbonate rocks, terrestrial sediments, and as fossilised organic material in the continental and oceanic crust exceeds that of the deep ocean by 3 orders of magnitude (Ciais et al. 2013). Anthropogenic CO_2 emissions through the burning of fossil fuels are unidirectionally short-circuiting the usually slow exchange between the vast pool of buried organic carbon and the atmosphere.

In this section, we describe the phases of carbon—and mechanisms for transitions between them—that comprise Earth's biogeochemical carbon cycle (see Falkowski et al. 2000; Hayes and Waldbauer 2006; Ciais et al. 2013, for reviews). Our focus is the mechanisms by which biology catalyzes transitions between phases of carbon, the isotopic fractionations associated with these changes, and the timescales over which these changes manifest in the geologic record. While there is a rich literature regarding the systematics and applications of radiocarbon (^{14}C), here we exclusively discuss the stable isotopes of carbon.

2.1 Organic Carbon

The term 'organic carbon' describes carbon in compounds containing at least one C–H bond. Carbon atoms exist in a wide variety of redox states (+4 to −4) depending on the electronegativity of available bonding partners. Some of the simplest organic molecules, such as methane (CH_4), formaldehyde (CH_2O) and methanol (CH_3OH), can form spontaneously by bombardment of high energy particles and are found in the interstellar medium and in the tails of comets (Snyder et al. 1969). More complex organic molecules are built upon chains and rings of carbon-carbon bonds. These 'carbon backbones' are versatile; the specific configuration of hetero-atoms (typically O, N, S, or P) and functional groups built off of the carbon backbone determine an organic molecule's physical properties (e.g., polarity, reactivity, solubility) and biologic function. For instance, the arrangement of hydrophobic and hydrophilic functional groups on opposite ends of lipids allows cells to build semi-permeable lipid membranes that selectively exclude the outside world. The rings of CH(OH) units of water-soluble carbohydrates are an efficient, flexible way to store and transport energy that can be polymerized to make larger, stabler compounds like cellulose and starch. The high density of functional groups in amino acids (each containing at least one amine (NH_2) and carboxyl (COOH) group) results in reactive molecules that can be linked into the proteins that catalyze reactions, replicate DNA, and perform tasks central to life.

The total carbon biomass of all living organisms on Earth is about 44 Pmol (Bar-On et al. 2018). Of this, about one third comprises metabolically active, high-turnover compounds in plants, animals, and vigorously-growing microorganisms. The remainder is in static tissues, such as the woody tissues of trees and deep biosphere microbes with exceptionally slow (superannual) turnover times (Bar-On et al. 2018; Magnabosco et al. 2018). The components of dead organisms are mostly respired (oxidized to CO_2) or assimilated by heterotrophs. Organic carbon compounds that escape consumption, either because they are difficult to degrade (e.g., lipids, lignin) or rapidly buried in inhospitable environments, become sequestered in soils and sediments. About 160 Pmol of organic carbon from recently-dead plants resides in soils (Batjes 2014), with an additional $\sim$ 160 Pmol of older carbon preserved in permafrost and anoxic wetlands (Bridgham et al. 2006; Tamocai et al. 2009). During further sedimentation and lithification, surviving organic compounds are transformed by polymerization and condensation reactions into recalcitrant, insoluble agglomerations of organic matter collectively known as kerogen. Globally, around $0.1 - 0.4\%$ of organic carbon that is cycled by the biosphere escapes remineralization and is buried in sediments (Berner and Canfield 1989; Middelburg 2019). Nonetheless, due to continued burial and sedimentation over billions of years, kerogen disseminated in the crust constitutes the largest organic carbon reservoir on Earth ($\sim 1.25 \times 10^6$ Pmol; Tissot and Welte 1984; Falkowski et al. 2000).

2.2 Methane

Methane (CH_4), the smallest organic molecule, plays an especially important role in the carbon cycle. As the most reduced form of carbon, it is an energetically-favorable endmember of abiotic and biotic redox reactions. In strongly reducing environments, organic carbon decomposition, microbial fermentation, and Fischer-Troph-type (FTT) reactions produce abundant methane (Reeburgh 2007). In relatively oxidizing environments, methane is oxidized either by aerobic methanotrophs or by microbial consortia performing anaerobic oxidation of methane (AOM) (Hinrichs et al. 2000). Despite its low abundance and short residence time in the modern atmosphere ($\sim$ 700 ppb in the pre-industrial era and less than 10 years respectively; Ciais et al. 2013; Prather et al. 2012), methane is a potent greenhouse gas, so methane that escapes oxidation contributes to the regulation of Earth's climate.

The sizes of the environmental sources and sinks of atmospheric methane are difficult to constrain, in no small part because they are dwarfed by anthropogenic methane emissions in the Industrial era (Kirschke et al. 2013; Schwietzke et al. 2016; Saunois et al. 2020). Modern abiotic methane fluxes are small compared to those involving biology, but were likely important for fertilizing early Earth with a viable reduced energy source in the pre-biotic world (Russell et al. 2010). Together, marine hydrothermal systems and terrestrial geothermal/volcanic areas contribute $\sim$0.004 Pmol/yr to the atmospheric methane budget (Etiope et al. 2008). Biotic fluxes are larger; obligately anaerobic archaea produce methane by either completely reducing CO_2 with H_2 or by adding a fourth hydrogen atom to a methyl-containing organic molecule (usually acetate, but also methanol, methylsulfides, or methylamines) produced by the fermentation of larger organic molecules (Ferry 2010). Worldwide, microbial methanogenesis in wetlands, peatlands, and lakes emits about 0.02 Pmol of methane to the atmosphere each year (Kirschke et al. 2013). These geologic methane fluxes would be larger were it not for microbial degradation; it is estimated that more than 90 % of methane produced in marine sediments is oxidized by cooperative consortia of methanotrophic archaea and sulfate reducing bacteria before it escapes the sediment column (Orphan et al. 2002; Knittel and Boetius 2009; Reeburgh 2007). Aerobic bacteria in soils oxidize and consume about 0.002 Pmol of atmospheric methane per year (Spahni et al. 2011). An additional methane reservoir of uncertain—but potentially significant—size exists in methane hydrates (a solid, high-pressure, cryslalline phase of methane bound to water) in ocean basins (Archer et al. 2009). The methane hydrate reservoir may be as large as 160 Pmol, and is susceptible to release in response to global warming (Archer et al. 2009). Such a feedback has been invoked as a major contributor to prominent disturbances to the ancient carbon cycle (Dickens 2003).

2.3 Carbon Isotopes in Organic Carbon and Methane

A range of factors control the carbon isotopic composition of organic matter. Broadly these factors fall into the following categories: *i*) processes dictating the isotopic composition of the primary organic matter, including environmental conditions and fixation pathway; *ii*) metabolic processes downstream of fixation that cause divergences in the isotopic compositions of various classes of organic compounds such as lipids, proteins and carbohydrates; and *iii*) heterotrophy and metabolic recycling, which lead to further alteration of the chemical and isotopic makeup of organic matter, largely through fermentation and preferential oxidation of certain classes of compounds.

2.3.1 Carbon Fixation Pathways

Almost all organic carbon on Earth is fixed through the Calvin cycle. The fixation reaction step, where an inorganic CO_2 molecule is bound, is catalysed by the enzyme ribulose-1,5-biphosphate oxygenase-carboxylase (RuBisCO). This fixation reaction is associated with a large negative fractionation of carbon isotopes ($^{13}\varepsilon_f$), causing the primary organic matter to be ^{13}C-depleted relative to the source CO_2 (Farquhar et al. 1982). Although once thought to carry a universal value of around -25 to $-28‰$ (e.g., Hayes 2001), the absolute magnitude of $^{13}\varepsilon_f$ has recently been shown to vary significantly between RuBisCOs extracted from different species of cyanobacteria, eukaryotic algae and higher plants (-11.1 to $\sim -30‰$; Boller et al. 2011, 2015; Tcherkez et al. 2006). In addition to the enzymatic fractionation factor, a multitude of factors influence the net (i.e., apparent) fractionation between CO_2 in the ambient environment and the isotopic composition of biomass produced ($^{13}\varepsilon_p$).

Reservoir effects tend to dampen the expression of the fractionation by the RuBisCO enzyme at higher utilization, so any process or infrastructure that increases separation of the substrate pool from the site of fixation will result in less-^{13}C-depleted organic products. In photosynthesizing microorganisms, physiological and environmental parameters that determine substrate availability—such as cell size, shape, and growth rate (Rau et al. 1996; Popp et al. 1998)—influence the magnitude of fractionation that is expressed (see: Wilkes and Pearson 2019; McClelland et al. 2017, for reviews). In plants that use the C3 photosynthetic pathway, $^{13}\varepsilon_p$ is sensitive to the relative rates of photosynthetic carbon fixation and diffusion of CO_2 into the leaf (Farquhar et al. 1989). Plants employing the C4-photosynthetic pathway or Crassulacean acid metabolism (CAM) separate the processes of CO_2 uptake and carbon fixation in space or time, so the expressed carbon isotope fractionations of these pathways are smaller than the full fractionation of RuBisCO ($^{13}\varepsilon_p \approx 10‰$; Farquhar et al. (1989)).

In addition to the Calvin cycle, at least five other carbon fixation pathways exist on Earth (Berg et al. 2010; Ward and Shih 2019). Of these, the reductive citric acid cycle (rTCA) and the reductive acetyl coenzyme A (Wood-Ljungdahl) pathway contribute most significantly to gross primary productivity today, though their importance has probably varied throughout Earth's history (Berg et al. 2010; Ward and Shih 2019). A handful of bacterial and archaeal clades fix carbon through the rTCA cycle, most notably green sulfur bacteria. The enzymes that catalyze the rTCA cycle are O_2-sensitive, limiting the habitat of the reliant organisms to sub-oxic or anoxic environments (Berg et al. 2010). When using H_2S, S^0, or H_2 as an electron donor for anoxygenic photosynthesis, these organisms express a relatively small fractionation against ^{13}C in the range of -2 to $-12‰$ (Hayes 2001). The flux of organic carbon fixed by the rTCA pathway today is estimated at ~ 0.08 Pmol/yr (about two orders of magnitude smaller than the Calvin cycle flux), but was probably more abundant in past periods when atmospheric O_2 levels were lower (Brocks et al. 2005; Ragsdale 2018). Due to its moderate carbon isotope fractionation, finding unambiguous carbon isotope evidence for ancient rTCA activity is problematic (Ward et al. 2019).

The Wood-Ljungdahl pathway is a strictly anaerobic carbon assimilation mechanism that enables the existence of autotrophic life in some of the planet's most extreme environments. Where environmental conditions make it possible (anoxic, H_2-replete, with abundant trace metals), the pathway is the least energetically costly of all carbon fixation schemes, and is used by (most notably) acetogens and methanogens to produce these environmentally significant biomolecules (Berg et al. 2010; Ward and Shih 2019). The pathway imparts exceptionally large carbon isotope fractionations; methanogenic archaea typically assimilate carbon with an $^{13}\varepsilon$ of -30 to $-36‰$ (Hayes 2001); the $^{13}\varepsilon$ of fixation by acetogenic bacteria can be as large as $-68‰$ (Blaser et al. 2013). The methane released by methanogens is also significantly depleted in $\delta^{13}C$. Depending on the substrate (CO_2, acetate, or another methyl-bearing compound), and the specific species, fractionations from substrate to methane range from -20 to $-80‰$ (Summons et al. 1998). The modern carbon fixation flux from this pathway is small (< 0.01 Pmol/yr), but its distinctive carbon isotope signature allows its presence to be recognized in past environments (Ward and Shih 2019). The widespread occurrence of kerogens with low-$\delta^{13}C$ values in late Archean sedimentary rocks, for instance, has been used to argue for the proliferation of reductive acetyl-CoA pathway at the time (Hayes and Waldbauer 2006; Slotznick and Fischer 2016).

2.3.2 Downstream Carbon Isotope Discrimination

Once fixed, primary organic carbon is converted into different classes of biochemicals including lipids, proteins, carbohydrates, and nucleic acids. Biosynthetic networks involve myriad branch-points, recycling processes, and enzymatic reactions with varying degrees of utilization and reversibility. These mechanisms all influence the carbon isotope compositions of metabolites and thus preclude accurate *de novo* predictions of the carbon isotope composition of any specific metabolic end-product (Hayes 2001). Broad-brush relationships between compound classes, however, are well established: within the same autotrophic organism utilizing the Calvin cycle, the $\delta^{13}C$ values of carbohydrates are canonically about 1‰ higher than proteins and nucleic acids, and about 6‰ higher than lipids (Blair et al. 1985; Hayes 2001). In general, this ordering is thought of as a function of metabolic path length; the larger number of metabolic steps between primary photosynthate and membrane lipids results in lipids that are relatively depleted in ^{13}C as compared to proteins and carbohydrates, which are metabolically 'closer' to primary photosynthate (Hayes 2001). In practice, however, specific metabolic steps—often, branch points—have outsized effects on the $\delta^{13}C$ values of metabolic products. For instance, the ^{13}C depletion of acetogenic lipids appears to arise specifically from the kinetic isotope effect associated with decarboxylation of pyruvate by pyruvate dehydrogenase, and the exact $\delta^{13}C$ value of such lipids is modulated by changes in the various metabolic demands for pyruvate (DeNiro and Epstein 1977; Monson and Hayes 1982). Environmental conditions and metabolic needs can also overwhelm the isotopic imprint of metabolic networks, such that the same set of biomolecules synthesized by the same organism will retain a different isotopic ordering depending on the conditions under which they are formed (e.g., Zhang et al. 2009; Bird et al. 2019). Some autotrophs, such as green sulfur bacteria, can even exhibit 'inverse' carbon isotope ordering ($\delta^{13}C_{lipids} > \delta^{13}C_{bulk\ biomass}$) due to utilization of the rTCA cycle for lipid synthesis (Van Der Meer et al.

1998). At the scale of the ocean, differences in the preservation potential of compounds from various community factions (e.g., prokaryotes vs. eukaryokes, autotrophs vs. heterotrophs) can overprint native relationships between compound classes to the extent that inverse carbon isotope ordering exists in sedimentary records from throughout the Phanerozoic (Close et al. 2011).

2.3.3 Heterotrophy

Heterotrophy alters the isotopic composition of fixed carbon in two ways: *i)* CO_2 respired by animals tends to have a lower $\delta^{13}C$ value than the food they eat (DeNiro and Epstein 1978). By mass balance, the biomass of animals is therefore enriched in ^{13}C compared to the biomass they consume by about 1‰ (DeNiro and Epstein 1978); heterotrophic microorganisms can also express this behavior (Mahmoudi et al. 2017). On an ecological scale, carbon is isotopically distilled over multiple trophic levels, and has been used as a proxy for position within the food chain. *ii)* Heterotrophic organisms preferentially consume some biomass components (i.e., proteins, carbohydrates) while leaving tough-to-metabolize compounds behind. In the oceans, the organic carbon that survives biodegradation and is buried in sediments is predominantly in the form of lipids and recalcitrant polymers (Tissot and Welte 1984). On land, the preservation potential of woody tissues exceed that of plant proteins and polysaccharides, leading to compositional bias in the material preserved in anoxic waters to become kerogen and coal (Tissot and Welte 1984). Due to the different $\delta^{13}C$ values of these different compound classes, selective degradation can, in principle, fractionate the carbon isotope composition of biomass by a few permil. In practice, variability in the ^{13}C content of biomass and sedimentary organic matter (Table 1) tends to outweigh such compound-based shifts in bulk $\delta^{13}C$ value.

2.4 Inorganic Carbon System

Whether in bodies of water, inside cells, or in a thin film of water on the surface of a cell, virtually all biologically mediated reactions occur in the aqueous phase. When CO_2 dissolves in water it reacts to form carbonic acid (H_2CO_3) and subsequently its conjugate bases, bicarbonate (HCO_3^-) and carbonate (CO_3^{2-}) ions. The two acid dissociation constants (pKas) of carbonic acid straddle the pH of most biologically relevant media such as seawater or the cellular cytosol. As a result, carbonate speciation in these systems is sensitive to pH, and, to a lesser extent, temperature and salinity (via the effect of these environmental parameters on the pKas). In modern seawater the total concentration of dissolved inorganic carbon (DIC) is around 2 mM, and CO_2 (CO_2 (aq) + carbonic acid), CO_3^{2-}, and HCO_3^- exist in an approximate ratio of 1:10:100. The pH, and thus the speciation of DIC, and other weak acids in solution, is controlled by the ratio of DIC to total concentration of conjugate bases of weak acids in solution (known as alkalinity; reviewed extensively in: Zeebe and Wolf-Gladrow 2001; Dickson and Goyet 1994). Seawater on the modern Earth is buffered dominantly by carbonic acid. There is a complex interplay between the system of reactions internal to the carbonate system, which are constantly relaxing towards a state of equilibrium, and any process involving a specific species (e.g., carbonate precipitation), which push the system out of equilibrium. Indeed, biological processes are often sufficiently fast that disequilibrium and interconversion kinetics need be considered. The interconversion of CO_2 and HCO_3^- is slow, but in some biological systems is catalyzed by an enzyme called carbonic anhydrase, with important isotopic implications (e.g., Chen et al. 2018). Since CO_2 is the only gaseous phase of the carbonate system, equilibration between a body of water and a gas requires all carbon to pass through the form CO_2. Therefore, as pH increases, which results in an increase in the fraction of DIC in forms other than CO_2, the equilibration timescale between the aqueous and gas phases also increases.

Carbon isotopes are fractionated among the different species of DIC. At equilibrium, HCO_3^- is the most ^{13}C-enriched species, followed by CO_3^{2-} and then $CO_2(aq)$. The sizes of these decrease with increasing temperature. In the range 0–35 °C, CO_3^{2-} and $CO_2(aq)$ are depleted in ^{13}C relative to HCO_3^- by $3 - 1‰$ and $12 - 8‰$, respectively. $CO_2(g)$ is consistently enriched compared to $CO_2(aq)$ by about $1‰$. Since the isotopes of carbon are distributed among all DIC species, the isotopic composition of each species depends on its relative abundance (see: Zeebe and Wolf-Gladrow 2001, for further reading). Various biogeochemically important reactions involve specific DIC species, so changes in these fluxes influence both the inorganic carbon speciation and isotope composition of the aqueous medium, on scales from the cellular cytosol to the global ocean. When the DIC system is disturbed by a rapid change, disequilibrium effects come into play (Sade and Halevy 2017; Zeebe 2014; Zeebe et al. 1999b). The rate of exchange between HCO_3^- and CO_3^{2-} is extremely rapid, and in the majority of cases these species can be assumed to comprise a single equilibrated pool. The comparatively sluggish interconversion of CO_2 and HCO_3^- can lead to isotopic disequilibrium in a variety of settings, including in the vicinity of phytoplankton (e.g., Rau et al. 1996), zooplankton (e.g., Zeebe et al. 1999a) and speleothems (e.g., Mickler et al. 2006).

2.5 Dynamics and Evolution of Earth's Carbon Cycle

On timescales of longer than a million years, Earth's exospheric carbon pools are approximately at steady state (Fig. 2). The principal inputs to the exogenic carbon cycle are CO_2 emissions from the crust and upper mantle, mostly from mid-ocean ridge degassing, arc volcanism, and metamorphism (Lee et al. 2019), and the outputs are the permanent burial and sedimentation (and ultimately subduction) of carbonate and organic carbon (Fig. 2). Once it enters the exosphere, CO_2 is partitioned with respect to chemical form, oxidation state, and isotopic composition into a number of interconnected reservoirs (Fig. 2). Organic and inorganic carbon move in parallel from more reactive surface reservoirs (e.g., the shallow ocean, the terrestrial biosphere) to progressively less reactive reservoirs with longer residence times (e.g., the deep ocean and sediments). Along the way, various processes interconvert organic and inorganic carbon within the exosphere before their ultimate removal (Fig. 2). The input of volcanic CO_2 to the exosphere is approximately constant today, but varies on multi-million year timescales with the cycles of continental rifting and collision. This has been widely invoked as the long-term driver of Earth's climatic state (Lee et al. 2016; McKenzie et al. 2016), but is an area of active debate (Macdonald et al. 2019; Jagoutz et al. 2016). On sufficiently long timescales, the silicate weathering feedback balances variations in the size of CO_2 inputs (Walker et al. 1981), but whether variations in Earth's climatic states are primarily source-driven or sink-driven remains uncertain (Lee et al. 2019).

Carbon appears in the geologic record in the form of carbonate and organic carbon in sediments and rocks. Most Phanerozoic carbonate sediment is biogenic in origin, and is precipitated in pelagic environments (dominantly by coccolithophores and planktonic foraminifera) and both deep and shallow benthic environments (mostly by benthic foraminifera, corals and molluscs). The carbon isotopic compositions of these carbonates represent that of DIC in the water in which they formed, with small offsets from equilibrium (typically $2 - 5‰$) known as "vital effects" (Zeebe et al. 1999a; McClelland et al. 2017; Chen et al. 2018). Unlike carbonate, organic carbon is more reduced than the volcanic influx. Any organic matter that escapes remineralization and is permanently buried therefore renders the exosphere increasingly oxidized. On long timescales (> 10^6 years), the fluxes of carbon entering and leaving the exosphere from the lithosphere are widely assumed to be maintained at steady state largely through the negative feedback of silicate weathering (Walker et al. 1981), although some models suggest that the surface-sediment carbon cycle may have a response time of as long as 10-100 million years (Rothman et al. 2003). Nonetheless, on sufficiently long timescales the weighted mean isotopic composition of the sequestered carbonates and organic carbon must equal that of the volcanic influx. Because organic matter is depleted in ^{13}C compared to carbonate, an increase in the fraction of carbon buried as organic matter (F_{org}) results in an increase in the isotopic enrichment of the buried carbonate.

Under the assumptions of *i*) steady state in the carbon cycle, *ii*) the applicability of a simple mass balance, and *iii*) the invariance of the isotopic composition of the volcanic influx through time (which may not be a good assumption, c.f., Mason et al. 2017), the $\delta^{13}C$ values of carbonate ($\delta^{13}C_{carb}$) and organic carbon ($\delta^{13}C_{org}$) in ancient rocks have been widely used to constrain F_{org}, and thus the oxidation state of the exosphere and (more tangibly) the concentration of oxygen in the atmosphere (e.g., Hayes and Waldbauer 2006; Krissansen-Totton et al. 2015). As reviewed in the preceding sections, however, the evolving picture is more nuanced: the $\delta^{13}C$ value of carbonate depends on variables influencing DIC speciation, such as temperature, pH, and pCO_2 (Sect. 2.4), while $\delta^{13}C_{org}$ depends on the magnitude of the carbon fixation fractionation that is expressed (Sect. 2.3). Both records may be strongly influenced by local effects (Swart 2008), and overprinted by post-depositional effects such as selective organic carbon compound degradation and authigenic carbonate precipitation (Banner and Hanson 1990; Des Marais 2001; Schrag et al. 2013). Use of a simple mass balance to reconstruct changes in organic carbon burial flux is only strictly valid at steady state (Hayes and Waldbauer 2006). Under conditions where the timescale of a perturbation is shorter than the residence time of the combined exospheric carbon pool, the system will evolve dynamically (e.g., Rothman et al. 2003). Given these complexities here we avoid describing quantitative changes in F_{org}. Instead, we emphasize the biogeochemical advances that best explain the qualitative changes observed.

2.5.1 Precambrian Carbon Cycling

Prior to the evolution of oxygenic photosynthesis, life was limited by the availability of viable terminal electron donors (Des Marais 2001; Ward and Shih 2019). In the Archean eon the most abundant reduced chemical species, which thus limited early primary production, was likely H_2 emitted from volcanic vents (Des Marais 2001). Recent estimates indicate a steady state atmospheric pH_2 of less than 0.01 bar (Kadoya and Catling 2019; Catling and Zahnle 2020). Even if seafloor spreading were more vigorous in the Archean due to higher radionuclide heat flux, and could have sustained H_2 emissions at an order of magnitude larger than today, the dependence on H_2 would have placed an upper bound on primary productivity of ~ 0.01 Pmol/yr (<0.1 % of modern GPP) (Des Marais 2001). This remains an active area of debate, with arguments made for stricter (Korenaga 2006) or more relaxed (Höink et al. 2013) limits on H_2. Regardless, due to imperfect utilization of volcanic reduced species, and with habitats likely limited to the photic zone (Tice and Lowe 2004), Archean primary productivity was likely orders of magnitude lower than this upper bound (Ward and Shih 2019; Canfield et al. 2006).

The $\delta^{13}C$ of early Archean organic matter is similar to that of the Phanerozoic ($\delta^{13}C_{org} \approx -28‰$), but with greater variability (≈ -50 to $-10‰$; Fig. 1). Carbonate $\delta^{13}C$ values were lower, however, by $2-4‰$. It has been argued that both the average ε_p of primary autotrophy and F_{org} were lower than today (e.g., Krissansen-Totton et al. 2015). The small ε_p of the rTCA cycle—canonically an ancient fixation pathway (e.g., Ragsdale 2018)—is compatible with isotopic evidence for a muted primary ^{13}C fractionation in the early Archean, but not uniquely so, given the overlap of this ε_p with those of other fixation pathways (Ward and Shih 2019). Mean organic carbon $\delta^{13}C$ values in the late Archean are lower than the early Archean by about 10‰; with some values as low as −60‰ (Strauss et al. 1992; Hayes and Waldbauer 2006) (Fig. 1). High variability—and possible bimodality—in $\delta^{13}C_{org}$ records is observed at various spatial scales at this time (Williford et al. 2016). This depletion in $\delta^{13}C_{org}$ has been canonically attributed to an 'age of methanotrophy' (Hayes 1994), but carbonate $\delta^{13}C$ records from the same sections are more consistent with the dominance of organisms employing a metabolism that does not produce such $\delta^{13}C$-depleted DIC (and thus $\delta^{13}C$-depleted authigenic carbonate) upon remineralization (Slotznick and Fischer 2016). Instead, the late Archean may have been an era when multiple primary carbon fixation metabolisms proliferated: a metabolism with a small ε_p (such as rTCA-utilizing anoxogenic photoautotrophs) and a metabolism with a large ε_p, such as the Wood-Ljungdahl pathway (by early acetogens and/or methanogens) (Des Marais 1997; Slotznick and Fischer 2016; Ward et al. 2019).

Multiple lines of evidence document a substantial rise in atmospheric O_2 levels about 2.4 Ga, in the so-called Great Oxidation Event (GOE) (Holland 2006; Gumsley et al. 2017). Although low levels of oxygen may have existed up to 50 million years prior (Anbar et al. 2007), the disappearance of sulfur mass-independent fractionations (S-MIF) in sedimentary pyrites and sulfates at this time requires that O_2 concentrations were higher than $10^{-5}\times$ present atmospheric levels (Farquhar et al. 2000). The appearance of Oxygen-MIF measured in contemporaneous sulfate evaporites further requires an oxygen concentration of greater than $10^{-3}\times$ present levels (Crockford et al. 2019), though it has been suggested to have risen higher still (potentially even greater than present levels; Bekker and Holland 2012). The introduction of O_2 is attributed to the proliferation—but necessarily the evolution (Anbar et al. 2007)—of oxygenic photosynthesis (Falkowski 2011; Fischer et al. 2016). Unlike their volcanism-dependent predecessors, oxygenic phototrophs had a comparably-limitless supply of terminal electron donors in the form of water. As a result, primary productivity may have reached 1% of modern levels (Crockford et al. 2019).

Despite the apparent increase in primary production across the GOE, a secular change in the carbon isotopic composition of organic matter is conspicuously absent (Fig. 1; Catling and Claire 2005). One explanation may be that an increase in primary production was compensated for by an increase in heterotrophy, and therefore there was no net change in the organic carbon exported from the exosphere (Catling and Claire 2005). Importantly, although localized 'oases' of the surface ocean may have been oxygenated by this time (Eigenbrode and Freeman 2006), these putative heterotrophs would have likely been anaerobes: aerobic heterotrophy appears to have postdated the evolution of oxygenic photosynthesis (Soo et al. 2019), and the deep ocean seems to have remained effectively anoxic for the remainder of the Proterozoic (Stolper and Keller 2018; Planavsky et al. 2011). An alternative explanation suggested that the appearance of atmospheric O_2 was due to a change in the *sinks* of O_2 instead of the sources (Catling and Claire 2005; Kump and Barley 2007). By decoupling a rise in O_2 from a rise in F_{org}, the GOE may have been a threshold response to a gradual (i.e., secular) change in the carbon cycle. Some primary pO_2 proxy evidence suggests that the GOE was characterized by a gradual rise (Murakami et al. 2011; Kump et al. 2011), while other evidence contends that it was a geologically stepwise event (Luo et al. 2016).

The oxygenation of Earth's atmosphere led to the removal of methane via photochemical oxidation reactions, scrubbing this potent greenhouse gas and cooling the exosphere (Pavlov et al. 2000; Kopp et al. 2005). The early Proterozoic saw at least two, and perhaps as many as four (Rasmussen et al. 2013), massive, global glaciations (so-called 'snowball Earth' events, Kirschvink and Kopp 2008; Hoffman 2013). Between and in the aftermath of these glacial intervals, major, globally-expressed perturbations to the carbon cycle are observed (Karhu and Holland 1996, Fig. 1). During one of these, the $\sim 2.2-2.0$ Ga Lomagundi-Jatuli Excursion, $\delta^{13}C_{carb}$ values exceed $+16‰$ (Karhu and Holland 1996; Martin et al. 2013). This excursion has been interpreted as a period of enhanced organic carbon burial (Karhu and Holland 1996; Des Marais 2001), but both the high scatter among $\delta^{13}C_{carb}$ values and the lack of a matching positive excursion in $\delta^{13}C_{org}$ values precludes a simple rise in F_{org} (Karhu and Holland 1996). Barring a diagenetic origin for these elevated $\delta^{13}C_{carb}$ values (Planavsky et al. 2012), some degree of basin-specific effects or non-steady state behavior may be required to account for this period. The Lomagundi-Jatuli Excursion is followed by a globally-synchronous negative carbon isotope excursion mirrored in both carbonate carbon and organic carbon at ca. 2.0 Ga (Kump et al. 2011). This event has been interpreted as a widespread oxidation of sedimentary organic carbon culminating in a reduction of global primary productivity of at least 80 % (Kump et al. 2011; Hodgskiss et al. 2019). Such a productivity crash may have ushered in an era in which nutrient-limited conditions prevented further substantial changes to the global carbon cycle (as observed by $\delta^{13}C_{carb}$ and $\delta^{13}C_{org}$ records) for much of the subsequent billion years (Buick et al. 1995; Hodgskiss et al. 2019; Laakso and Schrag 2019).

The apparent relative stability of the global carbon cycle for the duration of the mid-Proterozoic ($\sim$ 1.8 – 0.8 Ga) is terminated by a series of climatic and biogeochemical changes reminiscent of those seen in the Paleoproterozoic. The carbon isotope record from 0.8 to 0.54 Ga is interrupted by two 'snowball Earth' glaciations and one large regional glaciation (Kirschvink 1992; Hoffman et al. 1998). These glaciations are characterized by a gap in the sedimentary record (Partin and Sadler 2016). Large positive and negative excursions are seen throughout this interval, which may represent dramatic swings in the marine DIC reservoir, and may or may not have been productivity related. The timing and regional coherency of these excursions remains poorly constrained and the relationships to glaciations uncertain. Still, it is likely that the glaciations shut off marine autotrophy and the silicate weathering feedback until volcanogenic CO_2 inputs contributed sufficient atmospheric insulation to thaw the icehouse (Kaufman and Knoll 1995; Kirschvink et al. 2000). The terminations of these global glaciations are demarcated by 'cap carbonate' intervals. In the recovery from the Sturtian glaciation, eukaryotic algae may have replaced cyanobacteria as the dominant primary producers in the ocean (Brocks et al. 2017). This putative rise of eukaryotes was echoed by the proliferation of animals—and the restructuring of the carbon cycle that accompanied the evolution of predation and grazing—in the Ediacaran following the smaller Marinoan glaciation (Butterfield 2007).

2.5.2 *Phanerozoic Carbon Cycling*

The appearance of biomineralization in the late Neoproterozoic and its rapid convergent evolution throughout the tree of life in the Cambrian led to a step change in the way geochemical information was preserved in sediments. In sediments where carbonate is preserved in the form of fossils, taxonomic information is available—this can be used to infer habitat, and therefore the position in the water column where the carbonates formed. Through a comparative biology approach involving extant organisms, aspects of physiology can also be inferred, enabling 'vital effect' corrections to be applied. The start of the Phanerozoic saw a dramatic decrease in the amplitude of variation in the carbonate carbon isotope record compared to the Neoproterozoic (Hayes et al. 1999; Krissansen-Totton et al. 2015). This change suggests an increase in the relative stability of the Earth system (Bachan et al. 2017), possibly due to the emergence of multicellularity and complex, and relatively robust, ecosystems (Schobben and van de Schootbrugge 2019).

The marine sedimentary record prior to $\sim$ 180 Ma is biased towards continental shelf sediments, and is thus dominated by fossils produced by coastal, and shallow water species (Veizer et al. 1999; Müller et al. 2008). Acknowledging this caveat, the broad features of the marine Phanerozoic carbon isotope records begin with a secular trend in $\delta^{13}C_{carb}$ towards heavier values throughout the entire Paleozoic, followed by a relatively abrupt decline and stabilization of the long term mean at the present-day value for the duration of the Mesozoic and Cenozoic (Veizer et al. 1999). $\delta^{13}C_{org}$ approximately tracks $\delta^{13}C_{carb}$ throughout the Paleozoic and Early Mesozoic, probably representing an increase and subsequent decrease in the fraction of carbon buried as organic matter, and a negligible change in the fractionation of carbon into phytoplankton organic matter (Berner and Kothavala 2001; Krissansen-Totton et al. 2015). $\delta^{13}C_{org}$ undergoes a secular drift towards heavy values in the Cenozoic that is independent of $\delta^{13}C_{carb}$, which has been widely interpreted to reflect a global decrease in atmospheric CO_2 (Pagani et al. 1999; Witkowski et al. 2018).

Some of the largest perturbations to the global carbon cycle throughout the Phanerozoic are ocean anoxic events (OAEs). These $\sim$ million year intervals are characterized by high (relative) rates of organic carbon burial in thick shale deposits, and associated with widespread ocean anoxia, high temperatures and mass extinction of marine organisms (Schlanger and Jenkyns 1976; Jenkyns 2010, see Sect. 3.5 for further discussion on OAEs). Most of these organic matter burial events are associated with positive carbon isotope excursions and are thus consistent with the classic mechanism of global ocean ^{13}C enrichment via an increase in F_{org}. However, a few are preceded by large negative carbon isotope excursions, which suggest more nuanced mechanisms, such as recycling of isotopically light carbon from depth into the surface ocean (Schouten et al. 2000), or global methane release and oxidation (Hesselbo et al. 2000). The Phanerozoic also saw two massive glaciations, the Ordovician ($\sim$ 440 Ma) and Gondwanan ($\sim$ 250 – 320 Ma) glaciations, which were probably less intense than the Neoproterozoic glaciations, and not associated with the same decreases in apparent ε_p (Hayes et al. 1999).

The Cenozoic carbonate carbon isotope record is more complete than older intervals, owing largely to the ubiquitous presence of the calcitic shells of foraminifera, which are microscopic predators that have diversified to inhabit the surface (planktonic) ocean and benthic environments. The deep ocean carbonate carbon isotope record, generated by targeting benthic species of foraminifera, is rich in environmental information and low in noise due to the relative homogeneity and low energy of the environment (Zachos et al. 2001). The planktonic record is comparatively sparse and noisy. A group of planktonic calcifying algae called coccolithophores contribute similar amounts of calcite to the deep marine sedimentary record, but their carbonate exhibits large and variable vital effects, limiting their utility as compared to the foraminifera whose well-characterized vital effects are small and relatively constant. The Cenozoic benthic $\delta^{13}C_{carb}$ record contains a $\sim$ 2.5‰, 10 million year oscillation in the aftermath of the K-Pg mass extinction, followed by 30 million years of relative quiescence, terminated by an increase towards the heaviest values of the last 50 million years at the 'Middle Miocene Climatic Optimum' (16–14 Ma) (Zachos et al. 2001). Following this peak in $\delta^{13}C_{carb}$, values decrease towards the Plio-Pleistocene, likely driven in part by the expansion of C4 grasses (Cerling et al. 1997), which have a smaller isotopic fractionation than the previously dominant C3 metabolism (Farquhar et al. 1989), and a global decrease in CO_2, which modulates the net fractionation by plants and phytoplankton. Punctuating this curve are a number of hyperthermals, most notably the Paleocene/Eocene thermal maximum (PETM) (Zachos et al. 2001; Pagani et al. 2006). This event, thought to be the most rapid natural (i.e., not anthropogenic) CO_2 increase in the observable geologic record, is characterized by a sharp spike towards isotopically depleted values that represents a massive release of isotopically depleted organic matter, potentially supplemented by the mobilization of extremely isotopically depleted methane (Zeebe et al. 2009).

For at least the last 4 million years, the Earth system has oscillated between glacial periods, characterized by cold temperatures and high ice volume, and interglacial periods characterized by warmer temperatures and low ice volume (Broecker and van Donk 1970; Shackleton 1987). These cycles have a characteristic saw-tooth shaped oscillation: a gradual descent towards glacial maximum conditions and rapid emergence (Zachos et al. 2001; Lisiecki and Raymo 2005). The mixing ratio of CO_2 in the atmosphere, as measured in bubbles trapped in antarctic ice, has varied in concert with temperatures and ice volume over at least the last 800 thousand years (Lüthi et al. 2008; Neftel et al. 1982) and most likely beyond (Hönisch et al. 2009). During peak glacial periods, pCO_2 was around 180 ppm; the rapid end of glacial periods is associated with similarly rapid increases in CO_2, known as 'terminations', elevating pCO_2 to around 280 ppm (Lüthi et al. 2008). These glacial cycles appear to be paced by periodic variation in insolation (Hays et al. 1976), but the resonant frequency of the Earth system has shifted at least once, from a 41 kyr period prior to 1 Ma to 100 kyr period after 800 ka with no apparent change in insolation forcing (Lisiecki and Raymo 2005). During glacial periods, a third of the CO_2 that resides in the atmosphere at peak interglacial times is stored in the deep ocean (Sigman and Boyle 2000). These changes are reflected in the carbon isotopic composition of gaseous CO_2, possibly due to a smaller glacial terrestrial biosphere and a polar ocean with depressed productivity (Marino et al. 1992). Atmospheric pCO_2 can be estimated from the carbon isotope compositions of alkenones, a set of lipids that are specific to a single known family of coccolithophorid algae, in ocean sediments (Popp et al. 1998; Pagani et al. 2005, 2011). Although bulk $\delta^{13}C_{org}$ appears to respond to glacial cycles (Jasper and Hayes 1990), the carbon isotopic composition of alkenones do not vary predictably across glacial cycles (Badger et al. 2019), perhaps due to a trade-off between CO_2 concentration and growth rate of this phytoplankton group through glacial cycles.

The last glacial period reached a maximum around 20 ka, and the most recent interglacial period has persisted since 10 ka (Lisiecki and Raymo 2005). The recent period of increasing anthropogenic activity, which may become formally recognized as the Anthropocene epoch (Lewis and Maslin 2015), has been characterized by a rapid rise in atmospheric pCO_2 from less than 280 ppm before 1750 A.D. to exceeding 410 ppm by May, 2020 (Hartmann et al. 2013). The full impacts of this rise on Earth's biogeochemical carbon cycle are still being developed, but include a clear drop in the $\delta^{13}C$ of atmospheric CO_2 as a result of burning ^{13}C-depleted organic matter that constitutes fossil fuels (Keeling 1979), and a possible rise in terrestrial photosynthesis rates due to CO_2-fertilization (Keeling et al. 1996; Ciais et al. 2013).

3 Nitrogen

Nitrogen (N) is another biologically essential element that exists across a wide spectrum of redox states and speciations, reflecting its major involvement in both abiotic and biotic redox-driven transformations. In its most reduced form (–3), N exists as either ammonium (NH_4^+) or organic nitrogen, while in its most oxidized form (+5), N is found predominantly as nitrate (NO_3^-). In its zero-valent form nitrogen exists as the triple-bonded di-nitrogen molecule (N_2), which comprises the majority of Earth's atmosphere and constitutes the dominant form of N found in the exosphere (Gruber 2008; Johnson and Goldblatt 2015; Bebout et al. 2013). In general, N is most commonly found in one of these three oxidation states. Other more transient oxidation states are represented in several N species including aqueous species such as nitrite (NO_2^-), hydroxylamine (NH_2OH), and hydrazine (N_2H_4) as well as gaseous and aerosol phases including ammonia (NH_3), nitrous oxide (N_2O), nitric acid (HNO_3), nitrous acid (HNO_2), peroxynitrite ($ONOO^-$), and nitrogen oxides such as (NO_2, N_2O_3, N_2O_4, and N_2O_5). Many of these intermediate oxidation states arise as metabolic products or intermediates. For example, during the microbial oxidation of NH_4^+, NH_2OH is the first intermediate produced, followed by oxidation to NO and NO_2^- (Caranto and Lancaster 2017; Lehnert et al. 2018; Martens-Habbena et al. 2015)

Among these intermediate species, NO_2^- is a common redox 'crossroads species' with much evidence pointing to tight biological coupling of its oxidative and reductive production and consumption at sharp redox boundaries (e.g., sediment–water interfaces), with a pronounced relationship with ambient carbon cycling. Nitrite can be produced by both heterotrophic nitrate reduction or autotrophic ammonia oxidation, as well as be consumed by heterotrophic denitrification or autotrophic nitrite oxidation—highlighting the complexity of one of many key interfaces between the redox cycling of carbon and nitrogen.

On Earth's surface, after the atmospheric reservoir of N_2 (0.7809 mol fraction) and its dissolved counterpart in the global ocean, the most abundant form of N is organic N buried in ocean sediments and sedimentary rocks (Bebout et al. 2013). N_2 fixation into bio-available (reduced) forms is an energetically expensive process that can only be performed by a specialized group of prokaryotes (see below). As a result, more labile forms of dissolved inorganic nitrogen, such as NH_4^+ and NO_3^-, play the primary role in regulating levels of biological productivity in N-limited surface waters on Earth. Given the oxidizing nature of the modern Earth, however, it is perhaps not surprising that the primary pool of inorganic nitrogen in the ocean is NO_3^-. On land, most N is bound in soil organic matter or living biomass.

When taking into account the entirety of the Earth system, the geologic reservoirs of N dwarf those of the actively cycled surface Earth biosphere, similar to carbon. These include N in crustal rocks, the mantle and the Earth's core, with estimates of reservoir sizes varying over orders of magnitude (e.g., Johnson and Goldblatt 2015), and residence times of tens of millions to billions of years (Fig. 3).

3.1 Abiotic Processes in the N Cycle

Nitrogen transformations can occur in the absence of biological systems and would have played an important role in N cycling on a pre-biotic Earth. In particular, a wide range of photochemical and thermochemical processes are known to occur under Earth surface conditions that largely mimic those of the better-known biological nitrogen cycle (see review by: Doane 2017). For example, much attention has been given to the photochemical degradation of organic compounds producing free ammonium. Beyond this process, however, photochemical fixation of N_2 into the more bioavailable NH_3 has been demonstrated to occur in connection with surface reactions on transition metal oxides (especially those containing Fe, Ti, Mo, Ni and Zn; Medford and Hatzell 2017; Sclafani et al. 1993; Tennakone et al. 1991). At the same time, studies have also documented photochemical oxidation of

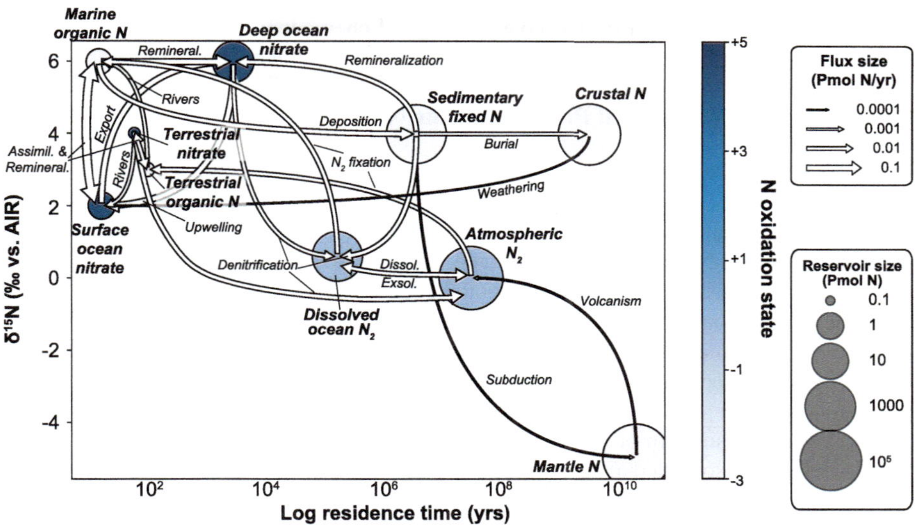

Fig. 3 A quantitative depiction of the modern (pre-Anthropogenic) nitrogen cycle. Sizes and positions of components are determined in a manner analogous to those in Fig. 2. Nitrogen reservoir sizes, flux sizes, and isotope compositions compiled from Galloway et al. (2004), Seitzinger et al. (2005), Gruber (2008), Sigman et al. (2009), Fowler et al. (2013), Johnson and Goldblatt (2015). $\delta^{15}N$ values of pools are estimates of mean modern values; the ranges in the $\delta^{15}N$ values of some pools are much larger due to the ranges in the relevant isotopic fractionations expressed, as listed in Table 1. All fluxes (and thus, residence times) have uncertainties in the range of $\pm 20 - 50\%$

N_2 into NO_x, NO_2^- and NO_3^-, highlighting their potential importance in desert-like, light-intensive settings (Bickley and Vishwanathan 1979; Medford and Hatzell 2017). In addition to surface-catalyzed fixation of N_2, transformation of N_2 into NO_x by lightning through Earth's atmosphere has also been shown, and argued to play a role in the transformation of N on prebiotic Earth (Kumar et al. 1995; Ridley et al. 1996; Rakov and Uman 2003). Meteorite impacts, volcanic lightning, and (potentially) cosmic rays may have been additional sources of fixed N in this prebiotic world (Laneuville et al. 2018; Navarro-González et al. 1998; Tabataba-Vakili et al. 2016).

Non-photochemical reactions with transition metals also play a key role in catalyzing abiotic transformations of N compounds with notable experimental and observational attention having been given to reactions involving Fe and Mn (e.g., Buchwald et al. 2016; Hansen et al. 1996; Heil et al. 2016; Luther et al. 1997; Ottley et al. 1997; Samarkin et al. 2010; Zhu-Barker et al. 2015). While approaches to teasing apart the relative environmental contribution of such transformations from those of enzyme-driven processes have proven challenging, many laboratory studies have established their feasibility under surface conditions. A number of factors influence reactivity, including: *i*) coordination environment of metal cations (e.g., surface bound, mineral form), *ii*) presence, abundance, and nature of organic molecules, *iii*) solution pH, *iv*) reactant concentrations, and *v*) presence of catalytic metals (e.g., Cu, Ni) (Buresh and Moraghan 1976; Dhakal et al. 2013; Grabb et al. 2017; Ottley et al. 1997; Postma 1990; Rakshit et al. 2008; Sørensen and Thorling 1991). Notably, many reactions involve the production and release of reactive intermediate and radiatively absorptive N species, such as NO and N_2O (e.g., Cavazos et al. 2018; Coby and Picardal 2005; Kampschreur et al. 2011; Klueglein and Kappler 2013; Nelson and Bremner 1970), prompting some speculation on the potential role of chemodenitrification in atmospheric N_2O budgets of Earth in the past (Robertson et al. 2011; Stanton et al. 2018).

3.2 Enzymatically Mediated Processes in the N Cycle

Similar to carbon, enzymatically catalyzed redox processes drive the short-term global nitrogen cycle. Biological nitrogen fixation comprises the enzymatic reduction of N_2 to NH_4^+ for nutritive N_2 assimilation and synthesis of biomass (Hoffman et al. 2014). Ecologically, this pathway of nutritive N assimilation is relatively restricted to specialized organisms known as diazotrophs, that are conventionally thought to be favored for lifestyles in environments otherwise devoid of bioavailable forms of N (e.g., the center of nutrient-poor—oligotrophic—ocean gyres) since catalysis by the metalloenzyme nitrogenase is energy intensive. More recent work, however, suggests that other more N-replete environments may also host nitrogen fixation (Bertics et al. 2010; Dekas et al. 2009; Knapp 2012; Mehta and Baross 2006). The most widespread form of nitrogenase contains an iron-molybdenum (Fe-Mo) reaction center, while other so-called alternative nitrogenases contain Fe-V and Fe-Fe reaction centers. The relative distribution and activity of these disparate enzyme forms may have important implications for isotope budgets under some conditions and remains an active area of research (McRose et al. 2017; Zhang et al. 2014).

Nitrification, the biological oxidation of NH_4^+ (the most reduced form of inorganic N) to the more oxidized forms NO_2^- and NO_3^- (also nitritation and nitratation), comprises the single most important oxidative flux of N on the Earth surface. This process is carried out by a group of specialized chemoautotrophic archaea and bacteria that use ammonia monooxygenase to harness the oxidizing power of molecular O_2 for transforming NH_3 into an intermediate hydroxylamine (NH_2OH). The conversion of NH_2OH to NO_2^- proceeds via the enzyme hydroxylamine oxidoreductase (HAO), with recent work suggesting an NO intermediate (Caranto et al. 2016; Caranto and Lancaster 2017; Lehnert et al. 2018). A separate group of chemoautotrophic nitrite-oxidizing bacteria (NOB) have been conventionally considered responsible for the ultimate oxidation of NO_2^- to NO_3^-, also through the use of molecular O_2. Recent cultures of a novel bacterial lineage have also demonstrated that conversion of NH_4^+ to NO_3^- by nitrification can also occur within a single species (Daims et al. 2015; Van Kessel et al. 2015).

From the large oxidized pool of NO_3^- in the modern global ocean and terrestrial biosphere, the largest return flux of N to the atmosphere occurs through reductive microbial processes known as denitrification and anammox (anaerobic ammonia oxidation) (Gruber 2008; Galloway et al. 2004; Alhaique et al. 1975). Denitrification comprises the stepwise reduction of nitrate and its loss to the atmosphere as gas phase N_2, proceeding through the intermediate species nitrite, nitric oxide and nitrous oxide. As a microbial respiration process, denitrification is frequently coupled to the oxidation of organic carbon (heterotrophy), and proceeds only in the absence of more thermodynamically viable oxidants (e.g., O_2, Mn^{4+}, Fe^{3+} at low pH). Although denitrification generally occurs only in the absence of O_2, the presence of O_2 in the biosphere is a necessary precursor for NO_2^- formation and hence for the advent of denitrification as a respiratory metabolic pathway. Notably, biological denitrification can also be coupled with oxidation of inorganic substrates (autotrophic denitrification) including reduced sulfur and iron species, and molecular hydrogen (Batchelor and Lawrence 1978; Frey et al. 2014; Melton et al. 2014; Sievert and Vetriani 2012).

Anammox, the production of N_2 through microbially coupled oxidation of NH_4^+ with nitrite reduction, is conducted by specialized chemoautotrophic bacteria. In contrast to denitrification, in which respiratory nitrate reduction is often facultatively engaged after O_2 levels are drawn down, anammox is very sensitive to the presence of O_2. Anammox has been detected in a wide range of anoxic environments including ocean oxygen minimum zones, soils, wastewater treatment reactors, and marine sediments. Stoichiometric examination of ocean carbon and nitrogen budgets suggests that the relative contribution of anammox likely does not exceed 30% of N loss in most cases (Babbin et al. 2014; Dalsgaard et al. 2012).

In addition to reductive transformation and return of inorganic forms of N to the atmosphere, dissimilatory nitrate reduction to ammonium (DNRA) represents an additional microbially catalyzed pathway in which N is retained in bioavailable form. Ecologically, DNRA therefore represents an important N retention pathway for maintenance of N in biological reservoirs. DNRA appears to generally be favored over denitrification when organic carbon is abundant relative to nitrate (Hardison et al. 2015; Koop-Jakobsen and Giblin 2010; Kraft et al. 2014; van den Berg et al. 2017). This trend could be tied to growth yields, where denitrification yields more energy per electron, but DNRA yields more energy per nitrate molecule (Strohm et al. 2007).

All known living organisms require nitrogen for biosynthesis and growth. Thus, biochemical pathways of N assimilation are, at least in part, central to the ecological success of all organisms. The biological assimilation of N for synthesis of organic biomolecules (amino acids, proteins, nucleic acids, chlorophyll, etc.) represents an important link between actively cycling N pools and those that may ultimately be stored in the geological record of sedimentary rocks. Heterotrophic organisms that consume complex organic molecules generally possess a host of metabolic pathways for funneling organic N moieties into cellular assimilatory machinery. In general, both single-celled prokaryotes as well as a wide array of eukaryotic organisms have evolved means of assimilating ambient forms of inorganic N, including NO_2^-, NO_3^- and NH_4^+. As biomolecules generally comprise N in its most reduced form, assimilative pathways typically involve the necessary reduction of oxidized forms (NO_2^- and NO_3^-) prior to shuttling of NH_4^+ into biosynthesis pathways.

3.3 Nitrogen Isotopes in the N Cycle

The majority of the biological processes driving the biogeochemical nitrogen cycle produce measurable fractionations between the $^{15}N/^{14}N$ ratios of products and substrates. For example, biological nitrogen fixation results in organic matter with $\delta^{15}N$ values of ~ -4 to $+2$‰ (relative to atmospheric N_2, which is also the $\delta^{15}N$ standard) (Bauersachs et al. 2009; Carpenter et al. 1997; Minagawa and Wada 1986; Delwiche and Steyn 1970). While it appears that the Mo form of nitrogenase dominates N-fixation budgets under the more oxidizing conditions of modern surface Earth, the so-called alternative nitrogenases (V and Fe forms) impart a stronger fractionation, yielding biomass that is up to $7-8$‰ lower in $\delta^{15}N$ than N_2 (Zhang et al. 2014). Denitrification and anammox both return ^{15}N-depleted N_2 to the atmosphere, leaving the residual nitrate enriched in ^{15}N by $10-25$‰ (Delwiche and Steyn 1970; Granger et al. 2008; Brunner et al. 2013, and references therein). Loss of low-$\delta^{15}N$ N_2 to the atmosphere via denitrification and anammox, therefore, exerts the dominant control on $\delta^{15}N$ values of nutrients and biomass in the modern oceans, resulting in sedimentary organic matter with average $\delta^{15}N$ values of +6 to $+7$‰ (Peters et al. 1978).

As with carbon, the extent to which nitrogen isotope effects associated with biological processes are expressed in the environment depends on the relative fraction of the N pool that is transformed or consumed. For example, uptake of dissolved inorganic nitrogen as nitrate or ammonium can also produce large fractionations in $\delta^{15}N$ (Hoch et al. 1992), but these isotope effects are not expressed in N-limited ecosystems where assimilation proceeds to completion. Similarly, in environments with low nitrate and high organic carbon flux, denitrification can consume the entirety of the nitrate pool, erasing any isotopic trace of its existence. This consideration becomes important when interpreting isotopic evidence for N cycling processes from ancient sediments on Earth, and would be similarly important when considering isotopic evidence for biological N cycling on other planets.

Nitrogen is preserved in the sedimentary rock record primarily as organic N or as ammonium in phyllosilicates, and generally records the δ^{15}N of the input biomass. Thus the sedimentary record of δ^{15}N has been used to infer the evolutionary history of N cycling metabolisms on Earth. Care must be taken, however, to consider that sedimentary δ^{15}N values can be modified by post-depositional processes, including diagenesis, metamorphism, and hydrothermal or hydrocarbon fluid migration. In particular, degradation of organic matter during early diagenesis, high-temperature metamorphism above greenschist facies and/or thermal maturation during hydrothermal fluid migration can all increase the absolute δ^{15}N values of the residue (Freudenthal et al. 2001; Bebout and Fogel 1992; Boyd and Philippot 1998). Low-temperature metamorphism can even cause mineral and organic δ^{15}N values to diverge within the same sample (Stüeken et al. 2017). The possible impacts of post-depositional alteration must therefore be considered before interpreting sedimentary δ^{15}N data as reflecting primary signals, particularly when applying these to Precambrian sediments. Additionally, different combinations of N cycling processes can produce identical or non-unique δ^{15}N patterns, meaning that additional depositional or redox context is often required for a convincing interpretation of ancient N cycling. We expand on these ideas in our following discussion of the evolution of Earth's early N cycle.

3.4 Evolution of the N Cycle on the Precambrian Earth

Abiotic sources of N to the early biosphere (e.g., from lightning or hydrothermal sources, as described above) would have been limited (Navarro-González et al. 2001), requiring biological N_2 fixation to have evolved very early in Earth history in order to keep pace with the N requirements of a rapidly evolving biosphere. Indeed, phylogenetic reconstructions suggest that the last universal common ancestor of extant life had proteins for N_2 fixation (Weiss et al. 2016), and δ^{15}N records have been interpreted as evidence of diazotrophy as far back as ~ 3.2 Ga (Stüeken et al. 2015a). Once biological N_2 fixation evolved, this enzymatic innovation would have provided a virtually unlimited source of nutrient N to an increasingly efficient biosphere, assuming no other nutrient (phosphorus or bioactive trace elements) became limiting.

Prior to $\sim$2.7 Ga, the early to mid-Archean δ^{15}N record exhibits a wide range of values, spanning from -6 to $+13$‰, and centered around 0‰ (Fig. 1). The most negative end of these δ^{15}N values are generally interpreted as reflecting a nitrogen cycle with inputs dominated by N_2 fixation and the efficient recycling of ammonium deriving from breakdown of diazotrophic biomass, with the higher δ^{15}N values attributed to the influence of metamorphism and other post-depositional effects (Beaumont and Robert 1999). However it has also been suggested that this range of δ^{15}N values in the pre-2.7 Ga world—which are too high to be buffered by primitive mantle N ($\delta^{15}N_{mantle} < -40$‰)—could reflect isotopic enrichment due to atmospheric escape (Cartigny and Marty 2013; Lammer et al. 2018). Regardless, rocks from around 2.7 Ga have a comparably large range of δ^{15}N values, spanning -11 to $+50$‰. The extremely high δ^{15}N values have been alternatively interpreted to reflect incomplete ammonium oxidation coupled to methanotrophy under low O_2 conditions (Thomazo et al. 2011) or ammonia volatilization from alkaline lakes (Stüeken et al. 2015b). The δ^{15}N values lower than 0‰ were recently interpreted as evidence for non-quantitative uptake of a pool of bioavailable ammonium available in excess sourced from enhanced primary production and export of organic matter following the proliferation of oxygenic photosynthesis (Yang et al. 2019). This ammonium would have provided an ample source of N to fuel the Late Archean biosphere and associated oxygen production, with its incomplete consumption implying that some other critical nutrient (e.g., phosphorus) must have been important in limiting primary productivity.

The timing of the evolution of the aerobic N cycle, driven by the oxidation of ammonium (nitrification), is highly debated. Nitrification would have been a prerequisite for the development of the major N loss pathways, since these require a supply of oxidized N in the form of nitrate (for denitrification) or nitrite (for anammox). Several studies have interpreted small increases in $\delta^{15}N$ in single rock sections (Garvin et al. 2009; Godfrey and Falkowski 2009) or statistical trends in the temporal $\delta^{15}N$ dataset (Stüeken et al. 2016) as evidence for episodic or localized aerobic N cycling as far back as 2.5 Ga. Interpretations of these data are complicated by the fact that these processes produce a non-unique signal in $\delta^{15}N$. For example, similar small and short-lived positive shifts in $\delta^{15}N$ could be produced by the incorporation of ^{15}N-enriched ammonium that has undergone partial nitrification or partial assimilation (Yang et al. 2019), or by nitrogen redox cycling independent of surface oxygenation, such as with Fe-oxides (Busigny et al. 2013).

Consideration of $\delta^{15}N$ data within depositional and redox context suggests that a clear isotopic fingerprint of nitrification/denitrification is not evident in the rock record until ~2.3 Ga, immediately following the Great Oxidation Event (Zerkle et al. 2017). However, as alluded to above, the isotopic signal of denitrification/anammox is lost if the reaction runs to completion in the environment. Therefore, the preservation of this signal in the rock record does not directly pinpoint the timing of the evolution of these metabolisms, but rather indicates when nitrate/nitrite availability became widespread enough for partial N loss processes to impart a long-lived isotopic imprint on the sedimentary $\delta^{15}N$ record.

Following the GOE, a hypothesized oxygen overshoot associated with the Lomagundi-Jatuli carbon isotope excursions could have increased the availability of nitrate, leading to enhanced denitrification rates and widespread loss of fixed N (Kump et al. 2011). The resulting nitrogen limitation has been implicated in hindering biological evolution, as well as the continued oxygenation of Earth surface environments, throughout much of the Proterozoic (Fennel et al. 2005). It is possible that N_2 fixation could have kept pace with N-loss (Luo et al. 2019), although biospheric N limitation could have been exacerbated by the drawdown of molybdenum (Mo) under more widespread sulfidic conditions, since Mo is a key cofactor of nitrogenase and is required for marine N_2 fixation (Anbar and Knoll 2002).

The remainder of the Proterozoic is characterized by apparent stasis in O_2, with concentrations held between the anoxic and "overshoot" conditions of the Archean, and less than near-modern levels realized in the Phanerozoic. Though data are limited, the $\delta^{15}N$ record is also muted, resting between +2 to +8‰, consistent with a mildly oxidized surface ocean (Ader et al. 2014), and perhaps reflecting prevailing N limitation (Koehler et al. 2017). The mechanisms for reduced N availability during the Mesoproterozoic may be linked to trace metals (Anbar and Knoll 2002) or redox related loss of N (Fennel et al. 2005) as O_2 increased. What remains unclear is whether higher O_2 in the later Neoproterozoic had little effect on the $\delta^{15}N$ record (Ader et al. 2014) and suggests that planetary oxygenation was not the sole control on the N cycle. One hypothesis states that O_2 concentrations may have been sufficiently high to inhibit nitrogenase expression, reducing N_2 fixation rates and limiting productivity until evolution of physiological changes in cyanobacteria allowed for N_2 fixation at higher O_2 concentrations in the Phanerozoic (Allen et al. 2019).

3.5 Phanerozoic N Cycling

Nitrogen cycling in the Phanerozoic stands in contrast with the Precambrian in that variability in the N-cycle and $\delta^{15}N$ record reflects changes in the redox state within the ocean as opposed to direct forcing by the oxygen content of the atmosphere. However, the primary issues ultimately still concern the balance between N_2 fixation, denitrification and anammox (Galbraith et al. 2013), and how these processes relate to changes in the physical and chemical structure of the ocean and nutrient delivery from land (phosphorus). An additional aspect that distinguishes the Phanerozoic from the Precambrian eras is the preservation of a broad range of primary organic phases within rocks and sediments, such as N-containing biomarkers (Chicarelli et al. 1993) and organic nitrogen preserved in the carbonate skeletons of invertebrates (e.g., Ren et al. 2009; Wang et al. 2015; Tornabene et al. 2017) that open a whole new range of potential areas of study.

Despite the relatively oxygenated atmosphere that characterized the Phanerozoic, its geologic record is punctuated by intervals of global-scale oceanic oxygen deprivation known as Oceanic Anoxic Events (OAEs) that are largely responses of the Earth system to global climate perturbations (Jenkyns 2010). Classically, OAEs are considered to be a Mesozoic phenomenon but recognition of similar depositional patterns and carbon cycle signatures have expanded to include intervals in the mid-Cambrian, Late Devonian, Early Triassic and during the Paleocene Eocene Thermal Maximum (Meyer and Kump 2008; Jenkyns 2010; Junium et al. 2018).

OAEs have been common targets for study into the role of nutrient cycling. Organic matter contents in black shales deposited during OAEs can be in excess of 30 wt.% and were, to some degree, the byproduct of increasing nutrient availability coupled with enhanced preservation of organic matter under anoxic conditions. Anomalous P geochemistry in black shales largely supports the hypothesis that OAEs were accompanied by perturbations to nutrient cycling (e.g., Van Cappellen and Ingall 1996; Rau et al. 1987). Reduced sedimentary retention of P in mineral-oxides and authigenic P minerals is indicated by high C:P values (Ingall and Jahnke 1994) and would have resulted in the accumulation of P in anoxic deep waters. If water-column oxygen deprivation stimulated nitrate reduction or anammox, N deficits may have resulted in overall N:P imbalance.

Nitrogen isotope records from OAEs are consistent with profound changes in water-column nitrogen cycling, with $\delta^{15}N$ values typically less than 0‰ and often as low as −5‰. The low $\delta^{15}N$ has been attributed to expanded N_2 fixation in response to fixed N-loss and P-excess (Rau et al. 1987; Kuypers et al. 2004). However, with $\delta^{15}N$ values below the average isotopic composition of diazotroph biomass (~ -1‰), a role for incomplete utilization of ammonium has become a favored model (Higgins et al. 2012; Junium and Arthur 2007). Alternatively, ^{15}N-depletion may reflect high Fe availability for diazotrophs (Zerkle et al. 2008) or utilization of alternative nitrogenases (Zhang et al. 2014). Regardless of the exact mechanism, compound-specific $\delta^{15}N$ of chlorophyll a-derived porphyrins confirm that ^{15}N-depletion was a shallow water signal, and not a byproduct of diagenesis or primary production by phototrophic sulfide oxidizers within the chemocline (Junium et al. 2015).

4 Sulfur

Sulfur (S) is a third versatile, bio-essential element that exists in solid (inorganic and organic), liquid and gaseous forms at Earth's surface, and spans redox states from −2 to +6.

Atmospheric sulfur sourced from volcanoes in the form of SO_2 ($\pm$ H_2S) constitutes a relatively minor component of the modern atmosphere, but can produce sulfate aerosols that have important climatic impacts. Sulfate (SO_4^{2-}) is the dominant form of sulfur in well-oxygenated aqueous environments, and forms a major anion in seawater, at concentrations of 28 mM (Canfield and Farquhar 2009). Aqueous sulfide (H_2S or HS^-) is less abundant but central to active biogeochemical cycling of S, and can build up in anoxic sediments and oxygen minimum zones. A variety of sulfur compounds with intermediate redox states, such as elemental sulfur (S^0), polysulfides (S_n^{2-}), thiosulfate ($S_2O_3^{2-}$), and sulfite (SO_3^{2-}) are additionally produced during abiotic and metabolic processes, but are generally recycled very quickly in most environments (Brimblecombe 2013). Sulfate can be incorporated into sediments and the rock record in the form of barite ($BaSO_4$), gypsum ($CaSO_4 \cdot 2H_2O$), anhydrite ($CaSO_4$), or bound in the lattice of carbonate minerals (carbonate associated sulfate, or CAS). Sulfide, elemental sulfur, and polysulfides can be preserved as Fe-monosulfides (e.g., pyrrhotite) and/or pyrite (FeS_2) (Fike et al. 2015).

Every living thing known to us requires sulfur. Sulfur is incorporated into organic compounds through both biological and abiological processes (Amrani 2014; Fike et al. 2015). Key organic sulfur compounds in living cells include the proteinogenic amino acids cysteine and methionine. The majority of important biological organosulfur compounds (e.g., coenzyme A, glutathione, S-adenosylmethionine) are biosynthesized from cysteine or methionine. However, some organosulfur compounds, such as sulfoquinovose (a sulfur-containg lipid important in photosynthesizing organisms) are produced from oxidized inorganic sulfur compounds like sulfite. Inorganic iron-sulfur clusters are required constituents of many electron-transfer proteins.

The abiotic reaction of organic compounds with reduced sulfur species is responsible for the formation of a wide range of organic sulfur compounds that can also build up in the environment (Amrani 2014). Hydrogen sulfide can react with hydrocarbons to produce thiols (mercaptans), acyclic and cyclic sulfides, polysulfides, and thiophenes (Sinninghe Damste and De Leeuw 1990). These reactions are widespread in sediments. The cross-linking of small organic compounds through sulfur bridges (natural vulcanization) is hypothesized to be one of the important processes in the formation of macromolecular carbon compounds with decreased lability, such as asphaltenes and kerogen (Vandenbroucke and Largeau 2007). Such macromolecular carbon compounds comprise the bulk of organic carbon in the exosphere.

4.1 Abiotic Redox Reactions in the S Cycle

As noted above, volcanic SO_2 forms an important flux of sulfur from the mantle to Earth's surface (Fig. 4). Volcanic SO_2 has a short residence time in the modern atmosphere, on the order of years, as it oxidizes to sulfate aerosols and deposits on the surface (Robock 2000). The dominant input of sulfate to the modern oceans is a riverine flux sourced from dissolution of sulfates and oxidative weathering of pyrite (Kump and Garrels 1986; Canfield 2004; Brimblecombe 2013; Fig. 4). Hydrothermal vents are thought to provide another important source of sulfur to the oceans, but this flux is largely unconstrained. Hydrothermal vent fluids can contain up to 12 mM sulfide, up to 89% of which could be derived by leaching of sulfur from seafloor basalt (Ono et al. 2007). The remaining sulfide, which may dominate in some hydrothermal vent fluids, is thought to come from thermochemical reduction of seawater sulfate, as seawater percolates into the oceanic crust and heats up. Given the ubiquitous nature of hydrothermal systems along mid-ocean ridges, these could therefore provide a significant source of sulfur to the deep oceans, and were potentially even more important earlier in Earth history prior to widespread oxidative weathering.

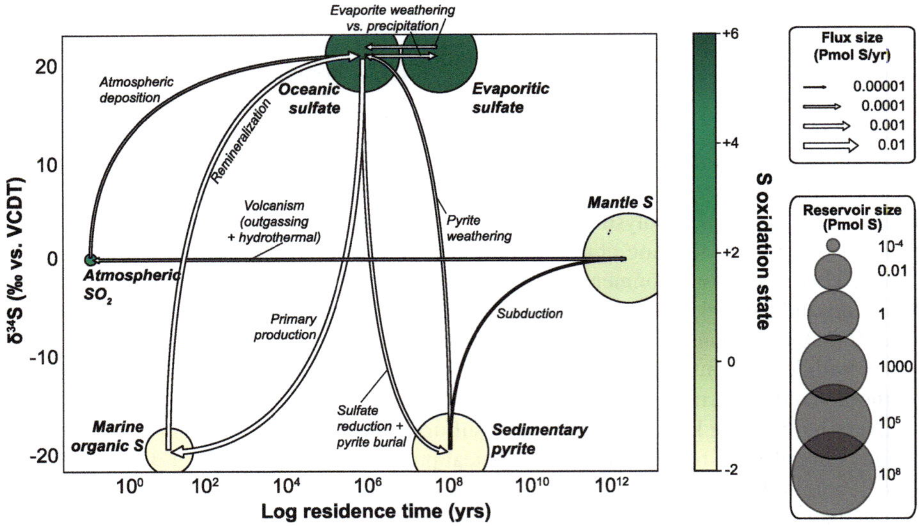

Fig. 4 A quantitative depiction of the modern (pre-industrial) sulfur cycle: construction of diagram is analogous to Figs. 2, 3. Sulfur reservoir and flux sizes after Canfield (2004), Palme and O'Neill (2013), Brimblecombe (2013), Moran and Durham (2019). For simplicity, less-abundant atmospheric species are omitted, and the marine organic sulfur network—which constitutes its own cycle of metabolites and fluxes (Amrani 2014; Moran and Durham 2019)—is subsumed into a single reservoir. Briefly, sulfur also exists as organic sulfur in sediments, deposited as a constituent of primary organic matter, or fixed via sulfurization reactions between sedimentary organic matter and sulfide in porefluids. This sedimentary organic sulfur pool is of order 0.1–1 Pmol S (for an S:C ratio of around 1–4%), and its isotopic composition is generally close to that of other reduced sulfur species (Raven et al. 2018, 2016; Sinninghe Damsté et al. 1998; Werne et al. 2000). Ordinate of each reservoir is an estimate of its modern mean δ^{34}S value; deviations from these mean δ^{34}S values are common, due to the variability in the underlying isotopic fractionations (Table 1)

The majority of processes that act to transform sulfur in marine and terrestrial systems are biological. However, several abiotic processes can also contribute to sulfur cycling in the environment. Sulfate can be reduced to sulfide abiotically at high temperatures by reaction with organic compounds, during thermochemical sulfate reduction (TSR). In modern marine sediments, most of this sulfide is subsequently oxidized mainly by interaction with molecular oxygen, and Fe- and Mn-oxides (Goldstein and Aizenshtat 1994).

4.2 Enzymatically Mediated Reactions in the S Cycle

A diverse set of enzymatically-catalyzed metabolic processes interconvert sulfur compounds among redox states (e.g., see review in Jørgensen et al. 2019). Most of these metabolic processes are dissimilatory in nature, i.e., they do not result in the direct incorporation of sulfur molecules into biological materials. However, all living cells require sulfur, and assimilatory processes incorporate abiotic sulfur into organic material, as mentioned previously.

Dissimilatory sulfate reduction (DSR) couples the reduction of sulfate to the oxidation of an electron donor, which may be an organic compound (typically the conjugate base of a small organic acid, e.g., lactate) or an inorganic compound (e.g., H_2). This process can be carried out by a diverse set of bacteria or archaea, all of which use the same set of enzymatic steps to catalyze the process (Pereira et al. 2011; Fike et al. 2015; Rabus et al. 2013; Oliveira et al. 2008). After sulfate is imported into the cell, it is converted to an organosulfate compound adenosine-5'-phosphosulfate. The sulfate moiety of this compound is subsequently reduced to sulfite, and then reduced again through a complex enzymatic cascade to hydrogen sulfide (Venceslau et al. 2014; Santos et al. 2015). Byproducts of this metabolic process can include thiosulfate and possibly trithionate (Drake and Akagi 1977; Bradley et al. 2011), though it is possible that these byproducts result from the reaction of excess sulfite with sulfide (Chambers and Trudinger 1975), and may not be relevant under most plausible environmental conditions. Of all the biological transformations that occur in the sulfur cycle, dissimilatory sulfate reduction exhibits the strongest discrimination against the heavier isotopes (Eldridge and Farquhar 2018). Observed fractionations during DSR can range from nil (when electron donors are abundant and DSR is fast or quantitative) to the equilibrium maximum of approximately 72‰ (Leavitt et al. 2013; Canfield et al. 2010; Sim et al. 2011, 2012; Zaarur et al. 2017).

The most reduced sulfur compound, H_2S, has the IUPAC-preselected name sulfane (Favre and Powell 2014). This reflects an updated systematic naming for parent hydrides, so that sulfane (H_2S) is consistent with other hydrides such as methane (CH_4), silane (SiH_4), phosphane (PH_3), etc. This term is easily confused with the traditional usage of 'sulfanes' to refer to (hydro)polysulfides and polysulfanes (Favre and Powell 2014). For consistency with the geochemical literature, we use the terms 'sulfide' and 'hydrogen sulfide' here. This compound can be oxidized by an electron acceptor with a suitably high redox potential, most commonly molecular oxygen, nitrate or Fe-oxides. Such oxidation of sulfide is common in marine sediments, where it is suggested that the majority ($\sim$90 %) of sulfide produced by microbial sulfate reduction is reoxidized to sulfate, often via compounds with intermediate oxidation states of sulfur (Jørgensen 1982, 1990; Thamdrup et al. 1994). The product of this reaction can include sulfur compounds at a variety of redox states, including various allotropes of elemental sulfur, thiosulfate, sulfite, and sulfate (Canfield 2001; Fike et al. 2015). Under permissive thermodynamic conditions, these intermediate compounds can also be oxidized by sulfide-oxidizing microbes (Jørgensen and Des Marais 1986). Reduced sulfur compounds such as S^0 and H_2S can also be oxidized by some photosynthetic organisms (Frigaard and Dahl 2008). Just as cyanobacteria split water to generate electrons for carbon fixation, releasing O_2 as a byproduct, anoxygenic phototrophs including green- and purple-sulfur bacteria use H_2S as an electron source and release more oxidized sulfur compounds such as S^0 or sulfate. Processes that oxidize reduced sulfur compounds are observed in laboratory cultures to involve only small discriminations against ^{34}S (Canfield 2001, and references therein), although this has been challenged recently, as discussed below.

Disproportionation of sulfur compounds involves the simultaneous oxidation and reduction of the same reactant, yielding two products (Bak and Pfennig 1987; Thamdrup et al. 1993; Finster 2008). For example, thiosulfate or S^0 can be disproportionated into sulfate and H_2S. The intermediate-valence compounds that are the substrates of disproportionation reactions are generally the products of partial oxidation of H_2S. Under many conditions, disproportionation is not thermodynamically viable, but it can be if H_2S concentrations are kept sufficiently low, e.g., by removal through environmental reactions producing iron sulfide.

4.3 Isotopic Records of the Sulfur Cycle

4.3.1 Sulfur Isotopes

The largest enzymatic fractionations of sulfur isotopes in the sulfur cycle occur during DSR (Canfield 2001). In principle, this process can generate fractionations of up to 72‰ in $\delta^{34}S$, the equilibrium sulfate-sulfide fractionation (e.g., Eldridge and Farquhar 2018). In practice, fractionations as large as 70‰ are rarely directly observed in laboratory settings (Canfield et al. 2010; Sim et al. 2011), but have been inferred in sedimentary pyrite records and in sulfate-limited environments (Crowe et al. 2014; Bryant et al. 2019). Observed fractionations during DSR depend on a range of biological and environmental factors. In general, large fractionations occur at higher ambient sulfate concentrations (Habicht et al. 2002), and when the sulfate-reduction rate by individual cells is low (Canfield 2001; Sim et al. 2011; Leavitt et al. 2013; Zaarur et al. 2017). These patterns are related to the energetics of the enzymatically-catalyzed reactions and the sizes of intracellular metabolite pools (Wing and Halevy 2014; Wenk et al. 2018) and to bulk cellular-level processes related to importing sulfate into cells (Bradley et al. 2011, 2016). Differences exist between different strains of sulfate-reducing microorganisms, so that under identical conditions different organisms will produce different fractionations (Sim et al. 2011; Leavitt et al. 2013; Zaarur et al. 2017). The origins of these strain-specific differences is not well understood.

In laboratory cultures, sulfide oxidation generally produces fractionations of only a few permil in $\delta^{34}S$ (see review in Zerkle et al. 2009). Both chemotrophic and phototrophic sulfide oxidation have usually been associated with fractionations of the order of $1-2$‰, although fractionations as large as 12‰ during chemotrophic sulfide oxidation have recently been observed in the laboratory (Pellerin et al. 2019) and in a natural environment (Zerkle et al. 2016).

Disproportionation of S^0 has been reported to produce H_2S that is depleted in ^{34}S by $4-9$‰ relative to the reactant (Habicht et al. 1998; Johnston et al. 2005). However, since oxidation of H_2S is not associated with a large fractionation, repeated cycles of the production of H_2S via disproportionation, its subsequent oxidation to ^{34}S-depleted sulfate, and ultimate reduction of this sulfate to even more ^{34}S-depleted H_2S have been invoked to explain large sulfate-sulfide sulfur isotope fractionations that have sometimes been observed in nature (Canfield and Teske 1996). In recent years, these large fractionations have often been attributed to DSR occuring at slow rates (Canfield et al. 2010; Zerkle et al. 2010; Wing and Halevy 2014).

In addition to producing large changes in $\delta^{34}S$, biological processes can produce changes in the minor isotopes of sulfur, ^{33}S (0.75%) and ^{36}S (0.02%) (Farquhar et al. 2003; Johnston et al. 2005; Zerkle et al. 2009, 2016). Isotope effects in the minor sulfur isotopes are expressed as deviations from $\delta^{33}S$ or $\delta^{36}S$ values predicted by a reference value, often with a theoretical basis, e.g., $\Delta^{33}S \equiv \delta^{33}S_{measured} - \delta^{33}S_{predicted}$. These differences, termed mass conservation effects, are generated during complex cycling of sulfur compounds between different S pools in the environment or within individual cellular metabolisms (e.g., Farquhar et al. 2007). Mass conservation effects are generally orders of magnitude smaller than mass-independent fractionations of ^{33}S and ^{36}S formed in the atmosphere (Sect. 4.4), but can be very useful in unraveling sulfur cycling processes in natural environments and at the cellular level (e.g., Johnston 2011).

4.3.2 Oxygen Isotopes in Sulfate

Exchange of oxygen between sulfate and water is extremely slow unless biologically mediated or associated with oxidation of compounds with sulfur of a lower oxidation state (Lloyd 1968; Kusakabe and Robinson 1977). Thus, the triple oxygen isotopic composition ($^{18}O/^{16}O$ and $^{17}O/^{16}O$, expressed in δ-notation and permil units, as $\delta^{18}O$ and $\delta^{17}O$) of atmospheric, dissolved and mineral-bound sulfate has been used to study both biotic and abiotic sulfur redox transformations. In sulfate aerosols, where redox transformations are strictly abiotic, the combined sulfur and the oxygen isotopic compositions of the sulfate have been extensively used to identify its source. The different oxygen isotopic compositions of the possible oxidants of atmospheric SO_2, some of which also display "mass-independent" oxygen isotope compositions, allow distinction among sources of atmospheric and riverine sulfate, both in the modern atmosphere (e.g., Holt et al. 1982; Lee and Thiemens 2001; Alexander et al. 2005, 2012) and hydrosphere (e.g., Waldeck et al. 2019), and in geologic repositories such as ice cores (e.g., Alexander et al. 2002; Sofen et al. 2011, 2014), soils, crusts and rock varnishes (e.g., Bao et al. 2001a,b; Bao and Marchant 2006), volcanic ash beds (e.g., Bindeman et al. 2007; Martin and Bindeman 2009; Bao et al. 2010) and massive sulfate deposits (Bao et al. 2000). The oxygen isotopic composition of sulfate has also been used to constrain the oxidants and the (biotic and abiotic) oxidation mechanisms of pyrite in laboratory experiments and natural environments (Balci et al. 2007). Using these insights, the triple oxygen isotopic compositions of seawater sulfate have been used to constrain the sources of sulfate to the ocean (Waldeck et al. 2019), whereas the isotopic composition of sulfate-bearing minerals in sedimentary rocks have been used to constrain the atmospheric abundance of CO_2 during the Neoproterozoic "snowball Earth" events (Bao et al. 2008, 2009; Bao 2015), the paleoproductivity of the biosphere over the Proterozoic Eon (Crockford et al. 2018), and the atmospheric chemistry on early Mars (Farquhar and Thiemens 2000).

The oxygen isotopic compositions of aqueous sulfate have also been used to study microbial redox cycling in laboratory and natural settings, where they are thought to record information complementary to that revealed by sulfur isotopes. The sulfate–water oxygen isotope fractionations associated with a subset of sulfur metabolisms, including DSR (e.g., Böttcher et al. 1998; Brunner and Bernasconi 2005; Turchyn et al. 2006; Wortmann et al. 2007), elemental sulfur disproportionation (e.g., Böttcher et al. 2001, 2005), and pyrite oxidation (Balci et al. 2007) have been studied in laboratory cultures and natural environments. On the basis of these studies, it is understood that the $\delta^{18}O$ values of sulfate in aqueous environments with microbial redox cycling reflect a combination of disequilibrium isotope effects associated with enzymatic reactions and equilibration of oxygen isotopes between sulfur species of intermediate valence state (e.g., sulfite, SO_3^{2-}) and water, both intracellularly and extracellularly (e.g., Böttcher et al. 1998; Brunner and Bernasconi 2005; Turchyn et al. 2006, 2010; Wortmann et al. 2007; Wankel et al. 2014; Bertran et al. 2020).

The relationship between sulfate $\delta^{18}O$ and $\delta^{34}S$ values has been shown to reflect the intensity of microbial sulfur redox cycling, whether intracellularly during the multi-step reduction of sulfate to sulfide, or in the environment by concurrent activity of DSR and oxidative sulfur metabolisms. The consensus from laboratory cultures, environmental observations and models is that the slope of an array of sulfate $\delta^{18}O$ vs. $\delta^{34}S$ values is sensitive to the rate of sulfate reduction vs. that of reoxidation of intermediate sulfur species. Specifically, under conditions where sulfate reduction is slow relative to the reoxidation of intermediates, the isotopic (Rayleigh) distillation of sulfur isotopes in the sulfate (i.e., ^{34}S enrichment of the sulfate) is outpaced by the exchange of oxygen isotopes between the sulfur intermediates and water, resulting in a steeper slope in sulfate $\delta^{18}O$ vs. $\delta^{34}S$ (e.g., Böttcher et al. 1998; Aharon and Fu 2000; Böttcher et al. 2001; Brunner and Bernasconi 2005; Wortmann et al. 2007; Antler et al. 2013; Wankel et al. 2014; Johnston et al. 2014; Mills et al. 2016; Antler et al. 2017). Thus, oxidative systems, such as sediments with a low content of organic matter and/or a high content of iron oxides, in which a large fraction of the sulfate reduced during DSR is reoxidized display a steep $\delta^{18}O$–$\delta^{34}S$ slope. In contrast, more reducing systems, such as gas seeps and estuaries, in which only a negligible fraction of sulfate is reoxidized, are characterized by shallower slopes of sulfate $\delta^{18}O$ vs. $\delta^{34}S$ (Aharon and Fu 2000, 2003; Antler et al. 2014; Antler and Pellerin 2018).

4.3.3 Preservation of S Isotope Records

In geological materials, sulfate is usually preserved in sulfate minerals (barite, gypsum, anhydrite), and carbonate-associated sulfate (CAS). Barite is a mineral of low solubility, and is generally considered to be a high-fidelity recorder of sulfur isotopes, particularly through the Cenozoic (Paytan et al. 1998; Griffith and Paytan 2012; Masterson et al. 2016). Gypsum and anhydrite are often deposited as massive beds that precipitate from seawater in evaporative basins. CAS is a more widespread record (Kampschulte et al. 2001; Kampschulte and Strauss 2004; Gill et al. 2007; Wotte et al. 2012; Wu et al. 2014), but is subject to recrystallization and variable concentration among various carbonate components (Swart 2015; Present et al. 2015; Fichtner et al. 2017; Rose et al. 2019). Recent work has shown that recrystallization of carbonate cements expels CAS from rocks, decreasing its bulk-rock concentrations (Fichtner et al. 2017; Rose et al. 2019). This has not been demonstrated to have an effect on the isotopic composition ($\delta^{34}S$) of CAS, but neither can this possibility be excluded (Gill et al. 2008; Rennie and Turchyn 2014; Fichtner et al. 2017), especially if recrystallization is accompanied by DSR and a change in the isotopic composition of porewater sulfate.

Pyrite is the most commonly interrogated phase of reduced sulfur in sedimentary rocks. The ratio of sulfur isotopes in pyrite is the outcome of a complex suite of early diagenetic processes, including DSR, various other sulfur metabolisms that may cycle sedimentary sulfur among its oxidation states, and interaction with sedimentary iron-bearing phases (Claypool 2004; Rickard and Luther 2007; Leavitt et al. 2013; Aller 2014; Fike et al. 2015; Shawar et al. 2018; Jørgensen et al. 2019). The early diagenetic ratio of sulfur isotopes in pyrite can be disturbed by late-stage processes, including pyrite overgrowth and, more generally, precipitation of iron sulfides when sulfide-bearing fluids flow through iron-bearing sedimentary rocks. A number of studies have also pointed out a relationship between the variability in $\delta^{34}S$ of pyrite and depositional environment (Fike et al. 2015, and references therein), where sediments that have undergone remobilization and/or bioturbation have more variable $\delta^{34}S$ values and pyrite that is on average more enriched in ^{34}S (Aller 2014; Fike et al. 2015; Pasquier et al. 2017). The mechanisms that result in this enrichment have not been elucidated, but suggest that the determination of $\delta^{34}S$ value of primary pyrite in sedimentary rocks is rarely simple.

4.4 Evolution of the Precambrian Sulfur Cycle

The record of sulfur isotopes preserved in pyrite and sulfate minerals over geologic time can provide a window into past sulfur cycling processes. This record shows a generally small spread in pyrite $\delta^{34}S$ values ($\sim 20‰$) in the Archean, increasing to a $\sim 40‰$ spread in the early Proterozoic (Canfield and Raiswell 1999; Halevy et al. 2010; Fike et al. 2015; Fig. 1). Even the small spread of Archean pyrite $\delta^{34}S$ values has been interpreted to represent active sulfur cycling and DSR at > 3.0 Ga (Shen et al. 2001, 2009; Zhelezinskaia et al. 2014). The increased spread in pyrite $\delta^{34}S$ values in the early Paleoproterozoic is roughly coincident with pervasive oxygenation of Earth's surface oceans and atmosphere during the Great Oxidation Event, ~ 2.4 Ga. This change has therefore been interpreted to reflect enhanced sulfate delivery via oxidative weathering of pyrites on land, allowing for preservation of larger fractionation effects during microbial sulfate reduction, and the spread of sulfide-rich conditions (euxinia) on continental shelves (Canfield 1998; Habicht et al. 2002). An additional increase in the spread of $\delta^{34}S$ values near the end of the Proterozoic, to values spanning more than 55‰, has been interpreted to reflect larger biological fractionation effects associated with an expansion in oxidative sulfur cycling and widespread oxygenation of the deep oceans in the Neoproterozoic (Canfield and Teske 1996; Canfield et al. 2010; Poulton et al. 2010). Trends in minor sulfur isotopes have pushed the environmental expression of oxidative sulfur cycling and disproportionation back even farther in Earth history (Johnston et al. 2005). However, these data have been challenged by recent evidence from experiments and natural environments suggesting that DSR alone can produce very large fractionations in $\delta^{34}S$ without the need for oxidative S cycling (Canfield et al. 2010; Sim et al. 2011). Furthermore, experiments, observations from natural environments and models all suggest that isotopic fractionation during DSR may be large even when sulfate concentrations are low (Canfield et al. 2010; Crowe et al. 2014; Wing and Halevy 2014), casting doubt on the inference of sulfate-limited microbial fractionation of sulfur isotopes prior to the oxygenation of the atmosphere.

A unique feature of the Archean sulfur isotope record is the existence of mass-independent sulfur isotope compositions, preserved mostly in sedimentary pyrite (Farquhar et al. 2000, 2003; Thiemens 2006; Johnston 2011; Ono 2017). Most physical and chemical processes fractionate the heavy isotopes of sulfur proportionally to their mass, such that an enrichment or depletion in ^{33}S is approximately half of the enrichment or depletion in ^{34}S. Sulfur isotope mass-independent fractionations (S-MIF) in the Archean deviate significantly from these canonical relationships (Thiemens 2006; Johnston 2011). Although S-MIF is observed in present-day atmospheric sulfate aerosols and preserved in glacial ice following stratospheric volcanic eruptions (Romero and Thiemens 2003; Savarino et al. 2003; Burke et al. 2019), it is absent from sedimentary sulfate minerals and pyrite younger than ~ 2.4 Ga (Farquhar et al. 2000, 2003; Bekker et al. 2004; Luo et al. 2016; Warke et al. 2020). The abrupt disappearance of S-MIF from the sedimentary pyrite record ~ 2.4 Ga coincides with abundant geological and geochemical evidence for the oxygenation of the atmosphere around this time (Holland 1984, 2006; Lyons et al. 2014). In contrast to other proxi es for the oxidation state of the Earth's surface that are temporally discontinuous, sedimentary pyrite is a relatively continuous record, offering (potentially) more refined insight into the redox evolution of the surface environment, provided the mechanism(s) governing the existence of S-MIF are understood. Although it is generally agreed that the disappearance of S-MIF is related to the increasing oxidation state of Earth's oceans and atmosphere, the exact mechanism(s) are a topic of ongoing debate. While laboratory photolysis experiments can broadly reproduce the direction of sulfur isotope fractionations observed in Archean sediments (Farquhar et al. 2001; Masterson et al. 2011; Whitehill and Ono 2012; Whitehill et al. 2013, 2015; Endo et al. 2016), and models can constrain the magnitude of atmospheric O_2 level at which the S-MIF signal would be suppressed (Pavlov and Kasting 2002; Zahnle et al. 2006; Claire et al. 2014), fundamental aspects of the Archean S-MIF record remain highly uncertain (e.g., see reviews by Ono 2017; Stüeken et al. 2020). For instance, there is ongoing debate as to the specific atmospheric sulfur species and reaction(s) responsible for the generation of S-MIF signals and the delivery to Archean sediments (e.g., Farquhar et al. 2001; Lyons 2007; Danielache et al. 2008, 2012, 2014; Ono et al. 2013; Whitehill et al. 2013; Endo et al. 2015; Babikov 2017; Harman et al. 2018), plus the role of non-sulfur-bearing gases (e.g., CH_4, O_2) and biological activity in regulating (or erasing) these effects (Zahnle et al. 2006; Domagal-Goldman et al. 2008; Zerkle et al. 2012; Halevy 2013; Izon et al. 2015, 2017; Liu et al. 2019). For a review of the latest understanding of the Archean S-MIF record, we direct the reader to Stüeken et al. (2020) in this volume.

4.5 Late Neoproterozoic and Phanerozoic Sulfur Cycles

Large differences between $\delta^{34}S$ of sedimentary sulfate and sulfide minerals persisted from the late Neoproterozoic into the Phanerozoic Eon (Fike et al. 2015). The ^{34}S-enrichments of sedimentary sulfate minerals reached a maximum in the late Neoproterozoic, with $\delta^{34}S$ values approaching +40‰ (Fig. 1). This has been interpreted to reflect the values of seawater sulfate during this time period. The origins of these ^{34}S-enrichments are not well understood. Generally, they are assumed to reflect steady-state conditions requiring either addition to the oceans of ^{34}S-enriched sources (e.g., sulfate evaporites) or the preferential removal of ^{34}S-depleted sinks (e.g., sedimentary sulfides) (Fike et al. 2015). However, an alternative to a radical change in the global sulfur cycle is that the extremely ^{34}S-enriched sulfate phases represent the action of sulfate reducing bacteria in partially-closed (likely, sedimentary) environments, which would leave sediment porewater (but not global seawater) sulfate concentrations low and sulfate $\delta^{34}S$ values enriched through isotopic (Rayleigh) distillation (Fike et al. 2015).

Sedimentary rocks deposited during the early Phanerozoic show a long-term secular decline in the $\delta^{34}S$ of marine sulfate from the Neoproterozoic maximum to a value of around +10‰ at 300 Ma (Canfield and Farquhar 2009; Fike et al. 2015; Fig. 1). This decline is followed by a secular increase through the mid to late Phanerozoic to modern values of +21‰. Sedimentary sulfide minerals also reach maximum values of $\delta^{34}S$ in deposits of the late Neoproterozoic, up to $\delta^{34}S$ of > +40‰. The Phanerozoic records a long-term secular decline in these values, with values of −40‰ or lower common through most of the Phanerozoic (Canfield and Farquhar 2009; Wu et al. 2010; Halevy et al. 2012). These changes are thought to reflect the increasing reservoir of marine sulfate over the course of the Phanerozoic (Canfield and Farquhar 2009; Algeo et al. 2015), and variations in the proportions of sulfate removal from the ocean as sulfate minerals or as pyrite (the product of microbial sulfate reduction) (Garrels and Lerman 1984; Berner 1987; Canfield 2004; Kampschulte and Strauss 2004; Halevy et al. 2012; Fike et al. 2015). Sulfur isotopic fractionations expressed during these biological processes may have changed in response to large-scale environmental shifts that were imposed during the onset of the Phanerozoic (Wu et al. 2010; Leavitt et al. 2013; Rennie et al. 2018). Relevant changes include secular changes, such as the long-term increase in marine sulfate concentrations, and evolutionary innovations, such as the advent of bioturbation or of organic carbon derived from terrestrial plants. The long-term changes in the isotopic records since the end of the Neoproterozoic may also reflect changes in the sulfur isotopic composition of the sulfate inputs to the ocean.

5 Synthesis: Carbon, Nitrogen, and Sulfur Cycles in the Absence of Life

One way to consider how isotopic evidence may be used to detect life on other planets is to compare how the (bio)geochemical C, N, and S cycles on Earth would look without it. Here we synthesize the above material to speculate on how these cycles could be structured.

At temperatures and pressures close to that of Earth's surface, the most thermodynamically stable heteronuclear phases of carbon, oxygen, and hydrogen are water, carbon dioxide, and/or methane, depending on the oxidation state of the system (Saleh and Oganov 2016; Krissansen-Totton et al. 2018). The equilibrium assemblages of N- and S-bearing species are similarly sparse (Lewis and Randall 1923). Where life is present, it acts to catalyse exergonic reactions that may otherwise be kinetically sluggish. These catabolic processes convert chemical potential energy to heat, accelerating the descent of the entire system towards the equilibrium state, despite often producing metastable intermediates in the process. On the other hand, life's anabolic reactions store chemical potential energy, and may contribute to prolonging a state of chemical disequilibrium in the system as a whole. For example, cell walls compartmentalize the chemicals of the Earth's surface against the flow of entropy, autotrophic metabolisms bolster the lifetimes of calorie-dense organic molecules while the organism is alive, and leave a recalcitrant form of reduced material after death. A continued input of chemical potential energy from an external energy source, such as hydrothermal vent H_2, which may have been short-lived in the absence of life, may therefore not only sustain, but also be sustained by, life itself. In other words, through kinetic inhibition, anabolic reactions can act to slow the inexorable conversion of chemical potential energy (or light energy, when phototrophy exists) to heat. Whether the net effect of life is to maintain disequilibrium in a system or accelerate a system in its entirety towards equilibrium is therefore a balance between the anabolic and catabolic reactions in a system. On Earth, by sequestering reduced carbon, nitrogen, and sulfur in recalcitrant kerogens and minerals, life separates electrons from their previous owners, which (today) allows O_2 to persist in the atmosphere.

The simultaneous presence of certain chemical combinations in a planetary atmosphere that represent chemical disequilibrium has been suggested to be a biosignature in extraterrestrial settings. For example the coexistence of atmospheric methane and CO_2 is thermodynamically unstable at Earth surface-like conditions, and would require large, possibly biogenic, fluxes of the unstable component (likely methane) in order to be sustained (Krissansen-Totton et al. 2018). Isotopic compositions of these gases could provide additional constraints on the degree of disequilibrium in the system.

An exploration of isotopic compositions of solid phases in the surface environment would provide an additional layer of information. If sediment cores were obtained and dated, and isotopic time series reconstructed, characteristic timescales may be diagnostic of the past existence of differentiated pools with contrasting residence times (i.e., a (bio)geochemical cycle). Comparing the structures of the C, N and S cycles in the modern Earth (Figs. 2, 3 and 4), it is likely that they would respond differently to a drastic reduction in biologic activity. The modern pre-industrial carbon cycle features connected oxidized and reduced reservoirs (C_{carb} and C_{org}) that occupy the same physical spaces (e.g., deep ocean, sediments, and crust) and have similar residence times (Fig. 2), so they often respond in complementary ways to the same perturbations. It is not uncommon for Phanerozoic carbon isotope excursions to be mirrored in $\delta^{13}C_{carb}$ and $\delta^{13}C_{org}$, and situations that induce non-steady-state behavior often do so in both records (Rothman et al. 2003). In the absence of a biosphere, the input fluxes of new organic carbon to these reservoirs would be reduced (at least to ca. Archean levels). Predicting the resulting changes to the size and residence time of the marine organic carbon reservoir is less straightforward because these would depend on the exact response of the output fluxes. Nonetheless, barring stark differences in the way organic carbon is buried in a biosphere-free world, we might expect the size of the marine organic carbon pool to be reduced as well. If the biosphere-free organic carbon pool were orders of magnitude smaller than the corresponding carbonate carbon pool, it would likely make it easier to decouple the two; perturbations to the former, such as a burst of hydrothermal abiotic organic synthesis, may not register if oxidized and added to the latter.

One key difference between the structures of the biogeochemical C and N cycles is that the latter is buffered by a well-mixed inorganic pool with a >Myr residence time (atmospheric N_2; residence time ca. 10^8 yr), while the former is not (Figs. 2, 3). In the existing world, the atmospheric N_2 pool buffers the exogenic N cycle against some types of continuous, coupled isotope excursions like those observed in the C cycle. These contrasting behaviors are reflected in the differences in the dynamics of the Phanerozoic C and N isotope records (Fig. 1). In the absence of life, with no N_2 fixers, the fixation flux would be smaller and more reliant on lightning and other minor abiotic fixation fluxes (reviewed in Sect. 3.1). However, in the absence of denitrifiers, there also would have been no mechanism to return N back to the atmosphere. Without the strong negative isotopic fractionation associated with the modern denitrification flux, the nitrate pool in the ocean may not be as isotopically enriched as it is today (note, however, that atmospheric escape of low-δ^{15}N N_2 could still have altered the nitrate record; Sect. 3.4).

Like the carbon cycle, the sulfur cycle is principally recorded in two sedimentary reservoirs that are distinguished in oxidation state and isotopic composition by (largely) a single, biologically mediated reaction (i.e., primary CO_2 fixation vs. dissimilatory sulfate reduction; Figs. 2, 4). Like the nitrogen cycle, the sulfur cycle is buffered against sub-Myr changes by a large reservoir with a long residence time (i.e., the atmospheric N_2 and marine sulfate pools). Unlike both C and N cycles, however, in the modern (oxidized) world, the principal metabolic reaction of the S cycle that introduces reduced sulfur to the sedimentary record—dissimilatory sulfate reduction—largely occurs in depositional environments that are in inconsistent communication with the global substrate pool. This variability in the degree of openness at the site of sulfur cycling is reflected in the high variability in the δ^{34}S composition of the Phanerozoic sedimentary pyrite record at any one time (Fig. 1). On an abiotic earth, in addition to a lack of microbial sulfate reduction, thermal sulfate reduction—which requires reaction with an electron donor (often, hydrocarbons)—would also have been reduced. The oxidation of volcanic sulfide would have been reduced, so the isotopic composition of any pyrite in the geologic record would more likely reflect its mantle source, rather than carry a metabolic signature of sulfate reduction. Gas-phase chemistry in an anoxic atmosphere is expected to produce S-MIF, which may be observable both in the atmospheric and the sedimentary components of the surface sulfur reservoir. Under an anoxic atmosphere, which is expected to produce S-MIF, the absence of this isotopic anomaly in sedimentary sulfur repositories (e.g., sulfate evaporites, sedimentary pyrite) may be indicative of intense microbial sulfur cycling.

Within all three of these cycles, the absence of phototrophy may all but remove the dependence of nutrient consumption on light penetration in the ocean. The lack of a biological pump would probably result in a more homogeneous ocean column due to a weaker decoupling between the surface and deep ocean. Furthermore, without nitrate utilization by organisms and burial in organic form, and without sulfate reduction in sediments, the concentration of nitrate and sulfate may have accumulated in the oceans over geologic time. In response to such accumulations, burial/subduction in the form of evaporitic salts may become an increasingly important means of export from the surface to the mantle.

Synthesizing the above discussion, in the absence of life, across the C, N and S cycles, we might expect co-localized pools to be dominantly either reduced or oxidized because their simultaneous existence is unstable. On the other hand, in the absence of enzymes to catalyze catabolic reactions, sluggish kinetics may nonetheless allow disequilibrated species to coexist. Where such oxidised and reduced pools do coexist, we might expect this disequilibrium to be reflected in non-equilibrium isotope fractionations. On a global scale, without similarly sized pools with significant partitioning of isotopes between phases, we might also expect isotopic excursions to be more modest in size. However, without the buffering capacity of differently sized organic reservoirs, we expect isotopic perturbations to be rapid with slow recoveries, and without a parallel expression in paired pools (such as $\delta^{13}C_{org}$ and $\delta^{13}C_{carb}$). Though limited by temporal resolution, there is some evidence for this behavior in the Archean (Fig. 1).

Our framework for interpreting geochemical observations as biosignatures on another planet is informed by—and limited by—our understanding of the imprint of biology on the geochemical records of our own. In this chapter, we have described present understanding of how life on Earth influences and is influenced by the exospheric cycles of carbon, nitrogen and sulfur. We have reviewed evidence for past changes in Earth's biosphere as told through the sedimentary isotopic record of these three elements. Clearly, many aspects of these records remain unresolved. In order to resolve these various uncertainties, we expect that research into the C, N, and S isotopic signatures in past terrestrial environments will continue, and continue to serve as guides for these elemental cycles beyond this planet.

Acknowledgements This paper is the product of the geobiology working group within the International Space Science Institute (ISSI) workshop 'Reading Terrestrial Planet Evolution in Isotopes and Element Measurements'. We would like to thank ISSI and Europlanet for their support. Thanks to Peter Crockford, who provided invaluable feedback on the manuscript. We would also like to thank Guest Editor Helmut Lammer and three anonymous reviewers for careful and constructive comments.

References

C. Achtnich, F. Bak, R. Conrad, Competition for electron donors among nitrate reducers, ferric iron reducers, sulfate reducers, and methanogens in anoxic paddy soil. Biol. Fertil. Soils **19**(1), 65–72 (1995)

M. Ader, P. Sansjofre, G.P. Halverson, V. Busigny, R.I. Trindade, M. Kunzmann, A.C. Nogueira, Ocean redox structure across the late neoproterozoic oxygenation event: a nitrogen isotope perspective. Earth Planet. Sci. Lett. **396**, 1–13 (2014)

P. Aharon, B. Fu, Microbial sulfate reduction rates and sulfur and oxygen isotope fractionations at oil and gas seeps in deepwater Gulf of Mexico. Geochim. Cosmochim. Acta **64**(2), 233–246 (2000)

P. Aharon, B. Fu, Sulfur and oxygen isotopes of coeval sulfate–sulfide in pore fluids of cold seep sediments with sharp redox gradients. Chem. Geol. **195**(1), 201–218 (2003)

B. Alexander, J. Savarino, N.I. Barkov, R.J. Delmas, M.H. Thiemens, Climate driven changes in the oxidation pathways of atmospheric sulfur. Geophys. Res. Lett. **29**(14), 30-1–30-4 (2002)

B. Alexander, R.J. Park, D.J. Jacob, Q.B. Li, R.M. Yantosca, J. Savarino, C.C.W. Lee, M.H. Thiemens, Sulfate formation in sea-salt aerosols: constraints from oxygen isotopes. J. Geophys. Res., Atmos. **110**(D10), D10307 (2005)

B. Alexander, D.J. Allman, H.M. Amos, T.D. Fairlie, J. Dachs, D.A. Hegg, R.S. Sletten, Isotopic constraints on the formation pathways of sulfate aerosol in the marine boundary layer of the subtropical northeast Atlantic Ocean. J. Geophys. Res., Atmos. **117**(D6), D06304 (2012)

T. Algeo, G. Luo, H. Song, T. Lyons, D. Canfield, Reconstruction of secular variation in seawater sulfate concentrations. Biogeosciences **12**(7), 2131–2151 (2015)

F. Alhaique, D. Giacchetti, M. Marchetti, F.M. Riccieri, Effect of surfactant monomers on chloramphenicol association to an albuminlecithin complex: a model for modified drug absorption. J. Pharm. Pharmacol. **27**(11), 811–817 (1975)

J.F. Allen, B. Thake, W.F. Martin, Nitrogenase inhibition limited oxygenation of Earth's proterozoic atmosphere. Trends Plant Sci. **24**(11), 1022–1031 (2019)

R. Aller, Sedimentary diagenesis, depositional environments, and benthic fluxes, in *Treatise on Geochemistry* (Elsevier, Amsterdam, 2014), pp. 293–334

A. Amrani, Organosulfur compounds: molecular and isotopic evolution from biota to oil and gas. Annu. Rev. Earth Planet. Sci. **42**(1), 733–768 (2014)

A.D. Anbar, A.H. Knoll, Proterozoic ocean chemistry and evolution: a bioinorganic bridge? Science **297**(5584), 1137–1142 (2002)

A.D. Anbar, Y. Duan, T.W. Lyons, G.L. Arnold, B. Kendall, R.A. Creaser, A.J. Kaufman, G.W. Gordon, C. Scott, J. Garvin, R. Buick, A whiff of oxygen before the great oxidation event? Science **317**(5846), 1903–1906 (2007)

G. Antler, A. Pellerin, A critical look at the combined use of sulfur and oxygen isotopes to study microbial metabolisms in methane-rich environments. Front. Microbiol. **9**, 519 (2018)

G. Antler, A.V. Turchyn, V. Rennie, B. Herut, O. Sivan, Coupled sulfur and oxygen isotope insight into bacterial sulfate reduction in the natural environment. Geochim. Cosmochim. Acta **118**, 98–117 (2013)

G. Antler, A.V. Turchyn, B. Herut, A. Davies, V.C.F. Rennie, O. Sivan, Sulfur and oxygen isotope tracing of sulfate driven anaerobic methane oxidation in estuarine sediments. Estuar. Coast. Shelf Sci. **142**, 4–11 (2014)

G. Antler, A.V. Turchyn, S. Ono, O. Sivan, T. Bosak, Combined ^{34}S, ^{33}S and ^{18}O isotope fractionations record different intracellular steps of microbial sulfate reduction. Geochim. Cosmochim. Acta **203**, 364–380 (2017)

D. Archer, B. Buffett, V. Brovkin, Ocean methane hydrates as a slow tipping point in the global carbon cycle. Proc. Natl. Acad. Sci. USA **106**(49), 20596–20601 (2009)

A.R. Babbin, R.G. Keil, A.H. Devol, B.B. Ward, Organic matter stoichiometry, flux, and oxygen control nitrogen loss in the ocean. Science **344**(6182), 406–408 (2014)

D. Babikov, Recombination reactions as a possible mechanism of mass-independent fractionation of sulfur isotopes in the Archean atmosphere of Earth. Proc. Natl. Acad. Sci. **114**(12), 3062–3067 (2017)

A. Bachan, K.V. Lau, M.R. Saltzman, E. Thomas, L.R. Kump, J.L. Payne, American journal of science: a model for the decrease in amplitude of carbon isotope excursions across the Phanerozoic. Am. J. Sci. **317**(6), 641–676 (2017)

M.P. Badger, T.B. Chalk, G.L. Foster, P.R. Bown, S.J. Gibbs, P.F. Sexton, D.N. Schmidt, H. Pälike, A. Mackensen, R.D. Pancost, Insensitivity of alkenone carbon isotopes to atmospheric CO_2 at low to moderate CO_2 levels. Clim. Past **15**(2), 539–554 (2019)

F. Bak, N. Pfennig, Chemolithotrophic growth of Desulfovibrio sulfodismutans sp. nov. by disproportionation of inorganic sulfur compounds. Arch. Microbiol. **147**(2), 184–189 (1987)

N. Balci, W.C. Shanks, B. Mayer, K.W. Mandernack, Oxygen and sulfur isotope systematics of sulfate produced by bacterial and abiotic oxidation of pyrite. Geochim. Cosmochim. Acta **71**(15), 3796–3811 (2007)

J.L. Banner, G.N. Hanson, Calculation of simultaneous isotopic and trace element variations during water-rock interaction with applications to carbonate diagenesis. Geochim. Cosmochim. Acta **54**(11), 3123–3137 (1990)

H. Bao, Sulfate: a time capsule for Earth's O_2, O_3, and H_2O. Chem. Geol. **395**, 108–118 (2015)

H. Bao, D.R. Marchant, Quantifying sulfate components and their variations in soils of the McMurdo Dry Valleys, Antarctica. J. Geophys. Res., Atmos. **111**(D16), D16301 (2006)

H. Bao, M.H. Thiemens, J. Farquhar, D.A. Campbell, C.C.-W. Lee, K. Heine, D.B. Loope, Anomalous ^{17}O compositions in massive sulphate deposits on the Earth. Nature **406**(6792), 176–178 (2000)

H. Bao, G.M. Michalski, M.H. Thiemens, Sulfate oxygen-17 anomalies in desert varnishes. Geochim. Cosmochim. Acta **65**(13), 2029–2036 (2001a)

H. Bao, M.H. Thiemens, K. Heine, Oxygen-17 excesses of the central namib gypcretes: spatial distribution. Earth Planet. Sci. Lett. **192**(2), 125–135 (2001b)

H. Bao, J.R. Lyons, C. Zhou, Triple oxygen isotope evidence for elevated CO_2 levels after a neoproterozoic glaciation. Nature **453**(7194), 504–506 (2008)

H. Bao, I.J. Fairchild, P.M. Wynn, C. Spötl, Stretching the envelope of past surface environments: neoproterozoic Glacial Lakes from Svalbard. Science **323**(5910), 119–122 (2009)

H. Bao, S. Yu, D.Q. Tong, Massive volcanic SO_2 oxidation and sulphate aerosol deposition in Cenozoic North America. Nature **465**(7300), 909–912 (2010)

Y.M. Bar-On, R. Phillips, R. Milo, The biomass distribution on Earth. Proc. Natl. Acad. Sci. USA **115**(25), 6506–6511 (2018)

B. Batchelor, A.W. Lawrence, A kinetic model for autotrophic denitrification using elemental sulfur. Water Res. **12**(12), 1075–1084 (1978)

N.H. Batjes, Total carbon and nitrogen in the soils of the world. Eur. J. Soil Sci. **65**(1), 10–21 (2014)

T. Bauersachs, S. Schouten, J. Compaore, U. Wollenzien, L.J. Stal, J.S. Sinninghe Damste, Nitrogen isotopic fractionation associated with growth on dinitrogen gas and nitrate by cyanobacteria. Limnol. Oceanogr. **54**(4), 1403–1411 (2009)

D.A. Beard, H. Qian, Relationship between thermodynamic driving force and one-way fluxes in reversible processes. PLoS ONE **2**(1), 1–4 (2007)

V. Beaumont, F. Robert, Nitrogen isotope ratios of kerogens in precambrian cherts: a record of the evolution of atmosphere chemistry? Precambrian Res. **96**(1–2), 63–82 (1999)

G.E. Bebout, M.L. Fogel, Nitrogen-isotope compositions of metasedimentary rocks in the Catalina Schist, California: implications for metamorphic devolatilization history. Geochim. Cosmochim. Acta **56**(7), 2839–2849 (1992)

G.E. Bebout, M.L. Fogel, P. Cartigny, Nitrogen: highly volatile yet surprisingly compatible. Elements **9**(5), 333–338 (2013)

A. Bekker, H.D. Holland, Oxygen overshoot and recovery during the early paleoproterozoic. Earth Planet. Sci. Lett. **317**(318), 295–304 (2012)

A. Bekker, H.D. Holland, P.L. Wang, D. Rumble, H.J. Stein, J.L. Hannah, L.L. Coetzee, N.J. Beukes, Dating the rise of atmospheric oxygen. Nature **427**(6970), 117–120 (2004)
I.A. Berg, D. Kockelkorn, W.H. Ramos-Vera, R.F. Say, J. Zarzycki, M. Hügler, B.E. Alber, G. Fuchs, Autotrophic carbon fixation in archaea. Nat. Rev. Microbiol. **8**(6), 447–460 (2010)
R.A. Berner, Models for carbon and sulfur cycles and atmospheric oxygen; application to Paleozoic geologic history. Am. J. Sci. **287**(3), 177–196 (1987)
R.A. Berner, D.E. Canfield, A new model for atmospheric oxygen over Phanerozoic time. Am. J. Sci. **289**(4), 333–361 (1989)
R.A. Berner, Z. Kothavala, Geocarb III: a revised model of atmospheric CO_2 over Phanerozoic time. Am. J. Sci. **301**(2), 182–204 (2001)
V.J. Bertics, J.A. Sohm, T. Treude, C.E.T. Chow, D.G. Capone, J.A. Fuhrman, W. Ziebis, Burrowing deeper into benthic nitrogen cycling: the impact of bioturbation on nitrogen fixation coupled to sulfate reduction. Mar. Ecol. Prog. Ser. **409**, 1–15 (2010)
E. Bertran, A. Waldeck, B.A. Wing, I. Halevy, W.D. Leavitt, A.S. Bradley, D.T. Johnston, Oxygen isotope effects during microbial sulfate reduction: applications to sediment cell abundances. ISME J. **14**(6), 1508–1519 (2020)
R.I. Bickley, V. Vishwanathan, Photocatalytically induced fixation of molecular nitrogen by near UV radiation. Nature **280**(5720), 306–308 (1979)
I.N. Bindeman, J.M. Eiler, B.A. Wing, J. Farquhar, Rare sulfur and triple oxygen isotope geochemistry of volcanogenic sulfate aerosols. Geochim. Cosmochim. Acta **71**(9), 2326–2343 (2007)
L.R. Bird, K.S. Dawson, G.L. Chadwick, J.M. Fulton, V.J. Orphan, K.H. Freeman, Carbon isotopic heterogeneity of coenzyme F430 and membrane lipids in methane-oxidizing archaea. Geobiology **17**(6), 611–627 (2019)
N. Blair, A. Leu, E. Munoz, J. Olsen, E. Kwong, D. Des Marais, Carbon isotopic fractionation in heterotrophic microbial metabolism. Appl. Environ. Microbiol. **50**(4), 996–1001 (1985)
C.C. Blair, S. D'Hondt, A.J. Spivack, R.H. Kingsley, Radiolytic hydrogen and microbial respiration in subsurface sediments. Astrobiology **7**(6), 951–970 (2007)
R.E. Blankenship, M.T. Madigan, C.E. Bauer (eds.), *Anoxygenic Photosynthetic Bacteria, Advances in Photosynthesis and Respiration* (Springer, Dordrecht, 2004)
M.B. Blaser, L.K. Dreisbach, R. Conrad, Carbon isotope fractionation of 11 acetogenic strains grown on H_2 and CO_2. Appl. Environ. Microbiol. **79**(6), 1787–1794 (2013)
A.J. Boller, P.J. Thomas, C.M. Cavanaugh, K.M. Scott, Low stable carbon isotope fractionation by coccolithophore RubisCO. Geochim. Cosmochim. Acta **75**(22), 7200–7207 (2011)
A.J. Boller, P.J. Thomas, C.M. Cavanaugh, K.M. Scott, Isotopic discrimination and kinetic parameters of RubisCO from the marine bloom-forming diatom, Skeletonema costatum. Geobiology **13**(1), 33–43 (2015)
M.E. Böttcher, B. Oelschläger, T. Höpner, H.-J. Brumsack, J. Rullkötter, Sulfate reduction related to the early diagenetic degradation of organic matter and "black spot" formation in tidal sandflats of the German Wadden Sea (southern North Sea): stable isotope (13C, 34S, 18O) and other geochemical results. Org. Geochem. **29**(5–7), 1517–1530 (1998)
M.E. Böttcher, B. Thamdrup, T. Vennemann, Oxygen and sulfur isotope fractionation during anaerobic bacterial disproportionation of elemental sulfur. Geochim. Cosmochim. Acta **65**(10), 1601–1609 (2001)
M.E. Böttcher, B. Thamdrup, M. Gehre, A. Theune, $^{34}S/^{32}S$ and $^{18}O/^{16}O$ fractionation during sulfur disproportionation by desulfobulbus propionicus. Geomicrobiol. J. **22**(5), 219–226 (2005)
S.R. Boyd, P. Philippot, Precambrian ammonium biogeochemistry: a study of the Moine metasediments, Scotland. Chem. Geol. **144**(3–4), 257–268 (1998)
A.S. Bradley, W.D. Leavitt, D.T. Johnston, Revisiting the dissimilatory sulfate reduction pathway. Geobiology **9**(5), 446–457 (2011)
A.S. Bradley, W.D. Leavitt, M. Schmidt, A.H. Knoll, P.R. Girguis, D.T. Johnston, Patterns of sulfur isotope fractionation during microbial sulfate reduction. Geobiology **14**(1), 91–101 (2016)
S.D. Bridgham, J.P. Megonigal, J.K. Keller, N.B. Bliss, C. Trettin, The carbon balance of North American wetlands. Wetlands **26**(4), 889–916 (2006)
P. Brimblecombe, The global sulfur cycle, in *Treatise on Geochemistry*, vol. 10, 2nd edn. (Elsevier, Amsterdam, 2013), pp. 559–591
H.G. Britton, The Ussing relationship and chemical reactions: possible application to enzymatic investigations. Nature **205**(4978), 1323–1324 (1965)
J.J. Brocks, G.D. Love, R.E. Summons, A.H. Knoll, G.A. Logan, S.A. Bowden, Biomarker evidence for green and purple sulphur bacteria in a stratified palaeoproterozoic sea. Nature **437**(7060), 866–870 (2005)
J.J. Brocks, A.J. Jarrett, E. Sirantoine, C. Hallmann, Y. Hoshino, T. Liyanage, The rise of algae in cryogenian oceans and the emergence of animals. Nature **548**(7669), 578–581 (2017)

W.S. Broecker, J. van Donk, Insolation Changes, Ice Volumes, and the O^{18} Record in Deep-Sea Cores Rev. Geophys. **8**(1), 169–198 (1970)
B. Brunner, S.M. Bernasconi, A revised isotope fractionation model for dissimilatory sulfate reduction in sulfate reducing bacteria. Geochim. Cosmochim. Acta **69**(20), 4759–4771 (2005)
B. Brunner, S. Contreras, M.F. Lehmann, O. Matantseva, M. Rollog, T. Kalvelage, G. Klockgether, G. Lavik, M.S.M. Jetten, B. Kartal, M.M.M. Kuypers, Nitrogen isotope effects induced by anammox bacteria. Proc. Natl. Acad. Sci. **110**(47), 18994–18999 (2013)
R.N. Bryant, C. Jones, M.R. Raven, M.L. Gomes, W.M. Berelson, A.S. Bradley, D.A. Fike, Sulfur isotope analysis of microcrystalline iron sulfides using secondary ion mass spectrometry imaging: extracting local paleo-environmental information from modern and ancient sediments. Rapid Commun. Mass Spectrom. **33**(5), 491–502 (2019)
C. Buchwald, K. Grabb, C.M. Hansel, S.D. Wankel, Constraining the role of iron in environmental nitrogen transformations: dual stable isotope systematics of abiotic NO_2^- reduction by Fe(II) and its production of N_2O. Geochim. Cosmochim. Acta **186**, 1–12 (2016)
R. Buick, D.J. Des Marais, A.H. Knoll, Stable isotopic compositions of carbonates from the mesoproterozoic bangemall group, northwestern Australia. Chem. Geol. **123**(1–4), 153–171 (1995)
R.J. Buresh, J.T. Moraghan, Chemical reduction of nitrate by ferrous iron. J. Environ. Qual. **5**(3), 320–325 (1976)
A. Burke, K.A. Moore, M. Sigl, D.C. Nita, J.R. McConnell, J.F. Adkins, Stratospheric eruptions from tropical and extra-tropical volcanoes constrained using high-resolution sulfur isotopes in ice cores. Earth Planet. Sci. Lett. **521**, 113–119 (2019)
V. Busigny, O. Lebeau, M. Ader, B. Krapež, A. Bekker, Nitrogen cycle in the late Archean ferruginous ocean. Chem. Geol. **362**, 115–130 (2013)
N.J. Butterfield, Macroevolution and macroecology through deep time. Palaeontology **50**(1), 41–55 (2007)
D.E. Canfield, A new model for proterozoic ocean chemistry. Nature **396**(6710), 450–453 (1998)
D.E. Canfield, Biogeochemistry of sulfur isotopes. Rev. Mineral. Geochem. **43**, 606–636 (2001)
D.E. Canfield, The evolution of the Earth surface sulfur reservoir. Am. J. Sci. **304**(10), 839–861 (2004)
D.E. Canfield, J. Farquhar, Animal evolution, bioturbation, and the sulfate concentration of the oceans. Proc. Natl. Acad. Sci. USA **106**(20), 8123–8127 (2009)
D.E. Canfield, R. Raiswell, The evolution of the sulfur cycle. Am. J. Sci. **299**(7–9), 697–723 (1999)
D.E. Canfield, A. Teske, Late proterozoic rise in atmospheric oxygen concentration inferred from phylogenetic and sulphur-isotope studies. Nature **382**(6587), 127–132 (1996)
D.E. Canfield, M.T. Rosing, C. Bjerrum, Early anaerobic metabolisms. Philos. Trans. R. Soc. Lond. B, Biol. Sci. **361**(1474), 1819–1834 (2006)
D.E. Canfield, J. Farquhar, A.L. Zerkle, High isotope fractionations during sulfate reduction in a low-sulfate euxinic ocean analog. Geology **38**(5), 415–418 (2010)
J.D. Caranto, K.M. Lancaster, Nitric oxide is an obligate bacterial nitrification intermediate produced by hydroxylamine oxidoreductase. Proc. Natl. Acad. Sci. USA **114**(31), 8217–8222 (2017)
J.D. Caranto, A.C. Vilbert, K.M. Lancaster, Nitrosomonas europaea cytochrome P460 is a direct link between nitrification and nitrous oxide emission. Proc. Natl. Acad. Sci. USA **113**(51), 14704–14709 (2016)
E.J. Carpenter, H.R. Harvey, F. Brian, D.G. Capone, Biogeochemical tracers of the marine cyanobacterium trichodesmium. Deep-Sea Res., Part 1, Oceanogr. Res. Pap. **44**(1), 27–38 (1997)
P. Cartigny, B. Marty, Nitrogen isotopes and mantle geodynamics: the emergence of life and the atmosphere–crust–mantle connection. Elements **9**(5), 359–366 (2013)
K.L. Casciotti, Inverse kinetic isotope fractionation during bacterial nitrite oxidation. Geochim. Cosmochim. Acta **73**(7), 2061–2076 (2009)
D.C. Catling, M.W. Claire, How Earth's atmosphere evolved to an oxic state: a status report. Earth Planet. Sci. Lett. **237**(1–2), 1–20 (2005)
D.C. Catling, K.J. Zahnle, The Archean atmosphere. Sci. Adv. **6**(9), eaax1420 (2020)
A.R. Cavazos, M. Taillefert, Y. Tang, J.B. Glass, Kinetics of nitrous oxide production from hydroxylamine oxidation by birnessite in seawater. Mar. Chem. **202**, 49–57 (2018)
T.E. Cerling, J.M. Harris, B.J. MacFadden, M.G. Leakey, J. Quade, V. Eisenmann, J.R. Ehieringer, Global vegetation change through the Miocene/Pliocene boundary. Nature **389**(6647), 153–158 (1997)
L.A. Chambers, P.A. Trudinger, Are thiosulfate and trithionate intermediates in dissimilatory sulfate reduction? J. Bacteriol. **123**(1), 36–40 (1975)
S. Chen, A.C. Gagnon, J.F. Adkins, Carbonic anhydrase, coral calcification and a new model of stable isotope vital effects. Geochim. Cosmochim. Acta **236**, 179–197 (2018)
M.I. Chicarelli, J.M. Hayes, B.N. Popp, C.B. Eckardt, J.R. Maxwell, Carbon and nitrogen isotopic compositions of alkyl porphyrins from the Triassic Serpiano oil shale. Geochim. Cosmochim. Acta **57**(6), 1307–1311 (1993)

P. Ciais, C. Sabine, G. Bala, L. Bopp, V. Brovkin, J. Canadell, A. Chhabra, R. DeFries, J. Galloway, M. Heimann, C. Jones, C.L. Quéré, R. Myneni, S. Piao, P. Thornton, Carbon and other biogeochemical cycles, in *Climate Change 2013 the Physical Science Basis: Working Group i Contribution to the Fifth Assessment Report of the Intergovernmental Panel on Climate Change*, vol. 9781107057 (2013), pp. 465–570

M.W. Claire, J.F. Kasting, S.D. Domagal-Goldman, E.E. Stüeken, R. Buick, V.S. Meadows, Modeling the signature of sulfur mass-independent fractionation produced in the Archean atmosphere. Geochim. Cosmochim. Acta **141**, 365–380 (2014)

G.E. Claypool, in *Ventilation of Marine Sediments Indicated by Depth Profiles of Pore Water Sulfate and $\delta^{34}S$.* The Geochemical Society Special Publications, vol. 9 (Elsevier, Amsterdam, 2004), pp. 59–65

H.G. Close, R. Bovee, A. Pearson, Inverse carbon isotope patterns of lipids and kerogen record heterogeneous primary biomass. Geobiology **9**(3), 250–265 (2011)

A.J. Coby, F.W. Picardal, Inhibition of NO_3^- and NO_2^- reduction by microbial Fe(III) reduction: evidence of a reaction between NO_2^- and cell surface-bound Fe^{2+}. Appl. Environ. Microbiol. **71**(9), 5267–5274 (2005)

R.E. Criss, *Principles of Stable Isotope Distribution* (Oxford University Press, London, 1999)

P.W. Crockford, J.A. Hayles, H. Bao, N.J. Planavsky, A. Bekker, P.W. Fralick, G.P. Halverson, T.H. Bui, Y. Peng, B.A. Wing, Triple oxygen isotope evidence for limited mid-Proterozoic primary productivity. Nature **559**(7715), 613–616 (2018)

P.W. Crockford, M. Kunzmann, A. Bekker, J. Hayles, H. Bao, G.P. Halverson, Y. Peng, T.H. Bui, G.M. Cox, T.M. Gibson, S. Wörndle, R. Rainbird, A. Lepland, N.L. Swanson-Hysell, S. Master, B. Sreenivas, A. Kuznetsov, V. Krupenik, B.A. Wing, *Claypool Continued: Extending the Isotopic Record of Sedimentary Sulfate* (2019)

S.A. Crowe, G. Paris, S. Katsev, C. Jones, S.-T. Kim, A.L. Zerkle, S. Nomosatryo, D.A. Fowle, J.F. Adkins, A.L. Sessions, J. Farquhar, D.E. Canfield, Sulfate was a trace constituent of Archean seawater. Science **346**(6210), 735–739 (2014)

H. Daims, E.V. Lebedeva, P. Pjevac, P. Han, C. Herbold, M. Albertsen, N. Jehmlich, M. Palatinszky, J. Vierheilig, A. Bulaev, R.H. Kirkegaard, M. Von Bergen, T. Rattei, B. Bendinger, P.H. Nielsen, M. Wagner, Complete nitrification by nitrospira bacteria. Nature **528**(7583), 504–509 (2015)

T. Dalsgaard, B. Thamdrup, L. Farías, N.P. Revsbech, Anammox and denitrification in the oxygen minimum zone of the eastern South Pacific. Limnol. Oceanogr. **57**(5), 1331–1346 (2012)

S.O. Danielache, C. Eskebjerg, M.S. Johnson, Y. Ueno, N. Yoshida, High-precision spectroscopy of ^{32}S, ^{33}S and ^{34}S sulfur dioxide: ultraviolet absorption cross sections and isotope effects. J. Geophys. Res., Atmos. **113**(D17), D17314 (2008)

S.O. Danielache, S. Hattori, M.S. Johnson, Y. Ueno, S. Nanbu, N. Yoshida, Photoabsorption cross-section measurements of ^{32}S, ^{33}S, ^{34}S, and ^{36}S sulfur dioxide for the B1B1-X1A1 absorption band. J. Geophys. Res., Atmos. **117**(D24), D24301 (2012)

S.O. Danielache, S. Tomoya, A. Kondorsky, I. Tokue, S. Nanbu, Nonadiabatic calculations of ultraviolet absorption cross section of sulfur monoxide: isotopic effects on the photodissociation reaction. J. Chem. Phys. **140**(4), 044319 (2014)

A.E. Dekas, R.S. Poretsky, V.J. Orphan, Deep-sea archaea fix and share nitrogen in methane-consuming microbial consortia. Science **326**(5951), 422–426 (2009)

C.C. Delwiche, P.L. Steyn, Nitrogen isotope fractionation in soils and microbial reactions. Environ. Sci. Technol. **4**(11), 929–935 (1970)

B. Demirel, P. Scherer, The roles of acetotrophic and hydrogenotrophic methanogens during anaerobic conversion of biomass to methane: a review. Rev. Environ. Sci. Biotechnol. **7**(2), 173–190 (2008)

M.J. DeNiro, S. Epstein, Mechanism of carbon isotope fractionation associated with lipid synthesis. Science **197**(4300), 261–263 (1977)

M.J. DeNiro, S. Epstein, Influence of diet on the distribution of carbon isotopes in animals. Geochim. Cosmochim. Acta **42**(5), 495–506 (1978)

D.J. Des Marais, Isotopic evolution of the biogeochemical carbon cycle during the Proterozoic Eon. Org. Geochem. **27**(5–6), 185–193 (1997)

D.J. Des Marais, Isotopic evolution of the biogeochemical carbon cycle during the precambrian. Rev. Mineral. Geochem. **43**(1), 555–578 (2001)

P. Dhakal, C.J. Matocha, F.E. Huggins, M.M. Vandiviere, Nitrite reactivity with magnetite. Environ. Sci. Technol. **47**(12), 6206–6213 (2013)

G.R. Dickens, Rethinking the global carbon cycle with a large, dynamic and microbially mediated gas hydrate capacitor. Earth Planet. Sci. Lett. **213**(3–4), 169–183 (2003)

A.G. Dickson, C. Goyet, Handbook of methods for the analysis of the various parameters of the carbon dioxide system in sea water. U. S. Department of Energy **1994(September):176**, 187 (1994)

T.A. Doane, The abiotic nitrogen cycle. ACS Earth Space Chem. **1**(7), 411–421 (2017)
S.D. Domagal-Goldman, J.F. Kasting, D.T. Johnston, J. Farquhar, Organic haze, glaciations and multiple sulfur isotopes in the mid-Archean era. Earth Planet. Sci. Lett. **269**(1–2), 29–40 (2008)
H.L. Drake, J.M. Akagi, Characterization of a novel thiosulfate-forming enzyme isolated from Desulfovibrio vulgaris. J. Bacteriol. **132**(1), 132–138 (1977)
J.L. Eigenbrode, K.H. Freeman, Late Archean rise of aerobic microbial ecosystems. Proc. Natl. Acad. Sci. USA **103**(43), 15759–15764 (2006)
D.L. Eldridge, J. Farquhar, Rates and multiple sulfur isotope fractionations associated with the oxidation of sulfide by oxygen in aqueous solution. Geochim. Cosmochim. Acta **237**, 240–260 (2018)
Y. Endo, S.O. Danielache, Y. Ueno, S. Hattori, M.S. Johnson, N. Yoshida, H.G. Kjaergaard, Photoabsorption cross-section measurements of ^{32}S, ^{33}S, ^{34}S and ^{36}S sulfur dioxide from 190 to 220 nm. J. Geophys. Res., Atmos. **120**(6), 2546–2557 (2015)
Y. Endo, Y. Ueno, S. Aoyama, S.O. Danielache, Sulfur isotope fractionation by broadband uv radiation to optically thin so2 under reducing atmosphere. Earth Planet. Sci. Lett. **453**, 9–22 (2016)
G. Etiope, K.R. Lassey, R.W. Klusman, E. Boschi, Reappraisal of the fossil methane budget and related emission from geologic sources. Geophys. Res. Lett. **35**(9), L09307 (2008)
P.G. Falkowski, *The Biological and Geological Contingencies for the Rise of Oxygen on Earth* (2011)
P. Falkowski, R.J. Scholes, E. Boyle, J. Canadell, D. Canfield, J. Elser, N. Gruber, K. Hibbard, P. Hogberg, S. Linder, F.T. Mackenzie, B. Moore, T. Pedersen, Y. Rosental, S. Seitzinger, V. Smetacek, W. Steffen, The global carbon cycle: a test of our knowledge of Earth as a system. Science **290**(5490), 291–296 (2000)
J. Farquhar, M.H. Thiemens, Oxygen cycle of the Martian atmosphere-regolith system: $\Delta^{17}O$ of secondary phases in Nakhla and Lafayette. J. Geophys. Res., Planets **105**(E5), 11991–11997 (2000)
G.D. Farquhar, M.H. O'Leary, J.A. Berry, On the relationship between carbon isotope discrimination and the intercellular carbon dioxide concentration in leaves. Aust. J. Plant Physiol. **9**(2), 121–137 (1982)
G.D. Farquhar, J.R. Ehleringer, K.T. Hubick, Carbon isotope discrimination and photosynthesis. Annu. Rev. Plant Physiol. Plant Mol. Biol. **40**(1), 503–537 (1989)
J. Farquhar, H. Bao, M. Thiemens, Atmospheric influence of Earth's earliest sulfur cycle. Science **289**(5480), 756–758 (2000)
J. Farquhar, J. Savarino, S. Airieau, M.H. Thiemens, Observation of wavelength-sensitive mass-independent sulfur isotope effects during SO_2 photolysis: implications for the early atmosphere. J. Geophys. Res., Planets **106**(E12), 32829–32839 (2001)
J. Farquhar, D.T. Johnston, B.A. Wing, K.S. Habicht, D.E. Canfield, S. Airieau, M.H. Thiemens, Multiple sulphur isotopic interpretations of biosynthetic pathways: implications for biological signatures in the sulphur isotope record. Geobiology **1**(1), 27–36 (2003)
J. Farquhar, D.T. Johnston, B.A. Wing, Implications of conservation of mass effects on mass-dependent isotope fractionations: influence of network structure on sulfur isotope phase space of dissimilatory sulfate reduction. Geochim. Cosmochim. Acta **71**(24), 5862–5875 (2007)
H.A. Favre, W.H. Powell, *Nomenclature of Organic Chemistry* (Royal Chem. Soc., London, 2014)
K. Fennel, M. Follows, P.G. Falkowski, The co-evolution of the nitrogen, carbon and oxygen cycles in the proterozoic ocean. Am. J. Sci. **305**(6-8 SPEC. ISS.), 526–545 (2005)
J.G. Ferry, How to make a living by exhaling methane. Annu. Rev. Microbiol. **64**(1), 453–473 (2010)
V. Fichtner, H. Strauss, A. Immenhauser, D. Buhl, R.D. Neuser, A. Niedermayr, Diagenesis of carbonate associated sulfate. Chem. Geol. **463**, 61–75 (2017)
D.A. Fike, A.S. Bradley, C.V. Rose, Rethinking the ancient sulfur cycle. Annu. Rev. Earth Planet. Sci. **43**(1), 593–622 (2015)
K. Finster, Microbiological disproportionation of inorganic sulfur compounds. J. Sulfur Chem. **29**(3–4), 281–292 (2008)
W.W. Fischer, J. Hemp, J.E. Johnson, Evolution of oxygenic photosynthesis. Annu. Rev. Earth Planet. Sci. **44**(1), 647–683 (2016)
D. Fowler, M. Coyle, U. Skiba, M.A. Sutton, J.N. Cape, S. Reis, L.J. Sheppard, A. Jenkins, B. Grizzetti, J.N. Galloway, P. Vitousek, A. Leach, A.F. Bouwman, K. Butterbach-Bahl, F. Dentener, D. Stevenson, M. Amann, M. Voss, The global nitrogen cycle in the twenty-first century. Philos. Trans. R. Soc. Lond. B, Biol. Sci. **368**(1621), 20130164 (2013)
H.B. Franz, A.C. McAdam, D.W. Ming, C. Freissinet, P.R. Mahaffy, D.L. Eldridge, W.W. Fischer, J.P. Grotzinger, C.H. House, J.A. Hurowitz, S.M. McLennan, S.P. Schwenzer, D.T. Vaniman, P.D. Archer, S.K. Atreya, P.G. Conrad, J.W. Dottin, J.L. Eigenbrode, K.A. Farley, D.P. Glavin, S.S. Johnson, C.A. Knudson, R.V. Morris, R. Navarro-González, A.A. Pavlov, R. Plummer, E.B. Rampe, J.C. Stern, A. Steele, R.E. Summons, B. Sutter, Large sulfur isotope fractionations in Martian sediments at Gale crater. Nat. Geosci. **10**(9), 658–662 (2017)

T. Freudenthal, T. Wagner, F. Wenzhöfer, M. Zabel, G. Wefer, Early diagenesis of organic matter from sediments of the Eastern subtropical Atlantic: evidence from stable nitrogen and carbon isotopes. Geochim. Cosmochim. Acta **65**(11), 1795–1808 (2001)

C. Frey, J.W. Dippner, M. Voss, Close coupling of N-cycling processes expressed in stable isotope data at the redoxcline of the Baltic Sea. Glob. Biogeochem. Cycles **28**(9), 974–991 (2014)

N.U. Frigaard, C. Dahl, Sulfur metabolism in phototrophic sulfur bacteria. Adv. Microb. Physiol. **54**, 103–200 (2008)

E.D. Galbraith, M. Kienast, A.L. Albuquerque, M.A. Altabet, F. Batista, D. Bianchi, S.E. Calvert, S. Contreras, X. Crosta, R. De Pol-Holz, N. Dubois, J. Etourneau, R. Francois, T.C. Hsu, T. Ivanochko, S.L. Jaccard, S.J. Kao, T. Kiefer, S. Kienast, M.F. Lehmann, P. Martinez, M. McCarthy, A.N. Meckler, A. Mix, J. Möbius, T.F. Pedersen, L. Pichevin, T.M. Quan, R.S. Robinson, E. Ryabenko, A. Schmittner, R. Schneider, A. Schneider-Mor, M. Shigemitsu, D. Sinclair, C. Somes, A.S. Studer, J.E. Tesdal, R. Thunell, J.Y. Terence Yang, The acceleration of oceanic denitrification during deglacial warming. Nat. Geosci. **6**(7), 579–584 (2013)

J.N. Galloway, D.G. Capone, E.W. Boyer, R.W. Howarth, S.P. Seitzinger, G.P. Asner, C.C. Cleveland, P.A. Green, E.A. Holland, D.M. Karl, A.F. Michales, J.H. Porter, A.R. Townsend, C.J. Vorosmarty, Nitrogen cycles: past, present, and future. Biogeochemistry **70**(153), 226 (2004)

R.M. Garrels, A. Lerman, Coupling of the sedimentary sulfur and carbon cycles; an improved model. Am. J. Sci. **284**(9), 989–1007 (1984)

J. Garvin, R. Buick, A.D. Anbar, G.L. Arnold, A.J. Kaufman, Isotopic evidence for an aerobic nitrogen cycle in the latest Archean. Science **323**(5917), 1045–1048 (2009)

B.C. Gill, T.W. Lyons, M.R. Saltzman, Parallel, high-resolution carbon and sulfur isotope records of the evolving Paleozoic marine sulfur reservoir. Palaeogeogr. Palaeoclimatol. Palaeoecol. **256**(3–4), 156–173 (2007)

B.C. Gill, T.W. Lyons, T.D. Frank, Behavior of carbonate-associated sulfate during meteoric diagenesis and implications for the sulfur isotope paleoproxy. Geochim. Cosmochim. Acta **72**(19), 4699–4711 (2008)

L.V. Godfrey, P.G. Falkowski, The cycling and redox state of nitrogen in the archaean ocean. Nat. Geosci. **2**(10), 725–729 (2009)

T.P. Goldstein, Z. Aizenshtat, Thermochemical sulfate reduction a review. J. Therm. Anal. **42**(1), 241–290 (1994)

K.C. Grabb, C. Buchwald, C.M. Hansel, S.D. Wankel, A dual nitrite isotopic investigation of chemodenitrification by mineral-associated Fe(II) and its production of nitrous oxide. Geochim. Cosmochim. Acta **196**, 388–402 (2017)

J. Granger, D.M. Sigman, M.F. Lehmann, P.D. Tortell, Nitrogen and oxygen isotope fractionation during dissimilatory nitrate reduction by denitrifying bacteria. Limnol. Oceanogr. **53**(6), 2533–2545 (2008)

E.M. Griffith, A. Paytan, Barite in the ocean - occurrence, geochemistry and palaeoceanographic applications. Sedimentology **59**(6), 1817–1835 (2012)

N. Gruber, The marine nitrogen cycle: overview and challenges, in *Nitrogen in the Marine Environment* (Elsevier, Amsterdam, 2008), pp. 1–50

A.P. Gumsley, K.R. Chamberlain, W. Bleeker, U. Söderlund, M.O. De Kock, E.R. Larsson, A. Bekker, Timing and tempo of the great oxidation event. Proc. Natl. Acad. Sci. USA **114**(8), 1811–1816 (2017)

K.S. Habicht, D.E. Canfield, J. Rethmeier, Sulfur isotope fractionation during bacterial reduction and disproportionation of thiosulfate and sulfite. Geochim. Cosmochim. Acta **62**(15), 2585–2595 (1998)

K.S. Habicht, M. Gade, B. Thamdrup, P. Berg, D.E. Canfield, Calibration of sulfate levels in the Archean ocean. Science **298**(5602), 2372–2374 (2002)

I. Halevy, Production, preservation, and biological processing of mass-independent sulfur isotope fractionation in the Archean surface environment. Proc. Natl. Acad. Sci. **110**(44), 17644–17649 (2013)

I. Halevy, D.T. Johnston, D.P. Schrag, Explaining the structure of the Archean mass-independent sulfur isotope record. Science **329**(5988), 204–207 (2010)

I. Halevy, S.E. Peters, W.W. Fischer, Sulfate burial constraints on the Phanerozoic sulfur cycle. Science **337**(6092), 331–334 (2012)

H.C.B. Hansen, C.B. Koch, H. Nancke-Krogh, O.K. Borggaard, J. Sørensen, Abiotic nitrate reduction to ammonium: key role of green rust. Environ. Sci. Technol. **30**(6), 2053–2056 (1996)

A.K. Hardison, C.K. Algar, A.E. Giblin, J.J. Rich, Influence of organic carbon and nitrate loading on partitioning between dissimilatory nitrate reduction to ammonium (DNRA) and N_2 production. Geochim. Cosmochim. Acta **164**, 146–160 (2015)

C.E. Harman, A.A. Pavlov, D. Babikov, J.F. Kasting, Chain formation as a mechanism for mass-independent fractionation of sulfur isotopes in the Archean atmosphere. Earth Planet. Sci. Lett. **496**, 238–247 (2018)

D.L. Hartmann, A.M. Klein Tank, M. Rusticucci, L.V. Alexander, S. Brönnimann, Y.A.R. Charabi, F.J. Dentener, E.J. Dlugokencky, D.R. Easterling, A. Kaplan, B.J. Soden, P.W. Thorne, M. Wild, P. Zhai, Observations: atmosphere and surface, in *Climate Change 2013 the Physical Science Basis: Working Group I*

Contribution to the Fifth Assessment Report of the Intergovernmental Panel on Climate Change (2013), pp. 159–254
J.M. Hayes, Global methanotrophy of the Archean-proterozoic transition, in *Early Life on Earth, Nobel Symposium No. 84* (Columbia University Press, New York, 1994), pp. 220–235
J.M. Hayes, Fractionation of carbon and hydrogen isotopes in biosynthetic processes, in *Stable Isotope Geochemistry*, vol. 43 (2001), pp. 225–277
J.M. Hayes, J.R. Waldbauer, The carbon cycle and associated redox processes through time. Philos. Trans. R. Soc. Lond. B, Biol. Sci. **361**(1470), 931–950 (2006)
J.M. Hayes, H. Strauss, A.J. Kaufman, The abundance of ^{13}C in marine organic matter and isotopic fractionation in the global biogeochemical cycle of carbon during the past 800 Ma. Chem. Geol. **161**(1), 103–125 (1999)
J.D. Hays, J. Imbrie, N.J. Shackleton, *Variations in the Earth's Orbit: Pacemaker of the Ice Ages* (1976)
J. Heil, H. Vereecken, N. Brüggemann, A review of chemical reactions of nitrification intermediates and their role in nitrogen cycling and nitrogen trace gas formation in soil. Eur. J. Soil Sci. **67**(1), 23–39 (2016)
S.P. Hesselbo, D.R. Gröcke, H.C. Jenkyns, C.J. Bjerrum, P. Farrimond, H.S. Morgans Bell, O.R. Green, Massive dissociation of gas hydrate during a Jurassic oceanic anoxic event. Nature **406**(6794), 392–395 (2000)
M.B. Higgins, R.S. Robinson, J.M. Husson, S.J. Carter, A. Pearson, Dominant eukaryotic export production during ocean anoxic events reflects the importance of recycled NH_4^+. Proc. Natl. Acad. Sci. USA **109**(7), 2269–2274 (2012)
K.U. Hinrichs, R.E. Summons, V. Orphan, S.P. Sylva, J.M. Hayes, Molecular and isotopic analysis of anaerobic methaneoxidizing communities in marine sediments. Org. Geochem. **31**(12), 1685–1701 (2000)
K.U. Hinrichs, J.M. Hayes, W. Bach, A.J. Spivackl, L.R. Hmelo, N.G. Holm, C.G. Johnson, S.P. Sylva, Biological formation of ethane and propane in the deep marine subsurface. Proc. Natl. Acad. Sci. USA **103**(40), 14684–14689 (2006)
M.P. Hoch, M.L. Fogel, D.L. Kirchman, Isotope fractionation associated with ammonium uptake by a marine bacterium. Limnol. Oceanogr. **37**(7), 1447–1459 (1992)
M.S. Hodgskiss, P.W. Crockford, Y. Peng, B.A. Wing, T.J. Horner, A productivity collapse to end Earth's great oxidation. Proc. Natl. Acad. Sci. USA **116**(35), 17207–17212 (2019)
P.F. Hoffman, The great oxidation and a siderian snowball Earth: MIF-S based correlation of paleoproterozoic glacial epochs. Chem. Geol. **362**, 143–156 (2013)
P.F. Hoffman, A.J. Kaufman, G.P. Halverson, D.P. Schrag, A neoproterozoic snowball Earth. Science **281**(5381), 1342–1346 (1998)
B.M. Hoffman, D. Lukoyanov, Z.Y. Yang, D.R. Dean, L.C. Seefeldt, Mechanism of nitrogen fixation by nitrogenase: the next stage. Chem. Rev. **114**(8), 4041–4062 (2014)
T. Höink, A. Lenardic, A.M. Jellinek, Earth's thermal evolution with multiple convection modes: a Monte-Carlo approach. Phys. Earth Planet. Inter. **221**, 22–26 (2013)
H.D. Holland, *The Chemical Evolution of the Atmosphere and Oceans* (Princeton University Press, Princeton, 1984)
H.D. Holland, The oxygenation of the atmosphere and oceans. Philos. Trans. R. Soc. Lond. B, Biol. Sci. **361**(1470), 903–915 (2006)
B.D. Holt, R. Kumar, P.T. Cunningham, Primary sulfates in atmospheric sulfates: estimation by oxygen isotope ratio measurements. Science **217**(4554), 51–53 (1982)
B. Hönisch, N.G. Hemming, D. Archer, M. Siddall, J.F. McManus, Atmospheric carbon dioxide concentration across the mid-Pleistocene transition. Science **324**(5934), 1551–1554 (2009)
C.H. House, J.W. Schopf, K.D. McKeegan, C.D. Coath, Carbon Isotopic Composition of Individual Precambrian Microfossils. Geology **28**(8), 707–710 (2000)
E. Ingall, R. Jahnke, Evidence for enhanced phosphorus regeneration from marine sediments overlain by oxygen depleted waters. Geochim. Cosmochim. Acta **58**(11), 2571–2575 (1994)
G. Izon, A.L. Zerkle, I. Zhelezinskaia, J. Farquhar, R.J. Newton, S.W. Poulton, J.L. Eigenbrode, M.W. Claire, Multiple oscillations in Neoarchaean atmospheric chemistry. Earth Planet. Sci. Lett. **431**, 264–273 (2015)
G. Izon, A.L. Zerkle, K.H. Williford, J. Farquhar, S.W. Poulton, M.W. Claire, Biological regulation of atmospheric chemistry en route to planetary oxygenation. Proc. Natl. Acad. Sci. **114**(13), E2571–E2579 (2017)
O. Jagoutz, F.A. Macdonald, L. Royden, Low-latitude arc-continent collision as a driver for global cooling. Proc. Natl. Acad. Sci. USA **113**(18), 4935–4940 (2016)
J.P. Jasper, J.M. Hayes, A carbon isotope record of CO_2 levels during the late quaternary. Nature **347**(6292), 462–464 (1990)
H.C. Jenkyns, Geochemistry of oceanic anoxic events. Geochem. Geophys. Geosyst. **11**(3), Q03004 (2010)
B. Johnson, C. Goldblatt, The nitrogen budget of Earth. Earth-Sci. Rev. **148**, 150–173 (2015)

D.T. Johnston, Multiple sulfur isotopes and the evolution of Earth's surface sulfur cycle. Earth-Sci. Rev. **106**(1–2), 161–183 (2011)
D.T. Johnston, J. Farquhar, B.A. Wing, A.J. Kaufman, D.E. Canfield, K.S. Habicht, Multiple sulfur isotope fractionations in biological systems: a case study with sulfate reducers and sulfur disproportionators. Am. J. Sci. **305**(6-8 SPEC. ISS.), 645–660 (2005)
D.T. Johnston, B.C. Gill, A. Masterson, E. Beirne, K.L. Casciotti, A.N. Knapp, W. Berelson, Placing an upper limit on cryptic marine sulphur cycling. Nature **513**(7519), 530–533 (2014)
B.B. Jørgensen, Mineralization of organic matter in the sea bed - the role of sulphate reduction. Nature **296**(5858), 643–645 (1982)
B.B. Jørgensen, A thiosulfate shunt in the sulfur cycle of marine sediments. Science **249**(4965), 152–154 (1990)
B.B. Jørgensen, D.J. Des Marais, Competition for sulfide among colorless and purple sulfur bacteria in cyanobacterial mats. FEMS Microbiol. Lett. **38**(3), 179–186 (1986)
B.B. Jørgensen, A.J. Findlay, A. Pellerin, The biogeochemical sulfur cycle of marine sediments. Front. Microbiol. **10**, 849 (2019)
C.K. Junium, M.A. Arthur, Nitrogen cycling during the Cretaceous, Cenomanian- Turonian Oceanic Anoxic Event II. Geochem. Geophys. Geosyst. **8**(3), Q03002 (2007)
C.K. Junium, M.A. Arthur, K.H. Freeman, Compound-specific δ15N and chlorin preservation in surface sediments of the Peru Margin with implications for ancient bulk δ^{15}N records. Geochim. Cosmochim. Acta **160**, 306–318 (2015)
C.K. Junium, A.J. Dickson, B.T. Uveges, Perturbation to the nitrogen cycle during rapid Early Eocene global warming. Nat. Commun. **9**, 1 (2018)
S. Kadoya, D.C. Catling, Constraints on hydrogen levels in the Archean atmosphere based on detrital magnetite. Geochim. Cosmochim. Acta **262**, 207–219 (2019)
M.J. Kampschreur, R. Kleerebezem, W.W. de Vet, M.C. Van Loosdrecht, Reduced iron induced nitric oxide and nitrous oxide emission. Water Res. **45**(18), 5945–5952 (2011)
A. Kampschulte, H. Strauss, The sulfur isotopic evolution of Phanerozoic seawater based on the analysis of structurally substituted sulfate in carbonates. Chem. Geol. **204**(3–4), 255–286 (2004)
A. Kampschulte, P. Bruckschen, H. Strauss, The sulphur isotopic composition of trace sulphates in carboniferous brachiopods: implications for coeval seawater, correlation with other geochemical cyles and isotope stratigraphy. Chem. Geol. **175**(1–2), 149–173 (2001)
J.A. Karhu, H.D. Holland, Carbon isotopes and the rise of atmospheric oxygen. Geology **24**(10), 867–870 (1996)
A.J. Kaufman, A.H. Knoll, Neoproterozoic variations in the C-isotopic composition of seawater: stratigraphic and biogeochemical implications. Precambrian Res. **73**(1–4), 27–49 (1995)
C.D. Keeling, The Suess effect: ^{13}Carbon-^{14}Carbon interrelations. Environ. Int. **2**(4–6), 229–300 (1979)
R.F. Keeling, S.C. Piper, M. Heimann, Global and hemispheric CO_2 sinks deduced from changes in atmospheric O_2 concentration. Nature **381**(6579), 218–221 (1996)
S. Kirschke, P. Bousquet, P. Ciais, M. Saunois, J.G. Canadell, E.J. Dlugokencky, P. Bergamaschi, D. Bergmann, D.R. Blake, L. Bruhwiler, P. Cameron-Smith, S. Castaldi, F. Chevallier, L. Feng, A. Fraser, M. Heimann, E.L. Hodson, S. Houweling, B. Josse, P.J. Fraser, P.B. Krummel, J.F. Lamarque, R.L. Langenfelds, C. Le Quéré, V. Naik, S. O'doherty, P.I. Palmer, I. Pison, D. Plummer, B. Poulter, R.G. Prinn, M. Rigby, B. Ringeval, M. Santini, M. Schmidt, D.T. Shindell, I.J. Simpson, R. Spahni, L.P. Steele, S.A. Strode, K. Sudo, S. Szopa, G.R. Van Der Werf, A. Voulgarakis, M. Van Weele, R.F. Weiss, J.E. Williams, G. Zeng, *Three Decades of Global Methane Sources and Sinks* (2013)
J.L. Kirschvink, Late proterozoic low-latitude global glaciation: the snowball Earth, in *The Proterozoic Biosphere: A Multidisciplinary Study*, vol. 52, ed. by J.W. Schopf, C. Klein (1992), pp. 51–52, Chap. 2.3
J.L. Kirschvink, R.E. Kopp, Palaeoproterozoic ice houses and the evolution of oxygen-mediating enzymes: the case for a late origin of photosystem II. Philos. Trans. R. Soc. Lond. B, Biol. Sci. **363**(1504), 2755–2765 (2008)
J.L. Kirschvink, E.J. Gaidos, L.E. Bertani, N.J. Beukes, J. Gutzmer, L.N. Maepa, R.E. Steinberger, Paleoproterozoic snowball Earth: extreme climatic and geochemical global change and its biological consequences. Proc. Natl. Acad. Sci. USA **97**(4), 1400–1405 (2000)
N. Klueglein, A. Kappler, Abiotic oxidation of Fe(II) by reactive nitrogen species in cultures of the nitrate-reducing Fe(II) oxidizer Acidovorax sp. BoFeN1 - questioning the existence of enzymatic Fe(II) oxidation. Geobiology **11**(2), 180–190 (2013)
A.N. Knapp, The sensitivity of marine N_2 fixation to dissolved inorganic nitrogen. Front. Microbiol. **3**, 374 (2012)
K. Knittel, A. Boetius, Anaerobic oxidation of methane: progress with an unknown process. Annu. Rev. Microbiol. **63**(1), 311–334 (2009)

M.C. Koehler, E.E. Stüeken, M.A. Kipp, R. Buick, A.H. Knoll, Spatial and temporal trends in precambrian nitrogen cycling: a mesoproterozoic offshore nitrate minimum. Geochim. Cosmochim. Acta **198**, 315–337 (2017)

K. Koop-Jakobsen, A.E. Giblin, The effect of increased nitrate loading on nitrate reduction via denitrification and DNRA in salt marsh sediments. Limnol. Oceanogr. **55**(2), 789–802 (2010)

R.E. Kopp, J.L. Kirschvink, I.A. Hilburn, C.Z. Nash, The paleoproterozoic snowball Earth: a climate disaster triggered by the evolution of oxygenic photosynthesis. Proc. Natl. Acad. Sci. USA **102**(32), 11131–11136 (2005)

J. Korenaga, in *Archean Geodynamics and the Thermal Evolution of Earth*. Geophysical Monograph Series, vol. 164 (Blackwell Publishing Ltd., Oxford, 2006), pp. 7–32

S. Kotelnikova, Microbial production and oxidation of methane in deep subsurface. Earth-Sci. Rev. **58**(3–4), 367–395 (2002)

B. Kraft, H.E. Tegetmeyer, R. Sharma, M.G. Klotz, T.G. Ferdelman, R.L. Hettich, J.S. Geelhoed, M. Strous, The environmental controls that govern the end product of bacterial nitrate respiration. Science **345**(6197), 676–679 (2014)

J. Krissansen-Totton, R. Buick, D.C. Catling, A statistical analysis of the carbon isotope record from the Archean to Phanerozoic and implications for the rise of oxygen. Am. J. Sci. **315**(4), 275–316 (2015)

J. Krissansen-Totton, S. Olson, D.C. Catling, Disequilibrium biosignatures over Earth history and implications for detecting exoplanet life. Sci. Adv. **4**(1), eaao5747 (2018)

P.P. Kumar, G.K. Manohar, S.S. Kandalgaonkar, Global distribution of nitric oxide produced by lightning and its seasonal variation. J. Geophys. Res. **100**(D6), 11203–11208 (1995)

L.R. Kump, M.E. Barley, Increased subaerial volcanism and the rise of atmospheric oxygen 2.5 billion years ago. Nature **448**(7157), 1033–1036 (2007)

L.R. Kump, R.M. Garrels, Modeling atmospheric O_2 in the global sedimentary redox cycle. Am. J. Sci. **286**(5), 337–360 (1986)

L.R. Kump, C. Junium, M.A. Arthur, A. Brasier, A. Fallick, V. Melezhik, A. Lepland, A.E. Črne, G. Luo, Isotopic evidence for massive oxidation of organic matter following the great oxidation event. Science **334**(6063), 1694–1696 (2011)

M. Kusakabe, B.W. Robinson, Oxygen and sulfur isotope equilibria in the $BaSO_4$-HSO_4-H_2O system from 110 to 350C and applications. Geochim. Cosmochim. Acta **41**(8), 1033–1040 (1977)

M.M. Kuypers, Y. van Breugel, S. Schouten, E. Erba, J.S. Damsté, N_2-fixing cyanobacteria supplied nutrient N for Cretaceous oceanic anoxic events. Geology **32**(10), 853–856 (2004)

T.A. Laakso, D.P. Schrag, A small marine biosphere in the proterozoic. Geobiology **17**(2), 161–171 (2019)

H. Lammer, A.L. Zerkle, S. Gebauer, N. Tosi, L. Noack, M. Scherf, E. Pilat-Lohinger, M. Güdel, J.L. Grenfell, M. Godolt, A. Nikolaou, Origin and evolution of the atmospheres of early Venus, Earth and Mars. Astron. Astrophys. Rev. **26**(1), 2 (2018)

M. Laneuville, M. Kameya, H.J. Cleaves, Earth without life: a systems model of a global abiotic nitrogen cycle. Astrobiology **18**(7), 897–914 (2018)

W.D. Leavitt, I. Halevy, A.S. Bradley, D.T. Johnston, Influence of sulfate reduction rates on the Phanerozoic sulfur isotope record. Proc. Natl. Acad. Sci. USA **110**(28), 11244–11249 (2013)

C.C.-W. Lee, M.H. Thiemens, The $\delta^{17}O$ and $\delta^{18}O$ measurements of atmospheric sulfate from a coastal and high Alpine region: a mass-independent isotopic anomaly. J. Geophys. Res., Atmos. **106**(D15), 17359–17373 (2001)

H. Lee, J.D. Muirhead, T.P. Fischer, C.J. Ebinger, S.A. Kattenhorn, Z.D. Sharp, G. Kianji, Massive and prolonged deep carbon emissions associated with continental rifting. Nat. Geosci. **9**(2), 145–149 (2016)

C.-T.A. Lee, H. Jiang, R. Dasgupta, M. Torres, A framework for understanding whole-Earth carbon cycling, in *Deep Carbon* (Cambridge University Press, Cambridge, 2019), pp. 313–357

N. Lehnert, H.T. Dong, J.B. Harland, A.P. Hunt, C.J. White, Reversing nitrogen fixation. Nat. Rev. Chem. **2**(10), 278–289 (2018)

S.L. Lewis, M.A. Maslin, Defining the Anthropocene. Nature **519**(7542), 171–180 (2015)

G.N. Lewis, M. Randall, *Thermodynamics and the Free Energy of Chemical Substances* (McGraw-Hill, New York, 1923)

L.E. Lisiecki, M.E. Raymo, A Pliocene-Pleistocene stack of 57 globally distributed benthic $\delta^{18}O$ records. Paleoceanography **20**(1), 1–17 (2005)

P. Liu, C.E. Harman, J.F. Kasting, Y. Hu, J. Wang, Can organic haze and O_2 plumes explain patterns of sulfur mass-independent fractionation during the Archean? Earth Planet. Sci. Lett. **526**, 115767 (2019)

R.M. Lloyd, Oxygen isotope behavior in the sulfate-water system. J. Geophys. Res. **73**(18), 6099–6110 (1968)

G. Luo, S. Ono, N.J. Beukes, D.T. Wang, S. Xie, R.E. Summons, Rapid oxygenation of Earth's atmosphere 2.33 billion years ago. Sci. Adv. **2**(5), 1–10 (2016)

M. Luo, Z.Q. Chen, G.R. Shi, X. Feng, H. Yang, Y. Fang, Y. Li, Microbially induced sedimentary structures (MISSs) from the Lower Triassic Kockatea Formation, northern Perth Basin, Western Australia: Palaeoenvironmental implications. Palaeogeogr. Palaeoclimatol. Palaeoecol. **519**, 236–247 (2019)

G.W. Luther, B. Sundby, B.L. Lewis, P.J. Brendel, N. Silverberg, Interactions of manganese with the nitrogen cycle: alternative pathways to dinitrogen. Geochim. Cosmochim. Acta **61**(19), 4043–4052 (1997)

D. Lüthi, M. Le Floch, B. Bereiter, T. Blunier, J.M. Barnola, U. Siegenthaler, D. Raynaud, J. Jouzel, H. Fischer, K. Kawamura, T.F. Stocker, High-resolution carbon dioxide concentration record 650,000-800,000 years before present. Nature **453**(7193), 379–382 (2008)

J.R. Lyons, Mass-independent fractionation of sulfur isotopes by isotope-selective photodissociation of so_2. Geophys. Res. Lett. **34**(22), L22811 (2007)

T.W. Lyons, C.T. Reinhard, N.J. Planavsky, The rise of oxygen in Earth's early ocean and atmosphere. Nature **506**(7488), 307–315 (2014)

F.A. Macdonald, N.L. Swanson-Hysell, Y. Park, L. Lisiecki, O. Jagoutz, Arc-continent collisions in the tropics set Earth's climate state. Science **364**(6436), 181–184 (2019)

S.A. Macko, M.L. Fogel, P. Hare, T. Hoering, Isotopic fractionation of nitrogen and carbon in the synthesis of amino acids by microorganisms. Chem. Geol., Isot. Geosci. Sect. **65**(1), 79–92 (1987)

C. Magnabosco, L.H. Lin, H. Dong, M. Bomberg, W. Ghiorse, H. Stan-Lotter, K. Pedersen, T.L. Kieft, E. van Heerden, T.C. Onstott, The biomass and biodiversity of the continental subsurface. Nat. Geosci. **11**(10), 707–717 (2018)

N. Mahmoudi, S.R. Beaupré, A.D. Steen, A. Pearson, Sequential bioavailability of sedimentary organic matter to heterotrophic bacteria. Environ. Microbiol. **19**(7), 2629–2644 (2017)

B.D. Marino, M.B. McElroy, R.J. Salawitch, W.G. Spaulding, Glacial-to-interglacial variations in the carbon isotopic composition of atmospheric CO_2. Nature **357**(6378), 461–466 (1992)

W. Martens-Habbena, W. Qin, R.E. Horak, H. Urakawa, A.J. Schauer, J.W. Moffett, E.V. Armbrust, A.E. Ingalls, A.H. Devol, D.A. Stahl, The production of nitric oxide by marine ammonia-oxidizing archaea and inhibition of archaeal ammonia oxidation by a nitric oxide scavenger. Environ. Microbiol. **17**(7), 2261–2274 (2015)

E. Martin, I. Bindeman, Mass-independent isotopic signatures of volcanic sulfate from three supereruption ash deposits in Lake Tecopa, California. Earth Planet. Sci. Lett. **282**(1), 102–114 (2009)

A.P. Martin, D.J. Condon, A.R. Prave, A. Lepland, A Review of Temporal Constraints for the Palaeoproterozoic Large, Positive Carbonate Carbon Isotope Excursion (the Lomagundi-Jatuli Event). Earth-Sci. Rev. **127**, 242–261 (2013)

E. Mason, M. Edmonds, A.V. Turchyn, Remobilization of crustal carbon may dominate volcanic arc emissions. Science **357**(6348), 290–294 (2017)

A.L. Masterson, J. Farquhar, B.A. Wing, Sulfur mass-independent fractionation patterns in the broadband uv photolysis of sulfur dioxide: pressure and third body effects. Earth Planet. Sci. Lett. **306**(3–4), 253–260 (2011)

A.L. Masterson, B.A. Wing, A. Paytan, J. Farquhar, D.T. Johnston, The minor sulfur isotope composition of Cretaceous and Cenozoic seawater sulfate. Paleoceanography **31**(6), 779–788 (2016)

H.L. McClelland, J. Bruggeman, M. Hermoso, R.E. Rickaby, The origin of carbon isotope vital effects in coccolith calcite. Nat. Commun. **8**, 1–16 (2017)

T.M. McCollom, E.L. Shock, Geochemical constraints on chemolithoautotrophic metabolism by microorganisms in seafloor hydrothermal systems. Geochim. Cosmochim. Acta **61**(20), 4375–4391 (1997)

W.F. McDonough, S.s. Sun, The composition of the Earth. Chem. Geol. **120**(3–4), 223–253 (1995)

N.R. McKenzie, B.K. Horton, S.E. Loomis, D.F. Stockli, N.J. Planavsky, C.T.A. Lee, Continental arc volcanism as the principal driver of icehouse-greenhouse variability. Science **352**(6284), 444–447 (2016)

D.L. McRose, X. Zhang, A.M. Kraepiel, F.M. Morel, Diversity and activity of alternative nitrogenases in sequenced genomes and coastal environments. Front. Microbiol. **8**, 267 (2017)

A.J. Medford, M.C. Hatzell, Photon-driven nitrogen fixation: current progress, thermodynamic considerations, and future outlook. ACS Catal. **7**(4), 2624–2643 (2017)

M.P. Mehta, J.A. Baross, Nitrogen fixation at 92°C by a hydrothermal vent archaeon. Science **314**(5806), 1783–1786 (2006)

E.D. Melton, E.D. Swanner, S. Behrens, C. Schmidt, A. Kappler, The interplay of microbially mediated and abiotic reactions in the biogeochemical Fe cycle. Nat. Rev. Microbiol. **12**(12), 797–808 (2014)

K.M. Meyer, L.R. Kump, Oceanic Euxinia in Earth history: causes and consequences. Annu. Rev. Earth Planet. Sci. **36**(1), 251–288 (2008)

P.J. Mickler, L.A. Stern, J.L. Banner, Large kinetic isotope effects in modern speleothems. Bull. Geol. Soc. Am. **118**(1–2), 65–81 (2006)

J.J. Middelburg, *Marine Carbon Biogeochemistry*. Springer Briefs in Earth System Sciences (Springer, Cham, 2019)

J.V. Mills, G. Antler, A.V. Turchyn, Geochemical evidence for cryptic sulfur cycling in salt Marsh sediments. Earth Planet. Sci. Lett. **453**, 23–32 (2016)
M. Minagawa, E. Wada, Nitrogen isotope ratios of red tide organisms in the East China Sea: a characterization of biological nitrogen fixation. Mar. Chem. **19**(3), 245–259 (1986)
K.D. Monson, J.M. Hayes, Biosynthetic control of the natural abundance of carbon 13 at specific positions within fatty acids in Saccharomyces cerevisiae. Isotopic fractionations in lipid synthesis as evidence for peroxisomal regulation. J. Biol. Chem. **257**(10), 5568–5575 (1982)
M.A. Moran, B.P. Durham, Sulfur metabolites in the pelagic ocean. Nat. Rev. Microbiol. **17**(11), 665–678 (2019)
R.D. Müller, M. Sdrolias, C. Gaina, W.R. Roest, Age, spreading rates, and spreading asymmetry of the world's ocean crust. Geochem. Geophys. Geosyst. **9**(4), 1–19 (2008)
T. Murakami, B. Sreenivas, S.D. Sharma, H. Sugimori, Quantification of atmospheric oxygen levels during the paleoproterozoic using paleosol compositions and iron oxidation kinetics. Geochim. Cosmochim. Acta **75**(14), 3982–4004 (2011)
R. Navarro-González, M.J. Molina, L.T. Molina, Nitrogen fixation by volcanic lightning in the early Earth. Geophys. Res. Lett. **25**(16), 3123–3126 (1998)
R. Navarro-González, C.P. McKay, D.N. Mvondo, A possible nitrogen crisis for archaean life due to reduced nitrogen fixation by lightning. Nature **412**(6842), 61–64 (2001)
A. Neftel, H. Oeschger, J. Schwander, B. Stauffer, R. Zumbrunn, Ice core sample measurements give atmospheric CO2 content during the past 40,000 yr. Nature **295**(5846), 220–223 (1982)
D.W. Nelson, J.M. Bremner, Gaseous products of nitrite decomposition in soils. Soil Biol. Biochem. **2**(3), 203–204 (1970)
E. Noor, A. Flamholz, W. Liebermeister, A. Bar-Even, R. Milo, A note on the kinetics of enzyme action: a decomposition that highlights thermodynamic effects. FEBS Lett. **587**(17), 2772–2777 (2013)
T.F. Oliveira, C. Vonrhein, P.M. Matias, S.S. Venceslau, I.A. Pereira, M. Archer, The crystal structure of Desulfovibrio vulgaris dissimilatory sulfite reductase bound to DsrC provides novel insights into the mechanism of sulfate respiration. J. Biol. Chem. **283**(49), 34141–34149 (2008)
S. Ono, Photochemistry of sulfur dioxide and the origin of mass-independent isotope fractionation in Earth's atmosphere. Annu. Rev. Earth Planet. Sci. **45**, 301–329 (2017)
S. Ono, W.C. Shanks, O.J. Rouxel, D. Rumble, S-33 constraints on the seawater sulfate contribution in modern seafloor hydrothermal vent sulfides. Geochim. Cosmochim. Acta **71**(5), 1170–1182 (2007)
S. Ono, A.R. Whitehill, J.R. Lyons, Contribution of isotopologue self-shielding to sulfur mass-independent fractionation during sulfur dioxide photolysis. J. Geophys. Res., Atmos. **118**(5), 2444–2454 (2013)
V.J. Orphan, C.H. House, K.U. Hinrichs, K.D. McKeegan, E.F. DeLong, Multiple archaeal groups mediate methane oxidation in anoxic cold seep sediments. Proc. Natl. Acad. Sci. USA **99**(11), 7663–7668 (2002)
C.J. Ottley, W. Davison, W.M. Edmunds, Chemical catalysis of nitrate reduction by iron(II). Geochim. Cosmochim. Acta **61**(9), 1819–1828 (1997)
M. Pagani, K.H. Freeman, M.A. Arthur, Late miocene atmospheric CO_2 concentrations and the expansion of C4 grasses. Science **285**(5429), 876–879 (1999)
M. Pagani, J.C. Zachos, K.H. Freeman, B. Tipple, S. Bohaty, Atmospheric science: marked decline in atmospheric carbon dioxide concentrations during the Paleogene. Science **309**(5734), 600–603 (2005)
M. Pagani, K. Caldeira, D. Archer, J.C. Zachos An Ancient Carbon Mystery. Science **341**(5805), 1556–1557 (2006)
M. Pagani, M. Huber, Z. Liu, S.M. Bohaty, J. Henderiks, W. Sijp, S. Krishnan, R.M. DeConto, The role of carbon dioxide during the onset of Antarctic glaciation. Science **334**(6060), 1261–1264 (2011)
H. Palme, H. O'Neill, Cosmochemical estimates of mantle composition, in *Treatise on Geochemistry*, vol. 3, 2nd edn. (2013), pp. 1–39
C.A. Partin, P.M. Sadler, Slow net sediment accumulation sets snowball Earth apart from all younger glacial episodes. Geology **44**(12), 1019–1022 (2016)
V. Pasquier, P. Sansjofre, M. Rabineau, S. Revillon, J. Houghton, D.A. Fike, Pyrite sulfur isotopes reveal glacial-interglacial environmental changes. Proc. Natl. Acad. Sci. USA **114**(23), 5941–5945 (2017)
A.A. Pavlov, J.F. Kasting, Mass-independent fractionation of sulfur isotopes in Archean sediments: strong evidence for an anoxic Archean atmosphere. Astrobiology **2**(1), 27–41 (2002)
A.A. Pavlov, J.F. Kasting, L.L. Brown, K.A. Rages, R. Freedman, Greenhouse warming by CH4 in the atmosphere of early Earth. J. Geophys. Res., Planets **105**(E5), 11981–11990 (2000)
A. Paytan, M. Kastner, D. Campbell, M.H. Thiemens, Sulfur isotopic composition of Cenozoic seawater sulfate. Science **282**(5393), 1459–1462 (1998)
A. Pellerin, G. Antler, S.A. Holm, A.J. Findlay, P.W. Crockford, A.V. Turchyn, B.B. Jørgensen, K. Finster, Large sulfur isotope fractionation by bacterial sulfide oxidation. Sci. Adv. **5**(7), eaaw1480 (2019)
I.A. Pereira, A.R. Ramos, F. Grein, M.C. Marques, S.M. da Silva, S.S. Venceslau, A comparative genomic analysis of energy metabolism in sulfate reducing bacteria and archaea. Front. Microbiol. **2**, 69 (2011)

K.E. Peters, R.E. Sweeney, I.R. Kaplan, Correlation of carbon and nitrogen stable isotope ratios in sedimentary organic matter. Limnol. Oceanogr. **23**(4), 598–604 (1978)

N.J. Planavsky, P. McGoldrick, C.T. Scott, C. Li, C.T. Reinhard, A.E. Kelly, X. Chu, A. Bekker, G.D. Love, T.W. Lyons, Widespread iron-rich conditions in the mid-Proterozoic ocean. Nature **477**(7365), 448–451 (2011)

N.J. Planavsky, A. Bekker, A. Hofmann, J.D. Owens, T.W. Lyons, Sulfur record of rising and falling marine oxygen and sulfate levels during the Lomagundi event. Proc. Natl. Acad. Sci. USA **109**(45), 18300–18305 (2012)

B.N. Popp, E.A. Laws, R.R. Bidigare, J.E. Dore, K.L. Hanson, S.G. Wakeham, Effect of phytoplankton cell geometry on carbon isotopic fractionation. Geochim. Cosmochim. Acta **62**(1), 69–77 (1998)

D. Postma, Kinetics of nitrate reduction by detrital Fe(II)-silicates. Geochim. Cosmochim. Acta **54**(3), 903–908 (1990)

S.W. Poulton, P.W. Fralick, D.E. Canfield, Spatial variability in oceanic redox structure 1.8 billion years ago. Nat. Geosci. **3**(7), 486–490 (2010)

M.J. Prather, C.D. Holmes, J. Hsu, Reactive greenhouse gas scenarios: systematic exploration of uncertainties and the role of atmospheric chemistry. Geophys. Res. Lett. **39**(9), L09803 (2012)

T.M. Present, G. Paris, A. Burke, W.W. Fischer, J.F. Adkins, Large carbonate associated sulfate isotopic variability between brachiopods, micrite, and other sedimentary components in Late Ordovician strata. Earth Planet. Sci. Lett. **432**, 187–198 (2015)

R. Rabus, T.A. Hansen, F. Widdel, Dissimilatory sulfate- and sulfur-reducing prokaryotes, in *The Prokaryotes: Prokaryotic Physiology and Biochemistry* (Springer, Berlin, 2013), pp. 309–404

S.W. Ragsdale, Stealth reactions driving carbon fixation new twists to bacterial metabolic pathways that contribute to the global carbon cycle. Science **359**(6375), 517–518 (2018)

V.A. Rakov, M.A. Uman, *Lightning: Physics and Effects* (Cambridge University Press, Cambridge, 2003)

S. Rakshit, C.J. Matocha, M.S. Coyne, Nitrite reduction by siderite. Soil Sci. Soc. Am. J. **72**(4), 1070–1077 (2008)

B. Rasmussen, A. Bekker, I.R. Fletcher, Correlation of Paleoproterozoic glaciations based on U-Pb zircon ages for tuff beds in the Transvaal and Huronian Supergroups. Earth Planet. Sci. Lett. **382**, 173–180 (2013)

G.H. Rau, M.A. Arthur, W.E. Dean, $^{15}N/^{14}N$ variations in Cretaceous Atlantic sedimentary sequences: implication for past changes in marine nitrogen biogeochemistry. Earth Planet. Sci. Lett. **82**(3–4), 269–279 (1987)

G.H. Rau, U. Riebesell, D. Wolf-Gladrow, A model of photosynthetic ^{13}C fractionation by marine phytoplankton based on diffusive molecular CO_2 uptake. Mar. Ecol. Prog. Ser. **133**(1–3), 275–285 (1996)

M.R. Raven, A.L. Sessions, J.F. Adkins, R.C. Thunell, Rapid organic matter sulfurization in sinking particles from the Cariaco Basin water column. Geochim. Cosmochim. Acta **190**, 175–190 (2016)

M.R. Raven, D.A. Fike, M.L. Gomes, S.M. Webb, A.S. Bradley, H.L.O. McClelland, Organic carbon burial during OAE2 driven by changes in the locus of organic matter sulfurization. Nat. Commun. **9**, 3409 (2018)

W.S. Reeburgh, *Oceanic Methane Biogeochemistry* (2007)

H. Ren, D.M. Sigman, A.N. Meckler, B. Plessen, R.S. Robinson, Y. Rosenthal, G.H. Haug, Foraminiferal isotope evidence of reduced nitrogen fixation in the ice age Atlantic Ocean. Science **323**(5911), 244–248 (2009)

V.C. Rennie, A.V. Turchyn, The preservation of $\delta^{34}S_{SO_4}$ and $\delta^{18}O_{SO_4}$ in carbonate-associated sulfate during marine diagenesis: a 25 Myr test case using marine sediments. Earth Planet. Sci. Lett. **395**, 13–23 (2014)

V.C. Rennie, G. Paris, A.L. Sessions, S. Abramovich, A.V. Turchyn, J.F. Adkins, Cenozoic record of $\delta^{34}S$ in foraminiferal calcite implies an early Eocene shift to deep-ocean sulfide burial. Nat. Geosci. **11**(10), 761–765 (2018)

D. Rickard, G.W. Luther, Chemistry of iron sulfides. Chem. Rev. **107**(2), 514–562 (2007)

B.A. Ridley, J.E. Dye, J.G. Walega, J. Zheng, F.E. Grahek, W. Rison, On the production of active nitrogen by thunderstorms over New Mexico. J. Geophys. Res., Atmos. **101**(15), 20985–21005 (1996)

A.L. Robertson, J. Roadt, I. Halevy, J.F. Kasting, Greenhouse warming by nitrous oxide and methane in the Proterozoic Eon. Geobiology **9**(4), 313–320 (2011)

A. Robock, Volcanic eruptions and climate. Rev. Geophys. **38**(2), 191–219 (2000)

A.B. Romero, M.H. Thiemens, Mass-independent sulfur isotopic compositions in present-day sulfate aerosols. J. Geophys. Res., Atmos. **108**(D16), 4524 (2003)

C.V. Rose, S.M. Webb, M. Newville, A. Lanzirotti, J.A. Richardson, N.J. Tosca, J.G. Catalano, A.S. Bradley, D.A. Fike, Insights into past ocean proxies from micron-scale mapping of sulfur species in carbonates. Geology **47**(9), 833–837 (2019)

D.H. Rothman, J.M. Hayes, R.E. Summons, Dynamics of the Neoproterozoic carbon cycle. Proc. Natl. Acad. Sci. USA **100**(14), 8124–8129 (2003)

M.J. Russell, A.J. Hall, W. Martin, Serpentinization as a source of energy at the origin of life. Geobiology **8**(5), 355–371 (2010)
Z. Sade, I. Halevy, New constraints on kinetic isotope effects during CO_2(aq) hydration and hydroxylation: revisiting theoretical and experimental data. Geochim. Cosmochim. Acta **214**, 246–265 (2017)
G. Saleh, A.R. Oganov, Novel stable compounds in the C-H-O ternary system at high pressure. Sci. Rep. **6**, 32486 (2016)
V.A. Samarkin, M.T. Madigan, M.W. Bowles, K.L. Casciotti, J.C. Priscu, C.P. McKay, S.B. Joye, Abiotic nitrous oxide emission from the hypersaline Don Juan Pond in Antarctica. Nat. Geosci. **3**(5), 341–344 (2010)
A.A. Santos, S.S. Venceslau, F. Grein, W.D. Leavitt, C. Dahl, D.T. Johnston, I.A.C. Pereira, A protein trisulfide couples dissimilatory sulfate reduction to energy conservation. Science **350**(6267), 1541–1545 (2015)
M. Saunois, A.R. Stavert, B. Poulter, P. Bousquet, J.G. Canadell, R.B. Jackson, P.A. Raymond, E.J. Dlugokencky, S. Houweling, P.K. Patra, P. Ciais, V.K. Arora, D. Bastviken, P. Bergamaschi, D.R. Blake, G. Brailsford, L. Bruhwiler, K.M. Carlson, M. Carrol, S. Castaldi, N. Chandra, C. Crevoisier, P.M. Crill, K. Covey, C.L. Curry, G. Etiope, C. Frankenberg, N. Gedney, M.I. Hegglin, L. Höglund-Isaksson, G. Hugelius, M. Ishizawa, A. Ito, G. Janssens-Maenhout, K.M. Jensen, F. Joos, T. Kleinen, P.B. Krummel, R.L. Langenfelds, G.G. Laruelle, L. Liu, T. Machida, S. Maksyutov, K.C. McDonald, J. McNorton, P.A. Miller, J.R. Melton, I. Morino, J. Müller, F. Murguia-Flores, V. Naik, Y. Niwa, S. Noce, S. O'Doherty, R.J. Parker, C. Peng, S. Peng, G.P. Peters, C. Prigent, R. Prinn, M. Ramonet, P. Regnier, W.J. Riley, J.A. Rosentreter, A. Segers, I.J. Simpson, H. Shi, S.J. Smith, L.P. Steele, B.F. Thornton, H. Tian, Y. Tohjima, F.N. Tubiello, A. Tsuruta, N. Viovy, A. Voulgarakis, T.S. Weber, M. van Weele, G.R. van der Werf, R.F. Weiss, D. Worthy, D. Wunch, Y. Yin, Y. Yoshida, W. Zhang, Z. Zhang, Y. Zhao, B. Zheng, Q. Zhu, Q. Zhu, Q. Zhuang, The global methane budget 2000–2017. Earth Syst. Sci. Data **12**(3), 1561–1623 (2020)
J. Savarino, A. Romero, J. Cole-Dai, S. Bekki, M. Thiemens, UV-induced mass-independent sulfur isotope fractionation in stratospheric volcanic sulfate. Geophys. Res. Lett. **30**(21), 2131 (2003)
E.A. Schauble, Applying stable isotope fractionation theory to new systems. Rev. Mineral. Geochem. **55**, 65–111 (2004)
S.O. Schlanger, H.C. Jenkyns, Cretaceous oceanic anoxic events: causes and consequences. Geol. Mijnb. **55**(3–4), 179–184 (1976)
M. Schobben, B. van de Schootbrugge, Increased stability in carbon isotope records reflects emerging complexity of the biosphere. Front. Earth Sci. **7**, 87 (2019)
S. Schouten, H.M. Van Kaam-Peters, W.I.C. Rijpstra, M. Schoell, J.S. Sinninghe Damste, Effects of an oceanic anoxic event on the stable carbon isotopic composition of early Toarcian carbon. Am. J. Sci. **300**(1), 1–22 (2000)
D.P. Schrag, J.A. Higgins, F.A. Macdonald, D.T. Johnston, Authigenic carbonate and the history of the global carbon cycle. Science **339**(6119), 540–543 (2013)
S. Schwietzke, O.A. Sherwood, L.M. Bruhwiler, J.B. Miller, G. Etiope, E.J. Dlugokencky, S.E. Michel, V.A. Arling, B.H. Vaughn, J.W. White, P.P. Tans, Upward revision of global fossil fuel methane emissions based on isotope database. Nature **538**(7623), 88–91 (2016)
A. Sclafani, L. Palmisano, M. Schiavello, N_2 photoreduction and phenol and nitrophenol isomers photooxidation as examples of heterogeneous photocatalytic reactions. Res. Chem. Intermed. **18**(2), 211–226 (1993)
S.P. Seitzinger, J.A. Harrison, E. Dumont, A.H.W. Beusen, A.F. Bouwman, Sources and delivery of carbon, nitrogen, and phosphorus to the coastal zone: an overview of global nutrient export from watersheds (NEWS) models and their application. Glob. Biogeochem. Cycles **19**(4), GB4S05 (2005)
N.J. Shackleton, Oxygen isotopes, ice volume and sea level. Quat. Sci. Rev. **6**(3–4), 183–190 (1987)
L. Shawar, I. Halevy, W. Said-Ahmad, S. Feinstein, V. Boyko, A. Kamyshny, A. Amrani, Dynamics of pyrite formation and organic matter sulfurization in organic-rich carbonate sediments. Geochim. Cosmochim. Acta **241**, 219–239 (2018)
Y. Shen, R. Buick, D.E. Canfield, Isotopic evidence for microbial sulphate reduction in the early archaean era. Nature **410**(6824), 77–81 (2001)
Y. Shen, J. Farquhar, A. Masterson, A.J. Kaufman, R. Buick, Evaluating the role of microbial sulfate reduction in the early Archean using quadruple isotope systematics. Earth Planet. Sci. Lett. **279**(3–4), 383–391 (2009)
S.M. Sievert, C. Vetriani, Chemoautotrophy at deep-sea vents, past, present, and future. Oceanography **25**(1), 218–233 (2012)
D.M. Sigman, E.A. Boyle, Glacial/interglacial variations in atmospheric carbon dioxide. Nature **407**(6806), 859–869 (2000)

D.M. Sigman, K.L. Karsh, K.L. Casciotti, Ocean process tracers: nitrogen isotopes in the ocean, in *Encyclopedia of Ocean Sciences*, ed. by S. Edition, J.H. Steele (Academic Press, Oxford, 2009), pp. 40–54
M.S. Sim, T. Bosak, S. Ono, Large sulfur isotope fractionation does not require disproportionation. Science **333**(6038), 74–77 (2011)
M.S. Sim, S. Ono, T. Bosak, Effects of iron and nitrogen limitation on sulfur isotope fractionation during microbial sulfate reduction. Appl. Environ. Microbiol. **78**(23), 8368–8376 (2012)
J.S. Sinninghe Damste, J.W. De Leeuw, Analysis, structure and geochemical significance of organically-bound sulphur in the geosphere: state of the art and future research. Org. Geochem. **16**(4–6), 1077–1101 (1990)
J.S. Sinninghe Damsté, M.D. Kok, J. Köster, S. Schouten, Sulfurized carbohydrates: an important sedimentary sink for organic carbon? Earth Planet. Sci. Lett. **164**(1–2), 7–13 (1998)
R. Sirevåg, B.B. Buchanan, J.A. Berry, J.H. Troughton, Mechanisms of CO_2 fixation in bacterial photosynthesis studied by the carbon isotope fractionation technique. Arch. Microbiol. **112**(1), 35–38 (1977)
S.P. Slotznick, W.W. Fischer, Examining Archean methanotrophy. Earth Planet. Sci. Lett. **441**, 52–59 (2016)
L.E. Snyder, D. Buhl, B. Zuckerman, P. Palmer, Microwave detection of interstellar formaldehyde. Phys. Rev. Lett. **22**(13), 679–681 (1969)
E. Sofen, B. Alexander, S. Kunasek, The impact of anthropogenic emissions on atmospheric sulfate production pathways, oxidants, and ice core $\Delta^{17}O(SO_4{}^{2-})$. Atmos. Chem. Phys. **11**, 3565–3578 (2011)
E.D. Sofen, B. Alexander, E.J. Steig, M.H. Thiemens, S.A. Kunasek, H.M. Amos, A.J. Schauer, M.G. Hastings, J. Bautista, T.L. Jackson, L.E. Vogel, J.R. Mcconnell, D.R. Pasteris, E.S. Saltzman, WAIS Divide ice core suggests sustained changes in the atmospheric formation pathways of sulfate and nitrate since the 19th century in the extratropical Southern Hemisphere. Atmos. Chem. Phys. **14**(11), 5749–5769 (2014)
R.M. Soo, J. Hemp, P. Hugenholtz, Evolution of photosynthesis and aerobic respiration in the cyanobacteria. Free Radic. Biol. Med. **140**, 200–205 (2019)
J. Sørensen, L. Thorling, Stimulation by lepidocrocite (7-FeOOH) of Fe(II)-dependent nitrite reduction. Geochim. Cosmochim. Acta **55**(5), 1289–1294 (1991)
R. Spahni, R. Wania, L. Neef, M. Van Weele, I. Pison, P. Bousquet, C. Frankenberg, P.N. Foster, F. Joos, I.C. Prentice, P. Van Velthoven, Constraining global methane emissions and uptake by ecosystems. Biogeosciences **8**(6), 1643–1665 (2011)
C.L. Stanton, C.T. Reinhard, J.F. Kasting, N.E. Ostrom, J.A. Haslun, T.W. Lyons, J.B. Glass, Nitrous oxide from chemodenitrification: a possible missing link in the proterozoic greenhouse and the evolution of aerobic respiration. Geobiology **16**(6), 597–609 (2018)
C.J. Still, J.A. Berry, G.J. Collatz, R.S. DeFries, Global distribution of C3 and C4 vegetation: carbon cycle implications. Glob. Biogeochem. Cycles **17**(1), 6-1–6-14 (2003)
D.A. Stolper, C.B. Keller, A record of deep-ocean dissolved O_2 from the oxidation state of iron in submarine basalts. Nature **553**(7688), 323–327 (2018)
K.L. Straub, M. Benz, B. Schink, F. Widdel, Anaerobic, nitrate-dependent microbial oxidation of ferrous iron. Appl. Environ. Microbiol. **62**(4), 1458–1460 (1996)
H. Strauss, D.J. Des Marais, J.M. Hayers, R.E. Summons, Concentrations of organic carbon and maturities and elemental compositions of kerogens, in *The Proterozoic Biosphere, a Multidisciplinary Study* (1992), pp. 95–101
T.O. Strohm, B. Griffin, W.G. Zumft, B. Schink, Growth yields in bacterial denitrification and nitrate ammonification. Appl. Environ. Microbiol. **73**(5), 1420–1424 (2007)
E.E. Stüeken, R. Buick, B.M. Guy, M.C. Koehler, Isotopic evidence for biological nitrogen fixation by molybdenum-nitrogenase from 3.2 Gyr. Nature **520**(7549), 666–669 (2015a)
E.E. Stüeken, R. Buick, A.J. Schauer, Nitrogen isotope evidence for alkaline lakes on late Archean continents. Earth Planet. Sci. Lett. **411**, 1–10 (2015b)
E.E. Stüeken, M.A. Kipp, M.C. Koehler, R. Buick, The evolution of Earth's biogeochemical nitrogen cycle. Earth-Sci. Rev. **160**, 220–239 (2016)
E.E. Stüeken, J. Zaloumis, J. Meixnerová, R. Buick, Differential metamorphic effects on nitrogen isotopes in kerogen extracts and bulk rocks. Geochim. Cosmochim. Acta **217**, 80–94 (2017)
E.E. Stüeken, S.M. Som, M. Claire, S. Rugheimer, M. Scherf, L. Sproß, N. Tosi, Y. Ueno, H. Lammer, Mission to planet Earth: the first two billion years. Space Sci. Rev. **216**(2), 31 (2020)
R.E. Summons, P.D. Franzmann, P.D. Nichols, Carbon isotopic fractionation associated with methylotrophic methanogenesis. Org. Geochem. **28**(7–8), 465–475 (1998)
P.K. Swart, Global synchronous changes in the carbon isotopic composition of carbonate sediments unrelated to changes in the global carbon cycle. Proc. Natl. Acad. Sci. USA **105**(37), 13741–13745 (2008)
P.K. Swart, The geochemistry of carbonate diagenesis: the past, present and future. Sedimentology **62**(5), 1233–1304 (2015)

F. Tabataba-Vakili, J.L. Grenfell, J.-M. Grießmeier, H. Rauer, Atmospheric effects of stellar cosmic rays on Earth-like exoplanets orbiting M-dwarfs. Astron. Astrophys. **585**, A96 (2016)

C. Tamocai, J.G. Canadell, E.A. Schuur, P. Kuhry, G. Mazhitova, S. Zimov, Soil organic carbon pools in the northern circumpolar permafrost region. Glob. Biogeochem. Cycles **23**(2), GB2023 (2009)

G.G. Tcherkez, G.D. Farquhar, T.J. Andrews, Despite slow catalysis and confused substrate specificity, all ribulose bisphosphate carboxylases may be nearly perfectly optimized. Proc. Natl. Acad. Sci. USA **103**(19), 7246–7251 (2006)

K. Tennakone, J.M. Bandara, C.T. Thaminimulla, W.D. Jayatilake, U.S. Ketipearachchi, O.A. Ileperuma, M.K. Priyadarshana, Photoreduction of dinitrogen to ammonia by ultrafine particles of Fe(O)OH formed by photohydrolysis of Iron(II) bicarbonate. Langmuir **7**(10), 2166–2168 (1991)

B. Thamdrup, Bacterial manganese and iron reduction in aquatic sediments, in *Advances in Microbial Ecology*, ed. by B. Schink (Springer, Berlin, 2000), pp. 41–84

B. Thamdrup, K. Finster, J.W. Hansen, F. Bak, Bacterial disproportionation of elemental sulfur coupled to chemical reduction of iron or manganese. Appl. Environ. Microbiol. **59**(1), 101–108 (1993)

B. Thamdrup, K. Finster, H. Fossing, J.W. Hansen, B.B. Jørgensen, Thiosulfate and sulfite distributions in porewater of marine sediments related to manganese, iron, and sulfur geochemistry. Geochim. Cosmochim. Acta **58**(1), 67–73 (1994)

M.H. Thiemens, History and applications of mass-independent isotope effects. Annu. Rev. Earth Planet. Sci. **34**, 217–262 (2006)

C. Thomazo, M. Ader, P. Philippot, Extreme ^{15}N-enrichments in 2.72-Gyr-old sediments: evidence for a turning point in the nitrogen cycle. Geobiology **9**(2), 107–120 (2011)

M.M. Tice, D.R. Lowe, Photosynthetic microbial mats in the 3,416-Myr-old ocean. Nature **431**(7008), 549–552 (2004)

B. Tissot, D. Welte, *Principles of Stable Isotope Distribution* (Springer, Berlin, 1984)

C. Tornabene, R.C. Martindale, X.T. Wang, M.F. Schaller, Detecting photosymbiosis in fossil scleractinian corals. Sci. Rep. **7**(1), 9465 (2017)

A.V. Turchyn, O. Sivan, D. Schrag, Oxygen isotopic composition of sulfate in deep sea pore fluid: evidence for rapid sulfur cycling. Geochim. Cosmochim. Acta **70**(18), A660 (2006)

A.V. Turchyn, V. Brüchert, T.W. Lyons, G.S. Engel, N. Balci, D.P. Schrag, B. Brunner, Kinetic oxygen isotope effects during dissimilatory sulfate reduction: a combined theoretical and experimental approach. Geochim. Cosmochim. Acta **74**(7), 2011–2024 (2010)

P. Van Cappellen, E.D. Ingall, Redox stabilization of the atmosphere and oceans by phosphorus-limited marine productivity. Science **271**(5248), 493–496 (1996)

E.M. van den Berg, J.L. Rombouts, J.G. Kuenen, R. Kleerebezem, M.C. van Loosdrecht, Role of nitrite in the competition between denitrification and DNRA in a chemostat enrichment culture. AMB Express **7**(1), 91 (2017)

M.T. Van Der Meer, S. Schouten, J.S. Sinninghe Damsté, The effect of the reversed tricarboxylic acid cycle on the ^{13}C contents of bacterial lipids. Org. Geochem. **28**(9–10), 527–533 (1998)

M.A. Van Kessel, D.R. Speth, M. Albertsen, P.H. Nielsen, H.J. Op Den Camp, B. Kartal, M.S. Jetten, S. Lücker, Complete nitrification by a single microorganism. Nature **528**(7583), 555–559 (2015)

M. Vandenbroucke, C. Largeau, Kerogen origin, evolution and structure. Org. Geochem. **38**(5), 719–833 (2007)

M. Vargas, K. Kashefi, E.L. Blunt-harris, D.R. Lovley, Fe (III) reduction on early Earth. Nature **395**, 65–67 (1998)

J. Veizer, D. Ala, K. Azmy, P. Bruckschen, D. Buhl, F. Bruhn, G.A. Garden, A. Diener, S. Ebneth, Y. Godderis, T. Jasper, C. Korte, F. Pawellek, O.G. Podlaha, H. Strauss, ^{87}Sr/^{86}Sr, δ^{13}C and δ^{18}O evolution of Phanerozoic seawater. Chem. Geol. **161**(1), 59–88 (1999)

S.S. Venceslau, Y. Stockdreher, C. Dahl, I.A. Pereira, The "bacterial heterodisulfide" DsrC is a key protein in dissimilatory sulfur metabolism. Biochim. Biophys. Acta, Bioenerg. **1837**(7), 1148–1164 (2014)

A.R. Waldeck, B.R. Cowie, E. Bertran, B.A. Wing, I. Halevy, D.T. Johnston, Deciphering the atmospheric signal in marine sulfate oxygen isotope composition. Earth Planet. Sci. Lett. **522**, 12–19 (2019)

J.C. Walker, P.B. Hays, J.F. Kasting, A negative feedback mechanism for the long-term stabilization of Earth's surface temperature. J. Geophys. Res. **86**(C10), 9776–9782 (1981)

X.T. Wang, D.M. Sigman, A.L. Cohen, D.J. Sinclair, R.M. Sherrell, M.A. Weigand, D.V. Erler, H. Ren, Isotopic composition of skeleton-bound organic nitrogen in reef-building symbiotic corals: a new method and proxy evaluation at Bermuda. Geochim. Cosmochim. Acta **148**, 179–190 (2015)

S.D. Wankel, A.S. Bradley, D.L. Eldridge, D.T. Johnston, Determination and application of the equilibrium oxygen isotope effect between water and sulfite. Geochim. Cosmochim. Acta **125**, 694–711 (2014)

L.M. Ward, P.M. Shih, The evolution and productivity of carbon fixation pathways in response to changes in oxygen concentration over geological time. Free Radic. Biol. Med. **140**, 188–199 (2019)

L.M. Ward, B. Rasmussen, W.W. Fischer, Primary productivity was limited by electron donors prior to the advent of oxygenic photosynthesis. J. Geophys. Res., Biogeosci. **124**(2), 211–226 (2019)

M.R. Warke, T.D. Rocco, A.L. Zerkle, A. Lepland, A.R. Prave, A.P. Martin, Y. Ueno, D.J. Condon, M.W. Claire, The great oxidation event preceded a paleoproterozoic "snowball Earth". Proc. Natl. Acad. Sci. **117**(24), 13314–13320 (2020)

M.C. Weiss, F.L. Sousa, N. Mrnjavac, S. Neukirchen, M. Roettger, S. Nelson-Sathi, W.F. Martin, The physiology and habitat of the last universal common ancestor. Nat. Microbiol. **1**(9), 16116 (2016)

C.B. Wenk, B.A. Wing, I. Halevy, Electron carriers in microbial sulfate reduction inferred from experimental and environmental sulfur isotope fractionations. ISME J. **12**(2), 495–507 (2018)

J.P. Werne, D.J. Hollander, A. Behrens, P. Schaeffer, P. Albrecht, J.S. Sinninghe Damsté, Timing of early diagenetic sulfurization of organic matter: a precursor-product relationship in Holocene sediments of the anoxic Cariaco Basin, Venezuela. Geochim. Cosmochim. Acta **64**(10), 1741–1751 (2000)

A.R. Whitehill, S. Ono, Excitation band dependence of sulfur isotope mass-independent fractionation during photochemistry of sulfur dioxide using broadband light sources. Geochim. Cosmochim. Acta **94**, 238–253 (2012)

A.R. Whitehill, C. Xie, X. Hu, D. Xie, H. Guo, S. Ono, Vibronic origin of sulfur mass-independent isotope effect in photoexcitation of SO_2 and the implications to the early Earth's atmosphere. Proc. Natl. Acad. Sci. **110**(44), 17697–17702 (2013)

A.R. Whitehill, B. Jiang, H. Guo, S. Ono, SO_2 photolysis as a source for sulfur mass-independent isotope signatures in stratospheric aerosols. Atmos. Chem. Phys. Discuss. **14**(16), 23499–23554 (2015)

E.B. Wilkes, A. Pearson, A general model for carbon isotopes in red-lineage phytoplankton: interplay between unidirectional processes and fractionation by RubisCO. Geochim. Cosmochim. Acta **265**, 163–181 (2019)

K.H. Williford, T. Ushikubo, K. Lepot, K. Kitajima, C. Hallmann, M.J. Spicuzza, R. Kozdon, J.L. Eigenbrode, R.E. Summons, J.W. Valley, Carbon and sulfur isotopic signatures of ancient life and environment at the microbial scale: neoArchean shales and carbonates. Geobiology **14**(2), 105–128 (2016)

B.A. Wing, I. Halevy, Intracellular metabolite levels shape sulfur isotope fractionation during microbial sulfate respiration. Proc. Natl. Acad. Sci. USA **111**(51), 18116–18125 (2014)

C.R. Witkowski, J.W. Weijers, B. Blais, S. Schouten, J.S. Sinninghe Damsté, Molecular fossils from phytoplankton reveal secular PCO2 trend over the Phanerozoic. Sci. Adv. **4**(11), eaat4556 (2018)

U.G. Wortmann, B. Chernyavsky, S.M. Bernasconi, B. Brunner, M.E. Böttcher, P.K. Swart, Oxygen isotope biogeochemistry of pore water sulfate in the deep biosphere: Dominance of isotope exchange reactions with ambient water during microbial sulfate reduction (ODP Site 1130). Geochim. Cosmochim. Acta **71**(17), 4221–4232 (2007)

T. Wotte, G.A. Shields-Zhou, H. Strauss, Carbonate-associated sulfate: experimental comparisons of common extraction methods and recommendations toward a standard analytical protocol. Chem. Geol. **326**(327), 132–144 (2012)

N. Wu, J. Farquhar, H. Strauss, S.-T. Kim, D.E. Canfield, Evaluating the S-isotope fractionation associated with Phanerozoic pyrite burial. Geochim. Cosmochim. Acta **74**(7), 2053–2071 (2010)

N. Wu, J. Farquhar, H. Strauss, δ^{34}S and δ^{33}S records of Paleozoic seawater sulfate based on the analysis of carbonate associated sulfate. Earth Planet. Sci. Lett. **399**, 44–51 (2014)

J. Yang, C.K. Junium, N.V. Grassineau, E.G. Nisbet, G. Izon, C. Mettam, A. Martin, A.L. Zerkle, Ammonium availability in the late archaean nitrogen cycle. Nat. Geosci. **12**(7), 553–557 (2019)

S. Zaarur, D.T. Wang, S. Ono, T. Bosak, Influence of phosphorus and cell geometry on the fractionation of sulfur isotopes by several species of Desulfovibrio during microbial sulfate reduction. Front. Microbiol. **8**, 890 (2017)

J. Zachos, H. Pagani, L. Sloan, E. Thomas, K. Billups, Trends, rhythms, and aberrations in global climate 65 Ma to present. Science **292**(5517), 686–693 (2001)

K. Zahnle, M. Claire, D. Catling, The loss of mass-independent fractionation in sulfur due to a palaeoproterozoic collapse of atmospheric methane. Geobiology **4**(4), 271–283 (2006)

R.E. Zeebe, Kinetic fractionation of carbon and oxygen isotopes during hydration of carbon dioxide. Geochim. Cosmochim. Acta **139**, 540–552 (2014)

R.E. Zeebe, D. Wolf-Gladrow, in *CO_2 in Seawater: Equilibrium, Kinerics, Isotopes*. Elsevier Oceanography Series, vol. 65 (2001), pp. 1–341

R.E. Zeebe, J. Bijma, D.A. Wolf-Gladrow, A diffusion-reaction model of carbon isotope fractionation in foraminifera. Mar. Chem. **64**(3), 199–227 (1999a)

R.E. Zeebe, D.A. Wolf-Gladrow, H. Jansen, On the time required to establish chemical and isotopic equilibrium in the carbon dioxide system in seawater. Mar. Chem. **65**(3–4), 135–153 (1999b)

R.E. Zeebe, J.C. Zachos, G.R. Dickens, Carbon dioxide forcing alone insufficient to explain palaeocene-Eocene thermal maximum warming. Nat. Geosci. **2**(8), 576–580 (2009)

A.L. Zerkle, C.K. Junium, D.E. Canfield, C.H. House, Production of ^{15}N-depleted biomass during cyanobacterial N_2-fixation at high Fe concentrations. J. Geophys. Res., Biogeosci. **113**(3), G03014 (2008)
A.L. Zerkle, J. Farquhar, D.T. Johnston, R.P. Cox, D.E. Canfield, Fractionation of multiple sulfur isotopes during phototrophic oxidation of sulfide and elemental sulfur by a green sulfur bacterium. Geochim. Cosmochim. Acta **73**(2), 291–306 (2009)
A.L. Zerkle, A. Kamyshny, L.R. Kump, J. Farquhar, H. Oduro, M.A. Arthur, Sulfur cycling in a stratified euxinic lake with moderately high sulfate: constraints from quadruple S isotopes. Geochim. Cosmochim. Acta **74**(17), 4953–4970 (2010)
A.L. Zerkle, M.W. Claire, S.D. Domagal-Goldman, J. Farquhar, S.W. Poulton, A bistable organic-rich atmosphere on the neoarchaean Earth. Nat. Geosci. **5**(5), 359–363 (2012)
A.L. Zerkle, D.S. Jones, J. Farquhar, J.L. Macalady, Sulfur isotope values in the sulfidic Frasassi cave system, central Italy: a case study of a chemolithotrophic S-based ecosystem. Geochim. Cosmochim. Acta **173**, 373–386 (2016)
A.L. Zerkle, S.W. Poulton, R.J. Newton, C. Mettam, M.W. Claire, A. Bekker, C.K. Junium, Onset of the aerobic nitrogen cycle during the Great Oxidation Event. Nature **542**(7642), 465–467 (2017)
J. Zhang, P.D. Quay, D.O. Wilbur, Carbon isotope fractionation during gas-water exchange and dissolution of CO_2. Geochim. Cosmochim. Acta **59**(1), 107–114 (1995)
X. Zhang, A.L. Gillespie, A.L. Sessions, Large D/H variations in bacterial lipids reflect central metabolic pathways. Proc. Natl. Acad. Sci. USA **106**(31), 12580–12586 (2009)
X. Zhang, D.M. Sigman, F.M. Morel, A.M. Kraepiel, Nitrogen isotope fractionation by alternative nitrogenases and past ocean anoxia. Proc. Natl. Acad. Sci. USA **111**(13), 4782–4787 (2014)
I. Zhelezinskaia, A.J. Kaufman, J. Farquhar, J. Cliff, Large sulfur isotope fractionations associated with neoArchean microbial sulfate reduction. Science **346**(6210), 742–744 (2014)
X. Zhu-Barker, A.R. Cavazos, N.E. Ostrom, W.R. Horwath, J.B. Glass, The importance of abiotic reactions for nitrous oxide production. Biogeochemistry **126**(3), 251–267 (2015)

Space Sci Rev (2020) 216:121
https://doi.org/10.1007/s11214-020-00736-0

Future Missions Related to the Determination of the Elemental and Isotopic Composition of Earth, Moon and the Terrestrial Planets

Iannis Dandouras[1] · Michel Blanc[1] · Luca Fossati[2] · Mikhail Gerasimov[3] · Eike W. Guenther[4] · Kristina G. Kislyakova[2,5] · Helmut Lammer[2] · Yangting Lin[6] · Bernard Marty[7] · Christian Mazelle[1] · Sarah Rugheimer[8] · Manuel Scherf[2] · Christophe Sotin[9] · Laurenz Sproß[2,10] · Shogo Tachibana[11] · Peter Wurz[12] · Masatoshi Yamauchi[13]

Received: 8 January 2020 / Accepted: 25 September 2020 / Published online: 22 October 2020

Abstract In this chapter, we review the contribution of space missions to the determination of the elemental and isotopic composition of Earth, Moon and the terrestrial planets, with special emphasis on currently planned and future missions. We show how these missions are going to significantly contribute to, or sometimes revolutionise, our understanding of planetary evolution, from formation to the possible emergence of life. We start with the Earth, which is a unique habitable body with actual life, and that is strongly related to its atmo-

Reading Terrestrial Planet Evolution in Isotopes and Element Measurements
Edited by Helmut Lammer, Bernard Marty, Aubrey L. Zerkle, Michel Blanc, Hugh O'Neill and Thorsten Kleine

✉ I. Dandouras
Iannis.Dandouras@irap.omp.eu

1 Institut de Recherche en Astrophysique et Planétologie, Université de Toulouse / CNRS / UPS / CNES, Toulouse, France

2 Space Research Institute, Austrian Academy of Sciences, Graz, Austria

3 Space Research Institute of RAS (IKI), Moscow, Russia

4 Thüringer Landessternwarte Tautenburg, Tautenburg, Germany

5 Department of Astrophysics, University of Vienna, Vienna, Austria

6 Institute of Geology and Geophysics, Chinese Academy of Sciences, Beijing, China

7 Nancy Université, Nancy, France

8 Department of Physics, Oxford University, Oxford, UK

9 Jet Propulsion Laboratory, NASA, Pasadena, CA, USA

10 Institute of Physics, University of Graz, Graz, Austria

11 Hokkaido University, Sapporo, Japan

12 Physikalisches Institut, University of Bern, Bern, Switzerland

13 Swedish Institute of Space Physics (IRF), Kiruna, Sweden

sphere. The new wave of missions to the Moon is then reviewed, which are going to study its formation history, the structure and dynamics of its tenuous exosphere and the interaction of the Moon's surface and exosphere with the different sources of plasma and radiation of its environment, including the solar wind and the escaping Earth's upper atmosphere. Missions to study the noble gas atmospheres of the terrestrial planets, Venus and Mars, are then examined. These missions are expected to trace the evolutionary paths of these two noble gas atmospheres, with a special emphasis on understanding the effect of atmospheric escape on the fate of water. Future missions to these planets will be key to help us establishing a comparative view of the evolution of climates and habitability at Earth, Venus and Mars, one of the most important and challenging open questions of planetary science. Finally, as the detection and characterisation of exoplanets is currently revolutionising the scope of planetary science, we review the missions aiming to characterise the internal structure and the atmospheres of these exoplanets.

Keywords Space missions · Planetary evolution · Planetary atmospheres · Planetary magnetospheres · Terrestrial planets · Exoplanets · Biosignatures

1 Introduction

The previous chapters of this topical series presented in considerable detail what we know on the way the evolution of a planet and its atmosphere, from formation to the possible emergence of habitable worlds, influences its elemental, chemical and isotopic composition. Indeed, the chemical and isotopic composition of the different constituents of a planet are shaped by initial conditions such as the composition and activity of its host star (Güdel 2020) and the structure and composition of the circumstellar nebula out of which it was born (Bermingham et al. 2020). In the course of its evolution, each planetary atmosphere is fed by additional material from its space environment and from the planet's degassing interior (Stüeken et al. 2020) while a fraction of its atmospheric chemical elements escape to outer space (Lammer et al. 2020a). A quantitative assessment of the effects of these mechanisms can be partly constrained in the chemical and isotopic composition of the different planetary layers, among them its atmosphere, and in the meteorite record (e.g., O'Neill et al. 2020) where they can be most easily measured. These measurements provide constraints on the formation history of Earth, Moon and the terrestrial planets (Lammer et al. 2020b), on the evolution of planetary atmospheres (Avice and Marty 2020; Stüeken et al. 2020) and on their current composition (Scherf et al. 2020). Last but not least, when life emerges, like on Earth, it also leaves a strong imprint on the planet's composition (Lloyd et al. 2020).

In this article, we review the contribution of space missions to Earth, Moon and the terrestrial planets, to the determination of their elemental and isotopic composition, with special emphasis on currently planned and future missions. We show how these missions are going to strongly contribute to, or sometimes revolutionise, our understanding of planetary evolution, from formation to the possible emergence of life. Our scope is not to provide an exhaustive list of missions, but to highlight some missions to the inner planets of our Solar System addressing these objectives, and then to expand our review to missions aiming to characterise planets around other stars.

We start this review with missions to the Earth-Moon system (Sect. 2). Building on the legacy of Apollo and of several other important missions to the Moon, we review the current plans for a new wave of Moon missions from a broad range of countries and from the public and private spheres. These missions, combining orbiter and lander elements, are aimed at studying the formation history of the Moon, the structure and dynamics of its tenuous exosphere, and the interaction of the Moon's surface and exosphere with the different sources of plasma and radiation of its environment, including the solar wind and the escaping Earth's upper atmosphere. We also review the current plans for missions in Earth orbit that will help better quantify the atmospheric losses processes and also their effects on the Earth's upper atmosphere and magnetosphere.

In Sect. 3 we review past studies and current plans to use noble gases in the atmospheres of the terrestrial planets, Venus and Mars, to trace the evolutionary paths of these atmospheres, with a special emphasis on understanding the effect of atmospheric escape on the fate of water. Future missions to these planets will be key to help us establish a comparative view of the evolution of climates and habitability at Earth Venus and Mars, one of the most important and challenging open questions of planetary science. In a similar way, past (Messenger) and current (BepiColombo) missions to Mercury are expected to provide a new view of the bulk and exosphere composition of this planet, the only example of a hot close-in terrestrial planet in our Solar System.

While remote sensing and in situ measurements are expected to provide new insights into these questions, return of samples of surface, sub-surface and atmospheres to our Earth laboratories is the most promising perspective to write an accurate evolutionary history of terrestrial planets, as illustrated by the outstanding scientific legacy of the Apollo and Lunakhod programs. Future, currently planned sample return missions to the Moon (Sects. 2.2.2 and 2.2.3) will give us access to the broad diversity of Moon terrains, spanning their different ages. They raise the hope of a far better understanding of the formation history of the Earth-Moon system and of a comprehensive inventory of the Moon's mineral and water resources. At Mars, sample return is the pillar of future exploration, bearing our hope to detect possible traces of extinct or extant life on the Red Planet (Sect. 3.2). Sample return from asteroids and comets, as the currently operating Hayabusa2 and OSIRIS-REx (Origins, Spectral Interpretation, Resource Identification, Security, Regolith Explorer) missions, will in parallel provide unique constraints on the initial composition of the building blocks of terrestrial planets and on their sources of water (Sect. 4). And while returning samples from the atmosphere or surface of Venus is probably one of the most challenging missions to fly, due to its very dense atmosphere, its huge scientific return on better understanding why Earth's sister planet is so different from its sibling fully justifies the required investment.

Finally, as the detection and characterisation of exoplanets is currently revolutionising the scope of planetary science, all means for assessing the composition of terrestrial exoplanets and their atmospheres will be extremely precious in order to place into a broader context the more accurate knowledge of the terrestrial planets composition in our Solar System (Sect. 5). Two methods are currently within the reach of past, planned and future exoplanet missions. The determination of the bulk composition of Ultra-short-period (USP) planets, in the range of Earth or super-Earth masses orbiting low-mass stars, provides precious constraints on their bulk composition and even on the existence of a significant atmosphere. In the longer term, missions dedicated to the spectroscopy of exoplanet atmospheres, using occultation or direct imaging, bear our greatest hopes to retrieve their origin and evolution. The most ambitious of these proposed missions (e.g., HabEx, LUVOIR and LIFE) open for us the perspective of detailed studies of habitable planets atmospheres, and hopes that inhabited ones might be discovered in a not-too-distant future.

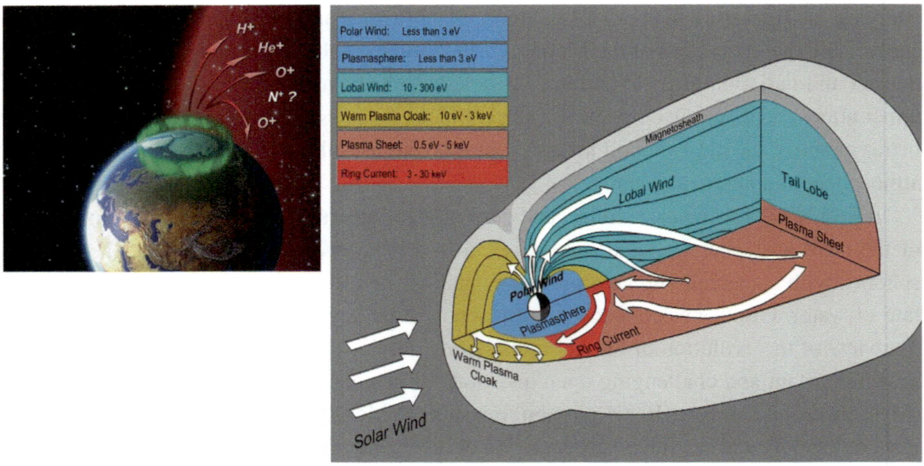

Fig. 1 Left: ion outflow from the terrestrial high-latitude ionosphere (adapted from Engwall 2009). Right: motion and energisation of outflowing low-energy ionospheric ions as they move through the magnetosphere. Depending on the outflow path, the interplanetary magnetic field conditions, the magnetospheric convection and the ion species, some of the ions are re-injected into the inner magnetosphere whereas some completely escape (from Chappell 2015)

2 The Earth-Moon System

2.1 Earth

The Earth is a unique habitable body with actual life, and that is strongly related to its atmosphere. However, the present atmospheric composition, following 4.5 billion years of evolution, is quite different from that of the ancient Earth (Stüeken et al. 2020). The evolution of the terrestrial atmosphere is driven by its interactions with the planet's crust and oceans, the biological activity, the influx from space (e.g., meteors) and the atmospheric escape to space (Yamauchi and Wahlund 2007; Lammer et al. 2008, 2011, 2018; Johnson and Goldblatt 2015; Airapetian et al. 2017a). The ions that outflow from the ionosphere (Fig. 1) are successively accelerated through a variety of energisation mechanisms and can eventually reach velocities above the gravitational escape velocity (e.g. Delcourt et al. 1993; Yau and André 1997; Huddleston et al. 2005; Nilsson et al. 2006, 2008). Depending on the outflow path, the interplanetary magnetic field conditions, the magnetospheric convection and the ion species, some of the outflowing ions are re-injected into the inner magnetosphere, whereas some completely escape (Ebihara et al. 2006; Haaland et al. 2012; Chappell 2015; Yamauchi 2019).

Ion outflow and atmospheric escape play an important role in the magnetospheric dynamics, by loading the magnetosphere with heavy ions (Kistler et al. 2005; Liu et al. 2013; Kronberg et al. 2014; Yamauchi and Slapak 2018), and in the evolution of the atmosphere on geological timescales, by slowly changing its composition (Ilie and Liemohn 2016; Lammer et al. 2018; Yamauchi 2019). They are highly dependent on the solar EUV flux and solar wind conditions and on the geomagnetic activity conditions, varying by 1 to 2 orders of magnitude (Slapak et al. 2017; Li et al. 2017; Schillings et al. 2019).

Earth being the most extensively explored body of our solar system, numerous missions since the beginning of the space era have been studying the upper atmosphere / ionosphere and the magnetosphere and have been collecting data on the ion and neutral sources, circulation within the magnetosphere and losses to outer space (e.g. Chappell 2015; Welling et al. 2015). However, several questions remain open, such as:

- What are the prevailing escape mechanisms and what are the escape rates, for ions and for neutrals, for different driver conditions (solar and geomagnetic activity), over geological time scales?
- How does the composition of the escaping populations change in response to different driver conditions, and in particular how does the nitrogen / oxygen escape ratio change as a function of the solar activity, over geological time scales?
- What is the isotope fractionation of the escaping fluxes and how it changes in response to different driver conditions?
- What is the degree of recirculation of plasma after it has left the ionosphere (recirculation versus escape)?

2.1.1 Thermospheric / Exospheric Missions

Table 1 provides a summary of the major past and present space missions, contributing to the study of the cold / thermal neutrals of the terrestrial thermosphere and exosphere (Dandouras et al. 2018).

It should be noted that for the extended exosphere these missions were mostly limited to hydrogen, whereas oxygen measurements were limited to altitudes below ~700 km and for nitrogen below ~600 km, and these measurements are quite old (Johnson 1969). However, nitrogen and oxygen are the two major constituents of the Earth's atmosphere.

Regarding recent missions, the GOLD (Global-scale Observations of the Limb and Disk) mission of NASA was launched in 2018. It consists of a UV imaging spectrograph on a geostationary satellite providing remotely measured densities and temperatures in Earth's thermosphere for O and N_2. Its objective is to study the global-scale response of the thermosphere and ionosphere to forcing in the integrated Sun-Earth system (https://gold.cs.ucf.edu/).

The ICON (Ionospheric Connection Explorer) mission of NASA was launched in October 2019 and will perform both in situ ion measurements, from its 575 km altitude orbit, i.e. close to the exobase, and remote-sensing measurements by pointing at the Earth's limb. These include an EUV and an FUV imager and a Michelson Interferometer for Global High-Resolution Thermospheric Imaging. Its objectives focus on the ionosphere response to space weather activity (http://icon.ssl.berkeley.edu/).

Daedalus is a proposed mission to ESA, which has been selected for a Phase-0 study within the Earth Explorer Program. Its objective is to quantify the key electrodynamic processes that determine the structure and composition of the lower thermosphere (called also the "ignorosphere", or "agnostosphere"), the gateway between Earth's atmosphere and space, and its response to solar and geomagnetic activity conditions (Sarris et al. 2020). It relies on an innovative mission concept, with a main satellite on an elliptical 140 × 2000 km altitude polar orbit, equipped with particle and wave instruments, including high-resolution ion and neutral mass spectrometers. The satellite will perform several short (day-long) excursions down to ~120 km. From there it could release each time an instrumented sub-satellite (e.g. cubesat) whose altitude would gradually decay, providing in this way measurements in the heart of the most under-explored region in the Earth's atmosphere, until the sub-satellite burns in the mesosphere (https://daedalus.earth). Although its focus is the 100–200 km altitude range, it will be also capable of measuring ions and neutrals beyond the exobase better than DE-2. Daedalus will be in competition with two other missions,

Table 1 Past and present thermospheric / exospheric missions

Mission	Altitude range	Remote sensing method	In situ method for cold neutrals	Species
DE-2 (1981–1983)	thermosphere (300–500 km) apogee: 1000 km		mass spectrometer	major species (N_2, N, O, H)
AE-C, D, E (1973–1978)	thermosphere (<400 km)	UV	mass spectrometer accelerometer	major species (N_2, N, O, H)
Air Force SCs (1968–)	thermosphere (<220 km)		accelerometer	total mass density
Castor (1975–1979)	thermosphere (250–600 km)		accelerometer	total mass density
San Marco 5 (1988, 2-weeks)	thermosphere (260–690 km)	UV (not used)	accelerometer	total mass density
MSS (1980–1989)	thermosphere (<250 km)	UV occultation		O_2 density
TIMED (2001–)	thermosphere <400 km (target: <180 km)	multi-wavelength UV (from 600 km)		major species (N_2, N, O, H)
DEMETER (2004–2010)	ionosphere			H^+, He^+, O^+
IMAGE (2000–2005)	exosphere	Lyman-α		H only
TWINS (2006–)	exosphere	Lyman-α		H only
GOES	exosphere	Lyman-α (from 6.6 R_E)		H only
GOLD (2018–)	thermosphere	UV		O, N_2
ICON (2019–)	thermosphere	EUV, FUV	RPA & drift meter	

STEREOID (Cryosphere, Oceanography, Geosphere) and G-CLASS:H_2O (rainfall, flooding landslides) for final selection within ESA's Earth Explorer Program. If selected, Daedalus launch is foreseen in 2027–2028.

AETHER (Aeronomy at Earth: Tools for Heliophysics Exploration and Research) is a mission proposed to NASA, which has been selected for a concept study. From a position on board the International Space Station, it would provide information on how the neutral, terrestrial-weather-driven thermosphere interacts with the ionosphere. AETHER will be in competition with two other missions, the Extreme Ultraviolet High-Throughput Spectroscopic Telescope (EUVST) and the Electrojet Zeeman Imaging Explorer (EZIE). If finally selected, AETHER would be launched in 2024.

2.1.2 Magnetospheric Missions

Table 2 provides a summary of the major past and present magnetospheric missions with composition measurements.

Table 2 Major past and present magnetospheric missions

Mission	Orbit	Key regions studied	Ion mass spectrometry
ISEE-1 & ISEE-2 (1977–1987)	6600 km × 23 R_E $i \approx 29°$	Equatorial magnetosphere	SHM Ion Composition Experiment
DE-1 (1981–1990)	490 km × 4.6 R_E $i \approx 90°$	Polar and Inner Magnetosphere	Retarding Ion Mass Spectrometer (RIMS)
AMPTE / CCE (1984–1989)	1100 km × 9 R_E $i \approx 5°$	Equatorial inner magnetosphere	Hot Plasma Composition Experiment (HPCE) & Charge-Energy-Mass Spectrometer (CHEM)
Akebono (1989–2015)	275 km × 10500 km $i \approx 75°$	Polar magnetosphere	Suprathermal Ion Mass Spectrometer (SMS)
Freja (1992–1996)	600 km × 1720 km $i \approx 63°$	Polar upper ionosphere	Three-dimensional Ion Composition Spectrometer (TICS)
SAMPEX (1992–2004)	512 km × 687 km $i \approx 82°$	Polar Inner magnetosphere, anomalous components of cosmic rays & SEPs	Low Energy Ion Composition Analyser (LEICA) & Mass Spectrometer Telescope (MAST)
FAST (1996–2009)	347 km × 3500 km $i \approx 83°$	Auroral region	Time-of-flight Energy Angle Mass Spectrograph (TEAMS)
Geotail (1992–)	9 R_E × 31 − 220 R_E $i \approx 10°$	Magnetotail	Comprehensive Plasma Investigation (CPI) & Energetic Particle and Ion Composition Experiment (EPIC)
Polar (1996–2008)	4200 km × 9 R_E $i \approx 79°$	Polar magnetosphere	Toroidal Imaging Mass Angle Spectrograph (TIMAS) & Charge and Mass Magnetospheric Ion Composition Experiment (CAMMICE)
Equator-S (1997–1998)	212 km × 35900 km $i \approx 4°$	Equatorial inner magnetosphere	Ion Composition Instrument (ICI)
Cluster-4 spacecraft (2000–)	4 R_E × 19.6 R_E $i \approx 90°$	Magnetosphere, magnetosheath, solar wind	Composition Distribution Function (CODIF) & Research with Adaptive Particle Imaging Detectors (RAPID)
IMAGE (2000–2005)	1000 km × 8.2 R_E $i \approx 90°$	Ring current & plasmasphere imaging	Energetic neutral atom imagers (LENA, MENA, HENA)
Double Star TC1 & TC2 (2004–2007)	570 km × 13.4 R_E $i \approx 28.5°$ (TC1) 700 km × 7.1 R_E $i \approx 90°$ (TC2)	Equatorial magnetosphere (TC1) & Polar magnetosphere (TC2)	Hot Ion Analyser (indirect composition)
TWINS A & B (2006–)	1000 km × 7.2 R_E $i \approx 63°$	Ring current & exosphere imaging	Energetic neutral atom imager (MENA)

Table 2 *(Continued)*

Mission	Orbit	Key regions studied	Ion mass spectrometry
THEMIS-5 spacecraft (2007–)	470 km × 12–16 R_E $i \approx 16°$	Magnetosphere, magnetosheath, solar wind	Electrostatic Analyser (indirect composition)
Van Allen Probes A & B (2012–2019)	618 km × 30400 km $i \approx 10°$	Inner Magnetosphere	Ion Composition Experiment (RBSPICE) & Energetic Particle, Composition, and Thermal Plasma (ECT)
MMS-4 spacecraft (2015–)	2550 km × 12–25 R_E $i \approx 63°$	Magnetosphere, magnetosheath, solar wind	Hot Plasma Composition Analyser (HPCA) & Energetic Ion Spectrometer (EIS)
ERG / Arase (2016–)	400 km × 32100 km $i \approx 31°$	Inner Magnetosphere	Low-Energy Particle Experiment—Ion Mass Analyser (LEP-i) & Medium-Energy Particle Experiment—Ion Mass Analyser (MEP-i)

The above list is not comprehensive, but focuses on the major missions equipped to supply ion composition information.

Currently, the Cluster mission of ESA, launched in 2000 (Escoubet et al. 2001, 2015), and the MMS (Magnetospheric Multiscale Mission) mission of NASA, launched in 2015 (Burch et al. 2016), are the two major missions supplying in situ magnetospheric ion composition measurements. Each of these missions involves four spacecraft, providing thus three-dimensional measurements of space plasma parameters. These missions have been very successful in analysing medium-scale (Cluster) and small-scale (MMS) plasma dynamics. They have been also supplying key information on plasma sources and losses (e.g. Engwall et al. 2009; Dandouras 2013; Parks et al. 2015; Nagai et al. 2016; Slapak et al. 2017; Xu et al. 2019). However, the moderate mass resolution of their instrumentation ($m/\Delta m \approx$ 5–7 for CODIF onboard Cluster (Rème et al. 2001) and $m/\Delta m \approx 4$ for HPCA on board MMS (Young et al. 2016)) does not allow distinguishing nitrogen from oxygen ions or any isotope ratios. The separation of nitrogen from oxygen was designed for DE-1, AMPTE, Akebono and Arase, but for limited energy ranges, and such separation is partially possible for the Van Allen Probes.

ESCAPE (European SpaceCraft for the study of Atmospheric Particle Escape) is a mission that has been proposed to ESA, in response to the M5 call, to quantitatively estimate the amount of escaping particles of the major atmospheric components (nitrogen and oxygen), as neutral and ionised species, escaping from the Earth as a magnetised planet. The goal is to understand the importance of each escape mechanism (thermal or non-thermal), its dependence on solar and geomagnetic activity, in order to infer the history of the Earth's atmosphere over a long (geological scale) time period. To achieve this goal a slowly spinning spacecraft on a high inclination 500 km × 33 000 km orbit has been proposed, equipped with a suite of instruments for both in situ measurements and for remote-sensing observations (Dandouras et al. 2018). This instrument suite would provide high-mass resolution measurements, providing nitrogen/oxygen separation and isotope fractionation over a wide energy range (http://cluster.irap.omp.eu/public/ESCAPE/ESCAPE_M5_Proposal_V1.1.pdf).

In March 2017 ESCAPE successfully passed the first technical and programmatic screening by ESA, but then in May 2018 ESCAPE was not selected for flight.

MEME-X (Mechanisms of Energetic Mass Ejection – eXplorer) is another proposed mission, that was selected in 2017 by NASA for a competitive study within the Small Explorer program (SMEX). Based on two spacecraft flying in twin 350 × 1200 km polar orbits, it would study how plasma escapes from the ionosphere into the magnetosphere (https://www.nasa.gov/feature/goddard/2017/new-nasa-mission-concept-aimed-at-studying-why-planets-lose-their-atmospheres). In June 2019 MEME-X was not selected for flight.

SMILE (Solar wind Magnetosphere Ionosphere Link Explorer) is a joint mission between ESA and the Chinese Academy of Sciences, currently under preparation and scheduled for launch in 2023 (http://sci.esa.int/smile/). Its objective is to investigate the dynamic response of the magnetosphere to the solar wind conditions in a global way. It will carry four instruments. Two of them will be for in situ measurements: a light ion analyser and a magnetometer. The other two will be imagers: an ultraviolet aurora imager and a novel soft X-ray imager. This imager will observe the magnetosheath, the magnetopause and the cusps, and their response to solar wind conditions, through the X-rays emitted from charge exchange between solar wind highly charged ions and exospheric neutrals (Sibeck et al. 2018). SMILE will be placed on a highly inclined elliptical orbit with an apogee at 19 R_E, enabling the spacecraft to spend most of its time at high altitudes, so as to collect continuous observations through imaging for long periods.

2.2 Moon

The Moon, during most part of its orbit around the Earth is directly exposed to the solar wind. Due to the absence of a substantial intrinsic magnetic field and of a collisional atmosphere, solar wind and solar energetic particles (SEPs) arrive almost without any deviation or absorption and impact directly on its surface, interacting with the lunar regolith and the tenuous lunar exosphere (e.g. Geiss et al. 2004; Wurz et al. 2007; Futaana et al. 2018). The same phenomenon occurs also with the galactic cosmic rays (GCRs), which present fluxes and energy spectra typical of the interplanetary space (e.g. Sohn et al. 2014).

During 5–6 days every orbit, however, the Moon crosses the tail of the terrestrial magnetosphere (Fig. 2). It is then exposed not to the solar wind but to the terrestrial magnetotail plasma environment, offering the possibility to study in situ magnetotail dynamics and its dependence on solar and geomagnetic activity (e.g. Kallio and Facskó 2015; Kallio et al. 2019). Phenomena such as plasmoids released from the near-Earth magnetotail and propagating anti-Sunward, bursty bulk flows (BBFs), magnetic reconnection and plasma sheet dynamics can be studied in situ (e.g. Nakamura 2006; Cao et al. 2013). The Moon is then also very well situated to study atmospheric escape from the Earth into space (Lammer et al. 2008), in the form of heavy ions upwelling from the terrestrial ionosphere and then transported and lost into the deep magnetotail, as shown from the wealth of data supplied from the THEMIS-ARTEMIS instruments (cf. Sect. 2.2.1) and from the Kaguya and Chandrayaan-1 spacecraft (Poppe et al. 2016; Terada et al. 2017; Dandouras et al. 2020; Li et al. 2020). The THEMIS-ARTEMIS data, however, did not include the crucial information on the plasma composition (Angelopoulos 2011), and the Kaguya plasma measurements where limited to a less than two-year mission and to low-energy plasma (Saito et al. 2010). Far magnetotail studies performed by the Geotail spacecraft supplied crucial information on the dynamics of ion beams streaming downtail (Christon et al. 1994;

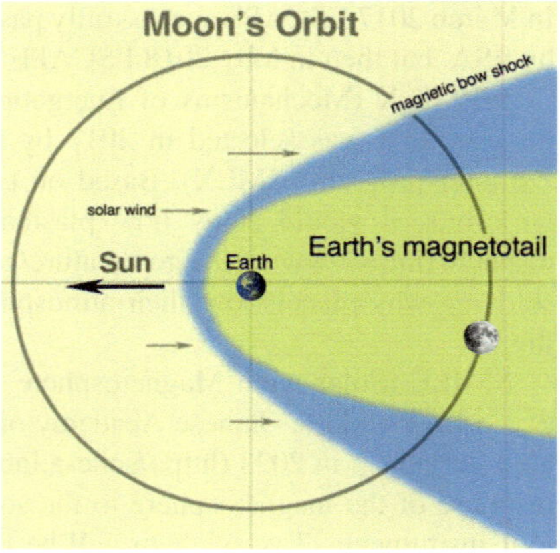

Fig. 2 Moon's orbit with respect to the Earth's magnetosphere. Earth's and Moon's sizes are not in scale (adapted from: *Tim Stubbs / University of Maryland / GSFC*

Seki et al. 1998), but lacked ion composition measurements at low energies (below $\sim$10 keV).

But, most important, the lunar environment offers a unique opportunity to study the Moon surface-bounded exosphere, its production mechanisms, its dynamics, its interaction with the solar wind and with the terrestrial magnetotail plasma, and its escape into space (Potter et al. 2000; Wurz et al. 2007; Leblanc and Chaufray 2011; Harada et al. 2014). The LADEE (Lunar Atmosphere and Dust Environment Explorer) observations have just shown the complexity of the lunar exosphere and of the associated physical mechanisms (Elphic et al. 2014; Benna et al. 2015; Hodges 2016).

The Moon surface also offers exciting possibilities for studying energetic ion implantation in the lunar regolith (Ozima et al. 2005; Ireland et al. 2006), solar wind ion implantation or neutralisation and reflection from the lunar regolith (Vorburger et al. 2015; Tucker et al. 2019), solar wind interaction with crustal magnetic anomalies (Vorburger et al. 2012; Poppe et al. 2015; Bamford et al. 2016), lunar pick-up ion generation (Poppe et al. 2012), or lunar surface electrostatic charging and dust levitation (Stubbs et al. 2007; Burinskaya 2015; Dyadechkin et al. 2015), just to mention few examples.

Table 3 provides a summary of the major past and present lunar missions with composition measurements.

Following the 50th anniversary of the Apollo missions to the Moon, and the more recent missions mentioned above (THEMIS-ARTEMIS, Kaguya, LADEE), a series of lunar missions is in preparation, or already operating, building on their outstanding legacy.

2.2.1 Orbiters

The THEMIS-ARTEMIS P1 and P2 NASA spacecraft entered lunar orbit on 27 June and 17 July of 2011 respectively, after they have been exploring the terrestrial magnetosphere since 2007 (Angelopoulos 2011). Each satellite carries identical instrumentation, including a fluxgate magnetometer (FGM), an electrostatic analyser (ESA), a solid state telescope (SST), a search-coil magnetometer (SCM) and an electric field instrument (EFI).

The Lunar Reconnaissance Orbiter (LRO) of NASA was launched on 18 June 2009 (Keller et al. 2016). The scientific payload consists of: (i) a Cosmic Ray Telescope for the Effects of Radiation (CRaTER) to characterise the effects of radiation in the lunar environment and its biological impacts; (ii) a Diviner Lunar Radiometer Experiment to measure the lunar

Table 3 Major past and present lunar missions with composition measurements

Mission	Mission type	Composition measurements
Luna 11 (1966)	Lunar orbiter	γ-ray spectrometer
Apollo 11–14 (1969–1971)	Human spaceflight: orbiter + lander	Lunar sample return & Solar wind composition experiment
Luna 16 (1970)	Lunar lander	Lunar sample return
Apollo 15–17 (1971–1972)	Human spaceflight: orbiter + lander + subsatellite	Lunar sample return & Solar wind composition experiment & Particles and fields subsatellite
Luna 19 & 22 (1971–1972, 1974–1975)	Lunar orbiter	γ-ray spectrometer
Luna 20 & 24 (1972, 1976)	Lunar lander	Lunar sample return
Geotail (1992–)	Earth orbiter + Lunar flybys	Comprehensive Plasma Investigation & Energetic Particle and Ion Composition Experiment
Clementine (1994–1995)	Lunar orbiter	Multi-spectral imaging of lunar surface
Lunar Prospector (1998–1999)	Lunar orbiter	γ-ray spectrometer & Neutron spectrometer & α particle spectrometer
SMART-1 (2003–2006)	Lunar orbiter	X-ray spectrometer
SELENE (Kaguya) (2007–2009)	Lunar orbiter	γ-ray spectrometer & X-ray spectrometer & Charged particle spectrometer
Chang'e 1 (2007–2009)	Lunar orbiter	γ-ray spectrometer & X-ray spectrometer & Imaging spectrometer
Chandrayaan-1 (2008–2009)	Lunar orbiter	High-Energy X-ray spectrometer & Hyper spectral imaging camera & Visible near-IR imaging spectrometer & ENA analyser
Lunar Reconnaissance Orbiter (LRO) (2009–)	Lunar orbiter	Lyman-α Mapping & Lunar Exploration Neutron Detector
Lunar Crater Observation and Sensing Satellite (LCROSS) (2009)	Lunar impactor	Multi-spectral cameras
Chang'e 2 (2010–2011)	Lunar orbiter	γ-ray spectrometer & X-ray spectrometer & Imaging spectrometer
THEMIS-ARTEMIS P1 & P2 (2011–)	Lunar orbiter	Electrostatic Analyser (indirect composition)
Lunar Atmosphere and Dust Environment Explorer (LADEE) (2013–2014)	Lunar orbiter	Neutral Mass Spectrometer & UV-Visible Spectrometer

Table 3 (*Continued*)

Mission	Mission type	Composition measurements
Chang'e 3 (2013–2015)	Lunar lander + rover	Ground-penetrating radar & α particle X-ray spectrometer
Chang'e 4 (2018–)	Lunar lander + rover + relay satellite	Ground-penetrating radar & α-particle X-ray spectrometer
Chandrayaan-2 (2019–)	Lunar orbiter	Soft X-ray spectrometer & Imaging IR Spectrometer & Atmospheric composition analyser

surface thermal emission and to provide information for future surface operations and exploration; (iii) the Lyman-Alpha Mapping Project to look into permanently shadowed craters in search of water ice, using ultraviolet light; (iv) the Lunar Exploration Neutron Detector to measure and map neutrons, in search of possible near-surface water ice deposits; (v) the Lunar Orbiter Laser Altimeter to produce a precise global lunar topographic model and geodetic grid; (vi) the Lunar Reconnaissance Orbiter Camera to address the requirements of landing site certification and polar illumination; (vii) the Miniature Radio Frequency radar to demonstrate new lightweight SAR and communications technologies and locate potential water-ice.

The Hakuto-R mission 1 of ispace inc. (Japan) is a commercial lunar exploration program (https://ispace-inc.com/hakuto-r/). Mission 1 is scheduled for launch in 2021 and will perform a soft landing on the Moon. Mission 2, scheduled for launch in 2023, will perform a soft landing and deploy a rover for surface exploration and data collection on the Moon.

The Deep Space Gateway (or Lunar Orbital Platform – Gateway, cf. Fig. 3) is a crewed platform that will be assembled and operated in the vicinity of the Moon by NASA, ESA, JAXA, Roscosmos and the CSA, and will offer new opportunities for fundamental and applied scientific research (Carpenter et al. 2018). Deployment of its different modules is planned within the 2022–2028 timeframe, and early utilisation for scientific payloads will be possible as of 2024 (Gateway Phase 1). In 2027 the ESPRIT (European System Providing Refuelling, Infrastructure and Tele-communication) module will also be available (Gateway Phase 2). Space plasma physics measurements could be performed either from the Deep Space Gateway platform, or from instrumented CubeSats released from the platform and placed into low-lunar orbits, or directly from the Moon surface (Dandouras et al. 2019).

2.2.2 Orbiters with Landers

The Chang'E-5 mission of CNSA is scheduled for launch in 2020, with a chosen landing site near Mons Rümker, an isolated volcanic formation in Oceanus Procellarum. Chang'E-5 has also a sample-return vehicle to return back to the Earth at least 2 kilograms of lunar soil and rock samples, including a 2-metre drill sample of the lunar regolith. The lander will carry landing cameras, a panoramic camera, a mineral spectrometer, a soil gas analytical instrument, a soil composition analytical instrument, a sampling sectional thermo-detector, and a ground-penetrating radar. For acquiring samples, it will be equipped with a robotic arm, a rotary-percussive drill, a scoop for sampling, and separation tubes to isolate individual samples. The planned landing area is one of the youngest regions on the Moon ($\sim$1.5 Ga), hence the returned samples will shed light on the final volcanic eruptions and thermal history of the Moon and determine the impact cratering rate between $\sim$1 to $\sim$3 Ga.

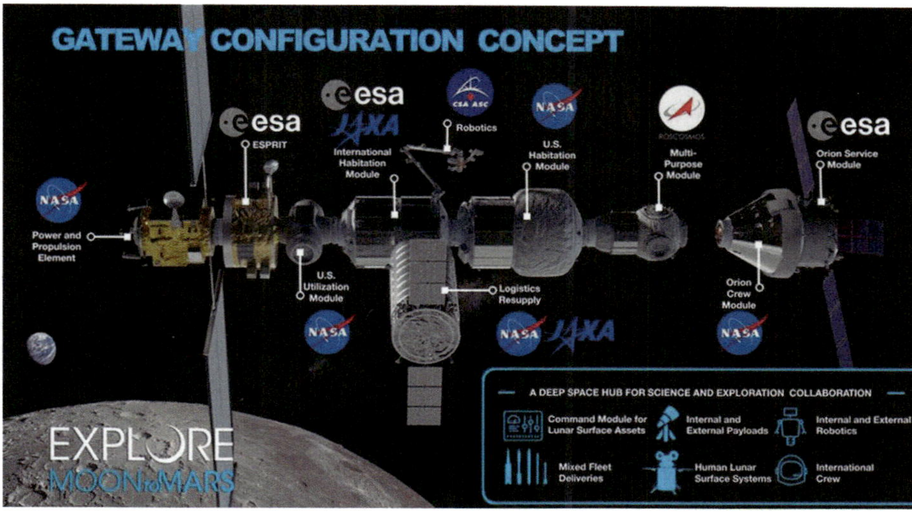

Fig. 3 Illustration showing the Deep Space Gateway (or Lunar Orbital Platform – Gateway) configuration, and the proposed international partners. Credit: NASA

The Chang'E-6 mission of CNSA is scheduled for launch in 2023–2024 and will perform China's second sample return mission. Chang'E-6 is the backup of Chang'E-5. If Chang'E-5 is successful, Chang'E-6 will be considered as a part of the 4th stage of the Chang'E lunar exploration program, which contains also two other landing missions (numbered as Chang'E-7 and 8). Chang'E-6 is planned to land in the south polar region of the Moon and bring back samples from there, including materials delivered from the SPA basin (cf. also Sect. 2.2.4).

2.2.3 Landers

The OMOTENASHI (Outstanding MOon exploration TEchnologies demonstrated by NAno Semi-Hard Impactor) mission of JAXA (via the Exploration Mission 1, or EM-1, of NASA) is a small spacecraft and semi-hard lander of the 6U CubeSat format that will demonstrate low-cost technologies to land and explore the lunar surface. The CubeSat will also take measurements of the radiation environment near the Moon as well as on the lunar surface.

The Peregrine mission of Astrobotic, USA (https://www.astrobotic.com/lunar-delivery), is scheduled to be launched in 2021 and to land at Lacus Mortis. The Peregrine Lander can accommodate a variety of payload types for science, exploration, resources etc.

The XL 1 mission of Masten Space System, USA (https://www.masten.aero/lunar-vehicles), is scheduled to be launched in 2021 and to land at the Aristarchus Crater. It has been designed in partnership with the NASA Lunar CATALYST program (public-private partnership) and is capable of placing a 100-kg payload on the lunar surface.

The Smart Lander for Investigating Moon (SLIM) mission of JAXA is scheduled to be launched in 2021–2022 and to land in the Mare Tranquillitatis. SLIM also has a sample-return vehicle to bring back a sample of about the size of the Hayabusa capability.

Luna-25 (Luna-Glob) and Luna-27 (Luna-Resurs) are two planned missions to send a lander to the South Pole-Aitken basin, under preparation by Roscosmos in collaboration with ESA. Their objectives include the detection and characterisation of lunar polar volatiles (Sanin and Mitrofanov 2016).

HERACLES (Human-Enhanced Robotic Architecture and Capability for Lunar Exploration and Science) is a robotic lunar transport system planned by ESA, JAXA and the CSA. It will feature the European Large Logistic Lander (EL3), designed to allow a series of different missions with different options for its payloads. The project is exploration-driven with a strong scientific content in a variety of domains, as samples return, polar ice and volatiles, internal geophysics, space plasma environment, radiation biological effects, radio astronomy, or identification of in situ resources (https://www.esa.int/Science_Exploration/Human_and_Robotic_Exploration/Exploration/European_Large_Logistics_Lander).

2.2.4 Landers and Rovers

In January 2019 the Chinese Chang' E-4 mission, with its relay satellite Queqiao, involving a robotic lander and the Yutu-2 rover, performed the first landing on the far-side of the Moon. The initial results of this ongoing mission include the spectroscopic identification of lunar far-side mantle-derived materials (Li et al. 2019) and the thick regolith and complex subsurface structures detected by the on-board Lunar Penetrating Radar (LPR).

Together with Chang'E-6, both Chang'E-7 and 8 missions of CNSA are the main parts of the 4th stage of the Chang'E program. All of these three missions aim to land in the south polar region of the Moon, and will establish a prototype lunar research station by the end of the 4th stage of Chang'E program. Chang'E-7 and 8 are scheduled for launch before and after Chang'E-6, respectively. The primary lunar science objectives of these missions are to explore in situ the possible water ice in the permanently shadowed craters, detect the lunar interior structures by delivering lunar seismometers, determine the volatile components in the lunar regolith in the polar region and explore the polar geology, geochemistry and space environment.

The Indian Chandrayaan-2 mission, launched on 22 July 2019, is composed of a lunar orbiter, a lander and a rover. Its scientific objectives include lunar topography, mineralogy, elemental abundance, the lunar exosphere, and signatures of hydroxyl and water ice (Goswami 2014). Communications with the lander were lost when touchdown was expected, on 7 September 2019, but the orbiter, equipped with eight scientific instruments, is operational and is expected to continue its seven-year mission to study the Moon.

The NASA VIPER (Volatiles Investigating Polar Exploration Rover) project consists of a mobile robot that will be sent to the South Pole of the Moon in December 2022. There it will get a close-up view of the location and concentration of water ice in the region and will, for the first time, actually sample the water ice at the same pole where the first woman and next man will land in 2024 under the Artemis program. VIPER will roam several km, using its four instruments, including a 1-meter drill, to sample various soil environments. It will collect about 100 days of data that will then be used to prepare the first global water resource maps of the Moon.

3 Noble Gas Atmosphere Missions

3.1 Venus

Our sister planet Venus attracts a large interest due to its very dense atmosphere, which shows drastically different evolutionary paths compared to Earth (Lammer et al. 2020c). Starting from the early 1960's multiple missions to Venus were accomplished by the USSR, USA and Europe, the most recent of which were Venus-Express (ESA, 2006–2015) and

Table 4 Major past and present Venus missions with composition measurements

Mission	Mission type	Composition measurements
Venera 4, 5, 6 (1967, 1969, 1969)	Venus orbiter + entry probe	Atmospheric composition analyser
Venera 8 (1972)	Venus orbiter + entry probe / lander	Atmospheric composition analyser & γ-ray spectrometer
Mariner 10 (1974)	Interplanetary probe + Venus flyby	UV spectrometer & IR radiometer
Venera 9, 11, 12 (1975, 1978, 1978)	Venus orbiter + entry probe / lander	Atmospheric composition analyser & γ-ray spectrometer & Imager
Pioneer Venus Orbiter (Pioneer Venus 1) (1978–1992)	Venus orbiter	UV spectrometer & Neutral mass spectrometer & Ion mass spectrometer
Pioneer Venus Multiprobe (Pioneer Venus 2) (1978)	Venus orbiter + entry probes	Ion mass spectrometer & Neutral mass spectrometer & Gas chromatograph
Venera 13, 14 (1982)	Interplanetary probe + Venus lander	Neutral mass spectrometer & Gas chromatograph & X-ray fluorescence spectrometer
Venera 15, 16 (1983–1985)	Venus orbiter	IR Fourier spectrometer
Vega 1, 2 (1985)	Interplanetary probe + Venus lander + Balloon	UV spectrometer & Aerosol mass spectrometer & Aerosol gas chromatograph & Aerosol X-ray spectrometer & γ-ray spectrometer
Venus Express (2006–2015)	Venus orbiter	Analyser of plasma and ENAs & IR spectrometer & Visible and IR imaging spectrometer
Akatsuki (Venus Climate Orbiter) (2015–)	Venus orbiter	UV imager

Akatsuki (JAXA, 2010–). Table 4 provides a summary of the major past and present missions to Venus with composition measurements.

There is still a question on how Venus could have lost its water and acquire a dense greenhouse-driven CO_2 atmosphere. Measurements of the abundance and of the isotopic composition of noble gases are a key element to understand the evolution of the atmosphere. This can be done only by direct measurements on board atmospheric probes. The latest successful landing on Venus was that of the Vega-2 lander in 1985 and many key parameters, including noble gases composition, need more precise measurements by modern instruments.

Russia is planning to launch a new mission to Venus in the late 2020s with the name Venera-D (D is an abbreviation of the Russian word "dolgozhivuschaya", that means long leaving). The mission is still at a level of planning and has a high probability to be included into the national space program. Following the intention of NASA to join the Venera-D mission, the space agencies of both countries established the JSDT (Joint Science Definition Team) bringing together 22 members for the formulation of the shape of the mission. Venera-D is planned to be a comprehensive investigation of Venus with three baseline elements: the orbiter, the lander and LLISSE (Long Lived In situ Solar System Explorer) (Report of the Venera-D JSDT, 2019).

The Russian-built orbiter is planned to have about 100 kg of science payload which includes a Fourier Spectrometer, an Ultraviolet mapping spectrometer, a Visible-to-Mid Infrared Spectrometer, an Ultraviolet–Near Infrared Imaging, a Solar and star occultation spectrometer, a Millimeter-radiometer, an Infrared heterodyne spectrometer, a Thermal Infrared Camera, an Infrared Imaging Camera, a Radio-package for two-frequency occultation in L- and X-bands, an Electromagnetic waves detector, and a Panoramic energy mass-analyser of ions. The scientific objectives of the orbiter are: a) study the dynamics and nature of superrotation, radiative balance, and the nature of the greenhouse effect; b) characterise the thermal structure of the atmosphere, winds, thermal tides, and solar-locked structures; c) measure the atmospheric composition, study the clouds, their structure, composition, microphysics, and chemistry; d) investigate the upper atmosphere, ionosphere, electrical activity, magnetosphere, atmospheric escape rate and solar wind interaction. The mission scenario would place the spacecraft in a highly inclined, near-polar orbit with a period of 24 hours. The expected operational lifetime of the orbiter would be more than 3 years.

The Russian-built Lander is planned to be based on the Vega heritage, with about 100–120 kg of science payload operating for about 1 hour during the descent and up to 3 hours at the surface. The scientific payload will include: an X-Ray Diffraction / X-Ray Fluorescent Analyser (XRD/XRF), a Mössbauer spectrometer with XRF mode, TV- cameras (landing, stereo, panoramic, high resolution up to 0.1 mm), a Chemical Analysis Package which will be an assemblage of Gas Chromatograph—Mass-Spectrometer—Laser-Induced Mass-Spectrometer (GCMS + LIMS), a Raman Spectrometer (with synergistic LIDAR), a Meteo Package, an Active Gamma and Neutron Spectrometers, a Multi-channel Tunable Diode Laser Absorption Spectrometer (MTDLAS), and an Infrared Radiometer & Ultraviolet-Visible Spectrometer. The scientific objectives of the lander are: a) perform a chemical analysis of the surface materials and study the elemental composition of the surface, including radiogenic elements; b) study the physical and chemical interactions between the surface and the atmosphere; c) investigate the structure and chemical composition of the atmosphere down to the surface, including the abundances and isotopic ratios of trace and noble gases; d) perform a direct chemical analysis of cloud aerosols and analyse the vertical structure of clouds during the lander descent; e) characterize the geology of local landforms at different scales.

LLISSE is a US-contributed element (attached to the lander) that is built on the basis of high-temperature silicon carbide electronics and will capture and transmit periodic science measurements for a minimum of 60 Earth days. LLISSE is equipped with a set of sensors to measure the 2D wind speed and direction, temperature, pressure, some specific chemical species abundance, and the Solar radiance.

The total mass of the Venera-D vehicle is about 7000 kg (depending on the launch window) and there is still room for additional elements, among which can be an aerial platform for detailed cloud layer investigation or additional long-lived stations for meteorological and/or seismic measurements.

Cupid's Arrow is a small satellite mission, proposed by JPL to measure noble gases in the atmosphere of Venus (Fig. 4). It was selected by the NASA PSDS3 Program and will be based on the VERITAS (Venus Emissivity, Radio Science, INSAR, Topography, and Spectroscopy) mission mothership (Sotin et al. 2018). The mission concept is to send an atmospheric probe that skims Venus atmosphere below the homopause to collect samples of Venus atmosphere, that are then analysed by a miniaturised quadrupole ion trap mass

Fig. 4 Picture of the Cupid's Arrow probe that will skim below Venus' homopause and will collect samples for analysis of its atmosphere (from Sotin et al. 2018)

spectrometer (QITMS) once the probe is outside of Venus' atmosphere. The goal is to understand the atmospheric formation, evolution, and climate history of Venus (Chassefière et al. 2012).

3.2 Mars

There is compelling evidence that liquid water was abundant on early Mars, despite Mars being too cold today to sustain significant amounts of liquid water. The most likely explanation, especially in the face of the Sun having been dimmer early in its history (Girardi et al. 2000), is that early Mars had a more-effective greenhouse atmosphere. Many of the key questions about Mars over the last three decades have been centred on the nature and evolution of this early atmosphere. Measurements from spacecraft in orbit and on the surface indicate that there are insufficient carbonate deposits on the Mars surface and in the subsurface to hold enough CO_2 from this early, thicker atmosphere to provide significant greenhouse warming. In that context, the Mars Atmosphere and Volatile Evolution (MAVEN) mission was designed to explore the loss of gas to space at the present epoch in detail. The goals of the mission are to understand the composition, structure, and variability of the upper-atmosphere system, the energetic drivers of the system, how they lead to loss of atmospheric gas to space, and the integrated amount of gas that has been lost to space through time (Fig. 5). The results on specific loss processes and their rates of loss are used to infer the integrated loss to space through time and the degree to which loss to space can explain the changes inferred for the Martian climate (Jakosky et al. 2015a, 2015b, 2015c; Jakosky et al. 2018 and references therein).

MAVEN has helped to determine the amount of gas lost to space through time by deriving the structure of $^{38}Ar/^{36}Ar$ between the homopause and exobase altitudes. Fractionation of argon occurs as a result of loss of gas to space by pickup-ion sputtering, which preferentially removes the lighter atom. The measurements require that 66% of the atmospheric argon has been lost to space. Thus, a large fraction of Mars' atmospheric gas has been lost to space, contributing to the transition in climate from an early, warm, wet environment to today's cold, dry atmosphere (Jakosky et al. 2017). After a full Martian year observation by MAVEN, the rate of loss of gas to space and the driving processes have been determined (Jakosky et al. 2018). Loss rates for H and O are sufficient to remove $\sim$2-3 kg/s to space. By itself, this loss would be significant over the history of the planet. In addition, loss rates would have been greater early in history due to a more-active Sun and enhanced solar EUV flux by a factor $\sim$100–1000 (Ribas et al. 2005; Tu et al. 2015;

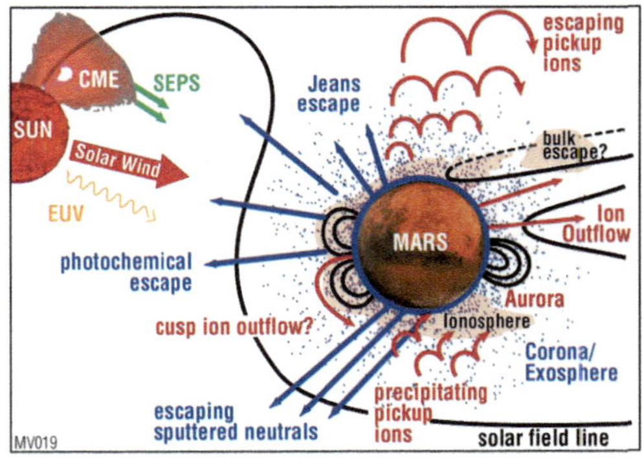

Fig. 5 Schematic showing the processes affecting the atmospheric escape on Mars. Credit: The Lunar and Planetary Institute and LASP

Güdel 2020), although the Sun was photometrically dimmer (Girardi et al. 2000). Integrated loss, based on current processes whose escape rates in the past are adjusted according to the expected solar evolution, would have been as much as 0.8 bar CO_2 or 23 m global equivalent layer of H_2O; these losses are likely to be lower limits due to the nature of the extrapolation of loss rates to the earliest times. Combined with the lack of surface or subsurface reservoirs for CO_2 that could hold remnants of an early, thick atmosphere (Edwards and Ehlmann 2015), these results suggest that loss of gas to space has been the dominant process responsible for changing the climate of Mars.

However, even if we understand the present-day H loss rate, it would be difficult to extrapolate to ancient times. On the one hand, loss could have been greater than it is today based on the increased solar EUV flux which would have increased the rate of photodissociation of water, and also would have provided additional heating of the upper atmosphere, both of which would increase the H escape rate. On the other hand, loss of H could have been lower in ancient times if, by having a more-Earth-like atmosphere, the Martian middle atmosphere had a more-efficient cold trap that kept the H_2O from getting high in the atmosphere regardless of the dust behaviour. This uncertainty, even in the indication of the change in loss rate at earlier times, makes extrapolation difficult. Integrated loss could have been as much as an order of magnitude greater than derived from the present loss rate. As a better understanding of the relative loss rates of deuterium and H is obtained utilising the MAVEN IUVS observations of D (Clarke et al. 2017), a better understanding of the integrated loss of D through time can be expected allowing to use the observed D/H as another way to infer the total amount of water lost through time.

Moreover, though the role of interplanetary coronal mass ejections (ICMEs) has been clearly shown (Jakosky et al. 2015b), the effects of flares and SEPs are still being analysed. In addition, the non-uniform solar wind contains what are known as stream interactive regions (SIRs), where low and high speed wind from different solar source regions collide. Observations of sun-like stars suggest that, in addition to more flares and SEPs, these events occurred more often early in the Sun's history, providing a space environment of essentially continuous ICMEs and SIRs. As these events individually result in an enhancement in loss at the present, it seems likely that they may have had similar effects early in Mars' history. There is no sufficient data yet to understand the cumulative contributions to atmospheric loss resulting from the largest events or how to extrapolate them back in time. However, because of their potential for enhancement of loss, the 0.8-bar loss described earlier is likely to be a conservative lower limit on total loss, conceivably by orders of magnitude.

Table 5 provides a summary of the major past and present missions to Mars with composition measurements.

Table 5 Major past and present Mars missions with composition measurements

Mission	Mission type	Composition measurements
Mariner 6, 7 (1969)	Interplanetary probes + Mars flybys	UV spectrometer & IR spectrometer
Mars 2, 3 (1971–1972)	Mars orbiter + lander	Lyman-α H/D spectrometer
Mariner 9 (1971–1972)	Mars orbiter	UV spectrometer & IR spectrometer
Viking 1, 2 (1976–1982)	Mars orbiter + lander	IR spectrometer & Gas chromatograph – mass spectrometer
Mars Global Surveyor (1997–2006)	Mars orbiter	Thermal emission imaging spectrometer
Mars Pathfinder (Sojourner) (1997)	Mars lander + rover	α proton X-ray spectrometer
2001 Mars Odyssey (2001–)	Mars orbiter	γ-ray spectrometer & Thermal emission imaging system
Mars Express (2003–)	Mars orbiter	Analyser of plasma and ENAs & IR spectrometer & Visible and IR mineralogical mapping spectrometer & Sub-surface sounding radar altimeter
Spirit (2004–2010)	Rover	α-particle X-ray spectrometer & Mössbauer spectrometer & Thermal emission spectrometer
Opportunity (2004–2018)	Rover	α-particle X-ray spectrometer & Mössbauer spectrometer & Thermal emission spectrometer
Mars Reconnaissance Orbiter (2006–)	Mars orbiter	Visible and near IR spectrometer
Phoenix (2008)	Mars lander	Thermal and evolved gas analyser & Wet chemistry laboratory
Mars Science Laboratory (Curiosity) (2012–)	Rover	Chemistry and camera complex (ChemCam) & α-particle X-ray spectrometer & Chemistry and mineralogy X-ray powder diffraction and fluorescence instrument & Sample analysis at Mars instrument & Dynamic albedo of neutrons instrument
Mars Orbiter Mission (2014–)	Mars orbiter	Lyman-α photometer & Exospheric neutral composition analyser
MAVEN (2014–)	Mars orbiter	Neutral gas and ion mass spectrometer & Suprathermal and thermal ion composition instrument & Imaging UV spectrometer
ExoMars Trace Gas Orbiter (2016–)	Mars orbiter	IR spectrometers & Epithermal neutron detector
Al Amal (Hope) Emirates Mars Mission (launched in 2020)	Mars orbiter	UV spectrometer & IR spectrometer
Tianwen-1 (Huoxing-1) (launched in 2020)	Mars orbiter + lander + rover	Mineralogy spectrometer & Ion and neutral particle analyser & Ground-penetrating radar
Mars 2020 (Perseverance) (launched in 2020)	Rover	Chemistry and camera complex (SuperCam) & X-ray fluorescence spectrometer & UV Raman spectrometer & Ground-penetrating radar

Since MAVEN is a single-spacecraft mission, it is not able to fully address some important plasma processes at Mars. When compared to the Earth, in the last 20 years, multi-spacecraft missions like Cluster, THEMIS-ARTEMIS, Van Allen Probes, and MMS have revolutionised our understanding of the causes, patterns, and variability of a wide array of space plasma phenomena in the Earth's magnetospheric environment (cf. Sect. 2.1). The ESCAPADE mission (Escape, Plasma and Acceleration Dynamics Explorers) is a twin-spacecraft Mars mission that has been selected for Phase A/B study by NASA as one of three finalists in the SIMPLEX-II program (Lillis et al. 2018). It aims to similarly revolutionise our understanding of how solar wind momentum and energy flows throughout Mars' magnetosphere to drive ion and sputtering escape, two processes which have helped shape Mars' climate evolution over solar system history. ESCAPADE will measure magnetic field strength and topology, ion plasma distributions (separated into light and heavy masses), as well as suprathermal electron flows and thermal electron and ion densities from elliptical, 200 km $\times$ $\sim$7000 km orbits. ESCAPADE consists of two small spacecraft (<90 kg), traveling to Mars via solar electric propulsion as a rideshare with the Psyche metal-asteroid mission in August 2022, matching Mars' heliocentric orbit until capture and spiral-down to science orbits. ESCAPADE's strategically-designed, 1-year, 2-part scientific campaign of temporally- and spatially-separated multipoint measurements in different parts of Mars' diverse plasma environment, will for the first time unravel the cause-and-effect of solar wind control of ion and sputtering escape.

For the following years, a series of orbiters and landers is also expected to study Mars' environment and surface. NASA launched on 30 July 2020 the Mars 2020 rover, Perseverance, derived from the Curiosity rover. Its payload includes the SuperCam laser-induced breakdown spectroscopy instrument, an X-ray spectrometer, a UV spectrometer, a ground-penetrating radar and a weather station (Fig. 6). In addition to the in situ analysis, it will also collect rock samples that will then be returned to Earth in two subsequent missions, in collaboration with ESA. Mars sample return, although presenting many challenges, including the issue of planetary protection (Haltigin et al. 2018; Raulin et al. 2019; Viso 2019; Chan et al. 2020), it will allow a range of analytical techniques to be applied to them, bearing also the hope to detect possible traces of extinct or extant life (Grady 2020). Mars 2020 features also a small detachable helicopter drone, Ingenuity.

China launched on 23 July 2020 its first orbiter with lander to Mars, Tianwen-1, known also as Huoxing-1, which will deploy a small instrumented rover. The United Arab Emirates Space Agency launched on 19 July 2020 the Al Amal (Hope) orbiter to Mars, which is equipped with an Ultraviolet Spectrometer, an Infrared Spectrometer and an Exploration Imager to study the atmospheric layers of Mars and the daily and seasonal weather cycles. And a Russian carrier spacecraft will deliver the ESA ExoMars rover (expected launch in August 2022), providing surface mobility, subsurface drilling and automatic sample collection, processing, and distribution to a suite of analytical instruments dedicated to exobiology and geochemistry research.

3.3 Mercury

Mercury being the closest planet to the Sun, it is subject to extreme environmental conditions that led to its unique evolutionary history. It lacks the collision- dominated atmosphere of the other terrestrial planets, but has instead a tenuous surface-bounded exosphere, composed primarily of atoms and molecules released from the planet's surface (e.g. Killen and Ip 1999; Wurz and Lammer 2003; Killen et al. 2007; Vervack et al. 2010; Leblanc and Johnson 2010). Production mechanisms for Mercury's exosphere include solar wind sputtering, micrometeorite impact and photon-stimulated desorption (Wurz et al. 2010; Pfleger et al. 2015; Killen

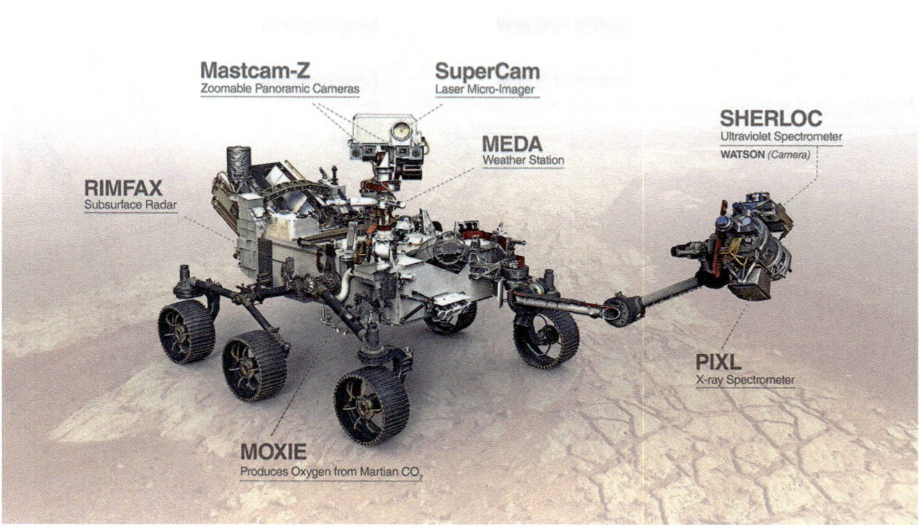

Fig. 6 Instruments onboard the Mars 2020 Perseverance rover. Credit: NASA

Table 6 Major Mercury missions with composition measurements

Mission	Mission type	Composition measurements
Mariner 10 (1974–1975)	Interplanetary probe + Mercury flybys	UV spectrometers & IR radiometer
MESSENGER (2008–2015)	Mercury orbiter	γ-ray spectrometer + X-ray spectrometer + UV and visible spectrometer IR spectrometer Neutron spectrometer
BepiColombo (launched in 2018)	Mercury planetary orbiter + magnetospheric orbiter	γ-ray spectrometer + Imaging X-ray spectrometer + UV spectrometer + IR spectrometer + Neutron spectrometer + Ion mass spectrometers + Neutral mass spectrometer + ENA imager + Sodium atmosphere spectral imager

and Hahn 2015), cf. Fig. 7. The weak planetary magnetic field and the high solar wind dynamic pressure result in a highly-compressed magnetosphere and a direct interaction between the solar wind and the surface-bounded exosphere (Massetti et al. 2003, 2007; Milillo et al. 2005; Slavin et al. 2009; Raines et al. 2015; Orsini et al. 2018).

Table 6 provides a summary of the major past and present missions to Mercury with composition measurements.

The MESSENGER (MErcury Surface, Space ENvironment, GEochemistry, and Ranging) NASA mission orbited Mercury between 2011 and 2015 (Solomon et al. 2001). It provided the first comprehensive observations of Mercury's magnetosphere (Benna et al. 2009; Slavin et al. 2009; Alexeev et al. 2010; Ho et al. 2011; Zurbuchen et al. 2011), global mag-

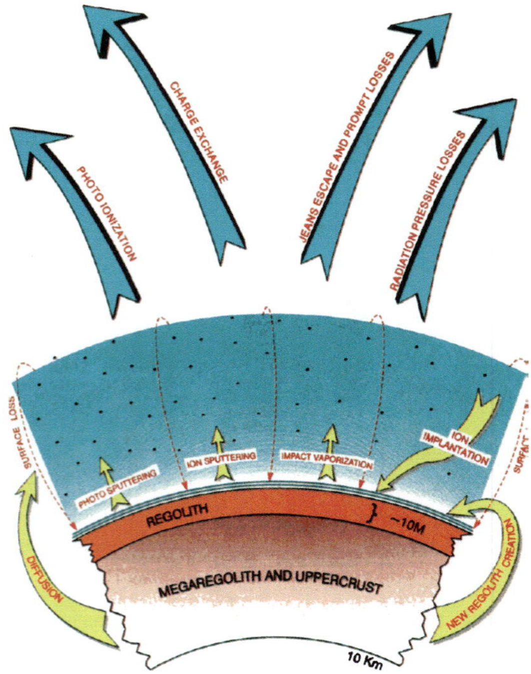

Fig. 7 Schematic showing sources and sinks of the Hermean exosphere (from Killen and Ip 1999)

netic field (Anderson et al. 2011), exosphere (McClintock et al. 2009; Burger et al. 2010; Killen et al. 2010; Vervack et al. 2010), surface chemical composition (Denevi et al. 2009; Nittler et al. 2011) and core size and state (Genova et al. 2019).

The BepiColombo mission has been prepared jointly by ESA and JAXA (Benkhoff et al. 2010). Launched in October 2018, after a series of flybys of Earth (April 2020) and Venus (October 2020 and August 2021), it will arrive at Mercury in 2025. It is an interdisciplinary mission that comprises a Mercury Planetary Orbiter (MPO), provided by ESA, that will study the exosphere, the surface and the internal composition of the planet, and a Mercury Magnetospheric Orbiter (MMO), provided by JAXA, that will study Mercury's magnetosphere and its interaction with the solar wind. MMO is equipped with five instruments, including a Mass Spectrum Analyser (Delcourt et al. 2009) and a Mercury Sodium Atmospheric Spectral Imager (Yoshikawa et al. 2009). MPO is equipped with eleven instruments, including the SERENA (Search for Exospheric Refilling and Emitted Natural Abundances) ion and neutral atoms instrument package (Orsini et al. 2010) and the PHEBUS (Probing of Hermean Exosphere By Ultraviolet Spectroscopy) ultraviolet spectrometer to observe Mercury's exosphere (Chassefière et al. 2010). The payload of both spacecraft will provide the detailed information necessary to understand the origin and evolution of the planet itself and its surrounding environment (Milillo et al. 2020). In addition, instrumentation onboard BepiColombo will be used to test Einstein's theory of general relativity.

Table 7 Major past and present asteroid and comet sample return missions

Mission	Sampled object
Stardust (1999–2006)	Coma of comet Wild 2
Hayabusa (2003–2010)	Asteroid 25143 Itokawa
Hayabusa2 (2014–)	Asteroid 162173 Ryugu
OSIRIS-REx (2016–)	Asteroid 101955 Bennu

4 Asteroid Sample Return Missions

4.1 Asteroid Ryugu Sample Return Mission: Hayabusa2

Hayabusa2 ("Peregrine falcon"-2) is an asteroid sample-return mission operated by JAXA. It follows on from the Hayabusa mission which in 2010 returned samples collected from the near-Earth asteroid 25143 Itokawa (Nakashima et al. 2013; Takashi et al. 2014; Fujiya et al. 2016).

Hayabusa2, launched in 2014, arrived at the near-Earth carbonaceous asteroid 162173 Ryugu in 2018. Such asteroids are assumed to have been produced from much older parent bodies through catastrophic disruption and re-accumulation of fragments during the evolution of the Solar System (O'Brien and Greenberg 2005; Sugita et al. 2019). Equipped with remote sensing instruments, sampling equipment and four small rovers, it acquired data characterising Ryugu as a spinning top-shaped rubble pile with a prominent circular equatorial ridge (Watanabe et al. 2019).

The first surface sampling by Hayabusa2 was performed in February 2019 and obtained a substantial amount of asteroid regolith material (Morota et al. 2020). The second sample collection was performed in July 2019 and used an impactor to collect sub-surface, pristine material. The sample-return capsule will re-enter Earth's atmosphere and land under a parachute in late 2020.

4.2 Asteroid Bennu Sample Return Mission: OSIRIS-REx

OSIRIS-REx (Origins, Spectral Interpretation, Resource Identification, Security, Regolith Explorer) is NASA's first asteroid sample return mission (Bierhaus et al. 2018). Its goal is to obtain a sample from 101955 Bennu, a carbonaceous near-Earth asteroid, and return the sample to Earth for a detailed analysis. Bennu is a primitive body that represents the objects that may have brought prebiotic molecules and volatiles such as water to Earth (Lauretta et al. 2015).

The OSIRIS-REx spacecraft carries a suite of instruments to characterise Bennu's global properties, support the selection of a sampling site and document that site at a sub-centimetre scale. Launched in 2016, the spacecraft reached the proximity of Bennu in December 2018, where it began analysing its surface for a target sample area. Preliminary surveys of the asteroid's surface indicated Bennu to be hydrated and volatile-rich, as expected, but presenting also an unexpected surficial diversity (Lauretta et al. 2019).

In April 2020 a sample collection rehearsal was successfully performed. The touch-and-go sample collection event is now scheduled for October 2020. The sample return capsule will re-enter Earth's atmosphere in September 2023 and land under a parachute.

Table 7 provides a summary of the major past and present asteroid and comet sample return missions.

5 Exoplanet Missions

5.1 The Internal Structure and the Atmospheres of Close-in Rocky Planets

One of the most exciting discoveries of the past decade was the detection of planets with masses smaller than 15 M_{Earth} having densities consistent with rocky compositions. A special class of such planets are transiting ultra-short period planets (USPs) that have orbital periods of one day or less (e.g. Howard et al. 2013; Dai et al. 2017). Because of the short orbital periods, their masses and radii can be determined with a high accuracy. The reason is that the K-amplitude of the radial-velocity variation along the orbit scales with $P^{-1/3}$, and the radius can be determined with high accuracy because the light-curves of many transits can be averaged. Many instruments have been used to detect low-mass planets by means of Radial-Velocity (RV) measurements. Today an accuracy of 1 m/s is achieved routinely with instruments like HARPS (Mayor et al. 2003; Bouchy et al. 2017), HARPS-N (Cosentino et al. 2012) and CARMENES (Quirrenbach et al. 2016). This accuracy allows to detect USPs with only a few M_{Earth}. The next step in accuracy is ESPRESSO (Pepe et al. 2010), which is designed to achieve even 0.1 m/s. ESPRESSO is also on a bigger telescope (8-m class) than the others (4-m class), which results in a significant gain in sensitivity. This allows determining the masses of USPs to a higher accuracy, and allows detecting planets of lower masses.

USPs do not have extended Hydrogen atmospheres. This is because a Hydrogen atmosphere would be quickly eroded due to the intense X-ray and extreme UV radiation from the host star (Fossati et al. 2017; Airapetian et al. 2017a; Kubyshkina et al. 2018). However, as explained below, they can have other types of atmospheres. USPs are thus ideal targets to study the internal structure and surface composition of rocky planets. USPs with masses in the range between 5 to 10 M_{Earth} are often called "super-Earths", although it is not so obvious how Earth-like they really are. In the following, we will discuss what we have learned from transiting USPs, and how we can use them to test and constrain the various scenarios of rocky planet formation.

Table 8 gives an overview of the transiting USPs with masses of 15 M_{Earth}, or less. We also include HD 213885 b and GJ 9827 b, although their orbital periods are a little longer than one day. At least eleven of the thirteen planets are members of systems which contain up to five known planets.

The densities of the planets can be divided into two broad categories: Planets with a density consistent with an Earth-like composition (30% metallic core, 70% silicate mantle) and planets where the density is consistent more with a Mercury-like composition (70% metallic core, 30% silicate mantle). There is possibly even a third category composed of planets with densities 10-20% lower than for an Earth-like composition. Examples are HD 219134 b, 55 Cancri e and WASP-47 e. As pointed out by Dorn et al. (2019), the relatively low densities of such planets can be explained with a different composition of key rocky species (Fe, Mg, Si, Ca, Al, Na). Alternatively, they could have an opaque but not very extended atmosphere. Figure 8 shows the mass-density diagram for USPs. Although the error bars for most planets are too large to put them definitely into one of the three categories, it is surprising that the majority of them fall below the line for an Earth-like composition. Perhaps most USPs belong into the third category, but the error bars of individual objects are usually too large to clearly separate them for the second category.

Up to now, the best cases for planets with a Mercury-like composition are K-106 b and K2-229 b. A third example is Kepler-107c, which has a density of 12.65±2.45 g cm^{-3} but it is not an USP (Bonomo et al. 2019). The Kepler-107-system is a remarkable one, because

Table 8 Overview of the transiting USPs (ultra-short period planets) with masses of 15 M_{Earth}, or less

Object	Semi-major axis	Orbital period [days]	Mass [M_{Earth}]	Radius [R_{Earth}]	Density [$g\,cm^{-3}$]	Multiple system	References
K2-141 b	0.00716	0.280	5.31 ± 0.46	$1.54^{+0.10}_{-0.09}$	$8.00^{+1.83}_{-1.45}$	YES	Barragán et al. (2018)
K2-131 b	0.0092	0.369	6.5 ± 1.6	$1.81^{+0.16}_{-0.12}$	$6.0^{+3.0}_{-2.7}$	YES	Dai et al. (2017)
Kepler-78 b	0.01	0.355	1.69 ± 0.41	1.20 ± 0.09	5.3 ± 1.8	NO	Howard et al. (2013)
K2-106 b	0.0116	0.571	$8.36^{+0.96}_{-0.94}$	1.52 ± 0.16	$13.1^{+5.4}_{-3.6}$	YES	Guenther et al. (2017)
K2-229 b	0.01289	0.584	2.59 ± 0.43	$1.164^{+0.066}_{-0.048}$	8.9 ± 2.1	YES	Santerne et al. (2018)
Kepler-1520 b	0.013	0.654	0.02–0.1	0.85	–	NO	Rappaport et al. (2013)
55 Cancri e	0.0154	0.737	8.0 ± 0.3	1.88 ± 0.03	6.7 ± 0.4	YES	Bourrier et al. (2018)
Kepler-10b	0.0169	0.837	3.33 ± 0.49	$1.47^{+0.03}_{-0.02}$	5.8 ± 0.8	YES	Dumusque et al. (2014)
CoRoT 7b	0.0172	0.854	4.73 ± 0.95	1.58 ± 0.09	6.61 ± 1.72	YES	Haywood et al. (2014)
WASP-47 e	0.0173	0.790	9.11 ± 1.17	1.87 ± 0.13	7.63 ± 1.90	YES	Sinukoff et al. (2017)
HD 3167 b	0.0182	0.960	5.69 ± 0.44	1.574 ± 0.054	$8.00^{+1.10}_{-0.98}$	YES	Gandolfi et al. (2017)
GJ 9827 b	0.020	1.209	$3.69^{+0.48}_{-0.46}$	$1.58^{+0.14}_{-0.13}$	$5.11^{+1.74}_{-1.27}$	YES	Prieto-Arranz et al. (2018)
HD 213885 b	0.0278	1.008	8.83 ± 0.66	1.745 ± 0.052	$9.15^{+1.38}_{-1.36}$	YES	Espinoza et al. (2019)

Fig. 8 Mass-density diagram of ultra-short period planets (adapted from Guenther et al. 2017, density calculation by R. Helled)

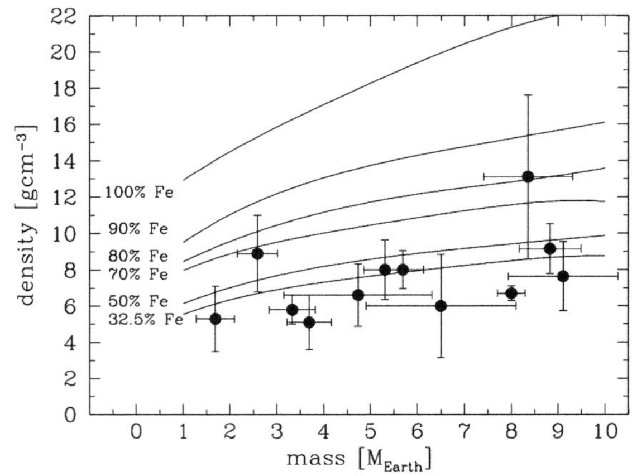

the inner planet, Kepler-107 b ($a = 0.0454$ AU), has a density that is consistent with an Earth-like composition, whereas Kepler-107 c ($a = 0.0606$ AU) has a much higher density (Bonomo et al. 2019). Because of this unusual architecture, Bonomo et al. (2019) argue that Kepler-107 c is the result of a giant impact that removed the outer layers of the planet.

The densest planets, such as K2-106 b and K2-229 b, might give us important clues about the formation of planets in general. Before calling all close-in and high-density planets Mercury-like, we should keep in mind that Mercury has a semi-major axis of 0.387 AU, a radius of 0.3829 R_{Earth}, a mass of 0.055 M_{Earth}, and a density of 5.427 $g\,cm^{-3}$. The USPs, instead, have a 50 to 170 times larger mass and they are 14 to 55 times closer to their host star than Mercury is. Such planets are not Mercury-like but far more extreme cases! It would be very interesting to know how the high-density USPs are formed. There are two possibilities: They are either the result of a giant impact (Leinhardt and Stewart 2012), or they are formed from metal rich material (Néri et al. 2019). The giant impact scenario requires a really dramatic event. Let us take K2-106 b and K2-229 b as examples. If these planets were the result of a giant impact they would have lost 40-50% of their masses, corresponding to 20-30 M_{Earth}. According to Valencia et al. (2019) such extreme events are rare. Thus, the more high-density planets are found, the less likely this giant impact hypothesis becomes.

Planets forming close to their stars are expected to have a higher metal content. The reasons are that planetesimals in the inner region of protoplanetary disks would preferentially be iron-rich, because the condensation temperatures of metals are higher than those of silicates (Wurm et al. 2013), and because the planetesimal formation is sensitive to the size of the largest aggregates, which have a larger iron content (Kruss and Wurm 2018). Thus, if USPs form close to the host stars, we expect them to be metal rich. This scenario explains thus the formation of high density USPs and also the iron gradient in the solar system.

Generally solid planets must have different compositions, because of the different condensation temperatures, chemical composition of the disks, interior dynamics of the planets, and outgasing histories. Not only the high-density planets can be explained with these mechanisms but also the low-density ones like 55 Cancri e, WASP-47 e, and HD219134 b (Elkins-Tanton 2008; Dorn et al. 2019).

Even better than the density measurements, it would be if we could determine the composition of the planets, at least the crust, directly. This would be important not only for the high-density planets but also for all solid planets. While USPs cannot have extended Hydrogen atmospheres (Fossati et al. 2017), they can have other types of atmospheres. In the following we will discuss volcanic atmospheres, mineral atmospheres, which very hot planets could have, and Mercury-like atmospheres. The chemical analysis of atmospheres that contain material released from the crust would be ideal to constrain the composition of the crust.

The currently preferred object for atmospheric studies of low-mass USPs is transiting planet 55 Cancri e. It is a small planet that orbits a nearby ($d = 12.590 \pm 0.012$ pc; Gaia Collaboration 2018), relatively bright ($V = 5.95$; SIMBAD) star. The mass and the radius of the planet can thus be determined with very high accuracy. Bourrier et al. (2018) and Crida et al. (2018) argue that, because of the relatively low density, 55 Cancri e has an atmosphere that is 0.08+/-0.05 R_{planet} thick. However, Madhusudhan et al. (2012) argue that there is no need for an atmosphere if it is assumed that the interior of the planet is carbon-rich, which is plausible given the carbon-rich nature of the host star. As discussed above, the density distribution of USPs does not scatter around the line for planets with Earth-like compositions, as there are many planets of smaller density. Perhaps this is a sign that all these planets host atmospheres that are not extended, perhaps containing molecules with a high mean molecular mass. In the case of 55 Cancri e, there is direct evidence for the presence of an atmosphere. For example, the infrared thermal emission phase curve has a pronounced eastward-asymmetry. The hot spot thus is not at the sub-stellar point but 41 ± 12 degrees east of it (Demory et al. 2016a). According to Angelo and Hu (2017), this requires an atmosphere with a pressure of about 1.4 bar. Repeated infrared observations have shown that the depth of the secondary eclipse varies with time, which has also been used as an argument in favour of an atmosphere (Deming et al. 2015; Demory et al. 2016b; Tamburo et al. 2018). The idea is that the substellar hemisphere is covered by material that has a high reflectivity, which varies in size. This material could be some kind of bright clouds, or haze, in an atmosphere. The most direct evidence for the presence of an atmosphere comes from the optical transmission spectroscopy. Combining the data from five transits, Ridden-Harper et al. (2016) presented hints of Na and the Ca^+ lines originating in the atmosphere of the planet. If confirmed, this signal would correspond to an optically thick Ca^+ atmosphere that is five times larger than the Roche lobe. In summary, there is some evidence that 55 Cancri e has an atmosphere, hence it is possible for USPs to host some sort of atmosphere.

Since a Hydrogen-rich atmosphere is most likely ruled out, what could it be? One possibility is an atmosphere of volcanic origin. Shortly after the discovery of CoRoT-7b (Léger et al. 2009), a number of authors came up with the idea of a lava ocean at the substellar point (Briot and Schneider 2010; Barnes et al. 2010; Rouan et al. 2011). The reason that such a planet should have a lava ocean is not only the intense radiation by the host star, but also tidal heating. If the host star is strongly magnetic and the planet along its orbit crosses the stellar magnetic field lines, the sub-surface regions of USPs can be subject to induction heating (Kislyakova et al. 2017, 2018) which can, in principle, be even stronger than the tidal heating. Since all stars with outer convection zones have magnetic fields, induction heating might be very common, particularly at young ages when stellar magnetic fields are stronger.

A lava-lake model would also explain the large albedo of Kepler-10b. Lava on Earth is dark, but it is expected to be as white as snow if the temperature is much higher. The reason is that the oxides with the highest fusion temperatures are MgO (3105 K), CaO (3200 K), UO_2 (3140 K), and particularly ThO_2 (3660 K). Thus, at the highest temperatures, a lava lake will contain ThO_2 particles, dispersed in the Al_2O_3 – CaO, and this material is white (Rouan et al. 2011). Kite et al. (2016) found that the surface composition of a magma ocean tracks bulk composition of the planet if the temperature is less than 2400 K, but at higher temperatures the composition of the surface of the magma ocean can differ from the bulk of the ocean. Nevertheless, it would be very important to detect the material released from

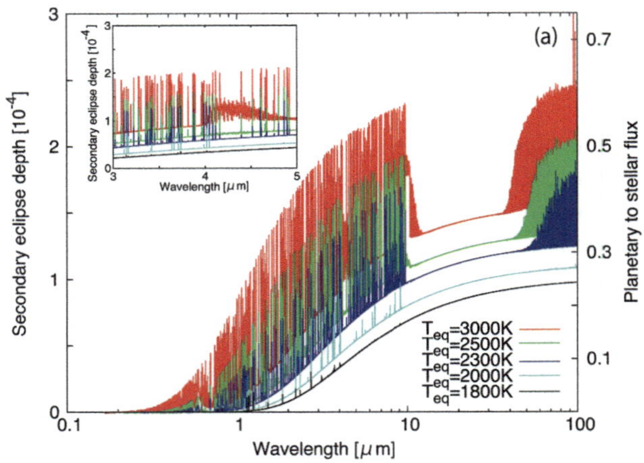

Fig. 9 Predicted dayside-averaged emission spectra of a hot rocky planet of 2 R_{Earth} in secondary eclipse, which has a mineral atmosphere in equilibrium with an underlying magma ocean (Ito et al. 2015)

a lava ocean to constrain the chemical composition of the crust. Volcanically active planets are expected to eject mostly SO_2, which would then dissociate into oxygen and sulphur atoms (Kislyakova et al. 2018). Planets with a high surface pressure are expected to have an atmosphere consisting of N_2-and CO_2-rich and dry volcanic gases, while low pressure ones will contain mostly sulphur-rich gases (Gaillard and Scaillet 2014). Possible tracers for such an atmosphere are the same lines that have been detected in the plasma torus of Jupiters moon Io. These are the [S III] 3722 Å, [O II] 3726 Å, [O II] 3729 Å, [S II] 4069 Å, [S II] 4076 Å, [O III] 5008 Å, [O I] 6300 Å, [S III] 6312 Å, [S II] 6716 Å, [S II] 6731 Å lines, as well as the Na D-lines (Brown and Chaffee 1974; Brown et al. 1975; Brown 1976; Brown and Shemansky 1982; Morgan and Pilcher 1982; Thomas 1993; Thomas 1996; Küppers and Jockers 1995, 1997; (Kupo et al. 1976)). Magma oceans are not only present in USPs but it is generally assumed that most, if not all, rocky planets had a magma ocean at young age. The outgassing of these oceans most likely leads to CO_2-rich atmospheres (Elkins-Tanton and Seager 2008; Lammer et al. 2018).

Very hot planets can have mineral atmospheres (Schaefer and Fegley 2009). The most abundant molecule, according to Miguel (2019), is expected to be N_2 followed by H_2 and CO. However, transmission spectra would be dominated by NH_3 and HCN, and CO and HCN. According to Ito et al. (2015) a planet with a rocky molten surface should have an atmosphere that is composed of gas-phase species containing Na, K, Fe, Si, SiO, O, and O_2. Figure 9 shows the predicted spectra for such planets at different temperatures (Ito et al. 2015).

Hot close-in planets could also have Mercury-like atmospheres (Vidotto et al. 2018). Mercury has an extremely thin atmosphere ($\sim 5 \times 10^{-12}$ mbar) containing Na, K and Ca, and other elements that are released from the crust through sputtering (Pfleger et al. 2015). Strictly speaking, this atmosphere is Mercury's exosphere (cf. also Sect. 3.3). The atmosphere of Mercury has already been detected 20 years ago using transit observations (e.g. Bida et al. 2000; Schleicher et al. 2004). Given the proximity of the USPs to their host stars, it is very well possible that they host a Mercury-like atmosphere. Since these planets are much bigger than Mercury and closer to their host stars, their atmospheres should be much thicker. For example, the surface area of CoRoT-7b is 17 times larger than Mercury and CoRoT-7b receives 250–370 times the amount of radiation. Mura et al. (2011) developed a model for Mercury-like atmosphere of CoRoT-7b. This model shows that the atmospheric loss rates of CoRoT-7b are expected to be 2–3 orders of magnitude higher than that of Mercury, forming an ionised tail of escaping particles. This tail should be observable in the

Na and Ca resonance lines in the optical band. A first attempt to detect the atmosphere of CoRoT-7b was carried out by Guenther et al. (2011) using spectra taken with UVES at the VLT. The observations could just provide an upper limit to CoRoT-7b's atmospheric column density, indicating the need of instruments with higher sensitivity, or similar objects that orbit closer and brighter stars. Guenther et al. (2011) selected CoRoT-7 b as a target because it was the only transiting USP known at the time (Léger et al. 2009).

Luckily we now know a number of USPs orbiting stars that are closer and brighter than CoRoT-7. We mentioned already 55 Cancri. Another interesting target is HD 3167, which is 13 times brighter than CoRoT-7. We recently observed one transit of HD 3167 b with UVES obtaining upper limits for the emission line-fluxes relative to the star of $(0.01 \pm 1.04) \times 10^{-3}$ and $(1.02 \pm 0.96) \times 10^{-3}$ for Na and Ca^{+}, respectively. For comparison, *Ridden-Harper et al.* (2016) detected the Na and Ca^{+} at a level of $\sim 2.3 \times 10^{-3}$ and $\sim 7 \times 10^{-2}$ respectively, relative to the stellar spectrum in 55 Cancri e. However, the Ca^{+} signal was only seen in one of the five observed transits. The difference between the two is that 55 Cancri e orbits a bit closer to its host star and it is a bit bigger than HD 3167 b. Perhaps HD 3167 b has weaker lines, or has a different crustal composition, or the absorption signal is time variable. Another interesting object is Kepler-1520 b, because it has a dusty tail containing material that is escaping from the planet (Schlawin et al. 2018). More observations of more objects are certainly required.

Although WASP-12 b is a hot Jupiter that orbits a main sequence G0 star at a distance of 0.0234 AU, it may also be interesting in the context of solid planets. The reason is that the host star, WASP-12, lacks emission components of the Mg II H and K and the Ca II H and K lines, though they are present in all stars with similar rotational velocity and spectra type. A possible explanation is that the star produces the line core emission but it is absorbed by a translucent torus of ionised Mg and Ca surrounding the system (Haswell et al. 2012; Fossati et al. 2013). The material composing the torus may originate from the escaping atmosphere of the hot Jupiter (Debrecht et al. 2018). However, Kislyakova et al. (2016) developed a model in which this material is released from the lava oceans of Trojan satellites of the planet. Trojan satellites of WASP-12b have not been found yet but Hippke and Angerhausen (2015) found statistical evidence for Trojans of Jupiter-like planets. After averaging all light-curves of all planets in the 50–350 days period range they found that there is a statistically significant dip corresponding to Trojans with radii between 400 and 1200 km. For comparison, the total transiting area equivalent of small Jupiter Trojans corresponds to one body of $\sim$600 km radius, and the largest 32 account for an additional radius equivalent of $\sim$300 km (Hippke and Angerhausen 2015; Fernández et al. 2003). There is thus some statistical evidence that at least longer period planets have Trojan satellites similar to Jupiter. Given that WASP-12b orbits at 0.023 AU (Hebb et al. 2009), Trojans would have a lava ocean, if they exist.

5.2 Exoplanet Missions

Table 9 provides a summary of the major past and present exoplanet space missions.

The new and upcoming exoplanet missions will dramatically change that situation of known exoplanets.

The TESS (Transiting Exoplanet Survey Satellite) mission of NASA is a big leap forward in exoplanet research because it will survey the whole sky to find planets around relatively bright stars (Ricker et al. 2015). Its main targets are 200 000 F5 to M5 stars (1.3 and 0.2 M_{Sun}) of $m_I = 8 - 12$ mag. TESS targets are thus about 100 times brighter than Kepler

Table 9 Major past and present exoplanet space missions

Mission	Mission type
CoRoT (*Convection, Rotation et Transits planétaires*) (2006–2013)	Space telescope: transit method
Kepler (2009–2018)	Space telescope: transit method
Gaia (2013–)	Astrometric space observatory
TESS (Transiting Exoplanet Survey Satellite) (2018–)	Space telescope: transit method
CHEOPS (CHaracterising ExOPlanets Satellite) (2019–)	Space telescope: transit method

targets. Detected planetary candidates can thus be followed-up with small- and mediums-sized ground-based telescopes. The satellite has been launched in April 2018. It was initially a two-year mission but it has already been extended for another two years. Further extensions are possible. It is estimated that TESS will detect about 49 Earth-size planets ($R_P < 1.25 R_{\mathrm{Earth}}$), 385 super-Earths (1.25 $\mathrm{R_{Earth}}$ $< R_P < 2R_{\mathrm{Earth}}$), 2049 sub-Neptunes (2.0 $\mathrm{R_{Earth}}$ $< R_P < 4R_{\mathrm{Earth}}$), and 2529 gas-giants (Sullivan et al. 2015; Bouma et al. 2017). TESS is an all-sky survey but only the areas located within 12° of the ecliptic poles are observed for 351 days. The area of the ecliptic pole fields are 905 deg^2, or only 2.2% of the sky. The rest of the sky is observed for a shorter time, mostly only for 27 days. Planets of small sizes and planets with long orbital periods will thus be found primarily in the ecliptic poles fields. Gas-giants with short orbital periods will be found in the whole sky.

Another new mission is ESA's CHEOPS (CHaracterising ExOPlanet Satellite) (Fortier et al. 2014; CHEOPS study team, 2013), launched in December 2019. CHEOPS is the first mission dedicated to searching for exoplanetary transits by performing ultra-high precision photometry on bright stars already known to host planets. It will thus determine the radii of planets found by other missions, for example those found by TESS. CHEOPS has a thermally stabilized f/8, Ritchey-Chrétien telescope with a diameter of 32 cm. It is thus much larger than TESS, which has cameras of only 10 cm aperture. CHEOPS can thus determine the radii of planets to a much higher accuracy and can also detect smaller ones than TESS. Combined with RV-measurements, the bulk densities can be determined. It is estimated that CHEOPS will be able to detect Earth-size planets transiting G5 dwarf stars (stellar radius of 0.9 $\mathrm{R}_\odot$) with V-band magnitudes in the range $6 \leq m_V \leq 9$ mag. It will thus provide a major improvement in the determination of planetary radii, with a strong impact on a better determination of their possible bulk composition.

The next step is ESA's PLATO mission (PLAnetary Transits and Oscillations of stars) (Rauer et al. 2014), which is planned to be launched in 2026. The payload consists of 24 'normal' cameras. These cameras will be read out with a cadence of 25 s and will monitor stars with $m_V > 8$. There are two additional cameras, which have a cadence 2.5 s. These fast cameras will be used for stars with $m_V \sim 4$ to 8. The 'normal' cameras are arranged in four groups of six. Each of them has a field-of view of 1100 deg^2. The central field were all cameras overlap has a size of 301 deg^2, and the whole field is 2250 deg^2. The core of the PLATO mission targets are 22000 FGK-stars with $m_v < 11$ mag. The light-curves of these PLATO will have a noise level of 34 ppm/h. For comparison, the transit depth of 2(1) $\mathrm{R_{Earth}}$

planet orbiting solar-like star is 334(84) ppm. Simulations show that PLATO will be capable of discovering 4000 super-Earth planets around stars with $m_v < 11$ mag. Because PLATO will stare at one field for at least two years, it will be possible to detect Earth-sized planets of long orbital period, e.g. planets in the habitable zone (HZ). It is expected that PLATO will find 40-70 super-Earths in the HZ of G-type stars. For detecting planets with periods of 90 d $< P <$ 500 d PLATO outperforms Kepler by a factor 10, and TESS by a factor 1000. PLATO will provide a huge increase in the number of small transiting planets, and many of them will be USPs.

Detailed studies of USPs will be possible with ESA's ARIEL (Atmospheric Remote-sensing Infrared Exoplanet Large-survey) mission (Puig et al. 2018; Tinetti et al. 2016, 2018; Ludovic 2018), planned to be launched in 2028. Many exoplanets will certainly be observed with the future James Webb Space Telescope (JWST) of NASA, but ARIEL will be the first dedicated space mission for atmospheric studies. It will measure the chemical composition and thermal structure of a large number ($\sim$1000) of exoplanets that have temperatures of at least 600 K. ARIEL will thus provide a truly representative picture of the chemical nature of exoplanets, including gas giants, Neptunes, super-Earths and Earth-size planets around a range of host star types using transit spectroscopy. In this way it will answer the following questions: what exoplanets are made of and how they form and have evolved. ARIEL has a three-mirror off-axis telescope with an elliptical primary of 1.1×0.7 m. The VIS channel has three photometric bands: 0.5-0.6 μm, 0.6-0.81 μm, and 0.81-0.1. μm. The NIR spectrographs cover the wavelength bands 1.1–1.95 μm ($R = 20$), 1.95-3.9 μm ($R = 100$), and 3.9-7.8 μm ($R = 30$). These wavelength ranges contain all the expected major atmospheric gases from e.g. H_2O, CO_2, CH_4, NH_3, HCN, H_2S through to the more exotic metallic compounds, such as TiO, VO, and condensed species.

The next generation of 30 and 40 m-class ground-based telescopes, for example the ELT (Extremely Large Telescope), for which the first light is planned for 2025, will also allow us to study the atmospheres in detail (Rodler 2018). They will even allow a much higher spectral resolution than it will be possible with ARIEL. Possibly the most interesting results can be obtained when the high-resolution spectra, obtained with ground-based telescopes, will be combined with spectra obtained with ARIEL.

5.3 η-Earth and the Relevance of N_2-O_2 Dominated Atmosphere Detection by Future Space Missions

Since the past two decades, the number of exoplanets continues to grow and an understanding of the frequency of habitable Earth-like exoplanets remains a major area of interest. The search for life beyond Earth and the Solar System focuses on finding rocky exoplanets orbiting a star at the orbit location within the so-called habitable zone that allows for the existence of liquid water (e.g. Hart 1979; Lammer et al. 2009; Kasting 2010; Wolf 2017). The mean number of rocky exoplanets per star that reside within their respective habitable zone is called eta-Earth, or ηE. Planets are thought to be rocky with radii less than 1.6 R_{Earth} (Rogers 2015). Usually, ηE is calculated separately for each stellar spectral type (Kasting et al. 1993; Kopparapu et al. 2013).

NASA's Kepler space telescope explored the parameter-space of Earth-sized exoplanets. One of the mission goals was to establish an estimate for ηE. However, a large number of possible ηE-values lying in the wide range between 0.01–0.35 were estimated by various researchers (Catanzarite and Shao 2011; Petigura et al. 2013; Silburt et al. 2015; Traub 2016; Barbato et al. 2018; Zink and Hansen 2019). A main reason for these significant differences in ηE values is that the climate models differ concerning the precise location annulus of these terrestrial planets that could potentially have liquid water on the surface, the maximum radius of a planet being "rocky" is still debated, and the lifetime of the Kepler mission was not enough to have complete statistics for FGK stars.

An additional problem with this size-related ηE definition is the lack of knowledge of the potential accretion histories of these ηE-like planets, their internal composition, tectonic regime, and more (see Bermingham et al. 2020; Lammer et al. 2020b). Each of the mentioned studies have chosen particular limits within possible ranges, but neglected potential relevant parameters such as the distribution and variation of radioactive heat producing elements (see O'Neill et al. 2020) within planetary building blocks and consequently within Earth-sized exoplanets. Moreover, the billion years lasting evolution of planetary atmospheres (see Avice and Marty 2020; Lammer et al. 2020a), including their shaping factors related to their host stars activity evolution (see Güdel 2020), are also neglected in such estimates. ηE is currently the next parameter in the Drake equation where we can get a robust estimate, that is $N_{\mathrm{civ}} = R \times f_p \times \eta E \times f_{\mathrm{l}} \times f_{\mathrm{i}} \times f_{\mathrm{c}} \times L$, which gives the number (N_{civ}) of "hypothetical" civilizations in the Galaxy that are able to communicate through electromagnetic waves (e.g. Sagan 1963; Carigi 2015). The other values in the Drake equation are R, the average rate of star formation in the galaxy, f_p, the fraction of stars that have planets, f_l, the fraction of these planets that could support and actually develop life at some point, f_i, the fraction of these planets with intelligent life, f_c, the fraction of civilizations that release detectable signs of their existence into space, and L, the timespan over which such civilizations actually release these signals into space. These last three parameters have the largest uncertainty and have no scientific estimate at this time.

Sara Seager (2017) proposed a revised Drake equation that aims towards biosignature gas detection in exoplanetary atmospheres. This equation can be written as $N_{\mathrm{bio}} = N_{\mathrm{star}} \times f_{\mathrm{q}} \times f_{\mathrm{HZ}} \times f_{\mathrm{o}} \times f_{\mathrm{live}} \times f_{\mathrm{s}}$, where N_{bio} is the number of planets with detectable signs of life, N_{star}, the number of stars observed, f_{q}, the fraction of stars that are quiet enough for terrestrial planets to sustain secondary atmospheres, f_{HZ}, the fraction of stars with rocky planets in the habitable zone, f_{o}, the fraction of these planets that can be observed, f_{live}, the fraction of these planets that have life, and f_{s}, the fraction on which life produces a detectable atmospheric biosignature.

In order to search for life and biospheres on Earth-like exoplanets, various molecular biosignatures have been studied and proposed (e.g. Lovelock 1975; Segura et al. 2003; Grenfell et al. 2007a, 2007b, 2010; Scalo et al. 2007; Cockell et al. 2009; Schwieterman et al. 2017). Besides biosignature molecules such as N_2O (Sagan et al. 1993; Segura et al. 2005; Rauer et al. 2011; Rugheimer et al. 2013, 2015; Airapetian et al. 2017b), CH_4 (Sagan et al. 1993; Krasnopolsky et al. 2004; Rugheimer et al. 2015; Airapetian et al. 2017b), CH_3Cl (Segura et al. 2005; Rugheimer et al. 2015), NH_3 (Seager et al. 2013a, 2013b), sulfur gases and C_2H_6 (Pilcher 2003; Domagal-Goldman et al. 2011), and organic hazes (Arney et al. 2016, 2017), O_2 has long been recognized as a key biosignature. This is because on Earth it is only produced by life and it has a high concentration in the modern terrestrial atmosphere and has a strong spectral feature at 0.76 microns. In the IR, O_2 can be detectable through its photochemical proxy, O_3 (Owen 1980; Léger et al. 1993; Sagan et al. 1993; Des Marais et al. 2002; Léger et al. 2011; Airapetian et al. 2017b).

Oxygen is thought to be a necessary ingredient for the evolution of complex life forms on habitable planets due to the large redox energy gradient pair of aerobic respiration (Catling et al. 2005; Meadows et al. 2018). There are known abiotic pathways to create O_2, CH_4 and N_2O (see e.g. Meadows et al. 2018; Guzmán-Marmolejo et al. 2013; Airapetian et al. 2016 and references therein). As such, no single biosignatures will confirm life by its own, as abiotic production scenarios under certain environmental conditions could always remain an alternative explanation for these gases.

As recently discussed by Stüeken et al. (2016) and Lammer et al. (2019), the two main atmospheric species in present-day Earth's atmosphere, N_2 and O_2, constitute a geo-biosignature for a biosphere that is populated by denitrifying life-forms (see Stüeken et al. 2020; Lloyd et al. 2020). According to the current state of knowledge, a long-time stable N_2-O_2 atmosphere, together with a considerably lower share of CO_2, seem to have strong correlations to the presence of aerobe lifeforms and even indicate an active tectonic regime (Stüeken et al. 2016; Lammer et al. 2019).

A physically untouched N_2 atmosphere is chemically exceedingly stable, due to its strong triple bond (e.g. Zumdahl and Zumdahl 2010). However, a surface totally or partly covered by liquid water moistens the air, providing conditions for efficient lightning fixation which converts N_2 into water soluble molecules (e.g. Chameides and Walker 1981; Navarro-González et al. 2001). This nitrogen is then dissolved in lakes/oceans, where it can form sediments (e.g. Voss et al. 2013). Impactor fixation (Fegley et al. 1986; Chyba and Sagan 1992; Parkos et al. 2018) and a productive photochemistry (Zahnle 1986; Tian et al. 2011; Airapetian et al. 2016) contribute to this depletion, especially on young planets. If the magnetospheric protection capability is low, cosmic rays may also play a role in nitrogen fixation (Navarro-González et al. 1998; Grenfell et al. 2012; Cooray 2015; Tabataba-Vakili et al. 2016). The strong return fluxes of nitrogen from the oceans into the atmosphere, observed on present Earth, are due to biological processes such as denitrification and anammox (e.g. Fowler et al. 2013; Zerkle and Mikhail 2017). In addition, these processes are effective especially in an oxidized environment, where nitric oxide is also a common reactant. In the later stages of a planet's evolution, it is not likely that volcanic nitrogen outgassing is strong enough to maintain more than a few mbar N_2 in an atmosphere; if one disregards biological influences on present Earth, one finds a net ingassing flux of N_2, since nitrogen fixation is stronger than nitrogen outgassing (Cartigny and Marty 2013; Lammer et al. 2019).

Plate tectonics remain a crucial factor for shaping a stable N_2-O_2 atmosphere. The subduction of oxidised material supports nitrogen outgassing through volatilisation of stored reduced nitrogen compounds, as well as it leads to a partly re-outgassing of sedimented nitrogen through metamorphism (Mikhail and Sverjensky 2014; Barry and Hilton 2016). Through gradually replacing oxidised surface, active tectonics help to remove CO_2 from the atmosphere through weathering (Walker et al. 1981; Krissansen-Totton et al. 2018). Combined with the outcome of the previous paragraph, a purely N_2-dominated atmosphere is possible only with an active tectonic regime that has led to CO_2 depletion and to a geologically-induced high volcanic outgassing rate. Here, converting of water-dissolved nitrogen into N_2 in volcanic environments at mid-ocean ridges, a process whose impact has not been quantified yet, could play a role (Kasting et al. 1993). Under such conditions, anaerobe life can exist undetectable with the here presented method. All in all, an N_2 atmosphere constitutes a geosignature, indicating some sort of tectonic activity.

In a pure N_2 atmosphere without any reaction partners (e.g. CO_2, CH_4), fixation is not possible; on the contrary, the more reaction partners, the more efficient lightning fixation works (Navarro-González et al. 2001). Therefore, when O_2 joins as a bulk gas, either N_2 or O_2 is expected to be depleted in the long-term. Generally, the atmospheric presence of O_2 in greater amounts needs an oxidized planet's surface and a lithosphere (e.g. Catling et al. 2005; Catling 2014). O_2 is required by complex life, not only in a chemical/metabolic sense but also as O_3 for shielding UV radiation (e.g. Levine 2014). Under the conditions discussed here, it can serve as indicator for complex lifeforms too. Both, the O_2 as well as the N_2 cycle, contain strong biological return fluxes to the atmosphere (e.g. Galloway 2003; Petsch 2014).

These deliberations lead to the conclusion that an N_2-O_2-atmosphere is most likely possible only under the influence of aerobic lifeforms. This makes the combination of N_2 and O_2 as bulk gases and a negligible amount of CO_2 the most likely geo-biosignature for life as we know it (Lammer et al. 2019). If such an atmosphere is additionally populated with O_3, H_2O, CH_4, and N_2O, then the corresponding planet can decisively indicate the presence of live on an Earth-like habitat (Airapetian et al. 2017b).

If one takes Drake's equation or the modified one by Seager (2017) and introduces the challenge of the detection of N_2-O_2 dominated atmospheres at rocky planets inside habitable zones as an updated $\eta E_{\mathrm{N2\text{-}O2}}$-constrain, then the Drake equation can be written as $N = R \times f_\mathrm{p} \times \eta E_{\mathrm{N2\text{-}O2}} \times f_\mathrm{i} \times f_\mathrm{c} \times L$. Seager's equation transforms then to an equation for the number of Earth-like habitats, equal here to the number of planets with an N_2-O_2-dominated atmosphere: $N_\mathrm{Earth} = N_\mathrm{star} \times f_\mathrm{q} \times f_\mathrm{HZ} \times \eta E_{\mathrm{N2\text{-}O2}}$.

One reason why N_2 atmospheres were not considered as an indirect biosignature in the past might be the very tricky detection of N_2 (and the detection of the domination of N_2), which is challenging even for future space telescopes. In an N_2-O_2-dominated atmosphere, N_2O molecules can be also deeply linked to biological and atmospheric processes (Muller 2013). From the Earth it is known that N_2O would be efficiently depleted by photodissociation in the troposphere, if O_3 would not protect it. Furthermore, the strength of molecular transition bands of N_2O and NO_2 molecules lie in the blue optical region and at near-UV wavelengths, but the much stronger lines of O_2 and O_3 are located at similar wavelengths (Bétrémieux and Kaltenegger 2013), which makes the detection of N_2O and NO_2 challenging. In addition, N_2O molecules have two absorption bands in the infrared at about 4.5 and 7.8 mm, but they are generally rather weak (Rauer et al. 2011; Hedelt et al. 2013), though could be detectable at higher concentrations around quiescent M dwarfs at 17 microns (Rugheimer et al. 2015).

Infrared absorption features of $O_2 \cdot N_2$ dimers at 1.26 mm should be detectable in the transmission spectrum of an Earth-like habitat, if the spectral resolution and signal-to-noise ratio are high enough (Pallé et al. 2009). Misra et al. (2014) carried out a study in which the dimers of $O_2 \cdot O_2$ were analysed and concluded that they would be extremely valuable for the characterisation of Earth-like planets, indicating perhaps an anti-biosignature with large O_2 concentrations produced by a run-away greenhouse or similar photochemical process.

Observationally, the $O_2 \cdot N_2$ dimer could be detected and measured from the ground using the ELTs, or from space by the proposed mission HabEX (Fig. 10), which will be able to observe the reflected light of exoplanets from ultraviolet to near-infrared wavelengths directly, through blocking out of the parent's light with a coronograph and a starshade (Gaudi et al. 2018, 2019). The upper atmosphere of an Earth-like habitat would be composed mainly of H, N, and O atoms. In such a case, charge exchange with stellar wind ions and resonant scattering would lead to the formation of a number of emission lines for these atoms, of which the strongest ones would be located between X-rays and optical wavelengths. Due to the low atmospheric density, these atoms can then be detected in the far-UV wavelengths (122–200 nm) of transmission spectra from transiting Earth-like habitats.

These absorption lines might be detectable by future large-aperture multipurpose telescopes, such as LUVOIR (Dalcanton et al. 2015), due to their unprecedented high far-UV sensitivity of the LUMOS (France et al. 2017) and POLLUX (Bouret et al. 2018) spectrographs, and HabEX. The LUVOIR (Fig. 11) and HabEx (Fig. 10) mission concepts, currently studied by NASA, or the ESA mission concept LIFE (Large Interferometer for Exoplanets) (Quanz et al. 2019), open the perspective of perhaps discovering inhabited worlds orbiting stars other than our Sun.

Fig. 10 Habitable Exoplanet Observatory HabEx (image by NASA)

Fig. 11 Large UV Optical Infrared Surveyor LUVOIR (image by NASA)

N_2-O_2 atmospheres are also transparent for starlight due to the oxidation of visible-light absorbers through O_2 molecules, which leads via Rayleigh scattering on these molecules to its famous "pale blue dot" appearance (Krissansen-Totton et al. 2016). By considering this process, an Earth-like Rayleigh absorption feature in the visible should then also be consistent with an N_2-O_2 dominated atmosphere and, therefore, could be considered as an indirect biosignature from an Earth-like habitat.

Finally, if N_2-dominated atmospheres are rare, due to the complex interplay of atmosphere, lithosphere, and biosphere (see O'Neill et al. 2020; Stüeken et al. 2020; Lloyd et al. 2020), then one may expect that the atmospheres of most Earth-sized exoplanets in the habitable zones would be CO_2-dominated (Lammer et al. 2019). Carbon dioxide molecules present a number of absorption bands in the IR, making them detectable from space with the future James Webb Space Telescope (JWST) and ARIEL space observatories, and from the ground with high-resolution spectrographs attached to the ELTs.

JWST is a visible to infrared mission planned to be launched by NASA, in cooperation with ESA (Ariane 5 launch) and the Canadian Space Agency, in 2021. JWST will be able to characterise atmospheres of exoplanets through transmission spectroscopy and secondary eclipse measurements. JWST will be the first mission capable of characterising an Earth-like exoplanet orbiting an M dwarf. However, it is expected that such a measurement will require hundreds of hours and JWST is a community-wide telescope, therefore such an observation will be limited to only a small number of terrestrial exoplanets (Kaltenegger and Traub 2009).

If CO_2-dominated atmospheres are common on Earth-sized exoplanets within habitable zones, these telescopes will detect such atmospheres by characterizing Earth-sized exoplanets discovered by the ongoing TESS and the future PLATO missions. For the far future (2030+), however, a realistic quantification of ηE that results in a less speculative occurrence rate of Earth-like habitats, would require the detection of N_2-O_2-dominated atmospheres and hence $\eta E_{\text{N2-O2}}$, by using the next generation ground- and space-based telescopes, such as the ground-based ELTs or the space-based LUVOIR.

Acknowledgements The authors thank the International Space Science Institute in Bern, Switzerland, for its support of the Workshop on "Reading Terrestrial Planet Evolution in Isotopes and Element Measurements". I. Dandouras, M. Blanc and C. Mazelle thank CNES for its support.

References

V.S. Airapetian, A. Glocer, G. Gronoff, E. Hébrard, W. Danchi, Prebiotic chemistry and atmospheric warming of early Earth by an active young Sun. Nat. Geosci. **9**, 452–455 (2016). https://doi.org/10.1038/ngeo2719

V.S. Airapetian, A. Glocer, G.V. Khazanov, R.O.P. Loyd, K. France, J. Sojka, W.C. Danchi, M.W. Liemohn, How hospitable are space weather affected habitable zones? The role of ion escape. Astrophys. J. Lett. (2017a). https://doi.org/10.3847/2041-8213/836/1/L3

V.S. Airapetian, C.H. Jackman, M. Mlynczak, W. Danchi, L. Hunt, Atmospheric beacons of life from exoplanets around G and K stars. Sci. Rep. **7**, 14141 (2017b). https://doi.org/10.1038/s41598-017-14192-4

I.I. Alexeev, E.S. Belenkaya, J.A. Slavin, H. Korth, B.J. Anderson, D.N. Baker, S.A. Boardsen, C.L. Johnson, M.E. Purucker, M. Sarantos, S.C. Solomon, Icarus (2010). https://doi.org/10.1016/j.icarus.2010.01.024

B.J. Anderson, C.L. Johnson, H. Korth, M.E. Purucker, R.M. Winslow, J.A. Slavin, S.C. Solomon, R.L. McNutt Jr., J.M. Raines, T.H. Zurbuchen, The global magnetic field of Mercury from MESSENGER orbital observations. Science (2011). https://doi.org/10.1126/science.1211001

I. Angelo, R. Hu, A case for an atmosphere on super-Earth 55 Cancri e. Astron. J. **154**, 232 (2017)

V. Angelopoulos, The ARTEMIS mission. Space Sci. Rev. (2011). https://doi.org/10.1007/s11214-010-9687-2

G.N. Arney, S.D. Domagal-Goldman, V.S. Meadows, E.T. Wolf, E. Schwieterman, B.D. Charnay, M.W. Claire, E. Hébrard, M.G. Trainer, The pale orange dot: the spectrum and habitability of hazy Archean Earth. Astrobiology **16**, 873–899 (2016). https://doi.org/10.1089/ast.2015.1422

G.N. Arney, V.S. Meadows, S.D. Domagal-Goldman, D. Deming, T.D. Robinson, G. Tovar, E.T. Wolf, E. Schwieterman, Pale orange dots: the impact of organic haze on the habitability and detectability of Earthlike exoplanets. Astrophys. J. **836**, 49 (2017). https://doi.org/10.3847/1538-4357/836/1/49

G. Avice, B. Marty, Perspectives on atmospheric evolution from noble gas and nitrogen isotopes. Space Sci. Rev. **216**, 36 (2020). https://doi.org/10.1007/s11214-020-00655-0

R.A. Bamford, E.P. Alves, F. Cruz, B.J. Kellett, R.A. Fonseca, L.O. Silva, R.M.G.M. Trines, J.S. Halekas, G. Kramer, E. Harnett, R.A. Cairns, R. Bingham, 3D PIC simulations of collisionless shocks at lunar magnetic anomalies and their role in forming lunar swirls. Astrophys. J. (2016). https://doi.org/10.3847/0004-637X/830/2/146

D. Barbato, A. Sozzetti, S. Desidera, M. Damasso, A.S. Bonomo, P. Giacobbe, L.S. Colombo, C. Lazzoni, R. Claudi, R. Gratton, G. Locurto, F. Marzari, C. Mordasini, Exploring the realm of scaled solar system analogues with HARPS. Astron. Astrophys. **615** (2018). https://doi.org/10.1051/0004-6361/201832791

R. Barnes, S.N. Raymond, R. Greenberg et al., CoRoT-7b: super-Earth or super-Io? Astrophys. J. **709**, 95 (2010)

O. Barragán, D. Gandolfi, F. Dai et al., K2-141 b. A 5-Mearth super-Earth transiting a K7 V star every 6.7 h. Astron. Astrophys. **612**, A95 (2018)

P.H. Barry, D.R. Hilton, Release of subducted sedimentary nitrogen throughout Earth's mantle. Geochem. Perspect. Lett. **2**, 148–159 (2016). https://doi.org/10.7185/geochemlet.1615

J. Benkhoff, J. van Casteren, H. Hayakawa, M. Fujimoto, H. Laakso, M. Novara, P. Ferri, H.R. Middleton, R. Ziethe, BepiColombo—comprehensive exploration of Mercury: mission overview and science goals. Planet. Space Sci. (2010). https://doi.org/10.1016/j.pss.2009.09.020

M. Benna et al., Modeling of the magnetosphere of Mercury at the time of the first MESSENGER flyby. Icarus (2009). https://doi.org/10.1016/j.icarus.2009.11.036

M. Benna, P.R. Mahaffy, J.S. Halekas, R.C. Elphic, G.T. Delory, Variability of helium, neon, and argon in the lunar exosphere as observed by the LADEE NMS instrument. Geophys. Res. Lett. (2015). https://doi.org/10.1002/2015GL064120

K.R. Bermingham, E. Füri, K. Lodders, B. Marty, The NC-CC isotope dichotomy: Implications for the chemical and isotopic evolution of the early Solar System. Space Sci. Rev. (2020). https://doi.org/10.1007/s11214-020-00748-w, in press

Y. Bétrémieux, L. Kaltenegger, Transmission spectrum of Earth as a transiting exoplanet—from the ultraviolet to the near-infrared. Astrophys. J. **772**, L31 (2013). https://doi.org/10.1088/2041-8205/772/2/L31

Th. Bida, R. Killen, T.H. Morgan, Discovery of calcium in Mercury's atmosphere. Nature **404**, 159 (2000)

E.B. Bierhaus, B.C. Clark, J.W. Harris, K.S. Payne, R.D. Dubisher, D.W. Wurts, R.A. Hund, R.M. Kuhns, T.M. Linn, J.L. Wood, A.J. May, J.P. Dworkin, E. Beshore, D.S. Lauretta (the OSIRIS-REx Team), The OSIRIS-REx spacecraft and the Touch-and-GoSample Acquisition Mechanism (TAGSAM). Space Sci. Rev. (2018). https://doi.org/10.1007/s11214-018-0521-6

A.S. Bonomo, L. Zeng, M. Damasso et al., A giant impact as the likely origin of different twins in the Kepler-107 exoplanet system. Nat. Astron. **3**, 416–423 (2019)

F. Bouchy et al., Near-InfraRed planet searcher to join HARPS on the near-InfraRed: ESO 3.6-metre telescope ESO 3.6-metre telescope. Messenger **169**, 21 (2017)

L.G. Bouma et al., Planet detection simulations for several possible TESS extended missions (2017). arXiv:1705.08891

J. Bouret, C. Neiner, A.I.G. de Castro, C. Evans, B. Gaensicke, S. Shore, L. Fossati, C. Gry, S. Charlot, F. Marin, P. Noterdaeme, J.-Y. Chaufray, The science case for POLLUX, a high-resolution UV spectropolarimeter onboard LUVOIR. Proceedings of the SPIE (2018). https://doi.org/10.1117/12.2312621

V. Bourrier, X. Dumusque, C. Dorn et al., The 55 Cancri system reassessed. Astron. Astrophys. **619**, A1 (2018)

D. Briot, J. Schneider, Occurrence, physical conditions, and observations of super-Ios and hyper-Ios, in *Pathways Towards Habitable Planets*, vol. 430 (2010), p. 409

R.A. Brown, A model of Jupiter's sulfur nebula. Astrophys. J. **206**, L179 (1976)

R.A. Brown, F.H. Chaffee Jr., High-resolution spectra of sodium emission from Io. Astrophys. J. **187**, L125 (1974)

R.A. Brown, D.E. Shemansky, On the nature of S II emission from Jupiter's hot plasma torus. Astrophys. J. **263**, 433 (1982)

R.A. Brown, R.M. Goody, F.J. Murcray, F.H. Chaffee Jr., Further studies of line emission from Io. Astrophys. J. **200**, L49 (1975)

J.L. Burch, T.E. Moore, R.B. Torbert, B.L. Giles, Magnetospheric multiscale overview and science objectives. Space Sci. Rev. (2016). https://doi.org/10.1007/s11214-015-0164-9

M.H. Burger, R.M. Killen, R.J. Vervack, E.T. Bradley, W.E. McClintock, M. Sarantos, M. Benna, N. Mouawad, Monte Carlo modeling of sodium in Mercury's exosphere during the first two MESSENGER flybys. Icarus (2010). https://doi.org/10.1016/j.icarus.2010.05.007

T.M. Burinskaya, Non-monotonic potentials above the day-side lunar surface exposed to the solar radiation. Planet. Space Sci. (2015). https://doi.org/10.1016/j.pss.2015.03.004

J. Cao, Y. Ma, G. Parks, H. Rème, I. Dandouras, T. Zhang, Kinetic analysis of the energy transport of bursty bulk flows in the plasma sheet. J. Geophys. Res. (2013). https://doi.org/10.1029/2012JA018351

L. Carigi, Solar neighborhood, in *Encyclopedia of Astrobiology* (Springer, Berlin, 2015), pp. 2286–2287. https://doi.org/10.1007/978-3-662-44185-5_1460

J. Carpenter, C. Speak, A. Chouker, M. Talevi, R. Nakamura, A. Santangelo, I. Crawford, D. Cullen, B. Bussey, J. Grenouilleau, Research Opportunities on the Deep Space Gateway: Findings from the Workshop and Call for Ideas (2018). ESA-HSO-K-RP-0284

P. Cartigny, B. Marty, Nitrogen isotopes and mantle geodynamics: the emergence of life and the atmosphere-crust-mantle connection. Elements **9**, 359–366 (2013). https://doi.org/10.2113/gselements.9.5.359

J. Catanzarite, M. Shao, The occurrence rate of Earth analog planets orbiting Sun-like stars. Astrophys. J. **738** (2011). https://doi.org/10.1088/0004-637X/738/2/151

D.C. Catling, The great oxidation event transition, in *Treatise on Geochemistry* (Elsevier, Amsterdam, 2014), pp. 177–195. https://doi.org/10.1016/B978-0-08-095975-7.01307-3

D.C. Catling, C.R. Glein, K.J. Zahnle, C.P. McKay, Why O_2 is required by complex life on habitable planets and the concept of planetary "oxygenation time". Astrobiology **8**, 377–395 (2005). https://doi.org/10.1089/ast.2005.5.415

W.L. Chameides, J.C.G. Walker, Rates of fixation by lightning of carbon and nitrogen in possible primitive atmospheres. Orig. Life Evol. Biosph. **11**, 291–302 (1981). https://doi.org/10.1007/BF00931483

Q.H.S. Chan, R. Stroud, Z. Martins, H. Yabuta, Concerns of organic contamination for sample return space missions. Space Sci. Rev. (2020). https://doi.org/10.1007/s11214-020-00678-7

C.R. Chappell, The role of the ionosphere in providing plasma to the terrestrial magnetosphere – an historical overview. Space Sci. Rev. (2015). https://doi.org/10.1007/s11214-015-0168-5

E. Chassefière et al., PHEBUS: a double ultraviolet spectrometer to observe Mercury's exosphere. Planet. Space Sci. (2010). https://doi.org/10.1016/j.pss.2008.05.018

E. Chassefière, R. Wieler, B. Marty, F. Leblanc, The evolution of Venus: present state of knowledge and future exploration. Planet. Space Sci. (2012). https://doi.org/10.1016/j.pss.2011.04.007

S.P. Christon, G. Gloeckler, D.J. Williams, T. Mukai, R.W. McEntire, C. Jacquey, V. Angelopoulos, A.T.Y. Lui, S. Kokubun, D.H. Fairfield, M. Hirahara, T. Yamamoto, Energetic atomic and molecular ions of ionospheric origin observed in distant magnetotail flow-reversal events. Geophys. Res. Lett. **21**(25), 3023–3026 (1994)

C. Chyba, C. Sagan, Endogenous production, exogenous delivery and impact-shock synthesis of organic molecules: an inventory for the origins of life. Nature **355**, 125–132 (1992). https://doi.org/10.1038/355125a0

J.T. Clarke et al., Variability of D and H in the Martian upper atmosphere observed with the MAVEN IUVS echelle channel. J. Geophys. Res. Space Phys. **122**, 2336–2344 (2017). https://doi.org/10.1002/2016JA023479

C.S. Cockell, L. Kaltenegger, J.A. Raven, Cryptic photosynthesis—extrasolar planetary oxygen without a surface biological signature. Astrobiology **9**, 623–636 (2009). https://doi.org/10.1089/ast.2008.0273

V. Cooray, 19. Interaction of lightning flashes with the Earth's atmosphere, in *An Introduction to Lightning* (Springer, Dordrecht, 2015), pp. 1–386. https://doi.org/10.1007/978-94-017-8938-7

R. Cosentino et al. Harps-N: the new planet hunter at TNG. Proc. SPIE **8446**, 84461V (2012)

A. Crida, R. Ligi, C. Dorn, Y. Lebreton, Mass, radius, and composition of the transiting planet 55 Cancri e: using interferometry and correlations. Astrophys. J. **860**, 122 (2018)

F. Dai, J.N. Winn, D. Gandolfi et al., The discovery and mass measurement of a new USP: K2-131b. Astron. J. **154**, 226 (2017)

J. Dalcanton, S. Seager, S. Aigrain, S. Battel, S. Brant, C. Conroy, L. Feinberg, S. Gezari, O. Guyon, W. Harris, C. Hirata, J. Mather, M. Postman, D. Redding, D. Schiminovich, H.P. Stahl, J. Tumlison, *From Cosmic Birth to Living Earths: The Future of UVOIR Space Astronomy. AURA Report* (2015). arXiv:1507.04779

I. Dandouras, Detection of a plasmaspheric wind in the Earth's magnetosphere by the Cluster spacecraft. Ann. Geophys. **31**, 1143–1153 (2013). https://doi.org/10.5194/angeo-31-1143-2013

I. Dandouras, M. Yamauchi, J. De Keyser, O. Marghitu, H. Rème, I. Yoshikawa, T. Sakanoi, (the ESCAPE proposal team), ESCAPE: a mission proposal for ESA-M5 to systematically study exosphere and atmospheric escape using European, Japanese, and US instruments, in *Proc. ISAS Symposium*, Japan (2018). https://repository.exst.jaxa.jp/dspace/handle/a-is/876320

I. Dandouras, R.A. Bamford, G. Branduardi-Raymont, D. Constantinescu, J. De Keyser, Y. Futaana, H. Lammer, F. Leblanc, A. Milillo, R. Nakamura, E. Roussos, M.G.G.T. Taylor, J. Carpenter, Space plasma physics science opportunities for the deep space gateway, in *EGU General Assembly*, vol. 2019 (2019), EGU2019-9690

I. Dandouras, A.R. Poppe, M.O. Fillingim, L.M. Kistler, C.G. Mouikis, H. Rème, P. Garnier, P. Pinet, G.K. Parks, First simultaneous detection of terrestrial ionospheric heavy ions in the Earth's inner magnetosphere and at the Moon. Nat. Commun. (2020), submitted. NCOMMS-18-24769-T

A. Debrecht, J. Carroll-Nellenback, A. Frank et al., Generation of a circumstellar gas disc by hot Jupiter WASP-12b. Astrophys. J. **478**, 2592 (2018)

D.C. Delcourt, J.A. Sauvaud, T.E. Moore, Polar wind ion dynamics in the magnetotail. J. Geophys. Res. **98**, 9155–9169 (1993). https://doi.org/10.1029/93JA00301

D. Delcourt, Y. Saito, J.-M. Illiano, N. Krupp et al., The mass spectrum analyzer (MSA) onboard BEPI COLOMBO MMO: scientific objectives and prototype results. Adv. Space Res. (2009). https://doi.org/10.1016/j.asr.2008.12.002

D. Deming, H. Knutson, J. Kammer et al., Spitzer secondary eclipses of the dense, modestly-irradiated, giant exoplanet HAT-P-20b using pixel-level decorrelation. Astrophys. J. **805**, 132 (2015)
B.-O. Demory, M. Gillon, J. de Wit et al., A map of the large day-night temperature gradient of a super-Earth exoplanet. Nature **532**, 207 (2016a)
B.-O. Demory, M. Gillon, N. Madhusudhan, D. Queloz, Variability in the super-Earth 55 Cancri e. Mon. Not. R. Astron. Soc. **455**, 2018 (2016b)
B.W. Denevi et al., The evolution of Mercury's crust: a global perspective from MESSENGER. Science **324**, 613–618 (2009)
D.J. Des Marais, M.O. Harwit, K.W. Jucks, J.F. Kasting, D.N.C. Lin, J.I. Lunine, J. Schneider, S. Seager, W.A. Traub, N.J. Woolf, Remote sensing of planetary properties and biosignatures on extrasolar terrestrial planets. Astrobiology **2**, 153–181 (2002). https://doi.org/10.1089/153110702601922 46
S.D. Domagal-Goldman, V.S. Meadows, M.W. Claire, J.F. Kasting, Using biogenic sulfur gases as remotely detectable biosignatures on anoxic planets. Astrobiology **11**, 419–441 (2011). https://doi.org/10.1089/ast.2010.0509
C. Dorn, J.H.D. Harrison, A. Bonsor, T.O. Hands, A new class of super-Earths formed from high-temperature condensates: HD219134 b, 55 Cancri e, WASP-47 e. Mon. Not. R. Astron. Soc. **484**, 712 (2019)
X. Dumusque, A.S. Bonomo, R.D. Haywood et al., The Kepler-10 planetary system revisited by HARPS-N: a hot rocky world and a solid Neptune-mass planet. Astrophys. J. **789**, 154 (2014)
S. Dyadechkin, E. Kallio, P. Wurz, New fully kinetic model for the study of electric potential, plasma and dust above lunar landscapes. J. Geophys. Res. **120**, 1589–1606 (2015). https://doi.org/10.1002/2014JA020511
Y. Ebihara, M. Yamada, S. Watanabe, M. Ejiri, Fate of outflowing suprathermal oxygen ions that originate in the polar ionosphere. J. Geophys. Res. **111**, A04219 (2006). https://doi.org/10.1029/2005JA011403
C.S. Edwards, B.L. Ehlmann, Carbon sequestration on Mars. Geology **43** (2015). https://doi.org/10.1130/G36983.1
L.T. Elkins-Tanton, Coreless terrestrial exoplanets. Astrophys. J. **688**, 628 (2008)
L.T. Elkins-Tanton, S. Seager, Ranges of atmospheric mass and composition of super-Earth exoplanets. Astrophys. J. **685**, 1237 (2008)
R.C. Elphic, G.T. Delory, B.P. Hine, P.R. Mahaffy, M. Horanyi, A. Colaprete, M. Benna, S.K. Noble, The lunar atmosphere and dust environment explorer mission. Space Sci. Rev. (2014). https://doi.org/10.1007/s11214-014-0113-z
E. Engwall, Low-Energy Ion Escape from the Terrestrial Polar Regions. Ph.D. thesis, Uppsala University, Department of Physics and Astronomy (2009)
E. Engwall, A.I. Eriksson, C.M. Cully, M. André, R. Torbert, H. Vaith, Earth's ionospheric outflow dominated by hidden cold plasma. Nat. Geosci. **2**, 24–27 (2009). https://doi.org/10.1038/ngeo387
C.P. Escoubet, M. Fehringer, M. Goldstein, The Cluster mission. Ann. Geophys. **19**, 1197–1200 (2001)
C.P. Escoubet, A. Masson, H. Laakso, M.L. Goldstein, Recent highlights from Cluster, the first 3-D magnetospheric mission. Ann. Geophys. (2015). https://doi.org/10.5194/angeo-33-1221-2015
N. Espinoza, R. Brahm, T. Henning et al., HD 213885b: a transiting 1-day-period super-Earth with an Earth-like composition around a bright (V=7.9) star unveiled by TESS (2019). arXiv:1903.07694
B. Fegley, R.G. Prinn, H. Hartman, G.H. Watkins, Chemical effects of large impacts on the Earth's primitive atmosphere. Nature **319**, 305–308 (1986). https://doi.org/10.1038/319305a0
Y.R. Fernández, S.S. Sheppard, D.C. Jewitt, The albedo distribution of Jovian Trojan asteroids. Astron. J. **126**, 1563 (2003)
A. Fortier, T. Beck, W. Benz, C. Broeg, V. Cessa, D. Ehrenreich, N. Thomas, Proc. SPIE **91432**, (2014)
L. Fossati, T.R. Ayres, C.A. Haswell et al., Absorbing gas around the WASP-12 planetary system. Astrophys. J. Lett. **766**, L20 (2013)
L. Fossati, N.V. Erkaev, H. Lammer et al., Aeronomical constraints to the minimum mass and maximum radius of hot low-mass planets. Astron. Astrophys. **598**, A90 (2017)
D. Fowler, J.A. Pyle, J.A. Raven, M.A. Sutton, The global nitrogen cycle in the twenty-first century: introduction. Philos. Trans. R. Soc. Lond. B, Biol. Sci. **368**, 20130165 (2013). https://doi.org/10.1098/rstb.2013.0165
K. France, B. Fleming, G. West, S.R. McCandliss, M.R. Bolcar, W. Harris, L. Moustakas, J.M. O'Meara, I. Pascucci, J. Rigby, D. Schiminovich, J. Tumlinson, J.-C. Bouret, C.J. Evans, M. Garcia, The LUVOIR Ultraviolet Multi-Object Spectrograph (LUMOS): instrument definition and design (2017)
W. Fujiya, P. Hoppe, U. Ott, Boron abundances and isotopic ratios of olivine grains on Itokawa returned by the Hayabusa spacecraft. Meteorit. Planet. Sci. **51**(9), 1721–1729 (2016). https://doi.org/10.1111/maps.12686
Y. Futaana et al., SELMA mission: how do airless bodies interact with space environment? The Moon as an accessible laboratory. Planet. Space Sci. (2018). https://doi.org/10.1016/j.pss.2017.11.002

Gaia Collaboration, *VizieR Online Data Catalog. SIMBAD I/345* (2018)

F. Gaillard, B. Scaillet, A theoretical framework for volcanic degassing chemistry in a comparative planetology perspective and implications for planetary atmospheres. Earth Planet. Sci. Lett. **403**, 307 (2014)

J.N. Galloway, The global nitrogen cycle, in *Treatise on Geochemistry* (Elsevier, Amsterdam, 2003), pp. 557–583. https://doi.org/10.1016/B0-08-043751-6/08160-3

D. Gandolfi, O. Barragán, A.P. Hatzes et al., The transiting multi-planet system HD 3167: a 5.7 MEarth super-Earth and an 8.3 M_{Earth} mini-Neptune. Astron. J. **154**, 123 (2017)

B.S. Gaudi, S. Seager, B. Mennesson, A. Kiessling, K. Warfield, G. Kuan, K. Cahoy, J.T. Clarke, S. Domagal-Goldman, L. Feinberg, O. Guyon, J. Kasdin, D. Mawet, T. Robinson, L. Rogers, P. Scowen, R. Somerville, K. Stapelfeldt, C. Stark, D. Stern, M. Turnbull, S. Martin, O. Alvarez-Salazar, R. Amini, W. Arnold, B. Balasubramanian, M. Baysinger, L. Blais, T. Brooks, R. Calvet, V. Cormarkovic, C. Cox, R. Danner, J. Davis, L. Dorsett, M. Effinger, R. Eng, J. Garcia, J. Gaskin, J. Harris, S. Howe, B. Knight, J. Krist, D. Levine, M. Li, D. Lisman, M. Mandic, L. Marchen, C. Marrese-Reading, J. McGowen, A. Miyaguchi, R. Morgan, B. Nemati, S. Nikzad, J. Nissen, M. Novicki, T. Perrine, D. Redding, M. Richards, M. Rud, D. Scharf, G. Serabyn, S. Shaklan, S. Smith, M. Stahl, P. Stahl, H. Tang, D. Van Buren, J. Villalvazo, S. Warwick, D. Webb, R. Wofford, J. Woo, M. Wood, J. Ziemer, E. Douglas, V. Faramaz, S. Hildebrandt, T. Meshkat, P. Plavchan, G. Ruane, N. Turner (2018). The Habitable Exoplanet Observatory (HabEx) Mission Concept Study Interim Report. https://www.jpl.nasa.gov/habex/pdf/HabEx_Interim_Report.pdf

S. Gaudi, S. Seager, B. Mennesson, A. Kiessling, K. Warfield, The Habitable Exoplanet Observatory (HabEx), in *SPIE-Intl. Soc. Optical Eng.*, p. 21 (2019). https://doi.org/10.1117/12.2530036

J. Geiss, F. Bühler, H. Cerutti, P. Eberhardt, Ch. Filleux, J. Meister, P. Signer, The Apollo SWC experiment: results, conclusions, consequences. Space Sci. Rev. (2004). https://doi.org/10.1023/B:SPAC.0000023409.54469.40

A. Genova et al., Geodetic evidence that Mercury has a solid inner core. Geophys. Res. Lett. (2019). https://doi.org/10.1029/2018GL081135

L. Girardi, A. Bressan, G. Bertelli, C. Chiosi, Evolutionary tracks and isochrones for low- and intermediate-mass stars: from 0.15 to 7 M_{sun}, and from Z=0.0004 to 0.03. Astron. Astrophys. Suppl. Ser. **141**, 371–383 (2000). https://doi.org/10.1051/aas:2000126

J.N. Goswami, Exploration of the Moon: Chandrayaan-1 and Chandrayaan-2, in *40th COSPAR Scientific Assembly* (2014). B0.1-7-14

M.M. Grady, Exploring Mars with returned samples. Space Sci. Rev. (2020). https://doi.org/10.1007/s11214-020-00676-9

J.L. Grenfell, J.-M. Grießmeier, B. Patzer, H. Rauer, A. Segura, A. Stadelmann, B. Stracke, R. Titz, P. Von Paris, Biomarker response to galactic cosmic ray-induced NOx and the methane greenhouse effect. Astrobiology **7**, 208–221 (2007a). https://doi.org/10.1089/ast.2006.0129. In *The Atmosphere of an Earth-Like Planet Orbiting an M Dwarf Star*

J.L. Grenfell, B. Stracke, P. von Paris, B. Patzer, R. Titz, A. Segura, H. Rauer, The response of atmospheric chemistry on earthlike planets around F, G and K Stars to small variations in orbital distance. Planet. Space Sci. **55**, 661–671 (2007b). https://doi.org/10.1016/j.pss.2006.09.002

J.L. Grenfell, H. Rauer, F. Selsis, L. Kaltenegger, C. Beichman, W. Danchi, C. Eiroa, M. Fridlund, T. Henning, T. Herbst, H. Lammer, A. Léger, R. Liseau, J.I. Lunine, F. Paresce, A. Penny, A. Quirrenbach, H. Röttgering, J. Schneider, D. Stam, G. Tinetti, G.J. White, Co-evolution of atmospheres, life, and climate. Astrobiology **10**, 77–88 (2010). https://doi.org/10.1089/ast.2009.0375

J.L. Grenfell, J.-M. Grießmeier, P. von Paris, A.B.C. Patzer, H. Lammer, B. Stracke, S. Gebauer, F. Schreier, H. Rauer, Response of Atmospheric Biomarkers to NOx-Induced Photochemistry Generated by Stellar Cosmic Rays for Earth-like Planets in the Habitable Zone of M Dwarf Stars. Astrobiology **12**, 1109–1122 (2012). https://doi.org/10.1089/ast.2011.0682,

M. Güdel, The Sun through time. Space Sci. Rev. (2020), accepted

E.W. Guenther, J. Cabrera, A. Erikson et al., Constraints on the exosphere of CoRoT-7b. Astron. Astrophys. **525**, A24 (2011)

E.W. Guenther, O. Barragán, F. Dai et al., K2-106, a system containing a metal-rich planet and a planet of lower density. Astron. Astrophys. **608**, A93 (2017)

A. Guzmán-Marmolejo, A. Segura, El. Escobar-Briones, Abiotic production of methane in terrestrial planets. Astrobiology **13**(6), 550–559 (2013). https://doi.org/10.1089/ast.2012.0817

S. Haaland, A. Eriksson, E. Engwall, B. Lybekk, H. Nilsson, A. Pedersen, K. Svenes, M. André, M. Förster, K. Li, C. Johnsen, N. Østgaard, Estimating the capture and loss of cold plasma from ionospheric outflow. J. Geophys. Res. **117**, A07311 (2012). https://doi.org/10.1029/2012JA017679

T. Haltigin et al., iMARS Phase 2: a draft mission architecture and science management plan for the return of samples from Mars. Astrobiology **18** (2018). https://doi.org/10.1089/ast.2018.29027.mars

Y. Harada, Y. Futaana, S. Barabash, M. Wieser, P. Wurz, A. Bhardwaj, K. Asamura, Y. Saito, S. Yokota, H. Tsunakawa, S. Machida, Backscattered energetic neutral atoms from the Moon in the Earth's plasma sheet observed by Chandarayaan-1/sub-keV atom reflecting analyzer instrument. J. Geophys. Res. **119**, 3573–3584 (2014). https://doi.org/10.1002/2013JA019682

M.H. Hart, Habitable zones about main sequence stars. Icarus **37**, 351–357 (1979). https://doi.org/10.1016/0019-1035(79)90141-6

C.A. Haswell, L. Fossati, T. Ayres et al., Near-ultraviolet absorption, chromospheric activity, and star-planet interactions in the WASP-12 system. Astrophys. J. **760**, 79 (2012)

R.D. Haywood, A. Collier Cameron, D. Queloz et al., Planets and stellar activity: hide and seek in the CoRoT-7 system. Mon. Not. R. Astron. Soc. **443**, 2517 (2014)

L. Hebb, A. Collier-Cameron, B. Loeillet et al., WASP-12b: the hottest transiting extrasolar planet yet discovered. Astrophys. J. **693**, 1920 (2009)

P. Hedelt, P. von Paris, M. Godolt, S. Gebauer, J.L. Grenfell, H. Rauer, F. Schreier, F. Selsis, T. Trautmann, Spectral features of Earth-like planets and their detectability at different orbital distances around F, G, and K-type stars. Astron. Astrophys. **553**, A9 (2013). https://doi.org/10.1051/0004-6361/201117723

M. Hippke, D. Angerhausen, A statistical search for a population of exo-trojans in the Kepler data set. Astrophys. J. **881**, 1 (2015)

G.C. Ho, S.M. Krimigis, R.E. Gold, D.N. Baker, J.A. Slavin, B.J. Anderson, H. Korth, R.D. Starr, D.J. Lawrence, R.L. McNutt, S.C. Solomon, MESSENGER observations of transient bursts of energetic electrons in Mercury's magnetosphere. Science (2011). https://doi.org/10.1126/science.1211141

R.R. Hodges Jr., Methane in the lunar exosphere: implications for solar wind carbon escape. Geophys. Res. Lett. (2016). https://doi.org/10.1002/2016GL068994

A.W. Howard, R. Sanchis-Ojeda, G.W. Marcy et al., A rocky composition for an Earth-sized exoplanet. Nature **503**, 381 (2013)

M.M. Huddleston, C.R. Chappell, D.C. Delcourt, T.E. Moore, B.L. Giles, M.O. Chandler, An examination of the process and magnitude of ionospheric plasma supply to the magnetosphere. J. Geophys. Res. (2005). https://doi.org/10.1029/2004JA010401

R. Ilie, M.W. Liemohn, The outflow of ionospheric nitrogen ions: a possible tracer for the altitude-dependent transport and energization processes of ionospheric plasma. J. Geophys. Res. (2016). https://doi.org/10.1002/2015JA022162

T.R. Ireland, P. Holden, M.D. Norman, J. Clarke, Isotopic enhancements of ^{17}O and ^{18}O from solar wind particles in the lunar regolith. Nature Lett. (2006). https://doi.org/10.1038/nature04611

Y. Ito, M. Ikoma, H. Kawahara et al., Theoretical emission spectra of atmospheres of hot rocky super-Earths. Astrophys. J. **801**, 144 (2015)

B.M. Jakosky, R.P. Lin, J.M. Grebowsky, J.G. Luhmann et al., The Mars Atmosphere and Volatile Evolution (MAVEN) mission. Space Sci. Rev. **195**, 3–48 (2015a). https://doi.org/10.1007/s11214-015-0139-x

B.M. Jakosky, J.M. Grebowsky, J.G. Luhmann, J. Connerney et al., MAVEN observations of the response of Mars to an interplanetary coronal mass ejection. Science **350**, aad0210 (2015b). https://doi.org/10.1126/science.aad0210

B.M. Jakosky, J.M. Grebowsky, J.G. Luhmann, D.A. Brain, Initial results from the MAVEN mission to Mars. Geophys. Res. Lett. **42**, 8791–8802 (2015c). https://doi.org/10.1002/2015GL065271

B.M. Jakosky, M. Slipski, M. Benna, P. Mahaffy, M. Elrod, R. Yelle, S. Stone, N. Alsaeed, Mars' atmospheric history derived from upper-atmosphere measurements of 38Ar/36Ar. Science **355**, 1408–1410 (2017)

B.M. Jakosky, D. Brain, M. Chaffin, S. Curry, J. Deighan, J. Grebowsky, J. Halekas, F. Leblanc, R. Lillis, J.G. Luhmann, L. Andersson, N. Andre, D. Andrews, D. Baird, D. Baker, J. Bell, M. Benna, D. Bhattacharyya, S. Bougher, C. Bowers, P. Chamberlin, J.-Y. Chaufray, J. Clarke, G. Collinson, M. Combi, J. Connerney, K. Connour, J. Correira, K. Crabb, F. Crary, T. Cravens, M. Crismani, G. Delory, R. Dewey, G. DiBraccio, C. Dong, Y. Dong, P. Dunn, H. Egan, M. Elrod, S. England, F. Eparvier, R. Ergun, A. Eriksson, T. Esman, J. Espley, S. Evans, K. Fallows, X. Fang, M. Fillingim, C. Flynn, A. Fogle, C. Fowler, J. Fox, M. Fujimoto, P. Garnier, Z. Girazian, H. Groeller, J. Gruesbeck, O. Hamil, K.G. Hanley, T. Hara, Y. Harada, J. Hermann, M. Holmberg, G. Holsclaw, S. Houston, S. Inui, S. Jain, R. Jolitz, A. Kotova, T. Kuroda, D. Larson, Y. Lee, C. Lee, F. Lefevre, C. Lentz, D. Lo, R. Lugo, Y.-J. Ma, P. Mahaffy, M.L. Marquette, Y. Matsumoto, M. Mayyasi, C. Mazelle, W. McClintock, J. McFadden, A. Medvedev, M. Mendillo, K. Meziane, Z. Milby, D. Mitchell, R. Modolo, F. Montmessin, A. Nagy, H. Nakagawa, C. Narvaez, K. Olsen, D. Pawlowski, W. Peterson, A. Rahmati, K. Roeten, N. Romanelli, S. Ruhunusiri, C. Russell, S. Sakai, N. Schneider, K. Seki, R. Sharrar, S. Shaver, D.E. Siskind, M. Slipski, Y. Soobiah, M. Steckiewicz, M.H. Stevens, I. Stewart, A. Stiepen, S. Stone, V. Tenishev, N. Terada, K. Terada, E. Thiemann, R. Tolson, G. Toth, J. Trovato, M. Vogt, T. Weber, P. Withers, S. Xu, R. Yelle, E. Yiğit, R. Zurek, Loss of the Martian atmosphere to space: present-day loss rates determined from MAVEN observations and integrated loss through time. Icarus (2018). https://doi.org/10.1016/j.icarus.2018.05.030

C.Y. Johnson, Ion and neutral composition of the ionosphere. Ann. IQSY **5**, 197–213 (1969)

B. Johnson, C. Goldblatt, The nitrogen budget of Earth. Earth-Sci. Rev. **148**, 150–173 (2015). https://doi.org/10.1016/j.earscirev.2015.05.006

E. Kallio, G. Facskó, Properties of plasma near the moon in the magnetotail. Planet. Space Sci. (2015). https://doi.org/10.1016/j.pss.2014.11.007

E. Kallio, S. Dyadechkin, P. Wurz, M. Khodachenko, Space weathering on the Moon: farside-nearside solar wind precipitation asymmetry. Planet. Space Sci. **166**, 9–22 (2019). https://doi.org/10.1016/j.pss.2018.07.013

L. Kaltenegger, W.A. Traub, Transits of Earth-like planets. Astrophys. J. (2009). https://doi.org/10.1088/0004-637X/698/1/519

J.F. Kasting, *How to Find a Habitable Planet* (Princeton University Press, Princeton, 2010). ISBN 978-0-691-13805-3

J.F. Kasting, D.P. Whitmire, R.T. Reynolds, Habitable zones around main sequence stars. Icarus **101**, 108–128 (1993). https://doi.org/10.1006/icar.1993.1010

J.W. Keller, N.E. Petro, R.R. Vondrak (the LRO team), The lunar reconnaissance orbiter mission – six years of science and exploration at the Moon. Icarus (2016). https://doi.org/10.1016/j.icarus.2015.11.024

R.M. Killen, J.M. Hahn, Impact vaporization as a possible source of Mercury's calcium exosphere. Icarus (2015). https://doi.org/10.1016/j.icarus.2014.11.035

R. Killen, W.-H. Ip, The surface-bounded atmospheres of Mercury and the Moon. Rev. Geophys. **37**(3), 361–406 (1999). https://doi.org/10.1029/1999RG900001

R. Killen, G. Cremonese, H. Lammer, S. Orsini, A.E. Potter, A.L. Sprague, P. Wurz, M.L. Khodachenko, H.I.M. Lichtenegger, A. Milillo, A. Mura, Processes that promote and deplete the exosphere of Mercury. Space Sci. Rev. **132**, 433–509 (2007)

R.M. Killen, A.E. Potter, R.J. Vervack Jr., E.T. Bradley, W.E. McClintock, C.M. Anderson, M.H. Burger, Observations of metallic species in Mercury's exosphere. Icarus (2010). https://doi.org/10.1016/j.icarus.2010.02.018

K.G. Kislyakova, E. Pilat-Lohinger, B. Funk et al., On the ultraviolet anomalies of the WASP-12 and HD 189733 systems: Trojan satellites as a plasma source. Mon. Not. R. Astron. Soc. **461**, 988 (2016)

K.G. Kislyakova, L. Noack, C.P. Johnstone et al., Magma oceans and enhanced volcanism on TRAPPIST-1 planets due to induction heating. Nat. Astron. **1**, 878 (2017)

K.G. Kislyakova, L. Fossati, C.P. Johnstone et al., Effective induction heating around strongly magnetized stars. Astrophys. J. **858**, 105 (2018)

L.M. Kistler, C. Mouikis, E. Möbius, B. Klecker, J.A. Sauvaud, H. Rème, A. Korth, M.F. Marcucci, R. Lundin, G.K. Parks, A. Balogh, Contribution of nonadiabatic ions to the cross-tail current in an O^+ dominated thin current sheet. J. Geophys. Res. (2005). https://doi.org/10.1029/2004JA010653

E.S. Kite, B. Fegley Jr., L. Schaefer, E. Gaidos, Atmosphere-interior exchange on hot, rocky exoplanets. Astrophys. J. **828**, 80 (2016)

R.K. Kopparapu, R. Ramirez, J.F. Kasting, V. Eymet, T.D. Robinson, S. Mahadevan, R.C. Terrien, S. Domagal-Goldman, V. Meadows, R. Deshpande, Habitable zones around main sequence stars: new estimates. Astrophys. J. **765**, 131 (2013). https://doi.org/10.1088/0004-637X/765/2/131

V.A. Krasnopolsky, J.P. Maillard, T.C. Owen, Detection of methane in the Martian atmosphere: evidence for life? Icarus **172**, 537–547 (2004). https://doi.org/10.1016/j.icarus.2004.07.004

J. Krissansen-Totton, E.W. Schwieterman, B.D. Charnay, G.N. Arney, T.D. Robinson, V. Meadows, D.C. Catling, Is the pale blue dot unique? Optimized photometric bands for identifying Earth-like exoplanets. Astrophys. J. **817**, 31 (2016). https://doi.org/10.3847/0004-637X/817/1/31

J. Krissansen-Totton, G.N. Arney, D.C. Catling, Constraining the climate and ocean pH of the early Earth with a geological carbon cycle model. Proc. Natl. Acad. Sci. **115**(16), 4105–4110 (2018). https://doi.org/10.1073/pnas.1721296115

E.A. Kronberg, M. Ashour-Abdalla, I. Dandouras, D.C. Delcourt, E.E. Grigorenko, L.M. Kistler, I.V. Kuzichev, J. Liao, R. Maggiolo, H.V. Malova, K.G. Orlova, V. Peroomian, D.R. Shklyar, Y.Y. Shprits, D.T. Welling, L.M. Zelenyi, Circulation of heavy ions and their dynamical effects in the magnetosphere: recent observations and models. Space Sci. Rev. (2014). https://doi.org/10.1007/s11214-014-0104-0

M. Kruss, G. Wurm, Seeding the formation of Mercurys: an iron-sensitive bouncing barrier in disk magnetic fields. Astrophys. J. **869**, 45 (2018)

D. Kubyshkina, L. Fossati, N.V. Erkaev et al., Grid of upper atmosphere models for 1-40 MEarth planets: application to CoRoT-7 b and HD 219134 b,c. Astron. Astrophys. **619**, A151 (2018)

I. Kupo, Y. Mekler, Y. Mekler, A. Eviatar, Detection of ionized sulfur in the Jovian magnetosphere. Astrophys. J. **205**, L51 (1976)

M. Küppers, K. Jockers, Fabry-Perot imaging of O III lambda = 5007 Angstrom Emission in the Io plasma torus. Astrophys. J. **441**, L101 (1995)

M. Küppers, K. Jockers, A multi-emission imaging study of the Io plasma torus. Icarus **129**, 48 (1997)

H. Lammer, J.F. Kasting, E. Chassefière, R.E. Johnson, Y.N. Kulikov, F. Tian, Atmospheric escape and evolution of terrestrial planets and satellites. Space Sci. Rev. (2008). https://doi.org/10.1007/s11214-008-9413-5

H. Lammer, J.H. Bredehöft, A. Coustenis, M.L. Khodachenko, L. Kaltenegger, O. Grasset, D. Prieur, F. Raulin, P. Ehrenfreund, M. Yamauchi, J.-E. Wahlund, J.-M. Grießmeier, G. Stangl, C.S. Cockell, Y.N. Kulikov, J.L. Grenfell, H. Rauer, What makes a planet habitable? Astron. Astrophys. Rev. **17**, 181–249 (2009). https://doi.org/10.1007/s00159-009-0019-z

H. Lammer, H.I.M. Lichtenegger, M.L. Khodachenko, The loss of nitrogen-rich atmospheres from Earth-like exoplanets within M-star habitable zones, in *Molecules in the Atmospheres of Extrasolar Planets*. ASP Conference Series, vol. 450 (2011), p. 139. Astronomical Society of the Pacific

H. Lammer, A.L. Zerkle, S. Gebauer et al., Origin and evolution of the atmospheres of early Venus, Earth and Mars. Astron. Astrophys. Rev. **26**, 2 (2018)

H. Lammer, L. Sproß, J.L. Grenfell, M. Scherf, L. Fossati, M. Lendl, P.E. Cubillos, The role of N_2 as a geo-biosignature for the detection and characterization of Earth-like habitats. Astrobiology **19**, 927–950 (2019). https://doi.org/10.1089/ast.2018.1914

H. Lammer, M. Scherf, H. Kurokawa, Y. Ueno, C. Burger, T. Maindl, C.P. Johnstone, M. Leizinger, M. Benedikt, L. Fossati, K.G. Kislyakova, B. Marty, G. Avice, B. Fegley, P. Odert, Loss and fractionation of noble gas isotopes and moderately volatile elements from planetary embryos and early Venus, Earth and Mars. Space Sci. Rev. **216**, 74 (2020a). https://doi.org/10.1007/s11214-020-00701-x

H. Lammer, R. Brasser, A. Johansen, M. Scherf, M. Leitzinger, Formation of Venus, Earth and Mars: constrained by isotopes. Space Sci. Rev. (2020b), in press

H. Lammer, M. Leitzinger, M. Scherf, P. Odert, C. Burger, D. Kubyshkina, C. Johnstone, T. Maindl, C.M. Schäfer, M. Güdel, N. Tosi, A. Nikolaou, E. Marcq, N.V. Erkaev, L. Noack, K.G. Kislyakova, L. Fossati, E. Pilat-Lohinger, F. Ragossnig, E.A. Dorfi, Constraining the early evolution of Venus and Earth through atmospheric Ar, Ne isotope and bulk K/U ratios. Icarus **339**, 113551 (2020c). https://doi.org/10.1016/j.icarus.2019.113551

D.S. Lauretta et al., The OSIRIS-REx target asteroid (101955) Bennu: constraints on its physical, geological, and dynamical nature from astronomical observations. Meteorit. Planet. Sci. **50**, 834–849 (2015)

D.S. Lauretta et al., The unexpected surface of asteroid (101955) Bennu. Nature (2019). https://doi.org/10.1038/s41586-019-1033-6

F. Leblanc, J.Y. Chaufray, Mercury and Moon He exospheres: analysis and modelling. Icarus (2011). https://doi.org/10.1016/j.icarus.2011.09.028

F. Leblanc, R.E. Johnson, Mercury exosphere. I. Global circulation model of its sodium component. Icarus (2010). https://doi.org/10.1016/j.icarus.2010.04.020

A. Léger, M. Pirre, F.J. Marceau, Search for primitive life on a distant planet: relevance of O_2 and O_3 detections. Astron. Astrophys. **277**, 309 (1993)

A. Léger, D. Rouan, J. Schneider, P. Barge, M. Fridlund, B. Samuel, M. Ollivier, E. Guenther, M. Deleuil, H.J. Deeg et al., Transiting exoplanets from the CoRoT space mission. VIII. CoRoT-7b: the first super-Earth with measured radius. Astron. Astrophys. **506**, 287 (2009)

A. Léger, M. Fontecave, A. Labeyrie, B. Samuel, O. Demangeon, D. Valencia, Is the presence of oxygen on an exoplanet a reliable biosignature? Astrobiology **11**, 335–341 (2011). https://doi.org/10.1089/ast.2010.0516

Z.M. Leinhardt, S.T. Stewart, Collisions between gravity-dominated bodies. I. Outcome regimes and scaling laws. Astrophys. J. **745**, 79 (2012). https://doi.org/10.1088/0004-637X/745/1/79

J.S. Levine, *The Photochemistry of Atmospheres: Earth, the Other Planets, and Comets* (Academic Press, San Diego, 2014)

K. Li, Y. Wei, M. André, A. Eriksson, S. Haaland, E.A. Kronberg, H. Nilsson, L. Maes, Z.J. Rong, W.X. Wan, Cold ion outflow modulated by the solar wind energy input and tilt of the geomagnetic dipole. J. Geophys. Res. **122**, 10,658–10,668 (2017). https://doi.org/10.1002/2017JA024642

C. Li, D. Liu, B. Liu, X. Ren, J. Liu, Z. He, W. Zuo, X. Zeng, R. Xu, X. Tan, X. Zhang, W. Chen, R. Shu, W. Wen, Y. Su, H. Zhang, Z. Ouyang, Chang'E-4 initial spectroscopic identification of lunar far-side mantle-derived materials. Nature (2019). https://doi.org/10.1038/s41586-019-1189-0

S. Li, P.G. Lucey, A.A. Fraeman, A.R. Poppe, V.Z. Sun, D.M. Hurley, P.H. Schultz, Widespread hematite at high latitudes of the Moon. Sci. Adv. **6**, eaba1940 (2020)

R.J. Lillis et al., *The Escape, Plasma and Acceleration Dynamics Explorers (ESCAPADE), a Response to NASA Science Mission Directorate (SMD) Small Innovative Missions for Planetary Exploration* (2018). (SIMPLEX-II) program

Y. Liu, L.M. Kistler, C.G. Mouikis, B. Klecker, I. Dandouras, Heavy ion effects on substorm loading and unloading in the Earth's magnetotail. J. Geophys. Res. (2013). https://doi.org/10.1002/jgra.50240

M.K. Lloyd, H.L.O. McClelland, G. Antler, A.S. Bradley, I. Halevy, C.K. Junium, S.D. Wankel, A.L. Zerkle, The isotopic imprint of life on an evolving planet. Space Sci. Rev. **216**, 112 (2020). https://doi.org/10.1007/s11214-020-00730-6

J.E. Lovelock, Thermodynamics and the recognition of alien biospheres. Proc. R. Soc. Lond. B, Biol. Sci. (1975). https://doi.org/10.1098/rspb.1975.0051
P. Ludovic, The phase a study of the ESA M4 mission candidate ARIEL. Exp. Astron. **46**(1), 211–239 (2018)
N. Madhusudhan, K.K.M. Lee, O. Mousis, A possible carbon-rich interior in super-Earth 55 Cancri e. Astrophys. J. **759**, L40 (2012)
S. Massetti, S. Orsini, A. Milillo, A. Mura, E. De Angelis, H. Lammer, P. Wurz, Mapping of the cusp plasma precipitation on the surface of Mercury. Icarus (2003). https://doi.org/10.1016/j.icarus.2003.08.005
S. Massetti, S. Orsini, A. Milillo, A. Mura, Modelling Mercury's magnetosphere and plasma entry through the dayside magnetopause. Planet. Space Sci. (2007). https://doi.org/10.1016/j.pss.2006.12.008
M. Mayor et al., Setting new standards with HARPS. Messenger **114**, 20 (2003)
W.E. McClintock, R.J. Vervack Jr., E.T. Bradley, R.M. Killen, N. Mouawad, A.L. Sprague, M.H. Burger, S.C. Solomon, N.R. Izenberg, MESSENGER observations of Mercury's exosphere: detection of magnesium and distribution of constituents. Science (2009). https://doi.org/10.1126/science.1172525
V.S. Meadows, C.T. Reinhard, G.N. Arney, M.N. Parenteau, E.W. Schwieterman, S.D. Domagal-Goldman, A.P. Lincowski, K.R. Stapelfeldt, H. Rauer, S. DasSarma, S. Hegde, N. Narita, R. Deitrick, J. Lustig-Yaeger, T.W. Lyons, N. Siegler, J.L. Grenfell, Exoplanet biosignatures: understanding oxygen as a biosignature in the context of its environment. Astrobiology **18**, 630–662 (2018). https://doi.org/10.1089/ast.2017.1727
Y. Miguel, Observability of molecular species in a nitrogen dominated atmosphere for 55 Cancri e. Mon. Not. R. Astron. Soc. **482**, 2893 (2019)
S. Mikhail, D.A. Sverjensky, Nitrogen speciation in upper mantle fluids and the origin of Earth's nitrogen-rich atmosphere. Nat. Geosci. **7**, 816–819 (2014). https://doi.org/10.1038/ngeo2271
A. Milillo, P. Wurz, S. Orsini, D. Delcourt, E. Kallio, R.M. Killen, H. Lammer, S. Massetti, A. Mura, S. Barabash, G. Cremonese, I.A. Daglis, E. Angelis, A.M. Lellis, S. Livi, V. Mangano, K. Torkar, Surface-exosphere-magnetosphere system of Mercury. Space Sci. Rev. (2005). https://doi.org/10.1007/s11214-005-3593-z
A. Milillo, M. Fujimoto, G. Murakami, J. Benkhoff et al., Investigating Mercury's environment with the two-spacecraft BepiColombo mission. Space Sci. Rev. (2020). https://doi.org/10.1007/s11214-020-00712-8
A. Misra, V. Meadows, M.W. Claire, D. Crisp, Using dimers to measure biosignatures and atmospheric pressure for terrestrial exoplanets. Astrobiology **14**, 67–86 (2014). https://doi.org/10.1089/ast.2013.0990
J.S. Morgan, C.B. Pilcher, Plasma characteristics of the Io torus. Astrophys. J. **253**, 406 (1982)
T. Morota et al., Sample collection from asteroid (162173) Ryugu by Hayabusa2: implications for surface evolution. Science **368**, 654–659 (2020). https://doi.org/10.1126/science.aaz6306
C. Muller, N_2O as a biomarker, from the Earth and solar system to exoplanets, in *The Early Evolution of the Atmospheres of Terrestrial Planets, Astrophysics and Space Science Proceedings* (Springer, New York, 2013), pp. 99–106. https://doi.org/10.1007/978-1-4614-5191-4_8
A. Mura, P. Wurz, J. Schneider et al., Comet-like tail-formation of exospheres of hot rocky exoplanets: possible implications for CoRoT-7b. Icarus **211**, 1 (2011)
T. Nagai, N. Kitamura, H. Hasegawa, I. Shinohara et al., Thick escaping magnetospheric ion layer in magnetopause reconnection with MMS observations. Geophys. Res. Lett. (2016). https://doi.org/10.1002/2016GL069085
R. Nakamura, Substorms and their solar wind causes. Space Sci. Rev. (2006). https://doi.org/10.1007/s11214-006-9131-9
D. Nakashima, N.T. Kita, T. Ushikubo, T. Noguchi, T. Nakamura, J.W. Valley, Oxygen three-isotope ratios of silicate particles returned from asteroid Itokawa by the Hayabusa spacecraft: a strong link with equilibrated LL chondrites. Earth Planet. Sci. Lett. **379**, 127–136 (2013). https://doi.org/10.1016/j.epsl.2013.08.009
R. Navarro-González, M.J. Molina, L.T. Molina, Nitrogen fixation by volcanic lightning in the early Earth. Geophys. Res. Lett. **25**, 3123–3126 (1998). https://doi.org/10.1029/98GL02423
R. Navarro-González, C.P. McKay, D.N. Mvondo, A possible nitrogen crisis for Archean life due to reduced nitrogen fixation by lighting. Nature **412**(412), 61–64 (2001)
A. Néri, J. Guignard, M. Monnereau, M.J. Toplis, G. Quitté, Metal segregation in planetesimals: constraints from experimentally determined interfacial energies. Earth Planet. Sci. Lett. **518**, 40–52 (2019). https://doi.org/10.1016/j.epsl.2019.04.049
H. Nilsson, M. Waara, S. Arvelius, O. Marghitu, M. Bouhram, Y. Hobara, M. Yamauchi, R. Lundin, H. Rème, J.-A. Sauvaud, I. Dandouras, A. Balogh, L.M. Kistler, B. Klecker, C.W. Carlson, M.B. Bavassano-Cattaneo, A. Korth, Characteristics of high altitude oxygen ion energization and outflow as observed by Cluster; a statistical study. Ann. Geophys. **24**, 1099–1112 (2006)
H. Nilsson, M. Waara, O. Marghitu, M. Yamauchi, R. Lundin, H. Rème, J.-A. Sauvaud, I. Dandouras, E. Lucek, L.M. Kistler, B. Klecker, C.W. Carlson, M.B. Bavassano-Cattaneo, A. Korth, An assessment of the role of the centrifugal acceleration mechanism in high altitude polar cap oxygen ion outflow. Ann. Geophys. **26**, 145–157 (2008)

L.R. Nittler et al., The major-element composition of Mercury's surface from MESSENGER X-ray spectrometry. Science **333**, 1847–1850 (2011)

D.P. O'Brien, R. Greenberg, The collisional and dynamical evolution of the main-belt and NEA size distributions. Icarus **178**, 179–212 (2005). https://doi.org/10.1016/j.icarus.2005.04.001

C. O'Neill, H.St.C. O'Neill, A.M. Jellinek, On the distribution and variation of radioactive heat producing elements within meteorites, the Earth, and planets. Space Sci. Rev. **216**, 37 (2020)

S. Orsini, S. Livi, K. Torkar, S. Barabash, A. Milillo, P. Wurz, A.M. Di Lellis, E. Kallio (the SERENA team), SERENA: a suite of four instruments (ELENA, STROFIO, PICAM and MIPA) on board BepiColombo-MPO for particle detection in the Hermean environment. Planet. Space Sci. (2010). https://doi.org/10.1016/j.pss.2008.09.012

S. Orsini, V. Mangano, A. Milillo, C. Plainaki, A. Mura, J.M. Raines, E. De Angelis, R. Rispoli, F. Lazzarotto, A. Aronica, Mercury sodium exospheric emission as a proxy for solar perturbations transit. Nat. Sci. Rep. (2018). https://doi.org/10.1038/s41598-018-19163-x

T.C. Owen, The search for early forms of life in other planetary systems: future possibilities afforded by spectroscopic techniques, in *Strategies for the Search for Life in the Universe*, ed. by M.D. Papagiannis (Springer, Dordrecht, 1980), pp. 177–185. https://doi.org/10.1007/978-94-009-9115-6_17

M. Ozima, K. Seki, N. Terada, Y.N. Miura, F.A. Podosek, H. Shinagawa, Terrestrial nitrogen and noble gases in lunar soils. Nature (2005). https://doi.org/10.1038/nature03929

E. Pallé, M.R.Z. Osorio, R. Barrena, P. Montañés-Rodríguez, E.L. Martín, Earth's transmission spectrum from lunar eclipse observations. Nature **459**, 814–816 (2009). https://doi.org/10.1038/nature08050

D. Parkos, A. Pikus, A. Alexeenko, H.J. Melosh, HCN production via impact ejecta reentry during the late heavy bombardment. J. Geophys. Res., Planets **123**, 892–909 (2018). https://doi.org/10.1002/2017JE005393

G.K. Parks, E. Lee, S. Fu, M. Fillingim, I. Dandouras, Y. Cui, J. Hong, H. Rème, Outflow of low-energy O^+ ion beams observed during periods without substorms. Ann. Geophys. **33**, 333 (2015)

F.A. Pepe et al., ESPRESSO: the Echelle spectrograph for rocky exoplanets and stable spectroscopic observations. Proc. SPIE **7735**, 77354Y (2010)

E.A. Petigura, A.W. Howard, G.W. Marcy, Prevalence of Earth-size planets orbiting Sun-like stars. Proc. Natl. Acad. Sci. USA **110**, 19273–19278 (2013). https://doi.org/10.1073/pnas.1319909110

S.T. Petsch, The global oxygen cycle, in *Treatise on Geochemistry* (2014), pp. 437–473. https://doi.org/10.1016/B978-0-08-095975-7.00811-1

M. Pfleger, H.I.M. Lichtenegger, P. Wurz et al., 3D-modeling of Mercury's solar wind sputtered surface-exosphere environment. Planet. Space Sci. **115**, 90 (2015)

C.B. Pilcher, Biosignatures of early Earths. Astrobiology **3**, 471–486 (2003). https://doi.org/10.1089/153110703322610582

A.R. Poppe, R. Samad, J.S. Halekas, M. Sarantos, G.T. Delory, W.M. Farrell, V. Angelopoulos, J.P. McFadden, ARTEMIS observations of lunar pick-up ions in the terrestrial magnetotail lobes. Geophys. Res. Lett. (2012). https://doi.org/10.1029/2012GL052909

A.R. Poppe, S. Fatemi, I. Garrick-Bethell, D. Hemingway, M. Holmström, Solar wind interaction with the Reiner Gamma crustal magnetic anomaly: connecting source magnetization to surface weathering. Icarus (2015). https://doi.org/10.1016/j.icarus.2015.11.005

A.R. Poppe, M.O. Fillingim, J.S. Halekas, J. Raeder, V. Angelopoulos, ARTEMIS observations of terrestrial ionospheric molecular ion outflow at the Moon. Geophys. Res. Lett. (2016). https://doi.org/10.1002/2016GL069715

A.E. Potter, R.M. Killen, T.H. Morgan, Variation of lunar sodium during passage of the Moon through the Earth's magnetotail. J. Geophys. Res. (2000). https://doi.org/10.1029/1999JE001213

J. Prieto-Arranz, E. Palle, D. Gandolfi et al., Mass determination of the 1:3:5 near-resonant planets transiting GJ 9827 (K2-135). Astron. Astrophys. **618**, A116 (2018)

L. Puig, G. Pilbratt, A. Heske et al., The phase a study of the ESA M4 mission ARIEL. Exp. Astron. **46**, 211 (2018)

S. Quanz et al., Atmospheric characterization of terrestrial exoplanets in the mid-infrared: biosignatures, habitability & diversity (2019). Submitted to ESA in response to the Call for White Papers for the Voyage 2050 long-term plan. arXiv:1908.01316

A. Quirrenbach et al., CARMENES: an overview six months after first light. Proc. SPIE **990812**, SPIE.9908 (2016)

J.M. Raines, G.A. DiBraccio, T.A. Cassidy, D. Delcourt, M. Fujimoto, X. Jia, V. Mangano, A. Milillo, M. Sarantos, J.A. Slavin, P. Wurz, Plasma sources in planetary magnetospheres: Mercury. Space Sci. Rev. (2015). https://doi.org/10.1007/s11214-015-0193-4

S. Rappaport, R. Sanchis-Ojeda, L.A. Rogers, A. Levine, J.N. Winn, The Roche limit for close-orbiting planets: minimum density, composition constraints, and application to the 4.2 hr planet KOI 1843.03. Astrophys. J. **773**, L15 (2013)

H. Rauer, S. Gebauer, P. Von Paris, J. Cabrera, M. Godolt, J.L. Grenfell, A. Belu, F. Selsis, P. Hedelt, F. Schreier, Potential biosignatures in super-Earth atmospheres. I. Spectral appearance of super-Earths around M dwarfs. Astron. Astrophys. **529**, A8 (2011). https://doi.org/10.1051/0004-6361/201014368

H. Rauer et al., The PLATO 2.0 mission. Exp. Astron. **38**(1–2), 249–330 (2014)

F. Raulin, A. Coustenis, G. Kminek, N. Hedman, Special issue: planetary protection: new aspects of policy and requirements. Life Sci. Space Res. **23**, 1–142 (2019). https://doi.org/10.1016/j.lssr.2019.09.003

H. Rème, C. Aoustin, J.M. Bosqued, I. Dandouras et al., First multispacecraft ion measurements in and near the Earth's magnetosphere with the identical Cluster ion spectrometry (CIS) experiment. Ann. Geophys. **19**, 1303–1354 (2001)

Report of the Venera-D Joint Science Definition Team, Venera-D: Expanding Our Horizon of Terrestrial Planet Climate and Geology Through the Comprehensive Exploration of Venus, January 31, 2019. http://www.iki.rssi.ru/events/2019/Venera-DPhaseIIFinalReport.pdf

I. Ribas, E.F. Guinan, M. Güdel, M. Audard, Evolution of the solar activity over time and effects on planetary atmospheres. I. High-energy irradiances (1-1700 Å). Astrophys. J. (2005). https://doi.org/10.1086/427977

G.R. Ricker et al., Transiting Exoplanet Survey Satellite (TESS). J. Astron. Telesc. Instrum. Syst. **1**, 014003 (2015)

A.R. Ridden-Harper, I.A.G. Snellen, C.U. Keller et al., Search for an exosphere in sodium and calcium in the transmission spectrum of exoplanet 55 Cancri e. Astron. Astrophys. **593**, A129 (2016)

F. Rodler, *Exoplanet Research in the Era of the Extremely Large Telescope (ELT)*. Handbook of Exoplanets (Springer, Berlin, 2018), p. 194. ISBN 978-3-319-55332-0

L. Rogers, Most 1.6 Earth radii planets are not rocky. Astrophys. J. **801**, 41 (2015). https://doi.org/10.1088/0004-637X/801/1/41

D. Rouan, H.J. Deeg, O. Demangeon et al., The orbital phases and secondary transits of Kepler-10b. A physical interpretation based on the lava-ocean planet model. Astrophys. J. Lett. **741**, L30 (2011)

S. Rugheimer, L. Kaltenegger, A. Zsom, A. Segura, D. Sasselov, Spectral fingerprints of Earth-like planets around FGK stars. Astrobiology **13**, 251–269 (2013). https://doi.org/10.1089/ast.2012.0888

S. Rugheimer, L. Kaltenegger, A. Segura, J. Linsky, S. Mohanty, Effect of UV radiation on the spectral fingerprints of Earth-like planets orbiting M stars. Astrophys. J. **809** (2015). https://doi.org/10.1088/0004-637X/809/1/57

C. Sagan, Direct contact among galactic civilizations by relativistic interstellar spaceflight. Planet. Space Sci. **11**, 485–498 (1963). https://doi.org/10.1016/0032-0633(63)90072-2

C. Sagan, W.R. Thompson, R. Carlson, D. Gurnett, C. Hord, A search for life on Earth from the Galileo spacecraft. Nature **365**, 715–721 (1993)

Y. Saito et al., In-flight performance and initial results of plasma energy angle and composition experiment (PACE) on SELENE (Kaguya). Space Sci. Rev. (2010). https://doi.org/10.1007/s11214-010-9647-x

A.B. Sanin, I.G. Mitrofanov, Selection of Luna-25 and Luna-27 landing sites based on LRO data, in *AGU Fall General Assembly*, vol. 2016, (2016), id. P53A-2182

A. Santerne, B. Brugger, D.J. Armstrong et al., An Earth-sized exoplanet with a Mercury-like composition. Nat. Astron. **2**, 393 (2018)

T.E. Sarris, E.R. Talaat, M. Palmroth, I. Dandouras, E. Armandillo, G. Kervalishvili et al., Daedalus: a low-flying spacecraft for the exploration of the lower thermosphere – ionosphere. Geosci. Instrum. Method. Data Syst. (2020). https://doi.org/10.5194/gi-9-153-2020

J. Scalo, L. Kaltenegger, A. Segura, M. Fridlund, I. Ribas, Y.N. Kulikov, J.L. Grenfell, H. Rauer, P. Odert, M. Leitzinger, F. Selsis, M.L. Khodachenko, C. Eiroa, J.F. Kasting, H. Lammer, M stars as targets for terrestrial exoplanet searches and biosignature detection. Astrobiology **7**, 85–166 (2007). https://doi.org/10.1089/ast.2006.0125

L. Schaefer, B. Fegley, Chemistry of silicate atmospheres of evaporating super-Earths. Astrophys. J. Lett. **703**, L113 (2009)

M. Scherf, H. Lammer, N.V. Erkaev, K.E. Mandt, B. Marty, Nitrogen atmospheres of the icy bodies in the solar system. Space Sci. Rev. (2020). https://doi.org/10.1007/s11214-020-00752-0. in press

A. Schillings, R. Slapak, H. Nilsson, M. Yamauchi, I. Dandouras, L.-G. Westerberg, Earth atmospheric loss through the plasma mantle and its dependence on solar wind parameters. Earth Planets Space (2019). https://doi.org/10.1186/s40623-019-1048-0

E. Schlawin, T. Hirano, H. Kawahara et al., Back to "normal" for the disintegrating planet candidate KIC 12557548 b. Astron. J. **156**, 28 (2018)

H. Schleicher, G. Wiedemann, H. Wöhl, T. Berkefeld, D. Soltau, Detection of neutral sodium above Mercury during the transit on 2003 May 7. Astron. Astrophys. **425**, 1119 (2004)

E.W. Schwieterman, N.Y. Kiang, M.N. Parenteau, C.E. Harman, S. DasSarma, T.M. Fisher, G.N. Arney, H.E. Hartnett, C.T. Reinhard, S.L. Olson, V.S. Meadows, C.S. Cockell, S.I. Walker, J.L. Grenfell, S. Hegde, S. Rugheimer, R. Hu, T.W. Lyons, Exoplanet biosignatures: a review of remotely detectable signs of life. Astrobiology **18** (2017). https://doi.org/10.1089/ast.2017.1729

S. Seager, The search for habitable planets with biosignature gases framed by a "biosignature drake equation.". Int. J. Astrobiol. **17**, 294–302 (2017). https://doi.org/10.1017/S1473550417000052

S. Seager, W. Bains, R. Hu, Biosignature gases in H_2-dominated atmospheres on rocky exoplanets. Astrophys. J. **777**, 95 (2013a). https://doi.org/10.1088/0004-637X/777/2/95

S. Seager, W. Bains, R. Hu, A biomass-based model to estimate the plausibility of exoplanet biosignature gases. Astrophys. J. **775**, 104 (2013b). https://doi.org/10.1088/0004-637X/775/2/104

A. Segura, K. Krelove, J.F. Kasting, D. Sommerlatt, V. Meadows, D. Crisp, M. Cohen, E. Mlawer, Ozone concentrations and ultraviolet fluxes on Earth-like planets around other stars. Astrobiology **3**, 689–708 (2003). https://doi.org/10.1089/153110703322736024

A. Segura, J.F. Kasting, V. Meadows, M. Cohen, J. Scalo, D. Crisp, R.A.H. Butler, G. Tinetti, Biosignatures from Earth-like planets around M dwarfs. Astrobiology **5**, 706–725 (2005). https://doi.org/10.1089/ast.2005.5.706

K. Seki, M. Hirahara, T. Terasawa, T. Mukai, Y. Saito, S. Machida, T. Yamamoto, S. Kokubun, Statistical properties and possible supply mechanisms of tailward cold O^+ beams in the lobe/mantle regions. J. Geophys. Res. (1998). https://doi.org/10.1029/97JA02137

D.G. Sibeck, R. Allen, H. Aryan, D. Bodewits et al., Imaging plasma density structures in the soft X-rays generated by SolarWind charge exchange with neutrals. Space Sci. Rev. (2018). https://doi.org/10.1007/s11214-018-0504-7

A. Silburt, E. Gaidos, Y. Wu, A statistical reconstruction of the planet population around Kepler solar-type stars. Astrophys. J. **799**, 180 (2015). https://doi.org/10.1088/0004-637X/799/2/180

E. Sinukoff, A.W. Howard, E.A. Petigura et al., Mass constraints of the WASP-47 planetary system from radial velocities. Astron. J. **153**, 70 (2017)

R. Slapak, A. Schillings, H. Nilsson, M. Yamauchi, L.-G. Westerberg, I. Dandouras, Atmospheric loss from the dayside open polar region and its dependence on geomagnetic activity: implications for atmospheric escape on evolutionary time scales. Ann. Geophys. (2017). https://doi.org/10.5194/angeo-35-721-2017

J.A. Slavin et al., MESSENGER observations of magnetic reconnection in Mercury's magnetosphere. Science (2009). https://doi.org/10.1126/science.1172011

J. Sohn, S. Oh, Y. Yi, Lunar cosmic ray radiation environments during Luna and Lunar Reconnaissance Orbiter missions. Adv. Space Res. (2014). https://doi.org/10.1016/j.asr.2014.05.011

S.C. Solomon et al., The MESSENGER mission to Mercury: scientific objectives and implementation. Planet. Space Sci. (2001). https://doi.org/10.1016/S0032-0633(01)00085-X

C. Sotin, G. Avice, J. Baker, A. Freeman, S. Madzunkov, T. Stevenson, N. Arora, M. Darrach, G. Lightsey, B. Marty, Cupid's arrow: a small satellite concept to measure noble gases in Venus' atmosphere, in *49th Lunar and Planetary Science Conference*, 2018, The Woodlands, Texas. LPI Contribution, vol. 2083, (2018), id. 1763

T.J. Stubbs, R.R. Vondrak, W.M. Farrell, A dynamic fountain model for dust in the lunar exosphere, in *Dust in Planetary Systems, ESA SP-643* (2007)

E.E. Stüeken, M.A. Kipp, M.C. Koehler, E.W. Schwieterman, B.W. Johnson, R. Buick, Modeling pN_2 through geological time: implications for planetary climates and atmospheric biosignatures. Astrobiology **16**, 949–963 (2016). https://doi.org/10.1089/ast.2016.1537

E.E. Stüeken, S.M. Som, M.W. Claire, S. Rugheimer, M. Scherf, L. Sproß, N. Tosi, Y. Ueno, H. Lammer, Mission to planet Earth: the first two billion years. Space Sci. Rev. **216**, 41 (2020). https://doi.org/10.1007/s11214-020-00667-w

S. Sugita et al., The geomorphology, color, and thermal properties of Ryugu: implications for parent-body processes. Science (2019). https://doi.org/10.1126/science.aaw0422

P.W. Sullivan et al., The transiting exoplanet survey satellite: simulations of planet detections and astrophysical false positives. Astrophys. J. **809**, 77 (2015)

F. Tabataba-Vakili, J.L. Grenfell, J.-M. Grießmeier, H. Rauer, Atmospheric effects of stellar cosmic rays on Earth-like exoplanets orbiting M-dwarfs. Astron Astrophys. **585** (2016). https://doi.org/10.1051/0004-6361/201425602

M. Takashi et al., Mineralogy and crystallography of some itokawa particles returned by the Hayabusa asteroidal sample return mission. Earth Planets Space **66**, 82 (2014). https://doi.org/10.1186/1880-5981-66-82

P. Tamburo, A. Mandell, D. Deming, E. Garhart, Confirming variability in the secondary eclipse depth of the super-Earth 55 Cancri e. Astron. J. **155**, 221 (2018)

K. Terada, S. Yokota, Y. Saito, N. Kitamura, K. Asamura, M.N. Nishino, Biogenic oxygen from Earth transported to the Moon by a wind of magnetospheric ions. Nat. Astron. (2017). https://doi.org/10.1038/s41550-016-0026

N. Thomas, Detection of O III 5007-A forbidden line emission from the Io plasma torus. Astrophys. J. **414**, L41 (1993)

N. Thomas, High resolution spectra of Io's neutral potassium and oxygen clouds. Astron. Astrophys. **313**, 306 (1996)
F. Tian, J.F. Kasting, K.J. Zahnle, Revisiting HCN formation in Earth's early atmosphere. Earth Planet. Sci. Lett. **308**, 417–423 (2011). https://doi.org/10.1016/j.epsl.2011.06.011
G. Tinetti et al., A chemical survey of exoplanets with ARIEL. Exp. Astron. (2018). https://doi.org/10.1007/s10686-018-9598-x
G. Tinetti et al., The science of ARIEL, in *Space Telescopes and Instrumentation 2016: Optical, Infrared, and Millimeter Wave*. Proc. SPIE, vol. 9904 (2016). 99041X
W.A. Traub, Kepler exoplanets: a new method of population analysis (2016). arXiv:1605.02255
L. Tu, C.P. Johnstone, M. Güdel, H. Lammer, The extreme ultraviolet and X-ray Sun in time: high-energy evolutionary tracks of a solar-like star. Astron. Astrophys. **577**, L3 (2015). https://doi.org/10.1051/0004-6361/201526146
O.J. Tucker, W.M. Farrell, R.M. Killen, D.M. Hurley, Solar wind implantation into the lunar regolith: Monte Carlo simulations of H retention in a surface with defects and the H_2 exosphere. J. Geophys. Res. (2019). https://doi.org/10.1029/2018JE005805
D. Valencia, E. Paracha, A.P. Jackson, Can a machine learn the outcome of planetary collisions? Astrophys. J. (2019). https://doi.org/10.3847/1538-4357/ab2bfb
R.J. Vervack Jr., W.E. McClintock, R.M. Killen, A.L. Sprague, B.J. Anderson, M.H. Burger, E.T. Bradley, N. Mouawad, S.C. Solomon, N.R. Izenberg, Mercury's complex exosphere: results from MESSENGER's third flyby. Science (2010). https://doi.org/10.1126/science.1188572
A.A. Vidotto, H. Lichtenegger, L. Fossati et al., Characterization of the HD 219134 multi-planet system II. Stellar-wind sputtered exospheres in rocky planets b & c. Mon. Not. R. Astron. Soc. **481**, 5296 (2018)
M. Viso, Mars sample receiving facility or facilities? That is the question. Life Sci. Space Res. **23**, 69–72 (2019). https://doi.org/10.1016/j.lssr.2019.06.004
A. Vorburger, P. Wurz, S. Barabash, M. Wieser, Y. Futaana, M. Holmström, A. Bhardwaj, K. Asamura, Energetic neutral atom observations of magnetic anomalies on the lunar surface. J. Geophys. Res. **117**, A07208 (2012). https://doi.org/10.1029/2012JA017553
A. Vorburger, P. Wurz, S. Barabash, M. Wieser, Y. Futaana, A. Bhardwaj, K. Asamura, Imaging the South Pole–Aitken basin in backscattered neutral hydrogen atoms. Planet. Space Sci. (2015). https://doi.org/10.1016/j.pss.2015.02.007
M. Voss, H.W. Bange, J.W. Dippner, J.J. Middelburg, J.P. Montoya, B. Ward, The marine nitrogen cycle: recent discoveries, uncertainties and the potential relevance of climate change. Philos. Trans. R. Soc. Lond. B, Biol. Sci. **368** (2013). https://doi.org/10.1098/rstb.2013.0121
J.C.G. Walker, P.B. Hays, J.F. Kasting, A negative feedback mechanism for the long-term stabilization of Earth's surface temperature. J. Geophys. Res. **86**, 9776–9782 (1981). https://doi.org/10.1029/JC086iC10p09776
S. Watanabe et al., Hayabusa2 arrives at the carbonaceous asteroid 162173 Ryugu—a spinning top–shaped rubble pile. Science **364**, 268–272 (2019). https://doi.org/10.1126/science.aav8032
D.T. Welling, M. André, I. Dandouras, D. Delcourt, A. Fazakerley, D. Fontaine, J. Foster, R. Ilie, L. Kistler, J. Lee, M.W. Liemohn, J.A. Slavin, C.-P. Wang, M. Wiltberger, A. Yau, The Earth: plasma sources, losses, and transport processes. Space Sci. Rev. (2015). https://doi.org/10.1007/s11214-015-0187-2
E.T. Wolf, Assessing the habitability of the TRAPPIST-1 system using a 3D climate model. Astrophys. J. **839**, L1 (2017). https://doi.org/10.3847/2041-8213/aa693a
G. Wurm, M. Trieloff, H. Rauer, Photophoretic separation of metals and silicates: the formation of Mercury-like planets and metal depletion in chondrites. Astrophys. J. **769**, 78 (2013)
P. Wurz, H. Lammer, Monte-Carlo simulation of Mercury's exosphere. Icarus (2003). https://doi.org/10.1016/S0019-1035(03)00123-4
P. Wurz, U. Rohner, J.A. Whitby, C. Kolb, H. Lammer, P. Dobnikar, J.A. Martín-Fernández, The lunar exosphere: the sputtering contribution. Icarus (2007). https://doi.org/10.1016/j.icarus.2007.04.034
P. Wurz, J.A. Whitby, U. Rohner, J.A. Martín-Fernández, H. Lammer, C. Kolb, Self-consistent modelling of Mercury's exosphere by sputtering, micro-meteorite impact and photon-stimulated desorption. Planet. Space Sci. **58**, 1599–1616 (2010)
Y. Xu, H.S. Fu, C. Norgren, S. Toledo-Redondo, C.M. Liu, X.C. Dong, Ionospheric cold ions detected by MMS behind dipolarization fronts. Geophys. Res. Lett. (2019). https://doi.org/10.1029/2019GL083885
M. Yamauchi, Terrestrial ion escape and relevant circulation in space. Ann. Geophys. (2019). https://doi.org/10.5194/angeo-37-1197-2019
M. Yamauchi, R. Slapak, Energy conversion through mass loading of escaping ionospheric ions for different Kp values. Ann. Geophys. (2018). https://doi.org/10.5194/angeo-36-1-2018
M. Yamauchi, J.-E. Wahlund, Role of the ionosphere for the atmospheric evolution of planets. Astrobiology **7**, 783–800 (2007). https://doi.org/10.1089/ast.2007.0140

A. Yau, M. André, Sources of ion outflow in the high latitude ionosphere. Space Sci. Rev. (1997). https://doi.org/10.1023/A:1004947203046
I. Yoshikawa, O. Korablev, S. Kameda, D. Rees, H. Nozawa, S. Okano, V. Gnedykh, V. Kottsov, K. Yoshioka, G. Murakami, F. Ezawa, G. Cremonese, The Mercury sodium atmospheric spectral imager for the MMO spacecraft of BepiColombo. Planet. Space Sci. **56**, 224–237 (2009)
D.T. Young, J.L. Burch, R.G. Gomez, A. De Los Santos et al., Hot plasma composition analyzer for the magnetospheric multiscale mission. Space Sci. Rev. (2016). https://doi.org/10.1007/s11214-014-0119-6
K.J. Zahnle, Photochemistry of methane and the formation of hydrocyanic acid (HCN) in the Earth's early atmosphere. J. Geophys. Res. **91**, 2819–2834 (1986). https://doi.org/10.1029/JD091iD02p02819
A.L. Zerkle, S. Mikhail, The geobiological nitrogen cycle: from microbes to the mantle. Geobiology **15**, 343–352 (2017). https://doi.org/10.1111/gbi.12228
J.K. Zink, B.M.S. Hansen, Accounting for multiplicity in calculating eta Earth. Mon. Not. R. Astron. Soc. **487**, 246–252 (2019). https://doi.org/10.1093/mnras/stz1246
S.S. Zumdahl, S.A. Zumdahl, *Chemistry*, 8th edn. (Cengage Learning, Belmont, 2010)
T.H. Zurbuchen et al., MESSENGER observations of the spatial distribution of planetary ions near Mercury. Science (2011). https://doi.org/10.1126/science.1211302

Space Science Reviews (2021) 217:45
https://doi.org/10.1007/s11214-021-00826-7

SPECIAL COMMUNICATION

Earth's Nitrogen and Carbon Cycles

Ralf Halama[1] · **Gray Bebout**[2]

Received: 22 February 2021 / Accepted: 24 March 2021 / Published online: 2 April 2021

Abstract
Understanding the Earth's geological nitrogen (N) and carbon (C) cycles is fundamental for assessing the distribution of these volatiles between solid Earth (core, mantle and crust), oceans and atmosphere. This Special Communication about the Earth's N and C cycles contains material that is relevant for researchers who are interested in the Topical Collection on planetary evolution "Reading Terrestrial Planet Evolution in Isotopes and Element Measurements". Variations in the fluxes of N and C between these major reservoirs through geological time influenced the evolution and determined the unique composition of the Earth's atmosphere. Here we review several key geological aspects of the N and C cycles of which our understanding has significantly advanced during the last decade through field-based, experimental and theoretical studies. Subduction zones are the most important pathway of both N and C from the Earth's surface into the deep Earth. A key question in the flux quantification is how much of the volatile elements is stored in the downgoing slab and introduced into the mantle and how much is returned back to the surface and the atmosphere through arc magmatism. For N, the retention of N as NH_4^+ in minerals has a major influence on fluxes between reservoirs. The temperature-dependent stability of NH_4^+-bearing minerals determines whether N is predominantly retained in the slab to mantle depths (in subduction zones with a low geothermal gradient) or devolatilized (in subduction zones with a high geothermal gradient). Several lines of evidence suggest that the mantle is regassing with respect to N due to a net influx of subducted N over time, but this issue is highly debated and evidence to the contrary also exists. Nevertheless, there is consensus that the majority of the planetary N budget is stored in the Earth's mantle, with the continental crust also constituting a significant N reservoir. For C, release from the subducting slab occurs through decarbonation reactions, dissolution and formation of carbonatitic liquids, but reprecipitation of C in the slab or the forearc mantle wedge may limit the effectiveness of direct return of C into the atmosphere. Carbon release through regional metamorphism in collision zone orogens also has potentially profound effects on C release into the atmosphere and consensus has emerged that such orogens are sources rather than sinks of atmospheric CO_2. On shorter timescales, contact metamorphism through interaction of mantle-derived magmas with C-bearing country rocks, and the resulting release of large quantities of CH_4 and/or CO_2, has been linked to global warming events.

Keywords Nitrogen cycle · Carbon cycle · Planetary atmospheres · Planetary evolution

Reading Terrestrial Planet Evolution in Isotopes and Element Measurements
Edited by Helmut Lammer, Bernard Marty, Aubrey L. Zerkle, Michel Blanc, Hugh O'Neill and Thorsten Kleine

Extended author information available on the last page of the article

1 The Deep Nitrogen Cycle on Earth

The Earth's N-cycle links to observations from planetary and extraterrestrial N-reservoirs and to near-surface nutrient cycling and palaeoenvironmental investigations of ancient Earth. Regarding the evolution of the Earth's atmosphere, a key aspect is whether the Earth's mantle is presently degassing or regassing and whether this balance has shifted in the Earth's history (Zhang and Zindler 1993). One approach to constrain the whole-Earth N distribution is based on using existing observations about the sizes of the Earth's N-reservoirs based on data in key rock types and geotectonic settings (Javoy 1997; Tolstikhin and Marty 1998; Johnson and Goldblatt 2015, 2017a,b; Bebout et al. 2016; Zerkle and Mikhail 2017; Hirschmann 2018). Our understanding of the geologic processes responsible for distributing N has further improved thanks to the experimental work on mineral-fluid-melt N partitioning (Roskosz et al. 2006; Mysen and Fogel 2010; Mallik et al. 2018; Mysen 2018) and theoretical calculations on the speciation of N as a function of P-T conditions and pH (Mikhail and Sverjensky 2014; Mikhail et al. 2017), one can focus here on the following key areas for the distribution and cycling of N on Earth:

1. Subduction zones, which play an essential role for the transport of N from the Earth's surface to the deep interior of the Earth and its subsequent partial return via a "short circuit" through arc magmatism or a "longer loop" through incorporation into mantle plumes.
2. Earth's continental crust, which it appears could contain a large fraction of whole-Earth N (perhaps $\sim$15%; Goldblatt et al. 2009; Johnson and Goldblatt 2017a,b).
3. The Earth's mantle and core, which together contain large amounts of N outweighing those of atmosphere and crust.

Much of the N in Earth's lithosphere owes its existence to its biogeochemical processing and fixation into organic matter, then transfer into low-temperature mineral phases such as the clay minerals (Boyd 2001; Bebout et al. 2013b; Busigny and Bebout 2013). Once it is stored as NH_4^+ in silicate phases (see Fig. 1), N can be retained at impressive levels even in rocks experiencing partial melting (Jia 2006; Palya et al. 2011) and this retention in NH_4^+-bearing phases (e.g., micas, feldspars) likely plays a key role in determining the sizes of deep-Earth reservoirs and cycling among these reservoirs (Yang et al. 2017; Mikhail et al. 2017). Of all the different lithologies in subduction zones, marine sediments have the highest concentrations of N with concentrations of up to several hundred ppm (Sadofsky and Bebout 2014). Early diagenetic breakdown of organic matter (Sadofsky and Bebout 2003; Li et al. 2007) leads to an increase in NH_4^+ in sediment pore waters, which is then at greater depths incorporated into clays (Schroeder and McLain 1998). Reworking of sediment and organic matter (Junium et al. 2015) and diagenesis (Williams et al. 1995) affect the distribution of N within the sediments, resulting in a large variability in N concentration and isotopic composition. Extremely heterogeneously distributed N is also present in hydrothermally altered oceanic crust and mantle rocks (Busigny et al. 2005; Li et al. 2007; Halama et al. 2014). Because of the very large volume of oceanic crust and mantle being subducted, even very small concentrations of N added during alteration can produce a subducting N inventory rivalling that in the overlying sediment sections (Li and Bebout 2005; Busigny et al. 2011; Halama et al. 2014).

Subduction zone thermal structure determines the stability of N-bearing mineral phases and hence the efficiency of the deep subduction of N (Bebout and Fogel 1992; Busigny et al. 2003; Bebout et al. 2013a, 2016; Halama et al. 2017). During subduction zone metamorphism, preservation of pre-subduction heterogeneities in high-P/T metamorphosed rocks

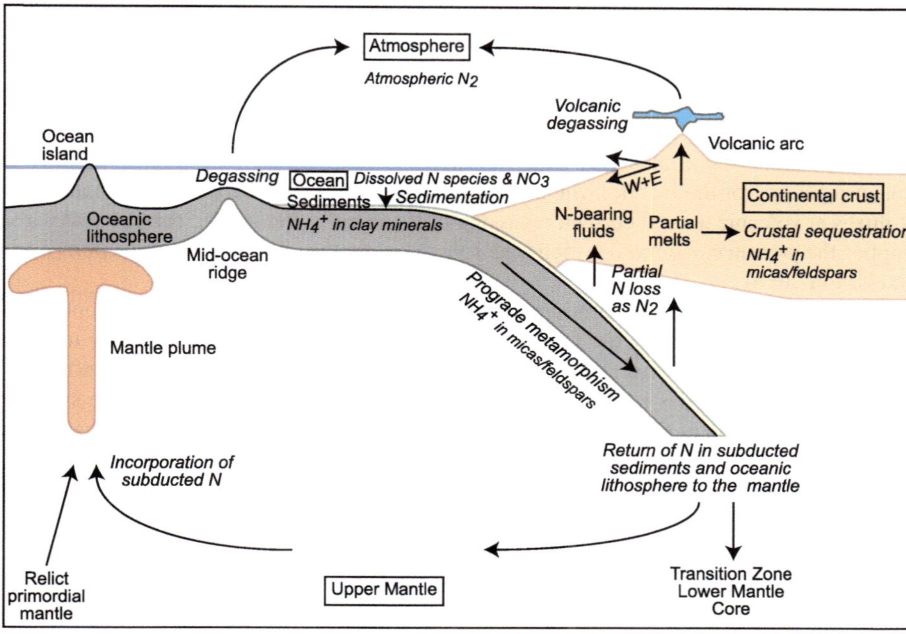

Fig. 1 Schematic representation of the geological nitrogen cycle (based on Bebout et al. 2013a,b; Cartigny and Marty 2013). W + E = Weathering and erosion

has been observed in some metasedimentary rocks (Sadofsky and Bebout 2003; Pitcairn et al. 2005; Bebout et al. 2013a), metabasaltic rocks (Halama et al. 2010; Busigny et al. 2011) and serpentinites (Philippot et al. 2007). In these rocks, the effects of devolatilization are variably masked by the initial heterogeneity related to seafloor alteration over wide ranges in temperature and at highly variable fluid/rock ratios. Only in subduction-related metamorphic rocks that formed under a relatively warm geothermal gradient has significant N devolatilization been observed, resulting in decreasing N contents with increasing grade (Bebout and Fogel 1992). The behaviour of N during partial melting is largely unknown (see Jia 2006; Palya et al. 2011), but for the mantle wedge above the subducting slab, thermodynamic calculations indicate relatively oxidising conditions under which N is likely to degas as N_2 in fluids (Mikhail and Sverjensky 2014). The behaviour of N during fractional crystallization of mantle-derived melts in the arc is also poorly constrained. However, studies of N fluxes in volcanic arcs can provide constraints on the efficiency of N recycling through the subduction cycle (see Hilton et al. 2002; Sano et al. 2001).

A study by Fischer et al. (2002) for the Central American margin concluded that almost 100% of sedimentary N is returned to the surface via arc volcanism. However, subsequent investigations updated this consideration employing data for the specific sediments subducting into that margin and demonstrated that, for some margins, the altered oceanic crust can be as important for the N input flux as the sediments (Li and Bebout 2005; Elkins et al. 2006; Li et al. 2007; Mitchell et al. 2010). If N stored in the altered oceanic crust is taken into consideration, the return of subducted N via arc volcanism is far less efficient, yielding values of 15-40% at the Central American margin and 4-17% at the Izu-Bonin arc. These calculations suggest that a large amount of N is transported to beyond sub-arc depths into mantle depths greater than 150-200 km.

Nitrogen-bearing material that has been subducted can be traced in plume-derived magmas using a combination of N elemental and isotopic systematic in combination with noble gas data (Marty and Dauphas 2003). Systematic differences in the N isotope composition of rocks derived from a plume mantle source compared to those from a depleted MORB mantle source suggest the presence of a recycled crustal component in the Iceland and Reunion plumes (Halldórsson et al. 2016; Barry and Hilton 2016). These data were interpreted to reflect regassing of the mantle, which can explain the imbalance between input and output N fluxes at volcanic arcs. If correct, regassing of the mantle over geological timescales also implies the presence of a more N-rich atmosphere before the initiation of subduction (Barry and Hilton 2016).

Estimation of the N concentration and isotopic composition of N in Earth's continental crust have been plagued by the fragmentary dataset for various materials likely to make up this highly heterogeneous reservoir (see Johnson and Goldblatt 2017a,b). The extremely wide range of lithologies in continental crust complicates any assessment of the size and isotopic composition of this reservoir (see Rudnick and Gao 2014) and estimates have been based on analyses of individual rock types (e.g., Palya et al. 2011) weighted by proportions in the crust. Johnson and Goldblatt (2017a) presented measurements of N concentration in glacial tills through time as a proxy for the upper continental crust and, building on arguments made earlier by Goldblatt et al. (2009), suggested that N in the early-Earth atmosphere was uptaken via biological processes and conveyed into (and stored in) continental crust. Johnson and Goldblatt (2017a) suggested that the isotopic composition of N shows no obvious secular variation.

Several recent experimental studies have investigated the distribution and behaviour of N in the deep mantle and the core (Cartigny and Marty 2013; Li et al. 2013; Li and Keppler 2014; Yoshioka et al. 2018; Li et al. 2016; Dalou et al. 2017). Despite the generally low N concentrations in mantle minerals (Yokochi et al. 2009), most of the planetary budget of N is stored in the Earth's mantle (see the review by Johnson and Goldblatt 2015, and papers cited therein). A high N storage capacity of mantle minerals indicates that N released during subduction would be redissolved in the mantle minerals (Watenphul et al. 2010; Li et al. 2013). Nitrogen solubility data for major minerals in the Earth's transition zone and lower mantle suggest that these regions can store at least 33 times the mass of N presently in the atmosphere (Yoshioka et al. 2018).

From this brief summary, one can see that the N-cycle is a complex system that is interlinked to the planet's tectonic regime and hence its initial radioactive heat producing elements (O'Neil et al.) and even life (see Stüeken et al.; Lloyd et al.). Three review articles within this book (Avice and Marty; Stüeken et al.; Lloyd et al.) summarize the latest knowledge of early Earth's nitrogen evolution based on constraints of a detailed analysis of early Earth's isotopic fingerprints.

2 A Tectonometamorphic Perspective on Carbon Cycling

For carbon cycling at the present day and in the Earth's geological past one can focus on three tectonic settings in which metamorphic processes play a key role in C cycling:

1. Subduction zones, which are fundamental for the return of material from the Earth's surface to the deep interior of the Earth and, largely (but not entirely) via arc volcanic degassing, into the atmosphere. The metamorphic processes in the subducting slab and the overlying lithosphere are essential to understand the balance between C input fluxes at trenches and the output via arc volcanism and forearc seeps.

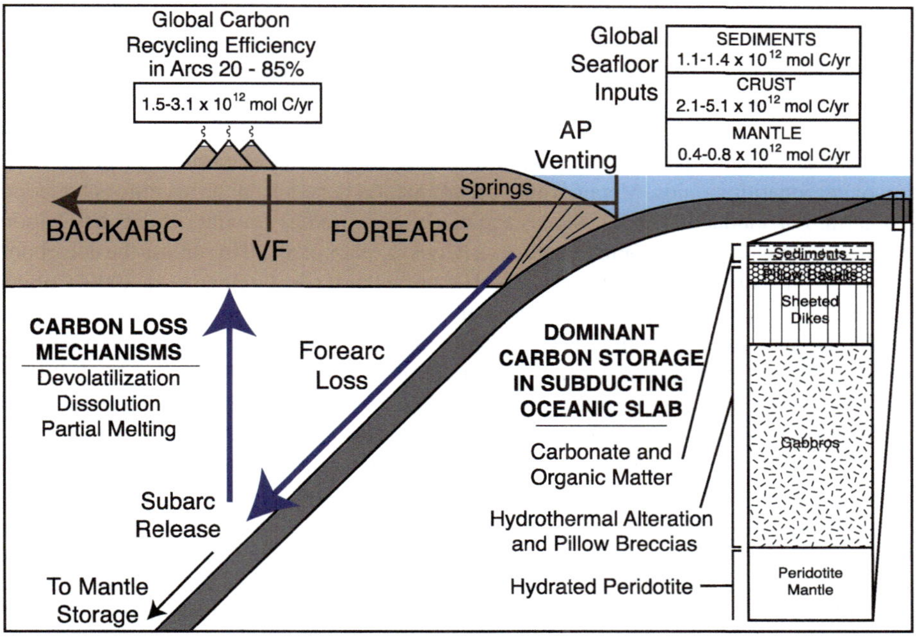

Fig. 2 Schematic illustration of an ocean-continent subduction zone, showing global-basis estimates of C fluxes discussed in the text (moles of C/year) and highlighting the uncertainties in the various flux estimates (from Cook-Kollars et al. 2014; other sources cited therein). Use of the fluxes shown here would suggest volcanic arc return efficiency (inputs/arc outputs) of ≈15 to 80%, indicating the need to examine C cycling in individual margins for which the input and output fluxes are better constrained. VF = volcanic front; AP = accretionary prism

2. Collision zone orogeneses, in which large quantities of C-bearing rocks undergo metamorphism with profound effects on C in the Earth's atmosphere.
3. Contact metamorphism, where the interaction of mantle-derived magmas with C in sedimentary layers is able to produce large amounts of CO_2 and/or CH_4.

The key question in assessing C cycling in subduction zones is how much of the C stored in the downgoing slab is directly transferred back to the Earth's surface (Fig. 2). Published global estimates vary widely from ≈ 15-80%, indicating that a significant proportion of C is not returned (see recent estimates by Dasgupta and Hirschmann 2010; Johnston et al. 2011; Collins et al. 2015). Detailed regional studies also show significant variations: In the Central American subduction zone, CO_2 output calculations based on petrological estimates as well as hydrothermal and gas emissions only account for a maximum of 24% of the input (Freundt et al. 2014). More CO_2 is released from arc volcanoes than the subducted C flux in sediments at the Sunda margin (Indonesia), requiring additional C sources in the subducting altered oceanic crust or perhaps the overriding plate (House et al. 2019). Aiuppa et al. (2017) suggested that, in some margins, decarbonation in the upper-plate crust (released by contact metamorphism) could contribute significantly to volcanic CO_2 output. A recent review of subduction zone C fluxes by Kelemen and Manning (2015) provided a contrasting perspective that little C is recycled into the convecting mantle as the lithosphere in the overlying plate represents an important reservoir for C (cf. Collins et al. 2015). In this scenario, the subduction zone C cycle is nearly balanced over geologically short timescales of 5-10 million years and outgassing at mid-ocean ridges and within-plate volcanic centres provides a net flux of C to the Earth's surface.

Understanding C release from subducting slabs is crucial for evaluating the important effects on the global C cycle, palaeoatmospheric CO_2 concentrations and arc magmatism (Kerrick and Connolly 2001). There are three main mechanisms for C release from the subducting slab: metamorphic decarbonation reactions, dissolution of carbonates, and formation of carbonatitic liquids. Metamorphic decarbonation depends on the specific subduction geotherm. For subduction of limestones and marls, only modest amounts of devolatilization of C occur along low-temperature geotherms (Cook-Kollars et al. 2014) unless the decarbonation is driven by infiltration of the rocks by H_2O-rich fluids (see Gorman et al. 2006; Collins et al. 2015). Similarly, a significant fraction of carbonate is retained to depths >200 km during the subduction of ophicarbonates (hydrothermally altered mantle rocks), providing a mechanism for carbonating the mantle (Kerrick and Connolly 1998; Collins et al. 2015). Hence, C is largely retained in the slab, in particular in relatively cool subduction zones (Gorman et al. 2006), explaining the discrepancy between input and output CO_2 fluxes. Even though the efficiency of C release from slab metamorphic decarbonation appears to be limited, dissolution of carbonate minerals by infiltrating fluids is a potentially voluminous pathway for slab decarbonation (Manning 2014). However, the magnitude of this release by carbonate dissolution, and whether it can impact margin-scale C cycling, have not yet been ascertained. In the exhumed Cycladic subduction complex, 60-90% of the initially present CO_2 was released adjacent to former fluid conduits (Ague and Nicolescu 2014) but extrapolation of this finding to scales of subducting slabs and overlying sediments will be problematic. Further evidence for fluid-mediated carbonate removal comes from diamond-bearing fluid inclusions with bicarbonate and carbonate ions in ultrahigh-pressure rocks from the Italian western Alps (Frezzotti et al. 2011). Theoretical modelling also shows that abundant dissolved organic and inorganic ionic C species are contained in subducted slab fluids, which can then be transported from the slab into the mantle wedge (Sverjensky et al. 2014). Ultimately, the removal of carbonate via dissolution is dependent upon the degree of infiltration of the rocks by fluids capable of dissolving and transporting the carbonate. Where studied in detail, carbonate-rich metasedimentary rocks that experienced high- or ultrahigh-pressure conditions can show only minimal petrologic and isotopic evidence for infiltration by far-traveled fluids (e.g., Cook-Kollars et al. 2014). An additional pathway for the release of C from subducted slabs is via the formation of carbonatitic liquids. Experimental data show that carbonatitic liquids can be produced at temperatures of 870-900 °C corresponding to ~120 km depth underneath arcs when the presence of water depresses the solidus for hydrous carbonate gabbro and limestone (Poli 2015). At even greater depths of 300-700 km, transport of C into the lower mantle is prevented by the formation of carbonatite melts when the slab geotherms intersects the melting curve of carbonated oceanic crust (Thomson et al. 2016). Whereas carbonatite melts at depths approaching the transition zone will metasomatise the overlying mantle and form diamonds, the sub-arc carbonatitic liquids provide a CO_2 source for subduction zone magmatism.

Even though there is widespread field evidence for mobilization of C in subduction-related metamorphic rocks via decarbonation and/or dissolution, this C is not necessarily recycled back into the atmosphere via arc magmatism. Precipitation of C can occur via vein formation or replacement mechanisms (Piccoli et al. 2016), enabling sequestration of C in the forearc and/or in crustal and mantle lithologies of the slab (Jaeckel et al. 2018; Scambelluri et al. 2016; van der Straaten et al. 2012). These high-pressure carbonates carry isotopic signals of mixed organic-inorganic C sources (van der Straaten et al. 2012). Carbonate destabilization in subduction zones may also be counterbalanced by carbonate reduction and the formation of graphite under reducing conditions (Galvez et al. 2013). Additional redistribution of C in the slab takes place via abiotic methanogenesis in ultramafic rocks during alteration (Vitale Brovarone et al. 2017). In summary, although there is a plethora of metamorphic processes operating in subduction zones that could impact carbon cycling, these processes have been demonstrated at only local (in some cases single outcrop) scales and their influence on regional (scales of individual margins) and global C recycling remains largely unconstrained.

The effects of C release during collisional orogeneses are less well studied than C cycling in subduction zones, but the magnitude of C release into the atmosphere via this pathway could be profound. For instance, CO_2 degassing from the Cordilleran belt has been linked to Eocene global warming (Kerrick and Caldeira 1998) and CO_2 flux for the Himalayan orogeny carrying large fluxes of CO_2 derived from metamorphic reactions was calculated to represent 7-60% of the global flux from volcanic arcs (Evans et al. 2008). Prograde metamorphism of carbonate-bearing sedimentary sequences will form calc-silicate rocks and release CO_2 in a wide range of orogenic settings during collisional metamorphism (Ague 2000; Groppo et al. 2013). The quantitative calculation of CO_2 release for the Himalayan orogeny through the study of calc-silicate rocks in the metamorphosed sedimentary sequences yielded a production of 1-3 Mt CO_2 (Groppo et al. 2017). Carbon fluxes during greenschist facies metamorphism alone were calculated to exceed C uptake by silicate weathering in orogenic belts (Skelton 2011). Whether the released CO_2 reaches the Earth's surface or is sequestered through carbonate or graphite precipitation is uncertain (Groppo et al. 2017), but the C sequestered by carbonate vein formation appears insignificant compared to the total amount of CO_2 released (Kerrick and Caldeira 1998) and cross-layer transport even allows for decarbonation during cooling (Ague 2000). There appears to be consensus that the orogenic belts are a source rather than a sink of atmospheric CO_2 (Skelton 2011; Evans et al. 2008). On geological timescales, metamorphic degassing during collisional metamorphism and orogeneses will create pulses of CO_2 relative to the slow silicate weathering that eventually sequesters CO_2 (Evans 2011).

A third metamorphic aspect of C release from crustal rocks relates to contact metamorphism, which can produce a high flux and volume of CH_4 and CO_2 around igneous intrusions (Svensen and Jamtveit 2010; Aiuppa et al. 2017). Addition of volatiles including CO_2 through shallow crustal sources via contact metamorphism has been observed for individual volcanoes such as Merapi (Indonesia) as they are located on thick sequences of limestones, marls and volcaniclastic deposits (Troll et al. 2012; Whitley et al. 2019). On a larger scale, the emplacement of voluminous mantle-derived melts in C-rich sedimentary strata (carbonates, evaporates, coal and organic-rich shales) generates large quantities of gases (CO_2, CH_4) that are vented directly into the atmosphere or rise as plumes through the water column into the atmosphere (Svensen et al. 2004; Ganino and Arndt 2009). For several large igneous provinces, this release of sediment-derived gases has been linked to global warming and mass extinctions (Ganino and Arndt 2009).

3 Conclusions

Field-based, experimental, and theoretical studies have greatly advanced our understanding of the Earth's geological N and C cycles and the distribution of these volatiles between the Earth's major reservoirs. Assessing key geological aspects of both the N and C cycles, we conclude:

1. The incorporation of N as NH_4^+ in minerals is crucial for N subduction and release upon breakdown of the host minerals. Cold subduction causes preferential N transport into the deep mantle, whereas warm subduction favours N recycling into the atmosphere through greater loss in forearcs and via arc magmatism.
2. Both the continental crust and the mantle are important N reservoirs. Long-term storage of N in continental crust appears to be largely as NH_4^+ in silicate minerals such as the micas. Despite low N contents in mantle minerals, N stored in the mantle outweighs N in the other Earth reservoirs and significantly more N can be stored in the mantle given its large N storage capacity.
3. Subducted C is released from the slab via decarbonation, dissolution of carbonates and carbonatite liquid formation. However, long-term carbonation of the mantle is indicated by the discrepancy between input and output CO_2 fluxes in subduction zones.
4. Release of CO_2 into the atmosphere also occurs during regional metamorphism in collisional orogens and by interaction with mantle-derived magmas with C-bearing country rocks. Periods when these fluxes appear to have been enhanced can be linked to specific past global warming events.

References

J.J. Ague, Release of CO_2 from carbonate rocks during regional metamorphism of lithologically heterogeneous crust. Geology **28**, 1123–1126 (2000). https://doi.org/10.1130/0091-7613(2000)28<1123:ROCFCR>2.0.CO;2

J. Ague, S. Nicolescu, Carbon dioxide released from subduction zones by fluid-mediated reactions. Nat. Geosci. **7**, 355–360 (2014). https://doi.org/10.1038/ngeo2143

A. Aiuppa, T.P. Fischer, T. Plank, P. Robidoux, R. Di Napoli, Along-arc, inter-arc, and arc-to-arc variations in volcanic gas CO_2/ST ratios reveal dual source of carbon in arc volcanism. Earth-Sci. Rev. **168**, 24–47 (2017). https://doi.org/10.1016/j.earscirev.2017.03.005

P. Barry, D. Hilton, Release of subducted sedimentary nitrogen throughout Earth's mantle. Geochem. Perspect. Lett. **2**, 148–159 (2016). https://doi.org/10.7185/geochemlet.1615

G.E. Bebout, M.L. Fogel, Nitrogen-isotope compositions of metasedimentary rocks in the Catalina Schist, California - implications for metamorphic devolatilization history. Geochim. Cosmochim. Acta **56**, 2839–2849 (1992). https://doi.org/10.1016/0016-7037(92)90363-N

G.E. Bebout, P. Agard, K. Kobayashi, T. Moriguti, E. Nakamura, Devolatilization history and trace element mobility in deeply subducted sedimentary rocks: evidence from Western Alps HP/UHP suites. Chem. Geol. **342**, 1–20 (2013a). https://doi.org/10.1016/j.chemgeo.2013.01.009

G.E. Bebout, M.L. Fogel, P. Cartigny, Nitrogen: highly volatile yet surprisingly compatible. Elements **9**, 333–338 (2013b). https://doi.org/10.2113/gselements.9.5.333

G.E. Bebout, K.E. Lazzeri, C.A. Geiger, Pathways for nitrogen cycling in Earth's crust and upper mantle: a review and new results for microporous beryl and cordierite. Am. Mineral. **101**, 7–24 (2016). https://doi.org/10.2138/am-2016-5363

S.R. Boyd, Nitrogen in future biosphere studies. Chem. Geol. **176**, 1–30 (2001). https://doi.org/10.1016/S0009-2541(00)00405-8

V. Busigny, G.E. Bebout, Nitrogen in the silicate Earth: speciation and isotopic behavior during mineral–fluid interactions. Elements **9**, 353–358 (2013). https://doi.org/10.2113/gselements.9.5.353

V. Busigny, P. Cartigny, P. Philippot, M. Ader, M. Javoy, Massive recycling of nitrogen and other fluid-mobile elements (K, Rb, Cs, H) in a cold slab environment: evidence from HP to UHP oceanic metasediments of the Schistes Lustrés nappe (Western Alps, Europe). Earth Planet. Sci. Lett. **215**, 27–42 (2003). https://doi.org/10.1016/S0012-821X(03)00453-9

V. Busigny, C. Laverne, M. Bonifacie, Nitrogen content and isotopic composition of oceanic crust at a superfast spreading ridge: a profile in altered basalts from ODP site 1256, Leg 206. Geochem. Geophys. Geosyst. **6**(12), Q12O01 (2005). https://doi.org/10.1029/2005GC001020

V. Busigny, P. Cartigny, P. Philippot, Nitrogen isotopes in ophiolitic metagabbros: a reevaluation of modern nitrogen fluxes in subduction zones and implication for the early Earth atmosphere. Geochim. Cosmochim. Acta **75**, 7502–7521 (2011). https://doi.org/10.1016/j.gca.2011.09.049

V. Busigny, J. Chen, P. Philippot, S. Borensztajn, F. Moynier, Insight into hydrothermal and subduction processes from copper and nitrogen isotopes in oceanic metagabbros. Earth Planet. Sci. Lett. **498**, 54–64 (2018). https://doi.org/10.1016/j.epsl.2018.06.030

P. Cartigny, B. Marty, Nitrogen isotopes and mantle geodynamics: the emergence of life and the atmosphere-crust-mantle connection. Elements **9**, 359–366 (2013). https://doi.org/10.2113/gselements.9.5.359

N.C. Collins, G.E. Bebout, S. Angiboust, P. Agard, M. Scambelluri, L. Crispini, T. John, Subduction zone metamorphic pathway for deep carbon cycling: II. Evidence from HP/UHP metabasaltic rocks and ophicarbonates. Chem. Geol. **412**, 132–150 (2015). https://doi.org/10.1016/j.chemgeo.2015.06.012

J. Cook-Kollars, G.E. Bebout, N.C. Collins, S. Angiboust, P. Agard, Subduction zone metamorphic pathway for deep carbon cycling: I. Evidence from HP/UHP metasedimentary rocks, Italian Alps. Chem. Geol. **386**, 31–48 (2014). https://doi.org/10.1016/j.chemgeo.2014.07.0123

C. Dalou, M.M. Hirschmann, A. von der Handt, J. Mosenfelder, L.S. Armstrong, Nitrogen and carbon fractionation during core-mantle differentiation at shallow depth. Earth Planet. Sci. Lett. **458**, 141–151 (2017). https://doi.org/10.1016/j.epsl.2016.10.026

R. Dasgupta, M.M. Hirschmann, The deep carbon cycle and melting in Earth's interior. Earth Planet. Sci. Lett. **298**, 1–13 (2010). https://doi.org/10.1016/j.epsl.2010.06.039

L.J. Elkins, T.P. Fischer, D.R. Hilton, Z.D. Sharp, S. McKnight, J. Walker, Tracing nitrogen in volcanic and geothermal volatiles from the Nicaraguan volcanic front. Geochim. Cosmochim. Acta **70**, 5215–5235 (2006). https://doi.org/10.1016/j.gca.2006.07.024

K.A. Evans, Metamorphic carbon fluxes: how much and how fast? Geology **39**, 95–96 (2011). https://doi.org/10.1130/focus012011.1

M.J. Evans, L.A. Derry, C. France-Lanord, Degassing of metamorphic carbon dioxide from the Nepal Himalaya. Geochem. Geophys. Geosyst. **9**, Q04021 (2008). https://doi.org/10.1029/2007GC001796

T.P. Fischer, D.R. Hilton, M.M. Zimmer, A.M. Shaw, Z.D. Sharp, J.A. Walker, Subduction and recycling of nitrogen along the Central American Margin. Science **297**, 1154–1157 (2002). https://doi.org/10.1126/science.1073995

A. Freundt, I. Grevemeyer, W. Rabbel, T.H. Hansteen, C. Hensen, H. Wehrmann, S. Kutterolf, R. Halama, M. Frische, Volatile (H_2O, CO_2, Cl, S) budget of the Central American subduction zone. Int. J. Earth Sci. **103**, 2101–2127 (2014). https://doi.org/10.1007/s00531-014-1001-1

M.L. Frezzotti, J. Selverstone, Z.D. Sharp, R. Compagnoni, Carbonate dissolution during subduction revealed by diamond-bearing rocks from the Alps. Nat. Geosci. **4**, 703–706 (2011). https://doi.org/10.1038/ngeo1246

M.E. Galvez, O. Beyssac, I. Martinez, K. Benzerara, C. Chaduteau, B. Malvoisin, J. Malavieille, Graphite formation by carbonate reduction during subduction. Nat. Geosci. **6**, 473–477 (2013). https://doi.org/10.1038/ngeo1827

C. Ganino, N.T. Arndt, Climate changes caused by degassing of sediments during the emplacement of large igneous provinces. Geology **37**, 323–326 (2009). https://doi.org/10.1130/G25325A.1

C. Goldblatt, M.W. Claire, T.M. Lenton, A.J. Matthews, A.J. Watson, K.J. Zahnle, Nitrogen-enhanced greenhouse warming on early Earth. Nat. Geosci. **2**, 891–896 (2009). https://doi.org/10.1038/ngeo692

P.J. Gorman, D.M. Kerrick, J.A.D. Connolly, Modeling open system metamorphic decarbonation of subducting slabs. Geochem. Geophys. Geosyst. **7**(4), Q04007 (2006). https://doi.org/10.1029/2005GC001125. J.A.D.

C. Groppo, F. Rolfo, D. Castelli, J.A.D. Connolly, Metamorphic CO_2 production from calc-silicate rocks via garnet-forming reactions in the CFAS-H_2O-CO_2 system. Contrib. Mineral. Petrol. **166**, 1655–1675 (2013). https://doi.org/10.1007/s00410-013-0947-5

C. Groppo, F. Rolfo, D. Castelli, P. Mosca, Metamorphic CO_2 production in collisional orogens: petrological constraints from phase diagram modeling of Himalayan, scapolite-bearing, calc-silicate rocks in the NKC(F)MAS(T)-HC system. J. Petrol. **58**, 53–84 (2017). https://doi.org/10.1093/petrology/egx005

R. Halama, G.E. Bebout, T. John, V. Schenk, Nitrogen recycling in subducted oceanic lithosphere: the record in high- and ultrahigh-pressure metabasaltic rocks. Geochim. Cosmochim. Acta **74**, 1636–1652 (2010). https://doi.org/10.1016/j.gca.2009.12.003

R. Halama, G.E. Bebout, T. John, M. Scambelluri, Nitrogen recycling in subducted mantle rocks and implications for the global nitrogen cycle. Int. J. Earth Sci. **103**, 2081–2099 (2014). https://doi.org/10.1007/s00531-012-0782-3

R. Halama, G.E. Bebout, H.R. Marschall, T. John, Fluid-induced breakdown of white mica controls nitrogen transfer during fluid-rock interaction in subduction zones. Int. Geol. Rev. **59**, 702–720 (2017). https://doi.org/10.1080/00206814.2016.1233834

S.A. Halldórsson, D.R. Hilton, P.H. Barry, E. Füri, K. Grönvold, Recycling of crustal material by the Iceland mantle plume: new evidence from nitrogen elemental and isotope systematics of subglacial basalts. Geochim. Cosmochim. Acta **176**, 206–226 (2016). https://doi.org/10.1016/j.gca.2015.12.021

D.R. Hilton, T.P. Fischer, B. Marty, Noble gases and volatile recycling at subduction zones. Rev. Mineral. Geochem. **47**, 319–370 (2002). https://doi.org/10.2138/rmg.2002.47.9

M.M. Hirschmann, Comparative deep Earth volatile cycles: The case for C recycling from exosphere/mantle fractionation of major (H_2O, C, N) volatiles and from H_2O/Ce, CO_2/Ba, and CO_2/Nb exosphere ratios. Earth Planet. Sci. Lett. **502**, 262–273 (2018). https://doi.org/10.1016/j.epsl.2018.08.023

B.M. House, G.E. Bebout, D.R. Hilton, Carbon cycling at the Sunda margin, Indonesia: a regional study with global implications. Geology **47**, 483–486 (2019). https://doi.org/10.1130/G45830.1

K. Jaeckel, G.E. Bebout, S. Angiboust, Deformation-enhanced fluid and mass transfer along Western and Central Alps paleo-subduction interfaces: significance for carbon cycling models. Geosphere **14**(6), 2355–2375 (2018). https://doi.org/10.1130/GES01587.1

M. Javoy, The major volatile elements of the Earth: their origin, behavior, and fate. Geophys. Res. Lett. **24**, 177–180 (1997). https://doi.org/10.1029/96GL03931

Y.F. Jia, Nitrogen isotope fractionations during progressive metamorphism: a case study from the Paleozoic Cooma metasedimentary complex, southeastern Australia. Geochim. Cosmochim. Acta **70**, 5201–5214 (2006). https://doi.org/10.1016/j.gca.2006.08.004

B. Johnson, C. Goldblatt, The nitrogen budget of Earth. Earth-Sci. Rev. **148**, 150–173 (2015). https://doi.org/10.1016/j.earscirev.2015.05.006

B. Johnson, C. Goldblatt, A secular increase in continental crust nitrogen during the Precambrian. Geochem. Perspect. Lett. **4**, 24–28 (2017a). https://doi.org/10.7185/geochemlet.1731

B.W. Johnson, C. Goldblatt, EarthN: a new Earth system nitrogen model. Geochem. Geophys. Geosyst. **19**(8), 2516–2542 (2017b). https://doi.org/10.1029/2017GC007392

F.K.B. Johnston, A.V. Turchyn, M. Edmonds, Decarbonation efficiency in subduction zones: implications for warm Cretaceous climates. Earth Planet. Sci. Lett. **303**, 143–152 (2011). https://doi.org/10.1016/j.epsl.2010.12.049

C.K. Junium, M.A. Arthur, K.H. Freeman, Compound-specific $\delta^{15}N$ and chlorin preservation in surface sediments of the Peru Margin with implications for ancient bulk $\delta^{15}N$ records. Geochim. Cosmochim. Acta **160**, 306–318 (2015). https://doi.org/10.1016/j.gca.2014.12.018

P.B. Kelemen, C.E. Manning, Reevaluating carbon fluxes in subduction zones, what goes down, mostly comes up. Proc. Natl. Acad. Sci. USA **112**(30), E3997–E4006 (2015). https://doi.org/10.1073/pnas.1507889112

D.M. Kerrick, K. Caldeira, Metamorphic CO_2 degassing from orogenic belts. Chem. Geol. **145**, 213–232 (1998). https://doi.org/10.1016/S0009-2541(97)00144-7

D.M. Kerrick, J.A.D. Connolly, Subduction of ophicarbonates and recycling of CO_2 and H_2O. Geology **26**, 375–378 (1998). https://doi.org/10.1130/0091-7613(1998)026<0375:SOOARO>2.3.CO;2

D.M. Kerrick, J.A.D. Connolly, Metamorphic devolatilization of subducted marine sediments and the transport of volatiles into the Earth's mantle. Nature **411**, 293–296 (2001). https://doi.org/10.1038/35077056

L. Li, G.E. Bebout, Carbon and nitrogen geochemistry of sediments in the Central American convergent margin: insights regarding subduction input fluxes, diagenesis, and paleoproductivity. J. Geophys. Res. **110**, B11202 (2005). https://doi.org/10.1029/2004JB003276

Y. Li, H. Keppler, Nitrogen speciation in mantle and crustal fluids. Geochim. Cosmochim. Acta **129**, 13–32 (2014). https://doi.org/10.1016/j.gca.2013.12.031

L. Li, G.E. Bebout, B.D. Idleman, Nitrogen concentration and $\delta^{15}N$ of altered oceanic crust obtained on ODP Legs 129 and 185: insights into alteration-related nitrogen enrichment and the nitrogen subduction budget. Geochim. Cosmochim. Acta **71**, 2344–2360 (2007). https://doi.org/10.1016/j.gca.2007.02.001

Y. Li, M. Wiedenbeck, S. Shcheka, H. Keppler, Nitrogen solubility in upper mantle minerals. Earth Planet. Sci. Lett. **377–378**, 311–323 (2013). https://doi.org/10.1016/j.epsl.2013.07.013

Y. Li, B. Marty, S. Shcheka, L. Zimmermann, H. Keppler, Nitrogen isotope fractionation during terrestrial core-mantle separation. Geochem. Perspect. Lett. **2**, 138–147 (2016). https://doi.org/10.7185/geochemlet.1614

A. Mallik, Y. Li, M. Wiedenbeck, Nitrogen evolution within the Earth's atmosphere-mantle system assessed by recycling in subduction zones. Earth Planet. Sci. Lett. **482**, 556–566 (2018). https://doi.org/10.1016/j.epsl.2017.11.045

C.E. Manning, A piece of the deep carbon puzzle. Nat. Geosci. **7**, 333–334 (2014). https://doi.org/10.1038/ngeo2152

B. Marty, N. Dauphas, The nitrogen record of crust-mantle interaction and mantle convection from Archean to present. Earth Planet. Sci. Lett. **206**, 397–410 (2003). https://doi.org/10.1016/S0012-821X(02)01108-1

S. Mikhail, D.A. Sverjensky, Nitrogen speciation in upper mantle fluids and the origin of Earth's nitrogen-rich atmosphere. Nat. Geosci. **7**, 816–819 (2014). https://doi.org/10.1038/NGEO2271

S. Mikhail, P.H. Barry, D.A. Sverjensky, The relationship between pH and the deep nitrogen cycle. Geochim. Cosmochim. Acta **209**, 149–160 (2017). https://doi.org/10.1016/j.gca.2017.04.007

C. Mitchell, T.P. Fischer, D.R. Hilton, E.H. Hauri, A.M. Shaw, J.M. de Moor, Z.D. Sharp, K. Kazahaya, Nitrogen sources and recycling at subduction zones: insights from the Izu–Bonin–Mariana arc. Geochem. Geophys. Geosyst. **11**(2), Q02X11 (2010). https://doi.org/10.1029/2009GC002783

B. Mysen, Mass transfer in the Earth's interior: fluid-melt interaction in aluminosilicate-C-O-H-N systems at high pressures and temperature under oxidizing conditions. Prog. Earth Planet. Sci. **5**, 6 (2018). https://doi.org/10.1186/s40645-017-0161-6

B. Mysen, M.L. Fogel, Nitrogen and hydrogen isotope compositions and solubility in silicate melts in equilibrium with reduced (N+H)-bearing fluids at high pressure and temperature: effects of melt structure. Am. Mineral. **95**, 987–999 (2010). https://doi.org/10.2138/am.2010.3364

A.P. Palya, I.S. Buick, G.E. Bebout, Storage and mobility of nitrogen in the continental crust: evidence from partially melted metasedimentary rocks, Mt. Stafford, Australia. Chem. Geol. **281**, 211–226 (2011). https://doi.org/10.1016/j.chemgeo.2010.12.009

P. Philippot, V. Busigny, M. Scambelluri, P. Cartigny, Oxygen and nitrogen isotopes as tracers of fluid activities in serpentinites and metasediments during subduction. Mineral. Petrol. **91**, 11–24 (2007). https://doi.org/10.1007/s00710-007-0183-7

F. Piccoli, A. Vitale Brovarone, O. Beyssac, I. Martinez, J.J. Ague, C. Chaduteau, Carbonation by fluid-rock interactions at high-pressure conditions: implications for carbon cycling in subduction zones. Earth Planet. Sci. Lett. **445**, 146–159 (2016). https://doi.org/10.1016/j.epsl.2016.03.045

I.K. Pitcairn, D.A.H. Teagle, R. Kerrich, D. Craw, T.S. Brewer, The behavior of nitrogen and nitrogen isotopes during metamorphism and mineralization: evidence from the Otago and Alpine Schists, New Zealand. Earth Planet. Sci. Lett. **233**, 229–246 (2005). https://doi.org/10.1016/j.epsl.2005.01.029

S. Poli, Carbon mobilized at shallow depths in subduction zones by carbonatitic liquids. Nat. Geosci. **8**, 633–636 (2015). https://doi.org/10.1038/ngeo2464

M. Roskosz, B.O. Mysen, G.D. Cody, Dual speciation of nitrogen in silicate melts at high pressure and temperature: an experimental study. Geochim. Cosmochim. Acta **70**, 2902–2918 (2006). https://doi.org/10.1016/j.gca.2006.03.001

R.L. Rudnick, S. Gao, Composition of the continental crust, in *Treatise on Geochemistry, Volume 4: The Crust*, ed. by H.D. Holland, K.K. Turekian (Elsevier, Amsterdam, 2014), pp. 1–51. https://doi.org/10.1016/B978-0-08-095975-7.00301-6

S.J. Sadofsky, G.E. Bebout, Record of forearc devolatilization in low-T, high-P/T metasedimentary suites: significance for models of convergent margin chemical cycling. Geochem. Geophys. Geosyst. **4**(4), 9903 (2003). https://doi.org/10.1029/2002GC000412

S.J. Sadofsky, G.E. Bebout, Nitrogen geochemistry of subducting sediments: new results from the Izu-Bonin-Mariana margin and insights regarding global nitrogen subduction. Geochem. Geophys. Geosyst. **5**(3), Q03115 (2014). https://doi.org/10.1029/2003GC000543

Y. Sano, N. Takahata, Y. Nishio, T.P. Fischer, S.N. Williams, Volcanic flux of nitrogen from the Earth. Chem. Geol. **171**, 263–271 (2001). https://doi.org/10.1016/S0009-2541(00)00252-7

M. Scambelluri, G.E. Bebout, D. Belmonte, M. Gilio, N. Campomenosi, N. Collins, L. Crispini, Carbonation of subduction-zone serpentinite (high-pressure ophicarbonate; Ligurian Western Alps) and implications for the deep carbon cycling. Earth Planet. Sci. Lett. **441**, 155–166 (2016). https://doi.org/10.1016/j.epsl.2016.02.034

P.A. Schroeder, A.A. McLain, Illite-smectites and the influence of burial diagenesis on the geochemical cycling of nitrogen. Clay Miner. **33**, 539–546 (1998). https://doi.org/10.1180/000985598545877

A. Skelton, Flux rates for water and carbon during greenschist facies metamorphism. Geology **39**, 43–46 (2011). https://doi.org/10.1130/G31328.1

H. Svensen, B. Jamtveit, Metamorphic fluids and global environmental changes. Elements **6**, 179–182 (2010). https://doi.org/10.2113/gselements.6.3.179

H. Svensen, S. Planke, A. Malthe-Sørenssen, B. Jamtveit, R. Myklebust, T. Rasmussen Eidem, S.S. Rey, Release of methane from a volcanic basin as a mechanism for initial Eocene global warming. Nature **429**, 542–545 (2004). https://doi.org/10.1038/nature02566

D.A. Sverjensky, V. Stagno, F. Huang, Important role for organic carbon in subduction-zone fluids in the deep carbon cycle. Nat. Geosci. **7**, 909–913 (2014). https://doi.org/10.1038/ngeo2291

A.R. Thomson, M.J. Walter, S.C. Kohn, R.A. Brooker, Slab melting as a barrier to deep carbon subduction. Nature **529**, 76–79 (2016). https://doi.org/10.1038/nature16174

I.N. Tolstikhin, B. Marty, The evolution of terrestrial volatiles: a view from helium, neon, argon and nitrogen isotope modeling. Chem. Geol. **147**, 27–52 (1998). https://doi.org/10.1016/S0009-2541(97)00170-8

V.R. Troll, D.R. Hilton, E.M. Jolis, J.P. Chadwick, L.S. Blythe, F.M. Deegan, L.M. Schwarzkopf, M. Zimmer, Crustal CO_2 liberation during the 2006 eruption and earthquake events at Merapi volcano, Indonesia. Geophys. Res. Lett. **39**, L11302 (2012). https://doi.org/10.1029/2012GL051307

F. van der Straaten, R. Halama, T. John, V. Schenk, F. Hauff, N. Andersen, Tracing the effects of high-pressure metasomatic fluids and seawater alteration in blueschist-facies overprinted eclogites: implications for subduction channel processes. Chem. Geol. **292–293**, 69–87 (2012). https://doi.org/10.1016/j.chemgeo.2011.11.008

A. Vitale Brovarone, I. Martinez, A. Elmaleh, R. Compagnoni, C. Chaduteau, C. Ferraris, I. Esteve, Massive production of abiotic methane during subduction evidenced in metamorphosed ophicarbonates from the Italian Alps. Nat. Commun. **8**, 14134 (2017). https://doi.org/10.1038/ncomms14134

A. Watenphul, B. Wunder, R. Wirth, W. Heinrich, Ammonium-bearing clinopyroxene: a potential nitrogen reservoir in the Earth's mantle. Chem. Geol. **270**, 240–248 (2010). https://doi.org/10.1016/j.chemgeo.2009.12.003

S. Whitley, R. Gertisser, R. Halama, K. Preece, V.R. Troll, F.M. Deegan, Crustal CO_2 contribution to subduction zone degassing recorded through calc-silicate xenoliths in arc lavas. Sci. Rep. **9**, 8803 (2019). https://doi.org/10.1038/s41598-019-44929-2

L.B. Williams, R.E. Ferrell Jr., I. Hutcheon, A.J. Bakel, M.D. Walsh, H.R. Krouse, Nitrogen isotope geochemistry of organic matter and minerals during diagenesis and hydrocarbon migration. Geochim. Cosmochim. Acta **59**, 765–779 (1995). https://doi.org/10.1016/0016-7037(95)00005-K

Y. Yang, V. Busigny, Z. Wang, Q. Xia, The fate of ammonium in phengite at high temperature. Am. Mineral. **102**, 2244–2253 (2017). https://doi.org/10.2138/am-2017-6094

R. Yokochi, B. Marty, G. Chazot, P. Burnard, Nitrogen in peridotite xenoliths: lithophile behavior and magmatic isotope fractionation. Geochim. Cosmochim. Acta **73**, 4843–4861 (2009). https://doi.org/10.1016/j.gca.2009.05.054

T. Yoshioka, M. Wiedenbeck, S. Shcheka, H. Keppler, Nitrogen solubility in the deep mantle and the origin of Earth's primordial nitrogen budget. Earth Planet. Sci. Lett. **488**, 134–143 (2018). https://doi.org/10.1016/j.epsl.2018.02.021

A.L. Zerkle, S. Mikhail, The geobiological nitrogen cycle: from microbes to the mantle. Geobiology **15**, 343–352 (2017). https://doi.org/10.1111/gbi.12228

Y. Zhang, A. Zindler, Distribution and evolution of carbon and nitrogen in Earth. Earth Planet. Sci. Lett. **117**, 331–345 (1993). https://doi.org/10.1016/0012-821X(93)90088-Q

Publisher's Note Springer Nature remains neutral with regard to jurisdictional claims in published maps and institutional affiliations.

Authors and Affiliations

Ralf Halama[1] · Gray Bebout[2]

✉ R. Halama
r.halama@keele.ac.uk

[1] School of Geography, Geology and the Environment, Keele University, Keele, UK

[2] Department of Earth and Environmental Sciences, Lehigh University, Bethlehem, USA

Space Science Reviews (2021) 217:44
https://doi.org/10.1007/s11214-021-00825-8

SPECIAL COMMUNICATION

Relative Atomic Solar System Abundances, Mass Fractions, and Atomic Masses of the Elements and Their Isotopes, Composition of the Solar Photosphere, and Compositions of the Major Chondritic Meteorite Groups

Katharina Lodders[1]

Received: 18 February 2021 / Accepted: 24 March 2021 / Published online: 1 April 2021

Abstract
This brief special communications article gives data for atomic abundances and mass fractions for the elemental and isotopic solar system composition, the atomic masses of the elements and their isotopes, the composition of the solar photosphere, and the compositions of the major chondritic meteorite groups. This additional material is relevant for researchers who are interested in this Topical Collection on planetary evolution.

Keywords Elemental abundances · Isotopic composition · Chondrites · Meteorites · Nuclides · Sun · Solar system

1 Introduction

This special communications article gives the elemental and isotopic composition of solar and meteoritic matter. A large review on the solar system composition is Lodders (2020) from which a portion of the material is drawn. The following briefly describes definitions and some data sources.

The solar system composition is the protosolar composition 4.567 Ga ago. Solar composition describes the current composition of the solar photosphere and the solar wind. The current photospheric composition is not the composition of the sun when it formed. The photospheric composition represents the composition of the underlying outer solar convection zone, which is a small portion of the entire sun. Heavy elements settled from the outer solar convection zone and the overlying photosphere into the sun's interior over the lifetime of the sun. Therefore, the element to hydrogen ratios are about 10-20% lower, which is model dependent, in the photosphere than they were originally. The settling effects on isotopic fractionations are not well known. In addition, the photospheric Li/H ratio is 170 times lower

Reading Terrestrial Planet Evolution in Isotopes and Element Measurements
Edited by Helmut Lammer, Bernard Marty, Aubrey L. Zerkle, Michel Blanc, Hugh O'Neill and Thorsten Kleine

✉ K. Lodders
lodders@wustl.edu

1 Dept. of Earth and Planetary Sciences and McDonnell Center of the Space Sciences, Washington University, One Brookings Drive Campus Box 1169, Saint Louis, MO 63130, USA

than in CI chondrites because of lithium destruction in the interior of the early and current sun; deuterium is essentially absent in the sun. The solar wind, which is derived from photospheric sources, has experienced further isotopic and elemental fractionation during solar wind acceleration.

2 Composition of Chondritic Meteorites

Table 1 gives the elemental composition by mass of the major chondrite groups in parts per million (μg/g). The CI-chondrite data are from Lodders (2020), and data for other chondrite groups are updated from Lodders and Fegley (1998, 2011). Several hundred individual analyses collected from the literature were used to calculate average compositions of observed meteorite falls in each chondrite group, although meteorite finds had to be considered for some elements if no analytical data were available for meteorite falls (particularly for CK and CR chondrites). The weighted chondrite group mean was obtained, if possible, by using the number of analyses for a given meteorite as statistical weight. Some of the data sources used here and other evaluations of the compositions of the chondrite groups can be found in Wasson and Kallemeyn (1988), Friedrich et al. (2003, 2004), Schaefer and Fegley (2007, 2010), Alexander (2019a, 2019b), Brauckmüller et al. (2018) and references therein. The average compositions should be taken as characteristic and could be taken as representative if it is kept in mind that natural concentration variations exist for mobile elements, and that there is variance in concentration determinations caused by differences in analytical methods. For meteorite finds, possible sources of terrestrial contamination or weathering losses of elements were considered. For the CI-chondrites, 2-sigma standard deviations ($\pm 2\sigma$) are listed. For other chondrite groups, typical uncertainties of element concentrations are below ten percent; however, for some elements (e.g., C, N, halogens, Cs, Cd, Hg, Ag, In, Tl, Pb, Bi), uncertainties can be considerably larger because of metamorphic redistribution and/or potential loss of these highly volatile elements as well as loss and redistribution of soluble salts (e.g., for halogens). Therefore, for several of these elements values characteristic for low-petrologic (least metamorphosed) types 3-4 are listed for ordinary (H, L, and LL) chondrites.

3 Solar Abundances: Composition of the Present Day Outer Solar Convection Zone, Mainly from the Photosphere

Table 2 gives the composition of the outer solar convection zone as mainly derived from spectroscopy of the solar photosphere. Many, but not all elements can be determined quantitatively in the solar photosphere. The abundance of He is determined by helioseismology (see Lodders 2020 for details). Currently there are two abundance sets to be considered, depending on which solar model atmospheres were used in the evaluation of spectroscopic observations. One set in Table 2 uses traditional 1D solar atmospheric models, the other set uses 3D atmospheric models, which contain more detailed descriptions of the physic in the solar atmosphere. The abundances are given on the logarithmic astronomical abundance scale:

$$A(X) = 12 + \log_{10}(N(X)/N(\mathrm{H})),$$

where the atomic abundance of hydrogen is set to $N(\mathrm{H}) = 1\mathrm{E}12$. The 3D abundances are generally lower than 1D abundances; in particular, the C, N, and O abundances dropped up to

Table 1 Characteristic elemental compositions of major chondritic meteorite groups (parts per million (ppm) by mass, μg/g)

Z	*E*	CI	±2*σ*	CM	CV	CO	CK	CR	H	L	LL	EH	EL
1	H	18600	3440	11500	2800	1000	4800	4200	800	1260	1700	630	...
2	He	0.00917	...	...	...	...	...	...	...	...	...	...	...
3	Li	1.51	0.12	1.63	1.8	1.8	1.2	...	1.7	1.9	2	1.7	2
4	Be	0.0220	0.0016	0.032	0.04	...	...	...	0.03	0.03	0.037	0.021	...
5	B	0.744	0.172	0.46	0.4	...	...	...	0.45	0.5	0.56	0.93	0.55
6	C	41300	8400	23200	4900	4900	1000	12600	2350	2500	3920	3800	4900
7	N	2500	660	1120	260	370	60	600	70	34	50	600	800
8	O	453840	20000	424100	367800	364200	382600	392700	337400	368400	383700	308000	332100
9	F	92	40	52	44	30	22	...	52	60	58	250	150
10	Ne	1.80E-4	...	...	...	...	...	...	...	...	...	...	...
11	Na	5100	500	4000	3400	4050	3200	2500	6100	6900	6900	7000	6000
12	Mg	95170	4000	119000	147000	145000	146000	136700	139400	148600	153000	108400	139000
13	Al	8370	600	11400	16900	14000	15300	11900	11100	11700	11800	8100	10200
14	Si	107740	7200	132000	159000	158000	158000	150000	170400	184600	189000	167300	190000
15	P	978	120	980	1100	1150	1100	840	1190	1030	980	2200	1200
16	S	53600	4400	30000	21200	21800	16000	16000	20000	22000	21000	56000	33000
17	Cl	717	270	470	280	290	230	...	87	99	126	670	210
18	Ar	0.00133	...	...	...	...	...	...	...	...	...	...	...
19	K	539	48	410	310	360	350	310	760	880	860	850	705
20	Ca	8840	700	12000	18000	15700	16300	13400	12400	13000	13200	8500	10000
21	Sc	5.83	0.40	8.2	11.1	9.5	11	8.3	8.3	8.4	8.3	5.7	7.5
22	Ti	450	30	610	850	750	770	600	630	690	680	480	510
23	V	53.6	4.0	72	92	90	92	74	72	75	76	54	61
24	Cr	2610	200	3090	3600	3550	3620	3640	3500	3660	3680	3140	3200
25	Mn	1896	160	1700	1470	1640	1440	1730	2340	2580	2600	2130	1600
26	Fe	185620	13000	211000	236000	248000	235000	237700	271500	218450	198000	311000	250000
27	Co	508	30	560	640	680	620	640	805	600	480	880	740
28	Ni	10950	700	12500	14000	14100	13200	13600	17100	12700	10200	18700	15400
29	Cu	130	20	129	109	134	100	86	90	95	83	210	120
30	Zn	311	20	177	113	101	90	65	47	54	55	320	16
31	Ga	9.45	0.70	7.5	6.1	7	5.8	4.2	6	5.4	5.2	17	11
32	Ge	33.4	3.0	24	16.5	20	16	11.6	10	11	11	43	30
33	As	1.77	0.16	1.8	1.57	2	1.6	1.5	2.2	1.4	1.3	3.3	2.3
34	Se	20.4	1.6	13	8.7	8	6.9	4.8	8.2	9.2	9.1	27	15
35	Br	3.77	1.80	2.5	1.5	1.6	1.1	1.1	0.7	0.9	1	3.3	1
36	Kr	5.22E-5	...	...	...	...	...	...	...	...	...	...	...
37	Rb	2.22	0.18	1.6	1.2	1.31	1.2	1.1	2.6	2.8	2.7	3.3	2.3
38	Sr	7.79	0.50	10	14.6	13.2	12.3	10	8.7	11	11	7.1	7.9
39	Y	1.50	0.10	2	2.5	2.4	2.7	...	2	2.1	2	1.43	1.6
40	Zr	3.79	0.28	5.2	6.8	6	6.7	5.2	5.7	5.8	6.8	3.8	4.8
41	Nb	0.279	0.015	0.38	0.51	0.43	0.49	[0.44]	0.38	0.4	0.4	0.3	0.25
42	Mo	0.976	0.050	1.2	1.45	1.6	1.4	1.3	1.4	1.2	1.1	1	0.730

Table 1 (*Continued*)

Z	*E*	CI	±2σ	CM	CV	CO	CK	CR	H	L	LL	EH	EL
43	Tc	...	...	...	...	...	...	...	...	...	...	...	...
44	Ru	0.666	0.04	0.86	1.07	1.03	1	0.91	1.1	0.75	0.55	0.91	0.920
45	Rh	0.133	0.008	0.16	0.183	0.17	0.19	0.18	0.22	0.155	0.11	0.17	0.17
46	Pd	0.558	0.030	0.63	0.664	0.76	0.67	0.73	0.845	0.62	0.51	0.91	0.7
47	Ag	0.204	0.008	0.13	0.097	0.1	0.055	0.076	0.06	0.08	0.065	0.29	0.023
48	Cd	0.679	0.024	0.4	0.245	0.25	0.21	0.16	0.04	0.035	0.02	0.83	0.03
49	In	0.0786	0.0040	0.048	0.0303	0.026	0.022	0.018	0.001	0.0005	0.004	0.11	0.005
50	Sn	1.63	0.16	0.93	0.59	0.69	0.47	0.43	0.39	0.4	0.33	1.6	0.68
51	Sb	0.169	0.018	0.13	0.0886	0.114	0.07	0.06	0.071	0.07	0.07	0.22	0.1
52	Te	2.31	0.18	1.6	1.02	0.94	0.8	0.63	0.35	0.4	0.39	2.5	1.1
53	I	0.77	0.62	0.39	0.23	0.19	0.22	...	0.054	0.07	0.16	0.26	0.1
54	Xe	1.74E-4	...	...	...	...	...	...	...	...	...	...	...
55	Cs	0.188	0.012	0.125	0.09	0.07	0.05	0.07	0.12	0.2	0.17	0.24	0.12
56	Ba	2.39	0.16	3.1	4.3	4.1	4.4	3	3.7	3.8	4	2.5	2.3
57	La	0.244	0.016	0.32	0.48	0.384	0.44	0.32	0.301	0.33	0.33	0.226	0.19
58	Ce	0.627	0.052	0.86	1.18	1.08	1.03	0.79	0.763	0.86	0.88	0.602	0.5
59	Pr	0.0951	0.0066	0.13	0.184	0.156	0.15	0.13	0.121	0.13	0.13	0.083	0.08
60	Nd	0.472	0.036	0.64	0.935	0.805	0.85	0.68	0.581	0.67	0.65	0.441	0.39
61	Pm	...	...	...	...	...	...	...	...	...	...	...	...
62	Sm	0.153	0.012	0.207	0.307	0.245	0.27	0.23	0.194	0.203	0.21	0.135	0.14
63	Eu	0.0577	0.0050	0.078	0.111	0.097	0.1	0.085	0.074	0.079	0.08	0.0535	0.051
64	Gd	0.208	0.018	0.287	0.384	0.372	0.36	0.32	0.27	0.3	0.29	0.198	0.185
65	Tb	0.0380	0.0030	0.052	0.069	0.064	0.064	0.057	0.049	0.054	0.055	0.0342	0.034
66	Dy	0.252	0.020	0.345	0.445	0.439	0.43	0.38	0.32	0.36	0.35	0.239	0.24
67	Ho	0.0563	0.0044	0.077	0.095	0.098	...	...	0.074	0.077	0.079	0.0513	0.053
68	Er	0.164	0.012	0.221	0.277	0.285	0.27	0.25	0.213	0.23	0.23	0.161	0.16
69	Tm	0.0259	0.0024	0.035	0.049	0.042	0.045	0.039	0.033	0.033	0.035	0.0245	0.027
70	Yb	0.167	0.014	0.225	0.313	0.272	0.3	0.25	0.203	0.226	0.23	0.162	0.18
71	Lu	0.0249	0.0020	0.035	0.045	0.042	0.044	0.036	0.033	0.034	0.034	0.0230	0.028
72	Hf	0.106	0.008	0.15	0.181	0.187	0.18	0.15	0.15	0.17	0.17	0.102	0.15
73	Ta	0.0148	0.0014	0.02	0.029	...	0.027	...	0.021	0.021	...	0.013	...
74	W	0.102	0.014	0.13	0.17	0.15	0.14	0.14	0.17	0.15	0.1	0.13	0.16
75	Re	0.0369	0.0028	0.049	0.065	0.056	0.054	0.048	0.074	0.055	0.032	0.052	0.06
76	Os	0.475	0.020	0.65	0.819	0.773	0.81	0.65	0.83	0.56	0.39	0.61	0.72
77	Ir	0.474	0.020	0.59	0.76	0.739	0.72	0.63	0.76	0.52	0.36	0.57	0.6
78	Pt	0.931	0.072	1.1	1.36	1.27	1.3	1.2	1.58	1.08	0.88	1.19	1.35
79	Au	0.147	0.024	0.152	0.146	0.186	0.12	0.15	0.22	0.16	0.14	0.34	0.26
80	Hg	0.288	0.140	...	0.035	0.05	0.007	...	...	0.002	0.022	0.06	...
81	Tl	0.141	0.014	0.088	0.061	0.04	...	0.042	0.004	0.002	0.002	0.095	0.007
82	Pb	2.64	0.16	1.76	1.08	0.84	0.75	0.47	0.14	0.06	0.06	2.24	0.24
83	Bi	0.113	0.016	0.07	0.05	0.041	0.031	0.044	0.009	0.004	0.012	0.084	0.012
90	Th	0.0298	0.0030	0.042	0.059	0.046	0.059	0.041	0.040	0.04	0.046	0.029	0.031
92	U	0.00816	0.00106	0.011	0.016	0.014	0.014	0.012	0.013	0.013	0.013	0.0095	0.008

30% from values considered 30-40 years ago. These lower abundances caused concerns for solar models, and this issue is currently still under investigation – see, e.g., Asplund et al. (2009), Grevesse et al. (2015), Scott et al. (2015a, 2015b), Lodders (2020). The (earlier) good agreement of 1D photospheric abundances with the meteoritic abundances of several non-volatile elements was no longer present with the early 3D abundances (Asplund et al. 2009), and disagreements persist for several elements in subsequent updates (Grevesse et al. 2015; Scott et al. 2015a, 2015b). There were also differences in some 3D results by different groups which were mainly due to different line selections (e.g., Asplund et al. 2009, Caffau et al. 2011). The latest 3D analyses employing important corrections for non-local thermodynamic equilibrium (NLTE) are moving closer to meteoritic values again (see references to individual elements in the following). The recommended 1D photospheric abundances are from Alexeeva and Mashonkina 2015; Holweger 2001; Grevesse et al. 2015, Lawler et al. 2014, 2017, 2018, 2019; Scott et al. 2015a, 2015b, Sneden et al. 2009, 2016, and Zhao et al. 2016. The recommended 3D abundances are largely from Caffau et al. 2011; Grevesse et al. 2015, and Scott et al. 2015a, 2015b, plus Amarsi et al. 2018 (O), Caffau et al. 2015 (O), Amarsi and Asplund 2017 (Si), Amarsi et al. 2019 (C), Caffau et al. 2011 (C), Bergemann et al. 2017 (Mg), Bergemann et al. 2019 (Mn), Lind et al. 2017 (Fe), Nordlander and Lind 2017 (Al). Sunspot data are from Maas et al. 2016 (Cl), Maiorca et al. 2014 (F), Vitas et al. 2008 (In). For more information, see Lodders (2020).

Indirect determinations provide data for the noble gases. The helium abundance is estimated from helioseismology (see Lodders 2020 and references therein). Related to the He abundance are the mass fractions of H, He, and the combined mass fraction of all other elements, called X, Y, and Z in the astronomical literature (see also Table 4 below). Mass fractions for the recommended 3D composition are: $X = 0.7389$ (Basu and Antia 2004), $Y = 0.2462$ for $Z = 0.0149$ and $Z/X = 0.0201$. Mass fractions for the recommended 1D composition are: $X = 0.7389$ (Basu and Antia 2004), $Y = 0.2458$ for $Z = 0.0153$ and $Z/X = 0.0207$. The neon abundance is estimated using solar wind and coronal Ne/O ratios from Young (2018) using oxygen abundances recommended here. The Ne abundance is still uncertain and debatable. The Ar abundance is from solar wind, B-stars and other considerations, see Lodders (2008). Krypton and xenon abundances are estimated from nuclear systematics, see Lodders (2003, 2020). Photospheric abundances of A(Ne) $= 8.060 \pm 0.033$ and A(Ar) $= 6.38 \pm 0.12$ from solar wind Genesis data (Huss et al. 2020) were not included when Table 2 was assembled.

4 Protosolar or Solar System Abundances from Solar and Meteoritic Data

Table 3 gives the atomic abundances of the solar photosphere, CI-chondrites, and the protosolar solar system composition. The Table is organized as follows. Columns (1) and (2) give atomic numbers and the element symbols. Columns (3)-(10) contain the atomic abundances of the elements on the cosmochemical abundance scale $N(X)$ where the amount of silicon is set to one million: $N(\mathrm{Si}) = 1\mathrm{E}+6$. Uncertainties are 1-sigma. The data for CI chondrites (columns (3) and (4)) are from Table 1 and are converted using the atomic weights listed in Table 4. The solar data in columns (5) and (6) are mainly from spectroscopic values of the photosphere, and are representative for the outer solar convection zone. Here the 3D photospheric results are adopted.

The present and protosolar compositions are in columns (7)-(8) and (9)-(10), respectively. Column (11) indicates how the values were selected: No entry means the values are based on

Table 2 Abundances in the Sun's outer convection zone mainly from photospheric and sunspot abundances, and solar wind and indirect data

Z	E	Recommended 3D		Recommended 1D	
		$A(E)$	$\pm a$	$A(E)$	$\pm a$
1	H	12		12	
2	He	[10.924]in	0.02	[10.923]in	0.02
3	Li	1.04	0.10	1.04	0.10
4	Be	1.38	0.09	1.38	0.09
5	B	2.70	0.20	2.70	0.20
6	C	8.47	0.06	8.58	0.11
7	N	7.85	0.12	8.00	0.11
8	O	8.71	0.04	8.76	0.06
9	F	[4.40]sp	0.25	[4.40]sp	0.25
10	Ne	[8.15]in	0.10	[8.10]in	0.10
11	Na	6.21	0.04	6.28	0.04
12	Mg	7.56	0.05	7.50	0.05
13	Al	6.43	0.04	6.41	0.03
14	Si	7.51	0.03	7.52	0.06
15	P	5.41	0.03	5.42	0.03
16	S	7.15	0.05	7.12	0.05
17	Cl	[5.25]sp	0.12	[5.25]sp	0.12
18	Ar	[6.50]in	0.10	[6.50]in	0.10
19	K	5.04	0.05	5.10	0.07
20	Ca	6.32	0.03	6.34	0.06
21	Sc	3.16	0.04	3.15	0.06
22	Ti	4.93	0.04	4.97	0.04
23	V	3.89	0.08	3.96	0.04
24	Cr	5.62	0.04	5.62	0.07
25	Mn	5.52	0.03	5.45	0.05
26	Fe	7.48	0.04	7.52	0.05
27	Co	4.93	0.05	4.96	0.06
28	Ni	6.20	0.04	6.28	0.06
29	Cu	4.18	0.05	4.21	0.03
30	Zn	4.56	0.05	4.61	0.09
31	Ga	3.02	0.05	3.09	0.05
32	Ge	3.63	0.07	3.62	0.07
33	As	...		...	
34	Se	...		...	
35	Br	...		...	
36	Kr	[3.25]in	0.08	[3.25]in	0.08
37	Rb	2.47	0.07	2.57	0.07
38	Sr	2.83	0.06	2.92	0.05
39	Y	2.21	0.05	2.20	0.05
40	Zr	2.59	0.04	2.59	0.06
41	Nb	1.47	0.06	1.49	0.06
42	Mo	1.88	0.09	2.04	0.05
43	Tc	...		...	

Table 2 *(Continued)*

Z	E	Recommended 3D		Recommended 1D	
		$A(E)$	$\pm a$	$A(E)$	$\pm a$
44	Ru	1.75	0.08	1.91	0.10
45	Rh	0.89	0.08	1.07	0.08
46	Pd	1.55	0.06	1.61	0.06
47	Ag	0.96	0.1	1.04	0.10
48	Cd	1.77	0.15	1.79	0.15
49	In	...		[0.8]sp	0.20
50	Sn	2.02	0.1	2.04	0.10
51	Sb	...		...	
52	Te	...		...	
53	I	...		...	
54	Xe	[2.25]in	0.08	[2.25]in	0.08
55	Cs	...		...	
56	Ba	2.25	0.07	2.19	0.07
57	La	1.11	0.04	1.14	0.03
58	Ce	1.58	0.04	1.61	0.06
59	Pr	0.72	0.04	0.76	0.04
60	Nd	1.42	0.04	1.45	0.05
61	Pm	...	...	...	...
62	Sm	0.95	0.04	1.00	0.05
63	Eu	0.52	0.04	0.52	0.04
64	Gd	1.08	0.04	1.11	0.05
65	Tb	0.31	0.1	0.28	0.10
66	Dy	1.1	0.04	1.13	0.06
67	Ho	0.48	0.11	0.51	0.10
68	Er	0.93	0.05	0.96	0.06
69	Tm	0.11	0.04	0.14	0.04
70	Yb	0.85	0.11	0.86	0.10
71	Lu	0.1	0.09	0.12	0.08
72	Hf	0.85	0.05	0.88	0.03
73	Ta	...		...	
74	W	0.83	0.11	1.03	0.11
75	Re	...		...	
76	Os	1.4	0.05	1.5	0.05
77	Ir	1.42	0.07	1.46	0.07
78	Pt	...		...	
79	Au	0.91	0.08	0.93	0.08
80	Hg	...		...	
81	Tl	...		[0.95]sp	[0.2]
82	Pb	1.92	0.08	2.05	0.08
83	Bi	...		...	
90	Th	$\leq$0.03	uncert.	$\leq$0.07	uncert.
92	U	...		...	

Abundances from photospheric data unless indicated by "in" for indirect determinations or by "sp" for sunspot data

Table 3 Recommended atomic solar system abundances: present and proto-solar 4.567 Ga ago (see text for data sources)

		CI-Chondrites		Sun, Outer Convection Zone, mainly photosphere, 3D		Solar system* (present)		Solar system* (proto-solar)†, ‡			Solar system* (present)		Solar system (proto-solar)†, ‡	
(1)	(2)	(3)	(4)	(5)	(6)	(7)	(8)	(9)	(10)	(11)	(12)	(13)	(14)	(15)
Z	E	$N(E)$	$\pm\sigma$	$N(E)$	$\pm\sigma$	$N(E)$	$\pm\sigma$	$N(E)$	$\pm\sigma$	Note	$A(E)$	$\pm a$	$A(E)$	$\pm a$
1	H	4.83E+6	4.5E+5	3.09E+10		3.09E+10		2.52E+10		s	12		12	
2	He	0.599		2.59E+9	1.2E+8	2.59E+9	1.2E+8	2.51E+9	1.2E+8	s	10.924	0.02	10.994	0.02
3	Li	56.9	3.4	0.339	0.088	0.339	0.088	56.9	3.4		3.27	0.03	3.35	0.03
4	Be	0.637	0.044	0.741	0.171	0.637	0.044	0.637	0.044		1.31	0.04	1.40	0.04
5	B	18	1.3	15.5	9.1	18	1.3	18	1.3		2.77	0.03	2.85	0.03
6	C	9.44E+5	1.79E+5	9.12E+6	1.35E+6	9.12E+6	1.8E+5	9.12E+6	1.8E+5	s	8.47	0.06	8.56	0.06
7	N	44830	9700	2.19E+6	7.0E+5	2.19E+6	9700	2.19E+6	9700	s	7.85	0.12	7.94	0.12
8	O	7.38E+6	1.3E+5	1.66E+7	2.9E+6	1.66E+7	1.3E+5	1.66E+7	1.3E+5	s	8.73	0.07	8.82	0.07
9	F	1270	275	780	600	1270	275	1270	275		4.61	0.09	4.79	0.09
10	Ne	0.00234		4.37E+6	1.13E+6	4.37E+6	1.13E+6	4.37E+6	1.13E+6	s,o	8.15	0.10	8.24	0.10
11	Na	57800	4700	50120	4840	57800	4700	57800	4700		6.27	0.03	6.36	0.03
12	Mg	1.03E+6	4.4E+4	1.12E+6	1.4E+5	1.03E+6	4.4E+4	1.03E+6	4.4E+4		7.52	0.02	7.61	0.02
13	Al	81820	6110	83180	8030	81820	6110	81820	6110		6.42	0.03	6.51	0.03
14	Si	1.00E+6	3.4E+4	1.00E+6	7E+4	1.00E+6	3.4E+4	1.00E+6	3.4E+4		7.51	0.01	7.60	0.01
15	P	8260	530	8430	810	8260	530	8260	530		5.43	0.03	5.52	0.03
16	S	4.37E+5	2.6E+4	4.37E+5	5.3E+4	4.37E+5	2.6E+4	4.37E+5	2.6E+4		7.15	0.03	7.24	0.03
17	Cl	5290	810	5495	1750	5290	810	5290	810		5.23	0.06	5.32	0.06
18	Ar	0.00959		...		97700	25300	97700	25300	s,o	6.50	0.10	6.59	0.10
19	K	3610	190	3685	850	3606	190	3611	190		5.07	0.02	5.16	0.02
20	Ca	57239	4500	64565	4620	57239	4500	57234	4500		6.27	0.03	6.36	0.03
21	Sc	33.7	2.2	44.7	4.3	33.7	2.2	33.7	2.2		3.04	0.03	3.13	0.03
22	Ti	2460	150	2630	254	2459	150	2459	150		4.90	0.03	4.99	0.03

Table 3 (*Continued*)

		CI-Chondrites		Sun, Outer Convection Zone, mainly photosphere, 3D		Solar system[*] (present)		Solar system[*] (proto-solar)[†, ‡]			Solar system[*] (present)		Solar system (proto-solar)[†, ‡]	
(1)	(2)	(3)	(4)	(5)	(6)	(7)	(8)	(9)	(10)	(11)	(12)	(13)	(14)	(15)
Z	E	$N(E)$	$\pm\sigma$	$N(E)$	$\pm\sigma$	$N(E)$	$\pm\sigma$	$N(E)$	$\pm\sigma$	Note	$A(E)$	$\pm a$	$A(E)$	$\pm a$
23	V	275	18	240	49	275	18	275	18		3.95	0.03	4.04	0.03
24	Cr	13130	460	12880	1240	13130	460	13130	460		5.63	0.02	5.72	0.02
25	Mn	9090	620	10230	730	9090	620	9090	620		5.47	0.03	5.56	0.03
26	Fe	8.72E+5	3.8E+4	9.33E+5	9.0E+4	8.72E+5	3.8E+4	8.72E+5	3.8E+4		7.45	0.02	7.54	0.02
27	Co	2260	100	2630	320	2260	100	2260	100		4.86	0.02	4.95	0.02
28	Ni	48670	2940	48980	4730	48670	2940	48670	2940		6.20	0.03	6.29	0.03
29	Cu	535	50	468	57	535	50	535	50		4.24	0.04	4.33	0.04
30	Zn	1260	180	1120	140	1260	180	1260	180		4.61	0.60	4.70	0.60
31	Ga	36.2	1.8	32.4	3.9	36.2	1.8	36.2	1.8		3.07	0.02	3.16	0.02
32	Ge	120	7	132	23	120	7	120	7		3.59	0.03	3.68	0.03
33	As	6.07	0.50	...		6.07	0.50	6.07	0.50		2.29	0.03	2.38	0.03
34	Se	67.6	5.0	...		67.6	5.0	67.6	5.0		3.34	0.03	3.43	0.03
35	Br	12.3	2.9	...		12.3	2.9	12.3	2.9		2.60	0.09	2.69	0.09
36	Kr	1.63E-04		...		51.3	10.4	51.3	10.4	o	3.22	0.08	3.31	0.08
37	Rb	7.04	0.46	9.12	1.60	7.04	0.46	7.17	0.47		2.36	0.03	2.45	0.03
38	Sr	23.4	1.3	20.9	3.1	23.4	1.3	23.3	1.3		2.88	0.02	2.97	0.02
39	Y	4.35	0.24	5.01	0.61	4.35	0.24	4.35	0.24		2.15	0.02	2.24	0.02
40	Zr	10.9	1.0	12.5	1.84	10.9	1.0	10.9	1.0		2.55	0.04	2.64	0.04
41	Nb	0.780	0.070	0.912	0.135	0.78	0.070	0.78	0.070		1.4	0.04	1.49	0.04
42	Mo	2.6	0.26	2.34	0.54	2.6	0.26	2.6	0.26		1.92	0.04	2.01	0.04
43	Tc	...		...		...		...				...		...
44	Ru	1.81	0.02	1.74	0.35	1.81	0.02	1.81	0.02		1.77	0.01	1.86	0.01

Table 3 (*Continued*)

		CI-Chondrites		Sun, Outer Convection Zone, mainly photosphere, 3D		Solar system* (present)		Solar system* (proto-solar)†, ‡			Solar system* (present)		Solar system (proto-solar)†, ‡	
(1)	(2)	(3)	(4)	(5)	(6)	(7)	(8)	(9)	(10)	(11)	(12)	(13)	(14)	(15)
Z	E	$N(E)$	$\pm\sigma$	$N(E)$	$\pm\sigma$	$N(E)$	$\pm\sigma$	$N(E)$	$\pm\sigma$	Note	$A(E)$	$\pm a$	$A(E)$	$\pm a$
45	Rh	0.338	0.015	0.240	0.049	0.338	0.015	0.338	0.015		1.04	0.02	1.13	0.02
46	Pd	1.38	0.07	1.10	0.16	1.38	0.07	1.38	0.07		1.65	0.02	1.74	0.02
47	Ag	0.497	0.022	0.282	0.073	0.497	0.022	0.497	0.022		1.21	0.02	1.29	0.02
48	Cd	1.58	0.06	1.82	0.75	1.58	0.06	1.58	0.06		1.71	0.02	1.80	0.02
49	In	0.179	0.008	0.195	0.114	0.179	0.008	0.179	0.008		0.76	0.02	0.85	0.02
50	Sn	3.59	0.22	3.24	0.84	3.59	0.22	3.59	0.22		2.07	0.03	2.15	0.03
51	Sb	0.359	0.045	...		0.359	0.045	0.359	0.045		1.06	0.05	1.15	0.05
52	Te	4.72	0.23	...		4.72	0.23	4.72	0.23		2.18	0.02	2.27	0.02
53	I	1.59	0.64	...		1.59	0.64	1.59	0.64		1.71	0.15	1.80	0.15
54	Xe	3.47E-04		...		5.50	1.11	5.50	1.11	o	2.25	0.08	2.34	0.08
55	Cs	0.368	0.043	...		0.368	0.043	0.368	0.043		1.08	0.02	1.16	0.02
56	Ba	4.55	0.27	5.5	0.92	4.55	0.27	4.55	0.27		2.17	0.02	2.26	0.02
57	La	0.459	0.024	0.398	0.038	0.459	0.024	0.459	0.024		1.17	0.02	1.26	0.02
58	Ce	1.16	0.05	1.17	0.11	1.16	0.05	1.16	0.05		1.58	0.02	1.66	0.02
59	Pr	0.175	0.012	0.162	0.016	0.175	0.012	0.175	0.012		0.75	0.01	0.84	0.01
60	Nd	0.865	0.022	0.813	0.078	0.865	0.022	0.864	0.022		1.45	0.01	1.53	0.01
61	Pm	...		...		...		...				...		...
62	Sm	0.271	0.012	0.275	0.027	0.271	0.012	0.273	0.012		0.94	0.02	1.03	0.02
63	Eu	0.100	0.005	0.102	0.010	0.1	0.005	0.1	0.005		0.51	0.02	0.60	0.02
64	Gd	0.346	0.013	0.372	0.036	0.346	0.013	0.346	0.013		1.05	0.02	1.14	0.02
65	Tb	0.0625	0.0025	0.0631	0.0163	0.0625	0.0025	0.0625	0.0025		0.31	0.02	0.39	0.02
66	Dy	0.407	0.016	0.389	0.038	0.407	0.016	0.407	0.016		1.12	0.02	1.21	0.02

Table 3 (*Continued*)

		CI-Chondrites		Sun, Outer Convection Zone, mainly photosphere, 3D		Solar system* (present)		Solar system* (proto-solar)†, ‡			Solar system* (present)		Solar system (proto-solar)†, ‡	
(1)	(2)	(3)	(4)	(5)	(6)	(7)	(8)	(9)	(10)	(11)	(12)	(13)	(14)	(15)
Z	E	$N(E)$	$\pm\sigma$	$N(E)$	$\pm\sigma$	$N(E)$	$\pm\sigma$	$N(E)$	$\pm\sigma$	Note	$A(E)$	$\pm a$	$A(E)$	$\pm a$
67	Ho	0.0891	0.0035	0.0933	0.0269	0.0891	0.0035	0.0891	0.0035		0.46	0.02	0.55	0.02
68	Er	0.256	0.009	0.263	0.032	0.256	0.009	0.256	0.009		0.92	0.02	1.01	0.02
69	Tm	0.0403	0.0014	0.0398	0.0038	0.0403	0.0014	0.0403	0.0014		0.11	0.02	0.20	0.02
70	Yb	0.252	0.009	0.219	0.063	0.252	0.009	0.252	0.009		0.91	0.02	1.00	0.02
71	Lu	0.038	0.0018	0.0389	0.0090	0.038	0.0018	0.0381	0.0018		0.09	0.02	0.18	0.02
72	Hf	0.155	0.011	0.219	0.027	0.155	0.011	0.155	0.011		0.70	0.03	0.79	0.03
73	Ta	0.0215	0.0010	…		0.0215	0.0010	0.0215	0.0010		-0.16	0.02	-0.07	0.02
74	W	0.144	0.013	0.209	0.060	0.144	0.013	0.144	0.013		0.67	0.04	0.76	0.04
75	Re	0.0521	0.0042	…		0.0521	0.0042	0.0547	0.0042		0.23	0.03	0.34	0.03
76	Os	0.655	0.040	0.776	0.095	0.655	0.040	0.652	0.040		1.33	0.03	1.41	0.03
77	Ir	0.633	0.029	0.813	0.142	0.633	0.029	0.633	0.029		1.31	0.02	1.40	0.02
78	Pt	1.24	0.10	…		1.24	0.10	1.24	0.10		1.60	0.03	1.69	0.03
79	Au	0.195	0.016	0.251	0.051	0.195	0.016	0.195	0.016		0.80	0.03	0.89	0.03
80	Hg	0.376	0.156	…		0.376	0.156	0.376	0.156		1.08	0.15	1.17	0.15
81	Tl	0.179	0.015	0.275	0.161	0.179	0.015	0.179	0.015		0.76	0.04	0.85	0.04
82	Pb	3.33	0.20	2.57	0.52	3.33	0.20	3.31	0.20		2.03	0.03	2.12	0.03
83	Bi	0.141	0.010	…		0.141	0.010	0.141	0.010		0.66	0.03	0.75	0.03
90	Th	0.0336	0.0017	> 0.033		0.0336	0.0017	0.0421	0.0021		0.04	0.02	0.22	0.02
92	U	0.00897	0.00064	…		0.00897	0.00064	0.02389	0.00170		-0.54	0.03	-0.02	0.03

Note. *based on CI-chondrites except as indicated: s = from sun, o = by other means

†corrected for radioactive decay. ‡present day photospheric values corrected for element settling from the outer convection zone to obtain protosolar values

CI-chondrites, "*s*" means data are from the Sun, and "*o*" indicates the data were derived by other means such as nucleosynthesis theory (see Lodders 2020 for details). The protosolar abundances (subscript "*o*" contain corrections for gain or loss of isotopes due to radioactive decay, and corrections to the present day photospheric values to take losses from continuous element settling from the outer convection zone into the solar interior into account (see also notes for Table 4). For convenience, the solar system data are also given on the astronomical abundance scale in columns (12)-(15). For practical purposes, the abundance scales are linked through the abundance of Si as:

$$A(E) = 1.510 + \log_{10} N(E) \text{ for the present-day abundances, and}$$

$$A(E)o = 1.598 + \log_{10} N(E) = 0.088 + A(E) \text{ for the proto-solar abundances except for H and He.}$$

For a detailed explanation and issues with the scaling see Lodders (2003, 2020), Lodders et al. (2009). The conversion of the uncertainties (U, in percent) from the linear cosmochemical scale to the logarithmic astronomical abundance scale (where the uncertainty corresponds to an uncertainty factor "a") can be done using:

$$U(\%) = \pm 100(10^{\pm a} - 1)$$

The percent uncertainty is smaller for $-a$ than for $+a$; or vice versa, a given percent uncertainty on the linear scale results in two different uncertainty factors on the logarithmic scale. Here the larger percentage U from a given uncertainty of "a" or the larger uncertainty "a" from a given U is taken (see Lodders 2003).

5 Atomic Abundances and Mass Fractions of the Isotopes in the Solar System and Atomic Weights of the Elements

Table 4 gives the proto-solar (solar system) isotopic composition of the elements, isotopic and elemental abundances relative to 1E6 silicon atoms, mass fractions of the isotopes in the overall proto-solar composition, atomic masses of the isotopes, and resulting protosolar and current mean atomic weights of the elements.

The first three columns of Table 2 list the element symbol, atomic number or charge number (Z), and mass number (A). Entries in column (1) marked with $*$ are long-lived radioactive nuclides with half-lives up to 10^{12} years and abundances are for 4.567 Ga ago. Isotopes with half-lives above 10^{12} years (marked with ^) can be regarded as stable compared to the age of the solar system and are of interest in studies of double-beta decay.

Column (4) gives the isotopic composition (in atom %) of each element for the protosolar composition 4.567 Ga ago. The isotopic compositions are mainly adopted from Meija et al. (2016). The sources for the isotopic compositions of H, C, N, O, and the noble gases are described below. Several elements (e.g., C, N, O, Ar) have different isotopic compositions in the Sun (taken as representative for the solar system composition) than on Earth. For radioactive isotopes and their stable decay products, the protosolar isotopic compositions differ from present-day values and the listed composition considers decay of radioactive isotopes and gain in stable isotopes over the past 4.567 Ga.

Column (5) gives the relative atomic *elemental* abundances on the cosmochemical abundance scale where the silicon abundance is fixed at $N(\text{Si}) = 1\text{E}6$ atoms. These are the same data as in Table 1 for the proto-solar abundances.

Table 4 Proto-Solar Isotopic Composition of the Elements, Isotopic and Elemental Abundances Relative to 1E6 Silicon Atoms, Mass Fractions of the Isotopes in the Overall Proto-Solar Composition, Atomic Masses of the Isotopes, and Resulting Protosolar and Current Mean Atomic Weights of the Elements (see text for data sources)

E	*Z*	*A*	Isotopic composition atom %	*N* (Elemental) Si = 1E+6	*N* (Isotopic) ΣSi = 1E+6	Elemental mass fractions	Isotope mass fractions	Atomic mass (Daltons, Da)	Protosolar mean atomic mass (Da)	Current mean atomic mass (Da)
(1)	(2)	(3)	(4)	(5)	(6)	(7)	(8)	(9)	(10)	(11)
H	1	1	99.9980		2.52E+10		7.057E-01	1.007825032		
H	1	2	0.00197		4.97E+5		2.781E-05	2.014101778		
			100	2.52E+10		7.057E-01			1.007844856	1.007825032
He	2	3	0.0166		4.13E+5		3.461E-05	3.01602932		
He	2	4	99.9834		2.49E+9		2.769E-01	4.002603254		
			100	2.49E+9		2.770E-01			4.002439456	4.002198759
Li	3	6	7.589		4.3		7.187E-10	6.015122885		
Li	3	7	92.411		52.6		1.025E-08	7.016003428		
			100	56.9		1.097E-08			6.940046604	6.940046604
Be	4	9	100	0.637	0.637	1.595E-10	1.595E-10	9.01218291	9.01218291	9.01218291
B	5	10	19.827		3.6		1.002E-09	10.01293696		
B	5	11	80.173		14.4		4.405E-09	11.00930537		
			100	18.0		5.406E-09			10.81175541	10.81175541
C	6	12	98.965		9.03E+6		3.011E-03	12.00000000		
C	6	13	1.035		9.44E+4		3.411E-05	13.00335484		
			100	9.12E+6		3.045E-03			12.01038885	12.01038885
N	7	14	99.774		2.18E+6		8.482E-04	14.003074		
N	7	15	0.226		4.94E+3		2.059E-06	15.0001089		
			100	2.19E+6		8.502E-04			14.00532465	14.00532465
O	8	16	99.777		1.66E+7		7.377E-03	15.99491462		
O	8	17	0.035		5.80E+3		2.739E-06	16.99913176		
O	8	18	0.188		3.13E+4		1.565E-05	17.99915961		
			100	1.66E+7		7.396E-03			15.99904151	15.99904151

Table 4 *(Continued)*

E	*Z*	*A*	Isotopic composition atom %	*N* (Elemental) Si = 1E+6	*N* (Isotopic) ΣSi = 1E+6	Elemental mass fractions	Isotope mass fractions	Atomic mass (Daltons, Da)	Protosolar mean atomic mass (Da)	Current mean atomic mass (Da)
(1)	(2)	(3)	(4)	(5)	(6)	(7)	(8)	(9)	(10)	(11)
F	9	19	100	1267	1267	6.688E-07	6.688E-07	18.99840317	18.99840317	18.99840317
Ne	10	20	93.1251		4.07E+6		2.261E-03	19.99244018		
Ne	10	21	0.2236		9.76E+3		5.693E-06	20.99384668		
Ne	10	22	6.6513		2.90E+5		1.772E-04	21.99138512		
			100	4.37E+6		2.444E-03			20.12763505	20.12763505
Na	11	23	100	57800	57800	3.692E-05	3.692E-05	22.98976928	22.98976928	22.98976928
Mg	12	24	78.992		8.10E+5		5.398E-04	23.9850417		
Mg	12	25	10.003		1.03E+5		7.151E-05	24.98583691		
Mg	12	26	11.005		1.13E+5		8.158E-05	25.98259295		
			100	1.03E+6		6.929E-04			24.30498176	24.30498176
Al	13	27	100	8.18E+4	8.18E+04	6.132E-05	6.132E-05	26.98153859		
Si	14	28	92.230		9.223E+5		7.167E-04	27.97692653		
Si	14	29	4.683		4.683E+4		3.768E-05	28.97649467		
Si	14	30	3.087		3.087E+4		2.573E-05	29.97377017		
			100	1.000E+6		7.801E-04			28.08538367	28.08538367
P	15	31	100		8260	7.109E-06	7.109E-06	30.973762	30.973762	30.973762
S	16	32	95.04074		415600		3.692E-04	31.97207117		
S	16	33	0.74869		3270		2.996E-06	32.97145569		
S	16	34	4.19599		18300		1.727E-05	33.9678669		
S	16	36	0.01458		64		6.396E-08	35.96708076		
			100	437300		3.895E-04			32.06387933	32.06387933
Cl	17	35	75.7647		4010		3.896E-06	34.96885268		

Table 4 (*Continued*)

E	*Z*	*A*	Isotopic composition atom %	*N* (Elemental) Si = 1E+6	*N* (Isotopic) ΣSi = 1E+6	Elemental mass fractions	Isotope mass fractions	Atomic mass (Daltons, Da)	Protosolar mean atomic mass (Da)	Current mean atomic mass (Da)
(1)	(2)	(3)	(4)	(5)	(6)	(7)	(8)	(9)	(10)	(11)
Cl	17	37	24.2353		1280		1.315E-06	36.96590259		
			100	5290		5.211E-06			35.45284372	35.45284372
Ar	18	36	84.596		82670		8.265E-05	35.96754511		
Ar	18	38	15.830		15030		1.582E-05	37.96273234		
Ar	18	40	0.025		23.5		2.609E-08	39.96238312		
			100	97724		9.850E-05			36.27537813	36.27535886
K	19	39	93.132		3363		3.638E-06	38.96370649		
K*	19	40	0.147		5		5.552E-09	39.96399848		
K	19	41	6.721		243		2.766E-07	40.96182526		
			100	3611		3.920E-06			39.09946879	39.09830181
Ca	20	40	96.941		55500		6.163E-05	39.96259086		
Ca	20	42	0.647		370		4.314E-07	41.95861801		
Ca	20	43	0.135		77		9.191E-08	42.95876667		
Ca	20	44	2.086		1194		1.458E-06	43.95548173		
Ca	20	46	0.004		2		2.554E-09	45.9536926		
Ca	20	48	0.187		107		1.426E-07	47.9525343		
			100	57200		6.375E-05			40.07802253	40.07802253
Sc	21	45	100	33.7	3.370E+01	4.209E-08	4.209E-08	44.9559119	44.9559119	44.9559119
Ti	22	46	8.249		203		2.592E-07	45.95262889		
Ti	22	47	7.437		183		2.387E-07	46.95176293		
Ti	22	48	73.72		1810		2.411E-06	47.94794631		
Ti	22	49	5.409		133		1.809E-07	48.94786998		

Table 4 *(Continued)*

E	*Z*	*A*	Isotopic composition atom %	*N* (Elemental) Si = 1E+6	*N* (Isotopic) ΣSi = 1E+6	Elemental mass fractions	Isotope mass fractions	Atomic mass (Daltons, Da)	Protosolar mean atomic mass (Da)	Current mean atomic mass (Da)
(1)	(2)	(3)	(4)	(5)	(6)	(7)	(8)	(9)	(10)	(11)
Ti	22	50	5.185		128		1.776E-07	49.94479117		
			100	2460		3.268E-06			47.8668887	47.8668887
V	23	50	0.2497		0.7		9.715E-10	49.9471585		
V	23	51	99.7503		274.5		3.886E-07	50.9439595		
			100	275.2		3.895E-07			50.94147049	50.94147049
Cr	24	50	4.3452		571		7.924E-07	49.9460442		
Cr	24	52	83.7895		11000		1.587E-05	51.94050751		
Cr	24	53	9.5006		1250		1.839E-06	52.94064943		
Cr	24	54	2.3647		310		4.646E-07	53.93888045		
			100	13100		1.897E-05			51.9961191	51.9961191
Mn	25	55	100	9090	9090	1.388E-05	1.388E-05	54.93804512	54.93804512	54.93804512
Fe	26	54	5.845		51000		7.643E-05	53.93961046		
Fe	26	56	91.754		8.01E+5		1.245E-03	55.93493745		
Fe	26	57	2.1191		18500		2.927E-05	56.93539427		
Fe	26	58	0.2819		2460		3.960E-06	57.93327558		
			100	8.73E+5		1.355E-03			55.84514458	55.84514458
Co	27	59	100	2260	2260	3.701E-06	3.701E-06	58.93319506	58.93319506	58.93319506
Ni	28	58	68.0769		33100		5.328E-05	57.9353435		
Ni	28	60	26.2231		12800		2.131E-05	59.93078635		
Ni	28	61	1.1399		555		9.396E-07	60.93105603		
Ni	28	62	3.6345		1770		3.046E-06	61.92834511		
Ni	28	64	0.9256		450		7.993E-07	63.92796594		
			100	48700		7.938E-05			58.69335196	58.69335196

Table 4 *(Continued)*

E (1)	*Z* (2)	*A* (3)	Isotopic composition atom % (4)	*N* (Elemental) Si = 1E+6 (5)	*N* (Isotopic) ΣSi = 1E+6 (6)	Elemental mass fractions (7)	Isotope mass fractions (8)	Atomic mass (Daltons, Da) (9)	Protosolar mean atomic mass (Da) (10)	Current mean atomic mass (Da) (11)
Cu	29	63	69.174		370		6.469E-07	62.92959751		
Cu	29	65	30.826		165		2.977E-07	64.92778945		
			100	535		9.446E-07			63.54556016	63.54556016
Zn	30	64	49.1704		620		1.101E-06	63.92914224		
Zn	30	66	27.7306		349		6.393E-07	65.92603345		
Zn	30	67	4.0401		51		9.484E-08	66.92712739		
Zn	30	68	18.4483		233		4.397E-07	67.9248442		
Zn	30	70	0.6106		8		1.554E-08	69.9253193		
			100	1260		2.291E-06			65.37776549	65.37776549
Ga	31	69	60.108		21.7		4.156E-08	68.9255735		
Ga	31	71	39.892		14.4		2.838E-08	70.9247026		
			100	36.2		6.994E-08			69.72306808	69.72306808
Ge	32	70	20.526		24.7		4.799E-08	69.9242474		
Ge	32	72	27.446		33.0		6.595E-08	71.9220758		
Ge	32	73	7.76		9.3		1.884E-08	72.9234589		
Ge	32	74	36.523		43.9		9.017E-08	73.92117777		
Ge	32	76	7.745		9.3		1.962E-08	75.92140273		
			100	120		2.426E-07			72.62958875	72.62958875
As	33	75	100	6.07	6.07	1.264E-08	1.264E-08	74.9215965	74.9215965	74.9215965
Se	34	74	0.86		0.58		1.191E-09	73.92247594		
Se	34	76	9.22		6.23		1.314E-08	75.91921372		
Se	34	77	7.59		5.13		1.096E-08	76.919914		

Table 4 (*Continued*)

E	*Z*	*A*	Isotopic composition atom %	*N* (Elemental) Si = 1E+6	*N* (Isotopic) ΣSi = 1E+6	Elemental mass fractions	Isotope mass fractions	Atomic mass (Daltons, Da)	Protosolar mean atomic mass (Da)	Current mean atomic mass (Da)
(1)	(2)	(3)	(4)	(5)	(6)	(7)	(8)	(9)	(10)	(11)
Se	34	78	23.69		16.00		3.464E-08	77.9173091		
Se	34	80	49.81		33.66		7.474E-08	79.9165213		
Se	34	82	8.83		5.96		1.357E-08	81.9166994		
			100	67.6		1.482E-07			78.97168088	78.97168088
Br	35	79	50.686		6.26		1.373E-08	78.9183371		
Br	35	81	49.314		6.09		1.369E-08	80.9162906		
			100	12.3		2.742E-08			79.90360789	79.90360789
Kr	36	78	0.362		0.19		4.114E-10	77.92036486		
Kr	36	80	2.326		1.20		2.665E-09	79.91637915		
Kr	36	82	11.655		5.99		1.363E-08	81.91348282		
Kr	36	83	11.546		5.94		1.368E-08	82.9141271		
Kr	36	84	56.903		29.18		6.803E-08	83.91149717		
Kr	36	86	17.208		8.79		2.098E-08	85.91061067		
			100	51.3		1.194E-07			83.79283539	83.7896357
Rb	37	85	70.844		5.080		1.199E-08	84.91178974		
Rb*	37	87	29.156		2.091		5.049E-09	86.90918054		
			100	7.17		1.703E-08			85.4941539	85.46775548
Sr	38	84	0.5584		0.13		3.031E-10	83.9134203		
Sr	38	86	9.8708		2.30		5.490E-09	85.9092602		
Sr	38	87	6.8982		1.61		3.888E-09	86.9088771		
Sr	38	88	82.6725		19.2		4.690E-08	87.9056122		
			100	23.3		5.658E-08			87.61750433	87.61352847

Table 4 (*Continued*)

E	*Z*	*A*	Isotopic composition atom %	*N* (Elemental) Si = 1E+6	*N* (Isotopic) ΣSi = 1E+6	Elemental mass fractions	Isotope mass fractions	Atomic mass (Daltons, Da)	Protosolar mean atomic mass (Da)	Current mean atomic mass (Da)
(1)	(2)	(3)	(4)	(5)	(6)	(7)	(8)	(9)	(10)	(11)
Y	39	89	100	4.35	4.35	1.075E-08	1.075E-08	88.9058483		
Zr	40	90	51.452		5.621		1.404E-08	89.9047044		
Zr	40	91	11.223		1.226		3.097E-09	90.9056458		
Zr	40	92	17.146		1.873		4.783E-09	91.9050408		
Zr	40	94	17.38		1.899		4.955E-09	93.9063152		
Zr^	40	96	2.799		0.306		8.154E-10	95.9082734		
			100	10.92		2.769E-08			91.22353759	91.22353759
Nb	41	93	100	0.780	0.780	2.014E-09	2.014E-09	92.9063781	92.9063781	92.9063781
Mo	42	92	14.525		0.380		9.704E-10	91.90680811		
Mo	42	94	9.151		0.238		6.210E-10	93.9050856		
Mo	42	95	15.838		0.412		1.086E-09	94.9058394		
Mo	42	96	16.672		0.433		1.154E-09	95.90467712		
Mo	42	97	9.599		0.249		6.704E-10	96.9060196		
Mo	42	98	24.391		0.630		1.714E-09	97.9054058		
Mo^	42	100	9.824		0.253		7.023E-10	99.9074724		
			100	2.60		6.918E-09			95.95978968	95.94866441
Ru	44	96	5.542		0.100		2.665E-10	95.9075939		
Ru	44	98	1.869		0.034		9.249E-11	97.9052876		
Ru	44	99	12.758		0.230		6.321E-10	98.9059393		
Ru	44	100	12.599		0.228		6.329E-10	99.9042195		
Ru	44	101	17.060		0.308		8.635E-10	100.9055821		
Ru	44	102	31.552		0.570		1.614E-09	101.9043493		

Table 4 (*Continued*)

E	*Z*	*A*	Isotopic composition atom %	*N* (Elemental) Si = 1E+6	*N* (Isotopic) ΣSi = 1E+6	Elemental mass fractions	Isotope mass fractions	Atomic mass (Daltons, Da)	Protosolar mean atomic mass (Da)	Current mean atomic mass (Da)
(1)	(2)	(3)	(4)	(5)	(6)	(7)	(8)	(9)	(10)	(11)
Ru	44	104	18.621		0.336		9.700E-10	103.9054326		
			100	1.81		5.071E-09			101.0650881	101.0650881
Rh	45	103	100	0.338	0.338	9.664E-10	9.664E-10	102.9055043	102.9055043	102.9055043
Pd	46	102	1.02		0.0141		3.992E-11	101.9056286		
Pd	46	104	11.14		0.1536		4.434E-10	103.9040359		
Pd	46	105	22.33		0.3079		8.975E-10	104.9050847		
Pd	46	106	27.33		0.377		1.109E-09	105.9034808		
Pd	46	108	26.46		0.365		1.094E-09	107.9038907		
Pd	46	110	11.72		0.162		4.947E-10	109.9051703		
			100	1.38		4.079E-09			106.4153292	106.4153292
Ag	47	107	51.839		0.258		7.664E-10	106.9050965		
Ag	47	109	48.161		0.239		7.232E-10	108.9047523		
			100	0.497		1.490E-09			107.8681467	107.8681467
Cd	48	106	1.249		0.020		5.885E-11	105.9064602		
Cd	48	108	0.89		0.014		4.197E-11	107.9041824		
Cd	48	110	12.485		0.197		6.016E-10	109.9030035		
Cd	48	111	12.804		0.202		6.225E-10	110.9041781		
Cd	48	112	24.117		0.381		1.185E-09	111.9027578		
Cd^	48	113	12.225		0.193		6.055E-10	112.9044026		
Cd	48	114	28.729		0.454		1.437E-09	113.9033595		
Cd^	48	116	7.501		0.119		3.832E-10	115.9047632		
			100	1.58		4.935E-09			112.4121436	112.4121436

Table 4 *(Continued)*

E	*Z*	*A*	Isotopic composition atom %	*N* (Elemental) Si = 1E+6	*N* (Isotopic) ΣSi = 1E+6	Elemental mass fractions	Isotope mass fractions	Atomic mass (Daltons, Da)	Protosolar mean atomic mass (Da)	Current mean atomic mass (Da)
(1)	(2)	(3)	(4)	(5)	(6)	(7)	(8)	(9)	(10)	(11)
In	49	113	4.281		0.008		2.510E-11	112.9040574		
In^	49	115	95.719		0.171		5.459E-10	114.9038788		
			100	0.179		5.710E-10			114.8182664	114.8182664
Sn	50	112	0.971		0.035		1.088E-10	111.9048218		
Sn	50	114	0.659		0.024		7.596E-11	113.9027788		
Sn	50	115	0.339		0.012		3.831E-11	114.9033424		
Sn	50	116	14.536		0.522		1.681E-09	115.9017405		
Sn	50	117	7.676		0.276		8.965E-10	116.9029516		
Sn	50	118	24.223		0.870		2.850E-09	117.9016031		
Sn	50	119	8.585		0.308		1.018E-09	118.9033076		
Sn	50	120	32.593		1.171		3.901E-09	119.9022002		
Sn	50	122	4.629		0.166		5.623E-10	121.9034391		
Sn	50	124	5.789		0.208		7.161E-10	123.9052761		
			100	3.59		1.185E-08			118.71035	118.71035
Sb	51	121	57.213		0.205		6.887E-10	120.9038157		
Sb	51	123	42.787		0.154		5.259E-10	122.904214		
			100	0.359		1.215E-09			121.7597261	121.7597261
Te	52	120	0.096		0.005		1.666E-11	119.9040452		
Te	52	122	2.603		0.123		4.166E-10	121.9030439		
Te^	52	123	0.908		0.043		1.468E-10	122.9042701		
Te	52	124	4.816		0.227		7.815E-10	123.9028176		
Te	52	125	7.139		0.337		1.170E-09	124.9044307		

Table 4 (*Continued*)

E	*Z*	*A*	Isotopic composition atom %	*N* (Elemental) Si = 1E+6	*N* (Isotopic) ΣSi = 1E+6	Elemental mass fractions	Isotope mass fractions	Atomic mass (Daltons, Da)	Protosolar mean atomic mass (Da)	Current mean atomic mass (Da)
(1)	(2)	(3)	(4)	(5)	(6)	(7)	(8)	(9)	(10)	(11)
Te	52	126	18.952		0.894		3.127E-09	125.9033117		
Te^	52	128	31.687		1.494		5.309E-09	127.9044621		
Te^	52	130	33.799		1.594		5.753E-09	129.9062228		
			100	4.72		1.672E-08			127.5855887	127.5855887
I	53	127	100	1.59	1.59	5.606E-09	5.606E-09	126.9044728	126.9044728	126.9044728
Xe^	54	124	0.129		0.007		2.410E-11	123.905893		
Xe	54	126	0.110		0.006		2.099E-11	125.9042912		
Xe	54	128	2.220		0.122		4.336E-10	127.9035313		
Xe	54	129	27.428		1.507		5.398E-09	128.9047809		
Xe	54	130	4.349		0.239		8.626E-10	129.9035094		
Xe	54	131	21.763		1.196		4.350E-09	130.9050524		
Xe	54	132	26.360		1.449		5.311E-09	131.9041551		
Xe	54	134	9.730		0.535		1.991E-09	133.9053945		
Xe^	54	136	7.911		0.435		1.643E-09	135.9072145		
			100	5.495		2.003E-08			131.1826916	131.1826916
Cs	55	133	100	0.368	0.368	1.359E-09	1.359E-09	132.905452	132.905452	132.905452
Ba^	56	130	0.106		0.005		1.805E-11	129.9063215		
Ba	56	132	0.101		0.005		1.833E-11	131.9050613		
Ba	56	134	2.417		0.110		4.093E-10	133.9045084		
Ba	56	135	6.592		0.300		1.125E-09	134.9056886		
Ba	56	136	7.853		0.357		1.348E-09	135.904576		
Ba	56	137	11.232		0.511		1.944E-09	136.9058274		

Table 4 *(Continued)*

E	*Z*	*A*	Isotopic composition atom %	*N* (Elemental) Si = 1E+6	*N* (Isotopic) ΣSi = 1E+6	Elemental mass fractions	Isotope mass fractions	Atomic mass (Daltons, Da)	Protosolar mean atomic mass (Da)	Current mean atomic mass (Da)
(1)	(2)	(3)	(4)	(5)	(6)	(7)	(8)	(9)	(10)	(11)
Ba	56	138	71.699		3.264		1.251E-08	137.9052473		
			100	4.552		1.737E-08			137.3269159	137.3269159
La*	57	138	0.092		4.000E-04		1.533E-12	137.907112		
La	57	139	99.908		0.459		1.772E-09	138.9063533		
			100	0.459		1.773E-09			138.905438	138.905466
Ce	58	136	0.186		0.002		7.552E-12	135.9071295		
Ce^	58	138	0.250		0.003		1.150E-11	137.905991		
Ce	58	140	88.450		1.030		4.004E-09	139.9054387		
Ce^	58	142	11.114		0.129		5.086E-10	141.9092442		
			100	1.165		4.532E-09			140.1157106	140.1156862
Pr	59	141	100	0.175	0.175	6.851E-10	6.851E-10	140.9076525	140.9076525	140.9076525
Nd	60	142	27.045		0.234		9.226E-10	141.9077233		
Nd	60	143	12.021		0.104		4.130E-10	142.9098143		
Nd^	60	144	23.729		0.205		8.197E-10	143.9100873		
Nd	60	145	8.763		0.076		3.060E-10	144.9125736		
Nd	60	146	17.130		0.148		6.000E-10	145.9131169		
Nd	60	148	5.716		0.049		2.014E-10	147.9168933		
Nd^	60	150	5.596		0.048		1.999E-10	149.9208949		
			100	0.864		3.463E-09			144.2446602	144.2427553
Sm	62	144	3.083		0.008		3.199E-11	143.9120046		
Sm*	62	147	15.017		0.042		1.714E-10	146.9148979		
Sm^	62	148	11.254		0.031		1.274E-10	147.9148227		

Table 4 (*Continued*)

E	*Z*	*A*	Isotopic composition atom %	*N* (Elemental) Si = 1E+6	*N* (Isotopic) ΣSi = 1E+6	Elemental mass fractions	Isotope mass fractions	Atomic mass (Daltons, Da)	Protosolar mean atomic mass (Da)	Current mean atomic mass (Da)
(1)	(2)	(3)	(4)	(5)	(6)	(7)	(8)	(9)	(10)	(11)
Sm	62	149	13.830		0.038		1.572E-10	148.9171847		
Sm	62	150	7.351		0.020		8.331E-11	149.9172755		
Sm	62	152	26.735		0.073		3.081E-10	151.9197324		
Sm	62	154	22.730		0.062		2.652E-10	153.9222093		
			100	0.273		1.145E-09			150.3632757	150.3649949
Eu^	63	151	47.81		0.0479		2.009E-10	150.9198502		
Eu	63	153	52.19		0.0523		2.222E-10	152.9212303		
			100	0.1002		4.231E-10			151.9643705	151.9643705
Gd^	64	152	0.2029		0.00013		5.357E-13	151.9197922		
Gd	64	154	2.1809		0.00755		3.227E-11	153.9208693		
Gd	64	155	14.7998		0.0512		2.204E-10	154.9226276		
Gd	64	156	20.4664		0.0708		3.068E-10	155.9221287		
Gd	64	157	15.6518		0.0542		2.361E-10	156.9239647		
Gd	64	158	24.8347		0.0859		3.770E-10	157.9241101		
Gd	64	160	21.8635		0.0756		3.361E-10	159.9270585		
			100	0.3454		1.509E-09			157.2520476	157.2520476
Tb	65	159	100	0.0625	6.250E-02	2.760E-10	2.760E-10	158.9253468	158.9253468	158.9253468
Dy	66	156	0.054		0.00005		2.166E-13	155.9242829		
Dy	66	158	0.095		0.0004		1.755E-12	157.9244096		
Dy	66	160	2.329		0.0095		4.221E-11	159.9251975		
Dy	66	161	18.889		0.0769		3.438E-10	160.9269334		
Dy	66	162	25.479		0.1038		4.670E-10	161.9267984		

Table 4 (*Continued*)

E (1)	*Z* (2)	*A* (3)	Isotopic composition atom % (4)	*N* (Elemental) Si = 1E+6 (5)	*N* (Isotopic) ΣSi = 1E+6 (6)	Elemental mass fractions (7)	Isotope mass fractions (8)	Atomic mass (Daltons, Da) (9)	Protosolar mean atomic mass (Da) (10)	Current mean atomic mass (Da) (11)
Dy	66	163	24.895		0.1014		4.590E-10	162.9287312		
Dy	66	164	28.260		0.1151		5.243E-10	163.9291748		
			100	0.407		1.838E-09			162.4997627	162.4997627
Ho	67	165	100	0.0891	8.910E-02	4.083E-10	4.083E-10	164.9303221	164.9303221	164.9303221
Er	68	162	0.139	0.0004			1.800E-12	161.9287799		
Er	68	164	1.601	0.0041			1.867E-11	163.9292065		
Er	68	166	33.503	0.0859			3.965E-10	165.9302931		
Er	68	167	22.869	0.0587			2.737E-10	166.9320482		
Er	68	168	26.978	0.0692			3.220E-10	167.9323702		
Er	68	170	14.910	0.0382			1.794E-10	169.9354643		
			100	0.256		1.192E-09			167.2590764	167.2590764
Tm	69	169	100	0.0403	4.030E-02	1.892E-10	1.892E-10	168.9342133		
Yb	70	168	0.12		0.0003		1.400E-12	167.9338869		
Yb	70	170	2.98		0.0075		3.541E-11	169.9347618		
Yb	70	171	14.09		0.0356		1.691E-10	170.9363258		
Yb	70	172	21.69		0.0547		2.613E-10	171.9363815		
Yb	70	173	16.10		0.0406		1.951E-10	172.9382108		
Yb	70	174	32.03		0.0808		3.905E-10	173.9388621		
Yb	70	176	13.00		0.0328		1.603E-10	175.9425717		
			100	0.252		1.213E-09			173.0544616	173.0544616
Lu	71	175	97.1795		0.0370		1.798E-10	174.9407712		

Table 4 *(Continued)*

E	*Z*	*A*	Isotopic composition atom %	*N* (Elemental) Si = 1E+6	*N* (Isotopic) ΣSi = 1E+6	Elemental mass fractions	Isotope mass fractions	Atomic mass (Daltons, Da)	Protosolar mean atomic mass (Da)	Current mean atomic mass (Da)
(1)	(2)	(3)	(4)	(5)	(6)	(7)	(8)	(9)	(10)	(11)
Lu*	71	176	2.8205		0.0011		5.377E-12	175.9426867		
			100	0.0381		1.852E-10			174.9690303	174.9668077
Hf^	72	174	0.161		0.0003		1.450E-12	173.9400462		
Hf	72	176	5.205		0.0081		3.960E-11	175.9414091		
Hf	72	177	18.604		0.0289		1.421E-10	176.9432224		
Hf	72	178	27.297		0.0424		2.096E-10	177.9437004		
Hf	72	179	13.627		0.0212		1.054E-10	178.945817		
Hf	72	180	35.105		0.0545		2.725E-10	179.9465512		
			100	0.155		7.707E-10			178.4865658	178.4851456
Ta*	73	180	0.0120		2.600E-06		1.300E-14	179.9474648		
Ta	73	181	99.9880		0.0215		1.081E-10	180.9479958		
			100	0.0215		1.081E-10			180.9478756	180.9478756
W^	74	180	0.120		0.0002		1.000E-12	179.9467091		
W	74	182	26.499		0.0381		1.926E-10	181.9482042		
W	74	183	14.314		0.0206		1.047E-10	182.9502223		
W	74	184	30.642		0.0440		2.249E-10	183.9509312		
W	74	186	28.426		0.0408		2.108E-10	185.9543641		
			100	0.144		7.340E-10			183.8416979	183.8416979
Re	75	185	35.662		0.0195		1.002E-10	184.9529549		
Re*	75	187	64.338		0.0352		1.828E-10	186.9557531		
			100	0.0547		2.831E-10			186.2415239	186.2067466
Os^	76	184	0.020		0.0001		5.111E-13	183.9524891		

Table 4 (*Continued*)

E	*Z*	*A*	Isotopic composition atom %	*N* (Elemental) Si = 1E+6	*N* (Isotopic) ΣSi = 1E+6	Elemental mass fractions	Isotope mass fractions	Atomic mass (Daltons, Da)	Protosolar mean atomic mass (Da)	Current mean atomic mass (Da)
(1)	(2)	(3)	(4)	(5)	(6)	(7)	(8)	(9)	(10)	(11)
Os^	76	186	1.597		0.0104		5.373E-11	185.9538382		
Os	76	187	1.298		0.0085		4.415E-11	186.9557505		
Os	76	188	13.325		0.0869		4.538E-10	187.9558382		
Os	76	189	16.252		0.106		5.565E-10	188.9581475		
Os	76	190	26.433		0.172		9.078E-10	189.9584471		
Os	76	192	41.076		0.268		1.429E-09	191.9614807		
			100	0.652		3.446E-09			190.2476945	190.2349378
Ir	77	191	37.272		0.236		1.252E-09	190.9605941		
Ir	77	193	62.728		0.397		2.129E-09	192.9629264		
			100	0.633		3.381E-09			192.2166171	192.2166171
Pt*	78	190	0.013		0.0004		1.056E-12	189.9599321		
Pt	78	192	0.794		0.01		5.334E-11	191.961038		
Pt	78	194	32.808		0.407		2.193E-09	193.962679		
Pt	78	195	33.787		0.419		2.270E-09	194.9647901		
Pt	78	196	25.290		0.314		1.710E-09	195.9649515		
Pt	78	198	7.308		0.091		5.006E-10	197.967891		
			100	1.24		6.728E-09			195.0839451	195.0839494
Au	79	197	100	0.195	0.195	1.067E-09	1.067E-09	196.9665687	196.9665687	196.9665687
Hg	80	196	0.16		0.001		5.445E-12	195.9658326		
Hg	80	198	10.04		0.038		2.090E-10	197.9667689		
Hg	80	199	16.94		0.064		3.538E-10	198.9682804		
Hg	80	200	23.14		0.087		4.834E-10	199.968326		

Table 4 (*Continued*)

E (1)	*Z* (2)	*A* (3)	Isotopic composition atom % (4)	*N* (Elemental) Si = 1E+6 (5)	*N* (Isotopic) ΣSi = 1E+6 (6)	Elemental mass fractions (7)	Isotope mass fractions (8)	Atomic mass (Daltons, Da) (9)	Protosolar mean atomic mass (Da) (10)	Current mean atomic mass (Da) (11)
Hg	80	201	13.17		0.049		2.736E-10	200.9703022		
Hg	80	202	29.74		0.112		6.285E-10	201.970643		
Hg	80	204	6.82		0.026		1.474E-10	203.9734941		
			100	0.376		2.101E-09			200.5923999	200.5923999
Tl	81	203	29.524		0.053		2.989E-10	202.9723442		
Tl	81	205	70.476		0.126		7.176E-10	204.9744275		
			100	0.179		1.017E-09			204.3833324	204.3833324
Pb^	82	204	1.997		0.066		3.740E-10	203.9730436		
Pb	82	206	18.582		0.615		3.520E-09	205.9744653		
Pb	82	207	20.563		0.680		3.911E-09	206.9758969		
Pb	82	208	58.858		1.947		1.125E-08	207.976652		
			100	3.308		1.906E-08			207.3188698	207.3162565
Bi^	83	209	100	0.1414	1.414E-01	8.210E-10	8.210E-10	208.9803987	208.9803987	208.9803987
Th*	90	232	100	0.0421	4.210E-02	2.714E-10	2.714E-10	232.0380553	232.0380553	232.0380553
U*	92	234	0.004		9.90E-07		6.438E-15	234.0409521		
U*	92	235	24.302		0.0058		3.788E-11	235.0439299		
U*	92	238	75.694		0.0181		1.197E-10	238.0507882		
			100	0.0239		1.576E-10			237.3199066	238.0289064

Column (6) gives the relative atomic *isotopic* composition on the same abundance scale as in column (5), hence the sum of all silicon atoms ("ΣSi" in header of column 6) must sum up to 1E6 atoms. These values are easily calculated from the abundance of an element (col5) times the isotopic composition of the respective element (col. 4).

Column (7) gives the mass fraction of each element in the overall protosolar composition, and column (8) the mass fraction of each isotope. Thus the sum of all elemental mass fractions = sum of all isotope mass fractions = 1.

Column (9) lists the atomic masses in atomic mass units (AMU) or Daltons (Da, which is 1/12 mass of a single ^{12}C atom) for each isotope, the data are from Audi et al. (2003), Wang et al. (2012, 2017), and Pfeiffer et al. (2014).

Column (10) lists the mean atomic masses (or mean atomic weights) for the elements with the given protosolar isotopic composition. For reference, column (11) present-day gives the atomic weights if the present-day isotopic compositions (not listed here) are used. The atomic weight or atomic mass of an element is generally calculated as the weighted mean of the individual atomic masses of the isotopes an element (column 9 in Table 4), where the statistical weight is the fractional isotopic atomic composition (column 4 in Table 4):

$$\text{Mean Atomic Mass Element [AMU]} = \frac{\sum \text{Isotopic Composition (atom\% of element)}}{100 \text{Isotope Atomic Mass [AMU]}}$$

The atomic weights derived from the adopted isotopic compositions of the elements must be used to be self-consistent. (The difference to the usual atomic weight tables is particularly noteworthy for Ar, which is mainly ^{40}Ar in the terrestrial atmosphere and thus dominates the atomic weight of Ar on Earth). Differences between the protosolar and present-day atomic weights are noticeable for elements with larger losses of radioactive isotopes over the past 4.567 Ga and for elements with larger gain of stable isotopes from decay of radioactive isotopes of other elements.

5.1 Calculation of the Overall Mass-Fractions of the Elements for the Solar-System Composition

The calculation of mass fractions requires knowledge of the abundances of all (major) elements and their isotopic composition. The calculation of the isotopic mass fractions from the abundances on the atomic scale are done by multiplying the abundances of the isotopes by their respective atomic masses and renormalizing the mass abundances to unity (or 100%). The mass fractions of an element (column 7) are just the sum of the mass fraction of its isotopes (col. 8). Another way to get the elemental mass fractions of the protosolar composition is multiplying the abundances of the elements (col. 5) by their respective mean atomic masses (col.10) and renormalizing the mass abundances to unity (or 100%). If the elemental mass fractions for the present-day composition is desired, the mean atomic weights in column (11) need to be used. In astronomy, often the mass fractions are designated as X for H, Y for He, and Z for heavy elements from Li to U (the Z for mass fraction of heavy elements should not be confused with Z for atomic number or charge).

5.2 Isotopic Composition of H, C, N, O, and the Noble Gases He, Ne, Ar, Kr, and Xe

The H isotopic composition (i.e., D/H ratio) of the Earth is not representative for the solar system as a whole because of secondary fractionations of the D/H ratio in the Earth's hydrological cycle. The D/H ratio cannot be derived from the solar photosphere or the solar wind because D was destroyed in the proto-Sun when it was fully convective. The proto-solar D/H ratio ($1.97[\pm 0.35] \times 10^{-5}$; Geiss and Gloeckler 2003) is thus derived from comparing the helium isotopic measurements of Jupiter's atmosphere with that of the solar wind and mass-balance calculations accounting for ^{3}He produced by deuterium burning. Direct measurements of the D/H ratio from Jupiter may not be representative according to Geiss and Gloeckler. However, Jupiter is the most massive planet in the solar system and has the largest proportion of captured solar gas, so it is the best proxy for the proto-solar D/H. But it should be kept in mind that the increase of D/H ratios in the giant planets' atmospheres with increasing distance from the Sun makes the assignment of the "correct" D/H ratio for the solar system more problematic.

The helium isotopic composition of ^{3}He/^{4}He of $1.66(\pm 0.05) \times 10^{-4}$ of Jupiter's atmosphere is adopted as proto-solar value (Geiss and Gloeckler 2003; Mahaffy et al. 1998). The current outer atmosphere of the sun has a higher ^{3}He/^{4}He because D-burning produced ^{3}He, therefore the ratio is above the respective proto-solar ratio. In addition, the solar wind ^{3}He/^{4}He from in-situ measurements shows that the various solar wind regimes are isotopically fractionated, and is seen with much higher precision in Genesis regime targets (Heber et al. 2012). Very ^{3}He-enriched solar energetic particles from solar flares have been observed (e.g., Mason et al. 2004), complicating the interpretation of the current "solar" ^{3}He/^{4}He ratio because the derivation of the proto-solar D/H ratio depends on the ^{3}He/^{4}He ratio of the present-day Sun. Hence differences in reported proto-solar D/H values exist (see Geiss and Gloeckler 2003; Trieloff 2018; Wieler 2016 for alternate choices).

The isotopic compositions of the other noble gases, C, N, and O are from solar wind measurements; a preferred choice for the solar system isotopic composition by several authors. One caveat is that the solar wind isotopic compositions may not be representative for the solar system as a whole because the solar wind is fractionated in elemental composition relative to the photosphere and meteorites, and corresponding mass-dependent isotopic fractionations are observed in the solar wind. Fractionation effects from the settling of elements during the sun's lifetime need to be considered in addition to the model-dependent corrections that are applied to account for fractionations effect during solar wind acceleration. Therefore the adopted values remain subject to change, however, the solar wind values appear more realistic choices for representative solar system values than terrestrial values that have been used in the past (see below).

The C, N, and O isotopic compositions in solar system materials are more variable and can reach 10 to 100 per-mil levels. The choices for representative solar system values are not necessarily straight-forward. Ideally, the solar isotopic composition should be the most representative choice, as it should be for the elemental composition. Spectroscopic measurements of the photospheric C and O isotopic compositions have large uncertainties. The recent solar wind measurements from Genesis samples give a light isotopic composition (i.e., the lower-mass isotope of an element is more abundant than heavier isotopes of the element relative to terrestrial standard materials). The C-isotopic composition is still relatively close to the terrestrial and meteoritic values.

The solar wind nitrogen is poor in ^{15}N when compared to Earth and most meteorites, and the adopted ratio of $^{14}N/^{15}N = 440$ is similar to that found for Jupiter). which seems to vindicates the adoption of the solar wind as isotopic standard (e.g., see Füri and Marty 2015; Marty et al. 2010; Avice and Marty 2020). The oxygen isotopic composition of the solar wind is depleted in ^{17}O and ^{18}O by about 7% relative to Earth and most meteoritic materials (Laming 2017; McKeegan et al. 2011).

The Ne, Ar, Kr, and Xe data are from soar wind measurements from the Genesis mission Crowther and Gilmour (2013), Heber et al. (2009), Heber et al. (2012), Pepin et al. (2012),

Vogel et al. (2011), and Meshik et al. (2014). Isotopic fractionations are less for the heavier Ne and Ar than for He. However, theoretical corrections to mass-dependent fractionations of the adopted solar wind isotopic compositions were not applied here because they are not yet well understood and vary with different authors (e.g., Heber et al. 2012; Ott 2015, see also Huss et al. 2020).

Acknowledgements This work was supported in part by NSF grant AST 1517541 and the McDonnell Center for the Space Sciences.

References

C.M.O'D.Alexander, Quantitative models for the elemental and isotopic fractionations in chondrites: the non-carbonaceous chondrites. Geochim. Cosmochim. Acta **254**, 246–276 (2019a)

C.M.O'D.Alexander, Quantitative models for the elemental and isotopic fractionations in chondrites: the carbonaceous chondrites. Geochim. Cosmochim. Acta **254**, 277–309 (2019b)

S.A. Alexeeva, L.I. Mashonkina, Carbon abundances of reference late-type stars from 1D analysis of atomic C I and molecular CH lines. Mon. Not. R. Astron. Soc. **453**, 1619–1631 (2015)

A.M. Amarsi, M. Asplund, The solar silicon abundance based on 3D non-LTE calculations. Mon. Not. R. Astron. Soc. **464**, 264–273 (2017)

A.M. Amarsi, P.S. Barklem, M. Asplund, R. Collet, O. Zatarinny, Inelastic O+H collisions and the O I 777 nm solar centre-to-limb variation. Astron. Astrophys. **616**, A89 (2018)

A.M. Amarsi, P.S. Barklem, R. Collet, N. Grevesse, M. Asplund, 3D non-LTE line formation of neutral carbon in the Sun. Astron. Astrophys. **624**, A111 (2019)

M. Asplund, N. Grevesse, A.J. Sauval, P. Scott, The chemical composition of the Sun. Annu. Rev. Astron. Astrophys. **47**, 481–522 (2009)

G. Avice, B. Marty, Perspectives on atmospheric evolution from noble gas and nitrogen isotopes on Earth, Mars & Venus. Space Sci. Rev. **216**, 36 (2020). https://doi.org/10.1007/s11214-020-00655-0

S. Basu, H.M. Antia, Constraining solar abundances using helioseismology. Astrophys. J. **606**, L85–L88 (2004)

M. Bergemann, R. Collet, A.M. Amarsi, M. Kovalev, G. Ruchti, Z. Magic, Non-local thermodynamic equilibrium stellar spectroscopy with 1D and <3D> models. I. Methods and application to magnesium abundances in standard stars. Astrophys. J. **847**, 15 (2017)

M. Bergemann, A.J. Gallagher, P. Eitner, M. Bautista, R. Collet, S.A. Yakovleva, A. Mayriedl, B. Plez, M. Carlsson, J. Leenaarts, A.K. Belyaev, C. Hansen, On the origin of the elements, I. 3D NLTE formation of Mn lines in late-type stars. Astron. Astrophys. **631**, A80 (2019)

N. Brauckmüller, F. Wombacher, D.C. Hezel, R. Escoube, C. Münker, The chemical composition of carbonaceous chondrites: implications for volatile element depletion, complementarity and alteration. Geochim. Cosmochim. Acta **239**, 17–48 (2018)

E. Caffau, H.G. Ludwig, M. Steffen, B. Freytag, P. Bonifacio, Solar chemical abundances determined with a CO5BOLD 3D model atmosphere. Sol. Phys. **268**, 255–269 (2011)

E. Caffau, H.-G. Ludwig, M. Steffen, W. Livingston, P. Bonifacio, J.-M. Malherbe, H.-P. Doerr, W. Schmidt, The photospheric solar oxygen project. III. Investigation of the centre-to-limb variation of the 630 nm [OI]-NiI blend. Astron. Astrophys. **579**, A88 (2015)

S.A. Crowther, J.D. Gilmour, The Genesis solar wind composition and its relationship to planetary xenon signatures. Geochim. Cosmochim. Acta **123**, 17–34 (2013)

J.M. Friedrich, M.S. Wang, M.E. Lipschutz, Chemical studies of L chondrites. V: compositional patterns for 49 trace elements in 14 L4–6 and 7 LL4–6 falls. Geochim. Cosmochim. Acta **67**, 2467–2479 (2003)

J.M. Friedrich, J.C. Bridges, M.S. Wang, M.E. Lipschutz, Chemical studies of L chondrites. VI: variations with petrographic type and shock-loading among equilibrated falls. Geochim. Cosmochim. Acta **68**, 2889–2904 (2004)

E. Füri, B. Marty, Nitrogen isotope variations in the solar system. Nat. Geosci. **8**, 515–522 (2015)

J. Geiss, G. Gloeckler, Isotopic composition of H, He and Ne in the protosolar cloud. Space Sci. Rev. **106**, 3–18 (2003)

N. Grevesse, P. Scott, M. Asplund, A.J. Sauval, The elemental composition of the Sun. III. The heavy elements Cu to Th. Astron. Astrophys. **573**, A27 (2015)

V.S. Heber, R. Wieler, H. Baur, C. Olinger, T.A. Friedmann, D.S. Burnett, Noble gas composition of the solar wind as collected by the Genesis mission. Geochim. Cosmochim. Acta **73**, 7414–7432 (2009)

V.S. Heber, H. Baur, P. Bochsler, K.D. McKeegan, M. Neugebauer, D.B. Reisenfeld, R. Wieler, R.C. Wiens, Isotopic mass fractionation of solar wind: evidence from fast and slow solar wind collected by the Genesis mission. Astrophys. J. **759**, 121 (2012)

H. Holweger, Photospheric abundances, problems, updates, implications. AIP Conf. Proc. **598**, 23–30 (2001)

G.R. Huss, E. Koeman-Shields, A.J. Jurewicz, D.S. Burnett, K. Nagashima, R. Ogliore, C.T. Olinger, Hydrogen fluence in Genesis collectors: implications for acceleration of solar wind and for solar metallicity. Meteorit. Planet. Sci. **55**, 326–351 (2020)

J.M. Laming, The first ionization potential effect from the ponderomotive force: on the polarization and coronal origin of Alfvén waves. Astrophys. J. **844**, 153 (2017)

J.E. Lawler, M.P. Wood, E.A. Den Hartog, T. Feigenson, C. Sneden, J.J. Cowan, Improved V I log(gf) values and abundance determinations in the photospheres of the Sun and metal-poor Star HD 84937. Astrophys. J. Suppl. Ser. **215**, 20 (2014)

J.E. Lawler, C. Sneden, G. Nave, E.A. Den Hartog, N. Emrahoğlu, J.J. Cowan, Improved Cr II log(gf) values and abundance determinations in the photospheres of the Sun and metal-poor Star HD 84937. Astrophys. J. Suppl. Ser. **228**, 10 (2017)

J.E. Lawler, T. Feigenson, C. Sneden, J.J. Cowan, G. Nave, Transition probabilities of Co II weak lines to the ground and low metastable levels. Astrophys. J. Suppl. Ser. **238**, 7 (2018)

J.E. Lawler, F.N. Hala, C. Sneden, G. Nave, M.P. Wood, J.J. Cowan, Transition probabilities of Sc I and Sc II and Scandium abundances in the Sun, Arcturus, and HD 84937. Astrophys. J. Suppl. Ser. **241**, 21 (2019)

K. Lind, A.M. Amarsi, M. Asplund, P.S. Barklem, M. Bautista, M. Bergemann, R. Collet, D. Kiselman, J. Leenaarts, T.M.D. Pereira, Non-LTE line formation of Fe in late-type stars—IV. Modelling of the solar centre-to-limb variation in 3D. Mon. Not. R. Astron. Soc. **468**, 4311–4322 (2017)

K. Lodders, Solar System abundances and condensation temperatures of the elements. Astrophys. J. **591**, 1220–1247 (2003)

K. Lodders, The solar argon abundance. Astrophys. J. **674**, 607–611 (2008)

K. Lodders, Solar elemental abundances, in *Online Oxford Research Encyclopedia of Planetary Science* (2020). https://doi.org/10.1093/acrefore/9780190647926.013.145. arXiv:1912.00844

K. Lodders, B. Fegley Jr., *The Planetary Scientist's Companion* (Oxford University Press, Oxford, 1998), 384 pp.

K. Lodders, B. Fegley Jr., *Chemistry of the Solar System* (Roy. Soc. Chemistry Publishing, Cambridge, 2011), 476 pp.

K. Lodders, H. Palme, H.P. Gail, Abundances of the elements in the solar system, in *Landolt-Börnstein*, ed. by J.E. Trümper. New Series, vol. VI/4B (Springer, Berlin, 2009), pp. 560–630. Chap. 4.4

Z.G. Maas, C.A. Pilachowski, K. Hinkle, Chlorine abundances in cool stars. Astrophys. J. **152**, id196 (2016)

P.R. Mahaffy, T.M. Donahue, S.K. Atreya, T.C. Owen, H.B. Niemann, Galileo probe measurements of D/H and He/ He in Jupiter's atmosphere. Space Sci. Rev. **84**, 251–263 (1998)

E. Maiorca, H. Uitenbroek, S. Uttenthaler, S. Randich, M. Busso, L. Magrini, A new solar fluorine abundance and a fluorine determination in the two open clusters M67 and NGC 6404. Astrophys. J. **788**, id149 (2014)

B. Marty, L. Zimmermann, P.G. Burnard, R. Wieler, V.S. Heber, D.S. Burnett, R.C. Wiens, P. Bochsler, Nitrogen isotopes in the recent solar wind from the analysis of Genesis targets: evidence for large scale isotope heterogeneity in the early solar system. Geochim. Cosmochim. Acta **74**, 340–355 (2010)

G.M. Mason, J.E. Mazur, J.R. Dwyer, J.R. Jokipii, R.E. Gold, S.M. Krimigis, Abundances of heavy and ultraheavy ions in 3He-rich solar flares. Astrophys. J. **606**, 555–564 (2004)

K.D. McKeegan, A.P.A. Kallio, V.S. Heber, G. Jarzebinski, P.H. Mao, C.D. Coath, T. Kunihiro, R.C. Wiens, J.E. Nordholt, R.W. Moses, D.B. Reisenfeld, A.J.G. Jurewicz, D.S. Burnett, The oxygen isotopic composition of the Sun inferred from captured solar wind. Science **332**, 1528–2011 (2011)

J. Meija, T.B. Coplen, M. Berglund, W.A. Brand, P. de Bievre, M. Groning, N.E. Holden, J. Irrgher, R.D. Loss, T. Walczyk, T. Prohaska, Isotopic compositions of the elements 2013 (IUPAC technical report). Pure Appl. Chem. **88**, 293–306 (2016)

A. Meshik, C. Hohenberg, O. Pravdivtseva, D. Burnett, Heavy noble gases in solar wind delivered by Genesis mission. Geochim. Cosmochim. Acta **127**, 326–347 (2014)

T. Nordlander, K. Lind, Non-LTE aluminium abundances in late-type stars. Astron. Astrophys. **607**, A75 (2017)

U. Ott, Planetary and pre-solar noble gases in meteorites. Chem. Erde **74**, 519–544 (2015)

R.O. Pepin, D.J. Schlutter, R.H. Becker, D.B. Reisenfeld, Helium, neon, and argon composition of the solar wind as recorded in gold and other Genesis collector materials. Geochim. Cosmochim. Acta **89**, 62–80 (2012)

B. Pfeiffer, K. Venkataramaniahc, U. Czoka, C. Scheidenberger, Atomic mass compilation 2012. At. Data Nucl. Data Tables **100**, 403–535 (2014)

L. Schaefer, B. Fegley, Outgassing of ordinary chondritic material and some of its implications for the chemistry of asteroids, planets, and satellites. Icarus **186**, 462–483 (2007)

L. Schaefer, B. Fegley, Volatile element chemistry during metamorphism of ordinary chondritic material. Icarus **205**, 483–496 (2010)

P. Scott, N. Asplund, N. Grevesse, M. Bergemann, A.J. Sauval, The elemental composition of the Sun. II. The iron group elements Sc to Ni. Astron. Astrophys. **573**, A26 (2015a)

P. Scott, N. Grevesse, M. Asplund, A.J. Sauval, K. Lind, Y. Takeda, R. Collet, R. Trampedach, W. Hayek, The elemental composition of the Sun. I. The intermediate mass elements Na to Ca. Astron. Astrophys. **573**, A25 (2015b)

C. Sneden, J.E. Lawler, J.J. Cowan, I.I. Ivans, E.A. Den Hartog, New rare Earth element abundance distributions for the Sun and five r-process-rich very metal-poor stars. Astrophys. J. Suppl. Ser. **182**, 80–96 (2009)

C. Sneden, J.J. Cowan, C. Kobayashi, M. Pignatari, J.E. Lawler, E.A. Den Hartog, M.P. Wood, Iron-group abundances in the metal-poor main-sequence turnoff star HD84937. Astrophys. J. **817**, 53 (2016)

M. Trieloff, Noble gases, in *Oxford Research Encyclopedia of Planetary Science*, vol. 30, ed. by P. Read et al. (Oxford University Press, London, 2018). 978-0-190-64792-6. https://doi.org/10.1093/acrefore/9780190647926.013.33

N. Vitas, I. Vince, M. Lugaro, O. Andriyenko, M. Gosic, R.J. Rutten, On the solar abundance of indium. Mon. Not. R. Astron. Soc. **384**, 370–375 (2008)

N. Vogel, V.S. Heber Veronika, H. Baur, D.S. Burnett, R. Wieler, Argon, krypton, and xenon in the bulk solar wind as collected by the Genesis mission. Geochim. Cosmochim. Acta **75**, 3057–3071 (2011)

M. Wang et al., The Ame2012 atomic mass evaluation. Chin. Phys. C **36**, 1603–2014 (2012)

M. Wang et al., The Ame2016 atomic mass evaluation. Chin. Phys. C **41**, 030003 (2017)

J.T. Wasson, G.W. Kallemeyn, Compositions of chondrites. Philos. Trans. R. Soc. Lond. A **325**, 535–544 (1988)

R. Wieler, Do lunar and meteoritic archives record temporal variations in the composition of solar wind noble gases and nitrogen? A reassessment in the light of Genesis data. Chem. Erde **76**, 463–480 (2016)

P.R. Young, Element abundance ratios in the quiet sun transition region. Astrophys. J. **855**, 15 (2018)

G. Zhao, L. Mashonkina, H.L. Yan, S. Alexeeva et al., Systematic non-LTE study of the −2.6 <=[Fe/H]<=0.2 F and G dwarfs in the solar neighborhood. II. Abundance patterns from Li- to Eu. Astrophys. J. **833**, 225 (2016)

GPSR Compliance

The European Union's (EU) General Product Safety Regulation (GPSR) is a set of rules that requires consumer products to be safe and our obligations to ensure this.

If you have any concerns about our products, you can contact us on ProductSafety@springernature.com

In case Publisher is established outside the EU, the EU authorized representative is:

Springer Nature Customer Service Center GmbH
Europaplatz 3
69115 Heidelberg, Germany

Batch number: 10732076

Printed by Printforce, the Netherlands